# Wolf Prize in Mathematics

Volume 3

# Wolf Prize in Mathematics

## Volume 3


Edited by

**Y Sinai**
*Princeton University, USA*

**E Stein**
*Princeton University, USA*



World Scientific

NEW JERSEY • LONDON • SINGAPORE • BEIJING • SHANGHAI • HONG KONG • TAIPEI • CHENNAI

*Published by*

World Scientific Publishing Co. Pte. Ltd.
5 Toh Tuck Link, Singapore 596224
*USA office:* 27 Warren Street, Suite 401-402, Hackensack, NJ 07601
*UK office:* 57 Shelton Street, Covent Garden, London WC2H 9HE

**British Library Cataloguing-in-Publication Data**
A catalogue record for this book is available from the British Library.

The editors and publisher would like to thank the following organisations and publishers of the various journals and books for their assistance and permission to reproduce the selected reprints found in this volume:

American Mathematical Society
Elsevier
Gautheir-Villars
International Mathematical Union
Kluwer
Plenum Publishing Corporation
Polish Scientific Publishers
Springer-Verlag
The Wolf Foundation

While every effort has been made to contact the publishers of reprinted papers prior to publication, we have not been successful in some cases. Where we cold not contact the publishers, we have acknowledged the source of the material. Proper credit will be accorded to the publishers in future editions of this work after permission is granted.

**WOLF PRIZE IN MATHEMATICS — Vol. 3**
**edited by Y. G. Sinai and E. M. Stein**


ISBN-13 978-981-4390-27-9
ISBN-10 981-4390-27-5

Printed in Singapore by B & Jo Enterprise Pte Ltd

## List of Wolf Prize Winners

| | | |
|---|---|---|
| 1978 | Israel M. Gel'fand | 1913–2009 |
| | Carl L. Siegel | 1896–1981 |
| 1979 | Jean Leray | 1906–1998 |
| | André Weil | 1906–1998 |
| 1980 | Henri Cartan | 1904–2008 |
| | Andrei N. Kolmogorov | 1903–1987 |
| 1981 | Lars V. Ahlfors | 1907–1996 |
| | Oscar Zariski | 1899–1986 |
| 1982 | Hassler Whitney | 1907–1989 |
| | Mark Grigor'evich Krein | 1907–1989 |
| 1983/4 | Shiing-Shen Chern | 1911–2004 |
| | Paul Erdős | 1913–1996 |
| 1984/5 | Kunihiko Kodaira | 1915–1997 |
| | Hans Lewy | 1904–1988 |
| 1986 | Samuel Eilenberg | 1913–1998 |
| | Atle Selberg | 1917–2007 |
| 1987 | Kiyoshi Itō | 1915–2008 |
| | Peter D. Lax | 1926 |
| 1988 | Friedrich Hirzebruch | 1927 |
| | Lars Hörmander | 1931 |
| 1989 | Alberto P. Calderón | 1920–1998 |
| | John W. Milnor | 1931 |
| 1990 | Ennio de Giorgi | 1928–1996 |
| | Ilya Piatetski-Shapiro | 1929–2009 |
| 1992 | Lennart A. E. Carleson | 1928 |
| | John G. Thompson | 1932 |

| | | |
|---|---|---|
| 1993 | Mikhail Gromov | 1943 |
| | Jacques Tits | 1930 |
| 1994/5 | Jürgen K. Moser | 1928–1999 |
| 1995/6 | Robert Langlands | 1936 |
| | Andrew J. Wiles | 1953 |
| 1996/7 | Joseph B. Keller | 1923 |
| | Yakov G. Sinai | 1935 |
| 1999 | László Lovász | 1948 |
| | Elias M. Stein | 1931 |
| 2000 | Raoul Bott | 1923–2005 |
| | Jean-Pierre Serre | 1926 |
| 2001 | Vladimir I. Arnold | 1937–2010 |
| | Saharon Shelah | 1945 |
| 2002/3 | Mikio Sato | 1928 |
| | John T. Tate | 1925 |
| 2005 | Gregory A. Margulis | 1946 |
| | Sergei P. Novikov | 1938 |
| 2006/7 | Stephen Smale | 1930 |
| | Hillel Fürstenberg | 1935 |
| 2008 | Pierre R. Deligne | 1944 |
| | Phillip A. Griffiths | 1938 |
| | David Mumford | 1937 |
| 2010 | Shing-Tung Yau | 1949 |
| | Dennis P. Sullivan | 1941 |
| 2012 | Michael G. Aschbacher | 1944 |
| | Luis A. Caffarelli | 1948 |

## Preface

This is the third volume in the series dealing with the winners of the Wolf Prize in mathematics. The present volume covers mostly the period of the first decade of this century. It also includes I. M. Gel'fand, who in 1978 with C. L. Siegel, was a first laureate of the prize, as well as Atle Selberg, J. Thompson, and M. Gromov, each of whom won the prize before 2000.

It may be well to recall some facts about the Wolf Prize. It was founded by Dr. Ricardo Wolf and his wife Francisca Subirana Wolf and is given each year in five of the six areas of Agriculture, Chemistry, Mathematics, Medicine, Physics, and the Arts. Laureates receive their awards from the President of the State of Israel in a ceremony at the Knesset (Israel's parliament) in Jerusalem. About this prize, S. S. Chern and F. Hirzebrach said it well when they wrote: "The Fields medal goes to young people … The Wolf prize often honors the achievements of a whole life."

As in the previous volumes, the documents collected here are varied in nature but are intended to sketch a picture of the work and scientific life of each of the laureates. As such, these include bibliographies and curricula vitae, particularly noteworthy papers or surveys, as well as articles written about the winners, whenever these were available.

It is our pleasure to record here our appreciation to the Wolf foundation. We also wish to express our thanks to World Scientific for their role in publishing this volume, and in particular to the help we have received from Ms. E. H. Chionh.

*Yakov G. Sinai*
*Elias M. Stein*

# Contents

**Mikhail Gromov**



**Gregory A. Margulis**

## Curriculum Vitae

Vladimir I. Arnold

Date of Birth: June 12, 1937, Odessa, USSR
Died: June 3, 2010, Paris, France

**Education:**

1954–1961 Student of Mathematics, Moscow State University, Moscow USSR
1959 M.S. Diploma work: "On mappings of circle to itself"; under the supervision of Prof. A. N. Kolmogorov
1961 Dr. Thesis: "On the representation of continuous functions of 3 variables by the superpositions of continuous functions of 2 variables", under the supervision of Prof. A. N. Kolmogorov; Referees: Prof. A. G. Vitushkin and Prof. L. V. Keldysh
1963 Ph.D. Thesis: "Small denominators and stability problems in classical and celestial mechanics"; referees: Prof. N. N. Bogoliubov, Prof. V. M. Volosov and Prof. G. N. Duboshin

**Employment History:**

1961–1986 Moscow State University, Mech. Math. faculty: assistant, dozent, professor
1986–2010 Steklov Mathematical Institute, Moscow
1993–2004 Universite' Paris 9, France

**Member of Academies:**

1976 Honorary Member of London Math. Soc.
1983 French Acad of Sci.
1984 Foreign Associate, National Academy of Sciences, USA
1986 USSR Academy, corresponding member
1987 Academy of Arts and Sciences, USA
1988 Royal Soc. of London
1988 Acad. Lincei Roma
1989 American Philosoph. Soc.
1990 Full member of the Soviet Academy of Sciences
1991 Russian Acad. of Natural Sciences

**Doctor Honoris Causa:**

1979 Univ. P. et M. Curie Paris
1988 Warwick (Coventry)
1991 Utrecht
1991 Bologna

**Awards:**

1958 Moscow Math. Soc. Prize for Young Mathematicians
1965 Lenin Prize (with *A. N. Kolmogorov*)
1982 Crafoord Prize of the Swedish Acad. (with *L. Nirenberg*)
2001 Wolf Prize in Mathematics
2007 State Prize of the Russian Federation
2008 Shaw Prize in Mathematical Science

## List of Publications


1. V. I. Arnold, Gibbs phenomenon for Fourier series on finite sets, *Funct. Anal. Other Math.* **3** (2010), No. 1, 91–96.
2. V. I. Arnold, Periods and Young's diagram of Fermat–Euler's geometrical progressions of residues, *Funct. Anal. Other Math.* **3** (2010), No. 1, 21–38.
3. V. I. Arnold, Topological classification of Morse polynomials (in Russian), translated *Mat. Inst. Steklova*, Vol. 268 (2010), Differentsialnye Uravneniya i Topologiya. I, pp. 40–55.
4. V. I. Arnold, Are quadratic residues random?, *Regul. Chaotic Dyn.* **15** (2010), No. 4–5, 425–430.
5. V. I. Arnold, Mathematics of chaos, *Mosc. Math. J.* **10** (2010), No. 2, 273–283; 478.
6. V. I. Arnold, Permutations (in Russian), *Usp. Mat. Nauk* **64** (2009), No. 4(388), 3–44; translation in *Russ. Math. Surv.* **64** (2009), No. 4, 583–624.
7. V. I. Arnold, Stochastic and deterministic statistics of orbits in chaotically looking dynamical systems (in Russian), *Tr. Mosk. Mat. Obs.* **70** (2009), 46–101; translation in (Trans. Moscow Math. Soc., 2009), pp. 31–69 (2009).
8. V. I. Arnold, Sur quelques problèmes de la théorie des systèmes dynamiques (in French) [On some problems of dynamical systems theory], in *Aspects des Systèmes Dynamiques*, pp. 205–221 (Éc. Polytech., 2009).
9. A. M. Abramov, V. I. Arnold and A. V. Bolsinov *et al.*, Nikolaĭ Nikolaevich Nekhoroshev (in Russian), *Usp. Mat. Nauk* **64** (2009), No. 3(387), 174–178; translation in *Russ. Math. Surv.* **64** (2009), No. 3, 561–566.
10. V. I. Arnold, Random and algebraic permutations' statistics, *Funct. Anal. Other Math.* **2** (2009), No. 2–4, 247–248.
11. V. I. Arnold, Lengths of periods of continued fractions of square roots of integers, *Funct. Anal. Other Math.* **2** (2009), No. 2–4, 151–164.
12. V. I. Arnold, Empirical study of stochasticity for deterministic chaotical dynamics of geometric progressions of residues, *Funct. Anal. Other Math.* **2** (2009), No. 2–4, 139–149.
13. V. I. Arnold, Geometry of continued fractions associated with Frobenius numbers, *Funct. Anal. Other Math.* **2** (2009), No. 2–4, 129–138.
14. V. I. Arnold, Uniform distribution of indivisible vectors in the space of integers (in Russian), *Izv. Ross. Akad. Nauk Ser. Mat.* **73** (2009), No. 1, 21–30; translation in *Izv. Math.* **73** (2009), No. 1, 21–29.

15. V. I. Arnold, To what extent are arithmetic progressions of fractional parts random? (in Russian), *Usp. Mat. Nauk* **63** (2008), No. 2(380), 5–20; translation in *Russ. Math. Surv.* **63** (2008), No. 2, 205–220.
16. V. I. Arnold, On additive semigroups starting parts, *Funct. Anal. Other Math.* **2** (2008), No. 1, 81–86.
17. V. I. Arnold, Statistics of the period lengths of the continued fractions for the eigenvalues of the integer matrices of order two, *Funct. Anal. Other Math.* **2** (2008), No. 1, 15–26.
18. V. I. Arnold, Orbits' statistics in chaotic dynamical systems, *Nonlinearity* **21** (2008), No. 7, T109–T112.
19. V. I. Arnold, Statistics of the periods of continued fractions for quadratic irrationals (in Russian), *Izv. Ross. Akad. Nauk Ser. Mat.* **72** (2008), No. 1, 3–38; translation in *Izv. Math.* **72** (2008), No. 1, 1–34.
20. V. I. Arnold, Topological classification of trigonometric polynomials of the affine Coxeter group $\tilde{A}_2$ (in Russian), *Tr. Mat. Inst. Steklova*, Vol. 258 (2007), Chaps. 1, 7–16; translation in *Proc. Steklov Inst. Math.*, Vol. 258 (2007), No. 1, 3–12.
21. V. I. Arnold, Topological classification of Morse functions and generalisations of Hilbert's 16th problem, *Math. Phys. Anal. Geom.* **10** (2007), No. 3, 227–236.
22. V. I. Arnold, Continued fractions of square roots of rational numbers and their statistics (in Russian), *Usp. Mat. Nauk* **62** (2007), No. 5(377), 3–14; translation in *Russ. Math. Surv.* **62** (2007), No. 5, 843–855.
23. V. I. Arnold *et al.*, Yuriĭ Alekseevich Mitropol'skiĭ (on the occasion of his ninetieth birthday) (in Russian), *Usp. Mat. Nauk* **62** (2007), No. 4(376), 179–185; translation in *Russ. Math. Surv.* **62** (2007), No. 4, 829–835.
24. D. V. Anosov, V. I. Arnold, V. A. Vladimirov, V. V. Kozlov and Ya. G. Sinaĭ, Viktor Iosifovich Yudovich (in Russian), *Usp. Mat. Nauk* **62** (2007), No. 2(374), 165–168; translation in *Russ. Math. Surv.* **62** (2007), No. 2, 375–378.
25. V. I. Arnold, Arithmetical turbulence of self-similar fluctuations statistics of large Frobenius numbers of additive semigroups of integers, *Mosc. Math. J.* **7** (2007), No. 2, 173–193.
26. V. I. Arnold *et al.*, Askold Georgievich Khovanskii, *Mosc. Math. J.* **7** (2007), No. 2, 169–171.
27. V. I. Arnold, Yesterday and long ago. Translated from the 2006 Russian original by Leonora P. Kotova and Owen L. deLange (Springer-Verlag, 2007).
28. V. I. Arnold, Smooth functions statistics, *Funct. Anal. Other Math.* **1** (2006), No. 2, 111–118.

29. V. I. Arnold, Complexity of finite sequences of zeros and ones and geometry of finite spaces of functions, *Funct. Anal. Other Math.* **1** (2006), No. 1, 1–15.
30. V. I. Arnold, Statistics of the symmetric group representations as a natural science question on asymptotics of Young diagrams, in *Representation Theory, Dynamical Systems, and Asymptotic Combinatorics*, pp. 1–7, Amer. Math. Soc. Transl. Ser. 2, Vol. 217 (Amer. Math. Soc., 2006).
31. V. I. Arnold and G. Iooss, Vladimirov, Valdimir Obituary [Victor Iosifovich Yudovich]. *J. Math. Fluid Mech.* **8** (2006), No. 4, 455.
32. V. I. Arnold, Geometry and growth rate of Frobenius numbers of additive semigroups, *Math. Phys. Anal. Geom.* **9** (2006), No. 2, 95–108.
33. V. I. Arnold, V. V. Kozlov and A. I. Neishtadt, Mathematical aspects of classical and celestial mechanics. [Dynamical systems. III]. Translated from the Russian original by E. Khukhro. Third edition. Encyclopaedia of Mathematical Sciences, Vol. 3 (Springer-Verlag, 2006).
34. V. I. Arnold, On the matricial version of Fermat–Euler congruences, *Jpn. J. Math.* **1** (2006), No. 1, 1–24.
35. V. I. Arnold, Statistics of Young diagrams of cycles of dynamical systems for finite tori automorphisms, *Mosc. Math. J.* **6** (2006), No. 1, 43–56; 221.
36. V. I. Arnold *et al.*, Victor A. Vassiliev, *Mosc. Math. J.* **6** (2006), No. 1, 1–3.
37. V. I. Arnold, Statistics and classification of topologies of periodic functions and trigonometric polynomials, Proc. Steklov Inst. Math., Dynamical Systems: Modeling, Optimization, and Control, Suppl. 1 (2006) S13–S23.
38. V. I. Arnold, *Ordinary Differential Equations*. Translated from the Russian by Roger Cooke. Second printing of the 1992 edition (Springer-Verlag, 2006).
39. V. I. Arnold *et al.*, Vladimir Mikhaĭlovich Tikhomirov (on the occasion of his seventieth birthday) (in Russian), *Usp. Mat. Nauk* **61** (2006), No. 1(367), 187–190; translation in *Russ. Math. Surv.* **61** (2006), No. 1, 189–192.
40. V. I. Arnold, The underestimated Poincaré (in Russian), *Usp. Mat. Nauk* **61** (2006), No. 1(367), 3–24; translation in *Russ. Math. Surv.* **61** (2006), No. 1, 1–18.
41. V. I. Arnold, From Hilbert's superposition problem to dynamical systems, in *Mathematical Events of the Twentieth Century*, pp. 19–47 (Springer, 2006).
42. V. I. Arnold, On the topology of the eigenfields, *Topol. Methods Nonlinear Anal.* **26** (2005), No. 1, 9–16.
43. V. I. Arnold, The principle of topological economy in algebraic geometry, in *Surveys in Modern Mathematics*, pp. 13–23, London Math. Soc. Lecture Note Ser., Vol. 321 (Cambridge Univ. Press, 2005).

44. V. I. Arnold, Mysterious mathematical trinities, in *Surveys in Modern Mathematics*, pp. 1–12, London Math. Soc. Lecture Note Ser., Vol. 321 (Cambridge Univ. Press, 2005).
45. V. I. Arnold, A. Demidov, Yu. Ilyashenko, G. Magaril-Ilyaev, E. Mishchenko, K. Osipenko, A. Sossinsky, M. Tsfasman, V. Uspensky, M. Vyalyi, I. Yaschenko, M. Zelikin and V. M. Tikhomirov, *Mosc. Math. J.* **5** (2005), No. 1, back matter.
46. V. I. Arnold, Ergodic and arithmetical properties of geometrical progression's dynamics and of its orbits, *Mosc. Math. J.* **5** (2005), No. 1, 5–22.
47. D. Anosov, V. I. Arnold, A. Glutsyuk, A. Gorodetski, V. Kaloshin, A. Katok, A. Khovanskii, S. Lando, A. Sossinsky, M. Tsfasman, S. Yakovenko and Y. S. Ilyashenko, *Mosc. Math. J.* **5** (2005), No. 1, front matter.
48. V. I. Arnold, Number-theoretical turbulence in Fermat–Euler arithmetics and large Young diagrams geometry statistics, *J. Math. Fluid Mech.* **7** (2005), Suppl. 1, S4–S50.
49. V. I. Arnold, Lobachevsky triangle altitudes theorem as the Jacobi identity in the Lie algebra of quadratic forms on symplectic plane, *J. Geom. Phys.* **53** (2005), No. 4, 421–427.
50. V. I. Arnold, Geometry and dynamics of Galois fields (in Russian), *Usp. Mat. Nauk* **59** (2004), No. 6(360), 23–40; translation in *Russ. Math. Surv.* **59** (2004), No. 6, 1029–1046.
51. V. I. Arnold, M. Sh. Birman and A. M. Vershik *et al.*, Olga Aleksandrovna Ladyzhenskaya (in Russian), *Usp. Mat. Nauk* **59** (2004), No. 3(357), 151–152; translation in *Russ. Math. Surv.* **59** (2004), No. 3, 553–555.
52. V. I. Arnold, The matrix Euler–Fermat theorem (in Russian), *Izv. Ross. Akad. Nauk Ser. Mat.* **68** (2004), No. 6, 61–70; translation in *Izv. Math.* **68** (2004), No. 6, 1119–1128.
53. V. I. Arnold, From Hilbert's superposition problem to dynamical systems, *Amer. Math. Monthly* **111** (2004), No. 7, 608–624.
54. V. I. Arnold, Arnold's problems. Translated and revised edition of the 2000 Russian original. With a preface by V. Philippov, A. Yakivchik and M. Peters (Springer-Verlag, 2004).
55. V. I. Arnold, A. N. Kolmogorov and the natural sciences (in Russian), *Usp. Mat. Nauk* **59** (2004), No. 1(355), 25–44; translation in *Russ. Math. Surv.* **59** (2004), No. 1, 27–46.
56. V. I. Arnold, Fermat dynamics, matrix arithmetic, finite circle, and the finite Lobachevskiĭ plane (in Russian), *Functional. Anal. i Prilozhen.* **38** (2004), No. 1, 1–15; 95; translation in *Funct. Anal. Appl.* **38** (2004), No. 1, 1–13.

57. V. I. Arnold, *Lectures on Partial Differential Equations*. Translated from the second Russian edition by Roger Cooke (Springer-Verlag, 2004).
58. J.-P. Kahane, P. Cartier, V. I. Arnold, J. Coates, D. Zagier, H. Brézis, F. Pham, B. Malgrange, H. Helson, Y. Meyer, Y. Colin de Verdière and P.-L. Lions, *Leçons de mathématiques d'aujourd'hui*, Vol. 1 (in French) [Lectures on Mathematics Today, Vol. 1] With a preface by Éric Charpentier and Nicolaï Nikolski. Second edition. Le Sel et le Fer [Salt and Iron], Vol. 4 (Cassini, 2003).
59. V. I. Arnold, Frequent representations, *Mosc. Math. J.* **3** (2003), No. 4, 1209–1221.
60. D. V. Anosov, V. I. Arnold and V. M. Bukhshtaber *et al.*, Andreĭ Andreevich Bolibrukh (in Russian), *Usp. Mat. Nauk* **58** (2003), No. 6(354), 139–142; translation in *Russ. Math. Surv.* **58** (2003), No. 6, 1185–1189.
61. V. I. Arnold, Topology and statistics of formulas of arithmetic (in Russian), *Usp. Mat. Nauk* **58** (2003), No. 4(352), 3–28; translation in *Russ. Math. Surv.* **58** (2003), No. 4, 637–664.
62. V. I. Arnold, The topology of algebra: Combinatorics of squaring (in Russian), *Functional. Anal. i Prilozhen.* **37** (2003), No. 3, 20–35; 95; translation in *Funct. Anal. Appl.* **37** (2003), No. 3, 177–190.
63. V. A. Aleksandrov, V. I. Arnold and A. A. Borisenko *et al.*, Alekseĭ Vasilevich Pogorelov (in Russian), *Usp. Mat. Nauk* **58** (2003), No. 3(351), 173–175; translation in *Russ. Math. Surv.* **58** (2003), No. 3, 593–596.
64. V. I. Arnold, Arithmetics of binary quadratic forms, symmetry of their continued fractions and geometry of their de Sitter world. Dedicated to the 50th anniversary of IMPA, *Bull. Braz. Math. Soc.* (*N.S.*) **34** (2003), No. 1, 1–42.
65. V. I. Arnold, The Fermat–Euler dynamical system and the statistics of the arithmetic of geometric progressions (in Russian), *Functional. Anal. i Prilozhen.* **37** (2003), No. 1, 1–18; 95; translation in *Funct. Anal. Appl.* **37** (2003), No. 1, 1–15.
66. V. I. Arnold, On a variational problem connected with phase transitions of means in controllable dynamical systems, in *Nonlinear Problems in Mathematical Physics and Related Topics*, I, pp. 23–34, Int. Math. Ser. (N.Y.), Vol. 1 (Kluwer/Plenum, 2002).
67. V. I. Arnold and I. G. Petrovskiĭ, Hilbert's topological problems, and modern mathematics (in Russian), *Usp. Mat. Nauk* **57** (2002), No. 4(346), 197–207; translation in *Russ. Math. Surv.* **57** (2002), No. 4, 833–845.
68. V. I. Arnold, The longest curves of given degree and the quasicrystallic Harnack theorem in pseudoparabolic topology (in Russian), *Functional.*

*Anal. i Prilozhen.* **36** (2002), No. 3, 1–8; 96; translation in *Funct. Anal. Appl.* **36** (2002), No. 3, 165–171.

69. V. I. Arnold, The mathematical duel over Bourbaki (in Russian), *Vestnik of Russian Academy of Science* **72** (2002), No. 3, 245–250.
70. V. I. Arnold, Optimization in the mean, and phase transitions in controlled dynamical systems (in Russian), *Functional. Anal. i Prilozhen.* **36** (2002), No. 2, 1–11; 95; translation in *Funct. Anal. Appl.* **36** (2002), No. 2, 83–92.
71. V. I. Arnold, Pseudoquaternion geometry (in Russian), *Functional. Anal. i Prilozhen.* **36** (2002), No. 1, 1–15; 96; translation in *Funct. Anal. Appl.* **36** (2002), No. 1, 1–12.
72. V. I. Arnold, Astroidal geometry of hypocycloids and the Hessian topology of hyperbolic polynomials (in Russian), *Usp. Mat. Nauk* **56** (2001), No. 6(342), 3–66; translation in *Russ. Math. Surv.* **56** (2001), No. 6, 1019–1083.
73. V. I. Arnold, On teaching mathematics (in Polish), translated from the Russian by Danuta Śledziewska-Błocka, *Wiadom. Mat.* **37** (2001), 17–26.
74. V. I. Arnold, Complexification of a tetrahedron, and pseudoprojective transformations (in Russian), *Functional. Anal. i Prilozhen.* **35** (2001), No. 4, 1–7; 95; translation in *Funct. Anal. Appl.* **35** (2001), No. 4, 241–246.
75. V. I. Arnold and A. B. Givental, Symplectic Geometry. Dynamical Systems, IV, pp. 1–138, Encyclopaedia Math. Sci., Vol. 4 (Springer, 2001).
76. V. I. Arnold, The Lagrangian Grassmannian of a quaternion hypersymplectic space (in Russian), *Functional. Anal. i Prilozhen.* **35** (2001), No. 1, 74–77; translation in *Funct. Anal. Appl.* **35** (2001), No. 1, 61–63.
77. V. I. Arnold, Zadachi Arnold (in Russian) [Arnold problems] With a preface by M. B. Sevryuk and V. B. Filippov (Izdatelstvo FAZIS, 2000).
78. V. I. Arnold, The complex Lagrangian Grassmannian (in Russian), *Functional. Anal. i Prilozhen.* **34** (2000), No. 3, 63–65; translation in *Funct. Anal. Appl.* **34** (2000), No. 3, 208–210.
79. V. I. Arnold, On A. N. Kolmogorov, in *Kolmogorov in Perspective*, pp. 89–108, Hist. Math., Vol. 20 (Amer. Math. Soc., 2000).
80. V. I. Arnold, Singularity theory, in *Development of Mathematics* 1950–2000, pp. 63–95 (Birkhäuser, 2000).
81. V. I. Arnold, Dynamical systems, in *Development of Mathematics* 1950–2000, pp. 33–61 (Birkhäuser, 2000).
82. V. I. Arnold, From averaging to statistical physics (in Russian), *Tr. Mat. Inst. Steklova*, Vol. 228 (2000), Probl. Sovrem. Mat. Fiz., pp. 196–202; translation in *Proc. Steklov Inst. Math.* (2000), No. 1(228), 184–190.
83. V. I. Arnold, Symplectic geometry and topology, *J. Math. Phys.* **41** (2000), No. 6, 3307–3343.

84. V. I. Arnold, Jürgen Moser (1928–1999). Déclin des mathématiques (après la mort de Jürgen Moser) (in French) [Jürgen Moser (1928–1999). The decline of mathematics (after the death of Jürgen Moser)] Gaz. Math. No. 84 (2000), 92–95.
85. V. I. Arnold, Polymathematics: Is mathematics a single science or a set of arts?, in *Mathematics: Frontiers and Perspectives*, pp. 403–416 (Amer. Math. Soc., 2000).
86. V. I. Arnold, Repartitioning the world, *Quantum* **10** (2000), No. 3, 34–37.
87. V. I. Arnold, Simple singularities of curves (in Russian), *Tr. Mat. Inst. Steklova*, Vol. 226 (1999), Mat. Fiz. Probl. Kvantovoi Teor. Polya, pp. 27–35; translation in *Proc. Steklov Inst. Math.* (1999), No. 3(226), 20–28.
88. V. I. Arnold, First steps of local contact algebra, dedicated to H. S. M. Coxeter on the occasion of his 90th birthday, *Canad. J. Math.* **51** (1999), No. 6, 1123–1134.
89. V. I. Arnold, The anti-science revolution and mathematics (in Russian), *Vestnik of Russian Academy of Science* **69** (1999), No. 6, 553–558.
90. V. I. Arnold, Weak asymptotics of the numbers of solutions of Diophantine equations (in Russian), *Functional. Anal. i Prilozhen.* **33** (1999), No. 4, 65–66; translation in *Funct. Anal. Appl.* **33** (1999), No. 4, 292–293.
91. M. S. Agranovich, V. I. Arnold and D. G. Vasil'ev *et al.*, Viktor Borisovich Lidskiĭ (on the occasion of his seventieth birthday) (in Russian), *Usp. Mat. Nauk* **54** (1999), No. 5(329), 197–203; translation in *Russ. Math. Surv.* **54** (1999), No. 5, 1077–1085.
92. V. I. Arnold, Topological problems in wave propagation theory and topological economy principle in algebraic geometry, in *The Arnoldfest* (Toronto, ON, 1997), pp. 39–54, Fields Inst. Commun., Vol. 24 (Amer. Math. Soc., 1999).
93. V. I. Arnold, Symplectization, complexification and mathematical trinities, in *The Arnoldfest* (Toronto, ON, 1997), pp. 23–37, Fields Inst. Commun., Vol. 24 (Amer. Math. Soc., 1999).
94. V. I. Arnold, From Hilbert's superposition problem to dynamical systems, in *The Arnoldfest* (Toronto, ON, 1997), pp. 1–18, Fields Inst. Commun., Vol. 24 (Amer. Math. Soc., 1999).
95. V. I. Arnold, Kepler's second law and the topology of abelian integrals (according to Newton) [Kvant 1987, No. 12, 17–21], in *Kvant Selecta: Algebra and Analysis*, II, pp. 131–140, Math. World, Vol. 15 (Amer. Math. Soc., 1999).
96. V. I. Arnold, Evolution processes and ordinary differential equations [Kvant 1986, No. 2, 13–20], in *Kvant Selecta: Algebra and Analysis*, II, pp. 73–85, Math. World, Vol. 15 (Amer. Math. Soc., 1999).

97. V. I. Arnold, V. S. Afrajmovich, Yu. S. Ilyashenko and L. P. Shilnikov, Bifurcation theory and catastrophe theory. Translated from the 1986 Russian original by N. D. Kazarinoff. Reprint of the 1994 English edition from the series Encyclopaedia of Mathematical Sciences [Dynamical Systems, Vol. V, Encyclopaedia Math. Sci., Vol. 5] (Springer-Verlag, 1999).
98. V. I. Arnold, First steps of local symplectic algebra, in *Differential Topology, Infinite-Dimensional Lie Algebras, and Applications*, pp. 1–8, Amer. Math. Soc. Transl. Ser. 2, Vol. 194 (Amer. Math. Soc., 1999).
99. V. I. Arnold, R. M. Borisyuk and I. M. Gel'fand *et al.*, Èmmanuil Èl'evich Shnol' (on the occasion of his seventieth birthday) (in Russian), *Usp. Mat. Nauk* **54** (1999), No. 3(327), 199–204; translation in *Russ. Math. Surv.* **54** (1999), No. 3, 677–683.
100. V. I. Arnold, Topological problems in the theory of asymptotic curves (in Russian), *Tr. Mat. Inst. Steklova*, Vol. 225 (1999), Solitony Geom. Topol. na Perekrest., pp. 11–20; translation in *Proc. Steklov Inst. Math.* (1999), No. 2(225), 5–15.
101. V. I. Arnold, A mathematical trivium (in Swedish), Translated from the Russian original [*Usp. Mat. Nauk* **46** (1991), No. 1(277), 225–232] by Jaak Peetre. *Normat* **47** (1999), No. 3, 111–121.
102. V. I. Arnold, Relatives of the quotient of the complex projective plane by complex conjugation (in Russian), *Tr. Mat. Inst. Steklova*, Vol. 224 (1999), Algebra. Topol. Differ. Uravn. i ikh Prilozh., 56–67; translation in *Proc. Steklov Inst. Math.* (1999), No. 1(224), 46–56.
103. V. I. Arnold, Topologically necessary singularities on moving wavefronts and caustics, in *Hamiltonian Systems with Three or More Degrees of Freedom* (S'Agaró, 1995), pp. 11–12, NATO Adv. Sci. Inst. Ser. C Math. Phys. Sci., Vol. 533 (Kluwer, 1999).
104. V. I. Arnold, International Mathematical Congress in Berlin (in Russian), *Vestnik of Russian Academy of Science* **69** (1999), No. 2, 163–172.
105. V. I. Arnold, Problèmes mathématiques de l'hydrodynamique et de la magnétohydrodynamique. [Mathematical problems of hydrodynamics and magnetohydrodynamics] *Trois Applications des Mathématiques*, pp. 39–50 (Soc. Math. France, 1998).
106. V. I. Arnold, Higher-dimensional continued fractions. J. Moser at 70 (in Russian), *Regul. Chaotic Dyn.* **3** (1998), No. 3, 10–17.
107. V. I. Arnold, Singularities of fractions and the behavior of polynomials at infinity (in Russian), *Tr. Mat. Inst. Steklova* **221** (1998), 48–68; translation in *Proc. Steklov Inst. Math.* (1998), No. 2(221), 40–59.
108. V. I. Arnold, On the problem of realization of a given Gaussian curvature function, *Topol. Methods Nonlinear Anal.* **11** (1998), No. 2, 199–206.

109. V. I. Arnold, V. V. Goryunov, O. V. Lyashko and V. A. Vasil'ev, *Singularity Theory*. I. Translated from the 1988 Russian original by A. Iacob. Reprint of the original English edition from the series Encyclopaedia of Mathematical Sciences [Dynamical Systems, Vol. VI, Encyclopaedia Math. Sci., Vol. 6] (Springer-Verlag, 1998).
110. V. I. Arnold, "Hard" and "soft" mathematical models. (Catalan) Translated from the Russian to Spanish by Rafael Ramirez and Natalia Sadowskaia and translated from the Spanish to Catalan by Llorenç Roselló. Butl. Soc. Catalana Mat. **13** (1998), No. 1, 7–26.
111. V. I. Arnold, On the Legendrian Sturm theory of space curves (in Russian), *Functional. Anal. i Prilozhen.* **32** (1998), No. 2, 1–7; 95; translation in *Funct. Anal. Appl.* **32** (1998), No. 2, 75–80.
112. V. I. Arnold, "Hard" and "soft" mathematical models (in Russian), *Priroda* (1998), No. 4, 3–14.
113. V. I. Arnold, On the teaching of mathematics (in Russian), *Usp. Mat. Nauk* **53** (1998), No. 1(319), 229–234; translation in *Russ. Math. Surv.* **53** (1998), No. 1, 229–236.
114. V. I. Arnold and B. A. Khesin, *Topological Methods in Hydrodynamics, Applied Mathematical Sciences*, Vol. 125 (Springer-Verlag, 1998).
115. V. I. Arnold, V. V. Kozlov and A. I. Neishtadt, Mathematical aspects of classical and celestial mechanics. Translated from the 1985 Russian original by A. Iacob. Reprint of the original English edition from Encyclopaedia Math. Sci., Vol. 3 (Springer, 1993).
116. V. I. Arnold, Izbrannoe-60 (in Russian) [Selecta-60] *Izdatel'stvo* (FAZIS, 1997).
117. V. I. Arnold, Lektsii ob uravneniyakh s chastnymi proizvodnymi (in Russian) [Lectures on partial differential equations] Second edition. Biblioteka Studenta-Matematika [Undergraduate Mathematics Library], 2. *Izdatel'stvo* (FAZIS, 1997).
118. V. I. Arnold, Remarks on the Morse theory of a divergence-free vector field, the averaging method, and the motion of a charged particle in a magnetic field (in Russian), *Tr. Mat. Inst. Steklova*, Vol. 216 (1997), Din. Sist. i Smezhnye Vopr., pp. 9–19; translation in *Proc. Steklov Inst. Math.* (1997), No. 1(216), 3–13
119. D. V. Anosov, S. Kh. Aranson, V. I. Arnold, I. U. Bronshtein, V. Z. Grines and Yu. S. Il'yashenko, Ordinary differential equations and smooth dynamical systems. Translated from the 1985 Russian original by E. R. Dawson and D. O'Shea. Third printing of the 1988 translation, Encyclopaedia Math. Sci., Vol. 1 (Springer, 1988).

120. V. I. Arnold, Remarks on parabolic curves on surfaces and on higher-dimensional Möbius–Sturm theory (in Russian), *Functional. Anal. i Prilozhen.* **31** (1997), No. 4, 3–18; 95; translation in *Funct. Anal. Appl.* **31** (1997), No. 4, 227–239.
121. V. I. Arnold, Yu. M. Berezanskiĭ, Yu. L. Daletskiĭ, M. M. Kutateladze, M. M. Lavrent'ev, Yu. G. Reshetnyak, E. M. Semenov and S. A. Sklyadnev, Selim Grigor'evich Kreĭn (on the occasion of his eightieth birthday) (in Russian), *Usp. Mat. Nauk* **52** (1997), No. 6(318), 203–204; translation in *Russ. Math. Surv.* **52** (1997), No. 6, 1349–1350.
122. V. I. Arnold, A. A. Bolibrukh, R. V. Gamkrelidze, V. P. Maslov, E. F. Mishchenko, S. P. Novikov, Yu. S. Osipov, Ya. G. Sinaĭ, A. M. Stepin and L. L. Faddeev, Dmitriĭ Viktorovich Anosov (on the occasion of his sixtieth birthday) (in Russian), *Usp. Mat. Nauk* **52** (1997), No. 2(314), 193–200; translation in *Russ. Math. Surv.* **52** (1997), No. 2, 437–445.
123. V. I. Arnold, Topological classification of real trigonometric polynomials and cyclic serpents polyhedron, in *The Arnold–Gelfand Mathematical Seminars*, pp. 101–106 (Birkhäuser, 1997).
124. V. I. Arnold, H. Kalf and A. Schneider, Separated Dirac operators and asymptotically constant linear systems, *Math. Proc. Cambridge Philos. Soc.* **121** (1997), No. 1, 141–146.
125. V. I. Arnold, M. I. Vishik, A. S. Kalashnikov, V. P. Maslov, S. M. Nikolskiĭ and S. P. Novikov, Olga Arsenevna Oleĭnik (on the occasion of her seventieth birthday) (in Russian), *Tr. Semin. im. I. G. Petrovskogo*, No. 19 (1996), 5–25; translation in *J. Math. Sci. (N.Y.)* **85** (1997), No. 6, 2249–2259.
126. V. I. Arnold, Osobennosti kaustik i volnovykh frontov (in Russian) [Singularities of caustics and wave fronts] Biblioteka Matematika [Mathematics Library], 1. Izdatel'stvo (FAZIS, 1996).
127. V. I. Arnold, Topological content of the Maxwell theorem on multipole representation of spherical functions, *Topol. Methods Nonlinear Anal.* **7** (1996), No. 2, 205–217.
128. V. I. Arnold, Will mathematics survive?, Report on the Zurich Congress [Math. Intelligencer **17** (1995), No. 3, 6–10] (in Czech), translated from the English by Jiří Fiala, *Pokroky Mat. Fyz. Astronom.* **41** (1996), No. 1, 38–44.
129. V. I. Arnold, Remarks on the extactic points of plane curves, in *The Gelfand Mathematical Seminars*, 1993–1995, pp. 11–22 (Birkhäuser, 1996).
130. V. I. Arnold, Topological problems in the theory of wave propagation (in Russian), *Usp. Mat. Nauk* **51** (1996), No. 1(307), 3–50; translation in *Russ. Math. Surv.* **51** (1996), No. 1, 1–47.

131. V. I. Arnold, Topological classification of complex trigonometric polynomials and the combinatorics of graphs with an identical number of vertices and edges (in Russian), *Functional. Anal. i Prilozhen.* **30** (1996), No. 1, 1–17; 96; translation in *Funct. Anal. Appl.* **30** (1996), No. 1, 1–14.
132. V. I. Arnold, Remarks on the enumeration of plane curves, in *Topology of Real Algebraic Varieties and Related Topics*, pp. 17–32, Amer. Math. Soc. Transl. Ser. 2, Vol. 173 (Amer. Math. Soc., 1996).
133. V. I. Arnold, On the number of flattening points on space curves, in *Sinaĭ's Moscow Seminar on Dynamical Systems*, pp. 11–22, Amer. Math. Soc. Transl. Ser. 2, Vol. 171 (Amer. Math. Soc., 1996).
134. V. I. Arnold, Invariants and perestroikas of fronts on a plane (in Russian), *Tr. Mat. Inst. Steklov.* **209** (1995), Osob. Gladkikh Otobrazh. s Dop. Strukt., pp. 14–64
135. V. I. Arnold, Some remarks on symplectic monodromy of Milnor fibrations, in *The Floer Memorial Volume*, pp. 99–103, Progr. Math., Vol. 133 (Birkhäuser, 1995).
136. V. I. Arnold, Sur les propriétés topologiques des projections lagrangiennes en géométrie symplectique des caustiques (in French) [Topological properties of Lagrangian projections in the symplectic geometry of caustics] *Rev. Mat. Univ. Complut. Madrid* **8** (1995), No. 1, 109–119.
137. V. I. Arnold, Les mathématiques vont-elles survivre? (in French) [Will mathematics survive?] Gaz. Math. No. 65 (1995), 3–10.
138. Yu. S. Osipov, A. A. Gonchar, S. P. Novikov, V. I. Arnold, G. I. Marchuk, P. P. Kulish, V. S. Vladimirov and E. F. Mishchenko, Lyudvig Dmitrievich Faddeev (on the occasion of his sixtieth birthday) (in Russian), *Usp. Mat. Nauk* **50** (1995), No. 3(303), 171–186; translation in *Russ. Math. Surv.* **50** (1995), No. 3, 643–659.
139. V. I. Arnold, Will mathematics survive?, Report on the Zurich Congress. *Math. Intelligencer* **17** (1995), No. 3, 6–10.
140. V. I. Arnold, The geometry of spherical curves and quaternion algebra (in Russian), *Usp. Mat. Nauk* **50** (1995), No. 1(301), 3–68; translation in *Russ. Math. Surv.* **50** (1995), No. 1, 1–68.
141. V. I. Arnold, Remarks on eigenvalues and eigenvectors of Hermitian matrices, Berry phase, adiabatic connections and quantum Hall effect, *Selecta Math. (N.S.)* **1** (1995), No. 1, 1–19.
142. V. I. Arnold, Sur quelques problèmes de la théorie des systèmes dynamiques (in French) [Some problems of the theory of dynamical systems], in *Topol. Methods Nonlinear Anal.* **4** (1994), No. 2, 209–225.

143. V. I. Arnold, The Vassiliev theory of discriminants and knots, in *First European Congress of Mathematics*, Vol. I (Paris, 1992), pp. 3–29, Progr. Math., Vol. 119 (Birkhäuser, 1994).
144. V. I. Arnold, Plane curves, their invariants, perestroikas and classifications. With an appendix by F. Aicardi. Adv. Soviet Math., Vol. 21, *Singularities and Bifurcations*, pp. 33–91 (Amer. Math. Soc., 1994).
145. V. I. Arnold, Topological properties of Legendre projections in contact geometry of wave fronts (in Russian), *Algebra i Analiz* **6** (1994), No. 3, 1–16; translation in *St. Petersburg Math. J.* **6** (1995), No. 3, 439–452.
146. V. I. Arnold, M. Sh. Birman and I. M. Gel'fand *et al.*, Anatoliĭ Moiseevich Vershik (on the occasion of his sixtieth birthday) (in Russian), *Usp. Mat. Nauk* **49** (1994), No. 3(297), 195–204; translation in *Russ. Math. Surv.* **49** (1994), No. 3, 207–221.
147. V. I. Arnold, Topological invariants of plane curves and caustics. Dean Jacqueline B. Lewis Memorial Lectures presented at Rutgers University, New Brunswick, University Lecture Series, Vol. 5 (Amer. Math. Soc., 1994).
148. V. I. Arnold, Mathematical problems in classical physics, in *Trends and Perspectives in Applied Mathematics*, pp. 1–20, Appl. Math. Sci., Vol. 100 (Springer, 1994).
149. V. I. Arnold, On A. N. Kolmogorov (in Russian), in *Reminiscences about Kolmogorov*, pp. 144–172 (Nauka, 1993).
150. V. I. Arnold, V. V. Kozlov and A. I. Neĭshtadt, Mathematical aspects of classical and celestial mechanics, in *Dynamical Systems*, III, pp. vii–xiv and pp. 1–291, Encyclopaedia Math. Sci., Vol. 3 (Springer, 1993).
151. V. I. Arnold, Problems on singularities and dynamical systems, in *Developments in Mathematics: The Moscow School*, pp. 251–274, (Chapman & Hall, 1993).
152. V. I. Arnold, On A. N. Kolmogorov, in *Golden Years of Moscow Mathematics*, pp. 129–153, Hist. Math., Vol. 6 (Amer. Math. Soc., 1993).
153. V. I. Arnold, N. S. Bakhvalov and K. V. Brushlinskiĭ *et al.*, Nikolaĭ Nikolaevich Chentsov (in Russian), *Usp. Mat. Nauk* **48** (1993), No. 2(290), 165–168; translation in *Russ. Math. Surv.* **48** (1993), No. 2, 161–166.
154. V. I. Arnold, Bounds for Milnor numbers of intersections in holomorphic dynamical systems, in *Topological Methods in Modern Mathematics* (Stony Brook, NY, 1991), pp. 379–390 (Publish or Perish, 1993).
155. V. I. Arnold, Will Russian mathematics survive?, Special issue on mathematics in the former Soviet Union, *Notices Amer. Math. Soc.* **40** (1993), No. 2, 104–107.

156. V. I. Arnold *et al.*, Obituary: Dmitriĭ Andreevich Gudkov (in Russian), *Usp. Mat. Nauk* **47** (1992), No. 6(288), 195–198; translation in *Russ. Math. Surv.* **47** (1992), No. 6, 193–197.
157. V. I. Arnold, Congruences for Euler, Bernoulli and Springer numbers of Coxeter groups (in Russian), *Izv. Ross. Akad. Nauk Ser. Mat.* **56** (1992), No. 5, 1129–1133; translation in *Russ. Acad. Sci. Izv. Math.* **41** (1993), No. 2, 389–393.
158. V. I. Arnold, Poly-integrable flows (in Russian), *Algebra i Analiz* **4** (1992), No. 6, 54–62; translation in *St. Petersburg Math. J.* **4** (1993), No. 6, 1103–1110.
159. V. I. Arnold, Catastrophe theory. Translated from the Russian by G. S. Wassermann. Based on a translation by R. K. Thomas. Third edition (Springer-Verlag, 1992).
160. V. I. Arnold, Snake calculus and the combinatorics of the Bernoulli, Euler and Springer numbers of Coxeter groups (in Russian), *Usp. Mat. Nauk* **47** (1992), No. 1(283), 3–45; 240; translation in *Russ. Math. Surv.* **47** (1992), No. 1, 1–51.
161. L. A. Sena, L. P. Feoktistov, Yu. N. Smirnov, S. S. Gershteĭn, V. I. Arnold and A. D. Dolgov, Master (in Russian), *Priroda* (1992), No. 2, 84–111.
162. V. I. Arnold, Ordinary differential equations. Translated from the third Russian edition by Roger Cooke (Springer-Verlag, 1992).
163. V. I. Arnold, Trivium mathématique (in French) [Mathematical trivium] Gaz. Math. No. 52 (1992), 87–96.
164. V. I. Arnold and B. A. Khesin, Topological methods in hydrodynamics, in *Annual Review of Fluid Mechanics*, Vol. 24, pp. 145–166 (Annual Reviews, 1992).
165. V. I. Arnold, Springer numbers and Morsification spaces, *J. Algebraic Geom.* **1** (1992), No. 2, 197–214.
166. V. I. Arnold, Topological and ergodic properties of closed 1-forms with incommensurable periods (in Russian), *Functional. Anal. i Prilozhen.* **25** (1991), No. 2, 1–12; 96; translation in *Funct. Anal. Appl.* **25** (1991), No. 2, 81–90.
167. V. I. Arnold, Cardiac arrhythmias and circle mappings, *Chaos* **1** (1991), No. 1, 20–24.
168. V. I. Arnold, B. R. Vaĭnberg and L. D. Kudryavtsev *et al.*, Mikhail Vasil'evich Fedoryuk (in Russian), *Usp. Mat. Nauk* **46** (1991), No. 2(278), 205–207; translation in *Russ. Math. Surv.* **46** (1991), No. 2, 237–240.
169. V. I. Arnold, Kolmogorov's hydrodynamic attractors. Turbulence and stochastic processes: Kolmogorov's ideas 50 years on, *Proc. Roy. Soc. London Ser. A* **434** (1991), No. 1890, 19–22.

170. V. I. Arnold, The theory of singularities and its applications, in *Lezioni Fermiane* [Fermi Lectures] (Accademia Nazionale dei Lincei, 1991).
171. B. R. Vaĭnberg, V. A. Il'in, L. D. Kudryavtsev, V. B. Lidskiĭ, V. I. Arnold, V. P. Maslov, E. F. Mishchenko and V. Yu. Prikhod'ko, Mikhail Vasil'evich Fedoryuk (in Russian), *Differentsialnye Uravneniya* **27** (1991), No. 5, 914–915; translation in *Differential Equations* **27** (1991), No. 5, 639–640.
172. V. I. Arnold, Bernoulli–Euler updown numbers associated with function singularities, their combinatorics and arithmetics, *Duke Math. J.* **63** (1991), No. 2, 537–555.
173. V. I. Arnold, Gewöhnliche Differentialgleichungen (in German) [Ordinary differential equations] Translated from the third Russian edition by B. Mai and W. Plischke. Second edition. Hochschulbücher für Mathematik [University Books for Mathematics], Vol. 83 (Deutscher Verlag der Wissenschaften, 1991).
174. V. I. Arnold, Mathematical methods of classical mechanics. Translated from the 1974 Russian original by K. Vogtmann and A. Weinstein. Corrected reprint of the second (1989) edition. Graduate Texts in Mathematics, Vol. 60 (Springer-Verlag, 1991).
175. V. I. Arnold, Singularities of caustics and wave fronts, in Mathematics and Its Applications (Soviet Series), Vol. 62 (Kluwer, 1990).
176. V. I. Arnold, Dynamics of complexity of intersections, *Bol. Soc. Brasil. Mat. (N.S.)* **21** (1990), No. 1, 1–10.
177. V. I. Arnold, Contact geometry and wave propagation, *Enseign. Math.* (2) **36** (1990), No. 3–4, 215–266.
178. V. I. Arnold, Contact geometry: The geometrical method of Gibbs's thermodynamics, in *Proceedings of the Gibbs Symposium* (New Haven, CT, 1989), pp. 163–179 (Amer. Math. Soc., 1990).
179. V. I. Arnold, Teoriya katastrof (in Russian) [Catastrophe theory] Third edition. With an English summary (Nauka, 1990).
180. V. I. Arnold, Ten problems, in *Theory of Singularities and Its Applications*, pp. 1–8, Adv. Soviet Math., Vol. 1 (Amer. Math. Soc., 1990).
181. V. I. Arnold, Huygens and Barrow, Newton and Hooke, in *Pioneers in Mathematical Analysis and Catastrophe Theory from Evolvents to Quasicrystals*. Translated from the Russian by Eric J. F. Primrose (Birkhäuser, 1990).
182. V. I. Arnold, A. G. Vitushkin and E. A. Gorin *et al.*, Vyacheslav Alekseevich Oleĭnikov (in Russian), *Usp. Mat. Nauk* **45** (1990), No. 1(271), 163–165; translation in *Russ. Math. Surv.* **45** (1990), No. 1, 191–194.
183. V. I. Arnold, Dynamics of intersections, in *Analysis*, etc., pp. 77–84 (Academic Press, 1990).

184. V. I. Arnold and V. A. Vasil'ev, Addendum to: Newton's Principia read 300 years later. *Notices Amer. Math. Soc.* **37** (1990), No. 2, 144.
185. V. I. Arnold, V. A. Vasil'ev, V. V. Goryunov and O. V. Lyashko, Singularities. II. Classification and applications (in Russian), With the collaboration of B. Z. Shapiro. Itogi Nauki i Tekhniki, Current Problems in Mathematics. Fundamental Directions, Vol. 39, pp. 5–256 (Akad. Nauk SSSR, 1989).
186. V. I. Arnold, Matematicheskie metody klassicheskoĭ mekhaniki (in Russian) [Mathematical methods of classical mechanics] Third edition (Nauka, 1989).
187. V. I. Arnold, Contact geometry and wave propagation, in Monographies de L'Enseignement Mathématique [Monographs of L'Enseignement Mathématique], Vol. 34. Série des Conférences de l'Union Mathématique Internationale [Lecture Series of the International Mathematics Union], Vol. 9 (L'Enseignement Mathématique, 1989).
188. V. I. Arnold, Bifurcations and singularities in mathematics and mechanics, in *Theoretical and Applied Mechanics* (Grenoble, 1988), pp. 1–25 (North-Holland, 1989).
189. V. I. Arnold, Spaces of functions with moderate singularities (in Russian), *Functional. Anal. i Prilozhen.* **23** (1989), No. 3, 1–10; 96; translation in *Funct. Anal. Appl.* **23** (1989), No. 3, 169–177.
190. V. I. Arnold and V. A. Vasilév, Newton's Principia read 300 years later, *Notices Amer. Math. Soc.* **36** (1989), No. 9, 1148–1154.
191. V. I. Arnold, M. I. Vishik, Yu. S. Il'yashenko, A. S. Kalashnikov, V. A. Kondrat'ev, S. N. Kruzhkov, E. M. Landis, V. M. Millionshchikov, O. A. Oleĭnik, A. F. Filippov and M. A. Shubin, Some unsolved problems in the theory of differential equations and mathematical physics (in Russian), *Usp. Mat. Nauk* **44** (1989), No. 4(268), 191–202; translation in *Russ. Math. Surv.* **44** (1989), No. 4, 157–171.
192. V. I. Arnold, A-graded algebras and continued fractions, *Comm. Pure Appl. Math.* **42** (1989), No. 7, 993–1000.
193. V. I. Arnold, Gyuĭgens i Barrou, N'yuton i Guk (in Russian) [Huygens and Barrow, Newton and Hooke], Pervye shagi matematicheskogo analiza i teorii katastrof, ot èvolvent do kvazikristallov. [First steps in mathematical analysis and catastrophe theory, from evolutes to quasicrystals] With an English summary. *Sovremennaya Matematika dlya Studentov* [Contemporary Mathematics for Students], Vol. 1 (Nauka, 1989).
194. V. I. Arnold, O. Ya. Viro and E. A. Leontovich-Andronova *et al.*, Dmitriĭ Andreevich Gudkov (on the occasion of his seventieth birthday) (in

Russian), *Usp. Mat. Nauk* **44** (1989), No. 1(265), 223–225; translation in *Russ. Math. Surv.* **44** (1989), No. 1, 271–273.

195. V. I. Arnold, Mathematical methods of classical mechanics. Translated from the Russian by K. Vogtmann and A. Weinstein. Second edition. Graduate Texts in Mathematics, Vol. 60 (Springer-Verlag, 1989).
196. V. I. Arnold, Topological proof of the transcendence of the abelian integrals in Newton's Principia (in Russian), Istor.-Mat. Issled. No. 31 (1989), 7–17.
197. V. I. Arnold, V. A. Vasil'ev, V. V. Goryunov and O. V. Lyashko, Singularities. I, in *Local and Global Theory*. Current Problems in Mathematics. Fundamental Directions, Vol. 6, pp. 5–257 (Akad. Nauk SSSR, 1988).
198. V. I. Arnold, On the interior scattering of waves, defined by hyperbolic variational principles, *J. Geom. Phys.* **5** (1988), No. 3, 305–315.
199. V. I. Arnold, Mathematische Methoden der klassischen Mechanik (in German) [Mathematical methods of classical mechanics] Translated from the second Russian edition by Peter Möbius (VEB Deutscher Verlag der Wissenschaften, 1988).
200. V. I. Arnold, *Équations Différentielles Ordinaires* (in French) [*Ordinary Differential Equations*] Translated from the Russian by Djilali Embarek. Fourth edition, *Traduit du Russe: Mathématiques*. [Translations of Russian Works: Mathematics] (Mir, 1988).
201. V. I. Arnold, L. M. Barkov, S. T. Belyaev, G. I. Dimov, B. B. Kadomtsev, E. P. Kruglyakov, I. N. Meshkov, D. D. Ryutov, V. A. Sidorov and A. N. Skrinskiĭ, Boris Valerianovich Chirikov (on his sixtieth birthday), *Sov. Phys. Usp.* **31** (1988), No. 7, 682–683 (1989), translated from *Usp. Fiz. Nauk* **155** (1988), No. 3, 543–544.
202. V. I. Arnold, Remarks on quasicrystallic symmetries, Progress in chaotic dynamics, *Phys. D* **33** (1988), No. 1–3, 21–25.
203. V. I. Arnold, Some thoughts about Andreĭ Nikolaevich Kolmogorov (in Russian), *Usp. Mat. Nauk* **43** (1988), No. 6(264), 37; translation in *Russ. Math. Surv.* **43** (1988), No. 6, 43–44.
204. V. I. Arnold, The branched covering CP2→S4, hyperbolicity and projective topology (in Russian), *Sibirsk. Mat. Zh.* **29** (1988), No. 5, 36–47; 237; translation in *Siberian Math. J.* **29** (1988), No. 5, 717–726.
205. V. I. Arnold and Yu. S. Ilyashenko, *Ordinary Differential Equations* [Current Problems in Mathematics. Fundamental Directions, Vol. 1, pp. 7–149 (Akad. Nauk SSSR, 1985)]. Translated from the Russian by E. R. Dawson and D. O'Shea. Encyclopaedia Math. Sci., Vol. 1, Dynamical systems, I, pp. 1–148 (Springer, 1988).

206. V. I. Arnold, On some problems in symplectic topology, in *Topology and Geometry — Rohlin Seminar*, pp. 1–5, Lecture Notes in Math., Vol. 1346 (Springer, 1988).
207. V. I. Arnold, Three hundred years of mathematical natural sciences and celestial mechanics (in Bulgarian), translated from the Russian by G. I. Chobanov, *Fiz.-Mat. Spis. Blgar. Akad. Nauk.* **30**(63) (1988), No. 3, 181–190.
208. V. I. Arnold, S. M. Guseĭn-Zade and A. N. Varchenko, Singularities of differentiable maps. Vol. II. Monodromy and asymptotics of integrals. Translated from the Russian by Hugh Porteous. Translation revised by the authors and James Montaldi. Monographs in Mathematics, Vol. 83 (Birkhäuser, 1988).
209. V. I. Arnold, Contact structure, relaxation oscillations and singular points of implicit differential equations, in *Global Analysis — Studies and Applications*, III, pp. 173–179, Lecture Notes in Math., Vol. 1334 (Springer, 1988).
210. V. I. Arnold, Surfaces defined by hyperbolic equations (in Russian), *Mat. Zametki* **44** (1988), No. 1, 3–18; 154; translation in *Math. Notes* **44** (1988), No. 1–2, 489–497.
211. V. I. Arnold, *Geometrical Methods in the Theory of Ordinary Differential Equations*. Translated from the Russian by Joseph Szücs [József M. Szűcs]. Second edition. Grundlehren der Mathematischen Wissenschaften [Fundamental Principles of Mathematical Sciences], Vol. 250 (Springer-Verlag, 1988).
212. V. I. Arnold, V. V. Kozlov and A. I. Neĭshtadt, Dynamical Systems. III. Translated from the Russian by A. Iacob. Encyclopaedia of Mathematical Sciences, Vol. 3 (Springer-Verlag, 1988).
213. V. I. Arnold, The tercentennial of mathematical sciences and celestial mechanics (in Russian), *Priroda* (1987), No. 8, 5–15.
214. V. I. Arnold, Remarks on Poisson structures on a plane and on other powers of volume elements (in Russian), *Tr. Sem. Petrovsk.* **12** (1987), 37–46; 242; translation in *J. Sov. Math.* **47** (1989), No. 3, 2509–2516.
215. V. I. Arnold, M. I. Vishik, Yu. V. Egorov, A. S. Kalashnikov, S. P. Novikov and S. L. Sobolev, Olga Arsenevna Oleĭnik (on the occasion of her sixtieth birthday) (in Russian), *Tr. Sem. Petrovsk.* No. 12 (1987).
216. V. I. Arnold, Contact structure, relaxation oscillations and singular points of implicit differential equations (in Russian), Geometry and the theory of singularities in nonlinear equations, pp. 3–8; 182, Novoe Global. Anal., Voronezh. Gos. Univ., Voronezh, 1987.

217. V. I. Arnold, *Geometrische Methoden in der Theorie der Gewöhnlichen Differentialgleichungen* (in German) [*Geometrical Methods in the Theory of Ordinary Differential Equations*] Translated from the Russian by Ernst Günter Giessmann, Bernd Graw and Horst Theel (Birkhäuser, 1987).
218. V. I. Arnold, First steps of symplectic topology, in *VIII International Congress on Mathematical Physics* (Marseille, 1986), pp. 1–16 (World Scientific, 1987).
219. V. I. Arnold, *Geometrische Methoden in der Theorie der Gewöhnlichen Differentialgleichungen* (in German) [*Geometrical Methods in the Theory of Ordinary Differential Equations*] Translated from the Russian by Ernst Günter Gießmann, Bernd Graw and Horst Theel. Hochschulbücher für Mathematik [University Books for Mathematics], Vol. 90 (VEB Deutscher Verlag der Wissenschaften, 1987).
220. V. I. Arnold, Convex hulls and the increase of efficiency of systems under impulse loading (in Russian), *Sibirsk. Mat. Zh.* **28** (1987), No. 4, 29–31; 224.
221. V. I. Arnold, Catastrophe theory (in Russian), Current Problems in Mathematics. Fundamental Directions, Vol. 5, pp, 219–277 (Itogi Nauki i Tekhniki, Akad. Nauk SSSR, 1986).
222. V. I. Arnold, V. S. Afrajmovich, Yu. S. Ilyashenko and L. P. Shil'nikov, *Bifurcation Theory* (in Russian), Current Problems in Mathematics. Fundamental Directions, Vol. 5, pp. 5–218 (Itogi Nauki i Tekhniki, Akad. Nauk SSSR, 1986).
223. V. I. Arnold, The asymptotic Hopf invariant and its applications, *Selected Translations. Selecta Math. Sov.* **5** (1986), No. 4, 327–345.
224. V. I. Arnold, The first steps of symplectic topology (in Russian), *Usp. Mat. Nauk* **41** (1986), No. 6(252), 3–18; 229.
225. V. I. Arnold, *Teória Katastrof* (in Slovak) [*Catastrophe Theory*] Translated from the second Russian edition and with a preface by Jaroslav Smítal. Edícia Matematicko-Fyzikálnej Literatúry. [Publications in Mathematics and Physics] (Vydavatelstvo Technickej a Ekonomickej Literatúry, 1986).
226. V. I. Arnold, A. M. Vershik, O. Ya. Viro, A. N. Kolmogorov, S. P. Novikov, Ya. G. Sinaĭ and D. B. Fuks, Vladimir Abramovich Rokhlin (in Russian), *Usp. Mat. Nauk* **41** (1986), No. 3(249), 159–163.
227. V. I. Arnold, Hyperbolic polynomials and Vandermonde mappings (in Russian), *Functional. Anal. i Prilozhen.* **20** (1986), No. 2, 52–53.
228. V. I. Arnold, *Catastrophe Theory*. Second edition. Translated from the Russian by G. S. Wassermann. Based on a translation by R. K. Thomas (Springer-Verlag, 1986).

229. V. I. Arnold, Matematicheski Metodi na Klasicheskata Mekhanika (in Bulgarian) [*Mathematical Methods of Classical Mechanics*] Second edition. Translated from the Russian by Ivan Dimovski (Nauka i Izkustvo, 1985).
230. V. I. Arnold and A. B. Givental, *Symplectic Geometry* (in Russian), Current Problems in Mathematics. Fundamental Directions, Vol. 4, pp. 5–139; 291, Itogi Nauki i Tekhniki (Akad. Nauk SSSR, 1985).
231. V. I. Arnold, V. V. Kozlov and A. I. Neĭshtadt, Sovremennye Problemy Matematiki. Fundamentalnye Napravleniya, Tome 3 (in Russian) [Current Problems in Mathematics. Fundamental Directions, Vol. 3] Dinamicheskie Sistemy, Vol. 3. [Dynamical systems, Vol. 3] *Matematicheskie Aspekty Klassicheskoĭ i Nebesnoĭ Mekhaniki* [*Mathematical Aspects of Classical and Celestial Mechanics*] Edited by R. V. Gamkrelidze, Itogi Nauki i Tekhniki. [Progress in Science and Technology] (Akad. Nauk SSSR, 1985).
232. V. I. Arnold, *Equações Diferenciais Ordinárias* (in Portuguese) [Ordinary differential equations] Translated from the Russian by M. Dombrovsky (Mir, 1985).
233. V. I. Arnold and Yu. S. Ilyashenko, *Ordinary Differential Equations* (in Russian), Current Problems in Mathematics. Fundamental Directions, Vol. 1, pp. 7–149; 244 (Akad. Nauk SSSR, 1985).
234. V. I. Arnold, Sturm theorems and symplectic geometry (in Russian), *Functional. Anal. i Prilozhen.* **19** (1985), No. 4, 1–10.
235. V. I. Arnold, M. I. Vishik, I. M. Gel'fand, Yu. V. Egorov, A. S. Kalashnikov, A. N. Kolmogorov, S. P. Novikov and S. L. Sobolev, Olga Arsenevna Oleĭnik (on the occasion of her sixtieth birthday) (in Russian), *Usp. Mat. Nauk* **40** (1985), No. 5(245), 279–293.
236. V. I. Arnold, Singularities, bifurcations and catastrophes (in Bulgarian), translated from the Russian by E. Khorozov, *Fiz.-Mat. Spis. Blgar. Akad. Nauk.* **27**(60) (1985), No. 1, 25–48.
237. V. I. Arnold, S. M. Guseĭn-Zade and A. N. Varchenko, *Singularities of Differentiable Maps*, Vol. I. The Classification of Critical Points, Caustics and Wave Fronts. Translated from the Russian by Ian Porteous and Mark Reynolds. Monographs in Mathematics, Vol. 82 (Birkhäuser, 1985).
238. V. I. Arnold, Chapitres supplémentaires de la théorie des équations différentielles ordinaires (in French) [Supplementary chapters to the theory of ordinary differential equations] Translated from the Russian by Djilali Embarek. Reprint of the 1980 edition (Mir, 1984).
239. V. I. Arnold, Evolution of a magnetic field under the action of drift and diffusion (in Russian), in Some Problems in Modern Analysis, pp. 8–21 (Moskov. Gos. Univ., 1984).

240. V. I. Arnold, Exponential scattering of trajectories and its hydrodynamic applications (in Russian), in *N. E. Kochin and the Development of Mechanics*, pp. 185–193; 254 (Nauka, 1984).
241. V. I. Arnold, Reversible systems, in *Nonlinear and Turbulent Processes in Physics*, Vol. 3 (Kiev, 1983), pp. 1161–1174 (Harwood Academic, 1984).
242. V. I. Arnold, Singularities of ray systems, in *Proceedings of the International Congress of Mathematicians*, Vols. 1, 2 (Warsaw, 1983), pp. 27–49 (PWN, 1984).
243. V. I. Arnold, *Obyknovennye Differentsialnye Uravneniya* (in Russian) [*Ordinary Differential Equations*] Third edition (Nauka, 1984).
244. V. I. Arnold, A. N. Varchenko, A. B. Givental and A. G. Khovanskiĭ, *Singularities of Functions, Wave Fronts, Caustics and Multidimensional Integrals*. Translated from the Russian. Sov. Sci. Rev. Sect. C Math. Phys. Rev., 4, Mathematical Physics Reviews, Vol. 4, pp. 1–92 (Harwood Academic, 1984).
245. V. I. Arnold, G. I. Barenblatt, V. L. Ginzburg, B. B. Kadomtsev, P. L. Kapitsa, L. B. Okun, L. P. Pitaevskiĭ, R. Z. Sagdeev, R. A. Syunyaev, L. D. Faddeev, V. Ya. Frenkel and Yu. B. Khariton, Yakov Borisovich Zeldovich (on his seventieth birthday) (in Russian), *Sov. Phys. Usp.* **27** (1984), No. 3, 230–232; translated from *Usp. Fiz. Nauk* **142** (1984), No. 3, 531–532.
246. V. I. Arnold, A. N. Varchenko and S. M. Guseĭn-Zade, *Osobennosti Differentsiruemykh Otobrazheniĭ*. II (in Russian) [*Singularities of Differentiable Mappings*. II] Monodromiya i asimptotiki integralov [Monodromy and the asymptotic behavior of integrals] (Nauka, 1984).
247. V. I. Arnold, Vanishing inflections (in Russian), *Functional. Anal. i Prilozhen.* **18** (1984), No. 2, 51–52.
248. V. I. Arnold, Some remarks on elliptic coordinates (in Russian), Differential geometry, Lie groups and mechanics, VI, *Zap. Nauchn. Sem. Leningrad. Otdel. Mat. Inst. Steklov.* (LOMI) **133** (1984), 38–50.
249. V. I. Arnold, *Catastrophe Theory*. Translated from the Russian by R. K. Thomas (Springer-Verlag, 1984).
250. V. I. Arnold, *Problems on the Frontier of Mathematical Research* (in Russian), Outlines of the development of mathematics in the USSR, pp. 421–426 (Naukova Dumka, 1983).
251. V. I. Arnold, Singularities, bifurcations, and catastrophes (in Russian), *Sov. Phys. Usp.* **26** (1983), No. 12, 1025–1037 (1984), translated from *Usp. Fiz. Nauk* **141** (1983), No. 4, 569–590.
252. V. I. Arnold, Singularities of ray systems (in Russian), *Usp. Mat. Nauk* **38** (1983), No. 2(230), 77–147.

253. V. I. Arnold, Singularities in the calculus of variations (in Russian), Current Problems in Mathematics, Vol. 22, pp. 3–55 (Akad. Nauk SSSR, 1983).
254. V. I. Arnold, *Teoria Równań Różniczkowych* (in Polish) [*Theory of Differential Equations*] Translated from the Russian by Maciej Wojtkowski (PWN, 1983).
255. V. I. Arnold, *Some Open Problems in the Theory of Singularities*. Translated from the Russian. Proc. Sympos. Pure Math., Vol. 40, Singularities, Part 1 (Arcata, Calif., 1981), pp. 57–69 (Amer. Math. Soc., 1983).
256. V. I. Arnold, Some algebro-geometrical aspects of the Newton attraction theory, in *Arithmetic and Geometry*, Vol. II, pp. 1–3, Progr. Math., Vol. 36 (Birkhäuser, 1983).
257. V. I. Arnold, Remarks on perturbation theory for problems of Mathieu type (in Russian), *Usp. Mat. Nauk* **38** (1983), No. 4(232), 189–203.
258. V. I. Arnold, *Geometrical Methods in the Theory of Ordinary Differential Equations*. Translated from the Russian by Joseph Szücs. Translation edited by Mark Levi. Grundlehren der Mathematischen Wissenschaften [Fundamental Principles of Mathematical Science], Vol. 250 (Springer-Verlag, 1983).
259. V. I. Arnold, The Newton potential of hyperbolic layers (in Russian), *Tr. Tbiliss. Univ.* **232/233** (1982), 23–29.
260. V. I. Arnold, A. N. Varchenko and S. M. Guseĭn-Zade, *Osobennosti Differentsiruemykh Otobrazheniĭ* (in Russian) [*Singularities of Differentiable Mappings*] Klassifikatsiya kriticheskikh tochek, kaustik i volnovykh frontov. [Classification of critical points, caustics and wave fronts] (Nauka, 1982).
261. V. I. Arnold, Reconstructions of singularities of potential flows in a collision-free medium and caustic metamorphoses in three-dimensional space (in Russian), *Tr. Sem. Petrovsk.* No. 8 (1982), 21–57.
262. V. I. Arnold, Ya. B. Zeldovich, A. A. Ruzmaĭkin and D. D. Sokolov, Stationary magnetic field in a periodic flow (in Russian), *Dokl. Akad. Nauk SSSR* **266** (1982), No. 6, 1357–1361.
263. V. I. Arnold, Some remarks on the antidynamo theorem (in Russian), *Vestnik Moskov. Univ. Ser. I Mat. Mekh.* (1982), No. 6, 50–57; 121.
264. V. I. Arnold, Singularities of Legendre varieties, of evolvents and of fronts at an obstacle, *Ergodic Theory Dynam. Systems* **2** (1982), No. 3–4, 301–309.
265. V. I. Arnold, On some problems in singularity theory, *Proc. Indian Acad. Sci. Math. Sci.* **90** (1981), No. 1, 1–9.

266. V. I. Arnold, Ya. B. Zeldovich and S. F. Shandarin, The large-scale structure of the universe. I. General properties. One-dimensional and two-dimensional models (in Russian), *Akad. Nauk SSSR Inst. Prikl. Mat.* Preprint 1981, No. 100.
267. D. V. Anosov, V. I. Arnold, M. I. Zelikin, A. N. Kolmogorov, O. V. Lokutsievskiĭ, Yu. S. Osipov, Ya. G. Sinaĭ, V. M. Tikhomirov and M. V. Yakobson, Vladimir Mikhaĭlovich Alekseev. Obituary (in Russian), *Usp. Mat. Nauk* **36** (1981), No. 4(220).
268. V. I. Arnold, *Singularity Theory, Selected Papers*. Translated from the Russian. With an introduction by C. T. C. Wall. London Mathematical Society Lecture Note Series, Vol. 53 (Cambridge Univ. Press, 1981).
269. V. I. Arnold, Lagrangian manifolds with singularities, asymptotic rays and the unfurled swallowtail (in Russian), *Functional. Anal. i Prilozhen.* **15** (1981), No. 4, 1–14.
270. V. I. Arnold, *Metody Matematyczne Mechaniki Klasycznej* (in Polish) [*Mathematical Methods of Classical Mechanics*] Translated from the Russian by Piotr Kucharczyk (PWN, 1981).
271. V. I. Arnold, On some problems in singularity theory, in *Geometry and Analysis*, pp. 1–9 (Indian Acad. Sci., 1980).
272. V. I. Arnold, Chapitres supplémentaires de la *Théorie des Équations Différentielles Ordinaires* (in French) [Supplementary chapters to the *Theory of Ordinary Differential Equations*] Translated from the Russian by Djilali Embarek (Mir, 1980).
273. V. I. Arnold, Lagrange and Legendre cobordisms. II (in Russian), *Functional. Anal. i Prilozhen.* **14** (1980), No. 4, 8–17; 95.
274. V. I. Arnold, Lagrange and Legendre cobordisms. I (in Russian), *Functional. Anal. i Prilozhen.* **14** (1980), No. 3, 1–13; 96.
275. V. I. Arnold, Statistics of integral convex polygons (in Russian), *Functional. Anal. i Prilozhen.* **14** (1980), No. 2, 1–3.
276. V. I. Arnold, *Gewöhnliche Differentialgleichungen* (in German) [*Ordinary Differential Equations*] Translated from the Russian by Brigitte Mai (Springer-Verlag, 1980).
277. V. I. Arnold, Steady oscillations with potential energy harmonic in space and periodic in time (in Russian), *J. Appl. Math. Mech.* **43** (1979), No. 2, 360–363.
278. V. I. Arnold and O. A. Oleĭnik, Topology of real algebraic varieties (in Russian), *Vestnik Moskov. Univ. Ser. I Mat. Mekh.* (1979), No. 6, 7–17.
279. V. I. Arnold, Indexes of singular points of 1-forms on manifolds with boundary, convolutions of invariants of groups generated by reflections, and

singular projections of smooth surfaces (in Russian), *Usp. Mat. Nauk* **34** (1979), No. 2(206), 3–38.

280. V. I. Arnold, *Matematicheskie Metody Klassicheskoĭ Mekhaniki* (in Russian) [*Mathematical Methods of Classical Mechanics*] Second edition (Nauka, 1979).
281. V. I. Arnold, Critical points of functions on a manifold with boundary, the simple Lie groups Bk, Ck, F4 and singularities of evolutes (in Russian), *Usp. Mat. Nauk* **33** (1978), No. 5(203), 91–105; 237.
282. V. I. Arnold, *Dopolnitelye Glavy Teorii Obyknovennykh Differentsialnykh Uravneniĭ* (in Russian) [Supplementary chapters to the *Theory of Ordinary Differential Equations*] (Nauka, 1978).
283. V. I. Arnold, *Ordinary Differential Equations*. Translated from the Russian and edited by Richard A. Silverman (MIT Press, 1978).
284. V. I. Arnold, The index of a singular point of a vector field, the Petrovskiĭ–Oleĭnik inequalities, and mixed Hodge structures (in Russian), *Functional. Anal. i Prilozhen.* **12** (1978), No. 1, 1–14.
285. V. I. Arnold, *Mathematical Methods of Classical Mechanics*. Translated from the Russian by K. Vogtmann and A. Weinstein. Graduate Texts in Mathematics, Vol. 60 (Springer-Verlag, 1978).
286. V. I. Arnold, Correction: "Wave front evolution and equivariant Morse lemma" [*Comm. Pure Appl. Math.* **29** (1976), No. 6, 557–582], *Comm. Pure Appl. Math.* **30** (1977), No. 6, 823.
287. V. I. Arnold, Loss of stability of self-induced oscillations near resonance, and versal deformations of equivariant vector fields (in Russian), *Functional. Anal. i Prilozhen.* **11** (1977), No. 2, 1–10; 95.
288. V. I. Arnold, Some unsolved problems of singularity theory (in Russian), in *Theory of Cubature Formulas and the Application of Functional Analysis to Problems of Mathematical Physics*, pp. 5–15; 165, Trudy Sem. S. L. Soboleva, No. 1, 1976 (Akad. Nauk SSSR Sibirsk. Otdel., 1976).
289. V. I. Arnold, Spectral Sequences for the Reduction of Functions to Normal Forms (in Russian) (*Problems in Mechanics and Mathematical Physics*), pp. 7–20; 297 (Nauka, 1976).
290. V. I. Arnold, *Les Méthodes Mathématiques de la Mécanique Classique* (in French), Traduit du russe par Djilali Embarek (Mir, 1976).
291. V. I. Arnold, Local normal forms of functions, *Invent. Math.* **35** (1976), 87–109.
292. V. I. Arnold, Wave front evolution and equivariant Morse lemma, *Comm. Pure Appl. Math.* **29** (1976), No. 6, 557–582.

293. V. I. Arnold, Bifurcations of invariant manifolds of differential equations, and normal forms of neighborhoods of elliptic curves (in Russian), *Functional. Anal. i Prilozhen.* **10** (1976), No. 4, 1–12.
294. V. I. Arnold, *Równania Różniczkowe Zwyczajne* (in Polish) [*Ordinary Differential Equations*] Translated from the first Russian edition by Alicja Derkowska and Gabriel Derkowski (PWN, 1975).
295. V. I. Arnold, Critical points of smooth functions, in *Proceedings of the International Congress of Mathematicians* (Vancouver, B. C., 1974), Vol. 1, pp. 19–39 (Canad. Math. Congress, 1975).
296. V. I. Arnold, Critical points of smooth functions, and their normal forms (in Russian), *Usp. Mat. Nauk* **30** (1975), No. 5(185), 3–65.
297. V. I. Arnold, A spectral sequence for the reduction of functions to normal form (in Russian), *Functional. Anal. i Prilozhen.* **9** (1975), No. 3, 81–82.
298. V. I. Arnold, *Obyknovennye Differentsialnye Uravneniya* (in Russian) [*Ordinary Differential Equations*] Second edition (Nauka, 1975).
299. V. I. Arnold, Classification of two-modal critical points of functions (in Russian), *Functional. Anal. i Prilozhen.* **9** (1975), No. 1, 49–50.
300. V. I. Arnold, Normal forms of functions in the neighborhood of degenerate critical points (in Russian), Collection of articles dedicated to the memory of Ivan Georgievič Petrovskiĭ (1901–1973), I. *Usp. Mat. Nauk* **29** (1974), No. 2(176), 11–49.
301. V. I. Arnold, *Matematicheskie Metody Klassicheskoĭ Mekhaniki* (in Russian) [*Mathematical Methods of Classical Mechanics*] (Nauka, 1974).
302. V. I. Arnold, *Équations Différentielles Ordinaires* (in French), Champs de vecteurs, groupes à un paramètre, difféomorphismes, flots, systèmes linéaires, stabilités des positions d'équilibre, théorie des oscillations, équations différentielles sur les variétés. Traduit du russe par Djilali Embarek (Mir, 1974).
303. V. I. Arnold, Remarks on the method of stationary phase and on the Coxeter numbers (in Russian), *Usp. Mat. Nauk* **28** (1973), No. 5(173), 17–44.
304. V. I. Arnold, *Ordinary Differential Equations*. Translated from the Russian and edited by Richard A. Silverman (MIT Press, 1973).
305. V. I. Arnold, A classification of the unimodal critical points of functions (in Russian), *Functional. Anal. i Prilozhen.* **7** (1973), No. 3, 75–76.
306. V. I. Arnold, Versal families and bifurcations of differential equations (in Russian), in *Ninth Mathematical Summer School* (Kaciveli, 1971), pp. 42–49 (Akad. Nauk Ukrain. SSR, 1972).
307. V. I. Arnold, *Lectures on Bifurcations and Versal Families* (in Russian), A series of articles on the theory of singularities of smooth mappings, *Usp. Mat. Nauk* **27** (1972), No. 5(167), 119–184.

308. V. I. Arnold, Normal forms of functions near degenerate critical points, the Weyl groups $A_k$, $D_k$, $E_k$ and Lagrangian singularities (in Russian), *Functional. Anal. i Prilozhen.* **6** (1972), No. 4, 3–25.
309. V. I. Arnold, Integrals of rapidly oscillating functions, and singularities of the projections of Lagrangian manifolds (in Russian), *Functional. Anal. i Prilozhen.* **6** (1972), No. 3, 61–62.
310. V. I. Arnold, Modes and quasimodes (in Russian), *Functional. Anal. i Prilozhen.* **6** (1972), No. 2, 12–20.
311. V. I. Arnold, *Obyknovennye Differentsialnye Uravneniya* (in Russian) [*Ordinary Differential Equations*] (Nauka, 1971).
312. V. I. Arnold, Matrices depending on parameters (in Russian), *Usp. Mat. Nauk* **26** (1971), No. 2(158), 101–114.
313. V. I. Arnold, I. M. Gel'fand, Yu. I. Manin, B. G. Moĭšezon, S. P. Novikov and I. R. Šafarevič, Galina Nikolaevna Tjurina. Obituary (in Russian), *Usp. Mat. Nauk* **26** (1971), No. 1, 207–211.
314. P. S. Aleksandrov, V. I. Arnold, I. M. Gel'fand, A. N. Kolmogorov, S. P. Novikov and O. A. Oleĭnik, Ivan Georgievič Petrovskiĭ (on his seventieth birthday) (in Russian), *Usp. Mat. Nauk* **26** (1971), No. 2, 3–24.
315. V. I. Arnold, Distribution of ovals of the real plane of algebraic curves, of involutions of four-dimensional smooth manifolds, and the arithmetic of integer-valued quadratic forms (in Russian), *Functional. Anal. i Prilozhen.* **5** (1971), No. 3, 1–9.
316. V. I. Arnold, Topological invariants of algebraic functions. II (in Russian), *Functional. Anal. i Prilozhen.* **4** (1970), No. 2, 1–9.
317. V. I. Arnold, The cohomology classes of algebraic functions that are preserved under Tschirnhausen transformations (in Russian), *Functional. Anal. i Prilozhen.* **4** (1970), No. 1, 84–85.
318. V. I. Arnold, Local problems of analysis (in Russian), *Vestnik Moskov. Univ. Ser. I Mat. Meh.* **25** (1970), No. 2, 52–56.
319. V. I. Arnold, Certain topological invariants of algebrac functions (in Russian), *Tr. Moskov. Mat. Obšč.* **21** (1970), 27–46.
320. V. I. Arnold, Algebraic unsolvability of the problem of Ljapunov stability and the problem of the topological classification of the singular points of an analytic system of differential equations (in Russian), *Functional. Anal. i Prilozhen.* **4** (1970), No. 3, 1–9.
321. V. I. Arnold, Algebraic unsolvability of the problem of stability and the problem of the topological classification of the singular points of analytic systems of differential equations (in Russian), *Usp. Mat. Nauk* **25** (1970), No. 2(152), 265–266.

322. V. I. Arnold, *Lektsii po Obyknovennym Differentsialnym Uravneniyam*, Chast 3 (in Russian) [*Lectures on Ordinary Differential Equations*, Part 3] (Moskov. Gosudarstv. Univ. Meh.-Mat. Fakultet, 1969).
323. V. I. Arnold, *Lektsii po Teorii Obyknovennykh Differentsialnykh Uravneniĭ*, Chast 2 (in Russian) [*Lectures on the Theory of Ordinary Differential Equations*, Part 2] (Moskov. Gosudarstv. Univ., 1969).
324. V. I. Arnold, *Lektsii po Teorii Obyknovennykh Differentsialnykh Uravneniĭ*, Chast 1 (in Russian) [*Lectures on the Theory of Oordinary Differential Equations*, Part 1] (Moskov. Gosudarstv. Univ., 1969).
325. V. I. Arnold, The Hamiltonian nature of the Euler equations in the dynamics of a rigid body and of an ideal fluid (in Russian), *Usp. Mat. Nauk* **24** (1969), No. 3(147), 225–226.
326. V. I. Arnold, The one-dimensional cohomologies of the Lie algebra of divergence-free vector fields, and the winding numbers of dynamical systems (in Russian), *Functional. Anal. i Prilozhen.* **3** (1969), No. 4, 77–78.
327. V. I. Arnold, Remarks on singularities of finite codimension in complex dynamical systems (in Russian), *Functional. Anal. i Prilozhen.* **3** (1969), No. 1, 1–6.
328. V. I. Arnold, The cohomology ring of the group of dyed braids (in Russian), *Mat. Zametki* **5** (1969), 227–231.
329. V. I. Arnold, A remark on the branching of hyperelliptic integrals as functions of the parameters (in Russian), *Functional. Anal. i Prilozhen.* **2** (1968), No. 3, 1–3.
330. V. I. Arnold, A stability problem and ergodic properties of classical dynamical systems (in Russian), in *1968 Proc. Internat. Congr. Math.* (Moscow, 1966), pp. 387–392 (Mir, 1966).
331. V. I. Arnold, A letter to the editors (in Russian), *Usp. Mat. Nauk* **23** (1968), No. 6(144).
332. V. I. Arnold and A. Avez, *Ergodic Problems of Classical Mechanics*. Translated from the French by A. Avez (W. A. Benjamin, 1968).
333. V. I. Arnold, Braids of algebraic functions and cohomologies of swallowtails (in Russian), *Usp. Mat. Nauk* **23** (1968), No. 4(142), 247–248.
334. V. I. Arnold, Singularities of smooth mappings (in Russian), *Usp. Mat. Nauk* **23** (1968), No. 1, 3–44.
335. V. I. Arnold, A remark on the Weierstrass preparation theorem (in Russian), *Functional. Anal. i Prilozhen.* **1** (1967), No. 3, 1–8.
336. V. I. Arnold, Characteristic class entering in quantization conditions (in Russian), *Functional. Anal. i Prilozhen.* **1** (1967), 1–13.

337. V. I. Arnold and A. Avez, Problèmes ergodiques de la mécanique classique (in French), Monographies Internationales de Mathématiques Modernes, No. 9 (Gauthier-Villars, 1967).
338. V. I. Arnold, On the topology of three-dimensional steady flows of an ideal fluid, *Prikl. Mat. Meh.* **30**, 183–185 (Russian); translated as *J. Appl. Math. Mech.* **30** (1966), 223–226.
339. V. I. Arnold, An *a priori* estimate in the theory of hydrodynamic stability (in Russian), *Izv. Vysš. Učebn. Zaved. Matematika* **1966** (1966), No. 5(54), 3–5.
340. V. I. Arnold, Sur la géométrie différentielle des groupes de Lie de dimension infinie et ses applications à l'hydrodynamique des fluides parfaits (in French), *Ann. Inst. Fourier* (*Grenoble*) **16** (1966), fasc. 1, 319–361.
341. V. I. Arnold, Stability and instability in classical mechanics (in Russian), 1965 Second Math. Summer School, Part II, pp. 85–119.
342. V. I. Arnold, Sur une propriété topologique des applications globalement canoniques de la mécanique classique (in French), *C. R. Acad. Sci. Paris* **261** (1965), 3719–3722.
343. V. I. Arnold, Sur la topologie des écoulements stationnaires des fluides parfaits (in French), *C. R. Acad. Sci. Paris* **261** (1965), 17–20.
344. V. I. Arnold, On conditions for non-linear stability of plane stationary curvilinear flows of an ideal fluid (in Russian), *Dokl. Akad. Nauk SSSR* **162** (1965), 975–978.
345. V. I. Arnold, Applicability conditions and an error bound for the averaging method for systems in the process of evolution through a resonance (in Russian), *Dokl. Akad. Nauk SSSR* **161** (1965), 9–12.
346. V. I. Arnold, Sur la courbure de Riemann des groupes de difféomorphismes (in French), *C. R. Acad. Sci. Paris* **260** (1965), 5668–5671.
347. V. I. Arnold, Correction to V. Arnold's paper: "Small denominators. I." (in Russian), *Izv. Akad. Nauk SSSR Ser. Mat.* **28** (1964), 479–480.
348. V. I. Arnold, Instability of dynamical systems with several degrees of freedom (in Russian), *Dokl. Akad. Nauk SSSR* **156** (1964), 9–12.
349. V. I. Arnold, Small denominators and problems of stability of motion in classical and celestial mechanics (in Russian), *Usp. Mat. Nauk* **18** (1963), No. 6(114), 91–192.
350. V. I. Arnold, Proof of a theorem of A. N. Kolmogorov on the preservation of conditionally periodic motions under a small perturbation of the Hamiltonian (in Russian), *Usp. Mat. Nauk* **18** (1963), No. 5(113), 13–40.
351. V. I. Arnold, Representation of continuous functions of three variables by the superposition of continuous functions of two variables, *Amer. Math. Soc. Transl.* (2) **28** (1963), 61–147.

352. V. I. Arnold, On functions of three variables, *Amer. Math. Soc. Transl.* (2) **28** (1963), 51–54.
353. V. I. Arnold and A. L. Krylov, Uniform distribution of points on a sphere and certain ergodic properties of solutions of linear ordinary differential equations in a complex domain (in Russian), *Dokl. Akad. Nauk SSSR* **148** (1963), 9–12.
354. V. I. Arnold, A theorem of Liouville concerning integrable problems of dynamics (in Russian), *Sibirsk. Mat. Ž.* **4** (1963), 471–474.
355. V. I. Arnold, On the behavior of an adiabatic invariant under slow periodic variation of the Hamiltonian (in Russian), *Dokl. Akad. Nauk SSSR* **142** (1963), 758–761, translated as *Sov. Math. Dokl.* **3** (1962), 136–140.
356. V. I. Arnold and Ya. G. Sinai, On small perturbations of the automorphisms of a torus (in Russian), *Dokl. Akad. Nauk SSSR* **144** (1962), 695–698.
357. V. I. Arnold, On the classical perturbation theory and the stability problem of planetary systems (in Russian), *Dokl. Akad. Nauk SSSR* **145** (1962), 487–490.
358. V. I. Arnold, Letter to the editor (in Russian), *Mat. Sb. (N.S.)* **56**(98) (1962).
359. V. I. Arnold, Remarks on winding numbers (in Russian), *Sibirsk. Mat. Ž.* **2** (1961), 807–813.
360. V. I. Arnold, Some remarks on flows of line elements and frames (in Russian), *Dokl. Akad. Nauk SSSR* **138** (1961), 255–257.
361. V. I. Arnold, Small denominators. I. Mapping the circle onto itself (in Russian), *Izv. Akad. Nauk SSSR Ser. Mat.* **25** (1961), 21–86.
362. V. I. Arnold, Test for nomographability using Descartes' rectilinear abacus (in Russian), *Usp. Mat. Nauk* **16** (1961), No. 4(100), 133–135.
363. V. I. Arnold, Generation of quasi-periodic motion from a family of periodic motions (in Russian), *Dokl. Akad. Nauk SSSR* **138** (1961), 13–15.
364. V. I. Arnold, The stability of the equilibrium position of a Hamiltonian system of ordinary differential equations in the general elliptic case (in Russian), *Dokl. Akad. Nauk SSSR* **137** (1961), 255–257.
365. V. I. Arnold, The representation of functions of several variables (in Czech), *Pokroky Mat. Fys. Astronom.* **5** (1960), 399–416.
366. V. I. Arnold, Some questions on approximation and representation of functions (in Russian), in *1960 Proc. Internat. Congress Math.* (1958), pp. 339–348 (Cambridge Univ. Press, 1960).
367. V. I. Arnold, On the representation of continuous functions of three variables by superpositions of continuous functions of two variables (in Russian), *Mat. Sb. (N.S.)* **48**(90) (1959), 3–74.

368. V. I. Arnold, On functions of three variables (in Russian), *Dokl. Akad. Nauk SSSR* **114** (1957), 679–681.
369. V. I. Arnold, On the representability of a function of two variables in the form $\chi[\varphi(x)+\psi(y)]$ (in Russian), *Usp. Mat. Nauk* (*N.S.*) **12** (1957), No. 2(74), 119–121.

## Work related to V. I. Arnold

V. I. Arnold, Vladimir I. Arnold — Collected Works, Vol. I. *Representations of Functions, Celestial Mechanics and KAM Theory*, 1957–1965. Edited by A. B. Givental, B. A. Khesin, J. E. Marsden, A. N. Varchenko, V. A. Vassilev, O. Ya. Viro and V. M. Zakalyukin (Springer-Verlag, 2009).

# INSTABILITY OF DYNAMICAL SYSTEMS WITH SEVERAL DEGREES OF FREEDOM

V. I. ARNOL'D

1. Recent progress in perturbation enables us to find many conditionally periodic motions in every nonlinear dynamical system which is close to an integrable system (see [1,2]. The stability of all the motions of the system follows from these results only when the dimension of the phase space is $\leq 4$. The purpose of the present note is to give *an example* (3) *of a system with a* 5*-dimensional phase space which satisfies all the conditions of* [1,2] *but is nonstable.** The secular changes $I_2$ in the system (3) have the velocity $\exp(-1/\sqrt{\epsilon})$ and consequently cannot be dealt with by any approximation of the classical theory of perturbations.

We first introduce some definitions.

2. **The whiskered torus.** By a *torus* $T^k$ we shall mean a direct product of $k$ circumferences, which can be described by the $k$ angular coordinates $\phi = \phi_1, \cdots, \phi_k \pmod{2\pi}$. *A conditionally periodic motion with frequencies* $\omega$ is defined by the equations $\dot{\phi} = \omega = \text{const}$ (where $\Sigma\, n_i\omega_i \neq 0$ for integral $n_i$, $\Sigma\, n_i^2 \neq 0$). Assume that in the phase space of the dynamical system there is an invariant torus $T$ and on it a conditionally periodic motion. We shall call $T$ a *whiskered torus* if $T$ is a component of the intersection of two invariant open manifolds $Y^-$, $Y^+$ where all the trajectories on the *arriving whisker* $Y^-$ approach $T$ as $t \to +\infty$, and on the *departing whisker* $Y^+$ all the trajectories approach $T$ as $t \to -\infty$.

**Example 1.** By the *standard* whiskered torus we shall mean the torus $x = y = z = 0$ in the system

$$\dot{x} = \lambda x, \quad \dot{y} = -\mu y, \quad \dot{z} = 0, \quad \dot{\phi} = \omega \tag{1}$$

defined in the $(l_+ + l_- + l_0 + k)$-dimensional space $x, y, z, \phi$ ($\phi$ taken mod $2\pi$).

Of essential importance below is the concept of an *obstructing set.* Let $M$ be a smooth submanifold of the space $X$. The tangent plane to $M$ at the point $x$ will be denoted by $TM_x$. The manifold $N$ *complements* $M$ *at the point* $x \in M \cap N$, if $TM_x + TN_x = TX_x$. We shall say that the set $\Omega$ *obstructs the manifold* $M$ *at the point* $x \in M$ if every manifold $N$ which *complements* $M$ at $x$ is intersected by $\Omega$.

**Example 2.** A spiral $\Omega$ which winds onto a closed curve $M$ obstructs the curve.**

Another example is given by the standard whiskered torus (1). Let $U$ be a neighborhood of the point $\xi$ of an arriving whisker $x = z = 0$. By $\Omega = \bigcup_{t>0} U(t)$ we denote the set of all points of all the trajectories which begin in $U$. Then it is easy to prove the following theorem.

**Theorem 1.** *The set* $\Omega$ *obstructs the departing whisker* $y = z = 0$ *at an arbitrary point* $\eta$ *of the whisker.*

3. **The transition chain.** If the whiskered torus $T$ has the property that the images of an arbitrary

* In contradistinction to stability, nonstability is itself stable. I believe that the mechanism of "transition chains" which guarantees that nonstability in our example is also applicable to the general case (for example, to the problem of three bodies).

** The articles [3,4] are based on this fact.

neighborhood of an arbitrary point $\xi$ of one of its arriving whiskers obstruct the departing whisker at an arbitrary point $\eta$ of the latter, then the torus will be said to be a *transition torus*. By Theorem 1, *the standard torus* (1) *is a transition torus.*

Assume that the dynamical system with phase space $X$ has certain transition toruses $T_1, \ldots, T_s, \ldots$. These toruses will be said to form a *transition chain* if the departing whisker $Y_s^+$ of every preceding torus $T_s$ complements the arriving whisker of the following torus $Y_{s+1}^-$ at some point of their intersection $x_s \in Y_s^+ \cap Y_{s+1}^-$.

Let $T_1, \ldots, T_s, \ldots$ be a transition chain. Then the following theorem is easily proved.

**Theorem 2.** *An arbitrary neighborhood of the torus* $T_1$ *is connected with an arbitrary neighborhood of the torus* $T_s$ *of trajectories of the given dynamical system.*

Consequently, for the proof of nonstability it is sufficient to find a transition chain which connects distant toruses $T_1$, $T_s$. The search for whiskered toruses and especially the investigation of their intersections in the general problem of the theory of perturbations demands very complicated calculations. We will confine ourselves here to an example in which a specially chosen perturbation vanishes on the toruses $T_s$.

4. **A nonstable system.** We consider a system with two degrees of freedom which is periodic in the time $t$ with period $2\pi$. "The phase space" $I_1, I_2; \phi_1, \phi_2; t$ is the direct product of the planes $I_1, I_2$ with the three-dimensional torus $\phi_1, \phi_2, t \pmod{2\pi}$. *The Hamiltonian, depending on the parameters* $\epsilon, \mu$, *will have the form* $H = H_0 + \epsilon H_1$, *where**

$$H_0 = \tfrac{1}{2}(I_1^2 + I_2^2), \quad \varepsilon H_1 = \varepsilon(\cos\varphi_1 - 1)[1 + \mu\beta], \quad B = \sin\varphi_2 + \cos t. \tag{2}$$

*In other words, we consider the system of differential equations*

$$\varphi_1 = I_1, \varphi_2 = I_2; \quad I_1 = \varepsilon \sin\varphi_1 [1 + \mu B], \ I_2 = \varepsilon(1 - \cos\varphi_1)\mu\cos\varphi_2; \\ B = \sin\varphi_2 + \cos t. \tag{3}$$

We first investigate the nonperturbed system ($\epsilon = 0$). Every three-dimensional torus $I_1 = \omega_1$, $I_2 = \omega_2$ is invariant. On it the three-frequency motion $\dot\phi_1 = \omega_1$, $\dot\phi_2 = \omega_2$, $\dot t = 1$ takes place. A torus is said to be *nonresonant*, if the frequencies on it are independent of one another (i.e., $n_1\omega_1 + n_2\omega_2 + n_0 \neq 0$ for integers in $n \neq 0$). The equation $I_1 = 0$ determines a family of resonant toruses (since $\omega_1 = 0$).

We now consider a perturbed system: assume $0 < \epsilon\mu \ll \epsilon \ll 1$. In [1,2] it is proved that for the majority of nonresonant initial conditions the quantities $I_1(t)$, $I_2(t)$ will change little in the course of the whole infinite interval of time $-\infty < t < +\infty$. It turns out, however, that *close to the resonant manifold* $I_1 = 0$ *there is a zone of nonstability*. More precisely, we have the following theorem.

**Theorem 3.** *Assume* $0 < A < B$. *For every* $\epsilon > 0$ *we can find a* $\mu_0 > 0$ *such that for* $0 < \mu < \mu_0$ *the system* (3) *is nonstable: there exists a trajectory which connects the region* $I_2 < A$ *with the region* $I_2 > B$.

5. **A proof of nonstability.** Choose a fixed $\epsilon > 0$.

A. First assume $\mu = 0$. Then the variables can be separated:

$$H = H^{(1)} + H^{(2)}, \quad H^{(1)} = \tfrac{1}{2} I_1^2 + \varepsilon(\cos\varphi_1 - 1), \quad H^{(2)} = \tfrac{1}{2} I_2^2. \tag{4}$$

Thus $\dot I_2 = 0$, $\dot\phi_2 = I_2 = \omega = \text{const}$, and the change of $I_1, \phi_1$ with time will be described by the

* It is easy to construct an actual mechanical system with the Hamiltonian (2).

Hamiltonian of the ordinary pendulum $H^{(1)}$. Let the number $\omega$ be irrational. Then the following assertion is easily proved.

**Assertion A.** *The manifold $T_\omega$ defined by the equations $I_1 = \phi_1 = I_2 - \omega = 0$ is a two-dimensional whiskered torus of the system* (3). *The whiskers are three-dimensional and have the equations*

$$H^{(1)} = 0, \quad H^{(2)} = {}^1/_2\,\omega^2 \quad \textit{or} \quad I_1 = \pm 2\sqrt{\varepsilon}\sin\frac{\varphi_1}{2}, \quad I_2 = \omega. \tag{5}$$

*The whiskers are supplemented by the asymptotic trajectories*

$$I_1(t) = \pm 2\sqrt{\varepsilon}\,\mathrm{ch}^{-1}\tau, \quad \varphi_1(t) = \pm\,\mathrm{arc\,ctg}\,(-\,\mathrm{sh}\,\tau), \quad \varphi_2(t) = \varphi_2^0 + \omega(t - t^0), \tag{6}$$

*where* $\tau = \sqrt{\epsilon}\,(t - t^0)$, $I_1(t^0) = \pm 2\sqrt{\epsilon}$, $\phi_1(t^0) = \pm\pi$, $\phi_2(t^0) = \phi_2^0$.

Thus for $t \to +\infty$ the point of the departing whisker of the torus $T_\omega$ falls again on the same torus $T_\omega$. In other words, the departing whisker forms one manifold together with the arriving whisker. Of course, for $\mu \neq 0$ this manifold splits into two whiskers, which intersect each other.* We shall see that (in distinction to the separatrices of systems with a phase space of dimension $\leq 4$, considered in [5,6] *these whiskers also intersect the whiskers of neighboring toruses $T_{\omega'}$.*

B. Now suppose $\mu \neq 0$. From (3) it is evident that the toruses $T_\omega$ remain invariant for all $\mu$. Assume that $\omega$ is irrational. By the standard method of contractive mappings we can prove the following assertion.**

**Assertion B.** *The manifold $T_\omega$ is a whiskered transition torus of the system* (3), *if $\mu$ is sufficiently small.*

Assume $\omega_1 < A < B < \omega_s$. For the proof of Theorem 3 it is sufficient to construct a transition chain of toruses $T_{\omega_1}, \cdots, T_{\omega_s}$ and to make use of Theorem 2. The construction of such a chain is based on a study of the perturbation of the whiskers (5) for small $\mu$. It turns out that the following lemma holds.

**Lemma 1.** *Assume $A < \omega < B$. Then the departing whisker $Y_\omega^+$ of the torus $T_\omega$ intersects with the arriving whiskers $Y_{\omega'}^-$ of all toruses $T_{\omega'}$, which are sufficiently close that $|\omega - \omega'| \leq \kappa$ (where $\kappa = \kappa(\epsilon, \mu, A, B) > 0$).*

The proof of Lemma 1 requires certain calculations. The nonperturbed whiskers have the equation (5): $H^{(1)} = 0$, $H^{(2)} = \omega^2/2$, where $H^{(k)}$ are the function of (4). Assume $\alpha > 0$ (for example, $\alpha = \pi/2$). It is easy to show that for $|\phi_1| < 2\pi - \alpha$ the equations of the perturbed departing whisker $Y_\omega^+$ can be written in the form

$$H^{(1)} = \Delta_1^+(\varphi_1; \varphi_2, t; \omega); \quad H^{(2)} = {}^1/_2\,\omega^2 + \Delta_2^+(\varphi_1; \varphi_2, t; \omega), \tag{7}$$

where the functions $\Delta_k^+ = O(\mu)$ have the period $2\pi$ with respect to $\phi_2$, $t$ and are equal to 0 for $\phi_1 = 0$. In exactly the same way the arriving whisker $Y_{\omega'}^-$ for $|\phi_1 - 2\pi| < 2\pi - \alpha$ has equations

$$H^{(1)} = \Delta_1^-(\varphi_1; \varphi_2, t; \omega'), \quad H^{(2)} = {}^1/_2\omega'^2 + \Delta_2^-(\varphi_1; \varphi_2, t; \omega'). \tag{8}$$

The intersection of the whisker $Y_\omega^+$ and $Y_{\omega'}^-$ will be sought on the plane $\phi_1 = \pi$. In the notations of (7) and (8) Lemma 1 is an assertion concerning the solvability with respect to $\phi_2$, $t$ of a system of

---

* The splitting of separatrices was studied by Poincaré in the final chapter of "New methods" [5]. The investigations of Poincaré were recently continued by Mel'nikov [6].

** It is convenient to use the conical metric $\|f(x)\| = \max|x^{-1} f(x)|$.

equations

$$\Delta_1^+(\pi; \varphi_2, t; \omega) = \Delta_1^-(\pi; \varphi_2, t; \omega'),$$
$$1/_2\omega^2 + \Delta_2^+(\pi; \varphi_2, t; \omega) = \Delta_2^-(\pi; \varphi_2, t; \omega') + 1/_2\omega'^2. \tag{9}$$

The solvability of the system (9) can be deduced from the following approximate expressions for $\Delta_k^\pm$

Lemma 2 (compare [6]). *The perturbations of the whiskers are* $\Delta_k^\pm = \mu\delta_k^\pm + O(\mu^2)$, *where*

$$\mu\delta_k^\pm(\pi; \varphi_2^0, t^0; \omega) = \int_{\mp\infty}^{0} \{H, H^{(k)}\}\, d(t - t^0)|_{(6)} \tag{10}$$

*(the Poisson bracket is integrated along the nonperturbed trajectory* (6)).

For in fact, in accordance with the definitions (7), (8), the quantities $\Delta_k^\pm$ represent the increments of $H^{(k)}$ in the perturbed motion (3). The derivative of the function $H^{(k)}$, in view of the system of equations (3), is exactly the Poisson bracket $\{H, H^{(k)}\}$. Consequently, $\Delta_k^\pm$ *is exactly equal to the integrals* (10) *extended over the perturbed trajectories.* Thus we easily derive the estimate $\Delta_\mu^\pm - \mu\delta_k^\pm = O(\mu^2)$, which completes the proof of Lemma 2.

From Lemma 2 it is obvious that the solvability of system (9) depends basically on the solvability with respect to $\phi_2^0$, $t^0$ of the approximate system

$$\delta_1 = 0, \quad \mu\delta_2 = 1/_2(\omega^2 - \omega'^2), \tag{11}$$

where

$$\delta_k = \delta_k^+(\pi; \varphi_2^0, t^0, \omega) - \delta_k^-(\pi; \varphi_2^0, t^0, \omega) = \int_{-\infty}^{+\infty} \{H, H^{(k)}\}\, d(t - t^0)|_{(6)}. \tag{12}$$

An easy calculation, based on formulas (2)–(6), gives the result

$$\delta_1 = -2\varepsilon \int_{-\infty}^{+\infty} u\frac{\partial B}{\partial t}\,dt, \quad \delta_2 = 2\varepsilon\omega \int_{-\infty}^{+\infty} u\frac{\partial B}{\partial \varphi_2}\,dt, \tag{13}$$

where $u = \mathrm{ch}^{-2}\tau$, $\tau = \sqrt{\epsilon}\,(t - t^0)$, $B = B(\phi_2, t)$, $\phi_2 = \phi_2^0 + \omega(t - t^0)$. For $B = \sin\phi_2 + \cos t$ the integrals (13) involve* the residues

$$\delta_1 = 2\pi\left(\mathrm{sh}^{-1}\frac{\pi}{2\sqrt{\varepsilon}}\right)\sin t^0, \quad \delta_2 = 2\pi\omega^2\left(\mathrm{sh}^{-1}\frac{\omega\pi}{2\sqrt{\varepsilon}}\right)\cos\varphi_2^0. \tag{14}$$

Setting $t^0 = 0$ in (14) we see that the system (11) is solvable for

$$|\omega^2 - \omega'^2| < 4\pi\mu\omega^2\,\mathrm{sh}^{-1}\frac{\omega\pi}{2\sqrt{\varepsilon}} \asymp \mu e^{-1/\sqrt{\varepsilon}} \tag{15}$$

From Lemma 2 it now follows that for sufficiently small $\mu$ the system (9) is also solvable. From the inequality (15) it is easy to obtain, uniformly for $A < \omega < B$, the estimate $\max\limits_{\omega'} |\omega - \omega'| = \kappa(\omega)$ from below, as required in Lemma 1. Thus the proof of Lemma 1 is complete. It allows us to construct a chain of transitional toruses $T_{\omega_1}, \cdots, T_{\omega_s}$ $(\omega_1 < A < B < \omega_s)$. From the formula (14) it is clear that for sufficiently small $\mu$ this chain can be so chosen that the consecutive intersecting whiskers lie in general position and complement one another in the sense of §2. Then the chain $T_{\omega_1}, \cdots, T_{\omega_s}$ will be

* The analogous integrals in [6], on page 32, are wrongly calculated

a transition chain. The application of Theorem 2 to the transition chain $T_{\omega_1}, \cdots, T_{\omega_s}$ completes the proof of Theorem 3.

Moscow State University Received 30/JAN/64

## BIBLIOGRAPHY

[1] A. N. Kolmogorov, Dokl. Akad. Nauk SSSR 98 (1954), 527. MR 16, 924.

[2] V. I. Arnol'd, Uspehi Mat. Nauk 18 (1963), no. 5(113), 13, 91.

[3] K. A. Sitnikov, Dokl. Akad. Nauk SSSR 133 (1960), 303 = Soviet Physics Dokl. 5 (1961), 647. MR 23 #B435.

[4] An. M. Leontovič, Dokl. Akad. Nauk SSSR 145 (1962), 523 = Soviet Math. Dokl. 3 (1962), 1049. MR 25 #2287.

[5] H. Poincaré, *Les méthodes nouvelles de la mécanique céleste*, 3, Paris, 1899.

[6] V. K. Mel'nikov, Trudy Moskov. Mat. Obšč. 12 (1963), 3 = Trans. Moscow Math. Soc. 12 (1963) (to appear). MR 27 #5981.

Translated by:
S. H. Gould

# CHARACTERISTIC CLASS ENTERING IN QUANTIZATION CONDITIONS

V. I. Arnol'd

Recently V. P. Maslov gave a mathematically rigorous treatment of multidimensional asymptotic methods of "quasiclassical" type in the large, i.e., for any number of conjugate points [1, 2]. It turned out that there appeared in the asymptotic formulas certain integers, reflecting homological properties of curves on surfaces of the phase space and closely related to the Morse indexes of the corresponding variational problems. In particular, Maslov defined a one-dimensional class of integer-valued cohomologies whose values on the basis cycles enter into the so-called "quantization conditions."

In this note we give new formulas for the calculation of this class of cohomologies. This class is characteristic in the category of real vector bundles, whose complexification is trivial and trivialized, and also in certain wider categories.

## § 1. NOTATION

### 1.1. Phase Space

Phase space will be 2n-dimensional real arithmetic space

$$\mathbf{R}^{2n} = \{x\}, \quad x = q,\ p; \quad q = q_1, \ldots, q_n; \quad p = p_1, \ldots, p_n.$$

In $\mathbf{R}^{2n}$ we shall consider the following three structures:

1. The Euclidean structure, given by the scalar quadratic

$$(x,\ x) = p^2 + q^2.$$

2. The complex structure, given by the operator

$$I\colon\ \mathbf{R}^{2n} \to \mathbf{R}^{2n},\ I(p,\ q) = (-q,\ p); \quad z = p + iq, \quad \mathbf{C}^n = \{z\}.$$

3. The simplectic structure, given by the skew-scalar product

$$[x,\ y] = (Ix,\ y) = -[y,\ x]. \tag{1}$$

The automorphism groups of $\mathbf{R}^{2n}$ preserving these structures are called the orthogonal group O(2n), the complex linear group GL(n, C), and the simplectic group Sp(n), respectively. From (1) there follows

LEMMA 1.1. An automorphism preserving two of these structures preserves the third also, so that

$$\mathbf{O}(2n) \cap \mathbf{GL}(n,\ \mathbf{C}) = \mathbf{GL}(n,\ \mathbf{C}) \cap \mathbf{Sp}(n) = \mathbf{Sp}(n) \cap \mathbf{O}(2n) = \mathbf{U}(n).$$

The automorphisms preserving two (and thus all three) structures form the unitary group U(n). The determinant det of a unitary automorphism is a complex number with modulus 1. Thus there arises a mapping of U(n) onto the circle

$$\mathbf{SU}(n) \to \mathbf{U}(n) \xrightarrow{\det} S^1, \tag{2}$$

which is obviously a fibering (the fiber is the group SU(n) of unitary automorphisms with determinant 1).

### 1.2. The Lagrangian Grassmanian Λ(n)

We consider an n-dimensional plane $\mathbf{R}^n \subset \mathbf{R}^{2n}$. It is called Lagrangian if the skew-scalar product of any two vectors of $\mathbf{R}^n$ equals zero. For example, the planes p = 0 and q = 0 are Lagrangian.*

* The name comes from the "Lagrange brackets" in classical mechanics.

Moscow State University. Translated from Funktsional'nyi Analiz i Ego Prilozheniya, Vol. 1, No. 1, pp.1-14, January-March, 1967. Original article submitted September 28, 1966.

The manifold of all (nonoriented) Lagrangian subspaces of $\mathbf{R}^{2n}$ is called the Lagrangian Grassmanian $\Lambda(n)$.

From the complex point of view Lagrangian planes can be called real-similar, since there holds

LEMMA 1.2. The unitary group $\mathbf{U}(n)$ acts on $\Lambda(n)$ transitively with stationary subgroup $\mathbf{O}(n)$.

Proof. Let $\lambda$ be a Lagrangian plane. By (1) this means that the plane $I\lambda$ is orthogonal to $\lambda$. Let $\lambda' \in \Lambda(n)$ and $\xi, \xi'$ be orthogonal frames in $\lambda, \lambda'$. Then the automorphism of $\mathbf{R}^{2n}$ carrying $\xi$ into $\xi'$ and $I\xi$ into $I\xi'$ is unitary.

From this lemma it follows that $\Lambda(n)$ is a manifold, $\Lambda(n) = \mathbf{U}(n)/\mathbf{O}(n)$; thus there is a fibering

$$\mathbf{O}(n) \rightarrowtail \mathbf{U}(n) \rightarrow \Lambda(n). \tag{3}$$

## 1.3 The Mapping $\mathrm{Det}^2$: $\Lambda(n) \rightarrow S^1$

The determinant of an orthogonal automorphism $A \in \mathbf{O}(n) \subset \mathbf{U}(n)$ equals $\pm 1$. Therefore the square of the determinant of a unitary automorphism carrying the plane $p = 0$ into the Lagrangian plane $\lambda$ depends only on $\lambda$. In this way a mapping is constructed

$$\mathrm{Det}^2\colon \Lambda(n) \rightarrow S^1.$$

Denote by $S\Lambda(n)$ the set of Lagrangian planes $\lambda \in \Lambda(n)$ with $\mathrm{Det}^2 \lambda = 1$. On this set the group $\mathbf{SU}(n)$ of unitary unimodular automorphisms acts transitively, and the stationary subgroup of any point is isomorphic to the rotation group $\mathbf{SO}(n)$. Therefore $S\Lambda(n) = \mathbf{SU}(n)/\mathbf{SO}(n)$ is a manifold.

Thus we obtain a diagram (obviously commutative) of six fiberings:

$$\begin{array}{ccccc} \mathbf{SO}(n) & \rightarrow & \mathbf{O}(n) & \xrightarrow{\det} & S^0, \\ \downarrow & & \downarrow & & \downarrow \\ \mathbf{SU}(n) & \rightarrow & \mathbf{U}(n) & \xrightarrow{\det} & S^1, \\ \downarrow & & \downarrow & & \downarrow z^2 \\ S\Lambda(n) & \rightarrow & \Lambda(n) & \xrightarrow{\mathrm{Det}^2} & S^1, \end{array}$$

where $z^2$ is the mapping of the circle ($z = e^{i\varphi} \rightarrow e^{2i\varphi} = z^2$).

## 1.4. The Cohomology Class $\alpha \in H^1(\Lambda(n), \mathbf{Z})$

LEMMA 1.4.1. The fundamental group $\Lambda(n)$ is free cyclic,

$$\pi_1(\Lambda(n)) = \mathbf{Z},$$

and its generator goes into the generator of $S^1$ under the mapping induced by $\mathrm{Det}^2$.

The proof is obtained from the exact homotopy sequences of the left column and lower row of the diagram of section 1.3.

COROLLARY 1.4.2. The one-dimensional homology and cohomology groups of $\Lambda(n)$ are free cyclic:

$$H_1(\Lambda(n), \mathbf{Z}) \simeq H^1(\Lambda(n), \mathbf{Z}) \simeq \pi_1(\Lambda(n)) \simeq \mathbf{Z}.$$

For the generator $\alpha$ of the cohomology group $H^1(\Lambda(n), \mathbf{Z})$ we take the number of rotations of $\mathrm{Det}^2$, i.e., the cocycle whose value on a closed curve $\gamma\colon S^1 \rightarrow \Lambda(n)$ is equal to the degree of the composition

$$S^1 \xrightarrow{\gamma} \Lambda(n) \xrightarrow{\mathrm{Det}^2} S^1.$$

(Here $S^1$ is the circle $e^{i\varphi}$, oriented on the side of increasing $\varphi$.)

Example 1.4.3. Let $\lambda$ be a Lagrangian plane: $\lambda \in \Lambda(n)$. Consider the automorphisms $e^{i\varphi}E \in \mathbf{U}(n)$. The Lagrangian planes $e^{i\varphi}\lambda$ $(0 \leq \varphi \leq \pi)$ form a closed curve $\gamma\colon S^1 \rightarrow \Lambda(n)$, since $e^{i\pi}E = -E$.

The value of the class $\alpha$ on the curve $\gamma$ equals n.

Indeed, $\det(e^{i\varphi}E) = e^{in\varphi}$, therefore $\mathrm{Det}^2 e^{i\varphi}\lambda = e^{2in\varphi}\mathrm{Det}^2\lambda$.

## 1.5. Lagrangian Manifolds

Let M be an n-dimensional submanifold of the phase space $R^{2n}$. The manifold M is called Lagrangian if its tangent plane at each point is Lagrangian. For example, in the case n = 1 every curve M on the phase plane $R^2$ is Lagrangian.

Let M be a Lagrangian manifold. We consider the tangential mapping

$$\tau: \ M \to \Lambda(n),$$

carrying each point $x \in M$ into the subspace $\tau x \in \Lambda(n)$ parallel to the tangent plane to M at x.

The cohomology class $\alpha \in H^1(\Lambda(n), \mathbf{Z})$ introduced above induces on M a one-dimensional cohomology class

$$\alpha^* = \tau^* \alpha \in H^1(M, \mathbf{Z}).$$

The value of $\alpha^*$ on an oriented closed curve $\gamma$: $S^1 \to M$ is defined as the number of rotations of the square of the determinant of the tangent plane, i.e., as the degree of the composition

$$S^1 \xrightarrow{\gamma} M \xrightarrow{\tau} \Lambda(n) \xrightarrow{\mathrm{Det}^2} S^1.$$

The aim of this note is the proof of the following assertion.

THEOREM 1.5. The cohomology class $\alpha^* \in H^1(M, Z)$ coincides with the "index of closed curves on the Lagrangian manifold M" introduced by Maslov in [1].

## § 2. PROOF OF THEOREM 1.5

The Maslov index is defined by him as the index of intersection of a certain two-sided (n-1)-dimensional cycle on $M^n$ – the singular cycle.

### 2.1. The Singular Cycle

Let M be an n-dimensional Lagrangian manifold. Consider the projection $f$: M $\to \mathbf{R}^n$ of the manifold M onto the plane p = 0; $f(p, q) = q$. The set $\Sigma$ of points of M where the rank of the differential of $f$ is less than n is called the singularity of the mapping $f$. Regarding the singularity $\Sigma$, Maslov formulates the following assertions 1-5 (proofs are given below in § 3 and 4).

THEOREM 2.1. By an arbitrarily small unitary rotation the manifold M can be brought into "general position" relative to the projection $f$, so that the following assertions are valid:

ASSERTION 1. The singularity $\Sigma$ consists of an open (n−1)-dimensional manifold $\Sigma^1$, where $df$ has rank n−1, and the boundary $(\Sigma - \Sigma^1)$ dimension strictly less than n−2, so that $\Sigma$ determines an (n−1)-dimensional (unoriented) cycle in M.

ASSERTION 2. This cycle is two-sided in M.

The choice of a positive side of $\Sigma$ can be carried out in the following way.

ASSERTION 3. In a neighborhood of the point $x \in \Sigma^1$ the Lagrangian manifold M is given by n equations of the form

$$q_k = q_k(p_k, q_{\hat{k}}), \ p_{\hat{k}}(p_k, q_{\hat{k}}), \tag{4}$$

where $\hat{k} = 1, 2, \ldots, k-1, k+1, \ldots, n$ for some k, $1 \le k \le n$.

Obviously, in a neighborhood of such a point x the singularity $\Sigma^1$ is given by an equation $\dfrac{\partial q_k}{\partial p_k} = 0.$

ASSERTION 4. On passing through $\Sigma^1$ the quantity $\partial q_k / \partial p_k$ changes sign.

It turns out that for the positive side of $\Sigma^1$ we can take the one for which $\partial q_1 / \partial p_1 > 0$.

ASSERTION 5. Such a definition of the positive side is correct, i.e., does not depend on which of the coordinate systems $p_k, q_{\hat{k}}$ (k = 1, . . ., n) we use.

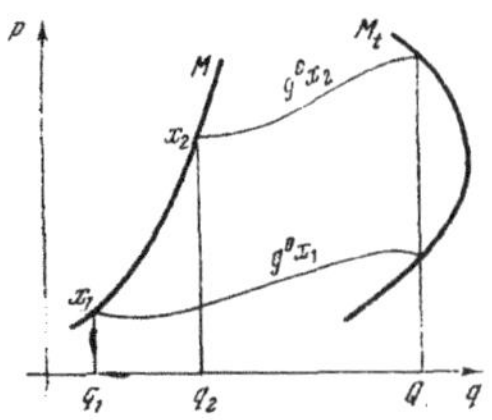


Fig. 1

## 2.2. The Maslov Index, ind $\in H^1(M, Z)$

Suppose there is given on the Lagrangian manifold M in "general position" in the sense of Theorem 2.1. a curve $\gamma$, transversal to the cycle $\Sigma$, with the initial and end points of $\gamma$ nonsingular:

$$\partial\gamma = x_1 - x_0, \; x_1 \overline{\in} \Sigma, \; x_0 \overline{\in} \Sigma.$$

Maslov calls the index ind $\gamma$ of the curve $\gamma$ its index of intersection with $\Sigma$, i.e., the number $\nu_+$ of points of passage from the negative side to the positive side minus the number $\nu_-$ of points of passage from the positive side to the negative:

$$\operatorname{ind}\gamma = \nu_+ - \nu_-.$$

Example. Let n = 1 and let M be a curve in the pq-plane (Fig. 1). For M in general position, $\Sigma$ consists of separate points a, b, c, . . . . The indexes of the curves $\gamma_i$ ($\partial\gamma_i = x_i - x_0$) are equal to 0, 1, 0, 1, 2. respectively.

THEOREM 2.2. The index of a closed oriented curve $\gamma$ on the Lagrangian manifold in general position M depends only on the homology class of $\gamma$ and is the value of a one-dimensional class of integer-valued cohomologies of M on the cycle $\gamma$: ind $\in H^1(M, Z)$.

## 2.3. Index of Curves on the Grassmanian $\Lambda(n)$

Proofs of the formulated theorems (1.5, 2.1, 2.2) are based on the following construction.

In the manifold of Lagrangian planes of $\Lambda(n)$ we single out the sets $\Lambda^k(n)$ of planes having a k-dimensional intersection with a fixed plane $\sigma \in \Lambda(n)$ (namely, the plane q = 0). It turns out that the closure $\overline{\Lambda^1(n)}$ determines a cycle (nonoriented) of codimension 1 (see 3.2.2).

In section 3.5 we prove

THE FUNDAMENTAL LEMMA. $\Lambda^1(n)$ is two-sidedly imbedded in $\Lambda(n)$ , i.e., there exists a continuous vector field transversal to $\Lambda^1(n)$ and tangent to $\Lambda(n)$.

Such a vector field is constructed by means of the orbits of action of $S^1 = \{e^{i\theta}\}$ on $\Lambda(n)$. In § 3 we prove

LEMMA 3.5.1. Every circle

$$\theta \to e^{i\theta}\lambda, \quad 0 \leqslant \theta \leqslant \pi, \quad \lambda \in \Lambda(n), \tag{4}$$

is transversal to $\Lambda^1(n)$.

For the positive side of $\Lambda^1(n)$ we choose the one toward which the velocity vectors of the curves (4) are directed.

The two-sidedness of $\Lambda^1(n)$ allows us to define

$$\operatorname{Ind} \in H^1(\Lambda(n), Z)$$

as the index of intersection of oriented closed curves on $\Lambda(n)$ with $\overline{\Lambda^1(n)}$ (Definition 3.6.1).

The index Ind is connected with the Maslov index ind and the cohomology class $\alpha$ of section 1.4. Namely, it turns out that the choice of a positive side of $\Lambda^1(n)$ by means of the curves (4) agrees with the definition of the positive side of $\Sigma^1$ from section 1.2. In § 4 we prove

LEMMA 4.3.1. The index Ind generates the Maslov index ind under the tangential mapping $\tau$: $M^n \to \Lambda(n)$; ind $= \tau^*$ Ind, i.e., for every curve $\gamma$: $S^1 \to M$ we have ind $\gamma$ = Ind $\tau\gamma$.

Proof of Theorem 1.5. Calculation of the index of the curves (4) (see Example 3.6.2) gives Ind $\gamma$ $= n = \alpha(\gamma)$ (Example 1.4.3). But $H^1(\Lambda(n), Z) = Z$ (Corollary 1.4.2). So, Ind $= \alpha$. By Lemma 4.3.1 ind $= \tau^*$ Ind and by Definition 1.5. $\alpha^* = \tau^* \alpha$. Thus ind $= \alpha^*$, which was to be proved.

## 3. PROOF OF THE FUNDAMENTAL LEMMA

In this section we prove the two-sidedness of the singular cycle $\overline{\Lambda^1(n)}$ and define the index Ind$\in H^1(\Lambda(n), Z)$.

### 3.1. Generating Functions

Let M be a manifold in phase space which is given in a simply connected neighborhood of the point $q = q_0$, $p = p_0$ by an equation of the form $p = p(q)$.

LEMMA 3.1.1. The manifold M is Lagrangian if and only if there is a "generating function" s(q) such that

$$p = \frac{\partial s}{\partial q}. \tag{5}$$

Proof. Let $s(q) = \int_{q_0}^{q} p(q)dq$. Independence of this integral of path is equivalent to the differential $d(p\,dq) = dp \wedge dq$ being 0 on M. But the value of $dp \wedge dq$ on the bivector $\xi \wedge \eta$ is exactly equal to the skew-scalar product $[\xi, \eta]$, so that equality of $dp \wedge dq$ to zero on M is equivalent to M being Lagrangian. The function s(q) satisfies (5), proving the lemma.

Remark 3.1.2. The function s(q) is determined up to a constant summand. In the particular case where M is a subspace, this summand can be chosed so that s(q) is a quadratic form. From this there follows

COROLLARY 3.1.3. The set of Lagrangian subspaces of the form $p = p(q)$ (i.e., transversal to the plane $q = 0$) make up in the manifold $\Lambda(n)$ an open cell $\Lambda^0(n)$, diffeomorphic to the linear space D of all real symmetric matrices of order n. The diffeomorphism is given by the mapping

$$\varphi\colon D \to \Lambda^0(n), \quad \varphi(S) = \lambda_S \quad (S \in D,\ \lambda_S \in \Lambda^0(n)),$$

where $\lambda_S$ denotes the plane $p = Sq$.

The proof is obtained from (5) by setting $s(q) = \frac{1}{2}(Sq, q)$.

The space of symmetric matrices D is $R^{n(n+1)/2}$. Thus we have proved

COROLLARY 3.1.4. The manifold $\Lambda(n)$ has dimension

$$\dim \Lambda(n) = n(n+1)/2.$$

### 3.2. The Singular Cycle $\overline{\Lambda^1(n)}$

Notation 3.2.0. Let $\sigma$ be the Lagrangian plane $q = 0$. We denote by $\Lambda^k(n)$ the set of all Lagrangian planes $\lambda \in \Lambda(n)$ whose intersection with the plane $\sigma$ is k-dimensional

LEMMA 3.2.1. The set $\Lambda^k(n)$ is an open manifold of codimension $k(k+1)/2$ in the Lagrangian Grassmanian $\Lambda(n)$.

Proof. We compare with each plane $\lambda \in \Lambda^k(n)$ its intersection with the plane $\sigma$. There arises a mapping of $\Lambda^k(n)$ on the Grassman manifold $G_{n,k}$ of all k-dimensional subspaces of the n-dimensional space $\sigma$. It is easily verified that this mapping determines a fibering

$$\Lambda^0(n-k) \to \Lambda^k(n) \to G_{n,k}.$$

By Corollary 3.1.4 $\dim \Lambda^0(n-k) = \frac{(n-k)(n-k+1)}{2}$. Since $\dim G_{n,k} = k(n-k)$, we find

$$\dim \Lambda^k(n) = \frac{(n-k)(n-k+1)}{2} + k(n-k) = \frac{n(n+1)}{2} - \frac{k(k+1)}{2} = \dim \Lambda(n) - \frac{k(k+1)}{2},$$

which was to be proved.

COROLLARY 3.2.2. $\overline{\Lambda^1(n)}$ determines an (unoriented) cycle of codimension 1 in $\Lambda(n)$.

Proof. The manifold $\Lambda(n)$ can be considered algebraic. The closure $\overline{\Lambda^1(n)} = \bigcup_{k \geq 1} \Lambda^k(n)$ is an algebraic submanifold of codimension $1(k(k+1)/2 \geq 1$ for $k \geq 1)$. Therefore $\overline{\Lambda^1(n)}$ determines a (nonoriented) chain. The singularity of the algebraic manifold $\overline{\Lambda^1(n)}$ is the algebraic submanifold $\overline{\Lambda^2(n)} = \bigcup_{k \geq 2} \Lambda^k(n)$ of codimension 3 in $\Lambda(n)$, since $k(k+1)/2 \geq 3$ for $k \geq 2$. Thus the homological boundary of the chain $\overline{\Lambda^1(n)}$ equals 0, which was to be proved.

## 3.3. Coordinates on $\Lambda(n)$

We consider a Lagrangian plane $\lambda \in \Lambda(n)$. Let $\lambda \in \Lambda^k(n)$, i.e., let the intersection $\lambda \cap \sigma$ be k-dimensional. We introduce coordinates on $\Lambda(n)$ in a neighborhood of $\lambda$.

Notation 3.3.0. Let K be a subset of the set 1,2, . . ., n. Denote by $\sigma_K$ the Lagrangian coordinate plane

$$\sigma_K = \{p,\ q:\ p_k = 0,\quad q_l = 0\quad \forall k \in K,\ \forall l \bar{\in} K\}.$$

LEMMA 3.3.1. The plane $\lambda \in \Lambda^k(n)$ is transversal to one of the $C_n^k$ coordinate planes $\sigma_K$, where K has k elements.

Proof. The intersection $\lambda \cap \sigma = \lambda_0$ is k-dimensional. Consequently, the plane $\lambda_0$ in $\sigma$ is transversal to one of the $C_n^k$ $(n-k)$-dimensional coordinate planes $\tau = \sigma_K \cap \sigma$, i.e., for some K we have $\lambda_0 \cap \sigma_K \cap \sigma = 0$. We shall show that the plane $\sigma_K$ is transversal to $\lambda$: $\sigma_K \cap \lambda = 0$.

By the condition, $\lambda_0 + \tau = \sigma$. From the Lagrangian property of $\lambda$ and $\sigma_K$ it follows that $[\lambda, \lambda_0] = 0$ (since $\lambda_0 \subset \lambda$) and $[\sigma_K, \tau] = 0$ (since $\tau \subset \sigma_K$). Thus, $[\lambda \cap \sigma_K, \lambda_0 + \tau] = 0$, i.e., $[\lambda \cap \sigma_K, \sigma] = 0$. But the largest number of pairwise skew-orthogonal independent vectors in $R^{2n}$ equals n. Therefore the n-dimensional plane $\sigma$ is itself a maximal skew-orthogonal plane, thus $(\lambda \cap \sigma_K) \subset \sigma$. So, $(\lambda \cap \sigma_K) \subseteq (\lambda \cap \sigma \cap \sigma_K) = (\lambda_0 \cap \tau) = 0$, which was to be proved.

From the lemma just proved it follows that every plane $\lambda \in \Lambda(n)$ is transversal to one of the $2^n$ coordinate planes $\sigma_K$. This allows us to set up an atlas of $\Lambda(n)$ of $2^n$ charts.

One of the maps was constructed in section 3.1.1: the region $\Lambda^0(n)$ is diffeomorphic to the space of symmetric matrices $D = R^{n(n+1)/2}$, where the diffeomorphism $\varphi: D \to \lambda^0(n)$ is defined as

$$\varphi(S) = \lambda_S = \{p,\ q: p = Sq\}\ \forall S \in D.$$

Notation 3.3.2. We denote by $I_K$ the operator of multiplication by i of the variables $z_\varkappa = p_\varkappa + iq_\varkappa$, $\varkappa \in K$:

$$I_K:\ R^{2n} \to R^{2n},$$

and for $\eta = I_K \xi$

$$q_\varkappa(\eta) = +p_\varkappa(\xi),\quad p_\varkappa(\eta) = -q_\varkappa(\xi),\quad \forall \varkappa \in K,$$

$$q_\nu(\eta) = q_\nu(\xi),\qquad p_\nu(\eta) = p_\nu(\xi),\qquad \forall \nu \bar{\in} K.$$

The operator $I_K$ is unitary, therefore it carries Lagrangian planes into Lagrangian planes. In particular, $I_K\sigma = \sigma_K$. Thus $I_K$ carries the set $\Lambda^0(n)$ of planes transversal to $\sigma$ into the set of planes $I_K\Lambda^0(n)$ transversal to $\sigma_K$. Thus the formula

$$\varphi_K(S) = I_K\lambda_S \in \Lambda(n),\quad S \in D \tag{6}$$

gives a diffeomorphism $\varphi_K$: $D \to I_K\Lambda^0(n)$, where $I_K\Lambda^0(n)$ is the set of all Lagrangian planes transversal to $\sigma_K$.

By Lemma 3.3.1 the $2^n$ regions $I_K\Lambda^0(n)$ cover $\Lambda(n)$ entirely, so that formula (6) gives an atlas of $\Lambda(n)$ of $2^n$ charts.

LEMMA 3.3.3. The set $\Lambda^k(n)$ is covered by $C_n^k$ charts $\varphi_K D$(K consisting of k elements) and in the coordinates $S = \varphi_K^{-1}\lambda$ is given by $k(k+1)/2$ linear equations $S_{\mu\nu} = 0$ $(\forall \mu \in K, \forall \nu \in K)$.

Proof. Let $\dim \lambda \cap \sigma = k$. By Lemma 3.3.1 $\lambda \cap \sigma_K = 0$ for some K of k elements. Therefore the plane $I_K\lambda = I_K^{-1}\lambda$ is transversal to $\sigma$ and has an equation $p = Sq$. The intersection $(I_K\lambda) \cap \sigma_K = I_K(\lambda \cap \sigma)$ has dimension k. But on $\sigma_K$ we have $q_l = 0$ $(\forall l \bar{\in} K)$, $p_m = 0$ $(\forall m \in K)$. Therefore on a k-dimensional subspace $q_l = 0$ $(l \bar{\in} K)$ of the plane $p = 0$, k of the functions $p_\mu$ $(p = Sq,\ \mu \in K)$ must vanish identically. This is equivalent to the equations $S_{\mu\nu} = 0$, as was to be proved.

## 3.4. Unitary Paramaterization

By means of the coordinates S introduced above it is possible to express the unitary transformations carrying the "purely imaginary" plane $p = 0$ into the plane $\lambda_S \in \Lambda^0(n)$.

It is obvious that we have

LEMMA 3.4.1. Let S, U be square $n \times n$ matrices with complex elements. Then

$$\left( U = \frac{E - iS}{E + iS} \right) \Leftrightarrow \left( S = -i \frac{E - U}{E + U} \right), \tag{7}$$

and for S, U, related by formulas (7),

S is self-adjoint if and only if U is unitary,

S is symmetric if and only if U is symmetric.

Thus, formulas (7) set up a diffeomorphism between the space D of real symmetric matrices S and the manifold of unitary symmetric nonsingular matrices U. (The unitary matrix U is nonsingular if $-1$ is not a proper value; for a real symmetric S we always have $\det(E + iS) \neq 0$).

It is always possible to take the square root of a nonsingular unitary matrix, defining it by continuity, beginning from $\sqrt{E} = E$.

LEMMA 3.4.2. Let $\lambda_S \in \Lambda^0(n)$ be a plane $p = Sq$. Then the matrix

$$\sqrt{U} = \frac{E - iS}{\sqrt{E + S^2}}$$

gives a unitary transformation carrying the plane $p = 0$ into the plane $\lambda_S$.

Proof. Since S is symmetric and real, $\sqrt{E + S^2}$ is real, and $\sqrt{U}$ carries the plane $p = 0$ into the same image as $E - iS$. The latter transformation carries the point

$$(0, q) \in \mathbf{R}^{2n}, \quad \text{i.e.}, \quad iq \in i\mathbf{R}^n \subset \mathbf{C}^n$$

into the point

$$iq + Sq \in \mathbf{C}^n, \quad \text{i.e.}, \quad (Sq, q) \in \lambda_S \subset \mathbf{R}^{2n},$$

which was to be proved.

COROLLARY 3.4.3. The mapping $\mathrm{Det}^2 \colon \Lambda(n) \to S^1$ of section 1.3 is given by the formula

$$\mathrm{Det}^2 \lambda_S = \det \frac{E - iS}{E + iS}.$$

COROLLARY 3.4.4. The symmetric nonsingular unitary matrix U for which $\sqrt{U}$ carries the plane $p = 0$ into the plane $\lambda$ is uniquely determined by this plane $\lambda \in \Lambda^0(n)$.

In fact, by 3.4.1 U is uniquely determined by S, and by 3.4.2 S is uniquely determined by $\lambda$.

## 3.5 Two-Sidedness of the Singular Cycle

Let $\lambda$ be a Lagrangian plane. Then each of the planes $e^{i\theta}\lambda$ is Lagrangian.

LEMMA 3.5.1. If $\lambda \in \Lambda^1(n)$, then the curve $\gamma \colon S^1 \to \Lambda(n), e^{i\theta} \to e^{i\theta}\lambda$ is transversal to the cycle $\Lambda^1(n)$ at the point $\theta = 0$.

Thus, the velocity vectors $v(\lambda) = \left.\frac{d}{d\theta}\right|_{\theta=0} (e^{i\theta}\lambda)$ form a transversal structure to $\Lambda^1(n)$, from which there follows the

FUNDAMENTAL LEMMA. The singular cycle $\overline{\Lambda^1}(n)$ is two-sidedly imbedded in $\Lambda(n)$.

We will carry out the proof of Lemma 3.5.1 in three stages.

A. First let $\lambda \in \Lambda^0(n)$, $\lambda = \lambda_S$, where $S \in D$ is a real symmetric matrix. We shall compute the coordinates of the velocity vector of the curve $e^{i\theta}\lambda$ in this coordinate system.

LEMMA 3.5.2. For any matrix $S \in D$

$$\left.\frac{d}{d\theta}\right|_{\theta=0} \varphi^{-1} e^{i\theta} \lambda_S = -(E + S^2).$$

Proof of Lemma 3.5.2. According to section 3.4 the plane $\lambda$ is in a unique correspondence with a nonsingular unitary symmetric matrix U so that

$$\lambda = \lambda_{S(U)}, \qquad S(U) = -i\frac{E-U}{E+U},$$

and $\sqrt{U}$ carries the plane $p = 0$ into $\lambda$.

Let $U(\theta) = e^{2i\theta}U$. The matrix $U(\theta)$ is unitary, symmetric, and for small $|\theta|$ nonsingular, so that $\sqrt{U(\theta)} = e^{i\theta}\sqrt{U}$. Therefore $\sqrt{U(\theta)}$ carries the plane $p = 0$ into $e^{i\theta}\lambda$, so that

$$\lambda_{S(U(\theta))} = e^{i\theta}\lambda_{S(U)}, \quad \text{or} \quad \varphi^{-1}e^{i\theta}\lambda_{S(U)} = S(U(\theta)). \tag{8}$$

The vector $\left.\frac{d}{d\theta}\right|_{\theta=0} \varphi^{-1} e^{i\theta}\lambda$ lies in the tangent space to the linear space of symmetric matrices D; this tangent space is naturally identified with D itself. With this identification, by the formula of 3.4.1,

$$-\left.\frac{d}{d\theta}\right|_{\theta=0} S(U(\theta)) = \left.\frac{d}{d\theta}\right|_{\theta=0} i\frac{E-e^{2i\theta}U}{E+e^{2i\theta}U} = \frac{4U}{(E+U)^2} = E + S^2,$$

which together with the formulas (8) proves Lemma 3.5.2.

B. Now let $\lambda \in \Lambda^1(n)$. By Lemma 3.3.3 the point $\lambda \in \Lambda^1(n)$ belongs to one of the n charts $\varphi_K D$, where K consists of one element $\varkappa$, $1 \leq \varkappa \leq n$. In other words, in the notation of 3.3.2

$$\lambda = I_K\lambda_S, \text{where } S \in D, \qquad \lambda_S \in \Lambda^0(n).$$

It is easy to calculate the velocity vector of the curve $\gamma$: $S^1 \to \Lambda(n)$, $e^{i\theta} \to e^{i\theta}\lambda$ for $\theta = 0$ in the coordinate system $\varphi_K^{-1}\lambda = S$.

LEMMA 3.5.3. For any matrix $S \in D$, $\lambda = I_K\lambda_S$,

$$\left.\frac{d}{d\theta}\right|_{\theta=0} \varphi_K^{-1}e^{i\theta}\lambda = -(E+S^2).$$

Indeed, by Definition 3.3.2, $\varphi_K = I_K\varphi$, and $I_K$ commutes with $e^{i\theta}$. Therefore

$$\varphi_K^{-1}e^{i\theta}\lambda = \varphi^{-1}I_K^{-1}e^{i\theta}I_K\lambda_S = \varphi^{-1}e^{i\theta}\lambda_S,$$

and Lemma 3.5.3 follows from Lemma 3.5.2.

C. The singular cycle $\Lambda^1(n)$ in the coordinates $S = \varphi_K^{-1}\lambda$ has an equation $S_{\varkappa\varkappa} = 0$ (Lemma 3.3.3). The velocity vector $\dot{v} = \left.\frac{d}{d\theta}\right|_{\theta=0} \varphi_K^{-1}e^{i\theta}\lambda$ by Lemma 3.5.3 is a negative definite matrix, $-v_{\varkappa\varkappa} \geq 1$. Thus, v and $\Lambda^1(n)$ are transversal, which proves Lemma 3.5.1.

Remark 3.5.4. At the same time we have shown that the vector v is directed to the side of $\Lambda^1(n)$ where $S_{\varkappa\varkappa} > 0$.

## 3.6. The Index Ind of Curves on $\Lambda(n)$

Let $\gamma$ be an oriented curve in $\Lambda(n)$, transversal to $\Lambda^1(n)$, and let $v(\lambda)$ be the velocity field of Lemma 3.5.1.

DEFINITION 3.6.1. By Ind $\gamma$ we shall denote the index of intersection of the curve $\gamma$ with the cycle $\Lambda^1(n)$, equipped with the field $v(\lambda)$.

In other words, Ind $\gamma = \nu_+ - \nu_-$, where $\nu_+$ is the number of points of intersection of $\gamma$ with $\Lambda^1(n)$ in which the vectors $\dot{\gamma}$ and v lie on the same side of $\Lambda^1(n)$, and $\nu_-$ is the number of those on which they lie on opposite sides.

The index of the closed curve $\gamma$, like every index of intersection, is determined by the homology class of $\gamma$ and can be considered as a one-dimensional cohomology class

$$\operatorname{Ind} \in H^1(\Lambda(n), \mathbf{Z}).$$

EXAMPLE 3.6.2. The index of the closed curve $\gamma$: $S^1 \to \Lambda(n)$, formed by $e^{i\theta}\lambda$, $0 \leq \theta \leq \pi$, is n:

$$\operatorname{Ind}\gamma = n.$$

Proof. We have that dim $\Lambda^2(n)$ = dim $\Lambda^1(n)$ is 2 (Lemma 3.2.1). Therefore for almost all $\lambda$ the curve $e^{i\theta}\lambda$ does not intersect $\Lambda^2(n)$. Such a curve is transversal to $\Lambda^1(n)$ at every point of intersection (Lemma 3.5.1). In this case Ind $\gamma$ is simply the number of these points of intersection (Definition 3.6.1).

Let $\lambda \in \Lambda^0(n)$. By Lemma 3.4.4 we have $\lambda = \lambda_{S(U)}$, where U is a unitary symmetric nonsingular matrix. We may consider the plane $\lambda$ to be such that all proper vales of the matrix U

$$e^{i\alpha_k},\ 1 \leqslant k \leqslant n,\ |\alpha_k| < \pi,$$

are distinct.

But

$$e^{i\theta}\lambda = \lambda_{S(e^{2i\theta}U)}$$

by formula (8) of 3.5.2, and

$$(\lambda_{S(e^{2i\theta}U)} \overline{\in} \Lambda^0(n)) \Leftrightarrow (\det(E + e^{2i\theta}U) = 0)$$

by Lemma 3.4.2. In other words, in the points of intersection of $\gamma$ with $\Lambda^1(n)$

$$\theta \equiv \frac{\pi - \alpha_k}{2} \pmod{\pi}.$$

There are precisely n such $\theta$ on the interval $0 \leq \theta \leq \pi$. Thus, Ind $\gamma$ = n, as was to be proved.

## § 4. PROOFS OF THE THEOREMS ON GENERAL POSITION

Here we prove Theorems 2.1 and 2.2 of § 2.

### 4.1. Transversality

Notation 4.1.1. Let A be a smooth manifold, and let $a \in A$. By $TA_a$ we denote the tangent space to A at the point a. If $f$: A → B is a smooth mapping, then by $f_*: TA_a \to TB_{f(a)}$ we denote the differential of $f$ at a.

Let $f$: A → B, h: C → B be two smooth mappings. The mappings $f$, h are called transversal at the point $b \in B$ if

$$f_* TA_a + h_* TC_c = TB_b$$

for every pair of points $a \in A$, $c \in C$, for which $f(a) = h(c) = b$. The mappings $f$ and h are transversal if they are transversal at every point $b \in B$.

In the particular case where $f$ or h is an imbedding, we may speak of the transversality of a mapping to a submanifold or of the transversality of two submanifolds.

The concept of transversality extends also to the case where A is the union of several manifolds, $A = \cup A_k$ [for example, $A^1(n) = \bigcup_{k \geq 1} \Lambda^k(n)$ in § 3] – in this case the restriction of $f$ to each $A_k$ must be transversal to h.

It is easy to prove (see, for example, [3]) the lemma of M. Morse and A. Sard

LEMMA 4.1.2. Let $f$: A → B be a smooth mapping. Then the measure of the set of points $b \in B$ not transversal to $f$ equals 0 (the point $b \subset B$ is a zero-dimensional submanifold of B).

From Lemma 4.1.2 there follows (see, for example, [4])

LEMMA 4.1.3. Let B be a homogeneous space on which a Lie group G acts transitively ($\forall g \in G$, g: B → B is a diffeomorphism). Let $C \subset B$ be a smooth submanifold of B and let $f$: A → B be a smooth mapping. Then the measure of the set of points $g \in G$ for which the mapping

$$f_g\colon A \to B,\quad f_g(a) = gf(a)$$

is not transversal to C is zero.

For completeness we carry out the proof of Lemma 4.1.3.

Remark 4.1.4. Since the union of a countable number of sets of measure zero has measure zero, it is sufficient to prove Lemma 4.1.3 for a neighborhood $A_0$ of a point $a_0 \in A$, a neighborhood $C_0$ of a point $c_0 \in C$ and a neighborhood of the identity e in the group G.

From the transitivity of the action of G there follows easily

ASSERTION 4.1.5. There exists a diffeomorphism of the product of spheres

$$u: D_1 \times D_2 \to G,$$

$$D_i = \{x \in R^{\nu_i}, |x| < 1\}, \quad \nu_1 = \dim B - \dim C, \quad \nu_2 = \dim G - \nu_1,$$

such that $u(0, 0) = e$, and the mapping

$$\beta: D_1 \times D_2 \times C_0 \to B \times D_2,$$

given by the formula

$$\beta(x, y, c) = (u(x, y)\, c, y), \forall x \in D_1, y \in D_2, c \in C_0,$$

is a diffeomorphism of $D_1 \times D_2 \times C_0$ onto some neighborhood E of the point $(c_0, 0)$ in $B \times D_2$.

Now define the projection of $E \subset B \times D_2$ on $D_1 \times D_2$

$$\Phi: E \to D_1 \times D_2 \text{ by the formula } \Phi(\beta(x, y, c)) = (x, y).$$

Further, define the mapping

$$\hat{f}: A \times D_2 \to B \times D_2 \text{ by the formula } \hat{f}(a, y) = (f(a), y).$$

We apply Lemma 4.1.2 to the composite mapping

$$\Theta = \Phi \circ \hat{f}: A_0 \times D_2 \to D_1 \times D_2.$$

ASSERTION 4.1.6. Let $x, y \in D_1 \times D_2$ be a point transversal to the mapping $\Theta$. Then the mapping

$$f_g: A_0 \to B, \; g = (u(x, y))^{-1},$$

is transversal to the imbedding $C_0 \subset B$.

Proof of Assertion 4.1.6. Consider $\Phi^{-1}(x, y)(x \in D_1, y \in D_2)$. Obviously, $\Phi^{-1}(x, y) = (C_{x,y}, y)$, where $C_{x,y} = u(x, y)C_0 \subset B$. The kernel of the differential $\Phi_*: T(B \times D_2)_{b,y} \to T(D_1 \times D_2)_{x,y}$ is exactly the tangent space to $(C_{x,y}, y)$:

$$\ker \Phi_* = T(C_{x,y}, y)_{b,y}.$$

Therefore the transversality of the mapping $\Theta = \Phi \circ \hat{f}$ to the point x,y implies

$$\hat{f}_* T(A_0 \times D_2)_{a,y} + T(C_{x,y}, y)_{f(a),y} = T(B \times D_2)_{b,y}$$

for all $a \in A_0$, for which $f(a) = b \in C_{x,y}$. Thus, the mapping $f: A_0 \to B$ is transversal to the imbedding $C_{x,y} \subset B$. Applying the diffeomorphism $g = (u(x, y))^{-1} \in G$, we see that $gf: A_0 \to B$ is transversal to $gC_{x,y} = C_0$, as required.

Proof of Lemma 4.1.3. We apply Lemma 4.1.2 to $\Theta$. The set of points $x, y \in D_1 \times D_2$ not transversal to $\Theta$ has measure zero. The corresponding set of points $g = (u(x, y))^{-1} \in G$ has measure zero in G. For the remaining g close to e the mapping $f_g$ is transversal to $C_0$ by 4.1.6. This proves Lemma 4.1.3 according to remark 4.1.4.

## 4.2. Proof of Theorem 2.1.

We apply Lemma 4.1.3 to the case where A is a Lagrangian manifold $M^n$, B is the Lagrangian Grassmanian $\Lambda(n)$, $f$ is the tangential mapping $\tau: M^n \to \Lambda(n)$, C is the submanifold $\Lambda^k(n) \subset \Lambda(n)$, and G is the unitary group $\mathbf{U}(n)$.

From Lemma 4.1.3 it follows that for almost all $u \in \mathbf{U}(n)$ the manifold $uM^n$ is such that its tangential mapping $\tau$ is transversal to every $\Lambda^k(n) \subset \Lambda(n)$, $k = 1, 2, \ldots$. Let us show that such a manifold $uM^n$ is in "general position" in the sense of Theorem 2.1.

Assertion 1 of Theorem 2.1 follows from the implicit functions theorem and Lemmas 3.2.1 and 3.2.2. Assertion 2 follows from the fundamental lemma of section 3.5. Assertion 3 is deduced from Lemma 3.3.1. Assertion 4 is obtained from Lemma 3.3.3 for k = 1. Finally, Assertion 5 follows from Lemma 3.5.1 and remark 3.5.4. Theorem 2.1 is thus proved.

### 4.3. Proof of Theorem 2.2

Let $M^n$ be a Lagrangian manifold in general position, and let $\gamma: S^1 \to M^n$ be an oriented closed curve, transversal to the singular cycle $\Sigma$.

LEMMA 4.3.1. Let $\tau \circ \gamma: S^1 \to \Lambda(n)$ be the tangential mapping of $M^n$ onto the curve $\gamma$. Then

$$\operatorname{ind} \gamma = \operatorname{Ind} \tau \circ \gamma.$$

In fact, $\Sigma^1 = \tau^{-1} \Lambda^1(n)$ (definitions of sections 2.1 and 3.2). Further, the positive (in the sense of section 2.1) side of $\Sigma^1$ is carried under the mapping $\tau$ into the positive (in the sense of Definition 3.6.1) side of $\Lambda^1(n)$ – this follows from remark 3.5.4. Thus, each point of intersection of $\gamma$ with $\Sigma^1$ gives to ind the same contribution as the corresponding point of intersection of $\tau \circ \gamma$ with $\Lambda^1(n)$ to Ind, which proves Lemma 4.3.1.

At the same time we have proved Theorem 2.2. also, since Ind $\tau \circ \gamma$ does not change under replacement of $\gamma$ by a homologous curve $\gamma'$ [this follows from the fact that $\dim \overline{\Lambda^2(n)} = \dim \Lambda^1(n) - 2$].

## § 5. A QUASICLASSICAL ASYMPTOTIC EXPRESSION

Here we give without proof the asymptotic formulas of Maslov, in which the index plays a role, for the simplest example.

### 5.1. Asymptotic Expression as $h \to 0$ of the Solution of the Schrödinger Equation

$$ih\frac{\partial \psi}{\partial t} = -\frac{h^2}{2}\Delta\psi + U(q)\psi, \quad \psi = \psi(q, t), \; q \in \mathbf{R}^n \tag{9}$$

with the initial condition

$$\psi|_{t=0} = \varphi(q) e^{\frac{i}{h} f(q)}, \text{ where } \varphi(q) \text{ is a finite function.} \tag{10}$$

To the Schrödinger equation there corresponds the classical dynamical system given in the 2n-dimensional phase space $\mathbf{R}^{2n}$ by the Hamilton equations

$$\dot{q} = \frac{\partial H}{\partial p}, \quad \dot{p} = -\frac{\partial H}{\partial q}, \qquad H = \frac{1}{2}p^2 + U(q).$$

The solutions of the Hamilton equation define a one-parameter group of canonical* diffeomorphisms of the phase space – the phase flow $g^t: \mathbf{R}^{2n} \to \mathbf{R}^{2n}$.

To the initial condition (10) there corresponds a function $\varphi(q)$ on a surface $M^n$, given in the phase space $\mathbf{R}^{2n}$ by the equations

$$M = \left\{p, q: \; p(q) = \frac{\partial f}{\partial q}\right\}.$$

The surface M is projected uniquely onto the q-plane. It is Lagrangian by Lemma 3.1.1. The phase flow $g^t$ carries M into another Lagrangian surface $g^t M = M_t$. The surface $M_t$ is no longer necessarily projected uniquely onto the q-plane. There arises a mapping $Q(q) = q(g^t(p(q), q))$ (Fig. 2).

Let $x_j = (p_j, q_j)$ be points of M such that $g^t x_j = (P_j, Q) \in M_t$. Assume that the Jacobian $\left.\frac{DQ}{Dq}\right|_{q = q_j} \neq 0.$

Maslov proved the following asymptotic formula for the solution of equation (9) with the condition (10).

THEOREM 5.1. As $h \to 0$

---

*A diffeomorphism $g: \mathbf{R}^{2n} \to \mathbf{R}^{2n}$ is called canonical, if for every closed curve $\gamma: S^1 \to \mathbf{R}^{2n}$ we have $\oint_\gamma p\,dq = \oint_{g\gamma} p\,dq$. The differential $g_*$ has then at each point a simplectic matrix: under a canonical transformation the 2-form $dp \wedge dq$ goes into itself. Therefore a Lagrangian manifold under a canonical diffeomorphism goes into a Lagrangian.

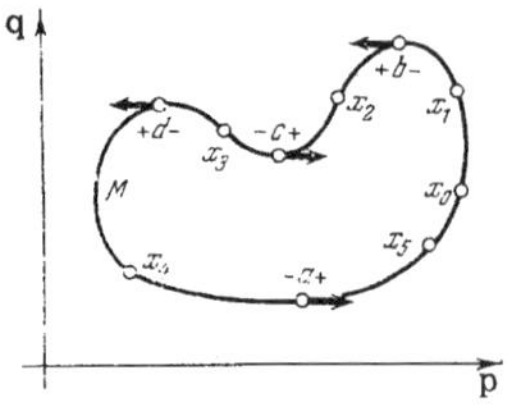


Fig. 2

$$\psi(Q, t) = \sum \varphi(q_j) \left| \frac{DQ}{Dq_j} \right|^{-1/2} e^{\frac{i}{h} S_j(Q,t) - \frac{i\pi}{2}\mu_j} + O(h),$$

where $S_j(Q, t)$ is the action along the trajectory $g^\theta x_j$:

$$S_j(Q, t) = f(q_j) + \int_0^t L\, d\theta, \quad L = \frac{q^2}{2} - U(q); \quad (p(\theta), q(\theta)) = g^\theta x_j,$$

and $\mu_j$ is the Morse index of the trajectory $g^\theta x_j$, i.e., the number of points $q(g^\theta x_j)$, $0 < \theta < t$, focal to M.

## 5.2. Connection between the Maslov and Morse Indexes

In Theorem 5.1 there appeared the Morse index $\mu$. The Morse index is a special case of the Maslov index; there holds

THEOREM 5.2. Consider in the $(2n + 2)$-dimensional phase space $\mathbf{R}^{2n+2} = \{\hat{p}, \hat{q}\}$, $\hat{p} = p_0$, p; $\hat{q} = q_0$, q; (p, q) $\in \mathbf{R}^{2n}$, the $(n + 1)$-dimensional manifold $\hat{M}$: (p, q) $\in M_t$, $q_0 = t$, $p_0 = -H(p, q)$. Then the manifold $\hat{M}$ is Lagrangian, and the Morse index of the trajectory $g^\theta x$, $0 < \theta < t$, is the Maslov index of the curve $(\theta, -H, g^\theta x)$ on the manifold $\hat{M}$.

The proof follows easily from the definitions of the indexes $\mu$ and ind: since $\left(\frac{\partial^2 H}{\partial p^2} \xi, \xi\right) > 0$, a simple focal point gives a contribution of + 1 to ind.

COROLLARY. For any curve $\gamma$ on M

$$\operatorname{ind} g^t\gamma - \operatorname{ind}\gamma = \mu(g^\theta\gamma^+) - \mu(g^\theta\gamma^-),$$

where $g^\theta\gamma^+$, $g^\theta\gamma^-$ $(0 \le \theta \le t)$ are trajectories with endpoints $\gamma$, $\partial\gamma = \gamma^+ - \gamma^-$.

For, the quadrilateral $\gamma$, $g^\theta\gamma^+$, $(g^t\gamma)^{-1}$, $(g^\theta\gamma^{-1})^{-1}$ on $\hat{M}$ is, obviously, homologous to zero; therefore its Maslov index equals zero (Theorem 2.2), which in view of Theorem 5.2 proves the desired relationship.

## 5.3. Quantization Conditions

In Theorem 5.1 there appear indexes of curves which are not closed. The indexes of closed curves enter into asymptotic formulas for stationary problems.

Let M be an invariant Lagrangian manifold of the phase flow $g^t$, lying on the hypersurface H = E (such invariant manifolds exist not only for integrable systems: see [5]).

Maslov proved

THEOREM 5.3. The equation

$$\frac{1}{2}\Delta\psi = \lambda^2 (U(q) - E)\psi$$

has a series of proper numbers $\lambda_N \to \infty$ with asymptotic expressions $\lambda_N = \mu_N + O(\mu_N^{-1})$ if for every $\gamma \in H_1(M, Z)$

$$\frac{2\mu_N}{\pi} \oint_\gamma p\, dq \equiv \operatorname{ind}\gamma \pmod 4. \tag{11}$$

In this case the characteristic functions $\psi_N$ are also related to the manifold M (in a sense defined precisely in [1] and under assumptions of the type of a simple spectrum).

In the particular case n = 1 the index of the circle equals 2, and formula (11) becomes the classical "quantization condition"

$$\mu_N \oint_\gamma p\, dq = 2\pi(N + 1/2).$$

## LITERATURE CITED

1. V. P. Maslov, Theory of Perturbations and Asymptotic Methods [in Russian], izd. MGU (1965).
2. V. P. Maslov, The WKB Method in the Multidimensional Case, Supplement to Heading's Book "An Introduction to Phase-Integral Methods " [in Russian], Biblioteka sb. "Matematika," Mir (1965).

3. L. S. Pontryagin, Smooth manifolds and their applications in homotopy theory, Trudy V. A. Steklov Matem. inst. AN SSSR, Vol. 45 (1955).
4. R. Abrakham, Supplement to Lang's Book "Differentiable Manifolds" [in Russian], Biblioteka sb. "Matematika," Mir (1966).
5. V. Arnold and A. Avez, Problèmes ergodiques de la mécanique classique, G.-V. (1966).

# DISTRIBUTION OF OVALS OF THE REAL PLANE OF ALGEBRAIC CURVES, OF INVOLUTIONS OF FOUR-DIMENSIONAL SMOOTH MANIFOLDS, AND THE ARITHMETIC OF INTEGER-VALUED QUADRATIC FORMS

V. I. Arnol'd

There is an interesting connection between the distribution of the branches of a real algebraic curve on the projective plane, on one hand, and, on the other hand, the topology of certain complex algebraic surfaces. In the present paper this connection is used in order to derive, from simple considerations of four-dimensional topology and the arithmetic of integer-valued quadratic forms, information as to the distribution of ovals of the real planes of algebraic curves.

## §1. Formulation of the Result

We consider a nonsingular real algebraic curve of degree n on the real projective plane $RP^2$. Such a curve is a compact smooth one-dimensional manifold without a boundary. Its connected components are diffeomorphic to circles and are called ovals.

A. Harnack [1] proved that the number of ovals does not exceed g + 1, where $g = (n-1)(n-2)/2$ is the genus of the curve. Curves with g + 1 ovals do exist, and are called M-curves.

The question as to how ovals can be distributed has been considered by many authors (in particular, D. Hilbert [2], K. Rohn [3], I. G. Petrovsky [4], and D. A. Gudkov [5]), but has been answered only for curves of degree 6 and less (see the survey in [5]).

To formulate our result we need to use the partition of ovals, introduced by I. G. Petrovsky, into positive and negative ovals. We assume that the degree of the curve is even: n = 2k (we retain this notation throughout the paper). Then, the ovals lie two-sidedly in $RP^2$, and each of them has an interior part (diffeomorphic to a circle) and an exterior (diffeomorphic to a Möbius sheet). We shall call an oval positive (or even) if it lies within an even number of others, and negative (or odd) if it lies within an odd number of other ovals. For example, the circle $x^2 + y^2 = 1$ is an even oval.

THEOREM 1. Let p be the number of positive, and m the number of negative, ovals of an M-curve of degree 2k. Then, the following congruence holds:

$$p - m \equiv k^2 \mod 4. \tag{1}$$

We note that congruence (1) does not exhaust all the constraints on the distribution of the ovals. For example, I. G. Petrovsky [4] proved the inequality

$$|2(p-m)-1| \leqslant 3k^2 - 3k + 1 \tag{2}$$

for any curve of degree 2k (with the necessarily maximal number of ovals), while D. A. Gudkov [5] proved, for M-curves of degree 6, a congruence of the form of (1) but modulo 8. For other constraints, see §9, paragraphs 4, 5, and 6.

Moscow State University. Translated from Funktsional'nyi Analiz i Ego Prilozheniya, Vol. 5, No. 3, pp. 1-9, July-September, 1971. Original article submitted March 22, 1971.

The proof given below of congruence (1) is based on a consideration of the actions of involutions of a certain four-dimensional manifold Y on its two-dimensional homology.

## §2. Structure of Four-Dimensional Manifold Y

Manifold Y is a two-sheeted covering of the complex projective plane, ramifying along the set of complex points of the curve in question.

Let $f(x, y) = 0$ be the affine equation of the curve in question. Here, x and y are real variables while $f$ is a nonhomogeneous polynomial of degree 2k with real coefficients. The corresponding homogeneous equation has the form $F(x_0, x_1, x_2) = 0$, where $f(x, y) = F(1, x, y)$.

Here, F is a homogeneous polynomial of degree 2k with real coefficients. Since the degree of F is even, the sign of F is the same at all points $(x_0, x_1, x_2)$ corresponding to one point $(x_0 : x_1 : x_2)$ of the projective plane $RP^2$ which does not lie on our curve. We can so choose F that, at the points corresponding to the nonorientable component of the complement to the curve in $RP^2$, the values of F will be negative. We fix such a polynomial F.

We now consider the equation

$$z^2 = F(x_0, x_1, x_2), \tag{3}$$

where $x_0$, $x_1$, and $x_2$ are complex variables not all simultaneously equal to zero. This equation gives a compact complex algebraic surface Y embedded in three-dimensional complex space E of the one-dimensional vector fibration over the complex projective plane

$$P' : E \to (CP^2 = \{(x_0 : x_1 : x_2)\}),$$

whose sections are homogeneous functions of degree k of the variables $x_0$, $x_1$, and $x_2$. From the real point of view, surface Y is a four-dimensional compact smooth orientable connected manifold without boundary.

In the affine map on E corresponding to the affine map (x, y) on $CP^2$, surface Y is given by the equation $z^2 = f(x, y)$.

The restriction of P to mapping P' on manifold Y gives a two-sheeted ramified covering of the complex projective plane. The manifold of the ramifications is the intersection of manifold Y with the zero fiber of fibration P'. We denote this manifold of ramification by A. It is clear from Eq. (3) that PA is precisely the set of complex points of our algebraic curve F = 0. Thus, A is a smooth orientable connected compact submanifold with boundary of the four-dimensional manifold Y. The real dimension (and codimension in Y) of manifold A equals 2.

## §3. Involution $\tau$ and Form $\Phi_\tau$

Multiplication of z by −1 gives a smooth involution $\tau$ of manifold Y. The set of fixed points is again our complex curve A.

On all our complex manifolds we choose an orientation in the natural way (by means of the form $\mathrm{Re}z_1 \wedge \mathrm{Im}z_1 \wedge \ldots \wedge \mathrm{Re}z_s \wedge \mathrm{Im}z_s$) such that the indices of intersection of the complex manifolds will be nonnegative. We remark that involution $\tau$ retains the orientation of manifold Y.

We denote by $H_2(Y) = H_2(Y, \underline{Z})$ mod Tors the group of two-dimensional integer-valued homologies of space reduced modulo the torsion. The index of intersection ( , ) of two-dimensional cycles gives, on $H_2(Y)$, a bilinear integer-valued nonsingular (Poincaré duality) form. The involution $\tau$: Y → Y induces the isomorphism $\tau_* : H_2(Y) \to H_2(Y)$. On $H_2(Y)$ we define the bilinear form $\Phi_\tau$ by the relationship

$$\Phi_\tau(a, b) = (\tau_* a, b), \quad a, b \in H_2(Y). \tag{4}$$

LEMMA 1. Form $\Phi_\tau$ is symmetric and nonsingular (det $\Phi_\tau = \pm 1$).

Proof. Since involution $\tau$ retains orientation on Y, $(\tau_* a, b) = \tau_*(\tau_* a, b) = (a, \tau_* b) = (\tau_* b, a)$. The second assertion follows from the Poincaré duality.

## §4. Arithmetic Lemma

Let $\Phi : Z^r \times Z^r \to Z$ be an integer-valued symmetric nonsingular (det $\Phi = \pm 1$) bilinear form. The following lemma is well known (see [6, 7, 8]).

LEMMA 2. There exists an element $w \in Z^r$ such that, for all $x \in Z^r$

$$\Phi(x, x) \equiv \Phi(w, x) \bmod 2. \tag{5}$$

Element w is defined univocally mod $2Z^r$. The magnitude of $\Phi(w, w)$ mod 8 does not depend on the choice of the representative w, and coincides, modulo 8, with the signature of the form.

In what follows we shall need only the fact that a consequence of (5) is

$$\Phi(w', w') \equiv \Phi(w, w) \bmod 8 \quad \text{for} \quad w' = w + 2x. \tag{6}$$

Proof of relationship (6). We have

$$\Phi(w', w') = \Phi(w, w) + 4\Phi(w, x) + 4\Phi(x, x),$$

where, according to (5), the numbers $\Phi(w, x)$ and $\Phi(x, x)$ have the same parity.

Element w (defined modulo 2) will be called the fundamental class of form $\Phi$.

## §5. Computing the Fundamental Class of Form $\Phi_\tau$

Let $\tau: Y^{4s} \to Y^{4s}$ be an orientation-retaining involution of the oriented compact smooth 4s-dimensional manifold without boundary with the 2s-dimensional oriented manifold of fixed points $A^{2s}$. We define form $\Phi_\tau$ on $H_{2s}(Y^{4s})$ by Eq. (4).

LEMMA 3. As the fundamental class of form $\Phi_\tau$ one can choose the homology class $[A^{2s}]$ represented by cycle $A^{2s}$.

Proof. To each point of intersection Q, not lying on $A^{2s}$, of the cycles $a$ and $\tau a$ there corresponds a second point $\tau Q$. At the points of intersection of $a$ and $A^{2s}$ one can achieve pairwise transversality of $A^{2s}$, $a$, and $\tau a$ by small displacements of $a$. Therefore, the indices of intersection of $a$ with $\tau a$ and with $A^{2s}$ coincide modulo 2.

Now, let $A \subset Y$ be the manifold of §3.

LEMMA 4. The index of self-intersection of cycle A in Y equals half the square of the degree of algebraic curve PA:

$$\Phi_\tau([A], [A]) = (A, A) = 2k^2.$$

Proof. Consider projection PA of curve A on $CP^2$. Let A' and A" be curves, homologous to PA, in $CP^2$ intersecting transversally, and not on PA. Then, the index of intersection of A' and A" in $CP^2$ equals the square of the degree of curve PA, i.e., $4k^2$. Consequently, the index of intersection of the cycles $P^{-1}A'$ and $P^{-1}A''$ in Y equals $8k^2$. But in Y we have $P^{-1}A' \sim 2A \sim P^{-1}A''$, so that $(A, A) = 2k^2$.

## §6. Real Part Π of Manifold Y

We now investigate the set of real points of manifold Y, i.e., the points to which there correspond real values of coordinates z and $x_0 : x_1 : x_2$.

Consider, on real projective plane $RP^2$, the set C of points to which there correspond values of $F \geq 0$. In $RP^2$ set C is a smooth two-dimensional submanifold whose boundaries are the ovals of interest to us. In general, manifold C is not connected; in view of the conventions made in §2 about the sign of F, it is orientable. We note that the Euler characteristic of manifold C equals the difference $p - m$ between the numbers of positive and of negative ovals.

LEMMA 5. Set Π of real points of manifold Y is a smooth compact two-dimensional orientable manifold without boundary, smoothly embedded in Y. Manifold Π is diffeomorphic to the doubled manifold with boundary C.

The proof follows from Eq. (3) and the triviality of fibration P over C; this latter is guaranteed by the choice of the sign of F.

LEMMA 6. The index of self-intersection of the real orientable compact smooth analytic manifold M of dimensionality 2s in its complexification equals $(-1)^s\chi$, where $\chi$ is the Euler characteristic of manifold M.

Proof. On M we construct the tangent vector field with simple singular points, and we multiply this by i. The sign of $(-1)^s$ corresponds to the difference between the two orientations of $C^{2s}$: we can initially take all the coordinate axes real, and then imaginary, and we can select the orientation as specified in §3.

LEMMA 7. The Euler characteristic of surface Π is expressed in terms of the numbers of positive and negative ovals by the formula

$$\chi = 2(p - m).$$

For the proof we note that $p-m$ is the Euler characteristic of the manifold with boundary C, and we then use Lemma 5.

Joining Lemmas 6 and 7, we obtain the following result.

LEMMA 8. The self-intersection index of surface Π in manifold Y equals $(\Pi, \Pi) = 2(m-p)$.

## §7. Homologies between the Cycles of A and Π

LEMMA 9. The homology classes represented by surfaces A and Π in Y coincide modulo 2; more precisely,

$$[A] = [\Pi] \in H_2(Y) \otimes \mathbf{Z}_2.$$

Proof. We denote by ∞ the "infinitely distant" complex line $x_0 = 0$ in $CP^2$ with coordinates $x_0 : x_1 : x_2$ from Eq. (3). We shall assume that this line intersects curve PA transversally in $n = 2k$ points. We can consider line ∞ as a cycle (with coefficients from Z or from $Z_2$).

We denote by [∞] the homology class of cycle ∞ in $H_2(CP^2)$. Each cycle c of $H_2(CP^2)$ is homologous to $(c, \infty)[\infty]$. For example, $PA \sim 2k[\infty]$.

Further, we denote by $\infty_Y$ the cycle $P^{-1}\infty$ in Y, and by $[\infty_Y]$ its homology class in $H_2(Y)$. Lemma 9 follows from the two relationships

$$[A] = k[\infty_Y] \in H_2(Y), \tag{7}$$

$$[\Pi] = k[\infty_Y] \in H_2(Y, \mathbf{Z}_2). \tag{8}$$

To prove relationship (7) we note that the integer-valued cycle $PA - 2k\infty$ is the boundary of some integer-valued three-dimensional chain $K_3$ in $CP^2$. We set $K_3' = P^{-1}K_3$. Then, $\partial K_3' = 2A - 2k\infty_Y$, i.e., $[A] - k[\infty_Y]$ is a second-order element in $H_2(Y, Z)$, which also proves Eq. (7) (we recall that $H_2(Y)$ is the torsion-factored homology group).

For the proof of relationship (8) we note that the ovals divide M-curve PA into two parts, the complex conjugate taking one part into the other (this is established in the proof of Harnack's Theorem). Let B be one of these parts. B is a connected compact oriented two-dimensional manifold with boundary. The boundaries of B are just the ovals of curve $F = 0$ on $RP^2$.

Consider the surface B + C, where C is the submanifold in $RP^2$ whose boundaries are the ovals of curve $F = 0$ (manifold C was defined in §6). The nonsmooth surface B + C is a combinatory, generally nonorientable, connected compact two-dimensional submanifold without boundary in $CP^2$. Therefore, surface B + C defines, in $CP^2$, cycles with coefficients in $Z_2$.

We now show that the index of intersection of cycles B + C and ∞ in $CP^2$ equals k modulo 2.

Indeed, if the ovals do not intersect the infinitely distant line, then exactly half of the 2k points of intersection of PA and ∞ fall in B (the complex conjugate interchanges the interiors of B and PA−B). If the ovals do intersect the infinitely distant line, then, by continuing this line up to B + C, one can achieve its intersection solely within B, transversally and at precisely k points.

For the proof we choose the real affine coordinates (x, y) so that line ∞ has the equation $x = 0$, and so that its infinitely distant point lies outside C. The line $x = i\varepsilon$, where $\varepsilon > 0$ is sufficiently small, is homologous to the line $x = 0$, does not intersect C, and transversally intersects B in precisely k points. Indeed, we orient the tangent field $\xi$ of each oval. Then, upon movement in the direction of along an oval in the (x, y) plane, we cross the line $x = 0$ from left to right as many times as from right to left. To intersections of one type correspond the points of intersection of PA with line $x = i\varepsilon$ lying in B and, to the other, on PA−B (this follows from the fact that $i\xi$ is a normal to an oval in B). The nonreal points of intersection of PA with $(x = i\varepsilon)$ lie in B and PA−B in equal parts if $\varepsilon$ is sufficiently small (because this is the case when $\varepsilon = 0$).

Thus, our earlier assertion is proven. It follows from it that $[B + C] = k[\infty] \in H_2(CP^2, Z_2)$. Therefore, the cycle $B + C - k\infty$ is homologous to 0 as a $Z_2$ chain: there exists a three-dimensional $Z_2$ chain in $L_3$ in $CP^2$ for which

$$\partial L_3 = B + C - k\infty.$$

Setting $L_3' = P^{-1}L_3$ we find

$$\partial L_3' = 2(P^{-1}B) + \Pi - k\infty_Y, \quad \text{i.e.,} \quad [\Pi] = k[\infty_Y] \in H_2(Y, Z_2).$$

Lemma 9 is proven.

## §8. Proof of Theorem 1

We apply Lemma 2 to form $\Phi_\tau$ and homology classes $w = [A] \in H_2(Y)$, $w' = [\Pi] \in H_2(Y)$.

According to Lemma 1, the form is symmetric and nonsingular; according to Lemma 3, class w is fundamental. By Lemma 9, class w' differs from w on even elements. Thus, Lemma 2 is applicable, and we find from Eq. (6) that

$$\Phi_\tau([A], [A]) \equiv \Phi_\tau([\Pi], (\Pi) \bmod 8.$$

We note that $\tau_* w = w$, $\tau_* w' = -w'$ (change of sign of z changes the orientation of $\Pi$). Thus, $\Phi_\tau([A], [A]) = (A, A)$, $\Phi_\tau([\Pi, \Pi]) = -(\Pi, \Pi)$. Therefore, $(A, A) + (\Pi, \Pi) \equiv 0 \bmod 8$. Substituting the values of the self-intersection indices $(A, A) = 2k^2$ from Lemma 4 and $(\Pi, \Pi) = 2(m-p)$ from Lemma 8, we find

$$2k^2 + 2(m-p) \equiv 0 \bmod 8,$$

q.e.d.

## §9. Remarks

1. Assertion (1) (modulo 8 rather than 4) was formulated by D. A. Gudkov in the form of a hypothesis supported by a large number of examples. Although the proof of congruence (1) does not use the results of D. A. Gudkov, the present paper could not have been produced had not D. A. Gudkov communicated his hypothesis to the author.

2. Manifold Y was studied by V. A. Rokhlin in a recent work [9] by means of the Hirzebruch-Atiyah-Zinger signature formulas ([10], § 6). By joining these computations with ours and with the Lefschetz-Dold-Atiyah-Bott formulas given by Hirzebruch in [11], we can obtain additional information on manifold Y and its involutions.

We note that the real differentiable type of manifold Y and its involution $\tau$ depends only on the degree of curve A, which may not even be real. However, the complex conjugate $\sigma$: $Y \to Y$ depends on the distribution of the real ovals of the curve.

Involutions $\sigma$ and $\tau$ commute, so that on Y there acts the group $H = Z_2 + Z_2$ of the four elements $1, \sigma, \tau, \sigma\tau$. We denote by $\Phi_h$ (where $h \in H$) the form $\Phi_h(x, y) = (h_* x, y)$ on $H_2(Y)$. We denote by $\Pi'$ the submanifold consisting of those points in Y for which the point $x_0 : x_1 : x_2$ is real while z is pure imaginary. Then, for any real curve $F = 0$ (not necessarily with maximal number of ovals), the relationships shown in Table 1 hold:

| $h$ | $1$ | $\sigma$ | $\tau$ | $\sigma\tau$ |
|---|---|---|---|---|
| Fixed points of involution h; | $Y$ | $\Pi$ | $A$ | $\Pi'$ |
| Fundamental class of form $\Phi_h$ | $(k+1)[\infty_Y]$ | $k[\infty_Y]$ | $k[\infty_Y]$ | $(k+1)[\infty_Y]$ |
| Trace of involution $h_*$ | $2+2g$ | $2(p-m)-2$ | $-2g$ | $2(m-p)$ |
| Signature of form $\Phi_h$ | $2-2k^2$ | $2(m-p)$ | $2k^2$ | $2(p-m)-2$ |

Not all these results were used for our proof of congruence (1), but without the computations of V. A. Rokhlin this proof would hardly have been found. To be more specific, the formulation of Lemma 9 resulted from a comparison of the data of Table 1 with the hypothesis of D. A. Gudkov.

3. The connection we have observed between the distribution of ovals and the involutions of four-dimensional manifold Y can also be used in the contrary direction, obtaining information on the involutions of manifold Y on the basis of data on real curves. For example, we find from relationships (1) and (2) that

$$2 + \operatorname{Tr}\sigma_* \equiv 2k^2 \bmod 8 \text{ (for M curves)}; \quad |1 + \operatorname{Tr}\sigma_*| \leqslant 3k^2 - 3k + 1.$$

It also follows from Table 1 that forms $\Phi_\sigma$ and $\Phi_T$ have the same parity as the number k, while forms $\Phi_1$ and $\Phi_{\sigma T}$ have that of k + 1. Indeed, for even k, the signature of form $\Phi_1$, and for odd k the signature of form $\Phi_T$, is not divisible by 8. Consequently, for all k there exists a cycle whose index of intersection with $\infty_Y$ is odd. By combining this information with the form of the fundamental classes, we obtain our assertion.

From Table 1, congruence (1), and inequality (2), it is clear that forms $\Phi_1$ and $\Phi_{\sigma T}$ are not sign-definite, while forms $\Phi_\sigma$, $\Phi_T$ are sign definite only when k = 1 and k = 2. Theorems on the structure of quadratic forms ([8], Chapter 5) therefore permit a complete reconstruction of the canonical form of forms $\Phi_1$, $\Phi_\sigma$, $\Phi_T$, $\Phi_{\sigma T}$ for all k. For example, when k = 2 (i.e., for curves of degree 4), the form $-\Phi_\sigma = \Phi_T$ is, in the notation of [8], the form $\Gamma_8$ corresponding to Lie algebra $E_8$.

4. It is of interest to note that our involution $\sigma$ allows us to provide a simple proof of the inequality of I. G. Petrovsk given in our (2). To this end we consider the linear space $E = H_2(Y, R)$ with scalar product given by the index of intersection. Involution $\sigma_*$ acts in E, retaining scalar products, so that E decomposes into the direct sum of two orthogonal proper subspaces $E_1$ and $E_{-1}$, corresponding to eigenvalues 1 and −1 of operator $\sigma_*$.

We mention that the scalar product is not degenerate. Therefore, each of the spaces $E_1$ and $E_{-1}$ can be presented in the form of the direct sum of orthogonal subspaces on which the scalar square is positive (negative) definite:

$$E_1 = E_1^+ + E_1^-, \quad E_{-1} = E_{-1}^+ + E_{-1}^-.$$

We have used here the following notation: $\dim E_1^+ = a$, $\dim E_1^- = b$, $\dim E_{-1}^+ = c$, $\dim E_{-1}^- = d$. In this notation,

$$\operatorname{Tr}\sigma_* = a + b - c - d, \qquad \operatorname{Tr} 1_* = a + b + c + d,$$
$$\operatorname{Sgn}\Phi_\sigma = a - b - c + d, \qquad \operatorname{Sgn}\Phi_1 = a - b + c - d.$$

We substitute into the left sides of these equations the values of traces and signatures from Table 1. Adding and subtracting the equations thus obtained, we find

$$a = (k-1)(k-2)/2, \; c = a + 1, \; b + d = 3k^2 - 3k + 1, \; b - d = 2(p - m) - 1.$$

Since b and d are non-negative, $|b - d| \leq b + d$, which also proves inequality (2).

5. Our constructions also lead to new constraints on the distribution of ovals. In order to formulate these constraints, we partition all ovals into three classes as a function of the sign of the Euler characteristic of that component of the complement to the curve forwhich the oval is an exterior boundary. We denote the numbers of positive ovals bounding domains with positive, zero, and negative Euler characteristic by $p_+$, $p_0$, and $p_-$ respectively, and the numbers of negative ovals by $m_+$, $m_0$, and $m_-$, so that

$$p = p_+ + p_0 + p_-, \quad m = m_+ + m_0 + m_-.$$

For example, $p_+$ is the number of positive ovals containing no other ovals inside themselves.

THEOREM 2. For any curve of degree 2k,

$$p_- \leqslant \frac{(k-1)(k-2)}{2}, \quad m_- \leqslant \frac{(k-1)(k-2)}{2}, \quad p_+ \leqslant b, \quad m_+ \leqslant d,$$

where numbers b and d are defined in paragraph 4.

Proof. Consider connected component $\Pi_i$ of surface $\Pi$, projected in the region of $RP^2$ bounded externally by the given oval $\gamma_i$. Of the p surfaces $\Pi_i$, the $p_-$ have a negative Euler characteristic.

The homology classes represented by these $p_-$ surfaces (of whatever orientation) $[\Pi_i] \in H_2(Y)$ are pairwise $\Phi_1$-orthogonal, and the quadratic form $\Phi_1$ on the $p_-$ classes $[\Pi_i]$ assumes positive values. Moreover, $\sigma_* [\Pi_i] = [\Pi_i]$.

It follows from this that the $p_-$ classes $[\Pi_i]$ are linearly independent, and that on the plane L spanning them, form $\Phi_1$ is positive definite, but $\sigma_* = 1$. Therefore,

$$p_- = \dim L \leqslant \dim E_1^+ = \frac{(k-1)(k-2)}{2},$$

which also proves our first inequality.

The other three inequalities are proven analogously (in considering $m_-$ it is necessary to take into account the nonorientable component of the complement).

6. From the linear independence of all the $\Pi_i$ would follow the stronger inequalities

$$p_- + p_0 \leqslant \frac{(k-1)(k-2)}{2}, \quad m_- + m_0 \leqslant \frac{(k-1)(k-2)}{2}, \quad p_+ + p_0 \leqslant b, \; m_+ + m_0 \leqslant d$$

and, for M-curves of degree 2k, the following lower bound on the number of empty ovals: $p_+ + m_+ \geq k^2$.

According to D. A. Gudkov, in all the interesting examples of M-curves, $p_+ + m_+ \geq k^2 + (k-1)(k-2)/2$, where, for any odd k, there exists an M-curve of degree 2k with $p_+ = k^2$, $p_0 = m_+ = (k-1)(k-2)/2$.

7. We also note that the factor space $X = Y/\tau\sigma$ from the naturally arising commutative diagram of the two-sheeted ramified coverings

$$\begin{array}{ccc} Y & \rightarrow & X \\ \downarrow & & \downarrow \\ CP^2 & \rightarrow & S^4 \end{array}$$

is a closed connected singly-connected oriented four-dimensional smooth manifold for which

$$\operatorname{Sgn} X = \frac{1}{2}(\operatorname{Sgn} Y + \operatorname{Sgn} \Phi_{\tau\sigma}) = p - m - k^2 = a - d,$$

where, for all known M-curves,

$$a \leqslant d, \quad c \leqslant b, \quad a - d \equiv 0 \pmod 8).$$

From the validity of these relationships for all M-curves would follow both the validity of the Gudkov hypothesis modulo 8 and the inequality

$$|p - m - 1| \leqslant k^2 - 1,$$

which is a strengthening of the Petrovsky inequality for M-curves.

8. For the first proof of Lemma 9 the author is indebted to A. N. Varchenko, to whom the author communicated this lemma in the form of a hypothesis. Although there were flaws in the proof of A. N. Varchenko, it convinced the author of the validity of the Lemma, without which the present paper could not have been produced. Our proof of Lemma 9 uses some ideas from the reasoning of A. N. Varchenko. Thus, the proof of congruence (1) is the result of the joint efforts of the author and A. N. Varchenko. Unfortunately, A. N. Varchenko would not agree to consider himself the co-author of this paper.

The author wishes to thank A. N. Varchenko, D. A. Gudkov, V. A. Rokhlin, and D. B. Fuks for numerous discussions and much support.

LITERATURE CITED

1. A. Harnack, "Über die Vielheitkeit der ebenen algebraischen Kurven," Math. Ann., 10, 189-199 (1876).
2. D. Hilbert, "Über die reelen Züge algebraischen Kurven," Math. Ann., 38, 115-138 (1891).
3. K. Rohn, "Die Maximalzahl und Anordnung der Ovale bei der ebenen Kurve 6 Ordnung und bei Fläche 4 Ordnung," Math. Ann., 73, 177-229 (1913).

4. I. Petrovsky, "On the topology of real plane algebraic curves," Ann. Math., 39, No. 1, 187-209 (1938).
5. D. A. Gudkov, "Topology of sixth-order curves and fourth-order surfaces," Uchenye Zapiski Gor'k. Un-ta, 87, 3-153 (1969).
6. F. Hirzebruch and K. H. Mayer, "O(n)-Manigfaltigkeiten, exotische Sphären und Singularitäten," Lecture Notes Math., 57, 89-91, J. Springer (1968).
7. J. Milnor, "On simply connected 4-manifold," in: Symposium International de Topologia Algebraica, 122-128, Mexico (1958).
8. J.-P. Serre, Cours d'arithmetique, Presses Universitaires de France, Paris (1970).
9. V. A. Rokhlin, "Two-dimensional submanifolds of four-dimensional manifolds," Funktsional.Analy.i Ego Prilozhen., 5, No. 1, 48-60 (1971).
10. M. F. Atiyah and I. M. Zinger, "Indices of elliptic operators. III," Usp. Matem. Nauk., 24, No. 1, 127-128 (1969).
11. F. Hirzebruch, "Elliptic differential operators on manifolds," Usp. Matem. Nauk, 23, No. 1, 191-209 (1968).

**Curriculum Vitae**
Pierre R. Deligne

Date of Birth: October 3, 1944, Etterbeek, Belgium

**Education:**

| | |
|---|---|
| 1950–1956 | Primary School at Schaerbeek, Belgium |
| 1956–1962 | Secondary School at Athénée Adolphe Max, Boulevard Clovis, Belgium |
| 1962–1966 | University of Brussels, Belgium |
| 1965–1966 | Foreign student (pensionnaire étranger) at Ecole Normale Supérieure |

At the University of Brussels, Pierre Deligne received degrees which are the equivalent of the American B.A. and Ph.D. in mathematics

| | |
|---|---|
| 1966 | B.A. Licence en Mathématiques |
| 1968 | Ph.D. Doctorat en mathématiques |

Deligne received the equivalent of the Ph.D. also from the University of Paris-Sud, France

| | |
|---|---|
| 1972 | Ph.D. Doctorat d'Etat des Sciences Mathématiques |

**Employment History:**

| | |
|---|---|
| September 1967 to August 1968 | Aspirant (=junior scientist) at F.N.R.S. (Fond National de la Recherche Scientifique, rue d'Egmont, Brussel, Belgium) and concurrently guest at IHES (Institut des Hautes Etudes Scientifiques, Bures-sur-Yvette, France) |
| September 1968 to January 1970 | Visiting member at IHES |
| Feburary 1970 to September 1984 | Permanent member at IHES |
| 1984 to present | Professor at the Institute for Advanced Study, Princeton, New Jersey |

**Awards and Honors:**

| | |
|---|---|
| 1974 | Francois Deruyts Prize, awarded by the Belgium Royal Academy (Period 70–74) |

1974 Henri Poincaré Medal, awarded by the Academy of Sciences (Paris)
1975 Quinquennal Prize "Doctor A. De Leeuw-Damry-Bourlart" for the period 1971–1975 (for mathematics, physics, and chemistry), awarded by the F.N.R.S. (the Belgian analogue of the National Science Foundation)
1978 Fields Medal, awarded at the International Congress of Mathematicians, Helsinki
1978 Elected Membre Associé Étranger, Académie des Sciences, Paris
1978 Elected Foreign Honorary Member, American Academy of Arts and Sciences
1988 Crafoord Prize (Stockholm). Awarded jointly to A. Grothendieck and P. Deligne
1989 Doctor Honoris Causa of the Vrije Universiteit Brussel.
1994 Elected Membre Associé, Académie Royale de Belgique
1995 Doctor Honoris Causa of Ecole Normale Supérieure
1995 Honorary Member of Moscow Mathematical Society
2003 Elected Honorary Member of the London Mathematical Society
2003 Elected Foreign Member, Accademia nazionale dei Lincei
2004 Recipient of Balzan Prize in Mathematics
2007 Elected Foreign Associate, National Academy of Sciences
2008 Recipient of Wolf Prize, Awarded jointly to P. Deligne, P. Griffiths and D. Mumford
2009 Foreign Member of the Royal Swedish Academy of Sciences
2009 Member of the American Philosophical Society

## List of Publications


1. P. Deligne, Voevodsky's lectures on motivic cohomology 2000/2001, in *Algebraic Topology*, pp. 355–409, Abel Symp., Vol. 4 (Springer, 2009).
2. P. Deligne, Equivalence numérique, équivalence cohomologique, et théorie de Lefschetz des variétés abéliennes sur les corps finis (in French) [Numerical equivalence, cohomological equivalence, and Lefschetz theory of abelian varieties on finite fields], edited by L. Clozel. *Pure Appl. Math. Q.* **5** (2009), No. 1, 495–506.
3. P. Deligne, Extended Euler congruence, *Funct. Anal. Other Math.* **2** (2009), No. 2–4, 249–250.
4. P. Deligne, B. Malgrange and J.-P. Ramis, Singularités irrégulières (in French) [Irregular singularities] Correspondance et documents. [Correspondence and documents] Documents Mathématiques (Paris) [Mathematical Documents (Paris)], Vol. 5 (Société Mathématique de France, 2007).
5. P. Deligne, La catégorie des représentations du groupe symétrique St, lorsque t n'est pas un entier naturel (in French) [The category of representations of the symmetric group St when t is not a natural number], in *Algebraic Groups and Homogeneous Spaces*, pp. 209–273 (Tata Inst. Fund. Res., 2007).
6. P. Deligne, The Hodge conjecture, in The Millennium Prize Problems, pp. 45–53 (Clay Math. Inst., 2006).
7. P. Deligne and A. B. Goncharov, Groupes fondamentaux motiviques de Tate mixte (in French) [Mixed Tate motivic fundamental groups], *Ann. Sci. École Norm. Sup.* (4) **38** (2005), No. 1, 1–56.
8. P. Deligne, Catégories tensorielles (in French) [Tensor categories], Dedicated to Yuri I. Manin on the occasion of his 65th birthday, *Mosc. Math. J.* **2** (2002), No. 2, 227–248.
9. P. Deligne and B. H. Gross, On the exceptional series, and its descendants, *C. R. Math. Acad. Sci. Paris* **335** (2002), No. 11, 877–881.
10. J.-L. Brylinski and P. Deligne, Central extensions of reductive groups by K2, *Publ. Math. Inst. Hautes Études Sci.* No. 94 (2001), 5–85.
11. P. Deligne, M. Goresky and R. MacPherson, L'algèbre de cohomologie du complément, dans un espace affine, d'une famille finie de sous-espaces affines (in French) [Cohomology algebra of the complement, in an affine space, of a finite family of affine subspaces], Dedicated to William Fulton on the occasion of his 60th birthday, *Michigan Math. J.* **48** (2000), 121–136.

12. P. Deligne, Super space descriptions of super gravity, in *Quantum Fields and Strings: A Course for Mathematicians*, Vol. 2 (Princeton, NJ, 1996/1997), pp. 1013–1016 (Amer. Math. Soc., 1999).
13. P. Deligne, Note on quantization, in *Quantum Fields and Strings: A Course for Mathematicians*, Vol. 1 (Princeton, NJ, 1996/1997), pp. 367–375 (Amer. Math. Soc., 1999).
14. P. Deligne and D. S. Freed, Sign manifesto, in *Quantum Fields and Strings: A Course for Mathematicians*, Vol. 1 (Princeton, NJ, 1996/1997), pp. 357–363 (Amer. Math. Soc., 1999).
15. P. Deligne and D. S. Freed, Supersolutions, in *Quantum Fields and Strings: A Course for Mathematicians*, Vol. 1 (Princeton, NJ, 1996/1997), pp. 227–355 (Amer. Math. Soc., 1999).
16. P. Deligne and D. S. Freed, Classical field theory, in *Quantum Fields and Strings: A Course for Mathematicians*, Vol. 1 (Princeton, NJ, 1996/1997), pp. 137–225 (Amer. Math. Soc., 1999).
17. P. Deligne, Notes on spinors, in *Quantum Fields and Strings: A Course for Mathematicians*, Vol. 1 (Princeton, NJ, 1996/1997), pp. 99–135 (Amer. Math. Soc., 1999).
18. P. Deligne and J. W. Morgan, Notes on supersymmetry (following Joseph Bernstein), in *Quantum Fields and Strings: A Course for Mathematicians*, Vol. 1 (Princeton, NJ, 1996/1997), pp. 41–97 (Amer. Math. Soc., 1999).
19. P. Deligne, Quelques idées maîtresses de l'œuvre de A. Grothendieck (in French) [Some fundamental ideas in the works of A. Grothendieck] *Matériaux pour l'histoire des Mathématiques au XXe Siècle* (Nice, 1996), pp. 11–19, Sémin. Congr., Vol. 3 (Soc. Math. France, 1998).
20. P. Deligne, Action du groupe des tresses sur une catégorie (in French) [Action of the braid group on a category], *Invent. Math.* **128** (1997), No. 1, 159–175.
21. P. Deligne, Local behavior of Hodge structures at infinity, in *Mirror Symmetry*, II, pp. 683–699, AMS/IP Stud. Adv. Math., Vol. 1 (Amer. Math. Soc., 1997).
22. P. Deligne, Extensions centrales de groupes algébriques simplement connexes et cohomologie galoisienne (in French) [Central extensions of simply connected algebraic groups and Galois cohomology], *Inst. Hautes Études Sci. Publ. Math.* No. 84 (1996), 35–89 (1997).
23. P. Deligne and R. de Man, La série exceptionnelle de groupes de Lie. II (in French) [The exceptional series of Lie groups. II], *C. R. Acad. Sci. Paris Sér. I Math.* **323** (1996), No. 6, 577–582.

24. P. Deligne, La série exceptionnelle de groupes de Lie (in French) [The exceptional series of Lie groups], *C. R. Acad. Sci. Paris Sér. I Math.* **322** (1996), No. 4, 321–326.
25. P. Deligne, Déformations de l'algèbre des fonctions d'une variété symplectique: comparaison entre Fedosov et De Wilde, Lecomte (in French) [Deformations of the algebra of functions of a symplectic manifold: comparison of the Fedosov and De Wilde-Lecomte methods] *Selecta Math. (N.S.)* **1** (1995), No. 4, 667–697.
26. E. Cattani, P. Deligne and A. Kaplan, On the locus of Hodge classes, *J. Amer. Math. Soc.* **8** (1995), No. 2, 483–506.
27. P. Deligne and G. Pappas, Singularités des espaces de modules de Hilbert, en les caractéristiques divisant le discriminant (in French) [Singularities of Hilbert modular varieties in characteristics dividing the discriminant] *Compositio Math.* **90** (1994), No. 1, 59–79.
28. A. Beĭlinson and P. Deligne, Interprétation motivique de la conjecture de Zagier reliant polylogarithmes et régulateurs (in French) [Motivic interpretation of the Zagier conjecture connecting polylogarithms and regulators], in *Motives* (Seattle, WA, 1991), pp. 97–121, Proc. Sympos. Pure Math., Vol. 55, Part 2 (Amer. Math. Soc., 1994).
29. P. Deligne, Structures de Hodge mixtes réelles (in French) [Real mixed Hodge structures], in *Motives* (Seattle, WA, 1991), pp. 509–514, Proc. Sympos. Pure Math., Vol. 55, Part 1 (Amer. Math. Soc., 1994).
30. P. Deligne, À quoi servent les motifs? (in French) [What is the use of motives?], in *Motives* (Seattle, WA, 1991), pp. 143–161, Proc. Sympos. Pure Math., Vol. 55, Part 1 (Amer. Math. Soc., 1994).
31. P. Deligne, Décompositions dans la catégorie dérivée (in French) [Decompositions in the derived category], in *Motives* (Seattle, WA, 1991), pp. 115–128, Proc. Sympos. Pure Math., Vol. 55, Part 1 (Amer. Math. Soc., 1994).
32. P. Deligne and G. D. Mostow, Commensurabilities among lattices in PU(1, $n$), in Annals of Mathematics Studies, Vol. 132 (Princeton Univ. Press, 1993).
33. P. Deligne, Le symbole modéré (in French) [The tame symbol], *Inst. Hautes Études Sci. Publ. Math.* No. 73 (1991), 147–181.
34. P. Deligne, Groupe de Heisenberg et réalité (in French) [The Heisenberg group and reality], *J. Amer. Math. Soc.* **4** (1991), No. 1, 197–206.
35. P. Deligne, Catégories tannakiennes (in French) [Tannakian categories], in *The Grothendieck Festschrift*, Vol. II, pp. 111–195, Progr. Math., Vol. 87 (Birkhäuser, 1990).

36. P. Deligne and A. Dimca, Filtrations de Hodge et par l'ordre du pôle pour les hypersurfaces singulières (in French) [Hodge filtration and filtration by the order of the pole for singular hypersurfaces], *Ann. Sci. École Norm. Sup.* (4) **23** (1990), No. 4, 645–656.
37. P. Deligne, Le groupe fondamental de la droite projective moins trois points (in French) [The fundamental group of the projective line minus three points], in *Galois Groups over Q* (Berkeley, CA, 1987), pp. 79–297, Math. Sci. Res. Inst. Publ., Vol. 16 (Springer, 1989).
38. P. Deligne, Le déterminant de la cohomologie (in French) [The determinant of the cohomology], in *Current Trends in Arithmetical Algebraic Geometry* (Arcata, Calif., 1985), pp. 93–177, Contemp. Math., Vol. 67 (Amer. Math. Soc., 1987).
39. P. Deligne and D. Husemoller, Survey of Drinfel'd modules, in *Current Trends in Arithmetical Algebraic Geometry* (Arcata, Calif., 1985), pp. 25–91, Contemp. Math., Vol. 67 (Amer. Math. Soc., 1987).
40. P. Deligne, Un théorème de finitude pour la monodromie (in French) [A finiteness theorem for monodromy], in *Discrete Groups in Geometry and Analysis* (New Haven, Conn., 1984), 1–19, Progr. Math., Vol. 67 (Birkhäuser, 1987).
41. P. Deligne and L. Illusie, Relèvements modulo $p_2$ et décomposition du complexe de de Rham (in French) [Liftings modulo and decomposition of the de Rham complex], *Invent. Math.* **89** (1987), No. 2, 247–270.
42. J. Bernstein, P. Deligne and D. Kazhdan, Trace Paley–Wiener theorem for reductive $p$-adic groups, *J. Anal. Math.* **47** (1986), 180–192.
43. P. Deligne and G. D. Mostow, Monodromy of hypergeometric functions and nonlattice integral monodromy, Inst. *Hautes Études Sci. Publ. Math.* No. 63 (1986), 5–89.
44. P. Deligne, Représentations $l$-adiques (in French) [$l$-adic representations], in *Seminar on Arithmetic Bundles: The Mordell Conjecture* (Paris, 1983/84). Astérisque No. 127 (1985), 249–255.
45. P. Deligne, Le lemme de Gabber (in French) [Gabber's lemma], in *Seminar on Arithmetic Bundles: The Mordell Conjecture* (Paris, 1983/84). Astérisque No. 127 (1985), 131–150.
46. P. Deligne, Preuve des conjectures de Tate et de Shafarevitch (d'après G. Faltings) (in French) [Proof of the Tate and Shafarevich conjectures (after G. Faltings)], in *Seminar Bourbaki*, Vol. 1983/84. Astérisque No. 121–122 (1985), 25–41.
47. P. Deligne, Les corps locaux de caractéristique, limites de corps locaux de caractéristique (in French) [Local fields of characteristic which are limits of

local fields of characteristic], in *Representations of Reductive Groups over a Local Field*, pp. 119–157 (Hermann, 1984).
48. P. Deligne, D. Kazhdan and M.-F. Vignéras, Représentations des algèbres centrales simples $p$-adiques (in French) [Representations of central simple $p$-adic algebras], in *Representations of Reductive Groups over a Local Field*, pp. 33–117 (Hermann, 1984).
49. P. Deligne, Intégration sur un cycle évanescent (in French) [Integration over a vanishing cycle], *Invent. Math.* **76** (1984), No. 1, 129–143.
50. P. Deligne and D. Sullivan, Division algebras and the Hausdorff–Banach–Tarski paradox, *Enseign. Math.* (2) **29** (1983), No. 1–2, 145–150.
51. P. Deligne and G. Lusztig, Duality for representations of a reductive group over a finite field, II, *J. Algebra* **81** (1983), No. 2, 540–545.
52. A. A. Beĭlinson, J. Bernstein and P. Deligne, Faisceaux pervers (in French) [Perverse sheaves], in *Analysis and Topology on Singular Spaces*, I (Luminy, 1981), pp. 5–171, Astérisque No. 100 (Soc. Math. France, 1982).
53. P. Deligne, J. S. Milne, A. Ogus and K.-Y. Shih, Hodge cycles, motives, and Shimura varieties, in Lecture Notes in Mathematics, Vol. 900 (Springer, 1982).
54. P. Deligne and G. Lusztig, Duality for representations of a reductive group over a finite field, *J. Algebra* **74** (1982), No. 1, 284–291.
55. P. Deligne, Cristaux ordinaires et coordonnées canoniques (in French) [Ordinary crystals and canonical coordinates], with the collaboration of L. Illusie, with an appendix by Nicholas M. Katz, in *Algebraic Surfaces* (Orsay, 1976–78), pp. 80–137, Lecture Notes in Math., Vol. 868 (Springer, 1981).
56. P. Deligne, Relèvement des surfaces en caractéristique nulle (in French) [Lifting of the surfaces with characteristic zero], prepared for publication by Luc Illusie, in *Algebraic Surfaces* (Orsay, 1976–1978), pp. 58–79, Lecture Notes in Math., Vol. 868 (Springer, 1981).
57. P. Deligne, Le groupe fondamental du complément d'une courbe plane n'ayant que des points doubles ordinaires est abélien (d'après W. Fulton) (in French) [The fundamental group of the complement of a plane curve having only ordinary double points is abelian (after W. Fulton)] in *Bourbaki Seminar*, Vol. 1979/80, pp. 1–10, Lecture Notes in Math., Vol. 842 (Springer, 1981).
58. P. Deligne and G. Henniart, Sur la variation, par torsion, des constantes locales d'équations fonctionnelles de fonctions (in French) [On the variation through torsion of the local constants of functional equations of functions] *Invent. Math.* **64** (1981), No. 1, 89–118.
59. P. Deligne, La conjecture de Weil. II (in French) [Weil's conjecture. II] *Inst. Hautes Études Sci. Publ. Math.* No. 52 (1980), 137–252.

60. P. Deligne, Cycles de Hodge absolus et périodes des intégrales des variétés abéliennes (in French) [*Abelian Functions and Transcendental Numbers*] (Colloq., École Polytech., Palaiseau, 1979) *Mém. Soc. Math. France* (*N.S.*) (1980/81), No. 2, pp. 23–33.
61. P. Deligne and K. A. Ribet, Values of abelian $L$-functions at negative integers over totally real fields, *Invent. Math.* **59** (1980), No. 3, 227–286.
62. P. Deligne, Sommes de Gauss cubiques et revêtements de [d'après S. J. Patterson]. (French), in *Séminaire Bourbaki* (1978/79), Exp. No. 539, pp. 244–277, Lecture Notes in Math., Vol. 770 (Springer, 1980).
63. P. Deligne, Variétés de Shimura: Interprétation modulaire, et techniques de construction de modèles canoniques (in French), in *Automorphic Forms, Representations and L-Functions* (Proc. Sympos. Pure Math., Oregon State Univ., Corvallis, Ore., 1977), Part 2, pp. 247–289, Proc. Sympos. Pure Math., XXXIII (Amer. Math. Soc., 1979).
64. P. Deligne, Valeurs de fonctions et périodes d'intégrales (in French), with an Appendix by N. Koblitz and A. Ogus. Proc. Sympos. Pure Math., XXXIII, *Automorphic Forms, Representations and L-Functions* (Proc. Sympos. Pure Math., Oregon State Univ., Corvallis, Ore., 1977), Part 2, pp. 313–346 (Amer. Math. Soc., 1979).
65. P. Deligne, Extensions centrales non résiduellement finies de groupes arithmétiques (in French) *C. R. Acad. Sci. Paris Sér. A-B* **287** (1978), No. 4, A203–A208.
66. P. Deligne, Cohomologie étale. Séminaire de Géométrie Algébrique du Bois-Marie SGA 4. Avec la collaboration de J. F. Boutot, A. Grothendieck, L. Illusie et J. L. Verdier, Lecture Notes in Mathematics, Vol. 569 (Springer, 1977).
67. P. A. Griffits, P. Deligne, J. Morgan and D. Sjullivan, Real homotopy theory of Kähler manifolds (in Russian) translated from the English by Ju. I. Manin, *Usp. Mat. Nauk* **32** (1977), No. 3(195), 119–152.
68. P. Deligne, Les difféomorphismes du cercle (d'après M. R. Herman) (in French), in *Séminaire Bourbaki*, Vol. 1975/76, 28ème année, Exp. No. 477, pp. 99–121, Lecture Notes in Math., Vol. 567 (Springer, 1977).
69. P. Deligne, Les constantes locales de l'équation fonctionnelle de la fonction d'Artin d'une représentation orthogonale (in French), *Invent. Math.* **35** (1976), 299–316.
70. P. Deligne, Le support du caractère d'une représentation supercuspidale (in French), *C. R. Acad. Sci. Paris Sér. A-B* **283** (1976), No. 4, Aii, A155–A157.
71. P. Deligne and G. Lusztig, Representations of reductive groups over finite fields, *Ann. of Math.* (2) **103** (1976), No. 1, 103–161.

72. P. Deligne, Poids dans la cohomologie des variétés algébriques, in *Proc. of the International Congress of Mathematicians* (Vancouver, B. C., 1974), Vol. 1, pp. 79–85 (Canad. Math. Congress, 1975).
73. P. Deligne and D. Sullivan, Fibrés vectoriels complexes à groupe structural discret, *C. R. Acad. Sci. Paris Sér. A-B* **281** (1975), No. 24, Ai, A1081–A1083.
74. P. Deligne, Correction to: "Formes modulaires et représentations de GL(2)", [*Modular Functions of One Variable*, II (Proc. Internat. Summer School, Univ. Antwerp, Antwerp, 1972), pp. 55–105, Lecture Notes in Math., Vol. 349 (Springer, 1973)], in *Modular Functions of One Variable*, IV (Proc. Internat. Summer School, Univ. Antwerp, Antwerp, 1972), p. 148, Lecture Notes in Math., Vol. 476 (Springer, 1975).
75. P. Deligne, Courbes elliptiques: Formulaire d'après J. Tate (in French), in *Modular Functions of One Variable*, IV (Proc. Internat. Summer School, Univ. Antwerp, Antwerp, 1972), pp. 53–73, Lecture Notes in Math., Vol. 476 (Springer, 1975).
76. P. Deligne, The Weil conjecture. I (in Russian), translated from the French (Inst. Hautes Études Sci. Publ. Math. No. 43 (1974), 273–307) by Ju. I. Manin, *Usp. Mat. Nauk* **30** (1975), No. 5(185), 159–196.
77. P. Deligne, Correction to: "Les constantes des équations fonctionelles des fonctions", [*Modular Functions of One Variable*, II (Proc. Internat. Summer School, Univ. Antwerp, Antwerp, 1972), pp. 501–597, Lecture Notes in Math., Vol. 349 (Springer, 1973)] (in French), in *Modular Functions of One Variable*, IV (Proc. Internat. Summer School, Univ. Antwerp, Antwerp, 1972), p. 149, Lecture Notes in Math., Vol. 476 (Springer, 1975).
78. P. Deligne, P. Griffiths, J. Morgan and D. Sullivan, Real homotopy theory of Kähler manifolds, *Invent. Math.* **29** (1975), No. 3, 245–274.
79. P. Deligne and M. Rapoport, Correction to: "Les schémas de modules de courbes elliptiques" [*Modular Functions of One Variable*, II (Proc. Internat. Summer School, Univ. Antwerp, Antwerp, 1972), pp. 143–316, Lecture Notes in Math., Vol. 349 (Springer, 1973)] (in French), in *Modular Functions of One Variable*, IV (Proc. Internat. Summer School, Univ. Antwerp, Antwerp, 1972), p. 149, Lecture Notes in Math., Vol. 476 (Springer, 1975).
80. P. Deligne, Théorie de Hodge. III (in French) *Inst. Hautes Études Sci. Publ. Math.* No. 44 (1974), 5–77.
81. P. Deligne and J.-P. Serre, Formes modulaires de poids (in French) *Ann. Sci. École Norm. Sup.* (4) **7** (1974), 507–530.
82. P. Deligne, La conjecture de Weil. I (in French), *Inst. Hautes Études Sci. Publ. Math.* No. 43 (1974), 273–307.

83. P. Deligne, Variétés unirationnelles non rationnelles [d'après M. Artin et D. Mumford] (in French), in *Séminaire Bourbaki*, 24ème année (1971/1972), Exp. No. 402, pp. 45–57. Lecture Notes in Math., Vol. 317 (Springer, 1973).
84. P. Deligne, Les constantes des équations fonctionnelles des fonctions (in French), in *Modular Functions of One Variable*, II (Proc. Internat. Summer School, Univ. Antwerp, Antwerp, 1972), pp. 501–597. Lecture Notes in Math., Vol. 349 (Springer, 1973).
85. P. Deligne, Formes modulaires et représentations de GL(2) (in French), in *Modular Functions of One Variable*, II (Proc. Internat. Summer School, Univ. Antwerp, Antwerp, 1972), pp. 55–105. Lecture Notes in Math., Vol. 349 (Springer, 1973).
86. P. Deligne and M. Rapoport, Les schémas de modules de courbes elliptiques (in French), *Modular Functions of One Variable*, II (Proc. Internat. Summer School, Univ. Antwerp, Antwerp, 1972), pp. 143–316. Lecture Notes in Math., Vol. 349 (Springer, 1973).
87. P. Deligne, Les immeubles des groupes de tresses généralisés (in French), *Invent. Math.* **17** (1972), 273–302.
88. P. Deligne, Les intersections complètes de niveau de Hodge un (in French), *Invent. Math.* **15** (1972), 237–250.
89. P. Deligne, La conjecture de Weil pour les surfaces (in French), *Invent. Math.* **15** (1972), 206–226.
90. P. Deligne, Travaux de Shimura (in French), in *Séminaire Bourbaki*, 23ème année (1970/71), Exp. No. 389, pp. 123–165. Lecture Notes in Math., Vol. 244 (Springer, 1971).
91. P. Deligne, Théorie de Hodge. II (in French), *Inst. Hautes Études Sci. Publ. Math.* No. **40** (1971), 5–57.
92. P. Deligne, Théorie de Hodge. I (in French), in *Actes du Congrès International des Mathématiciens* (Nice, 1970), Tome 1, pp. 425–430 (Gauthier-Villars, 1971).
93. P. Deligne, Équations différentielles à points singuliers réguliers (in French), in Lecture Notes in Mathematics, Vol. 163 (Springer-Verlag, 1970).
94. P. Deligne, Les constantes des équations fonctionnelles (in French), 1970 Séminaire Delange-Pisot-Poitou: 1969/70, Théorie des Nombres, Fasc. 2, Exp. 19 bis 13 pp. Secrétariat Mathématique, Paris.
95. P. Deligne and D. Mumford, The irreducibility of the space of curves of given genus, *Inst. Hautes Études Sci. Publ. Math.* No. 36 (1969), 75–109.
96. P. Deligne, Variétés abéliennes ordinaires sur un corps fini (in French), *Invent. Math.* **8** (1969), 238–243.

97. P. Deligne, Théorème de Lefschetz et critères de dégénérescence de suites spectrales (in French), *Inst. Hautes Études Sci. Publ. Math.* No. 35 (1968), 259–278.
98. P. Deligne, Congruences sur le nombre de sous-groupes d'ordre dans un groupe fini (in French), *Bull. Soc. Math. Belg.* **18** (1966), 129–132.

Proceedings of Symposia in Pure Mathematics
Volume 28, 1976

# An Overview of Deligne's Work on Hilbert's Twenty-First Problem

Nicholas M. Katz

Abstract Hilbert's twenty-first problem on the existence of differential equations with regular singular points and prescribed monodromy is interpreted as a GAGA-type problem of algebraic-analytic comparison. Deligne's solution is outlined. Several open questions are raised.

## The setting of the problem

Let X be a complete connected non-singular curve over $\mathbb{C}$, whose underlying complex manifold is thus a compact Riemann surface. Let U be a non-empty Zariski open set in X, the complement in X of a finite (possibly empty) set of closed points. The underlying complex manifold $U^{an}$ is thus a finitely punctured Riemann surface.

Consider a linear homogeneous differential equation of rank n on U. When $X = \mathbb{P}^1$ and $U \subset \mathbb{P}^1 - \{\infty\}$, this simply means an $n \times n$ system

$$\frac{d}{dz}\vec{f} + A(z)\cdot\vec{f} = 0$$

$$\begin{cases} \vec{f} = \begin{pmatrix} f_1(z) \\ \vdots \\ f_n(z) \end{pmatrix} \\ A(z) \text{ an } n \times n \text{ matrix of \underline{rational} functions of } z, \\ \quad \text{holomorphic on } U. \end{cases}$$

As we are so fond of emphasizing to engineering students, this includes the case of the n'th order linear homogeneous equation

$$\left(\frac{d}{dz}\right)^n f + a_1(z)\cdot\left(\frac{d}{dz}\right)^{n-1} f + \ldots + a_n(z)f = 0,$$

by taking for the matrix A(z) the particular choice

$$A(z) = \begin{pmatrix} 0 & -1 & 0 & \ldots & 0 \\ 0 & 0 & -1 & \ldots & 0 \\ \vdots & & & & \\ 0 & \ldots & \ldots & 0 & -1 \\ a_n(z) & \ldots & \ldots & \ldots & a_1(z) \end{pmatrix}$$

In case $X$ has higher genus, we have no global coordinate $z$ at our disposal, so we are led to define a differential equation on $U$ to be a pair $(M, \nabla)$ consisting of a locally free coherent algebraic sheaf $M$ on $U$ together with a connection $\nabla: M \longrightarrow M \otimes \Omega^1_{U/\mathbb{C}}$ (this simply means an additive mapping which satisfies the "product rule": for $f \in \mathcal{O}_U$ and $m \in M$, $\nabla(fm) = m \otimes df + f \otimes \nabla(m)$). In the example of the $n \times n$ system on an open set in $\mathbb{P}^1$, the locally free coherent algebraic sheaf $M$ is simply $(\mathcal{O}_U)^n$, and the connection $\nabla$ is the map $(\mathcal{O}_U)^n \longrightarrow (\Omega^1_{U/\mathbb{C}})^n$ given by

$$\nabla(\vec{f}\,) = d\vec{f} + A(z)\cdot\vec{f}\,dz$$

It is sometimes convenient to view a connection $\nabla$ as a way of having derivations of $\mathcal{O}_U$ act on $M$, viewing $\nabla$ as an $\mathcal{O}_U$-linear map $\underline{\mathrm{Der}}(U/\mathbb{C}) \longrightarrow \underline{\mathrm{End}}_{\mathbb{C}}(M)$, $D \longrightarrow \nabla(D)$ where $\nabla(D): M \longrightarrow M$ is the composite

$$\begin{array}{ccc} M & \xrightarrow{\ \nabla\ } & M \otimes \Omega^1_U \\ & \searrow^{\nabla(D)} & \downarrow{\scriptstyle \mathrm{id} \otimes \text{contract with } D} \\ & & M \end{array}$$

Thus, in the above example, $\nabla(\frac{d}{dz})$ is given by

$$\nabla(\frac{d}{dz})\vec{f} = \frac{d\vec{f}}{dz} + A(z)\vec{f}$$

Notice that the solutions of the differential equation become the sections of $M^{\nabla} \overset{\mathrm{dfn}}{=}$ the kernel of $\nabla: M \longrightarrow M \otimes \Omega^1_U$. If we fix a point $z_0 \in U$, then the space $S$ of germs of local holomorphic solutions near (in the analytic sense) $z_0$ is an $n = \mathrm{rank}\,(M)$-dimensional $\mathbb{C}$-space, by the local existence theorem for differential equations. Given any loop $\gamma$ in $U^{an}$ starting and ending at $z_0$, "analytic continuation of solutions along $\gamma$" defines an automorphism of $S$. It is immediate that this automorphism is "multiplicative" in $\gamma$, and that it depends only on the homotopy class of $\gamma$ in $\pi_1(U^{an}, z_0)$. In this way the fundamental group $\pi_1(U^{an}, z_0)$ acts on the $\mathbb{C}$-space S, defining the monodromy representation of the differential equation.

For example, if we take the differential equation

$$z \frac{df}{dz} = \alpha f \qquad \alpha \in \mathbb{C}$$

on the punctured complex "plane" $\mathbb{C} - \{0\}$, its solution is the function $z^{\alpha} = \exp(\alpha \log z)$, which under analytic continuation along a path $\gamma$ which loops once counterclockwise around the origin (e.g. $z \longrightarrow e^{2\pi i\theta} z$, $0 \leq \theta \leq 1$) is transformed into $e^{2\pi i \alpha} \cdot z^{\alpha}$. The fundamental group of $\mathbb{C} - \{0\}$ is isomorphic to $\mathbb{Z}$, with generator $\gamma$, and the monodromy representation in $\mathbb{C}^{x} = GL(1, \mathbb{C})$ is given by $\gamma \longrightarrow e^{2\pi i \alpha}$

Let us now recall the notion of a "regular singular point" of a differential equation. Classically, one considered an n'th order differential equation

$$\left(\frac{d}{dz}\right)^{n} f + a_1(z) \left(\frac{d}{dz}\right)^{n-1} f + \dots + a_n(z) f = 0$$

on an open set $U \subset \mathbb{P}^1$. A point $\alpha \in \mathbb{P}^1 - U$ is called a regular singular point of this equation if in any punctured angular sector around $\alpha$

the local holomorphic solutions satisfy a growth estimate

$$\begin{cases} \text{if } \alpha \text{ finite} & |f(z)| = O(|z-\alpha|^{-N}) \quad \text{for some } N \geq 0, \text{ as } z \longrightarrow \alpha \\ \text{if } \alpha = \infty & |f(z)| = O(|z|^{N}) \quad \text{for some } N \geq 0, \text{ as } z \longrightarrow \infty. \end{cases}$$

This notion is apparently analytic, rather than algebraic, in character. However, Fuchs discovered that in fact the notion is purely algebraic. He proved that a necessary and sufficient condition that $\alpha$ be a regular singular point is that the functions

$$\begin{cases} (z-\alpha)^i \cdot a_i(z), & i=1,\dots,n \quad \text{if } \alpha \text{ finite} \\ z^i a_i(z) & i=1,\ \ .,n \quad \text{if } \alpha = \infty \end{cases}$$

should all be holomorphic at $z = \alpha$.

We may rephrase Fuch's criterion as follows. Let $t$ be a uniformizing parameter at $\alpha$ (e.g. $t = z - \alpha$ if $\alpha$ finite, $t = 1/z$ if $\alpha = \infty$), and denote by D the differential operator $t\frac{d}{dt}$ Rewrite the equation as a monic equation in D:

$$D^n f + b_1(t)\, D^{n-1} f + \dots + b_n(t) \cdot f = 0.$$

Then $t=0$ is a regular singular point if and only if all of the functions $b_i(t)$ are holomorphic at $t=0$.

The notion of a regular singular point of a "fancy" differential equation $(M, \nabla)$ on U is defined as before in terms of the growth of local holomorphic solution vectors $\vec{f}$ near the singular point, where we measure the growth by expressing $\vec{f}$ in a local basis of any locally free coherent algebraic sheaf $\overline{M}$ on X which extends M (on a curve, such an $\overline{M}$ always exists, for any locally free M). In this more general setting, Fuch's algebraic criterion for regular singular points, as rephrased by Deligne, is that there exist a locally free coherent algebraic sheaf $\overline{M}$ on X extending M, such that near any singular point $\alpha$ of the equation, the derivation $D = t\frac{d}{dt}$ (t: uniformizing parameter at $\alpha$) should act (through $\nabla$) stably on $\overline{M}$, i.e. $\nabla(D)(\overline{M}) \subset \overline{M}$.

Notice the compatibility with Fuch's criterion for n'th order equations: If we convert the equation

$$D^n f + b_1(t)\, D^{n-1} f + \ \dots + b_n(t) f = 0$$

into system form, it becomes

$$\nabla(D)\,\vec{f} = D\vec{f} + \begin{pmatrix} 0 & -1 & 0 & \dots & & 0 \\ \vdots & & & & & \\ 0 & & & & 0 & -1 \\ b_n(t) & & & & & b_1(t) \end{pmatrix} \vec{f},$$

so that if the $b_i(t)$ are all holomorphic at $t=0$, then indeed $\nabla(D)$ acts stably on n-tuples of functions holomorphic at $t=0$.

Statement of the Problem

Hilbert's twenty-first problem asks whether any finite-dimensional complex representation of the fundamental group $\pi_1(U^{an})$ can be obtained as the monodromy representation of a differential equation on $U$ with regular singular points.

One could ask the same question without requiring regular singular points, but the simplest examples show that then there will be too many equations with given monodromy. For example, take $U = \mathbb{A}^1$ (thus $U^{an} = \mathbb{C}$), and the trivial one-dimensional representation of $\pi_1(U^{an}) = 0$. For any polynomial $P(z) \in \mathbb{C}[z]$, the differential equation

$$\frac{df}{dz} = P(z) \cdot f$$

has solution

$$f = \exp\left(\int_0^z P(t)\,dt\right)$$

which is an entire (singlevalued) function of $z$, so without monodromy. But as differential equations on the algebraic variety $\mathbb{A}^1$, these are pairwise non-isomorphic; only the choice $P \equiv 0$ gives regular singular points. For this reason, one insists on regular singular points.

In the case $U = \mathbb{P}^1 - \{0, 1, \infty\}$, the triply punctured Riemann sphere, the affirmative solution of Hilbert's problem for two-dimensional representations of $\pi_1(U^{an})$ goes back to Riemann, and amounts to the theory of the hypergeometric equation. Indeed, Hilbert's twenty-first problem is often referred to as the Riemann problem. Traditionally, it was viewed as a problem in "function theory," and in that setting has been solved repeatedly, by such men as Birkhoff, Plemelj, and most recently by Röhrl. But as Lipman

Bers remarked in his talk on uniformization, if a problem is worth solving, it's worth solving several times.

Algebro-Geometric Perspective

From the perspective of algebraic geometry, the point of the problem is to describe in purely algebraic terms on U (differential equations on U with regular singular points) a purely topological invariant of $U^{an}$ (the finite dimensional complex representations of its fundamental group).

The earliest result of this kind was the realization that complete connected nonsingular curves over $\mathbb{C}$ are identical with compact Riemann surfaces, the identification provided by the fact that a rational function on the curve is the same as a meromorphic function on the Riemann surface.

This fact was generalized by Chow and Kodaira, who proved that any compact complex connected surface with two algebraically independent (over $\mathbb{C}$) meromorphic functions was a nonsingular projective algebraic surface. The obvious generalization of this result to more than two dimensions is false, and its "counterexamples" have been studied these last few years by Artin and Moishezon as "algebraic spaces".

Around the same time, Chow proved that any closed analytic subset of $\mathbb{P}^n$ (i.e. defined locally by the vanishing of analytic functions) was in fact algebraic, and (by considering the graph) that any analytic map between such was in fact algebraic.

The final step in this direction was Serre's GAGA theorem according to which if X is any projective algebraic variety over $\mathbb{C}$, then the coherent algebraic sheaves $\mathcal{F}$ on X are identical with the coherent analytic sheaves $\mathcal{F}^{an}$ on $X^{an}$ (i.e. $\mathcal{F} \to \mathcal{F}^{an}$ is an equivalence of categories) and the cohomology groups of such sheaves are the "same" whether computed algebraically or analytically $(H^i(X, \mathcal{F}) \xrightarrow{\sim} H^i(X^{an}, \mathcal{F}^{an}))$.

For example, if we take $X$ to be non-singular, and $\mathcal{F} = \Omega^p_{X/\mathbb{C}}$ the sheaf of germs of holomorphic p-forms, then we get a purely algebraic description of the Hodge cohomology groups

$$H^q(X, \Omega^p_{X/\mathbb{C}}) \xrightarrow{\sim} H^q(X^{an}, \Omega^p_{X^{an}/\mathbb{C}})$$

$$\wr\| \text{ (Dolbeaut)}$$

$$\text{harmonic forms of type } (p,q)$$

If, instead of a single $\Omega^p$, we take the entire de Rham complex $\Omega^{\bullet}_{X/\mathbb{C}}$, then we get a purely algebraic description of the complex cohomology groups in terms of algebraic "hypercohomology."

$$\mathbb{H}^i(X, \Omega^{\bullet}_{X/\mathbb{C}}) \xrightarrow{\sim} \mathbb{H}^i(X^{an}, \Omega^{\bullet}_{X^{an}/\mathbb{C}})$$

$$\wr\uparrow \text{ (holomorphic Poincaré Lemma)}$$

$$H^i(X^{an}, \mathbb{C})$$

The general notion of a differential equation

Let $X$ be any non-singular connected algebraic variety over $\mathbb{C}$. A "differential equation" on $X$ is by definition a pair $(M, \nabla)$ consisting of a locally free coherent algebraic sheaf $M$ on $X$, together with an integrable connection

$$\nabla : M \longrightarrow M \otimes \Omega^1_{X/\mathbb{C}}$$

This means that $\nabla$ is an additive mapping which satisfies the product rule $\nabla(fm) = m \cdot df + f\nabla(m)$ for all $f \in \mathcal{O}_X$, $m \in M$, and is integrable in the sense that if we make $D \in \underline{\mathrm{Der}}(X/\mathbb{C})$ act on $M$ as the composite

$$\begin{array}{ccc} M & \xrightarrow{\ \nabla\ } & M \otimes \Omega^1_{X/\mathbb{C}} \\ & \underset{\nabla(D)}{\searrow} & \downarrow \text{ id} \otimes \text{ contract with } D \\ & & M \end{array}$$

then we have $\nabla([D_1, D_2]) = [\nabla(D_1), \nabla(D_2)]$. In local coordinates, this is a "completely integrable Pfaffian system."

Analogously, we define a differential equation on a complex manifold as an integrable connection on a locally free coherent analytic sheaf.

A holomorphic version of the problem, and its solution

Let $V$ be an arbitrary connected complex manifold. We may ask whether any finite dimensional complex representation of $\pi_1(V)$ arises as the monodromy representation associated to a differential equation on $V$. The answer is easily seen to be "yes," as follows.

On any complex manifold, the usual Frobenius existence theorem says that the functor "sheaf of germs of sections killed by $\nabla$" is an equivalence of categories between the category of analytic differential equations and the category of locally constant sheaves of complex vector spaces. (The inverse functor is to tensor a locally constant sheaf $E$ with the holomorphic structural sheaf, and to endow the tensor product $E \otimes \mathcal{O}^{an}$ with the connection $id \otimes d$.)

Finally, on any "reasonable" topological space, picking a base point defines an equivalence of categories between the category of locally constant sheaves of anythings (for example, of finite dimensional complex vector spaces) and the category of representations of the fundamental group on those anythings.

Combining these two equivalences, we see that the functor "associated monodromy representation" defines an equivalence of categories between the category of all differential equations on a connected complex manifold $V$ and the category of finite-dimensional complex representations of its fundamental group.

A compact form of the problem and its solution

Suppose that X is a projective non-singular connected algebraic variety over $\mathbb{C}$. Then Serre's GAGA theorem tells us that the functor $(M, \nabla) \longrightarrow (M^{an}, \nabla^{an})$ is an equivalence of categories between the algebraic differential equations (on X) and the analytic differential equations (on $X^{an}$). Combining this with the holomorphic solution to the problem, we see that the category of algebraic differential equations on X is equivalent to the category of finite dimensional complex representations of $\pi_1(X^{an})$. We should remark here that because X is supposed compact, there are no "missing points," and hence there are no regularity conditions to impose at them. Thus we have solved the compact case of the twenty-first problem "just" by using GAGA. Even in the case when X is one-dimensional (i.e. $X^{an}$ is a compact Riemann surface), this is of some interest (cf. open problem 1) below).

The non-compact case

As Grothendieck was the first to emphasize, the key result in all such questions is Hironaka's resolution of singularities, in the following form.

Let U be a smooth quasi-projective (Zariski-open in a projective) variety over $\mathbb{C}$. Then there exists a projective smooth variety $X/\mathbb{C}$ such that $U \subset X$ as an open dense set, and such that the closed set $D = X - U$ in X is a union of smooth divisors (subvarieties of codimension one) $D_i$ which cross transversely. Furthermore, if

$$\begin{array}{ccc} & & X_1 \\ & \nearrow & \\ U & & \\ & \searrow & \\ & & X_2 \end{array}$$

are two such compactifications, there always exists a third which "dominates" both:

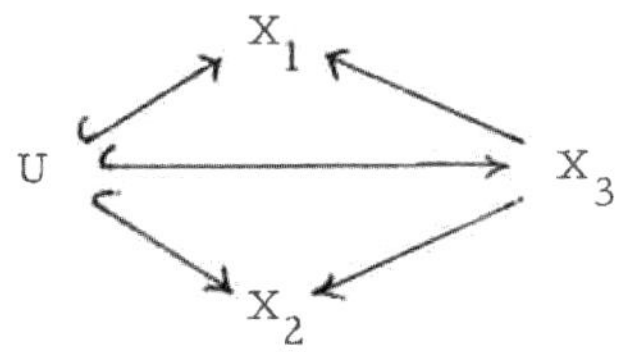


In this setup, we denote by $\underline{\mathrm{Der}}_D(X/\mathbb{C})$ the sub-sheaf of the tangent sheaf consisting of those derivations which preserve the ideals of each of the $D_i$. At a point on $X$ where $r$ of the divisors, say $D_1, \ldots, D_r$, cross, we may choose local coordinates $x_1, \ldots, x_n$ such that $D_i$ is defined by $x_i = 0$ for $i \leq r$. There, $\underline{\mathrm{Der}}_D(X/\mathbb{C})$ is free on $x_i \frac{\partial}{\partial x_i}$, $i \leq r$ and $\frac{\partial}{\partial x_j}$, $j > r$. The linear dual of $\underline{\mathrm{Der}}_D(X/\mathbb{C})$ is noted $\Omega^1_{X/\mathbb{C}}(\log D)$. It is locally free on

$$\frac{dx_1}{x_1}, \ldots, \frac{dx_r}{x_r}, \quad dx_{r+1}, \ldots, dx_n \quad \text{near this same point.}$$

In case $U$ is connected and <u>one-dimensional</u>, the $X$ is unique — it's just the complete non-singular model of the function field of $U$, the sheaf $\mathrm{Der}_D(X/\mathbb{C})$ is the subsheaf of $\underline{\mathrm{Der}}(X/\mathbb{C})$ of sections which <u>vanish</u> at the points of $X - U$, and $\Omega^1_{X/\mathbb{C}}(\log D)$ is the sheaf of one forms with at worst first order poles at the points of $D$.

<u>Deligne's definition of regular singular points in the general case</u>

In terms of the sheaves $\Omega^1_{X/\mathbb{C}}(\log D)$ and $\underline{\mathrm{Der}}_D(X/\mathbb{C})$, we may rephrase Deligne's version of Fuch's criterion for regular singular points in the one-variable case as follows: In order that an algebraic differential equation $(M, \nabla)$ on an open curve $U$ have regular singular points at all points of $D = X - U$, it is necessary and sufficient that there exist a locally free coherent algebraic sheaf $\overline{M}$ on $X$ extending $M$, such that the action of $\underline{\mathrm{Der}}(U/\mathbb{C})$ on $M$ extends to an action of $\underline{\mathrm{Der}}_D(X/\mathbb{C})$ on $\overline{M}$ (equivalently, that the arrow $\nabla: M \longrightarrow M \otimes \Omega^1_{U/\mathbb{C}}$ extends to an arrow $\overline{\nabla}: \overline{M} \longrightarrow \overline{M} \otimes \Omega^1_{X/\mathbb{C}}(\log D)$.

Deligne now takes over this definition verbatim for a differential equation on any connected non-singular quasi-projective variety $U$ over $\mathbb{C}$, where $X$ is some Hironaka compactification of $U$ of the type discussed above. We should remark that in the higher dimensional case, it is not a priori clear that there exists any locally free coherent algebraic sheaf $\overline{M}$ on $X$ which prolongs $M$ on $U$; we will prove this. And we will admit here that this definition does not depend on the auxiliary choice of $X$.

Deligne's solution

We can now prove that Hilbert's twenty-first problem has an affirmative solution for any quasi-projective smooth connected variety $U$ over $\mathbb{C}$. Suppose we're given a representation of the fundamental group of $U^{an}$. Then as explained above, it arises as the monodromy of a unique analytic differential equation $(M^{an}, \nabla^{an})$ on $U^{an}$. Let us admit for a moment the following analytic lemma:

**Key Lemma** Let $(M^{an}, \nabla^{an})$ be any analytic differential equation on $U^{an}$, and let $U \hookrightarrow X$ be a compactification as above. Then there exists a locally free coherent analytic sheaf $\overline{M^{an}}$ on $X^{an}$ which extends $M^{an}$, such that $\nabla^{an}$ extends to an arrow $\overline{\nabla^{an}}: \overline{M^{an}} \longrightarrow \overline{M^{an}} \otimes \left(\Omega^1_{X/\mathbb{C}}(\log D)\right)^{an}$.

Then we may apply GAGA to the pair $\left(\overline{M^{an}}, \overline{\nabla^{an}}\right)$, to conclude the existence of a locally free coherent algebraic sheaf $\overline{M}$ on $X$ and an arrow $\overline{\nabla}: \overline{M} \longrightarrow \overline{M} \otimes \Omega^1_{X/\mathbb{C}}(\log D)$ such that $\left(\overline{M^{an}}, \overline{\nabla^{an}}\right) = \left(\overline{M}, \overline{\nabla}\right)^{an}$. If we define $(M, \nabla)$ on $U$ to be the restriction to $U$ of $(\overline{M}, \overline{\nabla})$, it is immediate that $(M, \nabla)$ has regular singular points, and that $(M, \nabla)^{an} \xrightarrow{\sim} (M^{an}, \nabla^{an})$, so that $(M, \nabla)$ gives rise to the given representation of $\pi_1(U^{an})$.

Proof of the key lemma

The idea is to use the local monodromy of the equation to construct an extension of $M^{an}$ locally along $D$ which has the desired properties, and to do so in a sufficiently canonical way that these local extensions patch together to give the needed global extension $\overline{M^{an}}$ of $M^{an}$. [In the one-variable case, $D$ consists of isolated points, so the extension problem is local around each point of $D$, and there is no patching problem. But in several variables, where we can slide along $D$ from one point to another, there is a patching problem.]

Let us begin by constructing a local extension. At a point "0" of $D$ where $r$ of the divisors $D_i$ cross, say $D_1, \ldots, D_r$, we may choose local coordinates $x_1, \ldots, x_n$ such that $D_i$ is defined by $x_i = 0$ for $i=1,\ldots,r$. In a small coordinate polydisc $V$ around "0" defined by $|x_j| < \epsilon$ for $j=1,\ldots,n$, the open manifold $U^{an} \cap V$ is the product of $r$ punctured discs $0 < |x_i| < \epsilon$, $i=1,\ldots,r$, and of $n-r$ discs $|x_k| < \epsilon$, $k=r+1,\ldots,n$. The restriction of $(M^{an}, \nabla^{an})$ to $V \cap U^{an}$ is an analytic differential equation on $V \cap U^{an}$, so corresponds to a representation $\rho$ of $\pi_1(V \cap U^{an})$ in a finite dimensional complex vector space $L$.

The important thing about this local situation is that $\pi_1(V \cap U^{an})$ is the free Abelian group on the $r$ generators $\gamma_i$ = turning once counterclockwise around $D_i$. So the representation $\rho$ is specified by the $r$ commuting automorphisms $\rho(\gamma_i)$ of $L$, $i=1,\ldots,r$. A consideration of Jordan normal form shows that there are unique endomorphisms $B_j$ of the representation space $L$ such that

1) $\exp(2\pi i B_j) = \rho(\gamma_j)$, $j=1,\ldots,r$

2) the eigenvalues of $B_j$ have real parts in the strip $-1 < \mathrm{Re} \leq 0$

3) the $B_j$ mutually commute

In terms of this data, we may construct an extension to $V$ of $(M^{an}, \nabla^{an}) | V \cap U^{an}$ as follows. We define

$$\overline{M^{an}} \overset{\text{dfn}}{=} L \otimes_{\mathbb{C}} \mathcal{O}_V$$

and we define the "connection with logarithmic poles"

$$\begin{array}{ccc} \overline{\nabla^{an}} : \overline{M^{an}} & \longrightarrow & \overline{M^{an}} \otimes_{\mathcal{O}_V} \Omega^1_V(\log D) \\ \| & & \| \\ L \otimes_{\mathbb{C}} \mathcal{O}_V & & L \otimes \Omega^1_{V/\mathbb{C}}(\log D) \end{array}$$

by defining

$$\nabla^{an}(\ell \otimes f) = f\left(-\sum_{i=1}^{r} B_i \ell \otimes \frac{dx_i}{x_i}\right) + \ell \otimes df$$

To see that this pair $\left(\overline{M^{an}}, \overline{\nabla^{an}}\right)$ does in fact extend $(M^{an}, \nabla^{an})|V \cap U^{an}$, we need only check that its restriction to $V \cap U^{an}$ has the correct monodromy (by the "holomorphic case" of the Hilbert problem). But a fundamental solution matrix is given explicitly by

$$\prod_{i=1}^{r} x_i^{B_i} \overset{dfn}{=} \exp\left(\sum_{i=1}^{r} B_i \log x_i\right)$$

which does indeed have the correct monodromy.

This construction shows the <u>local</u> existence of locally free extensions of $M^{an}$ which are <u>stable</u> under $\underline{\mathrm{Der}}_D(X^{an}/\mathbb{C})$. Among <u>all</u> locally free extensions to $V$ of $M^{an}|V \cap U^{an}$, we may <u>characterize</u> the one constructed above by the following property:

* In any basis of it as an $\mathcal{O}_V$-module, a fundamental solution matrix has the form

$$H(x) \cdot \prod_{i=1}^{r} x_i^{B_i}$$

with $H(x) \in GL(n, \mathcal{O}_V)$ and with the $B_i$ commuting matrices, the real parts of whose eigenvalues lie in $-1 < \mathrm{Re} \leq 0$. To see that * indeed <u>characterizes</u> our extension $\overline{M^{an}}$, we argue as follows. Any <u>other</u> extension of $M^{an}$ is of the form $A(x) \cdot \overline{M^{an}}$ where $A(x)$ is an invertible $n \times n$ matrix of functions analytic in $V \cap U^{an}$, with possibly essential singularities along the $D_i$. The extension $A(x)\overline{M^{an}}$ depends only on the class of $A(x)$ in the double-coset space

$$GL(n, \mathcal{O}_V)\backslash GL(n, \mathcal{O}_{V\cap U^{an}})/GL(n, \mathcal{O}_V).$$

So suppose that in terms of such an extension $A(x)\overline{M^{an}}$, the fundamental solution matrix is of the form $K(x)\prod_i x_i^{C_i}$, with $K(x) \in GL(n, \mathcal{O}_V)$ and with the $C_i$ commuting matrices with all eigenvalues lying in $-1 < \mathrm{Re} \le 0$. If we equate the fundamental solution matrices, we obtain the equation

$$K(x)\prod_{i=1}^{r} x_i^{C_i} = A^{-1}(x)\cdot H(x)\cdot \prod x_i^{B_i}.$$

By considering the effect of analytic continuation along $\gamma_j$, we see that

$$\exp(2\pi i C_j) = \exp(2\pi i B_j).$$

By the unicity of the logarithm whose eigenvalues lie in $-1 < \mathrm{Re} \le 0$, we conclude that $C_j = B_j$ for $j=1,\ldots,r$, hence that $K(x) = A^{-1}(x)H(x)$. Thus $A(x) \in GL(n, \mathcal{O}_V)$, and our "other" extension $A(x)\overline{M^{an}}$ is just $\overline{M^{an}}$ itself.

It remains to check that these well-defined local extensions patch together, or equivalently that they are compatible with localization. If we restrict our extension to a small enough polydisc $V' \subset V$ around a point where only the divisors $D_1,\ldots,D_s$ ($s < r$ cross, and choose local coordinates $y_1,\ldots,y_n$ adopted to the situation ($D_i$ is defined in $V'$ by $y_i = 0$, $i=1,\ldots,s$), we must check that a fundamental solution matrix is of the form

$$(\text{matrix in } GL(n, \mathcal{O}_{V'})) \times \prod y_i^{B_i}.$$

In terms of the coordinates $x_1,\ldots,x_n$ on $V$, the fundamental solution matrix was

$$\prod_{i=1}^{r} x_i^{B_i}.$$

Since for $i=1,\ldots,s$, both $x_i = 0$ and $y_i = 0$ define $D_i$ near $P$, there are invertible functions $u_i$, $i=1,\ldots,s$ near $P$ such that $x_i = u_i y_i$, $i=1,\ldots,s$. As the divisors $D_{s+1},\ldots,D_r$ do not pass through $P$, the functions $x_{s+1}$, $\ldots,x_r$ are themselves invertible near $P$. So in a small enough polydisc $V'$ around $P$, we may take the logarithms of the invertible functions $u_1,\ldots,u_s$, $x_{s+1},\ldots,x_r$, i.e. there are functions $z_1,\ldots,z_r$ holomorphic on $V'$ such that

$$u_i = \exp(z_i) \qquad i=1,\ldots,s$$

$$x_i = \exp(z_i) \qquad i=s+1,\ldots,r.$$

Then

$$\prod_{i=1}^{r} x_i^{B_i} = \prod_{i=1}^{s} (u_i y_i)^{B_i} \prod_{i=s+1}^{r} x_i^{B_i} = \exp\left(\sum_{i=1}^{r} B_i z_i\right) \prod_{i=1}^{s} y_i^{B_i}$$

and $\exp(\sum z_i B_i) \in GL(n, \mathcal{O}_{V'})$ as desired. QED

Remark The proof shows clearly that the notion of regular singular points in several variables can be expressed as a growth condition on the local solutions in (angular sectors of) the punctured polydiscs at infinity, in complete analogy with the one-variable case (cf. [2] for more details).

## More of Deligne's results — Comparison theorems

Thus far we have proven that for any connected non-singular quasi-projective variety $U$ over $\mathbb{C}$, any finite-dimensional complex representation arises as the monodromy representation of an algebraic differential equation on $U$ with regular singular points. It is natural to ask if this differential equation is unique. In categorical terms, we have proven that the functor

$$\begin{array}{l}\text{Algebraic D.E.'s on } U \text{ with} \\ \text{regular singular points}\end{array} \longrightarrow \text{analytic D.E.'s on } U^{an} \simeq \text{rep'ns of } \pi_1(U^{an})$$

is essentially surjective, and we ask if it is an equivalence of categories (or equivalently, given its surjectivity, whether it is fully faithful). Concretely, given two algebraic differential equations $(M, \nabla)$ and $(M', \nabla')$ on $U$ with regular singular points, we are asking if the map

$$\operatorname{Hom}((M,\nabla),\ (M',\nabla')) \longrightarrow \operatorname{Hom}((M,\nabla)^{an},\ (M',\nabla')^{an})$$

is an isomorphism.

Consider the "internal hom" differential equation $(M'' = \underline{\operatorname{Hom}}(M,M'), \nabla'')$, defined by $(\nabla''(D)\varphi)(m) = \nabla'(D)(\varphi(m)) - \varphi(\nabla(D)(m))$ for $D \in \underline{\operatorname{Der}}(U/\mathbb{C})$, $\varphi \in \operatorname{Hom}(M,M')$, and $m \in M$; its horizontal (= killed by $\nabla''$) sections are exactly the D. E.-maps from $(M,\nabla)$ to $(M',\nabla')$, and it will have regular singular points if both $(M,\nabla)$ and $(M',\nabla')$ do. So we may rephrase the question as follows: given a differential equation $(M,\nabla)$ with regular singular points, is the map

$$\begin{matrix}\text{global horizontal} \\ \text{sections of } M \text{ on } U\end{matrix} \longrightarrow \begin{matrix}\text{global horizontal} \\ \text{sections of } M^{an} \text{ on } U^{an}\end{matrix}$$

an isomorphism?

The answer is easily seen to be yes, for in terms of the particular extension $\overline{M}$ we constructed, any global analytic horizontal section of $M^{an}$ extends to a holomorphic global section of $\overline{M}^{an}$, which by GAGA is a global section of $\overline{M}$, and so restricts to give a global section of M, as required.

In fact, there is something more that we can say. Let $(M,\nabla)$ be an algebraic differential equation on U with regular singular points, and let $(M^{an})^{\nabla}$ denote the locally constant sheaf of germs of local holomorphic horizontal sections. Because the connection is integrable, we can extend the map

$$\nabla: M \longrightarrow M \otimes \Omega^1_{V/\mathbb{C}}$$

to an entire complex

$$M \xrightarrow{\nabla} M \otimes \Omega^1_{U/\mathbb{C}} \xrightarrow{\nabla \otimes 1 + 1 \otimes d} M \otimes \Omega^2_{U/\mathbb{C}} \longrightarrow \cdots,$$

the "De Rham complex of $(M,\nabla)$," noted $(M \otimes \Omega^{\cdot}_{U/\mathbb{C}}, \nabla)$.

We can consider the hypercohomology groups

$$\mathbb{H}^i(U, (M\otimes\Omega^{\cdot}_{U/\mathbb{C}}, \nabla))$$

(for U affine, these are the cohomology groups of the complex of global sections of the De Rham complex). We have a canonical map

$$\mathbb{H}^i(U, (M\otimes\Omega^{\cdot}_{U/\mathbb{C}}, \nabla)) \longrightarrow \mathbb{H}^i(U^{an}, (M^{an}\otimes\Omega^{\cdot}_{U^{an}/\mathbb{C}}, \nabla^{an}))$$

$$\wr\Vert \quad \text{(an isomorphism by the holomorphic Poincaré lemma)}$$

$$H^i(U^{an}, (M^{an})^{\nabla})$$

which for $i=0$ is just the map comparing algebraic and analytic global horizontal sections. The "something more" is that this map is an isomorphism for all i. The simplest proof is to remark that for the extension $(\overline{M}, \overline{\nabla})$ we explicitly constructed, (the "quasi-canonical" extension in Deligne's terminology) we have isomorphisms

$$\mathbb{H}^i(X, (\overline{M}\otimes\Omega^{\cdot}_{X/\mathbb{C}}(\log D), \overline{\nabla})) \xrightarrow{\sim} \mathbb{H}^i(U, (M\otimes\Omega^{\cdot}_{U/\mathbb{C}}, \nabla))$$

and

$$\mathbb{H}^i(X^{an}, \overline{M}^{an}\otimes\Omega^{\cdot}_{X^{an}/\mathbb{C}}(\log D), \overline{\nabla}) \xrightarrow{\sim} \mathbb{H}^i(U^{an}, M^{an}\otimes\Omega^{\cdot}_{U/\mathbb{C}}, \nabla).$$

The desired result now follows by applying GAGA to see that the canonical map

$$\mathbb{H}^i(X, (\overline{M}\otimes\Omega^{\cdot}_{X/\mathbb{C}}(\log D), \overline{\nabla})) \longrightarrow \mathbb{H}^i(X^{an}, (\overline{M}^{an}\otimes\Omega^{\cdot}_{X^{an}/\mathbb{C}}, \overline{\nabla}))$$

is an isomorphism.

If we take the trivial representation of $\pi_1(U^{an})$, the corresponding locally constant sheaf is the constant sheaf $\mathbb{C}$ itself, and the corresponding algebraic differential equation with regular singular points is $\mathcal{O}_U$ with the "trivial" connection furnished by exterior differentiation $d: \mathcal{O}_U \longrightarrow \Omega^1_{U/\mathbb{C}}$. The comparison theorem in this special case asserts that the map

$$\begin{array}{ccc} \mathbb{H}^i(U, \Omega^{\cdot}_{U/\mathbb{C}}) & \longrightarrow & \mathbb{H}^i(U^{an}, \Omega^{\cdot}_{U^{an}/\mathbb{C}}) \\ & & \wr\downarrow \\ & & H^i(U^{an}, \mathbb{C}) \end{array}$$

is an isomorphism (e.g., on a smooth affine $U$, closed modulo exact global algebraic differential forms calculate the complex cohomology). This theorem, first proven by Grothendieck, was the starting point for the systematic application of Hironaka's resolution result to studying the cohomology of open varieties (classically, the problem of understanding "integrals of the second kind"). In fact, the first "modern" attempt to study such questions was that of Atiyah and Hodge [ 1 ], which essentially concerned itself with the cohomology of a variable Zariski open set $U$ in a fixed projective smooth $X$. Only after Hironaka's result, though, did it become apparent that one should instead study the cohomology of a fixed open smooth $U$ with the aid of an auxiliary Hironaka compactification $U \hookrightarrow X'$.

### The Moral

To an algebraic geometer, it is that so far as non-singular algebraic varieties $U$ over $\mathbb{C}$ are concerned, the apparently topological notions of locally constant sheaf on $U^{an}$, and of the cohomology of $U^{an}$ with coefficients in such a sheaf, are in fact the purely algebraic notions of differential equation on $U$ with regular singular points, and of the hypercohomology groups of $U$ with coefficients in their de Rham complexes.

### What we don't know

Here are three open problems that lie well within the scope of the traditional theory of differential equations. The first of them was raised by Lipman Bers in the course of a stimulating discussion at the Symposium. The second is a favorite of mine. The third was raised by Monsky.

1. Let $X^{an}$ be a compact Riemann surface of genus $g \geq 2$. Then its universal covering is the upper half-plane $\mathfrak{h}$, and $X^{an}$ is obtained as the quotient of $\mathfrak{h}$ by a fuchsian group $G \subset SL(2, R)/\pm 1$; $X^{an} = \mathfrak{h}/G$, and $G$ is the fundamental group $\pi_1(X^{an})$. As an abstract group, $\pi_1(X^{an})$ is generated by $2g$ elements $a_i, b_i, i = 1, \ldots, g$, with the single relation $\prod_{i=1}^{g} a_i b_i a_i^{-1} b_i^{-1} = 1$. Viewing $X^{an}$ as $\mathfrak{h}/G$ gives us an embedding of $\pi_1(X^{an})$ in $SL(2, \mathbb{R})/\pm 1$, which lifts in $2^{2g}$ ways to an embedding of $\pi_1(X^{an})$ in $SL(2, \mathbb{R})$. Thus we have $2^{2g}$ representations of $\pi_1(X^{an})$ in $SL(2, \mathbb{R})$, which may be realized as follows. There are $2^{2g}$ choices of line bundle $\mathcal{L}$ on $X$ such that $\mathcal{L}^{\otimes 2} \simeq \Omega^1_{X/\mathbb{C}}$. Each of these allows us to write $H^1(X, \Omega^1_{X/\mathbb{C}}) = H^1(X, \mathcal{L}^{\otimes 2}) = H^1(X, \underline{Hom}(\mathcal{L}^{-1}, \mathcal{L})) = Ext^1(X; \mathcal{L}^{-1}, \mathcal{L})$, under which isomorphism the canonical generator of $H^1(X, \Omega^1_{X/\mathbb{C}})$ ("the cohomology class of a point") gives rise to a rank two vector bundle $M(\mathcal{L})$ which sits in a short exact sequence

$$0 \longrightarrow \mathcal{L} \longrightarrow M(\mathcal{L}) \longrightarrow \mathcal{L}^{-1} \longrightarrow 0$$

It is classical that there is one and only one connection $\nabla$ on $M(\mathcal{L})$ such that the monodromy representation is a real representation; this representation gives an embedding $\pi_1(X^{an}) \longrightarrow SL(2, \mathbb{R})$, and as $\mathcal{L}$ varies we recover the $2^{2g}$ embeddings described above.

Let $X'^{an}$ be another compact Riemann surface of the same genus $g \geq 2$. Viewing it as $\mathfrak{h}/G'$, where $G' \subset SL(2, \mathbb{R})/\pm 1$ is another fuchsian group, we obtain as before $2^{2g}$ embeddings of $\pi_1(X'^{an})$ in $SL(2, \mathbb{R})$. But of course the groups $\pi_1(X^{an})$ and $\pi_1(X'^{an})$ are isomorphic as abstract groups, so we could compose with this group-isomorphism to produce $2^{2g}$ embeddings of $\pi_1(X^{an}) \longrightarrow SL(2, \mathbb{R})$, this time with image in $SL(2, \mathbb{R})/\pm 1$ the fuchsian group corresponding to $X'$ rather than to $X$. By the affirmative solution of the twenty-first problem, we know that these may be realized as the monodromy representations of certain rank two differential equations

$(M_i, \nabla_i)$, $i=1,\ldots,2^{2g}$.

Is it true that $\nabla_i$ is the unique connection on $M_i$ such that the monodromy representation is real? Can the bundles $M_i$ that arise in this way (for variable $X'$) be characterized algebraically? Can bundles which admit connections whose monodromy representation is real be characterized algebraically? (The "compact form" of this last question has been answered by Seshadri and Narasimhan, who show that it is precisely the "stable" bundles of degree zero (a purely algebraic notion) which admit connections with irreducible unitary monodromy, and that this unitary connection is unique.) If these questions could be answered affirmatively, one could try to resuscitate the original Poincaré "proof" of uniformization by the "continuity principle."

2. Can we characterize algebraically those differential equations with regular singular points $(M,\nabla)$ on an open curve $U$ whose monodromy representations factor through finite groups. This can be done rather strikingly for the hypergeometric equation, but already for second-order equations on $\mathbb{P}^1$ $-$ {4 or more points} the question is completely open and extremely interesting [cf. 7 ]. It should be remarked that this question is treated in Forsythe "in principle," but even in cases when one knows the answer ahead of time, it seems hopeless to ever carry out Forsythe's test procedure.

3. What is the role of equations with irregular singular points. (It would be absurd to ignore the equation $f' - f = 0$, which has an irregular singularity at $\infty$.) What is the meaning of the cohomology groups $H^i_{DR}(U, (M\otimes \Omega^{\cdot}_{U/\mathbb{C}}, \nabla))$ when $(M,\nabla)$ has irregular singular points? Are they finite-dimensional? The finite-dimensionality in the case when $U$ is a curve is known, and due essentially to Birkhoff (his theory of canonical forms for irregular singularities) $-$ Deligne gives a finally not-so-different proof, where he also shows that the algebraic "index" $\dim H^0 - \dim H^1$ is different from its analytic counterpart.

and that this difference measures the irregularity of the singularities.

References

1. Atiyah, M., Hodge, W.: Integrals of the second kind on an algebraic variety. Annals of Math. 62, 56-91 (1955).

2. Deligne, P.: Equations differentielles à points singuliers réguliers. Lecture Notes in Mathematics 163, Berlin-Heidelberg-New York: Springer 1970.

3. Forsythe, A.R.: Theory of Differential Equations, Vol. IV, Cambridge, 1900-1902.

4. Grothendieck, A.: On the de Rham cohomology of algebraic varieties. Publ. Math. I.H.E.S. 29 (1966).

5. Hironaka, H.: Resolution of singularities of an algebraic variety over a field of characteristic zero I, II, Annals of Math. 79, 109-326 (1964).

6. Ince, E.L.: Ordinary Differential Equations, New York, Dover, 1956 (esp. 356-372, 389-393).

7. Katz, N.M.: Algebraic Solutions of Differential Equations, Inventiones Math. 18, 1-118 (1972).

8. Manin, Y.: Moduli Fuchsiani, Ann. Scuola Norm. Sup. Pisa, Ser III, 19 (13-126 (1965).

9. Monsky, P.: Finiteness of de Rham Cohomology, Amer. J. Math., XCIV, 237-245 (1972).

10. Narasimhan, M.S., and Seshadri, C.S.: Stable and unitary vector bundles on a compact Riemann surface. Annals of Math., 82, 540-567 (1965).

11. Serre, J.-P.: Géometrie algébrique et géometrie analytique. Ann. Inst. Fourier Grenoble 6, 1-42 (1956).

Proceedings of the International Congress of Mathematicians
Helsinki, 1978

# The Work of Pierre Deligne

N. M. Katz

My purpose here is to convey to you some idea of the scope and the depth of the work for which we are today honoring Pierre Deligne with the Fields Medal. Deligne's work centers around the remarkable relations, first envisioned by Weil, which exist between the cohomological structure of algebraic varieties over the complex numbers, and the diophantine structure of algebraic varieties over finite fields.

**I. The Weil conjectures.** Let us first consider an algebraic variety $Y$ over a finite field $F_q$. For each integer $n \geqslant 1$ there is a unique field extension $F_{q^n}$ of degree $n$ over $F_q$. We denote by $Y(F_{q^n})$ the (finite) set of points of $Y$ with coordinates in $F_{q^n}$, and by $\# Y(F_{q^n})$ the cardinality of this set. The zeta function of $Y$ over $F_q$ is the formal series defined by

$$Z(Y/F_q, T) = \exp\left( \sum_{n \geqq 1} \frac{T^n}{n} \# Y(F_{q^n}) \right).$$

Knowledge of the zeta function is equivalent to knowledge of the numbers $\{\# Y(F_{q^n})\}$.

After the pioneering work of E. Artin, W. K. Schmid, H. Hasse, M. Deuring and A. Weil on the zeta functions of curves and abelian varieties, Weil in 1949 made the following conjectures about the zeta function of a projective non-singular $n$-dimensional variety $Y$ over a finite field $\mathbf{F}_q$.

(1) The zeta function is a rational function of $T$, i.e. it lies in $Q(T)$.

(2) There exists a factorization of the zeta function as an alternating product of polynomials $P_0(T), \ldots, P_{2n}(T)$,

$$Z(Y/F_q, T) = \frac{P_1(T)P_3(T) \ldots P_{2n-1}(T)}{P_0(T)P_2(T) \ldots P_{2n}(T)}$$

of the form

$$P_i(T) = \prod_{j=1}^{b_i} (1-\alpha_{i,j}T),$$

such that the map $\alpha \mapsto q^n/\alpha$ carries the $\alpha_{i,j}$ bijectively to the $\alpha_{2n-i,j}$.

(3) The polynomials $P_i(T)$ lie in $\boldsymbol{Z}[T]$, and their reciprocal roots $\alpha_{i,j}$ are algebraic integers which, together with all their conjugates, satisfy

$$|\alpha_{i,j}| = q^{i/2}.$$

This is the "Riemann Hypothesis" for varieties over finite fields.

(4) If $Y$ is the "reduction mod $p$" of a projective smooth variety $\boldsymbol{Y}$ in characteristic zero, then the degree $b_i$ of $P_i$ is the $i$th Betti number of $\boldsymbol{Y}$ as complex manifold.

Underlying these conjectures was Weil's belief in the existence of a "cohomology theory", with a coefficient field of characteristic zero, for varieties over finite fields. In this theory, the polynomial $P_i(T)$ would be the "inverse" characteristic polynomial det $(1-TF)$ of the "Frobenius endomorphism" acting on $H^i$. Conjectures (1) and (2) would then follow from a Lefschetz trace formula for $F$ and its iterates, and from a suitable form of Poincaré duality. Conjecture (4) would follow if the cohomology of the "reduction mod $p$" $Y$ of a projective smooth variety $\boldsymbol{Y}$ in characteristic zero were (essentially) equal to the topological cohomology of $\boldsymbol{Y}$ as complex manifold.

The next years saw the systematic introduction of sheaf-theoretic and cohomological methods into algebraic geometry. By the mid-1960s, M. Artin and A. Grothendieck had developed the étale cohomology theory of arbitrary schemes, along the lines foreseen in Grothendieck's 1958 Edinburgh address. For each prime number $l$, this gives a cohomology theory, "$l$-adic cohomology", with coefficients in the field $\boldsymbol{Q}_l$ of $l$-adic numbers, which is adequate to give parts (1), (2) and (4) of the Weil conjectures for projective smooth varieties over finite fields of characteristic $p \neq l$. In their theory, the cohomology of the "reduction mod $p$" $Y$ of a projective smooth $\boldsymbol{Y}$ in characteristic zero is just the singular cohomology, with coefficients in $\boldsymbol{Q}_l$, of "$\boldsymbol{Y}$ as complex manifold". In the case of curves and abelian varieties, these constructions agree with those already given by Weil.

For a given projective smooth $Y/\boldsymbol{F}_q$, we now have, for *each* $l \neq p$, a factorization of the zeta function as an alternating product of $l$-adic polynomials $P_{i,l}(T)$. There is, however, no assurance that the $P_{i,l}$ have coefficients in $\boldsymbol{Q}$ rather than in $\boldsymbol{Q}_l$, much less that their reciprocal zeros are algebraic integers with the predicted absolute values. Of course, if one could prove directly that the reciprocal zeros of $P_{i,l}$ were algebraic integers which, together with all their conjugates, had the correct absolute value $q^{i/2}$, then the polynomials $P_{i,l}$ could be described intrinsically in terms of the zeros and poles of the zeta function itself, and hence would have rational coefficients independent of $l$. But how could one even introduce archimedean considerations into the $l$-adic theory without first knowing the rationality of the cofficients of the $P_{i,l}$ ?

Let me now try to indicate the brilliant synthesis of ideas involved in Deligne's solution of these problems.

Initially, he tries to prove à priori that the $P_{i,l}$ have rational coefficients independent of $l$. The idea is to proceed by induction on the dimension of $Y$. If $Y$ is $n$-dimensional, then Poincaré duality and the fact that the zeta function is itself rational and independent of $l$ reduce us to treating the polynomials $P_{i,l}$ for $i \leq n-1$. Now let $Z$ be a smooth hyperplane section of $Y$. The "weak" Lefschetz theorem assures us that $Y$ and $Z$ have the same $P_{i,l}$ for $i \leq n-2$, and that the $P_{n-1,l}$ for $Y$ divides that for $Z$. This alone is enough to show inductively that the reciprocal zeroes of the $P_{i,l}$ are algebraic integers.

In order to go further, and show that the $P_{i,l}$ actually have rational coefficients independent of $l$, the idea is to show that $P_{n-1,l}$ for $Y$ is a generalized "greatest common divisor" of the $P_{n-1,l}$ of all possible smooth hyperplane sections. Unfortunately, this "g.c.d." argument, which itself depends on the full strength of the monodromy theory of Lefschetz pencils, works only when $Y$ satisfies the "hard" Lefschetz theorem (existence of the "primitive decomposition" on its cohomology), otherwise the "g.c.d." will be too big at some stage of the induction. But Deligne will later prove the hard Lefschetz theorem in arbitrary characteristic as a *consequence* of the Weil conjectures. What is to be done?

With characteristic daring, Deligne simply ignores the preliminary problem of establishing independence of $l$. Fixing one $l \neq p$, he turns to a direct attack on the absolute values of the algebraic integers which occur as the reciprocal roots of the $P_{i,l}$.

Consider a smooth projective even dimensional $Y$, and a Lefschetz pencil $Z_t$ of hyperplane sections, "fibering" $Y$ over the $t$-line. Factor the $P_{n-1,l}$ of each $Z_t$ as the product of the "g.c.d." of all of them, and of the "variable" part. Deligne shows à priori that these "variable" parts are each polynomials with rational coefficients whose reciprocal zeroes all satisfy the Riemann Hypothesis

$$|\alpha_{n-1,\,\text{variable}}| = q^{(n-1)/2}.$$

Deligne's proof of this is simply spectacular; no other word will do. He first uses a theorem of Kazdan–Margoulis, according to which the monodromy group of a Lefschetz pencil of odd fibre dimension is "as big as possible", to establish the rationality of the coefficients of the "variable" parts. Then he considers the $L$-function over the $t$-line whose Euler factors are the reciprocals of the "variable" parts. This $L$-function has rational Dirichlet coefficients. Deligne realizes that Rankin's method of estimating Ramanujan's function $\tau(n)$ by "squaring" might be applied in this context to estimate the reciprocal poles of the individual Euler factors (i.e. the reciprocal zeroes of the "variable" parts!). The problem is to control the *poles* of all the $L$-functions obtained from this one by passing to *even* tensor powers ("squaring"). Deligne gains this control by ingeniously combining Grothendieck's cohomological theory of such $L$-functions, the Kazdan–Margoulis theorem, and the classical invariant theory of the symplectic group!

Once he has this à priori estimate for the variable parts of the $P_{n-1,l}$ of the hyperplane sections $Z_t$, a Leray spectral sequence argument shows that in the $P_{n,l}$ for $Y$ itself, all the reciprocal zeroes are algebraic integers which, together with all their conjugates, satisfy the apparently too weak estimate

$$|\alpha_{n,j}| \leq q^{(n+1)/2} \quad \text{(instead of } q^{n/2}\text{)}.$$

But this estimate is valid for $Y$ of any *even* dimension $n$. The actual Riemann hypothesis for any projective smooth variety $X$ follows by applying this estimate to all the even cartesian powers of $X$.

**II. Consequences for number theory.** That there are many spectacular consequences for number theory comes as no surprise. Let us indicate a few of them.

(1) Estimation of $\# Y(\boldsymbol{F}_q)$ when $Y$ has a "simple" cohomological structure. For example, if $Y$ is a smooth $n$-dimensional hypersurface of degree $d$, we get

$$|\# Y(\boldsymbol{F}_q)-(1+q+\ldots+q^n)| \leq \left(\frac{(d-1)^{n+2}+(-1)^{n+2}(d-1)}{d}\right) q^{n/2}.$$

(2) Estimates for exponential sums in several variables, e.g

(a) if $f$ is a polynomial over $\boldsymbol{F}_p$ in $n$ variables of degree $d$ prime to $p$, whose part of highest degree defines a nonsingular projective hypersurface, then

$$\left|\sum_{x_i \in \boldsymbol{F}_p} \exp\left(\frac{2\pi i}{p} f(x_1, \ldots, x_n)\right)\right| \leq (d-1)^n p^{n/2};$$

(b) "multiple Kloosterman sums":

$$\left|\sum_{x_i \in \boldsymbol{F}_p^\times} \exp\left(\frac{2\pi i}{p}\left(x_1+\ldots+x_n+\frac{1}{x_1 \ldots x_n}\right)\right)\right| \leq (n+1) \cdot p^{n/2}.$$

(3) The Ramanujan–Petersson conjecture. Already in 1968 Deligne had combined techniques of $l$-adic cohomology and the arithmetic moduli of elliptic curves with earlier ideas of Kuga, Sato, Shimura and Ihara to reduce this conjecture to the Weil conjectures. Thus if $\sum a(n) q^n$ is the $q$-expansion of a normalized $(a(1)=1)$ cusp form on $\Gamma_1(N)$ of weight $k \geq 2$ which is a simultaneous eigenfunction of all Hecke operators, then

$$|a(p)| \leq 2p^{(k-1)/2} \quad \text{for all primes } p \nmid N.$$

**III. Cohomological consequences; weights.** As Grothendieck foresaw in the 1960s with his "yoga of weights", the truth of the Weil conjectures for varieties over finite fields would have important consequences for the cohomological structure of varieties over the complex numbers. The idea is that any reasonable algebro-geometric situation over $\boldsymbol{C}$ is actually defined over a subring of $\boldsymbol{C}$ which, as a ring, is finitely generated over $\boldsymbol{Z}$. Reducing modulo a maximal ideal $\mathfrak{m}$ of this ring, we find a situation over a finite field, and the corresponding Frobenius endomorphism $F(\mathfrak{m})$ operating on this situation. This Frobenius operates by functoriality on the $l$-adic cohomology, which is none other than the singular cohomology, with $\boldsymbol{Q}_l$-coefficients, of our original situation over $\boldsymbol{C}$. This natural operation of Frobenius imposes a previously unsuspected structure on the cohomology of complex algebraic varieties, the so-called "weight filtration", or filtration by the magnitude of the eigenvalues of Frobenius.

In a remarkable tour de force in the late 1960's and early 1970's, Deligne developed, independently of the Weil conjectures, a complete theory of the weight filtration of complex algebraic varieties, by making systematic use of Hironaka's resolution of singularities, of the notion of differential forms with "logarithmic poles" (i.e. products of $dt/t$'s), and of his own earlier work on cohomological descent. The resulting theory, which Deligne named "mixed Hodge theory", should be seen as a far-reaching generalization of the classical theory of "differentials of the second kind" on algebraic varieties, as well as of "usual" Hodge theory.

Consider, for example, a smooth affine variety $U$ over $\boldsymbol{C}$. By one of Hironaka's fundamental results, we can find a projective smooth variety $X$ and a collection of smooth divisors $D_i$ in $X$ which cross transversally, such that $U \cong X - UD_i$. Deligne shows that the Leray spectral sequence, in rational cohomology, of the inclusion map $U \hookrightarrow X$, degenerates at $E_3$, and that the filtration it defines on the cohomology of $U$ is independent of the choice of the compactification. This is the desired weight filtration; its smallest filtrant is the image of $H^{\bullet}(X)$ in $H^{\bullet}(U)$, i.e. the space of "differentials of the second kind on $U$".

One of the many applications of this theory is to the global monodromy of families of projective smooth varieties. Given a projective smooth map $X \to S$ of smooth complex varieties, and a point $s \in S$, the fundamental group $\pi_1(S, s)$ acts on each of the cohomology groups $H^i(X_s, \boldsymbol{C})$ of the fibre $X_s$. Deligne shows that these representations are all completely reducible, and that each of their isotypical components, especially the space of invariants, is stable under the Hodge decomposition into $(p, q)$-components.

By means of an extremely ingenious and difficult argument drawing upon Grothendieck's cohomological theory of $L$-functions and the ideas of the Hadamard–de la Vallée Poussin proof of the prime number theorem, Deligne later established an $l$-adic analogue of this theorem of complete reducibility for $l$-adic "local systems" in characteristic $p$ over open subsets of the projective $t$-line $\boldsymbol{P}^1$, provided that all of the "fibres" of the local system satisfy the Riemann Hypothesis (with a fixed power of $\sqrt{q}$). Once he had *proven* the Riemann Hypothesis for varieties over finite fields, he could apply this theorem to the local system coming from a Lefschetz pencil on a projective smooth variety over a finite field. The resulting complete reducibility is easily seen to imply the hard Lefschetz theorem. This theorem, previously known only over $\boldsymbol{C}$, and there by Hodge's theory of harmonic integrals, is thus established, in all characteristics as a consequence of the Weil conjectures.

Other outgrowths of Deligne's work on weights include his theory of differential equations with regular singular points on smooth complex varieties of arbitrary dimension, which yields a new solution of Hilbert's 21st problem, and his affirmative solution of the "local invariant cycle problem" in the local monodromy theory of families of projective smooth varieties.

Still another application of "weights" is to homotopy theory. Deligne, Griffiths, Morgan and Sullivan jointly apply mixed Hodge theory to prove that the homotopy theory ($\otimes Q$) of a projective smooth complex variety is a "formal consequence" of its cohomology.

**IV. Other work.** We have passed over in silence a considerable body of Deligne's work, which alone would be sufficient to mark him as a truly exceptional mathematician; duality in coherent cohomology, moduli of curves (jointly with Mumford), arithmetic moduli (jointly with Rapoport), the Ramanujan–Petersson conjecture for forms of weight one (jointly with Serre), the Macdonald conjecture (jointly with Lusztig), local "roots numbers", $p$-adic $L$-functions (jointly with Ribet), motivic $L$-functions, "Hodge cycles" on abelian varieties, and much more.

I hope that I have conveyed to you some sense, not only of Deligne's accomplishments, but also of the combination of incredible technical power, brilliant clarity, and sheer mathematical daring which so characterizes his work.

Princeton University
Princeton, N.J. 08540, U.S.A.

# LA CONJECTURE DE WEIL. I

*par* Pierre DELIGNE

SOMMAIRE



Dans cet article, je démontre la conjecture de Weil sur les valeurs propres des endomorphismes de Frobenius. Un énoncé précis est donné en (1.6). J'ai tenté de présenter la démonstration sous une forme aussi géométrique et élémentaire que possible et j'ai inclus force rappels : seuls sont originaux les résultats des §§ 3, 6, 7 et 8.

Dans un article faisant suite à celui-ci, je donnerai divers raffinements des résultats intermédiaires, et des applications, parmi lesquelles le théorème de Lefschetz « difficile » (sur les cup-produits itérés par la classe de cohomologie d'une section hyperplane).

Le texte suit fidèlement celui de six conférences données à Cambridge en juillet 1973. Je remercie N. Katz de m'avoir permis d'utiliser ses notes.

## 1. La théorie de Grothendieck : interprétation cohomologique de fonctions L.

**(1.1)** Soient X un schéma de type fini sur **Z**, $|X|$ l'ensemble des points fermés de X et, pour $x \in |X|$, soit $N(x)$ le nombre d'éléments du corps résiduel $k(x)$ de X en $x$. La fonction zêta de Hasse-Weil de X est

$$(1.1.1) \qquad \zeta_X(s) = \prod_{x \in |X|} (1 - N(x)^{-s})^{-1}$$

(ce produit converge absolument pour $\mathcal{R}(s)$ assez grand). Pour $X = \mathrm{Spec}(\mathbf{Z})$, $\zeta_X(s)$ est la fonction zêta de Riemann.

Nous considérerons exclusivement le cas où X est un schéma sur un corps fini $\mathbf{F}_q$.

Pour $x \in |X|$, nous écrirons $q_x$ plutôt que $N(x)$. Posant $\deg(x) = [k(x) : \mathbf{F}_q]$, on a $q_x = q^{\deg(x)}$. Il y a intérêt à introduire la variable $t = q^{-s}$. Posons

(1.1.2) $$Z(X; t) = \prod_{x \in |X|} (1 - t^{\deg(x)})^{-1};$$

ce produit infini converge pour $|t|$ assez petit, et on a

(1.1.3) $$\zeta_X(s) = Z(X; q^{-s}).$$

**(1.2)** Dwork (On the rationality of the zeta function of an algebraic variety, *Amer. J. Math.*, 82, 1960, p. 631-648) et Grothendieck ([1] et SGA 5) ont démontré que $Z(X, t)$ est une fonction rationnelle de $t$.

Pour Grothendieck, c'est là un corollaire de résultats généraux en cohomologie $\ell$-adique ($\ell$ désigne un nombre premier différent de la caractéristique $p$ de $\mathbf{F}_q$). Ceux-ci fournissent une interprétation cohomologique des zéros et des pôles de $Z(X; t)$, et une équation fonctionnelle lorsque X est propre et lisse. Les méthodes de Dwork sont $p$-adiques. Pour X une hypersurface non singulière dans l'espace projectif, elles lui fournissaient également une interprétation cohomologique des zéros et des pôles, et l'équation fonctionnelle. Elles ont inspiré la théorie cristalline de Grothendieck et Berthelot, qui pour X propre et lisse fournit une interprétation cohomologique $p$-adique des zéros et pôles, et l'équation fonctionnelle. Se basant sur des idées de Washnitzer, Lubkin a créé une variante de cette théorie, valable seulement pour X propre, lisse et relevable en caractéristique zéro (A $p$-adic proof of Weil's conjectures, *Ann. of Math.*, 87, 1968, p. 105-255).

Nous ferons un usage essentiel des résultats de Grothendieck, et les rappelons ci-dessous.

**(1.3)** Soit X une variété algébrique sur un corps algébriquement clos $k$ de caractéristique $p$, i.e. un schéma séparé de type fini sur $k$. On n'exclut pas le cas $p = 0$. Pour tout nombre premier $\ell \neq p$, Grothendieck a défini des groupes de cohomologie $\ell$-adiques $H^i(X, \mathbf{Q}_\ell)$. Il a aussi défini des groupes de cohomologie à support propre $H^i_c(X, \mathbf{Q}_\ell)$. Pour X propre, les deux coïncident. Les $H^i_c(X, \mathbf{Q}_\ell)$ sont des espaces vectoriels de dimension finie sur $\mathbf{Q}_\ell$, nuls pour $i > 2\dim(X)$.

**(1.4)** Soient $X_0$ une variété algébrique sur $\mathbf{F}_q$, $\overline{\mathbf{F}}_q$ une clôture algébrique de $\mathbf{F}_q$ et X la variété algébrique sur $\overline{\mathbf{F}}_q$ déduite de $X_0$ par extension des scalaires de $\mathbf{F}_q$ à $\overline{\mathbf{F}}_q$. Dans le langage de Weil ou Shimura, on exprimerait cette situation par : « Soit X une variété algébrique définie sur $\mathbf{F}_q$ ». Soit $F : X \to X$ le morphisme de Frobenius; il envoie le point de coordonnées $x$ vers le point de coordonnées $x^q$; en d'autres termes, pour $U_0$ un ouvert de Zariski de $X_0$, définissant un ouvert U de X, on a $F^{-1}(U) = U$; pour $x \in H^0(U_0, \mathcal{O})$, on a $F^* x = x^q$. Identifions l'ensemble $|X|$ des points fermés de X à $X_0(\overline{\mathbf{F}}_q)$ (l'ensemble $\mathrm{Hom}_{\mathbf{F}_q}(\mathrm{Spec}(\overline{\mathbf{F}}_q), X_0)$ des points de $X_0$ à coefficients dans $\overline{\mathbf{F}}_q$), et soit

$\varphi \in \mathrm{Gal}(\overline{\mathbf{F}}_q/\mathbf{F}_q)$ la substitution de Frobenius : $\varphi(x) = x^q$. L'action de F sur $|X|$ s'identifie à l'action de $\varphi$ sur $X_0(\overline{\mathbf{F}}_q)$. Dès lors :

*a)* L'ensemble $X^F$ des points fermés de X fixes sous F s'identifie à l'ensemble $X_0(\mathbf{F}_q) \subset X_0(\overline{\mathbf{F}}_q)$ des points de X définis sur $\mathbf{F}_q$. Ceci exprime simplement que, pour $x \in \overline{\mathbf{F}}_q$, on a $x \in \mathbf{F}_q \Leftrightarrow x^q = x$.

*b)* De même, l'ensemble $X^{F^n}$ des points fermés de X fixes sous le $n^{\text{ième}}$ itéré de F s'identifie à $X_0(\mathbf{F}_{q^n})$.

*c)* L'ensemble $|X_0|$ des points fermés de $X_0$ s'identifie à l'ensemble $|X|_F$ des orbites de F (ou de $\varphi$) dans $|X|$. Le degré $\deg(x)$ de $x \in |X_0|$ est le nombre d'éléments de l'orbite correspondante.

*d)* De *b)* et *c)* résulte la formule

$$\#X^{F^n} = \#X_0(\mathbf{F}_{q^n}) = \sum_{\deg(x)\mid n} \deg x \tag{1.4.1}$$

(pour $x \in |X_0|$ et $\deg(x) \mid n$, $x$ définit $\deg(x)$ points à coordonnées dans $\mathbf{F}_{q^n}$, tous conjugués sur $\mathbf{F}_q$).

**(1.5)** Le morphisme F est fini, en particulier propre. Il induit donc des applications

$$F^* : H^i_c(X, \mathbf{Q}_\ell) \to H^i_c(X, \mathbf{Q}_\ell).$$

Grothendieck a prouvé la formule de Lefschetz

$$\#X^F = \sum_i (-1)^i \operatorname{Tr}(F^*, H^i_c(X, \mathbf{Q}_\ell));$$

le membre de droite, *a priori* un nombre $\ell$-adique, est entier, et égal au membre de gauche. On notera qu'une telle formule n'est raisonnable que parce que $dF = 0$, même à l'infini (X n'est pas supposé propre); la relation $dF = 0$ implique que les points fixes de F sont de multiplicité un.

Une formule analogue vaut pour les itérés de F :

$$\#X^{F^n} = \#X_0(\mathbf{F}_{q^n}) = \sum_i (-1)^i \operatorname{Tr}(F^{*n}, H^i_c(X, \mathbf{Q}_\ell)). \tag{1.5.1}$$

Prenons la dérivée logarithmique de (1.1.2) :

$$t\frac{d}{dt}\log Z(X_0, t) = \frac{t\frac{d}{dt}Z(X_0, t)}{Z(X_0, t)} = \sum_{x \in |X_0|} -\frac{-\deg(x)t^{\deg(x)}}{1 - t^{\deg(x)}} \tag{1.5.2}$$

$$= \sum_{x \in |X_0|} \sum_{n>0} \deg(x) t^{n.\deg(x)} \underset{(1.4.1)}{=} \sum_n \#X_0(\mathbf{F}_{q^n}) . t^n.$$

Pour F un endomorphisme d'un espace vectoriel V, on a une identité de séries formelles

$$t\frac{d}{dt}\log(\det(1 - Ft, V)^{-1}) = \sum_{n>0} \operatorname{Tr}(F^n, V) t^n \tag{1.5.3}$$

(le vérifier pour $\dim(V)=1$, et observer que les deux membres sont additifs en V dans une suite exacte courte). Substituant (1.5.1) dans (1.5.2) et appliquant (1.5.3), on trouve

$$t\frac{d}{dt}\log Z(X_0, t)=\sum_i(-1)^i t\frac{d}{dt}\log\det(1-F^* t, H_c^i(X, \mathbf{Q}_\ell))^{-1},$$

soit

**(1.5.4)** $$Z(X_0, t)=\prod_i \det(1-F^* t, H_c^i(X, \mathbf{Q}_\ell))^{(-1)^{i+1}}.$$

Le membre de droite est un élément de $\mathbf{Q}_\ell(t)$. La formule affirme que son développement de Taylor en $t=0$, *a priori* une série formelle dans $\mathbf{Q}_\ell[[t]]$ de terme constant un, est dans $\mathbf{Z}[[t]]$, et est égal au membre de gauche, lui aussi considéré comme une série formelle en $t$. Cette formule est l'interprétation cohomologique de Grothendieck de la fonction Z.

Notre résultat principal est le suivant.

*Théorème* **(1.6)**. — *Soit* $X_0$ *une variété projective non singulière* (= *lisse*) *sur* $\mathbf{F}_q$. *Pour chaque i, le polynôme caractéristique* $\det(t.1-F^*, H^i(X, \mathbf{Q}_\ell))$ *est à coefficients entiers indépendants de* $\ell$ ($\ell \neq p$). *Les racines complexes* $\alpha$ *de ce polynôme (les conjugués complexes des valeurs propres de* $F^*$) *sont de valeur absolue* $|\alpha|=q^{i/2}$.

Montrons déjà que (1.6) résulte du résultat apparemment plus faible suivant.

*Lemme* **(1.7)**. — *Pour chaque i, et chaque* $\ell\neq p$, *les valeurs propres de l'endomorphisme* $F^*$ *de* $H^i(X, \mathbf{Q}_\ell)$ *sont des nombres algébriques dont tous les conjugués complexes* $\alpha$ *sont de valeur absolue* $|\alpha|=q^{i/2}$.

*Preuve de* (1.7) ⇒ (1.6). — Regardons $Z(X_0, t)$ comme une série formelle de terme constant 1, élément de $\mathbf{Z}[[t]]$ : $Z(X_0, t)=\sum_n a_n t^n$. D'après (1.5.3), l'image de $Z(X_0, t)$ dans $\mathbf{Q}_\ell[[t]]$ est le développement de Taylor d'une fraction rationnelle. Ceci signifie que pour N et M assez grands ($\geq$ les degrés des numérateurs et dénominateurs) les déterminants de Hankel

$$H_k=\det((a_{i+j+k})_{0\leq i,j\leq M}) \qquad (k>N)$$

sont nuls. Cette nullité est vraie dans $\mathbf{Q}_\ell$ si et seulement si elle l'est dans $\mathbf{Q}$; $Z(X_0, t)$ est donc le développement en série de Taylor d'un élément de $\mathbf{Q}(t)$. En d'autres termes,

$$Z(X_0, t)\in\mathbf{Z}[[t]]\cap\mathbf{Q}_\ell(t)\subset\mathbf{Q}(t).$$

Écrivons $Z(X_0, t)=P/Q$, avec $P, Q\in\mathbf{Z}[t]$, premiers entre eux, et de terme constant positif. D'après un lemme de Fatou, que $Z(X_0, t)$ soit dans $\mathbf{Z}[[t]]$ et de terme constant un implique que les termes constants de P et Q sont 1. Posons

$$P_i(t)=\det(1-F^* t, H^i(X, \mathbf{Q}_\ell)).$$

D'après l'hypothèse (1.7), les $P_i$ sont premiers entre eux. Le membre de droite de (1.5.4) est donc sous forme irréductible, et

$$P(t) = \prod_{i \text{ impair}} P_i(t)$$

$$Q(t) = \prod_{i \text{ pair}} P_i(t).$$

Soit K le sous-corps d'une clôture algébrique $\overline{\mathbf{Q}}_\ell$ de $\mathbf{Q}_\ell$ engendré sur $\mathbf{Q}$ par les racines de $R(t) = P(t)Q(t)$. Les racines de $P_i(t)$ sont celles des racines de $R(t)$ ayant la propriété que tous leurs conjugués complexes sont de valeur absolue $q^{-i/2}$. Cet ensemble est stable sous $\mathrm{Gal}(K/\mathbf{Q})$. Le polynôme $P_i(t)$ est donc à coefficients rationnels. D'après le lemme de Gauss (ou parce que les racines de $P_i$, étant racines de $R(t)$, sont des inverses d'entiers algébriques), il est même à coefficients entiers. La description ci-dessus des racines de $P_i(t)$ est indépendante de $\ell$; le polynôme $P_i(t)$ lui-même est donc indépendant de $\ell$.

La suite de cet article est consacrée à la démonstration de (1.7).

**(1.8)** La théorie de Grothendieck fournit une interprétation cohomologique non seulement de fonctions zêta, mais encore de fonctions L. Les résultats sont les suivants.

**(1.9)** Soit X une variété algébrique sur un corps $k$. Pour la définition d'un $\mathbf{Q}_\ell$-faisceau constructible sur X, je renvoie à SGA 5 VI. Qu'il suffise de dire que:

*a)* Si $\mathscr{F}$ est un $\mathbf{Q}_\ell$-faisceau constructible sur X, il existe une partition finie de X en parties localement fermées $X_i$ telles que $\mathscr{F}|X_i$ soit constant tordu.

*b)* Supposons X connexe et soit $\bar{x}$ un point géométrique de X. Pour $\mathscr{F}$ constant tordu, $\pi_1(X, \bar{x})$ agit sur la fibre $\mathscr{F}_{\bar{x}}$; le foncteur fibre en $\bar{x}$ est une équivalence de catégorie

($\mathbf{Q}_\ell$-faisceaux constructibles constants tordus sur X) $\mapsto$
$\mapsto$ (représentations continues de $\pi_1(X, \bar{x})$ sur un $\mathbf{Q}_\ell$-espace vectoriel de dimension finie).

Une telle représentation ne se factorise pas, en général, à travers un quotient fini de $\pi_1(X, \bar{x})$.

*c)* Si $k = \mathbf{C}$, les $\mathbf{Q}_\ell$-faisceaux constructibles sur X s'identifient aux faisceaux de $\mathbf{Q}_\ell$-espaces vectoriels $\mathscr{F}$ sur $X^{an}$, tels qu'il existe une partition finie de X en parties Zariski-localement fermées $X_i$, et pour chaque $i$ un système local de $\mathbf{Z}_\ell$-Modules libres de type fini $\mathscr{F}_i$ sur $X_i$, avec

$$\mathscr{F}|X_i = \mathscr{F}_i \otimes_{\mathbf{Z}_\ell} \mathbf{Q}_\ell.$$

Nous ne considérerons que des $\mathbf{Q}_\ell$-faisceaux constructibles, et les appellerons simplement $\mathbf{Q}_\ell$-faisceaux.

**(1.10)** Supposons $k$ algébriquement clos, et soit $\mathscr{F}$ un $\mathbf{Q}_\ell$-faisceau sur X. Grothendieck a défini des groupes de cohomologie $\ell$-adique $H^i(X, \mathscr{F})$ et $H^i_c(X, \mathscr{F})$. Les $H^i_c(X, \mathscr{F})$ sont des espaces vectoriels de dimension finie sur $\mathbf{Q}_\ell$, nuls pour $i > 2\dim(X)$.

Pour $k = \mathbf{C}$, les $\mathrm{H}^i(\mathrm{X}, \mathscr{F})$ et $\mathrm{H}^i_c(\mathrm{X}, \mathscr{F})$ sont les groupes de cohomologie usuels (resp. à support propre) de $\mathrm{X}^{\mathrm{an}}$, à coefficients dans $\mathscr{F}$.

**(1.11)** Soient $\mathrm{X}_0$ une variété algébrique sur $\mathbf{F}_q$, X la variété sur $\overline{\mathbf{F}}_q$ correspondante, et $\mathscr{F}_0$ un faisceau d'ensembles sur $\mathrm{X}_0$ (pour la topologie étale). On note $\mathscr{F}$ son image réciproque sur X. Outre le morphisme de Frobenius $\mathrm{F}: \mathrm{X} \to \mathrm{X}$, on dispose alors d'un isomorphisme canonique $\mathrm{F}^* : \mathrm{F}^*\mathscr{F} \xrightarrow{\sim} \mathscr{F}$. En voici une description. On regarde $\mathscr{F}_0$ comme un espace étalé sur $\mathrm{X}_0$, i.e. on identifie $\mathscr{F}_0$ à l'espace algébrique $[\mathscr{F}_0]$, muni d'un morphisme étale $f : [\mathscr{F}_0] \to \mathrm{X}_0$, tel que $\mathscr{F}_0$ soit le faisceau des sections locales de $[\mathscr{F}_0]$. L'espace analogue $[\mathscr{F}]$, étalé sur X, se déduit de $[\mathscr{F}_0]$ par extension des scalaires. On dispose donc d'un diagramme commutatif

$$\begin{array}{ccc} [\mathscr{F}] & \xrightarrow{\mathrm{F}} & [\mathscr{F}] \\ {\scriptstyle f}\downarrow & & \downarrow{\scriptstyle f} \\ \mathrm{X} & \xrightarrow{\mathrm{F}} & \mathrm{X} \end{array}$$

d'où un morphisme $[\mathscr{F}] \to \mathrm{X} \times_{(\mathrm{F},\mathrm{X},f)} [\mathscr{F}] = [\mathrm{F}^*\mathscr{F}]$, qui est un isomorphisme car $f$ est étale. Son inverse définit l'isomorphisme $\mathrm{F}^*\mathscr{F} \xrightarrow{\sim} \mathscr{F}$ cherché.

Cette construction se généralise aux $\mathbf{Q}_\ell$-faisceaux.

**(1.12)** Soient $\mathrm{X}_0$ une variété algébrique sur $\mathbf{F}_q$, $\mathscr{F}_0$ un $\mathbf{Q}_\ell$-faisceau sur $\mathrm{X}_0$, $(\mathrm{X}, \mathscr{F})$ déduit de $(\mathrm{X}_0, \mathscr{F}_0)$ par extension des scalaires de $\mathbf{F}_q$ à $\overline{\mathbf{F}}_q$, $\mathrm{F}: \mathrm{X} \to \mathrm{X}$ et $\mathrm{F}^* : \mathrm{F}^*\mathscr{F} \to \mathscr{F}$.

Le morphisme fini F et $\mathrm{F}^*$ définissent un endomorphisme

$$\mathrm{F}^* : \mathrm{H}^i_c(\mathrm{X}, \mathscr{F}) \to \mathrm{H}^i_c(\mathrm{X}, \mathrm{F}^*\mathscr{F}) \to \mathrm{H}^i_c(\mathrm{X}, \mathscr{F}).$$

Pour $x \in |\mathrm{X}|$, $\mathrm{F}^*$ définit un morphisme $\mathrm{F}^*_x : \mathscr{F}_{\mathrm{F}(x)} \to \mathscr{F}_x$. Pour $x \in \mathrm{X}^{\mathrm{F}}$, c'est un endomorphisme de $\mathscr{F}_x$. Grothendieck a démontré la formule de Lefschetz

$$\sum_{x \in \mathrm{X}^{\mathrm{F}}} \mathrm{Tr}(\mathrm{F}^*_x, \mathscr{F}_x) = \sum_i (-1)^i \, \mathrm{Tr}(\mathrm{F}^*, \mathrm{H}^i_c(\mathrm{X}, \mathscr{F})).$$

Une formule analogue vaut pour les itérés de F : le $n^{\text{ième}}$ itéré de $\mathrm{F}^*$ définit des morphismes $\mathrm{F}^{*n}_x : \mathscr{F}_{\mathrm{F}^n(x)} \to \mathscr{F}_x$ ; pour $x$ fixe sous $\mathrm{F}^n$, $\mathrm{F}^{*n}_x$ est un endomorphisme, et

**(1.12.1)** $$\sum_{x \in \mathrm{X}^{\mathrm{F}^n}} \mathrm{Tr}(\mathrm{F}^{*n}_x, \mathscr{F}_x) = \sum_i (-1)^i \, \mathrm{Tr}(\mathrm{F}^{*n}, \mathrm{H}^i_c(\mathrm{X}, \mathscr{F})).$$

**(1.13)** Soient $x_0 \in |\mathrm{X}_0|$, Z l'orbite correspondante de F dans $|\mathrm{X}|$ et $x \in \mathrm{Z}$. L'orbite Z a $\deg(x_0)$ éléments (1.4). On note $\mathrm{F}^*_{x_0}$ l'endomorphisme $\mathrm{F}^{*\deg(x_0)}_x$ de $\mathscr{F}_x$, et on pose

$$\det(1 - \mathrm{F}^*_{x_0} t, \mathscr{F}_0) = \det(1 - \mathrm{F}^*_{x_0} t, \mathscr{F}_x).$$

A isomorphisme près, $(\mathscr{F}_x, \mathrm{F}^*_{x_0})$ ne dépend pas du choix de $x$. Ceci justifie d'omettre $x$ de la notation. On utilisera une notation analogue pour d'autres fonctions de $(\mathscr{F}_x, \mathrm{F}^*_{x_0})$.

**(1.14)** Définissons $Z(X_0, \mathscr{F}_0, t) \in \mathbf{Q}_\ell[[t]]$ par le produit

**(1.14.1)** $$Z(X_0, \mathscr{F}_0, t) = \prod_{x \in |X_0|} \det(1 - F_x^* t^{\deg(x)}, \mathscr{F}_0)^{-1}.$$

Pour $\mathscr{F}$ le faisceau constant $\mathbf{Q}_\ell$, on retrouve (1.1.2). D'après (1.5.3), la dérivée logarithmique de Z est

**(1.14.2)** $$t\frac{d}{dt}\log Z(X_0, \mathscr{F}_0, t) \underset{\text{dfn}}{=} \frac{t\dfrac{d}{dt}Z(X_0, \mathscr{F}_0, t)}{Z(X_0, \mathscr{F}_0, t)} = \sum_n \sum_{x \in X^{F^n} = X_0(\mathbf{F}_{q^n})} \operatorname{Tr}(F_x^{*n}, \mathscr{F}_0) t^n.$$

Substituant (1.12.1) dans (1.14.2), on trouve par le même calcul qu'en (1.5) la généralisation suivante de (1.5.4)

**(1.14.3)** $$Z(X_0, \mathscr{F}_0, t) = \prod_i \det(1 - F^* t, H_c^i(X, \mathscr{F}))^{(-1)^{i+1}}$$

Cette formule est une identité dans $\mathbf{Q}_\ell[[t]]$.

**(1.15)** Il est parfois commode d'utiliser un langage galoisien plutôt que géométrique. Voici le dictionnaire.

Si $\overline{\mathbf{F}}_q^1$ et $\overline{\mathbf{F}}_q^2$ sont deux clôtures algébriques de $\mathbf{F}_q$, $(X_0, \mathscr{F}_0)$ sur $\mathbf{F}_q$ définit par extension des scalaires $(X_1, \mathscr{F}_1)$ sur $\overline{\mathbf{F}}_q^1$ et $(X_2, \mathscr{F}_2)$ sur $\overline{\mathbf{F}}_q^2$. Tout $\mathbf{F}_q$-isomorphisme $\sigma : \overline{\mathbf{F}}_q^1 \xrightarrow{\sim} \overline{\mathbf{F}}_q^2$ induit un isomorphisme

$$H_c^*(X_1, \mathscr{F}_1) \xrightarrow{\sim} H_c^*(X_2, \mathscr{F}_2).$$

En particulier, pour $\overline{\mathbf{F}}_q^1 = \overline{\mathbf{F}}_q^2$ (noté $\overline{\mathbf{F}}_q$), on trouve que $\operatorname{Gal}(\overline{\mathbf{F}}_q/\mathbf{F}_q)$ agit sur $H_c^*(X, \mathscr{F})$ (action par transport de structure). Soit $\varphi \in \operatorname{Gal}(\overline{\mathbf{F}}_q/\mathbf{F}_q)$ la substitution de Frobenius. On vérifie que

$$F^* = \varphi^{-1} \quad (\text{dans } \operatorname{End}(H_c^*(X, \mathscr{F}))).$$

Ceci amène à définir le *Frobenius géométrique* $F \in \operatorname{Gal}(\overline{\mathbf{F}}_q/\mathbf{F}_q)$ comme étant $\varphi^{-1}$. On a

**(1.15.1)** $$F^* = F.$$

Soit $x$ un point géométrique de $X_0$, localisé en $x_0 \in |X_0|$. Par transport de structure, le groupe $\operatorname{Gal}(k(x)/k(x_0))$ agit sur la fibre $(\mathscr{F}_0)_x$ de $\mathscr{F}_0$ en $x$; en particulier, le Frobenius géométrique relatif à $k(x_0)$ : $F_{x_0} \in \operatorname{Gal}(k(x)/k(x_0))$, agit. Pour $x$ défini par un point fermé, encore noté $x$, de X, on a $\mathscr{F}_x = (\mathscr{F}_0)_x$, et

**(1.15.2)** $$F_{x_0}^* \underset{\text{dfn}}{=} F_x^{*\deg(x_0)} = F_{x_0} \quad (\text{dans } \operatorname{End}(\mathscr{F}_x)).$$

En notations galoisiennes, (1.14.3) s'écrit

$$\prod_{x \in |X_0|} \det(1 - F_x t^{\deg(x)}, \mathscr{F}_0)^{-1} = \prod_i \det(1 - Ft, H_c^i(X, \mathscr{F}))^{(-1)^{i+1}}.$$

**2. La théorie de Grothendieck : dualité de Poincaré.**

**(2.1)** Pour expliquer la relation entre racines de l'unité et orientations, je vais au préalable redire dans un langage farfelu deux cas classiques.

a) *Variétés différentiables.* — Soit X une variété différentiable purement de dimension $n$. Le faisceau d'orientation $\mathbf{Z}'$ sur X est le faisceau localement isomorphe au faisceau constant $\mathbf{Z}$, dont les sections inversibles sur un ouvert U de X correspondent aux orientations de U. Une *orientation* de X est un isomorphisme de $\mathbf{Z}'$ sur le faisceau constant $\mathbf{Z}$. La *classe fondamentale* de X est un morphisme $\mathrm{Tr} : \mathrm{H}^n_c(\mathrm{X}, \mathbf{Z}') \to \mathbf{Z}$; si X est orientée, il s'identifie à un morphisme $\mathrm{Tr} : \mathrm{H}^n_c(\mathrm{X}, \mathbf{Z}) \to \mathbf{Z}$. La dualité de Poincaré s'exprime à l'aide de la classe fondamentale.

b) *Variétés complexes.* — Soit $\mathbf{C}$ une clôture algébrique de $\mathbf{R}$. Une variété algébrique complexe lisse, ou plutôt la variété différentiable sous-jacente, est toujours orientable. Pour l'orienter, il suffit d'orienter $\mathbf{C}$ lui-même. Ceci revient au choix :

*a)* à choisir l'une des deux racines de l'équation $\mathrm{X}^2 = -1$; on l'appelle $+i$;

*b)* à choisir un isomorphisme de $\mathbf{R}/\mathbf{Z}$ avec $\mathrm{U}^1 = \{z \in \mathbf{C} \mid |z| = 1\}$; $+i$ est l'image de $1/4$;

*c)* à choisir l'un des deux isomorphismes $x \mapsto \exp(\pm 2\pi i x)$ de $\mathbf{Q}/\mathbf{Z}$ sur le groupe des racines de l'unité de $\mathbf{C}$, qui se prolonge par continuité en un isomorphisme de $\mathbf{R}/\mathbf{Z}$ sur $\mathrm{U}^1$.

On note $\mathbf{Z}(1)$ un $\mathbf{Z}$-module libre de rang un dont l'ensemble à deux éléments des générateurs soit en correspondance canonique avec l'un des ensembles à deux éléments *a)*, *b)*, *c)*. Le plus simple est de prendre $\mathbf{Z}(1) = \mathrm{Ker}(\exp : \mathbf{C} \to \mathbf{C}^*)$. Au générateur $y = \pm 2\pi i$ correspond l'isomorphisme *c)* : $x \mapsto \exp(xy)$.

Soit $\mathbf{Z}(r)$ la puissance tensorielle $r^{\text{ième}}$ de $\mathbf{Z}(1)$. Si X est une variété algébrique complexe lisse purement de dimension complexe $r$, le faisceau d'orientation de X est le faisceau constant de valeur $\mathbf{Z}(r)$.

**(2.2)** Pour « orienter » les variétés algébriques sur $k$ algébriquement clos de caractéristique 0, il faut choisir un isomorphisme de $\mathbf{Q}/\mathbf{Z}$ sur le groupe des racines de l'unité de $k$. L'ensemble de ces isomorphismes est un espace principal homogène sous $\hat{\mathbf{Z}}^*$ (non plus sous $\mathbf{Z}^*$). Lorsqu'on s'intéresse seulement à la cohomologie $\ell$-adique, il suffit de considérer les racines de l'unité d'ordre une puissance de $\ell$, et de supposer la caractéristique $p$ de $k$ différente de $\ell$. On note $\mathbf{Z}/\ell^n(1)$ le groupe des racines de l'unité de $k$ d'ordre divisant $\ell^n$. Pour $n$ variable, les $\mathbf{Z}/\ell^n(1)$ forment un système projectif, d'applications de transition les

$$\sigma_{m,n} : \mathbf{Z}/\ell^m(1) \to \mathbf{Z}/\ell^n(1) : x \mapsto x^{\ell^{m-n}}.$$

On pose $\mathbf{Z}_\ell(1) = \mathrm{lim\ proj}\, \mathbf{Z}/\ell^n(1)$ et $\mathbf{Q}_\ell(1) = \mathbf{Z}_\ell(1) \otimes_{\mathbf{Z}_\ell} \mathbf{Q}_\ell$. On note $\mathbf{Q}_\ell(r)$ la puissance tensorielle $r^{\text{ième}}$ de $\mathbf{Q}_\ell(1)$; pour $r \in \mathbf{Z}$, $r$ négatif, on pose encore $\mathbf{Q}_\ell(r) = \mathbf{Q}_\ell(-r)^\vee$.

En tant qu'espace vectoriel sur $\mathbf{Q}_\ell$, $\mathbf{Q}_\ell(1)$ est isomorphe à $\mathbf{Q}_\ell$. Toutefois le groupe des automorphismes de $k$ agit non trivialement sur $\mathbf{Q}_\ell(1)$ : il agit *via* le caractère à valeurs dans $\mathbf{Z}_\ell^*$ qui donne son action sur les racines de l'unité. En particulier, si $k=\overline{\mathbf{F}}_q$, la substitution de Frobenius $\varphi : x \mapsto x^q$ agit par multiplication par $q$.

Soit X une variété algébrique lisse purement de dimension $n$ sur $k$. Le *faisceau d'orientation* de X en cohomologie $\ell$-adique est le $\mathbf{Q}_\ell$-faisceau constant $\mathbf{Q}_\ell(n)$. La *classe fondamentale* est un morphisme

$$\mathrm{Tr} : \mathrm{H}_c^{2n}(\mathrm{X}, \mathbf{Q}_\ell(n)) \to \mathbf{Q}_\ell,$$

soit encore

$$\mathrm{Tr} : \mathrm{H}_c^{2n}(\mathrm{X}, \mathbf{Q}_\ell) \to \mathbf{Q}_\ell(-n).$$

*Théorème* **(2.3)** (dualité de Poincaré). — *Pour* X *propre et lisse purement de dimension n, la forme bilinéaire*

$$\mathrm{Tr}(x \cup y) : \mathrm{H}^i(\mathrm{X}, \mathbf{Q}_\ell) \otimes \mathrm{H}^{2n-i}(\mathrm{X}, \mathbf{Q}_\ell) \to \mathbf{Q}_\ell(-n)$$

*est une dualité parfaite (elle identifie* $\mathrm{H}^i(\mathrm{X}, \mathbf{Q}_\ell)$ *au dual de* $\mathrm{H}^{2n-i}(\mathrm{X}, \mathbf{Q}_\ell(n))$*).*

**(2.4)** Soient $\mathrm{X}_0$ une variété algébrique propre et lisse sur $\mathbf{F}_q$, purement de dimension $n$, et X sur $\overline{\mathbf{F}}_q$ déduit de $\mathrm{X}_0$ par extension des scalaires. Le morphisme (2.3) est compatible à l'action de $\mathrm{Gal}(\overline{\mathbf{F}}_q/\mathbf{F}_q)$. Si les $(\alpha_j)$ sont les valeurs propres du Frobenius géométrique F agissant sur $\mathrm{H}^i(\mathrm{X}, \mathbf{Q}_\ell)$, les valeurs propres de F agissant sur $\mathrm{H}^{2n-i}(\mathrm{X}, \mathbf{Q}_\ell)$ sont donc les $(q^n \alpha_j^{-1})$.

**(2.5)** Supposons pour simplifier X connexe. La démonstration (2.4) se transpose comme suit dans un langage géométrique, plutôt que galoisien (cf. (1.15)).

*a)* Le cup-produit met $\mathrm{H}^i(\mathrm{X}, \mathbf{Q}_\ell)$ et $\mathrm{H}^{2n-i}(\mathrm{X}, \mathbf{Q}_\ell)$ en dualité parfaite à valeur dans $\mathrm{H}^{2n}(\mathrm{X}, \mathbf{Q}_\ell)$, qui est de dimension un.

*b)* Le cup-produit commute à l'image réciproque $\mathrm{F}^*$ par le morphisme de Frobenius $\mathrm{F} : \mathrm{X} \to \mathrm{X}$.

*c)* Le morphisme F est fini et de degré $q^n$ : sur $\mathrm{H}^{2n}(\mathrm{X}, \mathbf{Q}_\ell)$, $\mathrm{F}^*$ est la multiplication par $q^n$.

*d)* Les valeurs propres de $\mathrm{F}^*$ ont donc la propriété (2.4).

**(2.6)** On pose $\chi(\mathrm{X}) = \sum_i (-1)^i \dim \mathrm{H}^i(\mathrm{X}, \mathbf{Q}_\ell)$. Si $n$ est impair, la forme $\mathrm{Tr}(x \cup y)$ sur $\mathrm{H}^n(\mathrm{X}, \mathbf{Q}_\ell)$ est alternée; l'entier $n\chi(\mathrm{X})$ est donc toujours pair. On déduit facilement de (1.5.4) et de (2.3), (2.4) que

$$\mathrm{Z}(\mathrm{X}_0, t) = \varepsilon . q^{\frac{-n\chi(\mathrm{X})}{2}} . t^{-\chi(\mathrm{X})} . \mathrm{Z}(\mathrm{X}_0, q^{-n} t^{-1})$$

où $\varepsilon = \pm 1$. Si $n$ est pair, notons N la multiplicité de la valeur propre $q^{n/2}$ de $F^*$ agissant sur $H^n(X, \mathbf{Q}_\ell)$ (i.e. la dimension du sous-espace propre généralisé correspondant). On a

$$\varepsilon = \begin{cases} 1 & \text{si } n \text{ est impair} \\ (-1)^N & \text{si } n \text{ est pair.} \end{cases}$$

C'est là la formulation de Grothendieck de l'équation fonctionnelle des fonctions Z.

**(2.7)** Nous aurons besoin d'autres formes du théorème de dualité. Le cas des courbes nous suffirait dans celles-ci. Si $\mathscr{F}$ est un $\mathbf{Q}_\ell$-faisceau sur une variété algébrique X sur $k$ algébriquement clos, nous noterons $\mathscr{F}(r)$ le faisceau $\mathscr{F} \otimes \mathbf{Q}_\ell(r)$. Ce faisceau est (non canoniquement) isomorphe à $\mathscr{F}$.

*Théorème* **(2.8)**. — *Supposons* X *lisse purement de dimension n et* $\mathscr{F}$ *constant tordu. Soit* $\mathscr{F}^\vee$ *le dual de* $\mathscr{F}$. *La forme bilinéaire*

$$\mathrm{Tr}(x \cup y) : H^i(X, \mathscr{F}) \otimes H_c^{2n-i}(X, \mathscr{F}^\vee(n)) \to H_c^{2n}(X, \mathscr{F} \otimes \mathscr{F}^\vee(n)) \to H_c^{2n}(X, \mathbf{Q}_\ell(n)) \to \mathbf{Q}_\ell$$

*est une dualité parfaite.*

**(2.9)** Supposons X connexe et soit $x$ un point fermé de X. Le foncteur $\mathscr{F} \mapsto \mathscr{F}_x$ est une équivalence de la catégorie des $\mathbf{Q}_\ell$-faisceaux constants tordus avec celle des représentations $\ell$-adiques de $\pi_1(X, x)$. *Via* cette équivalence, $H^0(X, \mathscr{F})$ s'identifie aux invariants de $\pi_1(X, x)$ agissant dans $\mathscr{F}_x$ :

**(2.9.1)** $$H^0(X, \mathscr{F}) \xrightarrow{\sim} \mathscr{F}_x^{\pi_1(X, x)}.$$

D'après (2.8), pour X lisse et connexe de dimension $n$, on a donc

$$H_c^{2n}(X, \mathscr{F}) = H^0(X, \mathscr{F}^\vee(n))^\vee = ((\mathscr{F}_x^\vee(n))^{\pi_1(X, x)})^\vee.$$

La dualité échange invariants (plus grand sous-espace invariant) et coinvariants (plus grand quotient invariant). Cette formule se récrit donc

$$H_c^{2n}(X, \mathscr{F}) = (\mathscr{F}_x)_{\pi_1(X, x)}(-n).$$

Nous ne l'utiliserons que pour $n = 1$.

*Scholie* **(2.10)**. — *Soient* X *une courbe lisse et connexe sur k algébriquement clos, x un point fermé de* X *et* $\mathscr{F}$ *un* $\mathbf{Q}_\ell$*-faisceau constant tordu. On a*

(i) $H_c^0(X, \mathscr{F}) = 0$ *si* X *est affine.*

(ii) $H_c^2(X, \mathscr{F}) = (\mathscr{F}_x)_{\pi_1(X, x)}(-1)$.

L'assertion (i) signifie simplement que $\mathscr{F}$ n'a pas de section à support fini.

**(2.11)** Soient X une courbe projective, lisse et connexe sur $k$ algébriquement clos, U un ouvert de X, complément d'un ensemble fini S de points fermés de X, $j$ l'inclusion $U \hookrightarrow X$ et $\mathscr{F}$ un $\mathbf{Q}_\ell$-faisceau constant tordu sur U. Soit $j_*\mathscr{F}$ le $\mathbf{Q}_\ell$-faisceau constructible image directe de $\mathscr{F}$. Sa fibre en $x \in S$ est de rang inférieur ou égal au rang de sa fibre en un point général; c'est un espace d'invariants sous un groupe de monodromie locale.

*Théorème* (**2.12**). — *La forme bilinéaire*

$$\mathrm{Tr}(x \cup y) : \mathrm{H}^i(\mathrm{X}, j_*\mathscr{F}) \otimes \mathrm{H}^{2-i}(\mathrm{X}, j_*\mathscr{F}^\vee(1)) \to \mathrm{H}^2(\mathrm{X}, j_*\mathscr{F} \otimes j_*\mathscr{F}^\vee(1))$$
$$\to \mathrm{H}^2(\mathrm{X}, j_*(\mathscr{F} \otimes \mathscr{F}^\vee)(1)) \to \mathrm{H}^2(\mathrm{X}, j_*\mathbf{Q}_\ell(1)) = \mathrm{H}^2(\mathrm{X}, \mathbf{Q}_\ell(1)) \to \mathbf{Q}_\ell$$

*est une dualité parfaite.*

(**2.13**) Il nous sera commode de disposer des $\mathbf{Q}_\ell$-faisceaux $\mathbf{Q}_\ell(r)$ sur un quelconque schéma X où $\ell$ est inversible. Le tout est de définir les $\mathbf{Z}/\ell^n(1)$. Par définition, $\mathbf{Z}/\ell^n(1)$ est le faisceau étale des racines $(\ell^n)^{\text{ièmes}}$ de l'unité.

(**2.14**) *Indications bibliographiques sur les §§ 1 et 2.*

A) Tous les résultats importants en cohomologie étale se démontrent d'abord pour des faisceaux de torsion. L'extension aux $\mathbf{Q}_\ell$-faisceaux se fait par des passages à la limite formels. Dans ce qui suit, pour chaque théorème cité, je renverrai non point à une référence où il est démontré, mais à une référence où son analogue pour les faisceaux de torsion l'est.

B) A l'exception de la formule de Lefschetz et de (2.12), les résultats de cohomologie étale utilisés dans cet article sont tous démontrés dans SGA 4. Pour ceux déjà énoncés, les références sont : définition des $\mathrm{H}^i$ : VII; définition des $\mathrm{H}^i_c$ : XVII 5.1; théorème de finitude : XIV 1, complété en XVII 5.3; dimension cohomologique : X; dualité de Poincaré : XVIII.

C) La relation entre les divers Frobenius ((1.4), (1.11), (1.15)) est expliquée en détail dans SGA 5, XV, §§1, 2.

D) L'interprétation cohomologique des fonctions Z (1.14.3) est clairement exposée dans [1]; toutefois, la formule de Lefschetz (1.12), pour X une courbe projective et lisse, y est utilisée, mais non démontrée. Pour la démonstration, il faut, hélas, renvoyer à SGA 5.

E) La forme (2.12) du théorème de dualité de Poincaré résulte du résultat général SGA 4, XVIII (3.2.5) (pour $\mathrm{S} = \mathrm{Spec}(k)$, $\mathrm{X} = \mathrm{X}$, $\mathrm{K} = j_*\mathscr{F}$, $\mathrm{L} = \mathbf{Q}_\ell$) par un calcul local qui n'est pas difficile. L'énoncé figurera explicitement dans la version définitive de SGA 5. Dans le cas où nous l'utiliserons (ramification de $\mathscr{F}$ modérée), on pourrait l'obtenir par voie transcendante, en relevant X et $\mathscr{F}$ en caractéristique 0.

## 3. La majoration fondamentale.

Le résultat de ce paragraphe a été catalysé par la lecture de Rankin [3].

(**3.1**) Soient $\mathrm{U}_0$ une courbe sur $\mathbf{F}_q$, complément dans $\mathbf{P}^1$ d'un ensemble fini de points fermés, U la courbe sur $\overline{\mathbf{F}}_q$ qui s'en déduit, $u$ un point fermé de U, $\mathscr{F}_0$ un $\mathbf{Q}_\ell$-faisceau constant tordu sur $\mathrm{U}_0$ et $\mathscr{F}$ son image réciproque sur U.

Soit $\beta\in\mathbf{Q}$. Nous dirons que $\mathscr{F}_0$ est de *poids* $\beta$ si pour tout $x\in|U_0|$, les valeurs propres de $F_x$ agissant sur $\mathscr{F}_0$ (1.13) sont des nombres algébriques dont tous les conjugués complexes sont de valeur absolue $q_x^{\beta/2}$. Par exemple, $\mathbf{Q}_l(r)$ est de poids $-2r$.

*Théorème* (**3.2**). — *Faisons les hypothèses suivantes :*

(i) $\mathscr{F}_0$ *est muni d'une forme bilinéaire alternée non dégénérée*

$$\psi : \mathscr{F}_0\otimes\mathscr{F}_0 \to \mathbf{Q}_l(-\beta) \qquad (\beta\in\mathbf{Z}).$$

(ii) *L'image de* $\pi_1(U, u)$ *dans* $GL(\mathscr{F}_u)$ *est un sous-groupe ouvert du groupe symplectique* $Sp(\mathscr{F}_u, \psi_u)$.

(iii) *Pour tout* $x\in|U_0|$, *le polynôme* $\det(1-F_x t, \mathscr{F}_0)$ *est à coefficients rationnels.*

*Alors,* $\mathscr{F}$ *est de poids* $\beta$.

On peut supposer, et nous supposerons, que U est affine et que $\mathscr{F}\neq 0$.

*Lemme* (**3.3**). — *Soit* $2k$ *un entier pair et notons* $\overset{2k}{\otimes}\mathscr{F}_0$ *la puissance tensorielle* $(2k)^{\text{ième}}$ *de* $\mathscr{F}_0$. *Pour* $x\in|U_0|$, *la dérivée logarithmique*

$$t\frac{d}{dt}\log(\det(1-F_x t^{\deg(x)}, \overset{2k}{\otimes}\mathscr{F}_0)^{-1})$$

*est une série formelle à coefficients rationnels positifs.*

L'hypothèse (iii) assure que, pour tout $n$, $\mathrm{Tr}(F_x^n, \mathscr{F}_0)\in\mathbf{Q}$. Le nombre

$$\mathrm{Tr}(F_x^n, \overset{2k}{\otimes}\mathscr{F}_0) = \mathrm{Tr}(F_x^n, \mathscr{F}_0)^{2k}$$

est donc rationnel positif, et on applique (1.5.3).

*Lemme* (**3.4**). — *Les facteurs locaux* $\det(1-F_x t^{\deg(x)}, \overset{2k}{\otimes}\mathscr{F}_0)^{-1}$ *sont des séries formelles à coefficients rationnels positifs.*

La série formelle $\log\det(1-F_x t^{\deg(x)}, \overset{2k}{\otimes}\mathscr{F}_0)^{-1}$ est sans terme constant; d'après (3.3), ses coefficients sont $\geq 0$; les coefficients de son exponentielle sont donc également positifs.

*Lemme* (**3.5**). — *Soit* $f_i = \sum_n a_{i,n} t^n$ *une suite de séries formelles de terme constant un, et à coefficients réels positifs. On suppose que l'ordre de* $f_i - 1$ *tend vers l'infini avec* $i$, *et on pose* $f = \prod_i f_i$. *Alors, le rayon de convergence absolue de* $f_i$ *est au moins égal à celui de* $f$.

Si $f = \sum_n a_n t^n$, on a en effet $a_{i,n}\leq a_n$.

*Lemme* (**3.6**). — *Sous les hypothèses de* (3.5), *si* $f$ *et les* $f_i$ *sont les développements en série de Taylor de fonctions méromorphes, alors*

$$\inf\{|z|\,|\,f(z)=\infty\}\leq\inf\{|z|\,|\,f_i(z)=\infty\}.$$

Ces nombres sont en effet les rayons de convergence absolue.

**(3.7)** Pour chaque partition P de $[1, 2k]$ en parties à deux éléments $\{i_\alpha, j_\alpha\}$ $(i_\alpha < j_\alpha)$, on définit

$$\psi_P : \bigotimes^{2k} \mathscr{F}_0 \to \mathbf{Q}_\ell(-k\beta) : x_1 \otimes \ldots \otimes x_{2k} \mapsto \prod_\alpha \psi(x_{i_\alpha}, x_{j_\alpha}).$$

Soit $x$ un point fermé de X. L'hypothèse (ii) assure que les coinvariants de $\pi_1(U, u)$ dans $\bigotimes^{2k} \mathscr{F}_u$ sont les coinvariants dans $\bigotimes^{2k} \mathscr{F}_u$ du groupe symplectique tout entier ($\pi_1$ est Zariski-dense dans Sp). Soit $\mathscr{P}$ l'ensemble des partitions P. D'après H. Weyl (*The classical groups*, Princeton University Press, chap. VI, § 1), pour $\mathscr{P}' \subset \mathscr{P}$ convenable, dépendant de $\dim(\mathscr{F}_u)$, les $\psi_P$ pour $P \in \mathscr{P}'$ définissent un isomorphisme

$$(\bigotimes^{2k} \mathscr{F}_u)_{\pi_1} = (\bigotimes^{2k} \mathscr{F}_u)_{\mathrm{Sp}} \xrightarrow{\sim} \mathbf{Q}_\ell(-k\beta)^{\mathscr{P}'}.$$

Soit N le nombre d'éléments de $\mathscr{P}'$. D'après (2.10), la formule ci-dessus donne

$$H^2_c(U, \bigotimes^{2k} \mathscr{F}) \simeq \mathbf{Q}_\ell(-k\beta-1)^N.$$

Puisque $H^0_c(U, \bigotimes^{2k} \mathscr{F}) = 0$, la formule (1.14.3) se réduit à

$$Z(U_0, \bigotimes^{2k} \mathscr{F}_0, t) = \frac{\det(1 - F^* t, H^1(U, \bigotimes^{2k} \mathscr{F}))}{(1 - q^{k\beta+1} t)^N}.$$

Cette fonction Z est donc le développement en série de Taylor d'une fonction rationnelle n'ayant de pôle qu'en $t = 1/q^{k\beta+1}$. Nous utiliserons seulement le fait que les pôles sont de valeur absolue $1/q^{k\beta+1}$ dans **C**. Cela pourrait se déduire d'arguments généraux sur les groupes réductifs. Si $\alpha$ est une valeur propre de $F_x$ sur $\mathscr{F}_0$, alors $\alpha^{2k}$ est une valeur propre de $F_x$ sur $\bigotimes^{2k} \mathscr{F}_0$. Notons encore $\alpha$ un quelconque conjugué complexe de $\alpha$. La puissance inverse $1/\alpha^{2k/\deg(x)}$ est un pôle de $\det(1 - F_x t^{\deg(x)}, \bigotimes^{2k} \mathscr{F})^{-1}$. D'après (3.4) et (3.6), on a donc

$$|1/q^{k\beta+1}| \leq |1/\alpha^{2k/\deg(x)}|,$$

soit

$$|\alpha| \leq q_x^{\frac{\beta}{2} + \frac{1}{2k}}.$$

Faisant tendre $k$ vers l'infini, on trouve que

$$|\alpha| \leq q_x^{\beta/2}.$$

Par ailleurs, l'existence de $\psi$ assure que $q_x^\beta \alpha^{-1}$ est également valeur propre, d'où l'inégalité

$$|q_x^\beta \alpha^{-1}| \leq q_x^{\beta/2}$$

soit

$$q_x^{\beta/2} \leq |\alpha|.$$

Ceci achève la démonstration.

*Corollaire* (**3.8**). — *Soit* $\alpha$ *une valeur propre de* $F^*$ *agissant sur* $H^1_c(U, \mathscr{F})$. *Alors,* $\alpha$ *est un nombre algébrique, dont tous les conjugués complexes vérifient*

$$|\alpha| \leq q^{\frac{\beta+1}{2}+\frac{1}{2}}.$$

La formule (1.14.3) pour $\mathscr{F}_0$ se réduit à

$$Z(U_0, \mathscr{F}_0, t) = \det(1 - F^* t, H^1_c(U, \mathscr{F})).$$

Le premier membre est une série formelle à coefficients rationnels, vu son développement en produit et l'hypothèse (iii). Le second membre est donc un polynôme à coefficients rationnels; $1/\alpha$ en est racine. Ceci prouve déjà que $\alpha$ est algébrique. Pour achever la démonstration, il suffit de vérifier que le produit infini qui définit $Z(U_0, \mathscr{F}_0, t)$ converge absolument (donc est non nul) pour $|t| < q^{\frac{-\beta}{2}-1}$.

Soit N le rang de $\mathscr{F}$, et posons

$$\det(1 - F_x t, \mathscr{F}) = \prod_{i=1}^{N} (1 - \alpha_{i,x} t).$$

D'après (3.2), $|\alpha_{i,x}| = q_x^{\beta/2}$. La convergence du produit infini Z résulte de celle de la série

$$\sum_{i,x} |\alpha_{i,x} t^{\deg(x)}|.$$

Pour $|t| = q^{\frac{-\beta}{2}-1-\varepsilon}$ $(\varepsilon > 0)$, on a

$$\sum_{i,x} |\alpha_{i,x} t^{\deg(x)}| = N \sum_x q_x^{-1-\varepsilon}.$$

Sur la droite affine, il y a $q^n$ points à valeurs dans $\mathbf{F}_{q^n}$, donc au plus $q^n$ points fermés de degré $n$. On a donc

$$\sum_x q_x^{-1-\varepsilon} \leq \sum_n q^n . q^{n(-1-\varepsilon)} = \sum_n q^{-n\varepsilon} < \infty,$$

ce qui achève la démonstration.

*Corollaire* (**3.9**) — *Soient* $j_0$ *l'inclusion de* $U_0$ *dans* $\mathbf{P}^1_{\mathbf{F}_q}$, $j$ *celle de* U *dans* $\mathbf{P}^1$, *et* $\alpha$ *une valeur propre de* $F^*$ *agissant sur* $H^1(\mathbf{P}^1, j_* \mathscr{F})$. *Alors* $\alpha$ *est un nombre algébrique, dont tous les conjugués complexes vérifient*

$$q^{\frac{\beta+1}{2}-\frac{1}{2}} \leq |\alpha| \leq q^{\frac{\beta+1}{2}+\frac{1}{2}}.$$

Un segment de la suite longue de cohomologie définie par la suite exacte courte

$$0 \to j_! \mathscr{F} \to j_* \mathscr{F} \to j_* \mathscr{F} / j_! \mathscr{F} \to 0$$

($j_!$ := prolongement par 0) s'écrit

$$H^1_c(U, \mathscr{F}) \to H^1(\mathbf{P}^1, j_* \mathscr{F}) \to 0.$$

La valeur propre $\alpha$ apparaît donc déjà dans $H_c^1(U, \mathcal{F})$, et est justiciable de (3.8) :

$$|\alpha| \leq q^{\frac{\beta+1}{2}+\frac{1}{2}}.$$

La dualité de Poincaré (2.12) assure que $q^{\beta+1}\alpha^{-1}$ est également valeur propre, d'où l'inégalité

$$|q^{\beta+1}\alpha^{-1}| \leq q^{\frac{\beta+1}{2}+\frac{1}{2}}$$

et le corollaire.

## 4. La théorie de Lefschetz : théorie locale.

**(4.1)** Sur **C**, les résultats locaux de Lefschetz sont les suivants.

Soient $D = \{z \mid |z| < 1\}$ le disque unité, $D^* = D - \{0\}$, et $f : X \to D$ un morphisme d'espaces analytiques. On suppose que

*a)* X est non singulier, et purement de dimension $n+1$;

*b)* $f$ est propre;

*c)* $f$ est lisse en dehors d'un point $x$ de la fibre spéciale $X_0 = f^{-1}(0)$;

*d)* en $x$, $f$ présente un point quadratique non dégénéré.

Soient $t \neq 0$ dans D et $X_t = f^{-1}(t)$ « la » fibre générale. Aux données précédentes, on associe :

$\alpha$) des morphismes de spécialisation $sp : H^i(X_0, \mathbf{Z}) \to H^i(X_t, \mathbf{Z})$ : $X_0$ est un rétracte par déformation de X, et $sp$ est la flèche composée

$$H^i(X_0, \mathbf{Z}) \xleftarrow{\sim} H^i(X, \mathbf{Z}) \to H^i(X_t, \mathbf{Z});$$

$\beta$) des transformations de monodromie $T : H^i(X_t, \mathbf{Z}) \to H^i(X_t, \mathbf{Z})$, qui décrivent l'effet sur les cycles singuliers de $X_t$ de « faire tourner $t$ autour de 0 ». C'est encore l'action sur $H^i(X_t, \mathbf{Z})$, fibre en $t$ du système local $R^if_*\underline{\mathbf{Z}}|D^*$, du générateur positif de $\pi_1(D^*, t)$.

La théorie de Lefschetz décrit $\alpha$) et $\beta$) en terme du *cycle évanescent* $\delta \in H^n(X_t, \mathbf{Z})$. Ce cycle est bien défini au signe près. Pour $i \neq n$, $n+1$, on a

$$H^i(X_0, \mathbf{Z}) \xrightarrow{\sim} H^i(X_t, \mathbf{Z}) \quad (i \neq n,\ n+1).$$

Pour $i = n$, $n+1$, on a une suite exacte

$$0 \to H^n(X_0, \mathbf{Z}) \to H^n(X_t, \mathbf{Z}) \xrightarrow{x \mapsto (x, \delta)} \mathbf{Z} \to H^{n+1}(X_0, \mathbf{Z}) \to H^{n+1}(X_t, \mathbf{Z}) \to 0.$$

Pour $i \neq n$, la monodromie T est l'identité. Pour $i = n$, on a

$$Tx = x \pm (x, \delta)\delta.$$

Les valeurs de ce $\pm$, de $T\delta$, et de $(\delta, \delta)$ sont les suivantes :

| | | | | |
|---|---|---|---|---|
| $n$ mod 4 | 0 | 1 | 2 | 3 |
| $Tx = x \pm (x, \delta)\delta$ | $-$ | $-$ | $+$ | $+$ |
| $(\delta, \delta)$ | 2 | 0 | $-2$ | 0 |
| $T\delta$ | $-\delta$ | $\delta$ | $-\delta$ | $\delta$ |

La transformation de monodromie T respecte la forme d'intersection $\mathrm{Tr}(x \cup y)$ sur $H^n(X_t, \mathbf{Z})$. Pour $n$ impair, c'est une transvection symplectique. Pour $n$ pair, c'est une symétrie orthogonale.

**(4.2)** Voici l'analogue de (4.1) en géométrie algébrique abstraite. Le disque D est remplacé par le spectre d'un anneau de valuation discrète hensélien A à corps résiduel algébriquement clos. Soient S ce spectre, $\eta$ son point générique (spectre du corps des fractions de A), $s$ son point fermé (spectre du corps résiduel). Le rôle de $t$ est joué par un point générique géométrique $\bar{\eta}$ (spectre d'une clôture algébrique du corps des fractions de A).

Soit $f: X \to S$ un morphisme propre, avec X régulier purement de dimension $n+1$. On suppose $f$ lisse, sauf pour un point quadratique ordinaire $x$ dans la fibre spéciale $X_s$. Soit $\ell$ un nombre premier différent de la caractéristique résiduelle $p$ de S. Notant $X_{\bar{\eta}}$ la fibre générique géométrique, on dispose encore d'un morphisme de spécialisation

$$sp: H^i(X_s, \mathbf{Q}_\ell) \xleftarrow{\sim} H^i(X, \mathbf{Q}_\ell) \to H^i(X_{\bar{\eta}}, \mathbf{Q}_\ell). \qquad \textbf{(4.2.1)}$$

Le rôle de T est joué par l'action du groupe d'inertie $I = \mathrm{Gal}(\bar{\eta}/\eta)$, agissant sur $H^i(X_{\bar{\eta}}, \mathbf{Q}_\ell)$ par transport de structure (cf. (1.15)) :

$$I = \mathrm{Gal}(\bar{\eta}/\eta) \to GL(H^i(X_{\bar{\eta}}, \mathbf{Q}_\ell)). \qquad \textbf{(4.2.2)}$$

Les données (4.2.1), (4.2.2) décrivent entièrement les faisceaux $R^i f_* \mathbf{Q}_\ell$ sur S.

**(4.3)** Posons $n = 2m$ pour $n$ pair, et $n = 2m+1$ pour $n$ impair. (4.2.1) et (4.2.2) se décrivent encore en terme d'un cycle évanescent

$$\delta \in H^n(X_{\bar{\eta}}, \mathbf{Q}_\ell)(m). \qquad \textbf{(4.3.1)}$$

Ce cycle est bien défini au signe près.

Pour $i \neq n, n+1$, on a

$$H^i(X_s, \mathbf{Q}_\ell) \xrightarrow{\sim} H^i(X_{\bar{\eta}}, \mathbf{Q}_\ell) \quad (i \neq n, n+1). \qquad \textbf{(4.3.2)}$$

Pour $i = n, n+1$, on a une suite exacte

**(4.3.3)**

$$0 \to H^n(X_s, \mathbf{Q}_\ell) \to H^n(X_{\bar{\eta}}, \mathbf{Q}_\ell) \xrightarrow{x \mapsto \mathrm{Tr}(x \cup \delta)} \mathbf{Q}_\ell(m-n) \to H^{n+1}(X_s, \mathbf{Q}_\ell) \to H^{n+1}(X_{\bar{\eta}}, \mathbf{Q}_\ell) \to 0.$$

L'action (4.2.2) de I (la monodromie locale) est triviale si $i \neq n$. Pour $i = n$, elle se décrit comme suit.

A) *n impair.* — On dispose d'un homomorphisme canonique

$$t_\ell : I \to \mathbf{Z}_\ell(1),$$

et l'action de $\sigma \in I$ est

$$x \mapsto x \pm t_\ell(\sigma)(x, \delta)\delta.$$

B) *n pair.* — Ce cas ne nous servira pas. Disons seulement que, si $p \neq 2$, il existe un unique caractère d'ordre deux

$$\varepsilon : \mathrm{I} \rightarrow \{\pm 1\},$$

et qu'on a

$$\begin{aligned} \sigma x &= x && \text{si} \quad \varepsilon(\sigma) = 1 \\ \sigma x &= x \pm (x, \delta)\delta && \text{si} \quad \varepsilon(\sigma) = -1. \end{aligned}$$

Les signes $\pm$ dans A) et B) sont les mêmes qu'en (4.1).

**(4.4)** Ces résultats apportent les informations suivantes sur les $\mathrm{R}^i f_* \mathbf{Q}_\ell$.

a) *Si* $\delta \neq 0$ :

1) Pour $i \neq n$, le faisceau $\mathrm{R}^i f_* \mathbf{Q}_\ell$ est constant.
2) Soit $j$ l'inclusion de $\eta$ dans S. On a

$$\mathrm{R}^n f_* \mathbf{Q}_\ell = j_* j^* \mathrm{R}^n f_* \mathbf{Q}_\ell .$$

b) *Si* $\delta = 0$ : (C'est là un cas exceptionnel. Puisque $(\delta, \delta) = \pm 2$ pour $n$ pair, il ne peut se produire que pour $n$ impair.)

1) Pour $i \neq n+1$, le faisceau $\mathrm{R}^i f_* \mathbf{Q}_\ell$ est constant.
2) Soit $\mathbf{Q}_\ell(m-n)_s$ le faisceau $\mathbf{Q}_\ell(m-n)$ sur $\{s\}$, prolongé par zéro sur S. On a une suite exacte

$$0 \rightarrow \mathbf{Q}_\ell(m-n)_s \rightarrow \mathrm{R}^{n+1} f_* \mathbf{Q}_\ell \rightarrow j_* j^* \mathrm{R}^{n+1} f_* \mathbf{Q}_\ell \rightarrow 0,$$

où $j_* j^* \mathrm{R}^{n+1} f_* \mathbf{Q}_\ell$ est un faisceau constant.

## 5. La théorie de Lefschetz : théorie globale.

**(5.1)** Sur **C**, les résultats de Lefschetz sont les suivants. Soient **P** un espace projectif de dimension $\geq 1$, et $\check{\mathbf{P}}$ l'espace projectif dual; ses points paramétrisent les hyperplans de **P**, et on note $\mathrm{H}_t$ l'hyperplan défini par $t \in \check{\mathbf{P}}$. Si A est un sous-espace linéaire de codimension 2 de **P**, les hyperplans contenant A sont paramétrés par les points d'une droite $\mathrm{D} \subset \check{\mathbf{P}}$, la *duale* de A. Ces hyperplans $(\mathrm{H}_t)_{t \in \mathrm{D}}$ forment le *pinceau d'axe* A.

Soit $\mathrm{X} \subset \mathbf{P}$ une variété projective non singulière connexe et de dimension $n+1$. Soit $\tilde{\mathrm{X}} \subset \mathrm{X} \times \mathrm{D}$ l'ensemble des couples $(x, t)$ tels que $x \in \mathrm{H}_t$. Les applications première et seconde coordonnée forment un diagramme

$$\text{(5.1.1)} \qquad \begin{array}{ccc} \mathrm{X} & \xleftarrow{\pi} & \tilde{\mathrm{X}} \\ & & \downarrow \\ & & \mathrm{D} \end{array}$$

La fibre de $f$ en $t \in \mathrm{D}$ est la section hyperplane $\mathrm{X}_t = \mathrm{X} \cap \mathrm{H}_t$ de X.

Fixons X, et prenons A assez général. Alors :

A) A est transverse à X, et $\tilde{X}$ se déduit de X par éclatement de $A \cap X$. En particulier, $\tilde{X}$ est non singulier.

B) Il existe une partie finie S de D et pour chaque $s \in S$ un point $x_s \in X_s$, tels que $f$ soit lisse en dehors des $x_s$..

C) Les $x_s$ sont des points critiques non dégénérés de $f$.

Pour chaque $s \in S$, la théorie de Lefschetz locale (4.1) s'applique donc à un petit disque $D_s$ autour de $s$ et à $f^{-1}(D_s)$.

**(5.2)** On pose $U = D - S$. Soit $u \in U$, et choisissons des lacets disjoints $(\gamma_s)_{s \in S}$ partant de $u$, avec $\gamma_s$ tournant une fois autour de $s$ :

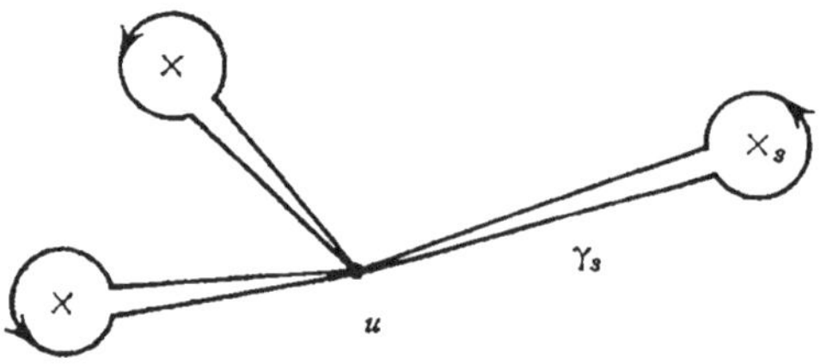


Ces lacets engendrent le groupe fondamental $\pi_1(U, u)$. Ce groupe agit sur $H^i(X_u, \mathbf{Z})$, fibre en $u$ du système local $R^i f_* \mathbf{Z} | U$. D'après la théorie locale (4.1), à chaque $s \in S$ correspond un cycle évanescent $\delta_s \in H^n(X_u, \mathbf{Z})$ ; ces cycles dépendent du choix des $\gamma_s$. Pour $i \neq n$, l'action de $\pi_1(U, u)$ sur $H^i(X_u, \mathbf{Z})$ est triviale. Pour $i = n$, on a

**(5.2.1)** $$\gamma_s x = x \pm (x, \delta_s)\delta_s.$$

Soit E le sous-espace de $H^n(X_u, \mathbf{Q})$ engendré par les $\delta_s$ (partie *évanescente* de la cohomologie).

*Proposition* **(5.3)**. — E *est stable sous l'action du groupe de monodromie* $\pi_1(U, u)$. *L'orthogonal* $E^\perp$ *de* E (*pour la forme d'intersection* $\mathrm{Tr}(x \cup y)$) *est le sous-espace des invariants de la monodromie dans* $H^n(X_u, \mathbf{Q})$.

Les $\gamma_s$ engendrant le groupe de monodromie, c'est clair sur (5.2.1).

*Théorème* **(5.4)**. — *Les cycles évanescents* $\pm \delta_s$ *(pris au signe près) sont conjugués sous l'action de* $\pi_1(U, u)$.

Soit $\check{X} \subset \check{\mathbf{P}}$ la variété duale de X : c'est l'ensemble des $t \in \check{\mathbf{P}}$ tels que $H_t$ soit tangent à X, i.e. tels que $X_t$ soit singulier, ou que $X \subset H_t$. La variété $\check{X}$ est irréductible. Soit $Y \subset X \times \check{\mathbf{P}}$ l'espace des couples $(x, t)$ tels que $x \in H_t$. On dispose d'un diagramme

$$\begin{array}{ccc} X & \longleftarrow & Y \\ & & \big\downarrow{\scriptstyle g} \\ & & \check{\mathbf{P}} \end{array}$$

La fibre de $g$ en $t \in \check{\mathbf{P}}$ est la section hyperplane $X_t = X \cap H_t$ de $X$, et $g$ est lisse en dehors de l'image réciproque de $\check{X}$.

On retrouve la situation de (5.1) en remplaçant $\check{\mathbf{P}}$ par une droite $D \subset \check{\mathbf{P}}$ et $Y$ par $g^{-1}(D)$. On a $S = D \cap \check{X}$. D'après un théorème de Lefschetz, pour $D$ assez générale, l'application

$$\pi_1(D - S, u) \to \pi_1(\check{\mathbf{P}} - \check{X}, u)$$

est surjective. Il suffit donc de montrer que les $\pm\delta_s$ sont conjugués sous $\pi_1(\check{\mathbf{P}} - \check{X})$.

Pour $x$ dans le lieu lisse de codimension 1 de $\check{X}$, soient *ch* un chemin de $t$ à $x$ dans $\check{\mathbf{P}} - \check{X}$, et $\gamma_x$ le lacet qui suit *ch* jusqu'au voisinage de $\check{X}$, tourne une fois autour de $\check{X}$, puis revient à $t$ par *ch*. Les lacets $\gamma_x$ (pour *ch* variable) sont conjugués entre eux. Puisque $\check{X}$ est irréductible, deux points du lieu lisse de $\check{X}$ peuvent toujours, dans $\check{X}$, être joints par un chemin qui ne quitte pas le lieu lisse. Il en résulte que la classe de conjugaison de $\gamma_x$ ne dépend pas de $x$. En particulier, les $\gamma_s$ sont conjugués entre eux. On lit sur (5.2.1) que ceci implique la conjugaison des $\pm\delta_s$.

*Corollaire* **(5.5)**. — *L'action de* $\pi_1(U, u)$ *sur* $E/(E \cap E^{\perp})$ *est absolument irréductible.*

Soit $F \subset E \otimes \mathbf{C}$ un sous-espace stable sous la monodromie. Si $F \not\subset (E \cap E^{\perp}) \otimes \mathbf{C}$, il existe $x \in F$ et $s \in S$ tels que $(x, \delta_s) \neq 0$. On a alors

$$\gamma_s x - x = \pm(x, \delta_s)\delta_s \in F,$$

et $\delta_s \in F$. D'après (5.4), tous les $\delta_s$ sont alors dans $F$, et $F = E$. Ceci prouve (5.5).

**(5.6)** Ces résultats se transposent comme suit en géométrie algébrique abstraite.

Soient $\mathbf{P}$ un espace projectif de dimension $> 1$ sur $k$ algébriquement clos de caractéristique $p$ et $X \subset \mathbf{P}$ une variété projective non singulière connexe et de dimension $n+1$. Pour $A$ un sous-espace linéaire de $\mathbf{P}$ de codimension 2, on définit comme en (5.1) $D$ et le pinceau $(H_t)_{t \in D}$, les $X_t$, $\tilde{X}$ et le diagramme (5.1.1). On dit que les $(X_t)_{t \in D}$ forment un *pinceau de Lefschetz* de sections hyperplanes si les conditions suivantes sont vérifiées :

A) L'axe $A$ est transverse à $X$. L'espace $\tilde{X}$ se déduit alors de $X$ par éclatement de $A \cap X$, et est lisse.

B) Il existe une partie finie S de D et pour chaque $s \in S$ un point $x_s \in X_s$, tels que $f$ soit lisse en dehors des $x_s$.

C) $x_s$ est un point singulier quadratique ordinaire de $X_s$.

Pour chaque $s \in S$, la théorie de Lefschetz locale du § 4 s'applique au spectre $D_s$ de l'hensélisé de l'anneau local de D en $s$, et à $\tilde{X}_{D_s} = \tilde{X} \times_D D_s$.

**(5.7)** Soient N la dimension de **P**, $r$ un entier $\geq 1$, et $\iota_{(r)}$ le plongement de **P** dans l'espace projectif de dimension $\binom{N+r}{N} - 1$, de coordonnées homogènes les monômes de degré $r$ en les coordonnées homogènes de **P**. Les sections hyperplanes de $\iota_{(r)}(\mathbf{P})$ sont les hypersurfaces de degré $r$ de **P**.

Si $p \neq 0$, il se peut qu'aucun pinceau de sections hyperplanes de X ne soit de Lefschetz. Toutefois, si $r \geq 2$ et qu'on remplace le plongement projectif donné $\iota_1 : X \hookrightarrow \mathbf{P}$ par $\iota_r = \iota_{(r)} \circ \iota_1$, alors, dans ce nouveau plongement, tout pinceau assez général est de Lefschetz. En d'autres termes, si $r \geq 2$, un pinceau assez général de sections hypersurfaces de degré $r$ de X est toujours de Lefschetz.

**(5.8)** Dans la suite de cette discussion, nous étudions un pinceau de Lefschetz de sections hyperplanes de X, *en excluant le cas* $p = 2$, *n pair*. Le cas où $n$ est impair nous suffirait pour la suite. On pose $U = D - S$. Soient $u \in U$ et $\ell$ un nombre premier $\neq p$. Les résultats locaux du § 4 montrent que $R^n f_* \mathbf{Q}_\ell$ est *modérément* ramifié en chaque $s \in S$. Le groupe fondamental modéré de U est un quotient du complété profini du groupe fondamental transcendant analogue (relèvement en caractéristique 0 des revêtements modérés, et théorème d'existence de Riemann). La situation algébrique est dès lors toute pareille à la situation transcendante, et la transposition de résultats de Lefschetz se fait par des arguments standards. Dans la démonstration de (5.4), le théorème de Lefschetz sur les $\pi_1$ devient le théorème de Bertini, et on doit invoquer le lemme d'Abhyankar pour contrôler la ramification de $R^\cdot g_* \mathbf{Q}_\ell$ le long du lieu lisse de codimension un de $\check{X}$.

Les résultats sont les suivants.

a) *Si les cycles évanescents sont non nuls :*

1) Pour $i \neq n$, le faisceau $R^i f_* \mathbf{Q}_\ell$ sur D est constant.

2) Soit $j$ l'inclusion de U dans D. On a

$$R^n f_* \mathbf{Q}_\ell = j_* j^* R^n f_* \mathbf{Q}_\ell.$$

3) Soit $E \subset H^n(X_u, \mathbf{Q}_\ell)$ le sous-espace de la cohomologie engendré par les cycles évanescents. Ce sous-espace est stable sous $\pi_1(U, u)$, et

$$E^\perp = H^n(X_u, \mathbf{Q}_\ell)^{\pi_1(U,u)}.$$

La représentation de $\pi_1(U, u)$ sur $E/(E \cap E^\perp)$ est absolument irréductible, et l'image de $\pi_1$ dans $GL(E/(E \cap E^\perp))$ est engendrée (topologiquement) par les $x \to x \pm (x, \delta_s)\delta_s$ $(s \in S)$ (le signe $\pm$ étant déterminé comme en (4.1)).

b) *Si les cycles évanescents sont nuls :* (C'est là un cas exceptionnel. Puisque $(\delta, \delta) = \pm 2$ pour $n$ pair, il ne peut se produire que pour $n$ impair : $n = 2m+1$. On notera que si un cycle évanescent est nul, ils le sont tous, car conjugués.)

1) Pour $i \neq n+1$, le faisceau $R^i f_* \mathbf{Q}_\ell$ est constant.

2) On a une suite exacte

$$0 \to \bigoplus_{s \in S} \mathbf{Q}_\ell(m-n)_s \to R^{n+1} f_* \mathbf{Q}_\ell \to \mathscr{F} \to 0$$

avec $\mathscr{F}$ constant.

3) $E = 0$.

**(5.9)** Le sous-espace $E \cap E^\perp$ de $E$ est le noyau de la restriction à $E$ de la forme d'intersection $\mathrm{Tr}(x \cup y)$. Cette forme induit donc une forme bilinéaire non dégénérée

$$\psi : E/(E \cap E^\perp) \otimes E/(E \cap E^\perp) \to \mathbf{Q}_\ell(-n),$$

alternée pour $n$ impair, et symétrique pour $n$ pair. Cette forme est respectée par la monodromie; pour $n$ impair, la représentation de monodromie induit donc

$$\rho : \pi_1(U, u) \to \mathrm{Sp}(E/(E \cap E^\perp), \psi).$$

*Théorème* **(5.10)** (Kajdan-Margulis). — *L'image de* $\rho$ *est ouverte.*

L'image de $\rho$ est un sous-groupe compact, donc analytique $\ell$-adique, de $\mathrm{Sp}(E/(E \cap E^\perp), \psi)$. Il suffit de montrer que son algèbre de Lie $\mathfrak{L}$ est égale à $\mathfrak{sp}(E/(E \cap E^\perp), \psi)$. L'analogue transcendant de cette algèbre de Lie est l'algèbre de Lie de l'adhérence de Zariski du groupe de monodromie.

On déduit de (5.8) que $\mathfrak{L}$ est engendrée par les transformations de carré nul

$$N_s : x \mapsto (x, \delta_s)\delta_s \qquad (s \in S)$$

et que $E/(E \cap E^\perp)$ est une représentation absolument irréductible de $\mathfrak{L}$. Le théorème résulte du lemme suivant.

*Lemme* **(5.11)**. — *Soient* V *un espace vectoriel de dimension finie sur un corps* $k$ *de caractéristique* 0, $\psi$ *une forme alternée non dégénérée et* $\mathfrak{L}$ *une sous-algèbre de Lie de* $\mathfrak{sp}(V, \psi)$. *On suppose que :*

(i) V *est une représentation simple de* $\mathfrak{L}$.

(ii) $\mathfrak{L}$ *est engendrée par une famille d'endomorphismes de* V *de la forme* $x \mapsto \psi(x, \delta)\delta$.

*Alors,* $\mathfrak{L} = \mathfrak{sp}(V, \psi)$.

On peut supposer et on supposera que V, donc $\mathfrak{L}$, est non nul. Soit $W \subset V$ l'ensemble des $\delta \in V$ tels que $N(\delta) : x \mapsto \psi(x, \delta)\delta$ soit dans $\mathfrak{L}$.

*a)* W est stable par homothétie (car $\mathfrak{L}$ est un sous-espace vectoriel de $\mathfrak{gl}(V)$).

*b)* Si $\delta \in W$, $\exp(\lambda N(\delta))$ est un automorphisme de $(V, \psi, \mathfrak{L})$, donc transforme W en lui-même. Si $\delta', \delta'' \in W$, on a donc $\exp(\lambda N(\delta')).\delta'' = \delta'' + \lambda\psi(\delta'', \delta')\delta' \in W$; si $\psi(\delta', \delta'') \neq 0$, le sous-espace vectoriel tendu par $\delta'$ et $\delta''$ est dans W.

*c)* Il en résulte que W est réunion de ses sous-espaces linéaires maximaux $W_\alpha$, et que ceux-ci sont deux à deux orthogonaux. Chaque $W_\alpha$ est dès lors stable sous les $N(\delta)$ $(\delta \in W)$, donc stable sous $\mathfrak{L}$. Vu l'hypothèse (i), $W_\alpha = V$ et $\mathfrak{L}$ contient tous les $N(\delta)$ pour $\delta \in V$. On conclut en notant que l'algèbre de Lie $\mathfrak{sp}(V, \psi)$ est engendrée par les $N(\delta)$ $(\delta \in V)$.

*Remarque* **(5.12)** (inutile pour la suite). — Il est maintenant facile de prouver (1.6) pour une hypersurface de dimension impaire $n$ dans $\mathbf{P}^{n+1}_{\mathbf{F}_q}$.

Soient $X_0$ une telle hypersurface, et $\overline{X}_0$ l'hypersurface sur $\overline{\mathbf{F}}_q$ qui s'en déduit par extension des scalaires. On a

$$H^{2i}(\overline{X}_0, \mathbf{Q}_\ell) = \mathbf{Q}_\ell(-i) \qquad (0 \leq i \leq n);$$

$H^{2i}(\overline{X}_0, \mathbf{Q}_\ell(i))$ est engendré par la $i^{\text{ème}}$ cup-puissance de $\eta$, la classe de cohomologie $c_1(\mathcal{O}(1))$ d'une section hyperplane. On a donc

$$Z(X_0, t) = \det(1 - F^* t, H^n(\overline{X}_0, \mathbf{Q}_\ell)) / \prod_{i=0}^{n} (1 - q^i t)$$

et $\det(1 - F^* t, H^n(\overline{X}_0, \mathbf{Q}_\ell))$ est un polynôme à coefficients entiers indépendants de $\ell$.

Faisons varier $X_0$ dans un pinceau de Lefschetz d'hypersurfaces, qui soit défini sur $\mathbf{F}_q$ (cf. (5.7) pour $X = \mathbf{P}^{n+1}$; l'existence d'un tel pinceau n'est pas claire; si on voulait compléter l'argument esquissé ici, il faudrait recourir aux arguments qui seront donnés en (7.1)). On vérifie que E coïncide ici avec le $H^n$ tout entier, et (3.2) fournit la conjecture de Weil pour toutes les hypersurfaces du pinceau, en particulier pour $X_0$.

**(5.13)** *Indications bibliographiques sur les §§ 4 et 5.*

A) Les résultats de Lefschetz (4.1) et (5.1) à (5.5) sont contenus dans son livre [2]. Pour la théorie locale (4.1), il peut être plus commode de consulter SGA 7, XIV (3.2).

B) Les résultats du § 4 sont démontrés dans les exposés XIII, XIV et XV de SGA 7.

C) (5.7) est démontré dans SGA 7, XVII.

D) (5.8) est démontré dans SGA 7, XVIII. Le théorème d'irréductibilité y est démontré pour E, mais seulement sous l'hypothèse que $E \cap E^\perp = \{0\}$. La démonstration dans le cas général (pour $E/(E \cap E^\perp)$) est pareille.

## 6. Un théorème de rationalité.

**(6.1)** Soient $\mathbf{P}_0$ un espace projectif de dimension $\geq 1$ sur $\mathbf{F}_q$, $X_0 \subset \mathbf{P}_0$ une variété projective non singulière, $A_0 \subset \mathbf{P}_0$ un sous-espace linéaire de codimension deux, $D_0 \subset \check{\mathbf{P}}_0$ la droite duale, $\overline{\mathbf{F}}_q$ une clôture algébrique de $\mathbf{F}_q$ et $\mathbf{P}$, X, A, D sur $\overline{\mathbf{F}}_q$ déduits de $\mathbf{P}_0$, $X_0$,

$A_0$, $D_0$ par extension des scalaires. Le diagramme (5.1.1) de (5.6) provient d'un diagramme analogue sur $\mathbf{F}_q$ :

$$\begin{array}{ccc} X_0 & \xleftarrow{\pi_*} & \widetilde{X}_0 \\ & & \downarrow f_* \\ & & D_0 \end{array} \qquad (6.1.1)$$

On suppose que X est connexe de dimension *paire* $n+1=2m+2$, et que le pinceau $(X_t)_{t\in D}$ de sections hyperplanes de X défini par D est un *pinceau de Lefschetz*. L'ensemble S des $t\in D$ tels que $X_t$ soit singulier est défini sur $\mathbf{F}_q$, i.e. provient de $S_0\subset D_0$. On pose $U_0=D_0-S_0$ et $U=D-S$.

Soit $u\in U$. La partie évanescente de la cohomologie $E\subset H^n(X_u, \mathbf{Q}_\ell)$ est stable sous $\pi_1(U, u)$, donc définit sur U un sous-système local $\mathscr{E}$ de $R^n f_* \mathbf{Q}_\ell$. Ce dernier est défini sur $\mathbf{F}_q$ : $R^i f_* \mathbf{Q}_\ell$ est l'image réciproque du $\mathbf{Q}_\ell$-faisceau $R^i f_{0*} \mathbf{Q}_\ell$ sur $D_0$, et, sur U, $\mathscr{E}$ est l'image réciproque d'un sous-système local

$$\mathscr{E}_0 \subset R^n f_{0*} \mathbf{Q}_\ell.$$

Le cup-produit est une forme alternée

$$\psi : R^n f_{0*} \mathbf{Q}_\ell \otimes R^n f_{0*} \mathbf{Q}_\ell \to \mathbf{Q}_\ell(-n).$$

Notant $\mathscr{E}_0^\perp$ l'orthogonal de $\mathscr{E}_0$ relativement à $\psi$, dans $R^n f_{0*} \mathbf{Q}_\ell | U_0$ on voit que $\psi$ induit une dualité parfaite

$$\psi : \mathscr{E}_0/(\mathscr{E}_0\cap\mathscr{E}_0^\perp)\otimes\mathscr{E}_0/(\mathscr{E}_0\cap\mathscr{E}_0^\perp) \to \mathbf{Q}_\ell(-n).$$

*Théorème* **(6.2)**. — *Pour tout* $x\in|U_0|$, *le polynôme* $\det(1-F_x^* t, \mathscr{E}_0/(\mathscr{E}_0\cap\mathscr{E}_0^\perp))$ *est à coefficients rationnels.*

*Corollaire* **(6.3)**. — *Soient* $j_0$ *l'inclusion de* $U_0$ *dans* $D_0$, *et* $j$ *celle de* U *dans* D. *Les valeurs propres de* $F^*$ *agissant sur* $H^1(D, j_*\mathscr{E}/(\mathscr{E}\cap\mathscr{E}^\perp))$ *sont des nombres algébriques dont tous les conjugués complexes* $\alpha$ *vérifient*

$$q^{\frac{n+1}{2}-\frac{1}{2}} \leq |\alpha| \leq q^{\frac{n+1}{2}+\frac{1}{2}}.$$

D'après (5.10) et (6.2), les hypothèses de (3.2) sont en effet vérifiées par $(U_0, \mathscr{E}_0/(\mathscr{E}_0\cap\mathscr{E}_0^\perp), \psi)$, pour $\beta=n$, et on applique (3.9).

*Lemme* **(6.4)**. — *Soit* $\mathscr{G}_0$ *un* $\mathbf{Q}_\ell$*-faisceau constant tordu sur* $U_0$, *tel que son image réciproque* $\mathscr{G}$ *sur* U *soit un faisceau constant. Il existe alors dans* $\overline{\mathbf{Q}}_\ell$ *des unités* $\alpha_i$ *telles que, pour tout* $x\in|U_0|$, *on ait*

$$\det(1-F_x^* t, \mathscr{G}_0)=\prod_i (1-\alpha_i^{\deg(x)} t).$$

Ce lemme exprime que $\mathscr{G}_0$ est l'image réciproque d'un faisceau sur $\mathrm{Spec}(\mathbf{F}_q)$, à savoir son image directe sur $\mathrm{Spec}(\mathbf{F}_q)$. Ce dernier s'identifie à une représentation $\ell$-adique $G_0$ de $\mathrm{Gal}(\overline{\mathbf{F}}_q/\mathbf{F}_q)$, et on prend

$$\det(1-\mathrm{F}t, G_0)=\prod_i(1-\alpha_i t).$$

Le lemme (6.4) s'applique aux $R^i f_{0*}\mathbf{Q}_\ell$ $(i\neq n)$, à $R^n f_{0*}\mathbf{Q}_\ell/\mathscr{E}_0$ et à $\mathscr{E}_0\cap\mathscr{E}_0^\perp$.

Pour $x\in|U_0|$ la fibre $X_x=f_0^{-1}(x)$ est une variété sur le corps fini $k(x)$. Si $\bar{x}$ est un point de U au-dessus de $x$, $X_{\bar{x}}$ se déduit de $X_x$ par extension des scalaires de $k(x)$ à sa clôture algébrique $k(\bar{x})=\overline{\mathbf{F}}_q$, et $H^i(X_{\bar{x}},\mathbf{Q}_\ell)$ est la fibre de $R^i f_*\mathbf{Q}_\ell$ en $\bar{x}$. La formule (1.5.4) pour la variété $X_x$ sur $k(x)$ s'écrit donc

$$Z(X_x,t)=\prod_i\det(1-F_x^*t, R^i f_{0*}\mathbf{Q}_\ell)^{(-1)^{i+1}}$$

et $Z(X_x,t)$ est le produit de

$$Z^f=\det(1-F_x^*t, R^n f_{0*}\mathbf{Q}_\ell/\mathscr{E}_0).\det(1-F_x^*t,\mathscr{E}_0\cap\mathscr{E}_0^\perp).\prod_{i\neq n}\det(1-F_x^*t, R^i f_{0*}\mathbf{Q}_\ell)^{(-1)^{i+1}}$$

par

$$Z^m=\det(1-F_x^*t,\mathscr{E}_0/(\mathscr{E}_0\cap\mathscr{E}_0^\perp)).$$

Posons $\mathscr{F}_0=\mathscr{E}_0/(\mathscr{E}_0\cap\mathscr{E}_0^\perp)$, $\mathscr{F}=\mathscr{E}/(\mathscr{E}\cap\mathscr{E}^\perp)$ et appliquons (6.4) aux facteurs de $Z^f$. On trouve qu'il existe des unités $\ell$-adiques $\alpha_i$ $(1\leq i\leq N)$ et $\beta_j$ $(1\leq j\leq M)$ dans $\overline{\mathbf{Q}}_\ell$ tels que pour tout $x\in|U_0|$

$$Z(X_x,t)=\frac{\prod_i(1-\alpha_i^{\deg x}t)}{\prod_j(1-\beta_j^{\deg x}t)}.\det(1-F_x^*t,\mathscr{F}_0)$$

et qu'en particulier le second membre est dans $\mathbf{Q}(t)$. Si un $\alpha_i$ coïncide avec un $\beta_j$, il est loisible de supprimer simultanément cet $\alpha_i$ de la liste des $\alpha$ et ce $\beta_j$ de la liste des $\beta$. On peut donc supposer, et nous supposerons, que $\alpha_i\neq\beta_j$ pour tout $i$ et tout $j$.

**(6.5)** Il suffit de prouver que les polynômes $\prod_i(1-\alpha_i t)$ et $\prod_j(1-\beta_j t)$ sont à coefficients rationnels, i.e. que la famille des $\alpha_i$ (resp. la famille des $\beta_j$) est définie sur $\mathbf{Q}$. Nous le déduirons des propositions suivantes.

*Proposition* **(6.6)**. — *Soient* $(\gamma_i)$ $(1\leq i\leq P)$ *et* $(\delta_j)$ $(1\leq j\leq Q)$ *deux familles d'unités* $\ell$*-adiques dans* $\overline{\mathbf{Q}}_\ell$. *On suppose que* $\gamma_i\neq\delta_j$. *Si* K *est un ensemble fini assez grand d'entiers* $\neq 1$, *et* L *une partie de densité* 0 *assez grande de* $|U_0|$, *alors, si* $x\in|U_0|$ *vérifie* $k\nmid\deg(x)$ (*pour tout* $k\in K$) *et* $x\notin L$, *le dénominateur de*

$$\textbf{(6.6.1)}\qquad \det(1-F_x^*t,\mathscr{F}_0)\prod_i(1-\gamma_i^{\deg(x)}t)/\prod_j(1-\delta_j^{\deg(x)}t),$$

*écrit sous forme irréductible, est* $\prod_j(1-\delta_j^{\deg(x)}t)$.

La démonstration sera donnée en (6.10-13). D'après (6.7) ci-dessous, (6.6) fournit une description intrinsèque de la famille des $\delta_j$ en terme de la famille des fractions rationnelles (6.6.1) pour $x \in |U_0|$.

*Lemme* **(6.7)**. — *Soient* K *un ensemble fini d'entiers* $\neq 1$ *et* $(\delta_j)$ $(1 \leq j \leq Q)$ *et* $(\varepsilon_j)$ $(1 \leq j \leq Q)$ *deux familles d'éléments d'un corps. Si, pour tout n assez grand, non divisible par aucun des* $k \in K$, *la famille des* $\delta_j^n$ *coïncide avec celle des* $\varepsilon_j^n$ *(à l'ordre près), alors la famille des* $\delta_j$ *coïncide avec celle des* $\varepsilon_j$ *(à l'ordre près).*

On procède par récurrence sur Q. L'ensemble des entiers $n$ tels que $\delta_Q^n = \varepsilon_j^n$ est un idéal $(n_j)$. Prouvons qu'il existe $j_0$ tel que $\delta_Q = \varepsilon_{j_0}$. Sinon les $n_j$ seraient distincts de 1, et il existerait des entiers arbitrairement grands $n$, non divisibles par aucun des $n_j$, ni par aucun des $k \in K$. On aurait $\delta_Q^n \neq \varepsilon_j^n$, et ceci contredirait l'hypothèse. Il existe donc $j_0$ tel que $\delta_Q = \varepsilon_{j_0}$. On conclut en appliquant l'hypothèse de récurrence aux familles $(\delta_j)$ $(j \neq Q)$ et $(\varepsilon_j)$ $(j \neq j_0)$.

*Proposition* **(6.8)**. — *Soient* $(\gamma_i)$ $(1 \leq i \leq P)$ *et* $(\delta_j)$ $(1 \leq j \leq Q)$ *deux familles d'unités p-adiques dans* $\overline{\mathbf{Q}}_\ell$, $R(t) = \prod_i (1 - \gamma_i t)$ *et* $S(t) = \prod_j (1 - \delta_j t)$. *Supposons que pour tout* $x \in |U_0|$, $\prod_i (1 - \delta_i^{\deg(x)} t)$ *divise*

$$\prod_i (1 - \gamma_i^{\deg(x)} t) . \det(1 - F_x^* t, \mathscr{F}_0).$$

*Alors* $S(t)$ *divise* $R(t)$.

Supprimons des familles $(\gamma_i)$ et $(\delta_j)$ des paires d'éléments communs jusqu'à vérifier l'hypothèse de (6.6). Appliquons (6.6). Par hypothèse, les fractions rationnelles (6.6.1) sont des polynômes. Aucun $\delta$ n'a donc subsisté, ce qui signifie que $S(t)$ divise $R(t)$.

Cette proposition fournit une caractérisation intrinsèque de $R(t)$ en terme de la famille de polynômes

$$\prod_i (1 - \gamma_i^{\deg(x)} t) . \det(1 - F_x^* t, \mathscr{F}_0).$$

C'est le ppcm des polynômes $S(t) = \prod_j (1 - \delta_j t)$ vérifiant l'hypothèse de (6.8).

**(6.9)** Prouvons (6.5) et donc (6.2) (modulo (6.6)). Faisons dans (6.6) $(\gamma_i) = (\alpha_i)$ et $(\delta_j) = (\beta_j)$. On trouve une caractérisation intrinsèque de la famille des $\beta_j$ en terme de la famille de fractions rationnelles $Z(X_x, t)$ $(x \in |U_0|)$. Celles-ci étant dans $\mathbf{Q}(t)$, la famille des $\beta_j$ est définie sur $\mathbf{Q}$.

Les polynômes $\prod_i (1 - \alpha_i^{\deg(x)} t) . \det(1 - F_x^* t, \mathscr{F}_0)$ sont donc dans $\mathbf{Q}[t]$. La proposition (6.8) fournit une description intrinsèque de la famille des $\alpha_i$ en terme de cette famille de polynômes. La famille des $\alpha_i$ est donc définie sur $\mathbf{Q}$.

**(6.10)** *Préliminaires.* — Soient $u \in U$ et $\mathscr{F}_u$ la fibre de $\mathscr{F}$ en $u$. Le groupe fondamental arithmétique $\pi_1(U_0, u)$, extension de $\hat{\mathbf{Z}} = \mathrm{Gal}(\overline{\mathbf{F}}_q/\mathbf{F}_q)$ (générateur : $\varphi$) par le groupe fondamental géométrique $\pi_1(U, u)$, agit sur $\mathscr{F}_u$ par similitudes symplectiques :

$$\rho : \pi_1(U_0, u) \to \mathrm{CSp}(\mathscr{F}_u, \psi).$$

Nous notons $\mu(g)$ le multiplicateur d'une similitude symplectique $g$. Soit

$$H \subset \hat{\mathbf{Z}} \times \mathrm{CSp}(\mathscr{F}_u, \psi)$$

le sous-groupe défini par l'équation

$$q^{-n} = \mu(g)$$

($q$ étant une unité $\ell$-adique, $q^n \in \mathbf{Q}_\ell^*$ est défini pour tout $n \in \hat{\mathbf{Z}}$). Le fait que $\psi$ soit à valeurs dans $\mathbf{Q}_\ell(-n)$ s'exprime en disant que l'application de $\pi_1$ dans $\hat{\mathbf{Z}} \times \mathrm{CSp}$, de coordonnées la projection canonique sur $\hat{\mathbf{Z}}$ et $\rho$, se factorise par

$$\rho_1 : \pi_1(U_0, u) \to H.$$

*Lemme* **(6.11)**. — *L'image* $H_1$ *de* $\rho_1$ *est ouverte dans* H.

En effet, $\pi_1(U_0, u)$ se projette *sur* $\hat{\mathbf{Z}}$, et l'image de $\pi_1(U, u) = \mathrm{Ker}(\pi_1(U_0, u) \to \hat{\mathbf{Z}})$ dans $\mathrm{Sp}(\mathscr{F}_u, \psi) = \mathrm{Ker}(H \to \hat{\mathbf{Z}})$ est ouverte (5.10).

*Lemme* **(6.12)**. — *Pour* $\delta \in \overline{\mathbf{Q}}_\ell$ *une unité $\ell$-adique, l'ensemble* Z *des* $(n, g) \in H_1$ *tels que* $\delta^n$ *soit valeur propre de* $g$ *est un fermé de mesure nulle.*

Il est clair que Z est fermé. Pour chaque $n \in \hat{\mathbf{Z}}$, soit $\mathrm{CSp}_n$ l'ensemble des $g \in \mathrm{CSp}(\mathscr{F}_u, \psi)$ tels que $\mu(g) = q^{-n}$, et soit $Z_n$ l'ensemble des $g \in \mathrm{CSp}_n$ tels que $\delta^n$ soit valeur propre de $g$. Alors, $\mathrm{CSp}_n$ est un espace homogène sous Sp, et on vérifie que $Z_n$ en est un sous-espace algébrique propre, donc de mesure o. D'après (6.11), $H_1 \cap (\{n\} \times Z_n)$ est donc de mesure o dans l'image inverse dans $H_1$ de $n$, et on applique Fubini à la projection $H_1 \to \hat{\mathbf{Z}}$.

**(6.13)** Prouvons (6.6). Pour chaque $i$ et $j$, l'ensemble des entiers $n$ tels que $\gamma_i^n = \delta_j^n$ est l'ensemble des multiples d'un entier fixe $n_{ij}$ (on n'exclut pas $n_{ij} = 0$). Par hypothèse, $n_{ij} \neq 1$.

D'après (6.12) et le théorème de densité de Čebotarev, l'ensemble des $x \in |U_0|$ tels qu'un $\beta_j^{\deg(x)}$ soit valeur propre de $F_x^*$ agissant sur $\mathscr{F}_0$ est de densité o. On prend pour K l'ensemble des $n_{ij}$ et pour L l'ensemble des $x$ ci-dessus.

## 7. Fin de la démonstration de 1.7.

*Lemme* **(7.1)**. — *Soit* $X_0$ *une variété projective non singulière absolument irréductible de dimension* paire $d$ *sur* $\mathbf{F}_q$. *Soient* X *sur* $\overline{\mathbf{F}}_q$ *déduit de* $X_0$ *par extension des scalaires et* $\alpha$ *une valeur propre de* $F^*$ *agissant sur* $H^d(X, \mathbf{Q}_\ell)$. *Alors,* $\alpha$ *est un nombre algébrique dont tous les conjugués complexes, encore notés* $\alpha$, *vérifient*

$$q^{\frac{d}{2}-\frac{1}{2}} \leq |\alpha| \leq q^{\frac{d}{2}+\frac{1}{2}}. \tag{7.1.1}$$

On procède par récurrence sur $d$ (toujours supposé pair). Le cas $d = 0$ est trivial, même sans supposer $X_0$ absolument irréductible; on suppose dorénavant $d \geq 2$. On pose $d = n + 1 = 2m + 2$.

Si $\mathbf{F}_{q^r}$ est une extension de degré $r$ de $\mathbf{F}_q$ et que $X'_0/\mathbf{F}_{q^r}$ se déduit de $X_0/\mathbf{F}_q$ par extension des scalaires, l'assertion (7.1) pour $X_0/\mathbf{F}_q$ équivaut à (7.1) pour $X'_0/\mathbf{F}_{q^r}$ : de même que $q$ est remplacé par $q^r$, les valeurs propres de $F^*$ sont remplacées par leurs puissances $r^{\text{ièmes}}$.

D'après (5.7), dans un plongement projectif convenable $i : X \to \mathbf{P}$, X admet un pinceau de Lefschetz de sections hyperplanes. La remarque précédente nous permet de supposer $i$ et ce pinceau définis sur $\mathbf{F}_q$ (quitte à remplacer $\mathbf{F}_q$ par une extension finie).

Supposons donc qu'il existe sur $\mathbf{F}_q$ un plongement projectif $X_0 \to \mathbf{P}_0$ et un sous-espace de codimension deux $A_0$ de $\mathbf{P}_0$ qui définisse un pinceau de Lefschetz. On reprend les notations de (6.1) et (6.3). Une nouvelle extension des scalaires nous ramène à supposer que :

*a)* Les points de S sont définis sur $\mathbf{F}_q$.

*b)* Les cycles évanescents en $x_s$ ($s \in S$) sont définis sur $\mathbf{F}_q$ (puisque seul $\pm\delta$ est intrinsèque, ils pourraient n'être définis que sur une extension quadratique).

*c)* Il existe un point rationnel $u_0 \in U_0$. On prend le point correspondant $u$ de U comme point base.

*d)* $X_{u_0} = f_0^{-1}(u_0)$ admet une section hyperplane lisse $Y_0$, définie sur $\mathbf{F}_q$. On pose $Y = Y_0 \otimes_{\mathbf{F}_q} \overline{\mathbf{F}}_q$.

Puisque $\tilde{X}$ se déduit de X par éclatement de la sous-variété lisse de codimension deux $A \cap X$, on a

$$H^i(X, \mathbf{Q}_\ell) \hookrightarrow H^i(\tilde{X}, \mathbf{Q}_\ell)$$

(en fait, $H^i(\tilde{X}, \mathbf{Q}_\ell) = H^i(X, \mathbf{Q}_\ell) \oplus H^{i-2}(A \cap X, \mathbf{Q}_\ell)(-1)$). Il suffit donc de prouver (7.1.1) pour les valeurs propres $\alpha$ de $F^*$ agissant sur $H^d(\tilde{X}, \mathbf{Q}_\ell)$.

La suite spectrale de Leray pour $f$ s'écrit

$$E_2^{pq} = H^p(D, R^q f_* \mathbf{Q}_\ell) \Rightarrow H^{p+q}(\tilde{X}, \mathbf{Q}_\ell).$$

Il suffit de prouver (7.1.1) pour les valeurs propres $\alpha$ de $F^*$ agissant sur les $E_2^{pq}$ pour $p+q=d=n+1$. Ce sont :

A) $E_2^{2,n-1}$. D'après (5.8), $R^{n-1} f_* \mathbf{Q}_\ell$ est constant. D'après (2.10), on a donc

$$E_2^{2,n-1} = H^{n-1}(X_u, \mathbf{Q}_\ell)(-1).$$

D'après le théorème de Lefschetz faible (corollaire de SGA 4, XIV (3.2), et de la dualité de Poincaré, SGA 4, XVIII), on a

$$H^{n-1}(X_u, \mathbf{Q}_\ell)(-1) \hookrightarrow H^{n-1}(Y, \mathbf{Q}_\ell)(-1),$$

et on applique l'hypothèse de récurrence à $Y_0$.

B) $E_2^{0,n+1}$. Si les cycles évanescents sont non nuls, $R^{n+1} f_* \mathbf{Q}_\ell$ est constant et

$$E_2^{0,n+1} = H^{n+1}(X_u, \mathbf{Q}_\ell).$$

Le morphisme de Gysin

$$H^{n-1}(Y, \mathbf{Q}_\ell)(-1) \to H^{n+1}(X_u, \mathbf{Q}_\ell)$$

est surjectif (cet argument est dual de A)), et on applique l'hypothèse de récurrence à $Y_0$.

Si les cycles évanescents sont nuls, la suite exacte de (5.8) b) fournit une suite exacte

$$\bigoplus_{s \in S} \mathbf{Q}_\ell(m-n) \to E_2^{0,n+1} \to H^{n+1}(X_u, \mathbf{Q}_\ell).$$

La valeur propre de F agissant sur $\mathbf{Q}_\ell(m-n)$ est $q^{d/2}$, et $H^{n+1}$ se traite comme plus haut.

C) $E_2^{1,n}$. Si on disposait du théorème de Lefschetz « difficile », on saurait que $\mathscr{E} \cap \mathscr{E}^\perp$ est nul, et que $R^n f_* \mathbf{Q}_\ell$ est somme directe de $j_* \mathscr{E}$ et d'un faisceau constant. Le $H^1$ d'un faisceau constant sur $\mathbf{P}^1$ est nul, et il suffirait d'appliquer (6.3).

Faute d'avoir déjà démontré Lefschetz « difficile », il nous va falloir dévisser. Si les cycles évanescents sont nuls, $R^n f_* \mathbf{Q}_\ell$ est constant ((5.8) b)) et $E_2^{1,n} = 0$. On peut donc supposer, et on supposera, les cycles évanescents non nuls. Filtrons $R^n f_* \mathbf{Q}_\ell = j_* j^* R^n f_* \mathbf{Q}_\ell$ (5.8) par les sous-faisceaux $j_* \mathscr{E}$ et $j_*(\mathscr{E} \cap \mathscr{E}^\perp)$. Si les cycles évanescents $\delta$ ne sont pas dans $\mathscr{E} \cap \mathscr{E}^\perp$, on dispose de suites exactes :

**(7.1.2)** $0 \to j_* \mathscr{E} \to R^n f_* \mathbf{Q}_\ell \to$ faisceau constant $\to 0$.

**(7.1.3)** $0 \to$ faisceau constant $j_*(\mathscr{E} \cap \mathscr{E}^\perp) \to j_* \mathscr{E} \to j_*(\mathscr{E}/(\mathscr{E} \cap \mathscr{E}^\perp)) \to 0$.

Si, à Dieu ne plaise, les $\delta$ sont dans $\mathscr{E} \cap \mathscr{E}^\perp$, on a $\mathscr{E} \subset \mathscr{E}^\perp$, et des suites exactes :

**(7.1.4)** $0 \to$ le faisceau constant $j_* \mathscr{E}^\perp \to R^n f_* \mathbf{Q}_\ell \to$ un faisceau $\mathscr{F} \to 0$

**(7.1.5)** $0 \to \mathscr{F} \to$ le faisceau constant $j_* j^* \mathscr{F} \to \bigoplus_{s \in S} \mathbf{Q}_\ell(n-m)_s \to 0$.

Dans le premier cas, les suites exactes longues de cohomologie fournissent

**(7.1.2′)** $H^1(D, j_* \mathscr{E}) \to H^1(R^n f_* \mathbf{Q}_\ell) \to 0$,

**(7.1.3′)** $0 \to H^1(D, j_* \mathscr{E}) \to H^1(D, j_*(\mathscr{E}/(\mathscr{E} \cap \mathscr{E}^\perp)))$

et on applique (6.3).

Dans le second cas, elles fournissent

**(7.1.4′)** $0 \to H^1(D, R^n f_* \mathbf{Q}_\ell) \to H^1(D, \mathscr{F})$,

**(7.1.5′)** $\bigoplus_{s \in S} \mathbf{Q}_\ell(n-m) \to H^1(D, \mathscr{F}) \to 0$

et on remarque que F agit sur $\mathbf{Q}_\ell(n-m)$ par multiplication par $q^{d/2}$.

*Lemme* **(7.2)**. — *Soit* $X_0$ *une variété projective non singulière absolument irréductible de dimension* $d$ *sur* $\mathbf{F}_q$. *Soient* X *sur* $\overline{\mathbf{F}}_q$ *déduit de* $X_0$ *par extension des scalaires et* $\alpha$ *une valeur propre de* $F^*$ *agissant sur* $H^d(X, \mathbf{Q}_\ell)$. *Alors,* $\alpha$ *est un nombre algébrique dont tous les conjugués complexes, encore notés* $\alpha$, *vérifient*

$$|\alpha| = q^{d/2}.$$

Prouvons déjà que (7.2)⇒(1.7). Pour $X_0$ projective non singulière sur $\mathbf{F}_q$ et $i$ un entier, il faut prouver l'assertion suivante :

$W(X_{0,i})$. *Soit* X *déduit de* $X_0$ *par extension des scalaires de* $\mathbf{F}_q$ *à* $\overline{\mathbf{F}}_q$. *Si* $\alpha$ *est une valeur propre de* $F^*$ *agissant sur* $H^i(X, \mathbf{Q}_\ell)$, *alors* $\alpha$ *est un nombre algébrique dont tous les conjugués complexes, encore notés* $\alpha$, *vérifient* $|\alpha| = q^{i/2}$.

*a)* Si $\mathbf{F}_{q^n}$ est une extension de degré $n$ de $\mathbf{F}_q$, et que $X'_0/\mathbf{F}_{q^n}$ se déduit de $X_0/\mathbf{F}_q$ par extension des scalaires, alors $W(X_0, i)$ équivaut à $W(X'_0, i)$ : étendre les scalaires revient à remplacer $\alpha$ par $\alpha^n$ et $q$ par $q^n$.

*b)* Si $X_0$ est purement de dimension $n$, $W(X_0, i)$ équivaut à $W(X_0, 2n-i)$ : ceci résulte de la dualité de Poincaré.

*c)* Si $X_0$ est somme de variétés $X_0^\alpha$, $W(X_0, i)$ équivaut à la conjonction des $W(X_0^\alpha, i)$.

*d)* Si $X_0$ est purement de dimension $n$, $Y_0$ une section hyperplane lisse de $X_0$, et que $i<n$, alors $W(Y_0, i) \Rightarrow W(X_0, i)$ : ceci résulte du théorème de Lefschetz faible.

Pour prouver les assertions $W(X_0, i)$, on se ramène successivement :

— par *c)*, à supposer $X_0$ purement d'une dimension $n$;
— par *b)*, à supposer de plus $0 \leq i \leq n$;
— par *a)* et *d)*, à supposer de plus $i = n$;
— par *a)* et *c)*, à supposer de plus $X_0$ absolument irréductible.

Ce cas est justiciable de (7.2).

**(7.3)** Prouvons (7.2). Pour tout entier $k$, $\alpha^k$ est valeur propre de $F^*$ agissant sur $H^{kd}(X^k, \mathbf{Q}_\ell)$ (formule de Künneth). Pour $k$ pair, $X^k$ est justiciable de (7.1), d'où

$$q^{\frac{kd}{2}-\frac{1}{2}} \leq |\alpha^k| \leq q^{\frac{kd}{2}-\frac{1}{2}}$$

et

$$q^{\frac{d}{2}-\frac{1}{2k}} \leq |\alpha| \leq q^{\frac{d}{2}+\frac{1}{2k}}.$$

Faisant tendre $k$ vers l'infini, on trouve (7.2).

## 8. Premières applications.

*Théorème* **(8.1)**. — *Soit* $X_0 \subset \mathbf{P}_0^{n+r}$ *une intersection complète non singulière sur* $\mathbf{F}_q$, *de dimension n et de multidegré* $(d_1, \ldots, d_r)$. *Soit* $b'$ *le* $n^{\text{ième}}$ *nombre de Betti des intersections complètes non singulières complexes, de mêmes dimension et multidegré. Posons* $b = b'$ *pour n impair, et* $b = b' - 1$ *pour n pair. Alors*

$$|\# X_0(\mathbf{F}_q) - \# \mathbf{P}^n(\mathbf{F}_q)| \leq b \cdot q^{n/2}.$$

Soient $X/\overline{\mathbf{F}}_q$ déduit de $X_0$, et $\mathbf{Q}_\ell \cdot \eta^i$ la droite dans $H^{2i}(X, \mathbf{Q}_\ell)$ engendrée par la $i^{\text{ème}}$ cup-puissance de la classe de cohomologie d'une section hyperplane. Sur cette droite, $F^*$ agit par multiplication par $q^i$. La cohomologie de X est somme des $\mathbf{Q}_\ell \cdot \eta^i$

$(0 \leq i \leq n)$ et de la partie primitive de $H^n(X, \mathbf{Q}_\ell)$, de dimension $b$. D'après (1.5), il existe donc $b$ nombres algébriques $\alpha_j$, les valeurs propres de $F^*$ agissant sur cette cohomologie primitive, tels que

$$\# X_0(\mathbf{F}_q) = \sum_{i=0}^{n} q^i + (-1)^n \sum_j \alpha_j.$$

D'après (1.7), $|\alpha_j| = q^{n/2}$ et

$$|\# X_0(\mathbf{F}_q) - \# \mathbf{P}(\mathbf{F}_q)| = |\# X_0(\mathbf{F}_q) - \sum_{i=0}^{n} q^i| = |\sum_j \alpha_j| \leq \sum_j |\alpha_j| = b \cdot q^{n/2}.$$

*Théorème* **(8.2)**. — *Soient* N *un entier* $\geqslant 1$, $\varepsilon : (\mathbf{Z}/N)^* \to \mathbf{C}^*$ *un caractère, $k$ un entier* $\geqslant 2$ *et $f$ une forme modulaire holomorphe sur* $\Gamma_0(N)$, *de poids $k$ et de caractère* $\varepsilon$ : *$f$ est une fonction holomorphe sur le demi-plan de Poincaré* X, *telle que pour* $\begin{pmatrix} a & b \\ c & d \end{pmatrix} \in SL(2, \mathbf{Z})$, *avec* $c \equiv 0$ (N), *on ait*

$$f\left(\frac{az+b}{cz+d}\right) = \varepsilon(a)^{-1}(cz+d)^k f(z).$$

*On suppose $f$* cuspidale *et* primitive *(« new » au sens d'Atkin-Lehner et de Miyake), en particulier vecteur propre des opérateurs de Hecke* $T_p$ $(p \nmid N)$. *Posons* $f = \sum_{n=1}^{\infty} a_n q^n$, *avec* $q = e^{2\pi i z}$ *(et* $a_1 = 1$*). Alors, pour $p$ premier ne divisant pas* N

$$|a_p| \leq 2 \cdot p^{\frac{k-1}{2}}.$$

*En d'autres termes, les racines de l'équation*

$$T^2 - a_p T + \varepsilon(p) \cdot p^{k-1}$$

*sont de valeur absolue* $p^{\frac{k-1}{2}}$.

Ces racines sont en effet des valeurs propres de Frobenius agissant sur le $H^{k-1}$ d'une variété projective non singulière de dimension $k-1$ définie sur $\mathbf{F}_p$.

Sous des hypothèses restrictives, ce fait est démontré dans mon exposé Bourbaki (Formes modulaires et représentations $\ell$-adiques, exposé 355, février 1969, dans : *Lecture Notes in Mathematics*, 179). Le cas général n'est pas beaucoup plus difficile.

*Remarque* **(8.3)**. — J. P. Serre et moi-même avons récemment démontré que (8.2) reste vrai pour $k=1$. La démonstration est toute différente.

L'application suivante m'a été suggérée par E. Bombieri.

*Théorème* **(8.4)**. — *Soient* Q *un polynôme à $n$ variables et de degré $d$ sur* $\mathbf{F}_q$, $Q_d$ *la partie homogène de degré $d$ de* Q *et* $\psi : \mathbf{F}_q \to \mathbf{C}^*$ *un caractère additif non trivial sur* $\mathbf{F}_q$. *On suppose que :*

(i) *$d$ est premier à $p$;*

(ii) *l'hypersurface* $H_0$ *dans* $\mathbf{P}^{n-1}_{\mathbf{F}_q}$ *définie par* $Q_d$ *est lisse.*

*Alors*

$$\Big|\sum_{x_1,\ldots,x_n\in\mathbf{F}_q}\psi(Q(x_1,\ldots,x_n))\Big|\leq(d-1)^n q^{n/2}.$$

Quitte à remplacer Q par un multiple scalaire, on peut supposer (et on supposera) que

**(8.4.1)** $$\psi(x)=\exp(2\pi i\,\mathrm{Tr}_{\mathbf{F}_q/\mathbf{F}_p}(x)/p).$$

Soit $X_0$ le revêtement étale de l'espace affine $A_0$ de dimension $n$ sur $\mathbf{F}_q$ d'équation $T^p-T=Q$, et soit $\sigma$ la projection de $X_0$ sur $A_0$ :

$$\sigma : X_0\to A_0$$
$$X_0=\mathrm{Spec}(\mathbf{F}_q[x_1,\ldots,x_n,T]/(T^p-T-Q)).$$

Le revêtement $X_0$ est galoisien, de groupe de Galois $\mathbf{Z}/p$; $i\in\mathbf{Z}/p=\mathbf{F}_p$ agit par $T\mapsto T+i$.

Soit $x\in A_0(\mathbf{F}_q)$ et calculons l'endomorphisme de Frobenius de la fibre de $X_0/A_0$ en $x$. Posons $q=p^f$ et soit $\overline{\mathbf{F}}_q$ une clôture algébrique de $\mathbf{F}_q$. Pour $(x,T)\in X_0(\overline{\mathbf{F}}_q)$ au-dessus de $x$, on a $F((x,T))=(x,T^q)$, et

$$T^q=T+\sum_{i=1}^{f}(T^{p^i}-T^{p^{i-1}})=T+\sum_i Q(x)^{p^{i-1}}=T+\mathrm{Tr}_{\mathbf{F}_q/\mathbf{F}_p}(Q(x)).$$

C'est l'action de l'élément $\mathrm{Tr}_{\mathbf{F}_q/\mathbf{F}_p}(Q(x))$ du groupe de Galois.

Soient E le corps des racines $p^{\text{ièmes}}$ de l'unité et $\lambda$ une place finie de E première à $p$. Nous travaillerons en cohomologie $\lambda$-adique. Pour $j\in\mathbf{Z}/p$, soit $\mathcal{F}_{j,0}$ le $E_\lambda$-système local de rang un sur $A_0$ défini par $X_0$ et $\psi(-jx):\mathbf{Z}/p\to E^*\to E_\lambda^*$ : on dispose de $\iota: X_0\to\mathcal{F}_{j,0}$ et $\iota(i*x)=\psi(-ji)\iota(x)$. Notons sans $_0$ les objets déduits de $A_0$, $X_0$, $\mathcal{F}_{j,0}$ par extension des scalaires à $\overline{\mathbf{F}}_q$. La formule des traces (1.12.1) pour $\mathcal{F}_{j,0}$ s'écrit

**(8.4.2)** $$\sum_{x_1,\ldots,x_n\in\mathbf{F}_q}\psi(Q(x_1,\ldots,x_n))=\sum_i(-1)^i\,\mathrm{Tr}(F^*,H_c^i(A,\mathcal{F}_1)).$$

On a $\sigma_*E_\lambda=\bigoplus_j\mathcal{F}_j$, et donc

**(8.4.3)** $$H_c^\bullet(X,\mathbf{Q}_\ell)\otimes_{\mathbf{Q}_\ell}E_\lambda=\bigoplus_j H_c^\bullet(A,\mathcal{F}_j).$$

Pour $j=0$, $\mathcal{F}_j$ est le faisceau constant $E_\lambda$; ce facteur correspond à l'inclusion, par image réciproque, de la cohomologie de A dans celle de X.

*Lemme* **(8.5)**. — (i) *Pour* $j\neq 0$, $H_c^i(A,\mathcal{F}_j)$ *est nul pour* $i\neq n$; *pour* $i=n$, *cet espace de cohomologie est de dimension* $(d-1)^n$.

(ii) *Pour* $j\neq 0$, *le cup-produit*

$$H_c^n(A,\mathcal{F}_j)\otimes H_c^n(A,\mathcal{F}_{-j})\to H_c^{2n}(A,E_\lambda)\xrightarrow{\mathrm{Tr}}E_\lambda(-n)$$

*est une dualité parfaite.*

(iii) $X_0$ *est un ouvert d'une variété projective non singulière* $Z_0$.

Déduisons (8.4) de (8.5). Soient $j_0 : X_0 \hookrightarrow Z_0$ et $j : X \hookrightarrow Z$ déduit par extension des scalaires à $\overline{\mathbf{F}}_q$. D'après (8.4.2), (i), et (1.7) pour $Z_0$, il suffit de prouver l'injectivité de

$$H^n_c(A, \mathscr{F}_1) \overset{\sigma^*}{\to} H^n_c(X, \mathscr{F}_1) = H^n_c(X, E_\lambda) \overset{j_!}{\to} H^n(Z, E_\lambda).$$

On a $\mathrm{Tr}(a \cup b) = \frac{1}{p}\,\mathrm{Tr}(j_! \sigma^* a \cup j_! \sigma^* b)$, donc cette injectivité résulte de (ii).

**(8.6)** Prouvons (8.5) (iii). Soient $P_0$ l'espace projectif sur $\mathbf{F}_q$ complété de $A_0$ par adjonction d'un hyperplan à l'infini $P_0^\infty$, $H_0 \subset P_0^\infty$ d'équation $Q_d = 0$ et $Y_0$ le revêtement de $P_0$ normalisé de $P_0$ dans $X_0$.

**(8.6.1)**
$$\begin{array}{ccccccc} X_0 & \hookrightarrow & Y_0 & & & & \\ \sigma\downarrow & & \downarrow & & & & \\ A_0 & \hookrightarrow & P_0 & \hookleftarrow & P_0^\infty & \hookleftarrow & H_0 \end{array}$$

Étudions $Y_0/P_0$ près de l'infini, localement pour la topologie étale.

*Lemme* **(8.7)**. — $Y_0$ *est lisse en dehors de l'image réciproque de* $H_0$.

Le diviseur de la fonction rationnelle $Q$ sur $P_0$ est somme de sa partie finie $\mathrm{div}(Q)_f$ et de $(-d)$ fois l'hyperplan à l'infini. On a :

**(8.7.1)**
$$\mathrm{div}(Q) = \mathrm{div}(Q)_f - d P_0^\infty$$
$$\mathrm{div}(Q)_f \cap P_0^\infty = H_0.$$

A distance finie, $Y_0 = X_0$ est étale sur $A_0$, donc lisse. A l'infini, mais en dehors de l'image réciproque de $H_0$, il existe des coordonnées locales $(z_1, \ldots, z_n)$ telles que $Q = z_1^{-d}$ (ceci utilise que $(d, p) = 1$). Dans ces coordonnées, $Y_0$ apparaît comme produit d'une courbe et d'un espace lisse (correspondant aux coordonnées $z_2, \ldots, z_n$). Par normalité, il est lisse.

*Lemme* **(8.8)**. — *Au voisinage étale d'un point au-dessus de* $H_0$, $Y_0$ *est lisse sur une surface singulière normale, toujours la même.*

On peut cette fois trouver des coordonnées locales telles que $Q = z_1^{-d} z_2$. En effet, puisque $H_0$ est lisse, $\mathrm{div}(Q)_f$ est lisse au voisinage de l'infini et coupe $P_0^\infty$ transversalement. Cette forme est indépendante du point choisi, et n'utilise que deux coordonnées, d'où l'assertion.

**(8.9)** On sait que le procédé suivant (dû à Zariski) permet de résoudre les singularités de surfaces : alternativement, on normalise et on éclate le lieu singulier (réduit). Les opérateurs en jeu commutent à la localisation étale et au produit par un espace lisse. Le procédé de Zariski permet donc encore de résoudre les singularités d'un espace qui, comme $Y_0$, est, localement pour la topologie étale, lisse sur une surface. La résolution obtenue de $Y_0$ est le $Z_0$ cherché.

Si T est une courbe sur une surface S, contenant le lieu singulier, et que T′ est l'image réciproque de T dans la résolution de Zariski S′ de S, on sait que si on éclate de façon itérée dans S′ le lieu singulier (réduit) de $(T')_{red}$, on obtient une surface S″ telle que l'image réciproque réduite $(T'')_{red}$ de T dans S″ soit un diviseur à croisements normaux. A nouveau, les opérations en jeu commutent à la localisation étale et au produit par un espace lisse. Raisonnant comme plus haut, et observant que ($Y_0$, l'infini) est localement lisse sur un (S, T), on peut même trouver $Z_0$ tel que $Z_0-X_0$ soit un diviseur à croisements normaux.

**(8.10)** Prouvons (8.5) (i), (ii). Ce sont des assertions géométriques; ceci nous permet de travailler dorénavant sur $\overline{\mathbf{F}}_q$. Soit S′ l'espace affine sur $\overline{\mathbf{F}}_q$ qui paramétrise les polynômes à $n$ variables de degré $\leq d$, et soit S l'ouvert de S′ correspondant aux polynômes dont la partie homogène de degré $d$ est de discriminant non nul. On note $Q_S \in H^0(S, \mathcal{O}_S[x_1, \ldots, x_n])$ le polynôme universel sur S, et $X_S$ le revêtement galoisien étale de $A_S = A^n \times S$ d'équation $T^p - T = Q_S$, de groupe de Galois $\mathbf{Z}/p$. Soient $P_S = P^n \times S$ l'espace projectif sur S complété de $A_S$ et $Y_S$ le normalisé de $P_S$ dans $X_S$. On dispose, sur S, d'un diagramme analogue à (8.6.1).

Les expressions de Q en coordonnées locales données en (8.7) et (8.8) restent valables dans la situation présente, avec paramètres, de sorte que, localement pour la topologie étale sur $Y_S$, $Y_S/S$ est isomorphe au produit de S (qui est lisse) avec une fibre. La méthode canonique de résolution utilisée en (8.9) nous fournit une compactification relative $Z_S/S$ de $X_S/S$, avec $Z_S - X_S$ un diviseur à croisements normaux relatifs sur S

$$\begin{array}{ccc} X_S & \overset{u}{\hookrightarrow} & Z_S \\ \Big\downarrow{\scriptstyle \sigma} & & \Big\downarrow{\scriptstyle f} \\ A_S & \overset{a}{\longrightarrow} & S \end{array}$$

($f$ propre et lisse, $u$ plongement ouvert, $Z_S - X_S$ diviseur à croisements normaux relatifs).

Soit $\mathscr{F}_{j,S}$ le $E_\lambda$-faisceau sur $A_S$ déduit comme en (8.4) de $X_S/A_S$. On a $\sigma_* E_\lambda = \bigoplus \mathscr{F}_{j,S}$, donc

$$R^\bullet(fu)_!(E_\lambda) = \bigoplus_j R^\bullet a_! \mathscr{F}_{j,S}.$$

Les propriétés de $Z_S$ assurent que les $R^i(fu)_! E_\lambda = R^i f_*(u_! E_\lambda)$ sont des faisceaux localement constants sur S. Dès lors, les $R^i a_! \mathscr{F}_{j,S}$ également sont localement constants. Puisque S est connexe, il suffit de prouver (8.5) (i), (ii) pour *un* polynôme Q particulier. On prendra $Q = \sum_i x_i^d$. Ce polynôme vérifie la condition de non-singularité car $(d, p) = 1$. Pour ce polynôme, les variables se séparent dans la somme exponentielle de (8.4). Ceci correspond au fait que $\mathscr{F}_j$ est le produit tensoriel des images réciproques de faisceaux analogues $\mathcal{F}_j^1$ sur les facteurs de dimension un $A^1$ de $A = A^n$. Par la formule de Künneth

$$H^\bullet(A, \mathcal{F}_j) = \bigotimes H^\bullet(A^1, \mathcal{F}_j^1).$$

Ceci ramène la preuve de (8.5) (i), (ii) au cas où $n = 1$ et où Q est $x^d$.

**(8.11)** Traitons ce cas particulier. Le revêtement X de A est irréductible, de sorte que pour $i = 0, 2$

$$H_c^i(A, E_\lambda) \xrightarrow{\sim} H_c^i(X, E_\lambda).$$

Pour $i \neq 1$ et $j \neq 0$, on a donc

$$H_c^i(A, \mathcal{F}_j) = 0.$$

L'assertion (ii) résulte de (2.8) ou (2.12) et de ce que $u_! \mathcal{F}_j = u_* \mathcal{F}_j$. Pour prouver (i), il reste à vérifier que

$$\chi_c(A, \mathcal{F}_j) = 1 - d.$$

D'après la formule d'Euler-Poincaré (voir l'exposé Bourbaki 286 de février 1965, par M. Raynaud), ceci équivaut au lemme suivant.

*Lemme* **(8.12)**. — *Le conducteur de Swan de $\mathcal{F}_j$ à l'infini est égal à d.*

Cet énoncé équivaut au suivant.

*Lemme* **(8.13)**. — *Soient k un corps fini de caractéristique p, $y \in k[[x]]$ un élément de valuation d première à p,* L *l'extension de* $K = k((x))$ *engendré par les racines de* $T^p - T = y^{-1}$ *et $\chi$ le caractère suivant de* Gal(L/K) *à valeurs dans* $\mathbf{Z}/p$ :

$$\chi(\sigma) = \sigma T - T.$$

*Alors, $\chi$ est de conducteur $d+1$.*

Par extension du corps résiduel, on se ramène à supposer $k$ algébriquement clos plutôt que fini, et on applique : J. P. Serre, Sur les corps locaux à corps résiduel algébriquement clos, *Bull. Soc. Math. France*, 89 (1961), p. 105-154, n° 4.4.

## BIBLIOGRAPHIE

[1] A. Grothendieck, Formule de Lefschetz et rationalité des fonctions L, *Séminaire Bourbaki*, 279, décembre 1964 (Benjamin).

[2] S. Lefschetz, *L'analysis situs et la géométrie algébrique* (Gauthier-Villars), 1924. Reproduit dans : *Selected papers* (Chelsea Publ. Co.).

[3] R. A. Rankin, Contributions to the theory of Ramanujan's function $\tau(n)$ and similar arithmetical functions. II, *Proc. Camb. Phil. Soc.*, **35** (1939), 351-372.

[4] A. Weil, Numbers of solutions of equations in finite fields, *Bull. Am. Math. Soc.*, 55 (1949), p. 497-508.

SGA, *Séminaire de Géométrie Algébrique du Bois-Marie* (IHES) :

SGA 4, Théorie des topos et cohomologie étale des schémas (dirigé par M. Artin, A. Grothendieck et J.-L. Verdier), *Lecture Notes in Math.*, 269, 270, 305.

SGA 5, *Cohomologie ℓ-adique et fonctions* L, diffusé par l'IHES.

SGA 7, Groupes de monodromie en géométrie algébrique.

1re partie : dirigé par A. Grothendieck, *Lecture Notes in Math.*, 288.

2e partie : par P. Deligne et N. Katz, *Lecture Notes in Math.*, 340.

*Manuscrit reçu le 20 septembre 1973.*

## Curriculum Vitae

Hillel Fürstenberg

Date of Birth: September 29, 1935, Berlin, Germany

**Education:**

1955 Graduated with a B.A. and an M.Sc. from Yeshiva College
1958 Graduated with a Ph.D. from Princeton University

**Employment History:**

1958–1959 Instructor at Massachusetts Institute of Technology
1959–1965 University of Minnesota
1965–2003 Full Professor at the Hebrew University of Jerusalem

**Awards:**

1993 Israel Prize
1993 Harvey Prize from Technion, Haifa, Israel
2006/7 Wolf Prize in Mathematics

## List of Publications

1. V. Bergelson and H. Fürstenberg, WM groups and Ramsey theory, *Topology Appl.* **156** (2009), No. 16, 2572–2580.
2. H. Fürstenberg, Ergodic fractal measures and dimension conservation, *Ergodic Theory Dynam. Syst.* **28** (2008), No. 2, 405–422.
3. V. Bergelson, H. Fürstenberg and B. Weiss, Piecewise-Bohr sets of integers and combinatorial number theory, in *Topics in Discrete Mathematics*, pp. 13–37, Algorithms Combin., Vol. 26 (Springer, 2006).
4. H. Fürstenberg and Y. Katznelson, Eigenmeasures, equidistribution, and the multiplicity of $\beta$-expansions, in *Fractal Geometry and Applications: A Jubilee of Benoît Mandelbrot*, Part 1, pp. 97–116, Proc. Sympos. Pure Math., Vol. 72, Part 1 (Amer. Math. Soc., 2004).
5. H. Fürstenberg and B. Weiss, Markov processes and Ramsey theory for trees, Special issue on Ramsey theory, *Combin. Probab. Comput.* **12** (2003), No. 5–6, 547–563.
6. H. Fürstenberg, From the Erdős–Turán conjecture to ergodic theory — The contribution of combinatorial number theory to dynamics, in *Paul Erdös and his Mathematics, I* (Budapest, 1999), pp. 261–277, Bolyai Soc. Math. Stud., Vol. 11 (János Bolyai Math. Soc., 2002).
7. H. Fürstenberg, Stiffness of group actions, in *Lie Groups and Ergodic Theory* (Mumbai, 1996), pp. 105–117, Tata Inst. Fund. Res. Stud. Math., Vol. 14 (Tata Inst. Fund. Res., 1998).
8. H. Fürstenberg and E. Glasner, Subset dynamics and van der Waerden's theorem, in *Topological Dynamics and Applications* (Minneapolis, MN, 1995), pp. 197–203, Contemp. Math., Vol. 215 (Amer. Math. Soc., 1998).
9. H. Fürstenberg and E. Glasner, Robert Ellis and the algebra of dynamical systems, in *Topological Dynamics and Applications* (Minneapolis, MN, 1995), pp. 3–12, Contemp. Math., Vol. 215 (Amer. Math. Soc., 1998).
10. V. Bergelson, H. Fürstenberg and R. McCutcheon, IP-sets and polynomial recurrence, *Ergodic Theory Dynam. Syst* **16** (1996), No. 5, 963–974.
11. H. Fürstenberg and B. Weiss, A mean ergodic theorem for $(1/N)\sum_{n=1}^{N} f(T_x^n)g(T_x^{n^2})$, in *Convergence in Ergodic Theory and Probability* (Columbus, OH, 1993), pp. 193–227, Ohio State Univ. Math. Res. Inst. Publ., Vol. 5 (de Gruyter, 1996).
12. H. Fürstenberg, A polynomial Szemerédi theorem, in *Combinatorics, Paul Erdős is Eighty*, Vol. 2 (Keszthely, 1993), pp. 253–269, Bolyai Soc. Math. Stud., Vol. 2 (János Bolyai Math. Soc., 1996).

13. H. Fürstenberg, Y. Peres and B. Weiss, Perfect filtering and double disjointness, *Ann. Inst. H. Poincaré Probab. Statist.* **31** (1995), No. 3, 453–465.
14. J. Auslander and H. Fürstenberg, Product recurrence and distal points, *Trans. Amer. Math. Soc.* **343** (1994), No. 1, 221–232.
15. H. Fürstenberg, H. B. Keynes, N. G. Markley and M. Sears, Topological properties of $\mathbf{R}_n$ suspensions and growth properties of $\mathbf{Z}_n$ cocycles, *Proc. London Math. Soc.* (3) **66** (1993), No. 2, 431–448.
16, H. Fürstenberg and Y. Katznelson, A density version of the Hales–Jewett theorem, *J. Anal. Math.* **57** (1991), 64–119.
17. H. Fürstenberg, Recurrent ergodic structures and Ramsey theory, in *Proc. of the International Congress of Mathematicians*, Vol. I, II (Kyoto, 1990), pp. 1057–1069 (Math. Soc. Japan, 1991).
18. H. Fürstenberg, Y. Katznelson and B. Weiss, Ergodic theory and configurations in sets of positive density, in *Mathematics of Ramsey Theory*, pp. 184–198, Algorithms Combin., Vol. 5 (Springer, 1990).
19. H. Fürstenberg, Nonconventional ergodic averages, in *The Legacy of John von Neumann* (Hempstead, NY, 1988), pp. 43–56, Proc. Sympos. Pure Math., Vol. 50 (Amer. Math. Soc., 1990).
20. V. Bergelson, H. Fürstenberg, N. Hindman and Y. Katznelson, An algebraic proof of van der Waerden's theorem, *Enseign. Math.* (2) **35** (1989), No. 3–4, 209–215.
21. H. Fürstenberg and Y. Katznelson, Idempotents in compact semigroups and Ramsey theory, *Israel J. Math.* **68** (1989), No. 3, 257–270.
22. H. Fürstenberg and B. Weiss, On almost 1-1 extensions, *Israel J. Math.* **65** (1989), No. 3, 311–322.
23. H. Fürstenberg and Y. Katznelson, A density version of the Hales–Jewett theorem for $k = 3$. Graph theory and combinatorics (Cambridge, 1988). *Discre. Math.* **75** (1989), No. 1–3, 227–241.
24. H. Fürstenberg and B. Weiss, Simultaneous Diophantine approximation and IP-sets, *Acta Arith.* **49** (1988), No. 4, 413–426.
25. H. Fürstenberg and Y. Katznelson, IP$_r$-sets, Szemerédi's theorem, and Ramsey theory, *Bull. Amer. Math. Soc. (N.S.)* **14** (1986), No. 2, 275–278.
26. H. Fürstenberg and Y. Katznelson, An ergodic Szemerédi theorem for IP-systems and combinatorial theory, *J. Anal. Math.* **45** (1985), 117–168.
27. H. Fürstenberg, IP-systems in ergodic theory, in *Conference in Modern Analysis and Probability* (New Haven, Conn., 1982), pp. 131–148, Contemp. Math., Vol. 26 (Amer. Math. Soc., 1984).
28. H. Fürstenberg and Y. Kifer, Random matrix products and measures on projective spaces, *Israel J. Math.* **46** (1983), No. 1–2, 12–32.

29. H. Fürstenberg, Y. Katznelson and D. Ornstein, The ergodic theoretical proof of Szemerédi's theorem, *Bull. Amer. Math. Soc.* (*N.S.*) **7** (1982), No. 3, 527–552.
30. H. Fürstenberg, Rigidity and cocycles for ergodic actions of semisimple Lie groups (after G. A. Margulis and R. Zimmer), in *Bourbaki Seminar*, Vol. 1979/80, pp. 273–292, Lecture Notes in Math., Vol. 842 (Springer, 1981).
31. H. Fürstenberg, Poincaré recurrence and number theory, *Bull. Amer. Math. Soc.* (*N.S.*) **5** (1981), No. 3, 211–234.
32. H. Fürstenberg, Recurrence in ergodic theory and combinatorial number theory, in *M. B. Porter Lectures* (Princeton Univ. Press, 1981).
33. A. Bellow and H. Fürstenberg, An application of number theory to ergodic theory and the construction of uniquely ergodic models, A collection of invited papers on ergodic theory, *Israel J. Math.* **33** (1979), No. 3–4, 231–240.
34. H. Fürstenberg and Y. Katznelson, An ergodic Szemerédi theorem for commuting transformations, *J. Anal. Math.* **34** (1978), 275–291.
35. H. Fürstenberg and B. Weiss, Topological dynamics and combinatorial number theory, *J. Anal. Math.* **34** (1978), 61–85.
36. H. Fürstenberg and S. Glasner, On the existence of isometric extensions, *Amer. J. Math.* **100** (1978), No. 6, 1185–1212.
37. H. Fürstenberg and B. Weiss, The finite multipliers of infinite ergodic transformations, in *The Structure of Attractors in Dynamical Systems* (Proc. Conf., North Dakota State Univ., Fargo, N.D., 1977), pp. 127–132, Lecture Notes in Math., Vol. 668 (Springer, 1978).
38. H. Fürstenberg, Ergodic behavior of diagonal measures and a theorem of Szemerédi on arithmetic progressions, *J. Anal. Math.* **31** (1977), 204–256.
39. H. Fürstenberg, A note on Borel's density theorem, *Proc. Amer. Math. Soc.* **55** (1976), No. 1, 209–212.
40. H. Fürstenberg, The unique ergodicity of the horocycle flow, in *Recent Advances in Topological Dynamics* (Proc. Conf., Yale Univ., New Haven, Conn., 1972; In Honor of Gustav Arnold Hedlund), pp. 95–115, Lecture Notes in Math., Vol. 318 (Springer, 1973).
41. H. Fürstenberg, Boundary theory and stochastic processes on homogeneous spaces, in *Harmonic Analysis on Homogeneous Spaces* (Proc. Sympos. Pure Math., XXVI, Williams Coll., Williamstown, Mass., 1972), pp. 193–229 (Amer. Math. Soc., 1973).
42. H. Fürstenberg, H. Keynes and L. Shapiro, Prime flows in topological dynamics, *Israel J. Math.* **14** (1973), 26–38.

43. H. Fürstenberg, Boundaries of Riemannian symmetric spaces, in *Symmetric Spaces* (Short Courses, Washington Univ., St. Louis, Mo., 1969–1970), pp. 359–377. Pure and Appl. Math., Vol. 8 (Dekker, 1972).
44. H. Fürstenberg, Boundaries of Lie groups and discrete subgroups, in *Actes du Congrès International des Mathématiciens* (Nice, 1970), Tome 2, pp. 301–306 (Gauthier-Villars, 1971).
45. H. Fürstenberg and I. Tzkoni, Spherical functions and integral geometry. *Israel J. Math.* **10** (1971), 327–338.
46. H. Fürstenberg, Random walks and discrete subgroups of Lie groups, in *Advances in Probability and Related Topics*, Vol. 1, pp. 1–63 (Dekker, 1971).
47. H. Fürstenberg, Intersections of Cantor sets and transversality of semigroups, in *Problems in Analysis* (Sympos. Salomon Bochner, Princeton Univ., Princeton, N.J., 1969), pp. 41–59 (Princeton Univ. Press, 1970).
48. H. Fürstenberg, Algebraic functions over finite fields, *J. Algebra* **7** (1967), 271–277.
49. H. Fürstenberg, Disjointness in ergodic theory, minimal sets, and a problem in Diophantine approximation, *Math. Systems Theory* **1** (1967), 1–49.
50. H. Fürstenberg, Poisson boundaries and envelopes of discrete groups, *Bull. Amer. Math. Soc.* **73** (1967), 350–356.
51. H. Fürstenberg, Translation-invariant cones of functions on semi-simple Lie groups, *Bull. Amer. Math. Soc.* **71** (1965), 271–326.
52. H. Fürstenberg, Noncommuting random products, *Trans. Amer. Math. Soc.* **108** (1963), 377–428.
53. H. Fürstenberg, The structure of distal flows, *Amer. J. Math.* **85** (1963), 477–515.
54. H. Fürstenberg, A Poisson formula for semi-simple Lie groups, *Ann. of Math.* (2) **77** (1963), 335–386.
55. H. Fürstenberg, Strict ergodicity and transformation of the torus, *Amer. J. Math.* **83** (1961), 573–601.
56. H. Fürstenberg, *Stationary Processes and Prediction Theory*, Ann. Mathematics Studies, No. 44 (Princeton Univ. Press, 1960).
57. H. Fürstenberg and H. Kesten, Products of random matrices, *Ann. Math. Statist.* **31** (1960), 457–469.
58. H. Fürstenberg, The inverse operation in groups, *Proc. Amer. Math Soc.* **6** (1955), 991–997.
59. H. Fürstenberg, On the infinitude of primes, *Amer. Math. Monthly* **62** (1955), 353.
60. H. Fürstenberg, Note on one type of indeterminate form, *Amer. Math. Monthly* **60** (1953), 700–703.

# THE UNIQUE ERGODICITY OF THE HOROCYCLE FLOW

HARRY FURSTENBERG

Introduction. Professor Hedlund's name is associated with two of the classical dynamical systems of topological dynamics, the geodesic flow and the horocycle flow on surfaces of constant negative curvature ([4], [5], [6]). Both flows arise in a similar way. Geometrically, they both represent motion of unit tangent vectors along lines of constant geodesic curvature on compact surfaces of constant negative curvature. The geodesics have 0 geodesic curvature and the horocycles have maximal constant curvature without being closed. Both have similar group-theoretic descriptions: both are actions of non-compact one-parameter subgroups of $G = SL(2, \mathbb{R})$ on a compact quotient of $G$ by a discrete subgroup $\Gamma$. The principal results for these flows were discovered in the 30's; namely the ergodicity of the geodesic flow (which implies that "almost all" orbits are dense) and the minimality of the horocycle flow (which is equivalent to the assertion that all orbits are dense) ([4], [5], [6], [7]). In the meantime, however, the two flows have fared differently. The geodesic flow is very well understood and has become the prototype of a large class of flows (Anosov flows), whereas the horocycle flow seems to stand in class by itself. Even confining ones attention to homogeneous flows, the phenomena associated with the geodesic flow, namely, ergodicity and mixing, are better understood than that associated with the horocycle flow, namely, minimality. To make this contention precise, let $G$ be a simple Lie group and $\Gamma$ a uniform subgroup (discrete and cocompact) and let $T$ be a one-parameter subgroup of $G$. A reasonable condition on $T$ (and in fact a necessary condition) for the ergodicity of the action of $T$ on $G/\Gamma$ turns out to be sufficient ($T$ not have compact closure in $G$; see [8]). Now an obvious necessary condition for the minimality of the action of $T$ on $G/\Gamma$ is that there are no periodic cycles ($\Longleftrightarrow gTg^{-1} \cap \Gamma = \{e\}$, $g \in G$). Is this condition sufficient? So far one knows the answer only for $G = SL(2,\mathbb{R})$ when under these hypotheses the flow is the horocycle flow and the result is Hedlund's theorem. Another more general question is the following. Is every minimal homogeneous flow on a compact homogeneous space of the form $G/\Gamma$ also uniquely ergodic? Here $G$ is a Lie group and $\Gamma$ is a uniform subgroup so that there is a unique probability measure $m_{G/\Gamma}$ on $G/\Gamma$ preserved by $G$. The unique ergodicity of the action of a subgroup $T$ on $G/\Gamma$ means that $m_{G/\Gamma}$ is the only probability measure on $G/\Gamma$ preserved by the action of $T$. When $T$ is abelian this implies that $T$ is minimal on $G/\Gamma$ (because $T$ then leaves invariant some probability measure on any closed $T$-invariant subset) and unique ergodicity implies both ergodicity and minimality. The example of nilflows which have been thoroughly studied bears out the conjecture that minimality does imply unique ergodicity for homogeneous flows. (See [2] for a counterexample in the general case.) In this paper we show that the horocycle flow also supports this conjecture. But the general question remains open.

We hope that the method developed here enables one to make headway with some of these questions. While we shall deal here exclusively with the classical horocycle flow, a similar method seems to yield at least the following: if $T \subset SL(m, \mathbb{R})$ is the subgroup leaving pointwise fixed an $r$-dimensional subspace of $\mathbb{R}^m$, where $2r < m$, then $T$ acts uniquely ergodically on $SL(m, \mathbb{R})/\Gamma$. For $m > 2$ this overlaps the minimality result of L. Greenberg in [1, appendix].

1. Measures on Locally Compact Spaces. We shall be discussing borel measures on a locally compact, $\sigma$- compact, metrizable space $X$, by which we shall mean $\sigma$- additive measures taking values in $[0,\infty]$ for all borel subsets of $X$, but finite for compact sets. We denote the space of these by $\mathcal{M}(X)$. $\mathcal{M}(X)$ may be identified with the space of positive linear functionals on the vector space $\mathcal{C}_o(X)$ of continuous functions having compact support in $X$. If $g$ is a borel measurable map from one space $X$ to another $Y$, we define $g: \mathcal{M}(X) \longrightarrow \mathcal{M}(Y)$ by $g\mu(A) = \mu(g^{-1}A)$, $A \subset Y$. If $G$ is a group of such transformations of $X$ to itself, we say a measure $\mu \in \mathcal{M}(X)$ is G-invariant if $g\mu = \mu$ for all $g \in G$. We denote by $\mu_1 \prec \mu_2$ the absolute continuity of $\mu_1$ with respect to $\mu_2$, and we say $\mu_1$ is equivalent to $\mu_2$ if $\mu_1 \prec \mu_2$ and $\mu_2 \prec \mu_1$. Note that under our hypotheses on $X$, every measure is equivalent to a finite measure (i.e., a measure $\mu$ with $\mu(X) < \infty$).

The following proposition will be indispensable.

Proposition 1.1. If $Z = X \times Y$ where $X$, $Y$ are locally compact, $\sigma$- compact, and metrizable, then every measure $\mu \in \mathcal{M}(Z)$ admits a decomposition $d\mu(x,y) = d\lambda_y(x)\,d\nu(y)$ in the sense that

$$\int_Z f(x,y)\,d\mu(x,y) = \int_Y \{ \int_X (x,y)\,d\lambda_y(x)\}\,d\nu(y)$$

for $f \in \mathcal{C}_o(Z)$. Here $\lambda_y$ is a borel measurable function from $Y$ to $\mathcal{M}(X)$ and $v \in \mathcal{M}(Y)$. This decomposition is unique in the sense that if $d\lambda_y'(x)\,d\nu'(y)$ is another such decomposition then $\nu$ and $\nu'$ are equivalent and $\lambda_y' = \frac{d\nu}{d\nu'}(y)\,\lambda_y$ for almost all $y$ (relative to $\nu$). Finally, if $\mu$ is a finite measure on $Z$ then the measure $\nu$ can be taken as the projection of $\mu$ on $Y$. In particular, the equivalence class of $\nu$ is determined by that of $\mu$.

We shall sketch a proof of this proposition which may be considered a "folk theorem". One sees readily that it suffices to prove the statements of the proposition in the case that $\mu$ is a finite, or even probability, measure. In this case we choose an appropriate countable dense subset of $\mathcal{C}_0(X)$, say, $\{f_n(x)\}$. We let $P$ denote the projection from $L^2(Z, \mu)$ to $L^2(Y, \nu)$ where $\nu \in \mathcal{M}(Y)$ is the image of $\mu \in \mathcal{M}(Z)$, so that $L^2(Y, \nu)$ may be regarded as a closed subspace of $L^2(Z, \mu)$, then $(Pf_n)(y)$ is defined almost everywhere in $Y$ and fixing $y \in Y$ one finds that for almost all $y$, $Pf_n(y)$ extends to a positive linear functional on $\mathcal{C}_0(X)$. This yields a measurable map $Y \longrightarrow \mathcal{M}(X)$ and it is not difficult to show that this map $y \longrightarrow \lambda_y$ together with $\nu$ yield a decomposition of $\mu$ as desired. One should observe that the operator $P$ here is essentially the <u>conditional expectation</u> for the borel field of $Z$ with respect to the subfield defined by sets of $Y$. For more details we refer the reader to [2].

One application of this proposition is to homogeneous spaces of Lie groups. Recall that a locally compact group is called <u>unimodular</u> if there exists a measure on the group which is invariant under both left and right translations.

<u>Proposition 1.2</u>. Let $H$ be a closed unimodular subgroup of the Lie group $G$. There is a one-one correspondence between $\mathcal{M}(G/H)$ and measures on $G$ invariant with respect to the action of right translation by elements of $H$. The correspondence is given by $\omega \longleftrightarrow \pi$ where

$$\int_G f(g)\,d\,\omega(g) = \int_{G/H}\int_H f(gh)\,dh\,d\pi(gH).$$

<u>Proof</u>. It is clear that the integral to the right defines an $H$-invariant measure on $G$. We have to show that every such measure arises in this way. Under our hypotheses it is known that there exists a borel measurable crossection $\sigma : G/H \longrightarrow G$ where we re-

gard G as a fibre bundle over G/H with fibre H. So, for measure-theoretic purposes we may consider G as a product $G/H \times H$ where $g \longleftrightarrow (x, h)$ with $x = gH$ and $g = \sigma(gH)h$. We then write $d\omega(g) = d\omega(x, h) = d\lambda_x(h)d\pi'(x)$. Since $\omega$ is right H-invariant, we have (by the uniqueness in Proposition 1.1) that for almost all $x, \lambda_x$ is right invariant on H. (This argument uses the fact that the measure $\nu$ in Proposition 1.1 can obviously be chosen arbitrarily in its equivalence class, and so if $\omega$ and a translate of $\omega$ by $h \in H$ are identical, both may be represented with the same $\pi' \in \mathcal{M}(G/H)$). Since H is unimodular we have $d\lambda_x(h) = \varphi(x)dh$, dh being haar measure on H, and so, setting $d\pi(x) = \varphi(x)d\pi'(x)$ we obtain the desired result. Once again the uniqueness in Proposition 1.1 shows that $\pi$ is uniquely determined by $\omega$ once the haar measure of H is fixed.

From this one deduces the following.

Proposition 1.3. If H, L are both closed unimodular subgroups of the Lie group G, there is a one-one correspondence between L-invariant measures on G/H and measures on G that are invariant under left translations by elements of L and right translations by elements of H.

Finally we have

Theorem 1.4. If H, L are both closed unimodular subgroups of the Lie group G, there is a one-one correspondence between L-invariant measures on G/H and H-invariant measures on G/L.

For the proof one lifts measures of both types to the group G using the foregoing proposition, and associates to a measure $\omega \in \mathcal{M}(G)$ the measure $\omega^*$ defined by $\omega^*(A) = \omega(A^{-1})$.

2. Homogeneous spaces of $SL(2, \mathbb{R})$. For the remainder of this paper $G$ will denote the $2 \times 2$ unimodular group $SL(2, \mathbb{R})$. $\Gamma$ will denote a fixed discrete subgroup of $G$ having compact quotient $G/\Gamma$. In this section we shall discuss some other homogeneous spaces of $G$ which play an important role in our subsequent discussion. We denote by $K$, $A$, $N$, respectively, the subgroups of matrices of the form

$$k_\theta = \begin{pmatrix} \cos\theta & \sin\theta \\ -\sin\theta & \cos\theta \end{pmatrix} \quad a_s = \begin{pmatrix} s & o \\ o & s^{-1} \end{pmatrix}, \ s > o \quad n_t = \begin{pmatrix} 1 & t \\ o & 1 \end{pmatrix}.$$

the horocycle flow is the flow on $G/\Gamma$ determined by the subgroup $N$ : $\tau_t(g\Gamma) = n_t g\Gamma$. The center of $G$ consists of $\{\pm 1\}$ and we denote by $T$ the subgroup of upper triangular matrices: $T = \{\pm 1\} \cdot A \cdot N$.

With the natural action of $G$ on the vector space $\mathbb{R}^2$, we find that $\mathbb{R}^2 - \{0\}$ is a homogeneous G-space which may be identified with $G/N$. $G$ also acts on the complex upper half plane $\mathcal{H} = \{x+iy \mid y>0\}$ by

$$\begin{pmatrix} a & b \\ c & d \end{pmatrix}(w) = \frac{aw+b}{cw+d}.$$

This action is transitive and we recognize the stability group of $i \in \mathcal{H}$ to be just $K$, whence we may identify $G/K$ with $\mathcal{H}$. It will also be useful to identify $\mathcal{H}$ with the unit disc $\mathcal{D}$ by

$$z = \sigma(w) = \frac{w-i}{w+i}.$$

This determines an action of $G$ on $\mathcal{D}$ which we again denote $z \longrightarrow g(z)$, but with a different formula than that given above. We frequently switch between the three spaces $G/K$, $\mathcal{H}$, and $\mathcal{D}$, using whichever suits our purposes best.

Since $T$ is the subgroup of $G$ taking the X-axis in $\mathbb{R}^2$ to itself, we may identify $G/T$ by the set of lines through the origin of $\mathbb{R}^2$; i.e., $G/T \cong P'$, the projective line. Note that it also corresponds to the boundaries $\partial\mathcal{D}$ and $\mathbb{R} \cup \{\infty\}$ of $\mathcal{D}$ and $\mathcal{H}$ respectively. If $x \in \mathbb{R}^2 - \{0\}$ we denote by $\langle x \rangle$ the point of $P'$ corresponding to the line in $\mathbb{R}^2$ joining $x$ to the origin. One

One may then identify $G/K$, $P'$, $\mathbb{R} \cup \{\infty\}$ and $\partial\mathcal{D}$ as follows

(2.1) $$\begin{pmatrix} a & b \\ c & d \end{pmatrix} T \quad <\!\!-\!\!> \quad \left\langle \begin{pmatrix} a \\ c \end{pmatrix} \right\rangle \quad <\!\!-\!\!> \quad \frac{c}{a} \quad <\!\!-\!\!> \quad \frac{c-ai}{c+ai} \, .$$

Note that $G$ preserves lebesgue measure on $\mathbb{R}^2$ and hence on $\mathbb{R}^2 - \{0\} = G/N$. The unique ergodicity of the horocycle flow is the statement that $N$ preserves a unique probability measure on $G/\Gamma$. Since $N$ and $\Gamma$ are unimodular we may apply Theorem 1.4 to obtain

Theorem 2.1. The horocycle flow is uniquely ergodic iff the only $\Gamma$- invariant measures on $\mathbb{R}^2 - \{0\}$ are the constant multiples of lebesgue measure.

3. Positive Harmonic Functions on $G/K$. We shall consider harmonic functions on $\mathcal{D}$ and $\mathcal{H}$, or, equivalently, on $G/K$. Note that a function on the coset space $G/K$ can be thought of as a function $f(g)$ on $G$ satisfying $f(gk) = f(g)$ for $k \in K$.

The following is basic for our considerations.

Proposition 3.1

(a) the functions

$$P(\zeta, z) = \frac{1 - |z|^2}{|\zeta - z|^2} \, ,$$

$\zeta \in \partial\mathcal{D}$, $z \in \mathcal{D}$, are harmonic functions in $\mathcal{D}$ for each $\zeta$.

(b) If $h(z)$ is a positive harmonic function in $\mathcal{D}$ it may be represented uniquely in the form

$$h(z) = \int_{\partial\mathcal{D}} P(\zeta, z)\, d\nu(\zeta)$$

where $\nu$ is a non-negative measure (clearly finite) on $\partial\mathcal{D}$.

(c) The integrals

$$I_p(r) = \int_0^{2\pi} |h(re^{i\theta})|^p d\theta \qquad 1 < p < \infty$$

increase with $r$ for any harmonic function $h(z)$ on $\mathcal{D}$. A necessary and sufficient condition for the measure $\nu$ in (b) to be of the form $d\nu(\zeta) = \psi(\zeta)dm(\zeta)$, $\psi \in L^p(\partial\mathcal{D}, m)$, $m$ lebesgue measure on $\partial\mathcal{D}$, is that $I_p(r)$ be bounded as $r \nearrow 1$. In this case we say $h(z)$ is $L^p$- bounded.

(d) For $x \in \mathbb{R}^2 - \{0\}$, let $\|x\|$ denote the euclidean norm. Let $\zeta_x$ correspond to $\langle x\rangle$ in the identifications (2.1). Then we can write

$$P(\zeta_x, g(0)) = \|x\|^2/\|g^{-1}x\|^2 \ , \ g(0) \in \mathcal{D}. \tag{3.1}$$

Proof: (a), (b), and (c) are classical (see, e.g., [9]). The identity in (d) can be (presumably) verified by direct computation, but we present a more transparent proof. If we write the Möbius transformation $g$ of $\mathcal{D}$ as

$$g(z) = \frac{\alpha z + \beta}{\bar\beta z + \bar\alpha}, \quad |\alpha|^2 - |\beta|^2 = 1$$

we find that

$$|g^{-1\prime}(\zeta)| = \frac{1}{|\alpha - \beta\zeta|^2} = \frac{|1 - g(o)|^2}{|\zeta - g(o)|^2} = P(\zeta, z)$$

with $g(0) = z$. Now if $m$ denotes lebesgue measure on $\partial\mathcal{D}$ we can write

$$|g^{-1\prime}(\zeta)| = \frac{dgm}{dm}(\zeta),$$

so that the Poisson kernel is represented as a Rodon-Nikodym derivative. To transfer this from $\partial\mathcal{D}$ to $P'$, we must replace $m$ by the K-invariant probability measure on $P'$. Denoting this measure by $\mu$, we may write

$$P(\zeta_x, g(0)) = \frac{dg\mu}{d\mu}(\langle x\rangle).$$

Finally a simple geometric argument will convince the reader that

$$\frac{dg\mu}{d\mu}(\langle x\rangle) = \frac{\|x\|^2}{\|g^{-1}x\|^2}\,.$$

(See [3] for a general framework for this type of formula.) This proves the proposition.

We remark that the form of the function of $g$ occurring in (3.1) is clearly dependent only on the coset $gK$.

The proposition shows that $\|g^{-1}x\|^{-2}$ defines a harmonic function on $G/K$. More generally, if $\nu$ is a measure with compact support on $\mathbb{R}^2 - \{0\}$, the integral

$$\int \|g^{-1}x\|^{-2} d\nu(x) \tag{3.2}$$

defines a positive harmonic function on $G/K$. Furthermore in view of (b) and (d) of the proposition, we find that every positive harmonic function on $G/K$ has such a representation (say, with $\nu$ concentrated on the unit circle of $\mathbb{R}^2$). Without any restriction on $\nu$, this representation will not be unique, since for $x_2 = \lambda x_1$, the functions $\|g^{-1}x_1\|^{-2}$ and $\|g^{-1}x_2\|^{-2}$ are proportional.

To determine the extent to which a representation of a positive harmonic function in the form (3.2) is unique, we use the fact that $\mathbb{R}^2 - \{0\}$ is a fibre bundle over $P'$ with fibre $\mathbb{R} - \{0\}$. With this structure each finite measure $\nu \in \mathcal{M}(\mathbb{R}^2 - \{0\})$ projects down to a measure $\bar{\nu} \in \mathcal{M}(P')$. Inasmuch as we can regard $\mathbb{R}^2 - \{0\}$ as the product, in a borel measurable fashion, of $P'$ and $\mathbb{R} - \{0\}$, we may apply Proposition 1.1 to obtain

$$d\nu(x) = d\lambda_{\langle x\rangle}(t)\, d\bar{\nu}(\langle x\rangle). \tag{3.3}$$

(Here $t \in \mathbb{R} - \{0\}$. The measures $\lambda_{\langle x\rangle}$ are not uniquely determined by $\nu$ but depend also on the choice of a borel-measurable crossection $P' \longrightarrow \mathbb{R}^2 - \{0\}$.) We now find in evaluating the integral (3.2) in accordance with (3.3) and comparing the result with the representation of harmonic functions in Proposition 3.1 (b):

Theorem 3.2. Every measure $\nu$ of compact support on $\mathbb{R}^2 - \{0\}$ defines a positive harmonic function by

(3.2 bis) $$h(gK) = \int \|g^{-1}x\|^{-2} d\nu(x).$$

Every positive harmonic function admits such a representation and for given $h(gK)$ the projection of the measure $\nu$ on $P^1$ is determined up to equivalence. This class is that of boundary measure occurring in the representation of Proposition 3.1.

4. Sketch of the Proof of Unique Ergodicity. According to Theorem 2.1 what must be shown is that if $\pi$ is a $\Gamma$- invariant measure on $\mathbb{R}^2 - \{0\}$, then $\pi$ is proportional to $m$, where $m$ will heretofore denote lebesgue measure on $\mathbb{R}^2 - \{0\}$. We write $d\pi(x) = d\lambda_\xi(t)\, d\tilde{\pi}(\xi)$ in accordance with Proposition 1.1. Our first step will be to show that without loss of generality we may suppose all the measures $\lambda_\xi$ absolutely continuous. We next show that it suffices to prove that $\pi$ is absolutely continuous with respect to $m$. Decomposing $m$ in the same way we find it sufficient to prove that $\tilde{\pi} \prec \tilde{m}$, the latter being, of course, lebesgue measure on $P^1$. We then show that the class of $\tilde{\pi}$ is determined by projecting onto $P^1$ the restriction $\pi_Q$ of $\pi$ to a sufficiently large compact set $Q \subset \mathbb{R}^2 - \{0\}$. Since this projection occurs in Theorem 3.2 in connection with the harmonic function

$$h_Q(gK) = \int_Q \|g^{-1}x\|^{-2} d\pi(x)$$

we find that we are called upon to study such functions. In particular it will follow from Proposition 3.1(c) that the measure class $\tilde{\pi}$ will be absolutely continuous iff $h_Q(gK)$ is $L^2$- bounded. The reason for choosing $p = 2$ will be apparent. (Once the result has been proven it will follow that $\pi = cm$ and $h_Q$ is in fact bounded!)

It will now develop that to determine if $h_Q(gK)$ is $L^2$- bounded it will suffice to examine $h_Q(\gamma K)$ for $\gamma \in \Gamma$. Here we apply the $\Gamma$-invariance of to write

(4.1) $$h_Q(\gamma K) = \int_{\gamma^{-1}Q} \|x\|^{-2} d\pi(x).$$

The crux of our argument will be that for evaluating averages of expressions of the form (4.1) we can replace the measure $\pi$ by m. Here the result will have to be true since m defines bounded harmonic functions. The details follow.

5. Preliminaries. We begin with

Lemma 5.1. If $\pi$ is a $\Gamma$- invariant measure on $\mathbb{R}^2 - \{0\}$ and $\pi$ is absolutely continuous (with respect to lebesgue measure), then $\pi$ is proportional to lebesgue measure.

This amounts to the ergodicity of the action of $\Gamma$ on $\mathbb{R}^2 - \{0\}$ which by the duality principle in [8] is equivalent to the ergodicity flow. (See [1] or [8].)

Lemma 5.1 reduces our problem to studying the equivalence class of the invariant measure $\pi$. We wish to reduce this further to the determination of the class of the "angular" part of the measure $\pi$.

Lemma 5.2. For $\ell > 0$ let $^{\ell}\pi$ denote the measure $^{\ell}\pi(A) = \pi(\ell^{-1}A)$. If $\psi \in C_0(0,\infty))$ we may form the measure $\pi_\psi = \int_0^\infty {}^{\ell}\pi\, \psi(\ell)\, d\ell$. If each measure $\pi_\psi$ is proportional to lebesgue measure on $\mathbb{R}^2 - \{0\}$ then so is $\pi$.

Proof: For we can choose $\psi_n \in C_0((0,\infty))$ with $\pi_{\psi_n} \longrightarrow \pi$ weakly in $\mathcal{M}(\mathbb{R}^2 - \{0\})$.

We can now enunciate

Proposition 5.3. For the proof of the unique ergodicity of the horocycle flow it suffices to show that if $\pi \in \mathcal{M}(\mathbb{R}^2 - \{0\})$ is $\Gamma$-invariant and $\pi$ has a decomposition

$$d\pi(x) = d\lambda_{<x>}(t)\,d\tilde{\pi}\ <x>)$$

then $\tilde{\pi}$ is absolutely continuous on $P^1$ .

Proof: For it suffices to show that $\pi$ is absolutly continuous. Now by Lemma 5.2, it suffices to restrict our attention to measures of the form $(\pi')_\psi$ . But if $\pi = (\pi')_\psi$ , the measures $\lambda_{<x>}(t)$ are clearly absolutely continuous. Thus it suffices to show that the "angular" part, $\tilde{\pi}$ is absolutely continuous.

Lemma 5.4. There exists a compact set $Q$ in $\mathbb{R}^2 - \{0\}$ with $\Gamma Q = \mathbb{R}^2 - \{0\}$.

This is clear from the cocompactness of $\Gamma$ in $G$.

Now let $\pi_Q$ denote the restriction of $\pi$ to $Q$. We then have

Lemma 5.5. If $\pi$ is a $\Gamma$-invariant measure in $\mathbb{R}^2 - \{0\}$, $\tilde{\pi}$ will be absolutely continuous iff the projection $\bar{\pi}_Q$ of $\pi_Q$ on $P^1$ is absolutely continuous.

Proof: One direction is clear since $\pi_Q < \pi$. For the other direction assume that $\bar{\pi}_Q$ is absolutely continuous. Then so is $\gamma\bar{\pi}_Q = \bar{\pi}_{\gamma Q}$ for $\gamma \in \Gamma$. Now if a set in $\mathbb{R}^2 - \{0\}$ is a null set for all $\pi_{\gamma Q}$ then it must be a null set for $\pi$, since $\Gamma Q = \mathbb{R}^2 - \{0\}$. It follows that if a set is a null set for all $\bar{\pi}_{\gamma Q}$, it is also a null set for $\tilde{\pi}$. The result now follows.

Combining the last lemma with Proposition 5.3, Theorem 3.2, and Proposition 3.1(c) we find

Theorem 5.6. To prove the unique ergodicity of the horocycle flow it suffices to prove that the harmonic function

$$h_Q(gK) = \int_Q \|g^{-1}x\|^{-2}\,d\pi(x)$$

is $L^2$- bounded whenever $\pi$ is a $\Gamma$- invariant measure.

6. $L^2$ - boundedness of the Functions $h_Q$. Let $\pi$ be a $\Gamma$- invariant measure on $\mathbb{R}^2 - \{0\}$.

Lemma 6.1. There exists a (finite) measure $\mu$ on $G$ with compact support such that for each $\varphi \in \mathcal{C}_0(\mathbb{R}^2 - \{0\})$

$$\int \varphi(x)\,dm(x) = \iint \varphi(gx)\,d\pi(x)\,d\mu(g).$$

Proof: Let $\tilde{\mu}$ be the G-invariant measure on $G/\Gamma$. Map $G/\Gamma$ to $G$ by some measurable crossection $\sigma$ and let $\mu = \sigma(\tilde{\mu})$. Let $\pi'$ be defined by

$$\int \varphi\,d\pi' = \int\int \varphi(gx)\,d\pi(x)\,d\mu(g).$$

We claim $\pi'$ is G-invariant. This will imply that $\pi'$ is proportional to m and this yields the lemma. Fix $g_0 \in G$. Since $g_0\tilde{\mu} = \tilde{\mu}$ it follows that $g_0\mu$ and $\mu$ project onto the same measure on $G/\Gamma$. It is not hard to see that this means that we can decompose $\mu$ as $\mu = \Sigma\,\mu_\gamma = \Sigma\,\mu'_\gamma$ where each part $\mu_\gamma$ satisfies $g_0\mu_\gamma = \mu'_\gamma\gamma$, the right translate of $\mu'_\gamma$ by $\gamma$. Then

$$\int \varphi(g_0x)\,d\pi'(x) = \int\int \varphi(g_0gx)\,d\pi(x)\,d\mu(g) =$$

$$\sum_\Gamma \iint \varphi(g_0gx)\,d\pi(x)\,d\mu_\gamma(g) = \sum_\Gamma \iint \varphi(gx)\,d\pi(x)\,dg_0\mu_\gamma(g) =$$

$$\sum_\Gamma \iint \varphi(gx)\,d\pi(x)\,d\mu'\ (g) = \sum_\Gamma \iint \varphi(g\gamma x)\,d\pi(x)\,d\mu'_\gamma(g) =$$

$$\sum_\Gamma \iint \varphi(gx)\,d\gamma\pi(x)\,d\mu'_\gamma(g) = \sum_\Gamma \iint \varphi(gx)\,d\pi(x)\,d\mu'_\gamma(g) =$$

$$\iint \varphi(gx)\,d\pi(x)\,d\mu(g) = \int \varphi(x)\,d\pi'(x)$$

This completes the proof.

In the remainder of our discussion $c_1$, $c_2$, $c_3$, etc. will denote various constants. For any given index i, the constant $c_i$ will vary according to the context.

Lemma 6.2. Let $\pi$ be a $\Gamma$- invariant measure on $\mathbb{R}^2 - \{0\}$ and let $f(x) = f_0(\|x\|)$ where $f_0$ is a monotonic function on $(0,\infty)$. There exist constants $c_1, c_2, c_3$ such that

$$\int_{r_1 < \|x\| < r_2} f(x)\,d\pi(x) < c_3 \int_{c_1 r_1 < \|x\| < c_2 r_2} f(x)\,dm(x) = 2\pi c_3 \int_{c_1 r_1}^{c_2 r_2} r f_o(r)\,dr.$$

$c_1, c_2, c_3$ depend only on $\pi$ and not on $f(x)$ or $r_1, r_2$.

Proof: If $M$ is a fixed compact set of $G$ then clearly $c_4\|x\| \leq \|gx\| \leq c_5\|x\|$ for all $x \in \mathbb{R}^2 - \{0\}$ and $g \in M$. Take $M$ to be the support of $\mu$ in the foregoing lemma. We shall give the argument for $f_o$ a monotone decreasing function, the idea being similar in the reverse case. Denote by $\chi_{r_1,r_2}$ the characteristic function of $[r_1, r_2] \subset (0,\infty)$. We have, for $g \in M$,

$$\int_{r_1<\|x\|<r_2} f(x)\,d\pi(x) = \int \chi_{r_1,r_2}(\|x\|)\,f_o(\|x\|)\,d\pi(x) \leq$$

$$\int \chi_{c_4 r_1, c_5 r_2}(\|gx\|)\,f_o(c_5^{-1}\|gx\|)\,d\pi(x).$$

Hence, integrating over $g \in M$

$$\int_{r_1<\|x\|< r_2} f(x)\,d\pi(x) \leq \mu(M)^{-1}\iint \chi_{c_4 r_1, c_5 r_2}(\|gx\|)\,f_o(c_5^{-1}\|gx\|)\,d\pi(x)\mu(g) =$$

$$\mu(M)^{-1}\int \chi_{c_4 r_1, c_5 r_2}(\|x\|)\,f_o(c_5^{-1}\|x\|)\,dm(x)$$

by Lemma 6.1. The latter integral may be written

$$c_6 \int_{c_4 r_1}^{c_5 r_2} f_o(c_5^{-1} r)\,r\,dr = c_7 \int_{c_5^{-1} c_4 r_1}^{r_2} f_o(r)\,r\,dr$$

which yields the assertion of the lemma.

Lemma 6.3. Let $g_N = \begin{pmatrix} N & 0 \\ 0 & N^{-1} \end{pmatrix}$ for $N \geq 1$ and let $Q$ denote the region $c_1 \leq \|x\| \leq c_2$ in $\mathbb{R}^2 - \{0\}$ with $\chi_Q$ denoting its characteristic function. Set

$$H_N(x) = \int_0^{2\pi} \chi_Q(g_N k_\theta x)\,d\theta.$$

then $H_N(x) = 0$ for $\|x\| < c_1 N^{-1}$ and for $\|x\| > c_2 N$. For all $x$

we have $H_N(x) \leq C\, N^{-1}\|x\|^{-1}$ for some c.

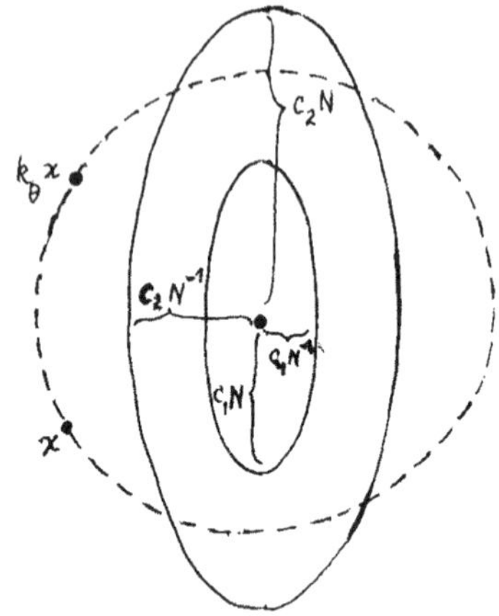


Proof: The proof is apparent from the diagram.

Lemma 6.4. With $g_N$ and Q as in the foregoing lemma, assume $\varphi$ is a bounded non-negative function with support in Q. Then

$$(6.1) \qquad \iiint \varphi(g_N k_\theta x)\, \varphi(g_N k_\theta y)\, d\theta \; \frac{d\pi(x)}{\|x\|^2} \frac{d\pi(y)}{\|y\|^2}$$

is bounded independently of N.

Proof: To begin with

$$\frac{1}{2\pi}\int_0^{2\pi} \varphi(g_N k_\theta x)\varphi(g_N k_\theta y)\, d\theta \leq \{\frac{1}{2\pi}\int_0^{2\pi} \varphi^2(g_N k_\theta x)\, d\theta\}^{1/2}\{\frac{1}{2\pi}\int_0^{2\pi} \varphi^2(g_N k_\theta y) d\theta\}^{1/2}$$

$$\leq c_3 \{\int_0^{2\pi} \chi_Q(g_N k_\theta x)\, d\theta\}^{1/2}\{\int_0^{2\pi} \chi_Q(g_N k_\theta y)\, d\theta\}^{1/2}$$

$$= c_3 \sqrt{H_N(x) H_N(y)}.$$

this vanishes unless both $\|x\|$ and $\|y\|$ exceed $c_1 N^{-1}$ and in any case it is less than $c_4 N^{-1}\sqrt{\|x\|^{-1}\|y\|^{-1}}$. Hence the expression in (6.1) is bounded by

$$c_4 N^{-1} \iint_{c_1 N^{-1} < \|x\|, \|y\| < c_2 N} \|x\|^{-5/2} \|y\|^{-5/2}\, d\pi(x)\, d\pi(y)$$

Thus it suffices to show that

$$\int_{c_1 N^{-1} < \|x\| < c_2 N} \|x\|^{-5/2}\, d\pi(x)$$

But by Lemma 6.2, this integral is comparable with

$$\int_{c_5N^{-1}}^{c_6N} r^{-5/2}, \ rdr = O(\sqrt{N})$$

this proves the lemma.

Let $g = \begin{pmatrix} a & b \\ c & d \end{pmatrix}$. We shall denote by $N(g)$ the number $(a^2+b^2+c^2+d^2)^{1/2}$. Clearly $N(k_\theta g k_{\theta'}) = N(g)$.

Lemma 6.5. Let $\varphi$ be a non-negative function of compact support on $\mathbb{R}^2 - \{0\}$. For $x, y \in \mathbb{R}^2 - \{0\}$ set

$$H_u(x,y) = \int_{N(g)<u} \varphi(gx)\,\varphi(gy)\,dg$$

and

$$V_u = \int_{N(g)<u} dg.$$

Then

$$\iint H_u(x,y)\,\frac{d\pi(x)}{\|x\|^2}\,\frac{d\pi(y)}{\|y\|^2} = O(V_u) \tag{6.2}$$

as $u \longrightarrow \infty$.

Proof: Let $F(g)$ be any bounded borel function on $G$. The region $\{N(g) < u\} \subset G$ is invariant under right and left multiplication by $K$. Since every group element can be expressed in precisely two ways in the form $k_\theta a_s k_{\theta'}$, $k_\theta, k_{\theta'} \in K$, $a_s \in A$, we can express

$$\int_{N(g)<u} F(g)\,dg = c_1 \int_{N(a_s)<u} \int_0^{2\pi}\int_0^{2\pi} F(k_\theta a_s k_{\theta'})\,d\theta\,d\theta'\,d\eta(t)$$

for some measure $\eta$ on $(0,\infty)$. Now both $V_u$ and $H_u(x,y)$ are integrals of the above type and for both $F(k_\theta g) = F(g)$. Hence (6.2) will follow if it is known that

$$\iiint \varphi(gk_\theta x)\,\varphi(gk_\theta y)\,\frac{d\pi(x)}{\|x\|^2}\,\frac{d\pi(y)}{\|y\|^2} < c_2$$

for all diagonal matrices $g$. But this is the content of Lemma 6.4.

The next lemma deals with G-invariant area on $\mathcal{D}$. We denote by $D(z, z')$ the G-invariant metric on $\mathcal{D}$ and by $\mathcal{D}_R$ the disc of D-radius R about O. It can be shown that the area of $\mathcal{D}_R$ is given by

$$A_R = \sinh^2 R/2 .$$

We use this in

Lemma 6.6. For fixed $a$ there exists $c$ with $A_{R+a} < cA_R$ for all R.

Lemma 6.6 can be used to prove

Lemma 6.7. Let $V_u$ be as in Lemma 6.5 and let $a$ be fixed. There exists $c$ with $V_{au} < c\, V_u$.

Proof: Let W be a subset of G-invariant under right multiplication by K. Because of the uniqueness, up to constant multiple, of the G-invariant measure on $\mathcal{D}$, it follows that the volume of G in W can be taken to equal to the area of $W(0) \subset \mathcal{D}$. For $W = \{g \mid N(g) < u\}$ we find $W(0)$ equal to a disc of radius $R(u)$. (It is not necessary to compute this.) Now any matrix with $N(g) < ua$ can be obtained by multiplying a matrix with $N(g) < u$ by one in a fixed compact set, and since

$$\begin{aligned} D(g_1g_2(0),0) &\le D(g_1g_2(0),\, g_1(0)) + D(g_1(0),0) \\ &= D(g_2(0),0) + D(g_1(0),0) \end{aligned}$$

we find that $R(ua) < R(u) + a'$. So the lemma follows from Lemma 6.6.

Lemma 6.8. With $x, y \in \mathbb{R}^2 - \{0\}$, $\varphi$ a fixed function of compact support in $\mathbb{R}^2 - \{0\}$, set

$$H_u'(x,y) = \sum_{\substack{\gamma \in \Gamma \\ N(\gamma) < u}} \varphi(\gamma x)\varphi(\gamma y)$$

then

$$\iint H_u'(x,y) \frac{d\pi(x)}{\|x\|^2} \frac{d\pi(y)}{\|y\|^2} = O(V_u)$$

Proof: Let M be a fundamental domain for $\Gamma$ in G; i.e., a borel set with $M_\gamma \cap M = \emptyset$ for $\gamma \neq e$, $r \in \Gamma$, and $\bigcup_\Gamma M\gamma = G$. We may assume that $\overline{M}$ is compact. Set $\Phi(x) = \max\{|\varphi(q^{-1}x)| \mid q \in \overline{M}\}$ so that

$\Phi$ is bounded and has compact support. If $g \in q\gamma$, $q \in M$, $\gamma \in M$ and if $N(\gamma) < u$, then $N(g) < cu$ for some constant $c$. It follows that if $F(g)$ is a non-negative borel measurable function on $G$, then

$$\sum_{\substack{N(\gamma)<u \\ \gamma\in\Gamma}} \int_M F(q\gamma)\,dq \le \int_{N(g)<cu} F(g)\,dg.$$

Also $|\varphi(\gamma x)\varphi(\gamma y)| \le \Phi(q\gamma x)\Phi(q\gamma y)$ for $q \in M$.

Hence

$$m_G(M)\left|\sum_{N(\gamma)<u} \varphi(\gamma x)\varphi(\gamma y)\right| \le \int_{N(g)<cu} \Phi(gx)\Phi(gy)\,dg$$

and we may apply Lemma 6.5 to obtain the desired result.

Lemma 6.9. Let the harmonic function $h_\varphi(z)$ be defined by

$$h_\varphi(gK) = \int \varphi(x)\,\|g^{-1}x\|^{-2}d\pi(x)$$

where $\varphi$ is a non-negative function of compact support in $\mathbb{R}^2 - \{0\}$.

Then

$$\sum_{\gamma(0)\in\mathcal{D}_R} h_\varphi(\gamma(0))^2 = O(A_R)$$

where $A_R$ is the non-euclidean area of the disc $\mathcal{D}_R$.

Proof: The condition $g(0) \in \mathcal{D}_R$ is equivalent to a condition of the form $N(g) < u$. We use the $\Gamma$-invariance of $\pi$ to write

$$h_\varphi(\gamma(0)) = \int \varphi(\gamma x)\,\|x\|^{-2}d\pi(x)$$

so that

$$\sum_{\gamma(0)\in\mathcal{D}_R} h_\varphi(\gamma(0))^2 = \iint H_u'(x,y)\,\frac{d\pi(x)}{\|x\|^2}\,\frac{d\pi(y)}{\|y\|^2}$$

for appropriate $u$, with $H_u'$ as in Lemma 6.8. Bearing in mind the connection between areas in $\mathcal{D}$ and volumes in $G$ we see that this lemma follows from Lemma 6.8.

Lemma 6.10. Let $M$ be a compact set in $G$. There exists a constant $c = c(M)$ such that for any positive harmonic function on $G/K$

$$h(gg_0K) \le c\,h(gK)$$

for $g \in G$, $g_0 \in M$.

Proof: It suffices to prove this for extremal harmonic functions $P(\zeta, z)$, or using the form in Proposition 3.1, it suffices to consider the functions $\|g^{-1}x\|^{-2}$. But

$$\|g_o^{-1}g^{-1}x\|^{-2} \leq \|g_o\|^2 \|g^{-1}x\|^{-2}$$

and $\|g_o\|$ is bounded in $M$.

Lemma 6.11. With $h_\varphi$ as in Lemma 6.9 we have

$$\int_{\mathcal{D}_R} h_\varphi(z)^2 dm(z) = O(A_R)$$

Proof: We let $Q$ be a bounded fundamental domain for $\Gamma$ in $\mathcal{D}$ with $o \in Q$. $\bigcup_\Gamma \gamma Q$ covers $\mathcal{D}$ and since $Q$ is bounded we find that each $z \in \mathcal{D}_R$ can be expressed as $\gamma z'$, $\gamma \in \Gamma$, $z' \in Q$ and

$$D(\gamma(0), 0) \leq D(\gamma z', \gamma(0)) + D(\gamma z', 0) \leq D(z', 0) + R.$$

Hence $\gamma(0) \in \mathcal{D}_{R+c}$ for appropriate $c$ independent of $R$. So

$$\int_{\mathcal{D}_R} h_\varphi(z)^2\, dm(z) \leq \sum_{\substack{\gamma \in \Gamma \\ \gamma(0) \in \mathcal{D}_{R+c}}} \int_Q h_\varphi(\gamma z')^2 dm(z').$$

Now $z' \in Q$ can be expressed $z' = g'(0)$ with $g'$ inside some compact set. Hence by the foregoing lemma

$$h_\varphi(\gamma z') = h_\varphi(\gamma g'(0)) < c_1\, h_\varphi(\gamma(0)).$$

This gives

$$\int_{\mathcal{D}_R} h_\varphi(z)^2 dm(z) < c_2 \sum_{\substack{\gamma \in \Gamma \\ \gamma(0) \in \mathcal{D}_{R+c}}} h_\varphi(\gamma(0))^2 = O(A_{R+c}) = O(A_R).$$

Finally we argue that the integral in the foregoing lemma is a weighted linear combination of integrals $I_r = \int h_\varphi(re^{i\theta})^2\, d\theta$ for various $r$. Since the weight given to $I_r$ for $r_1 < r < r_2$ depends on the non-euclidean area of this annulus, it follows that the total weight given to circles with $r < r_1 < 1$ tends to $0$ in the integral over $\mathcal{D}_R$ as $R \longrightarrow \infty$ (since $A_R \longrightarrow \infty$). In view of the fact that the $I_r$ are increasing as $r \nearrow 1$ it follows from Lemma 6.11 that $I_r$ must be bounded. We have proven

Proposition 6.12. If $\pi$ is a $\Gamma$-invariant measure on $\mathbb{R}^2 - \{0\}$ and $\varphi(x)$ is a function of compact support on $\mathbb{R}^2 - \{0\}$ then

$$h_{\varphi}(gK) = \int \varphi(x) \| g^{-1}x \|^{-2} d\pi(x)$$

is an $L^2$ - bounded harmonic function.

Corollary. The horocycle flow is uniquely ergodic.

BIBLIOGRAPHY

[1] L. Auslander, L. Green, and F. Hahn, Flows on Homogeneous Spaces. Annals of Mathematics Studies No. 53 (1963), Princeton.

[2] H. Furstenberg, "Strict ergodicity and transformations of the torus", American Journal of Mathematics 83 (1961) 573-601

[3] H. Furstenberg and I. Tzkoni, "Spherical functions and integral geometry", Israel Journal of Mathematics, 10 (1971) 327-338.

[4] G. A. Hedlund, "On the metrical transitivity of the geodesics on closed surfaces of constant negative curvature", Annals of Mathematics 35 (1934) 787-808.

[5] G. A. Hedlund, "A metrically transitive group defined by the modular group", Americal Journal of Mathematics, vol. 57 (1935) 668-678.

[6] G. A. Hedlund, "Fuchsian groups and transitive horocycles", Duke Mathematical Journal 2 (1936) 530-542.

[7] E. Hopf, Ergodentheorie, Ergebnisse der Mathematik und ihrer Grenzgebiete 5 (1937).

[8] C. C. Moore, "Ergodicity of flows on homogeneous spaces", American Journal of Mathematics 88 (1965) 154-178.

[9] A. Zygmund, Trigonometric Series , vol.I second edition, 1959, Cambridge.

Berkeley, California
and
Jerusalem, Israel

# POISSON BOUNDARIES AND ENVELOPES OF DISCRETE GROUPS

BY HARRY FURSTENBERG[1]

Communicated by G. D. Mostow, January 5, 1967

In [4] we defined the Poisson boundaries for semisimple Lie groups. These spaces play a role in the theory of generalized harmonic functions on the Lie group similar to that played by the boundary of the unit disc in the classical theory of harmonic functions on the unit disc. It is not hard to extend these notions to all separable, locally compact groups, and, in particular, they make sense for countable discrete groups. In this form we shall show that these ideas provide a useful tool for answering certain purely algebraic questions. Namely, we raise the following question. Let $G$ be a connected Lie group, $\Gamma$ a discrete subgroup for which $G/\Gamma$ has finite (left-) invariant measure. To what extent is $G$ determined by a knowledge of $\Gamma$ as an abstract group, and conversely, what is the influence of $G$ on the structure of $\Gamma$?

To make this question precise, let us say that $G$ is an *envelope* of $\Gamma$ if an isomorphic copy $\Gamma'$ of $\Gamma$ occurs as a discrete subgroup of $G$, and $G=D\Gamma'$, where $D$ is a subset of $G$ with finite left-invariant Haar measure. Our question may now be stated in this way. How different can two connected Lie groups $G_1$ and $G_2$ be if they both envelop the same countable group $\Gamma$?

We shall be discussing a rather restricted version of this question. We suppose that $G_1$ and $G_2$ are semisimple and have no compact components, and that $G_1$ and $G_2$ envelop the same group $\Gamma$. Does it follow that $G_1$ and $G_2$ are isomorphic? (Without the hypothesis that $G_1$ and $G_2$ have no compact components we could always take $G_2=G_1\times$a compact group.) Our guess is that this is the case. However all we can prove is the following:

THEOREM. *Let $H_r$, $r=1, 2, 3, \cdots$, denote the hyperbolic group of motions of the r-sphere $S^r$: $H_r$ consists of the $(r+2)\times(r+2)$ real matrices that leave the form $x_0^2+x_1^2+\cdots+x_r^2-t^2$ invariant. $SL(s, R)$ denotes the group of $s\times s$ real unimodular matrices. If $G_1$ is one of the groups $H_r$, $r\geqq 1$ and $G_2$ is one of the groups $SL(s, R)$, $s\geqq 3$, then $G_1$ and $G_2$ cannot simultaneously envelop the same countable group.*

[1] This research was partially supported by the Office of Scientific Research of the Office of Aerospace Research of the USAF under Grant #AF-AFOSR-381-63.

Note that $SL(2, \boldsymbol{R})$ is exceptional here. This is because $SL(2, \boldsymbol{R})$ is isomorphic to the hyperbolic group $H_1$. (Consider the adjoint representation of $SL(2, \boldsymbol{R})$.) Thus, as a special case of the theorems, $SL(2, \boldsymbol{R})$ and $SL(s, \boldsymbol{R})$ cannot envelop the same countable group, if $s \geqq 3$.

Our theorem is rather special, but it implies, for example, that the free group on two generators which occurs with finite index in $SL(2, \boldsymbol{Z})$, cannot be enveloped by any $SL(s, \boldsymbol{R})$, $s \geqq 3$. Actually we can say more: $SL(s, \boldsymbol{R})$, $s \geqq 3$, cannot envelop any free group. In particular, no subgroup of $SL(s, \boldsymbol{Z})$, $s \geqq 3$, of finite index is free. This last result can also be deduced from [1] and [6].

We present here the main ideas of the proof of this theorem. A more detailed discussion of a partial result is given in [5].

1. **The Poisson boundary.** Let $G$ be a separable, locally compact group, $M$ a $G$-space, $\mu$ a probability (Borel) measure on $G$, and $\nu$ a probability measure on $M$. The application $G \times M \rightarrow M$ defines the convolution $\mu * \nu$, and we say that $\nu$ is a *$\mu$-stationary* measure if $\mu * \nu = \nu$. A bounded Borel function $f(g)$ on $G$ is *$\mu$-harmonic* if

$$f(g) = \int f(gg')d\mu(g').$$

If $\nu$ is a $\mu$-stationary measure on $M$, $\phi$ a bounded Borel function on $M$, then

$$f(g) = \int \phi(g\xi)d\nu(\xi) = \int \phi(\xi)dg\nu(\xi) \tag{1}$$

defines a $\mu$-harmonic function on $G$.

THEOREM 1. *Let $\mu$ be a fixed probability measure on $G$. There exists a $G$-space $B$ and a $\mu$-stationary measure $\nu$ on $B$ such that every bounded $\mu$-harmonic function on $G$ admits the Poisson representation* (1) *for some function $\phi$ on $B$. The pair $(B, \nu)$ is called the Poisson boundary of $(G, \mu)$ and we write $(B, \nu) = P(G, \mu)$.*

Let $X_1, X_2, \cdots, X_n, \cdots$ denote a sequence of independent $G$-valued random variables, each with distribution $\mu$. If $M$ is a $G$-space and $\nu$ a $\mu$-stationary measure on $M$, it may be shown that with probability 1, the sequence of measures $\{X_1X_2 \cdots X_n\nu\}$ converges weakly on $M$, as $n \rightarrow \infty$.

DEFINITION 1. The pair $(M, \nu)$ is a *boundary* of $(G, \mu)$ if, with probability 1, $\lim X_1X_2 \cdots X_n\nu$ is a point measure (i.e. its support consists of a single point).

THEOREM 2. *Every boundary of $(G, \mu)$ is a measurable equivariant image of the Poisson boundary $P(G, \mu)$.*

That is, if $P(G, \mu) = (B, \nu_0)$, then there is an equivariant, measurable map $\rho: B \to M$ such that $\rho(\nu_0) = \nu$.

In case $G$ is a semisimple Lie group and $\mu$ is an absolutely continuous measure on $G$, we found in [4] that the underlying space $B$ of $P(G, \mu)$ is a compact homogeneous space of $G$. In fact, it must be one of finitely many covering spaces of the homogeneous space $B(G)$ which can be explicitly described: If $G = K \cdot A \cdot N$ is an Iwasawa decomposition of $G$ relative to the maximal compact subgroup $K$, then $B(G) = G/T$, where $T$ is the normalizer in $G$ of $A \cdot N$. Note that $KT = G$, so that $K$ is transitive on $B(G)$.

In certain cases we can be even more specific. We say $\mu$ is *spherical* if it is invariant with respect to left and right multiplication by elements of $K$. If $\mu$ is spherical, then $P(G, \mu) = (B(G), m_B)$, where $m_B$ denotes the unique $K$-invariant probability measure on $B(G)$. By Theorem 1, it follows that all spherical measures $\mu$ lead to the same class of $\mu$-harmonic functions. We call these simply harmonic functions on $G$. (They do depend, however, on the choice of $K$.)

2. **Measures on discrete subgroups.**

THEOREM 3. *If $G$ is a connected semisimple Lie group, $\Gamma$ a countable discrete subgroup enveloped by $G$, then there exists a measure $\mu$ whose support coincides with $\Gamma$ for which $P(\Gamma, \mu) = (B(G), m_B)$.*

According to Theorem 3, the Poisson representation of a $\mu$-harmonic function on $\Gamma$ coincides with that of a harmonic function on $G$. In fact, the theorem implies that the $\mu$-harmonic functions on $\Gamma$ are precisely the restrictions of harmonic functions from $G$ to $\Gamma$.

Theorem 3 provides the main tool for our subsequent analysis. It shows in what form one can obtain information about an envelope of $\Gamma$ from $\Gamma$ itself. For, the possible Poisson boundaries $P(\Gamma, \mu)$, $\mu$ a measure on $\Gamma$, depend only on the structure of $\Gamma$.

3. **Dynkin spaces and $\mathfrak{D}$-groups.** Let $H$ be a locally compact topological group and $M$ an $H$-space which is compact and metrizable.

DEFINITION 2. $M$ is a *Dynkin* space of $H$ if, for every $\epsilon > 0$, there is a compact subset of $H$ such that each $h \in H$ outside of this set maps all of $M$ except for some $\epsilon$-neighborhood into an $\epsilon$-neighborhood:

For example, the projective line $P^1$ is a Dynkin space of $SL(2, \boldsymbol{R})$. More generally,

PROPOSITION 1. *The r-dimensional sphere is a Dynkin space of the hyperbolic group $H_r$, $r = 1, 2, 3, \cdots$.*

DEFINITION 3. A group $H$ is a $\mathfrak{D}$-group if it possesses a Dynkin space $M$ with the property that no point of $M$ is a fixed-point of the group.

$H_r$ is a $D$-group. The same is true of any subgroup of $H_r$ that has no fixed-point on the $r$-sphere. Since the subgroup that leaves a given point of $S^r$ fixed is connected, it follows from [2] that a closed subgroup $H \subset H_r$ for which $H_r/H$ has finite measure cannot have a fixed-point on $S^r$. Hence

PROPOSITION 2. *If* $\Gamma$ *is enveloped by a hyperbolic group, then* $\Gamma$ *is a* $\mathfrak{D}$*-group.*

For example, $SL(2, \boldsymbol{Z})$ and its subgroups of finite index are $\mathfrak{D}$-groups. The fundamental groups of compact orientable 2-dimensional surfaces of genus $\geqq 2$ are $\mathfrak{D}$-groups. The *Picard group* of $2 \times 2$ unimodular matrices whose entries are Gaussian integers is enveloped by $SL(2, \boldsymbol{C})$ which is isomorphic to the Lorentz group $H_2$. Hence the Picard group is a $\mathfrak{D}$-group. We can obtain many other examples using the following proposition.

PROPOSITION 3. *$H$ is a $\mathfrak{D}$-group if some homomorphic image of $H$ is a $\mathfrak{D}$-group.*

For example, the commutator subgroup of $SL(2, \boldsymbol{Z})$ is of finite index and is a free group with two generators. Now any free group on $\geqq 2$ generators maps homomorphically onto the latter. This gives us

PROPOSITION 4. *The free group on $r$ generators, $2 \leqq r \leqq \infty$, is a $\mathfrak{D}$-group.*

Actually we may construct a Dynkin space for the free group $F_2$ on 2 generators directly as follows. Denote the generators by $a$ and $b$. Let $M$ be the set of all infinite "words," $w = w_1 w_2 \cdots w_n \cdots$ whose letters are chosen from $\{a, a^{-1}, b, b^{-1}\}$ and subject to the condition that no consecutive pair $(w_i, w_{i+1})$ is of the form $(a, a^{-1})$, $(b, b^{-1})$, $(a^{-1}, a)$, or $(b^{-1}, b)$. $M$ is a $F_2$-space if we define the action of $F_2$ by juxtaposition, cancelling where necessary. With the usual topology of a sequence space, $M$ is compact and metrizable. Moreover one sees quite easily that $M$ is a Dynkin space for $F_2$. This construction is essentially due to Dynkin and Malyutov [3] who use it to construct a Martin boundary for a class of harmonic functions on the free group.

A similar construction may be applied to other instances of groups presented in terms of generators and relations. The following is an example of this.

PROPOSITION 5. *A free product of nontrivial finite groups, $G_1, G_2, \cdots, G_m$ $(2 \leqq m \leqq \infty)$ is a $\mathfrak{D}$-group unless $m=2$ and $G_1$ and $G_2$ are both of order* 2.

Here the Dynkin space consists of words $w_1 w_2 \cdots w_n \cdots$ where the $w_i$ are nontrivial elements of the groups $G_j$ and where neighboring $w_i$ never come from the same group.

4. **$\mu$-harmonic functions on a $\mathfrak{D}$-group.** Let $\mu$ be a measure on a $\mathfrak{D}$-group $H$ whose support coincides with $H$, and suppose that $M$ is a Dynkin space for $H$. Since $M$ is compact, there exists a $\mu$-stationary measure $\nu$ on $M$. $\nu$ cannot concentrate on a single point of $M$, for then $H$ would have a fixed-point in $M$. Let us assume that, in fact, $\nu$ is entirely nonatomic. Then $(M, \nu)$ is a boundary of $(H, \mu)$. This follows from Definition 2, since, if $h_n\nu$ converges to a measure on $M$ and $h_n \to \infty$ in $H$, then $\lim h_n\nu$ must be a point measure. Now it can be shown that, with probability 1, the sequence $\{X_1X_2 \cdots X_n\}$ possesses a subsequence $\to \infty$ unless $\mu$ is concentrated on a compact subgroup of $H$. Thus, in our case, $\lim X_1X_2 \cdots X_n\nu$ is a point measure, and $(M, \nu)$ is a boundary of $(H, \mu)$.

Now suppose $A_1$ and $A_2$ are two disjoint closed subsets of $M$ with $\nu(A_i) > \frac{1}{2} - \epsilon$, where $\epsilon$ is a positive number. The functions

$$f_i(h) = \nu(h^{-1}(A_i)) = h\nu(A_i)$$

are $\mu$-harmonic and $f_i(e) > \frac{1}{2} - \epsilon$. Now as $h \to \infty$, the measure $h\nu$ tends to become a point measure, and since $A_1$ and $A_2$ are separated by a positive distance, we find

$$\min\{f_1(h), f_2(h)\} \to 0 \tag{2}$$

as $h \to \infty$. This gives us the following result.

LEMMA 1. *If $H$ is a $\mathfrak{D}$-group and $\mu$ a probability measure on $H$ with support all of $H$, then either there exists an atomic $\mu$-stationary measure on the Dynkin space of $H$, or, for every $\epsilon > 0$, there exist $\mu$-harmonic functions $f_1(h), f_2(h)$ on $H$ with*

(a) $0 \leqq f_1(h), f_2(h) \leqq 1$

(b) $f_1(e) > \frac{1}{2} - \epsilon, f_2(e) > \frac{1}{2} - \epsilon$

(c) $\min\{f_1(h), f_2(h)\} \to 0$ as $h \to \infty$.

5. **Boundary behavior of harmonic functions on $SL(s, \boldsymbol{R})$.** Let $s \geqq 3$ be fixed, set $G = SL(s, \boldsymbol{R})$, $B = B(G)$. $B$ can be explicitly determined; it is the "flag space" of subspaces of all dimensions of $\boldsymbol{R}^s$. $K$ will denote the orthogonal subgroup of $G$, and $m_B$ denotes the $K$-invariant measure on $B$. Let $\{g_n\}$ be a sequence in $G$ such that $g_n m_B$

converges to a measure $\pi$ on $B$. We then have the following version of the Fatou theorem.

PROPOSITION 6. *If $f(g)$ is a bounded harmonic function on $G$ corresponding to a boundary function $\hat{f}$ on $B$, then, as functions on $K$, the sequence $f(kg_n)$ converges in measure to a function $\hat{f}_\pi(k)$ which satisfies*
(a) $\hat{f}_\pi(k)=\int_M\hat{f}(k\xi)d\pi(\xi)$ *a.e.*,
(b) $\int_K\hat{f}_\pi(k)dk=f(e)$.

If $\pi$ is a point measure, $\pi=\delta_{\xi_0}$, then $\hat{f}_\pi$ simply "lifts" $\hat{f}$ from $B$ to $K$: $\hat{f}_\pi(k)=\hat{f}(k\xi_0)$. For $s\geqq 3$ however, there always are limit measures $\pi$ which are not point measures, and Proposition 6 describes the boundary behavior near these points as well. The values of $\hat{f}_\pi$ are averages of values of $\hat{f}(\xi)$; we may expect, therefore, that even if $\hat{f}$ takes on only the two values 0 and 1, the function $\hat{f}_\pi$ will take on intermediate values as well. This is made precise in the following.

LEMMA 2. *There exist two limit measures $\pi'$ and $\pi''$ on $B$ and a positive constant $\eta$ such that if $f(g)$ is a harmonic function on $G$ with $0\leqq f(g)\leqq 1$, and also $1/4\leqq f(e)\leqq 3/4$, then*

$$\int \min(\hat{f}_\pi(k),\, 1-\hat{f}_\pi(k))dk > \eta$$

*for either $\pi=\pi'$ or $\pi=\pi''$.*

In other words, unless $f(e)$ is close to either 0 or 1, it is not possible for *both* $\hat{f}_{\pi'}$ and $\hat{f}_{\pi''}$ to approximate characteristic functions.

A comparison of Lemmas 1 and 2 enables us to prove our main result.

THEOREM 4. *A unimodular group $SL(s, R)$, $s\geqq 3$, cannot envelop a $\mathfrak{D}$-group.*

Suppose $G=SL(s, \boldsymbol{R})$ envelopes the $\mathfrak{D}$-group $\Gamma$. Let $\mu$ be a measure on $\Gamma$ for which $P(\Gamma, \mu)=(B, m_B)$, let $M$ be the Dynkin space of $\Gamma$ and let $\nu$ be a $\mu$-stationary measure on $M$. Then $(M, \nu)$ is a boundary, and by Theorem 2, there is an equivariant map from $(B, m_B)$ to $(M, \nu)$. This may be seen to imply that $\nu$ is nonatomic. Applying Lemma 1, we find that the second alternative takes place, and we may find two $\mu$-harmonic functions $f_1$, $f_2$ on $\Gamma$ satisfying (a), (b), and (c). Since $P(\Gamma, \mu)=(B, m_B)$, $f_1$ and $f_2$ extend to harmonic functions on $G$. Now we can no longer assert that, in general, min $(f_1(g), f_2(g))\to 0$ as $g\to\infty$ in $G$, but this will be the case if $g$ stays "sufficiently close" to $\Gamma$. Using a result of [7] regarding the ergodicity of flows in $G/\Gamma$, we may show that each of the measures $\pi'$, $\pi''$ is the limit of a sequence of the form $\{g_n m_B\}$ such that

$$\min(f_1(kg_n), f_2(kg_n)) \to 0 \tag{3}$$

in measure (of $K$) as $n \to \infty$. But then $\min(\hat{f}_{1,\pi}, \hat{f}_{2,\pi}) = 0$ for $\pi = \pi'$ and $\pi = \pi''$. Now

$$\int \hat{f}_{1,\pi}(k)dk + \int \hat{f}_{2,\pi}(k)dk = f_1(e) + f_2(e) > 1 - 2\epsilon$$

so that $\int |(1 - \hat{f}_{1,\pi}) - \hat{f}_{2,\pi}| dk < 2\epsilon$. This gives

$$\int \min(\hat{f}_{1,\pi}(k),\ 1 - \hat{f}_{1,\pi}(k))dk < 2\epsilon$$

for both $\pi = \pi', \pi''$. This contradicts Lemma 2 if $2\epsilon < \eta$ and $\epsilon < 1/4$.

As a direct consequence of Theorem 4 we have

THEOREM 5. *If $\Gamma$ is a countable group enveloped by a unimodular group $SL(s, \boldsymbol{R})$, $s \geqq 3$, then $\Gamma$ is not a free group, nor is it a free product of finite groups, nor is it enveloped by a hyperbolic group, nor does it have a homomorphic image with any of these properties.*

We should remark that whereas free groups cannot occur as discrete subgroups of a group $SL(s, \boldsymbol{R})$, $s \geqq 3$, in such a way that the quotient space has finite measure, if we remove either the finiteness condition or the discreteness condition, they certainly do occur. In fact, free groups occur as dense subgroups in any connected Lie group.

## REFERENCES

1. H. Bass, M. Lazard and J.-P. Serre, *Sous-groupes d'indice fini dans $SL(n, Z)$*, Bull. Amer. Math. Soc. **70** (1964), 385.

2. A. Borel, *Density properties for certain subgroups of semi-simple groups without compact components*, Ann. of Math. **72** (1960), 179–188.

3. E. B. Dynkin and M. B. Malyutov, *Random walk on groups with finitely many generators*, Soviet Math. Dokl. **2** (1961), 399–402.

4. H. Furstenberg, *A Poisson Formula for semi-simple Lie groups*, Ann. of Math. **77** (1963), 335–386.

5. ———, "Random walks and discrete subgroups of Lie groups," to appear in *Advances in probability and related topics*.

6. J. L. Mennicke, *Finite factor groups of the unimodular groups*, Ann. of Math. **81** (1965), 31–37.

7. C. C. Moore, *Ergodicity of flows on homogeneous spaces*, Amer. J. Math. **88** (1966), 154–178.

THE HEBREW UNIVERSITY, JERUSALEM

# Recurrent Ergodic Structures and Ramsey Theory

*Hillel Furstenberg*

Institute of Mathematics, Hebrew University, Givat Ram, Jerusalem, Israel

## Introduction

Ramsey theory is best defined by example and the classic example of a Ramsey type theorem is the result of van der Waerden: if the integers $\mathbb{Z}$ are partitioned into finitely many sets, one of these contains arbitrarily long arithmetic progressions. An equivalent version states: given natural numbers $r, l$, $\exists N(r,l)$ so that if $N \geq N(r,l)$ and the integers $\{1,2,\ldots,N\}$ are partitioned into $r$ sets, one of these contains an $l$-term arithmetic progression. Erdös and Turán realized that this result would follow if it were the case that *any* subset of $\{1,2,\ldots,N\}$ comprising $\delta N$ elements contains an $l$-term arithmetic progression provided $N$ is sufficiently large. This result, conjectured in 1936 [ET1] was proved by E. Szemerédi in 1973 [Sz1], and was reproved using ergodic theoretic methods in 1976 [Fu1]. The theorems of van den Waerden and Szemerédi illustrate the two sides of Ramsey theory: coloring – or partition – results, and density results. The latter are clearly stronger than the former, and the proofs are typically more recondite. The role of ergodic theory in density theorems of this type stems from the fact that in a number of situations theorems about patterns found in sets having "density" bounded from below are equivalent to theorems about the "return", or "recurrence", patterns for measure preserving transformations acting on a measure space. It would appear that the ubiquity of certain patterns in the combinatorics of large sets reflects the phenomenon of recurrence in ergodic theoretic contexts and the latter has to be studied to gain insight into the former.

We shall be examining three different contexts for recurrence results in ergodic theory. We shall mention the combinatorial (Ramsey theoretic) equivalents of these results, although we shall have to refer the reader elsewhere for the proof of the equivalence. Our purpose here is to display the common features of these recurrence phenomena. We shall find that in each setting there is a notion of *rigidity* and the notion of a special system constructed from scratch by a finite succession of rigid extension. For these systems (*distal* and *quasi-distal*) it will be possible to deduce recurrence properties directly, the key tool being Ramsey theory of the van der Waerden sort. For arbitrary systems we shall find a method to "factor out" the "independent" component, reducing the recurrence property

Proceedings of the International Congress
of Mathematicians, Kyoto, Japan, 1990

for a general system to that of a factor system, which will be one of the special systems for which the result has been established.

To motivate the ergodic theoretic discussion we list here some of the combinatorial consequences. For more details, particularly as regards the connection between the ergodic theoretic formulation and the Ramsey theoretic results, we refer the reader to [Fu2, FK2, FK4, and GRS1].

**Theorem A.** *There is a function $N(\delta, l)$ for $\delta > 0$ and $l \in \mathbb{N}$ so that if $N \geq N(\delta, l)$ and $S \subset \{1, 2, \ldots, N\}$ with $|S| \geq \delta N$ then $S$ contains an $l$-term arithmetic progression. ( $|S|$ denotes cardinality of $S$.)*

**Theorem B.** *Let $P$ denote a finite subset of $\mathbb{R}^d$. There is a function $R(\delta, P)$ so that if $R \geq R(\delta, P)$ and $B$ is a ball of radius $R$ in $\mathbb{R}^d$ and $S \subset B$ is a measurable subset with $m(S) \geq \delta m(B)$, then $S$ contains an integral dilation of $P$, i.e., a set of the form $u + nP$, $u \in \mathbb{R}^d, n \in \mathbb{N}$.*

**Theorem C.** *There exists a function $D(\delta, d, q)$ for $\delta > 0, d \in \mathbb{N}$, $q =$ a prime power, so that if $V$ is a vector space over $\mathbb{F}_q$ of dimension $\geq D(\delta, d, q)$ and $S \subset V$ is a subset with $|S| \geq \delta |V|$, then $S$ contains a $d$-flat (translate of $d$-dimensional subspace).*

**Theorem D.** *If $\Lambda$ is a finite set (alphabet) let $W_N(\Lambda)$ denote words of length $N$ in $\Lambda$. Let $W_N^*(\Lambda) = W_N(\Lambda \cup \{t\}) \setminus W_N(\Lambda)$ so that $\varphi(t) \in W_N^*(\Lambda)$ is a word in which the variable $t$ appears. There exists a function $L(\delta, |\Lambda|)$ so that if $N \geq L(\delta, |\Lambda|)$ and $S \subset W_N(\Lambda)$ with $|S| \geq \delta |W_N(\Lambda)|$, then there exists $\varphi(t) \in W_N^*(\Lambda)$ so that the words $\varphi(a), \varphi(b), \ldots, \varphi(s)$, $a, b, \ldots, s \in \Lambda$ are all in $S$.*

Theorem A is Szemerédi's theorem. Theorem B is equivalent to a multidimensional version of Szemerédi's theorem. It was proved first in [FK1]. Theorem C is a density theoretic version of a coloring theorem first conjectured by Roth and proved by R.L. Graham and B. Rothschild [GR1]. The latter also follows from the theorem of A.W. Hales and R.I. Jewett [HJ1]. A special case of Theorem C was proved by T.C. Brown and J.P. Buhler [BB1], and in generality it was proved in [FK2]. Theorem D was conjectured by R.L. Graham and proved in [FK4].

By interpreting the letters of $\Lambda$ as the digits for writing numbers to a base $b$, we can see that $D \Rightarrow A$. On the other hand, interpreting $\Lambda$ as a vector space over a finite field, we can see that $D \Rightarrow C$. It is also not difficult to show that $D \Rightarrow B$, so that Theorem D holds the key to a number of results of Ramsey theory.

We remark that the major part of what is presented here is joint work with Y. Katznelson.

## 1. Classical Systems

In the classical setting for ergodic theory we speak of a *measure preserving system* (m.p.s.) $(X, \mathcal{B}, \mu, T)$ where $T : X \to X$ is a measurable, measure preserving map, and $\mu$ is a probability measure on the sets of $\mathcal{B}$. It is useful to distinguish between *ergodic* and *non-ergodic* systems. $(X,\mathcal{B},\mu,T)$ is *ergodic* if $T^{-1}A = A$ implies $\mu(A) = 0$ or 1. There is a rather general decomposition theorem for measure preserving systems to ergodic components, and the recurrence phenomena we shall study here follow for arbitrary systems once they are established for ergodic systems. We also can assume without loss of generality that $T$ is invertible.

If $T$ is a measure preserving transformation on $X$ then $T$ induces an operator on functions on $X$: $T^{-1}f(x) = f(Tx)$. (The notation corresponds to the fact that $S^{-1}(T^{-1}f) = (TS)^{-1}f$.) $T^{-1}$ defines a unitary operator on $L^2(X,\mathcal{B},\mu)$ and we can study spectral properties of $T^{-1}$. In particular ergodicity is equivalent to 1 being a simple eigenvalue, and if, in addition, there are no other eigenvalues, we say that $T$, or $(X,\mathcal{B},\mu,T)$, is *weak mixing*.

The mean ergodic theorem for $L^2$-functions is a special case of a general theorem for contraction operators in Hilbert space:

**Theorem 1.** *If $(X,\mathcal{B},\mu,T)$ is an ergodic system and $f \in L^2(X,\mathcal{B},\mu)$, then*

$$\frac{f + T^{-1}f + T^{-2}f + \cdots + T^{-N}f}{N+1} \longrightarrow \int f d\mu$$

*in the norm of $L^2(X,\mathcal{B},\mu)$.*

V. Bergelson has given a far reaching generalization of this in the case of a weak mixing system [Be1].

**Theorem 2.** *Let $(X,\mathcal{B},\mu,T)$ be a weak mixing system. Let $p_1(t), p_2(t), \ldots, p_k(t)$ be polynomials with integer coefficients with $\deg p_i > 0$ and $\deg(p_i - p_j) > 0$ for $i \neq j$. Then if $f_1, f_2, \ldots, f_k \in L^\infty(X,\mathcal{B},\mu)$ we will have*

$$\frac{1}{N+1}\sum_{n=0}^{N} T^{-p_1(n)}f_1 T^{-p_2(n)}f_2 \cdots T^{-p_k(n)}f_k \to \int f_1 d\mu \int f_2 d\mu \cdots \int f_k d\mu \tag{1}$$

*in the norm of $L^2(X,\mathcal{B},\mu)$.*

One consequence of this theorem is that if $(X,\mathcal{B},\mu,T)$ is weak mixing and $A_1, A_2, \ldots, A_k$ are any $k$ subsets of positive measure, we will have for some (in fact, most) $n$:

$$\mu(A_1 \cap T^{-n}A_2 \cap \cdots \cap T^{-(k-1)n}A_k) > 0\ , \tag{2}$$

and, in particular, if $\mu(A) > 0$,

$$\mu(A \cap T^{-n}A \cap \cdots \cap T^{-(k-1)n}A) > 0\ . \tag{3}$$

The foregoing property, *that points of A return to A along arithmetic progressions,* is the recurrence phenomenon we have in mind in the case of a classical system. In the weak mixing case we have found that it follows from a generalized ergodic theorem.

The proof of Theorem 2 is based on the following lemma for convergence of averages in Hilbert space to 0.

**Lemma 3.** *Let $\{u_n\}$ be a bounded sequence of vectors in a Hilbert space. Assume for each m, that*

$$\gamma_m = \lim_{N\to\infty} \frac{1}{N}\sum_{n=1}^{N}\langle u_n, u_{n+m}\rangle$$

*exists, and that*

$$\frac{1}{M}\sum_{m=1}^{M}\gamma_m \longrightarrow 0 \ .$$

*then*

$$\left\| \frac{1}{N}\sum_{n=1}^{N} u_n \right\| \longrightarrow 0 \ .$$

Using the lemma and the ergodic theorem, one proves (1.1) for progressively more complex situations, beginning with

$$\frac{1}{N+1}\sum_{n=0}^{N} T^{-an}fT^{-bn}g, \ \frac{1}{N+1}\sum_{n=0}^{N} T^{-an}fT^{-bn}gT^{-cn}h, \cdots$$

We refer to [Be1] for details.

When $(X,\mathscr{B},\mu,T)$ is not weak mixing the result of Theorem 2 cannot be expected to be valid. For example assume $\varphi$ is an eigenfunction, $T^{-1}\varphi = \lambda\varphi$, and set $f=\varphi^2, g=\varphi^{-1}$, then $T^{-n}fT^{-2n}g = \left(\lambda^{2n}\varphi^2\right)\left(\lambda^{-2n}\varphi^{-1}\right) = \varphi$ and

$$\frac{1}{N+1}\sum_{0}^{N} T^{-n}fT^{-2n}g \to \varphi$$

which is not constant. In the weak mixing case we would have obtained $\int f d\mu \int g d\mu$ as the limit.

## 2. Isometric Extensions and Distal Factors

If $(X,\mathscr{B},\mu,T)$ and $(Y,\mathscr{D},\nu,T)$ are two systems ( it is convenient to use the same letter for the transformation), we say that a map $\pi : X \to Y$ defines $(Y,\mathscr{D},\nu,T)$ as a *factor* of $(X,\mathscr{B},\mu,T)$ if $\pi$ is measurable, measure preserving, and $\pi(Tx) = T(\pi x)$. We lift functions from $Y$ to $X$, $f \to f\circ\pi$ and since this imbeds $L^2(Y,\mathscr{D},\nu)$ isometrically into $L^2(X,\mathscr{B},\mu)$, we shall identify $f \in L^2(Y,\mathscr{D},\nu)$ with $f\circ\pi \in L^2(X,\mathscr{B},\mu)$. It will be convenient to denote the conditional expectation operator $f \mapsto E(f \mid \pi^{-1}\mathscr{D})$ by $f \mapsto E(f \mid Y)$ and we regard $E(f \mid Y)$ either as a function on $Y$ or on $X$. Note that $T^{-1}$ commutes with $E(\mid Y)$.

We call a m.p.s. *Kronecker* if it has the form $(Z,\mathscr{D},m,T)$ where $(Z,+)$ is a compact abelian group with a dense cyclic subgroup $\mathbb{Z}\alpha$, $\mathscr{D}$ = Borel sets, $m$ = Haar measure, and $Tz = z+\alpha$. For a Kronecker system, $L^2(Z,\mathscr{D},m)$ is spanned by eigenfunctions (each character is an eigenfunction). It follows that if a system $(X,\mathscr{B},\mu,T)$ possesses a non-trivial Kronecker factor then it cannot be weak mixing. The converse is also true: if $(X, \mathscr{B}, \mu, T)$ is not weak mixing it has a non-trivial Kronecker factor.

We describe a relativized notion of a Kronecker factor, representing a certain relationship between a system and a factor.

**Definition 1.** $(X, \mathscr{B}, \mu, T)$ is an *isometric extension* of a factor $(Y, \mathscr{D}, \nu, T)$ if it can be represented – up to isomorphism – as follows: $X = Y \times M$, $M$ is a homogeneous space of a compact group $G$, $\mu = \nu \times m$ where $m$ is the $G$-invariant measure on $M$, and there exists a measurable function $\varrho : Y \to G$ so that for $y \in Y, u \in M$,

$$T(y, u) = (Ty, \varrho(y)u) \ .$$

An ergodic isometric extension of the trivial one-point system is a Kronecker system.

**Definition 2.** A m.p.s. $(X, \mathscr{B}, \mu, T)$ is ($n$*-step*) *distal* if it has a succession of factors

$$(X, \mathscr{B}, \mu, T) = (X_1, \mathscr{B}_1, \mu_1, T) \to \cdots \to (X_n, \mathscr{B}_n, \mu_n, T) \to \text{1-point} \tag{4}$$

where each extension is isometric.

The terminology comes from topological dynamics where distal systems are defined by the property that for $x \neq x'$, the distance $d(T^n x, T^n x')$ is bounded from below. It is easily seen that for metric spaces, a succession of isometric extensions leads to a distal system in this sense.

We now formulate the generalization of V. Bergelson's theorem 1.2 to the general ergodic situation. Note that weak mixing systems have only trivial distal factors.

**Theorem 3.** *Let* $(X, \mathscr{B}, \mu, T)$ *be an ergodic m.p.s. Let* $p_1(t), p_2(t), \ldots, p_k(t)$ *be polynomials with integer coefficients with* $\deg p_i > 0$ *and* $\deg(p_i - p_j) > 0$ *for* $i \neq j$. *There exists a distal factor* $\pi : X \to Y$ *so that for any* $f_1, f_2, \ldots f_k \in L^\infty(X, \mathscr{B}, \mu)$

$$\frac{1}{N+1} \sum_{n=0}^{N} T^{-p_1(n)} f_1 T^{-p_2(n)} f_2 \cdots T^{-p_k(n)} f_k - \frac{1}{N+1} \sum_{n=0}^{N} T^{-p_1(n)} E(f_1 \mid Y) \cdots T^{-p_k(n)} E(f_k \mid Y) \to 0 \tag{5}$$

*in* $L^2(X, \mathscr{B}, \mu)$.

We call $(Y, \mathscr{D}, \nu, T)$ a *characteristic factor for* $(T^{-p_1(n)}, \ldots, T^{-p_k(n)})$. We use the indefinite article here because any extension of a characteristic factor is again a characteristic factor. We do not know if there exists a "smallest" one.

For special cases the information is more specific [Fu3]:

**Theorem 4.** *For any ergodic m.p.s.* $(X, \mathscr{B}, \mu, T)$ *there exists a "largest" Kronecker factor* $(Z, \mathscr{D}, m, T)$ *and for* $f, g \in L^\infty(X, \mathscr{B}, \mu)$, $a, b \in \mathbb{Z}$, $a \neq b$,

$$\lim_{N\to\infty} \frac{1}{N+1} \sum_{n=0}^{N} T^{-an} f T^{-bn} g = \lim_{N\to\infty} \frac{1}{N+1} \sum_{n=0}^{N} T^{-an} E(f \mid Z) T^{-bn} E(g \mid Z) \ . \quad (6)$$

Here we are also guaranteed existence of the limits in question (in $L^2(X, \mathscr{B}, \mu)$). If we write $\bar{f} = E(f \mid Z), \bar{g} = E(g \mid Z)$ and regard the right hand side of (2.3) as a function $\psi(\zeta)$ on $Z$, then it can be rewritten as

$$\psi(\zeta) = \int \bar{f}(\zeta + a\theta)\bar{g}(\zeta + b\theta)d\theta$$

We can also be more explicit regarding characteristic factors for ($T^{-an}$, $T^{-bn}$, $T^{-cn}$). To formulate this result we need the notion of a 2-*step nilpotent system.* This is a system for which the underlying space has the form $N/\Gamma$ where $N$ is a 2-step nilpotent group, $\Gamma$ is a closed cocompact subgroup and $N$ possesses a sequence of closed subgroups $N_i$ so that $N/N_i$ is locally compact, $\Gamma/\Gamma \cap N_i$ is discrete and cocompact in $N/N_i$, the measure on $N/\Gamma$ is the (unique) $N$-invariant probability measure (lifted from $(N/N_i) \ / \ (\Gamma/\Gamma \cap N_i)$) and the measure preserving transformation is $T(g\Gamma) = (\tau g)\Gamma$ for some $\tau \in N$.

**Theorem 5.** *For any ergodic* $(X, \mathscr{B}, \mu, T)$ *there exists a 2-step nilpotent factor* $(Y, \mathscr{D}, \nu, T)$ *so that for* $f, g, h \in L^\infty(X, \mathscr{B}, \mu)$ *and* $a, b, c$ *distinct integers, we have*

$$\lim_{N\to\infty} \frac{1}{N+1} \sum_{n=0}^{N} T^{-an} f T^{-bn} g T^{-cn} h =$$

$$\lim_{N\to\infty} \frac{1}{N+1} \sum_{n=0}^{N} T^{-an} E(f \mid Y) T^{-bn} E(g \mid Y) T^{-cn} E(h \mid Y)$$

Theorem 5 was obtained jointly with B. Weiss and overlaps results of Conze and Lesigne [CL1]. It may be conjectured that similar results are valid for more general expressions with 2-step nilpotent systems replaced by arbitrary nilpotent systems.

## 3. Multiple Recurrence

We begin this section by formulating a version of (1.3) for distal systems.

**Theorem 1.** *Let* $(Y, \mathscr{D}, \nu, T)$ *be a distal system, let* $f \in L^\infty(Y, \mathscr{D}, \nu)$ *be non-negative but not vanishing a.e. Then for any* $l \in \mathbb{N}$,

$$\liminf_{N\to\infty} \frac{1}{N+1} \sum_{n=0}^{N} \int f \, T^{-n} f \, T^{-2n} f \cdots T^{-(l-1)n} f \, d\mu > 0 \ . \quad (7)$$

To explain the role of distality in this result we shall prove here by way of illustration that for 1-step distal systems one has the weaker statement: With $(Y, \mathscr{D}, \nu, T)$ and $f, l$ as above, $\exists n$ such that

$$\int fT^{-n}fT^{-2n}f \cdots T^{-(l-1)n}f d\mu > 0 \ . \tag{8}$$

A refinement of the argument will show that (7) is valid. Moreover a relativized version of this argument can be reapplied successively to handle the general distal case.

If $M$ is a homogeneous space of the compact group $G$, then $G$ acts continuously on $L^2(M,m) : \psi(u) \longrightarrow \psi^g(u) = \psi(gu)$. In particular, if $\psi \in L^\infty(M,m)$, its orbit under $G$, $\Psi = \{\psi^g \mid g \in G\}$, is compact in $L^2(M,m)$. Suppose $\psi \geq 0$ and that $\psi$ is not 0 a.e. For any $l$, $\exists \varepsilon > 0$ so that if $\psi_1, \psi_2, \ldots, \psi_l \in L^\infty(M,m)$ satisfy $\|\psi_j - \psi\| < \varepsilon$ then $\int \psi_1\psi_2 \ldots \psi_l dm > 0$. Since $\Psi$ is compact we can partition $\Psi$ into finitely many subsets of diameter $< \varepsilon$. So we can write $\Psi = \Psi_1 \cup \cdots \cup \Psi_r$ so that if $\psi_1, \psi_2, \ldots, \psi_l \in \Psi_j$ then $\int \psi_1\psi_2 \cdots \psi_l dm > 0$ .

Now consider a 1-step distal system, i.e., a Kronecker system, $(Z, \mathscr{D}, \nu, T)$. $Z$ plays the role of $G$ as well as $M$. Take $f$ as in the theorem and set $\psi = f$. By van der Waerden's theorem there is an $l$-term arithmetic progression, $a, a+n, a+2n, \ldots, a+(l-1)n$ such that each $T^{-(a+in)}f, i = 0, 1, \ldots, l-1$, belongs to the same $\Psi_j$. Then

$$\int T^{-a}fT^{-(a+n)}f \cdots T^{-(a+(l-1)n)}f \, d\mu > 0$$

or

$$\int f\,T^{-n}f \cdots T^{-(l-1)n}f \, d\mu > 0$$

as claimed.

We now combine Theorem 1 with Theorem 2.1 to obtain

**Theorem 2.** *If $(X, \mathscr{B}, \mu, T)$ is any ergodic m.p.s., $f \in L^\infty(X, \mathscr{B}, \mu)$ is a non-negative function not vanishing a.e., then*

$$\liminf_{N\to\infty} \frac{1}{N+1} \sum_{n=0}^{N} \int fT^{-n}fT^{-2n}f \cdots T^{-(l-1)n}f \, d\mu > 0 \ .$$

Decomposition to ergodic components gives the result for arbitrary measure preserving systems. Finally taking $f = 1_A$ we obtain the

**Theorem (Multiple Recurrence).** *Let $(X, \mathscr{B}, \mu, T)$ be any m.p.s., let $A \in \mathscr{B}$ with $\mu(A) > 0$. Then for any $l$ there exists $n$ with*

$$\mu(A \cap T^{-n}A \cap T^{-2n}A \cap \cdots \cap T^{-(l-1)n}A) > 0 \ . \tag{9}$$

As we show in [Fu1] and [Fu2] this is equivalent to Theorem A of the Introduction.

## 4. IP-Systems

We denote by $\mathscr{F}$ the collection of all finite subsets of $\mathbb{N} = \{1,2,3,\dots\}$. If $\alpha, \beta \in \mathscr{F}$ we write $\alpha < \beta$ if $\max \alpha < \min \beta$.

**Definition 1.** An *IP-subset* of $\mathscr{F}$ consists of a sequence $\alpha_1 < \alpha_2 < \dots$ in $\mathscr{F}$ together with all finite unions of these.

The elements of an IP-subset of $\mathscr{F}$ may be placed in one-one correspondence with $\mathscr{F}$: $\beta = \{i_1, i_2, \dots i_k\} \leftrightarrow \alpha_{i_1} \cup \alpha_{i_2} \cup \cdots \alpha_{i_k} = \alpha_\beta$.

**Definition 2.** An *$\mathscr{F}$-sequence* is any sequence $\{x_\alpha\}$ labelled by $\alpha \in \mathscr{F}$. An *$\mathscr{F}$-subsequence* of $\{x_\alpha\}$ is the subsequence corresponding to an IP-subset of $\mathscr{F}$: $x'_\beta = x_{\alpha_\beta}$.

**Definition 3.** If $X$ is a topological space, $\{x_\alpha\}$ an $\mathscr{F}$-sequence in $X$, we say that $x_\alpha \to x_0 \in X$, or $\lim\limits_{\alpha\to\infty} x_\alpha = x_0$, if for any neighborhood $V$ of $x_0$, $\exists \alpha_0$ so that for $\alpha > \alpha_0$ we have $x_\alpha \in V$.

The following theorem is equivalent to N. Hindman's theorem on finite sum sets [Hi1]:

**Theorem 4.** *If $X$ is a compact metric space and $\{x_\alpha\}$ is an $\mathscr{F}$-sequence in $X$, then $\{x_\alpha\}$ has a convergent $\mathscr{F}$-subsequence.*

If $X$ is finite this can be reformulated as: if $\mathscr{F} = \mathscr{C}_1 \cup \cdots \mathscr{C}_r$ then some $\mathscr{C}_j$ contains an IP-subset of $\mathscr{F}$. This latter formulation is due to Hindman.

The next definition describes the principal concept of this section.

**Definition 5.** An $\mathscr{F}$-sequence $\{\sigma_\alpha\}$ with values in a semigroup is an *IP-system* if $\alpha < \beta$ implies that

$$\sigma_{\alpha\cup\beta} = \sigma_\alpha \sigma_\beta \ . \tag{10}$$

Writing $\sigma_i$ for $\sigma_{\{i\}}$ we see that $\sigma_{\{i_1,i_2,\dots,i_k\}} = \sigma_{i_1}\sigma_{i_2}\cdots\sigma_{i_k}$ when $i_1 < i_2 < \cdots < i_k$. We note that in the commutative case (4.1) is valid whenever $\alpha \cap \beta = \emptyset$.

**Lemma 6.** *An $\mathscr{F}$-subsequence of an IP-system is an IP-system.*

We call this an *IP-subsystem.*

The classical notion of a m.p.s. will now be replaced by that of an IP-system of measure preserving transformations (m.p.t.). Inside a classical system we can form numerous IP-systems, taking $T_\alpha = T^{n_\alpha}$ where $n_\alpha = \sum_{i\in\alpha} n_i$ is an IP-system in $(\mathbb{Z},+)$. In addition to providing more information regarding classical systems, the analysis of IP-systems provides recurrence results for situations where the classical setup doesn't prevail – or is trivial, because as might be the case $T^p$ could equal the identity for some $p$ and all $T$ under consideration.

The main theorem of this section is

**Theorem (IP-Recurrence).** *Let $(X, \mathscr{B}, \mu)$ be a measure space and let $\{T_\alpha^{(1)}\}, \cdots, \{T_\alpha^{(k)}\}$ be k IP-systems of m.p.t. of $(X\ell, \mathscr{B}, \mu)$ where all the $T_\alpha^i$ commute with one another. Then if $A \in \mathscr{B}, \mu(A) > 0$, $\exists \alpha \in \mathscr{F}$ so that*

$$\mu(T_\alpha^{(1)-1}A \cap T_\alpha^{(2)-1}A \cap \cdots \cap T_\alpha^{(k)-1}A) > 0 \ . \tag{11}$$

We remark that this theorem implies Theorems B and C of the Introduction. We refer the reader to [FK2] for details.

The proof of the IP-recurrence theorem depends on a development of ergodic theory for IP-systems analogous to classical theory. We proceed to outline the salient points of such a theory. One important point should be made at the outset. Since in the recurrence theorem we are looking for one index $\alpha$ for which we have a multiple intersection, we may pass from the IP-systems to IP-subsystems that may be more convenient. Since on account of Theorem 4, an $\mathscr{F}$-subsequence of an $\mathscr{F}$-sequence in a compact metric space converges, we shall often find it convenient to deal with an appropriately chosen $\mathscr{F}$-subsequence, which means we are really dealing with IP-subsystems of the initially given systems.

If $\{T_\alpha\}$ is an IP-system of m.p.t. of $(X, \mathscr{B}, \mu)$ then writing as before $T_\alpha^{-1}f(x) = f(T_\alpha x)$ we obtain, in the commutative case, an IP-system of isometry operators $\{T_\alpha^{-1}\}$ on $L^2(X, \mathscr{B}, \mu)$. Let $\mathscr{K}$ be the semigroup of linear operators on $L^2$ with norm $\leq 1$, endowed with the weak operator topology. $\mathscr{K}$ is a compact metrizable space. Using Theorem 4 we can find a subsystem so that $T_\alpha^{-1}$ converges in $\mathscr{K}$. We now have

**Proposition 7.** *If $\{K_\alpha\} \subset \mathscr{K}$ is an IP-system and $K_\alpha \to Q$ then $Q$ is a self-adjoint projection operator. When $K_\alpha = T_\alpha^{-1}$ for m.p.t. $\{T_\alpha\}$, then $Q$ is the conditional expectation operator relative to a subalgebra $\mathscr{D} \subset \mathscr{B}$.*

*Proof.* If $Q = \lim_{\alpha \to \infty} K_\alpha$ then $Q^2 = \lim_{\alpha \to \infty} \lim_{\beta \to \infty} K_\alpha K_\beta = \lim_{\alpha \to \infty} \lim_{\beta \to \infty} K_{\alpha \cup \beta} = Q$. The functions in the range of $Q$ are characterized by: $K_\alpha f \to f$ weakly. Since $\|f\| \geq \|K_\alpha f\|$, weak convergence implies strong convergence. For strong convergence $K_\alpha f \to f, K_\alpha g \to g \Rightarrow K_\alpha f \vee K_\alpha g \to f \vee g$. Now $T_\alpha^{-1} f \vee T_\alpha^{-1} g = T_\alpha^{-1}(f \vee g)$ and so if $f, g \in$ range $Q$ also $f \vee g \in$ range $Q$. This characterizes conditional expectation operators. □

For an IP-system $\{T_\alpha\}$ we now suppose having passed to a subsystem for which $\lim T_\alpha^{-1} = Q_T$ exists. For IP-systems this operator $Q_T$ plays the role of the ergodic average. Since $Q_T$ is self-adjoint and the $T_\alpha^{-1}$ are isometric this convergence will generally not be in the strong topology, and certainly not pointwise. In the extreme case when $Q_T =$ identity then the convergence is strong and we speak of $\{T_\alpha\}$ as a *rigid* system. An example of a rigid system is one obtained from the Kronecker system $(Z, \mathscr{B}, \mu, T)$ setting $T_\alpha = T^{n_\alpha}$ for $\{n_\alpha\}$ an IP-system in $\mathbb{Z}$. if $T_\alpha \to Q_T$ it is not hard to show, applying $T_\alpha$ to eigenfunctions of $T$, that $Q_T =$identity. In IP-theory rigid systems play the role of Kronecker systems.

The opposite extreme to rigidity occurs when $Q_T f = E(f)$. This means that $\int f T_\alpha^{-1} g \, d\mu \to \int f d\mu \int g d\mu$ so that $f$ and $T_\alpha^{-1} g$ are asymptotically independent. We call this case *mixing*.

Let $(X, \mathscr{B}, \mu)$ be a measure space, $(Y, \mathscr{D}, \nu)$ a factor of $(X, \mathscr{B}, \mu)$ such that a system $\{T_\alpha\}$ is defined both on $X$ and on $Y$, that is to say, we have $\pi : X \to Y$ with $\pi T_\alpha = T_\alpha \pi$. A subspace $\mathscr{M} \subset L^2(X, \mathscr{B}, \mu)$ is called a *$Y$-module* if it is closed under multiplication by $L^\infty(Y, \mathscr{D}, \nu)$. A $Y$-module is of *finite rank* if it is spanned by finitely many functions over $L^\infty(Y, \mathscr{D}, \nu)$. Finally $\mathscr{M}$ is called $\{T_\alpha\}$ *quasi-invariant* if for each $g \in \mathscr{M}$ and $\varepsilon > 0$, $\exists \alpha_0$ so that for $\alpha > \alpha_0$ the $L^2$ distance satisfies

$$\operatorname{dist}(T_\alpha^{-1} g, \mathscr{M}) < \varepsilon \ .$$

**Definition 8.** $(X, \mathscr{B}, \mu, \{T_\alpha\})$ is a *rigid extension* of $(Y, \mathscr{D}, \nu, \{T_\alpha\})$ if $L^2(X, \mathscr{B}, \mu)$ is spanned by $\{T_\alpha\}$ quasi-invariant $Y$-modules of finite rank.

Now consider commuting IP-systems $\{T_\alpha^{(1)}\}, \cdots \{T_\alpha^{(k)}\}$ of m.p.t. on $(X, \mathscr{B}, \mu)$. We call $(X, \mathscr{B}, \mu, \{T_\alpha^{(1)}\}, \cdots \{T_\alpha^{(k)}\})$ a *$k$-fold IP-system.*

**Definition 9.** $(X, \mathscr{B}, \mu, \{T_\alpha^{(1)}\}, \cdots \{T_\alpha^{(k)}\})$ is *quasi-distal* if there is a sequence of $k$-fold IP-systems $(X_l, \mathscr{B}_l, \mu_l, \{T_\alpha^{(1)}\}, \cdots \{T_\alpha^{(k)}\})$, $l = 0, 1, 2, \ldots, n$, with $(X_0, \mathscr{B}_0, \mu_0) = (X, \mathscr{B}, \mu)$, and with $X_n$ =point, and with maps $\pi_l : X_l \to X_{l+1}$ which are measure preserving and satisfy $\pi_l T_\alpha = T_\alpha \pi_l$ and such that for each $l$, $(X_l, \mathscr{B}_l, \mu_l, \{T_\alpha^{(i)} T_\alpha^{(j)-1}\})$ is a rigid extension of $(X_{l+1}, \mathscr{B}_{l+1}, \mu_{l+1}, \{T_\alpha^{(i)} T_\alpha^{(j)-1}\})$ for some $i \neq j$.

We now have

**Theorem 10.** *Given a $k$-fold IP-system $(X, \mathscr{B}, \mu, \{T_\alpha^{(1)}\}, \cdots \{T_\alpha^{(k)}\})$ there is a subsystem and a factor $(Y, \mathscr{D}, \nu, \{T_\alpha^{(1)}\}, \cdots \{T_\alpha^{(k)}\})$ such that the latter is quasi-distal and for any $f_1, f_2, \ldots, f_k \in L^\infty(X, \mathscr{B}, \mu)$ we have*

$$\lim_\alpha \int T_\alpha^{(1)-1} f_1 \cdots T_\alpha^{(k)-1} f_k d\mu = \lim_\alpha \int T_\alpha^{(1)-1} E(f_1 \mid Y) \cdots T_\alpha^{(k)-1} E(f_k \mid Y) d\nu \ .$$

The implication of Theorem 10 for us is that the IP-recurrence theorem will follow in general if we can show that for quasi-distal systems and $f \geq 0$ with $f$ not vanishing a.e.,

$$\lim_{\alpha \to \infty} \int T_\alpha^{(1)^{-1}} f T_\alpha^{(2)^{-1}} f \cdots T_\alpha^{(k)^{-1}} f d\mu > 0 \ .$$

This can indeed be shown in a manner analogous to the proof of Theorem 3.1, by successively lifting the desired property via the rigid extensions that constitute the links of the quasi-distal system. This outlines the proof of the IP-recurrence theorem.

## 5. $W(\Lambda)$-Systems

In this section we describe the ergodic structure whose recurrence properties lead to Theorem D, the density version of the Hales-Jewett theorem. The structure will be similar to that of the previous section except that now the operators are no longer assumed to commute.

We take $\Lambda = \{1, 2, \ldots, k\}$. Suppose we are given $k$ sequences of m.p.t. $\{T_n^{(1)}, \{T_n^{(2)}\}, \ldots, \{T_n^{(k)}\}$ of a measure space $(X, \mathscr{B}, \mu)$. Given a word $w \in W_l(\Lambda)$ we can form the transformation

$$T(w) = T_1^{w(1)} T_2^{w(2)} \cdots T_l^{w(l)} \ . \tag{12}$$

**Definition 1.** A $W(\Lambda)$-system consists of a family of m.p.t. of a measure space $(X, \mathscr{B}, \mu)$ corresponding to $w \in W(\Lambda)$ and formed in accordance with (12). We denote the system $(X, \mathscr{B}, \mu, \{T(w)\}_{w \in W(\Lambda)})$.

We can now state the

**Theorem ($W(\Lambda)$-Recurrence).** *Let $(X, \mathscr{B}, \mu, \{T(w)\}_{w \in W(\Lambda)})$ be a $W(\Lambda)$-system. If $f \in L^\infty(X, \mathscr{B}, \mu)$ is non-negative and not 0 a.e. then $\exists \varphi(t) \in W^*(\Lambda)$ so that*

$$\int T(\varphi(1))^{-1} f \, T(\varphi(2))^{-1} f \cdots T(\varphi(k))^{-1} f \, d\mu > 0 \ . \tag{13}$$

The proof of this theorem is patterned after the proof of the IP-recurrence theorem of §4. There is a new feature which arises on account of the non-commutativity. This feature also reflects an aspect of Ramsey theory for $W^*(\Lambda)$ as opposed to $\mathscr{F}$.

We use the following notation. If $w \in W_N(\Lambda)$ and $\alpha \subset \{1, 2, \ldots, N\}$, we designate by $w_\alpha$ the word of $W^*(\Lambda)$ in which the letters at positions of $\alpha$ are replaced by the variable $t$. If we write $\varphi(t)$ for $w_\alpha$ then we denote $\varphi(i)$ by $w_\alpha^i$, $i \in \Lambda$. Finally set $\Omega = \Lambda^{\mathbb{N}}$. We regard $\Omega$ as part of the "boundary" of $W^*(\Lambda)$; namely for sequences $w^{(n)} \in W(\Lambda)$ of increasing length, and $\alpha^{(n)} \in \mathscr{F}$, we say $w_{\alpha_n}^{(n)} \to \omega \in \Omega$ if $\alpha_n \to \infty$ and $w^{(n)}(p) = \omega(p)$ for each $p \in \mathbb{N}$ and $n > n(p)$.

We next define *subspaces* of $W(\Lambda)$ and $W^*(\Lambda)$. Let $\Sigma = \{\varphi_1(t), \varphi_2(t), \ldots\}$ be a sequence in $W^*(\Lambda)$. We define $W_\Sigma^*(\Lambda)$ as consisting of words of the form $\varphi_1(u_1)\varphi_2(u_2) \cdots \varphi_h(u_h)$, $h$ arbitrary, with $u_i \in \Lambda$ or $u_i \in \Lambda \cup \{t\}$ respectively. It is clear how to define $\Omega_\Sigma$.

By analogy with $\mathscr{F}$-sequences we have

**Definition 2.** A function $x(w_\alpha)$ is called a *$W^*(\Lambda)$-sequence*. Its restriction to a subspace is a $W^*(\Lambda)$-subsequence.

**Definition 3.** If $x(w_\alpha)$ takes values in a metric space $M$. Then we say $x(w_\alpha)$ is *coherent* if there exists a function $x^* : \Omega \to M$ such that $x(w_\alpha) \to x^*(\omega)$ uniformly as $w_\alpha \to \omega \in \Omega$.

The following theorem is a far reaching extension of the Hales-Jewett theorem. It includes Hindman's theorem. See [Ca1, FK3].

**Theorem 4.** *A $W^*(\Lambda)$-sequence with values in compact metric space has a coherent $W^*(\Lambda)$-subsequence.*

Given a $W(\Lambda)$-system we will restrict to appropriate subspaces in order to achieve coherence of certain expressions. Note that the expression in (5.2) is a $W^*(\Lambda)$-sequence.

The next lemma shows how IP-systems occur within $W(\Lambda)$-systems.

**Lemma 5.** *Let $w \in W_N(\Lambda), \alpha, \beta \subset \{1, 2, \cdots, N\}, \alpha < \beta$ and suppose that for $n \in \alpha \cup \beta, w(n) = i$. Then for any $W(\Lambda)$-system $\{T(w)\}$*

$$T(w_\alpha^j)T(w_\alpha^i)^{-1}T(w_\beta^j)T(w_\beta^i)^{-1} = T(w_{\alpha\cup\beta}^j)T(w_{\alpha\cup\beta}^i)^{-1} \ . \tag{14}$$

Using Theorem 4 we pass to a subsystem for which all

$$Q_{ij}(\omega) = \lim_{\substack{\alpha\to\infty \\ w\to\omega}} T(w_\alpha^i)T(w_\alpha^j)^{-1}$$

exist uniformly. On account of Lemma 5 and Proposition 4.7 these define conditional expectation operators $E(\ |\ \mathscr{D}_{ij}(\omega))$ provided each $i \in \Lambda$ occurs infinitely often in $\omega$. These $Q_{ij}(\omega)$ represent the "ergodic averaging" operator in the $W(\Lambda)$ context. Note that we obtain a "field" of operators. In a similar way we can form rigid extensions, and quasi-distal factor systems, except that the entire structure will vary (continuously) with $\omega \in \Omega$. The $\sigma$-algebras would not be invariant but for $\alpha$ far out the operators

$$T_j^i(w_\alpha) = T(w_\alpha^i)T(w_\alpha^j)^{-1}$$

will move $\mathscr{D}_{ij}(\omega)$ to $\mathscr{D}_{ij}(\omega')$ with $\omega'$ close to $\omega$. With this apparatus we can proceed as in Sect. 4.

## References

[BB1] Brown, T.C., Buhler, J.P.: A density version of a geometric Ramsey theorem. J. Comb. Theory Ser. A **32** (1982) 20–34

[Be1] Bergelson, V.: Weakly mixing PET. Ergodic Theory Dyn. Systems **7** (1987) 337–349

[Ca1] Carlson, T.J.: Some unifying principles in Ramsey theory. Discrete Math. **68** (1988) 117–169

[CL1] Conze, J.P., Lesigne, E.: Sur un théorème ergodique pour les mesure diagonales. Preprint

[ET1] Erdös, P., Turán, P.: On some sequences of integers. J. London Math. Soc. **11** (1936) 261–264

[Fu1] Furstenberg, H.: Ergodic behavior of diagonal measures and a theorem of Szemerédi on arithmetic progressions. J. d'Analyse Math. **31** (1977) 204–256

[Fu2] Furstenberg, H.: Recurrence in Ergodic Theory and Combinatorial Number Theory. Princeton University Press (1981)

[Fu3] Furstenberg, H.: Nonconventional ergodic theorems. In: Symposium on Legacy of von Neumann. Proceedings of Symposia in Pure Mathematics, Amer. Math. Soc. **50** (1990) 43–56

[FK1] Furstenberg, H., Katznelson, Y.: An ergodic Szemerédi theorem for commuting transformations. J. d'Analyse Math. **34** (1978) 275–291

[FK2] Furstenberg, H., Katznelson, Y.: An ergodic Szemerédi theorem for IP-systems and combinatorial theory. J. d'Analyse Math. **45** (1985) 117–168

[FK3] Furstenberg, H., Katznelson, Y.: Idempotents in compact semigroups and Ramsey theory. Israel J. Math. **68** (1989) 257–270

[FK4] Furstenberg, H., Katznelson, Y.: A density version of the Hales-Jewett theorem. J. D'Analyse Math. (to appear)

[GR1] Graham, R.L., Rothschild, B.L.: Ramsey's theorem for $n$-parameter sets. Trans. Amer. Math. Soc. **159** (1971) 257–292

[GRS1] Graham, R.L., Rothschild, B.L., Spencer, J.H.: Ramsey theory. Wiley, New York 1980

[HJ1] Hales, A.W., Jewett, R.I.: Regularity and positional games. Trans. Amer. Math. Soc. **106** (1963) 222–229

[Hi1] Hindman, N.: Ultrafilters and combinatorial number theory. In: Number theory Carbondale 1979. Lecture Notes in Mathematics, vol. 751. Springer, Berlin Heidelberg New York 1979, pp. 119–184

[Sz1] Szemerédi, E.: On sets of integers containing no $k$ elements in arithmetic progression. Acta Arith. **27** (1975) 199–245

**Curriculum Vitae**

Israel M. Gel'fand

| | |
|---|---|
| Date of Birth: | September 2, 1913, Krasnye Okny, in Ukraine |
| Died: | October 5, 2009, New Brunswick, NJ, USA |

### Education:

1940 (Ph.D.) Doctor of Science at Moscow State University

### Employment History:

| | |
|---|---|
| 1935–1940 | Associate Professor at Moscow State University |
| 1941–1990 | Professor at Moscow State University |
| 1953 | Became head of the Department at the Institute of Applied Mathematics of the Academy of Sciences |
| 1989–1990 | Visiting Professor at Harvard University and at MIT |
| 1990–2009 | Distinguished Professor at the Department of Mathematics at Rutgers University, in Piscataway. The State University of New Jersey, USA |

### Awards and Honors:

| | |
|---|---|
| 1935 | Member, Moscow Mathematical Society |
| 1951 and 1953 | The State Prize of the USSR |
| 1961 | The Lenin Prize |
| 1978 | Wolf Prize in Mathematics |
| 1979 | Wigner Medal |
| 1989 | The Kyoto Prize |
| 1994 | MacArthur Foundation Fellowship |
| 1999 | The State Award of the Russian Federation |
| 2005 | The Leroy P. Steele Prize for Lifetime Achievement from American Mathematical Society |

A member of academies:

| | |
|---|---|
| 1964 | The American Academy of Arts and Sciences |
| 1970 | The Royal Irish Academy |
| 1970 | The National Academy of Sciences of the USA |
| 1974 | The Royal Swedish Academy |
| 1976 | The Académie des Sciences of France |

1977 The Royal Society of London
1988 The Academia dei Lincei of Italy
1989 The Imperial Academie of Sciences of Japan
1999 A Lifetime Member of the New York Academy of Sciences
2004 The European Academy of Sciences
Academy of Sciences of the USSR (now Russian Academy of Sciences)
Gelfand was a corresponding member of the USSR Academy of Sciences since 1953, and was elected as a full member in 1984.

The following honorary degrees:

Honorary doctorate from the University of Oxford (England)
Harvard University (USA)
Sorbonne University (France)
University of Lyons (France)
Uppsala University (Sweden)
University of Pennsylvania (USA)
New York University (USA)
Scuola Normale de Pisa (Italy)
University of Kyoto (Japan)
The Order of Lenin three times (1954, 1956, 1973) (USSR)
President of the Moscow Math Society, 1968–1970 (USSR)
An honorary member of the American Mathematical Society (USA)
An honorary member of the London Mathematical Society (England)

## List of Publications

1. I. M. Gel'fand, M. M. Kapranov and A. V. Zelevinsky, Discriminants, resultants and multidimensional determinants. Reprint of the 1994 edition. *Modern Birkhäuser Classics* (Birkhäuser, 2008).
2. I. M. Gel'fand, Mathematics as an adequate language, in *The Unity of Mathematics*, pp. xv–xxii, Progr. Math., Vol. 244 (Birkhäuser, 2006).
3. I. M. Gel'fand, M. I. Graev and V. S. Retakh, Gauss $q$-hypergeometric equation and its solutions as series and integrals, *Russ. J. Math. Phys.* **12** (2005), No. 2, 168–179.
4. I. M. Gel'fand, S. Gelfand, V. Retakh and R. L. Wilson, Factorizations of polynomials over noncommutative algebras and sufficient sets of edges in directed graphs, *Lett. Math. Phys.* **74** (2005), No. 2, 153–167.
5. I. M. Gel'fand, V. Retakh, S. Serconek and R. L. Wilson, On a class of algebras associated to directed graphs, *Selecta Math.* (*N.S.*) **11** (2005), No. 2, 281–295.
6. J. Barros-Neto and I. M. Gel'fand, Fundamental solutions of the Tricomi operator. III, *Duke Math. J.* (2005), No. 1, 119–140.
7. I. M. Gel'fand, S. Gelfand, V. Retakh and R. L. Wilson, Quasideterminants, *Adv. Math.* **193** (2005), No. 1, 56–141.
8. I. M. Gel'fand, M. I. Graev and V. S. Retakh, Hypergeometric functions over an arbitrary field (in Russian), *Usp. Mat. Nauk* **59** (2004), No. 5(359), 29–100; translation in *Russ. Math. Surv.* **59** (2004), No. 5, 831–905.
9. A. S. Fokas, I. M. Gel'fand and A. E. Kister, Prediction of the structural motifs of sandwich proteins, *Proc. Natl. Acad. Sci. USA* **101** (2004), No. 48, 16780–16783 (electronic).
10. I. M. Gel'fand, V. Retakh and R. L. Wilson, Quaternionic quasideterminants and determinants, in *Lie Groups and Symmetric Spaces*, pp. 111–123, *Amer. Math. Soc. Transl.* Ser. 2, Vol. 210 (Amer. Math. Soc., 2003).
11. I. M. Gel'fand, S. G. Gindikin and M. I. Graev, Selected topics in integral geometry, translated from the 2000 Russian original by A. Shtern. Translations of Mathematical Monographs, Vol. 220 (Amer. Math. Soc., 2003).
12. A. V. Borovik, I. M. Gel'fand and N. White, Coxeter matroids, in Progress in Mathematics, Vol. 216 (Birkhäuser, 2003).
13. J. Barros-Neto and I. M. Gel'fand, Correction to: "Fundamental solutions for the Tricomi operator. II" [*Duke Math. J.* **111** (2002), No. 3, 561–584], *Duke Math. J.* **117** (2003), No. 2, 385–387.

14. I. M. Gel'fand, E. G. Glagoleva and E. E. Shnol, Functions and graphs, translated and adapted from the Russian second edition by Thomas Walsh and Randell Magee. Reprint of the 1969 English edition (Dover, 2002).
15. I. M. Gel'fand, S. Gelfand, V. Retakh, S. Serconek and R. L. Wilson, Hilbert series of quadratic algebras associated with pseudo-roots of noncommutative polynomials, *J. Algebra* **254** (2002), No. 2, 279–299.
16. I. M. Gel'fand, E. G. Glagoleva and A. A. Kirillov, The method of coordinates, translated from the Russian by Leslie Cohn and David Sookne. Reprint of the 1967 translation (Dover, 2002).
17. V. L. Bodneva, V. G. Boltyanskiĭ, I. M. Gel'fand *et al.*, Alekseĭ Alekseevich Milyutin [1925–2001] (in Russian), *Usp. Mat. Nauk* **57** (2002), No. 3(345), 137–140; translation in *Russ. Math. Surv.* **57** (2002), No. 3, 577–580.
18. J. Barros-Neto and I. M. Gel'fand, Fundamental solutions for the Tricomi operator. II, *Duke Math. J.* **111** (2002), No. 3, 561–584.
19. A. V. Borovik, I. M. Gel'fand and D. A. Stone, On the topology of the combinatorial flag varieties, *Disc. Comput. Geom.* **27** (2002), No. 2, 195–214.
20. I. M. Gel'fand and M. I. Graev, Hypergeometric functions over finite fields (in Russian), *Dokl. Akad. Nauk* **381** (2001), No. 6, 732–737.
21. I. M. Gel'fand, S. Gelfand and V. Retakh, Noncommutative algebras associated to complexes and graphs, *Selecta Math.* (*N.S.*) **7** (2001), No. 4, 525–531.
22. I. M. Gel'fand, V. Retakh and R. L. Wilson, Quadratic linear algebras associated with factorizations of noncommutative polynomials and non-commutative differential polynomials, *Selecta Math.* (*N.S.*) **7** (2001), No. 4, 493–523.
23. I. M. Gel'fand and M. I. Graev, Hypergeometric *N*-functions (in Russian), *Dokl. Akad. Nauk* **379** (2001), No. 1, 12–15.
24. I. M. Gel'fand and M. I. Graev, *N*-functions and their relation to solutions of general hypergeometric systems and *GG*-systems (in Russian), *Usp. Mat. Nauk* **56** (2001), No. 4(340), 3–34; translation in *Russ. Math. Surv.* **56** (2001), No. 4, 615–647.
25. R. F. Booth, A. V. Borovik, I. M. Gel'fand and N. White, Oriented Lagrangian matroids, Combinatorial geometries (Luminy, 1999), *European J. Combin.* **22** (2001), No. 5, 639–656.
26. A. V. Borovik, I. M. Gel'fand and N. White, Representations of matroids in semimodular lattices, *European J. Combin.* **22** (2001), No. 6, 789–799.
27. È. B. Vinberg, I. M. Gel'fand and S. G. Gindikin *et al.*, Fridrikh Izrailevich Karpelevich (in Russian), *Usp. Mat. Nauk* **56** (2001), No. 1(337), 147–152; translation in *Russ. Math. Surv.* **56** (2001), No. 1, 141–147.

28. A. S. Fokas, I. M. Gel'fand, F. Finkel and Q. M. Liu, A formula for constructing infinitely many surfaces on Lie algebras and integrable equations, *Selecta Math.* (*N.S.*) **6** (2000), No. 4, 347–375.
29. I. M. Gel'fand and I. W. Zakharevich, Lenard schemes, and the local geometry of bi-Hamiltonian Toda and Lax structures, *Selecta Math.* (*N.S.*) **6** (2000), No. 2, 131–183.
30. I. M. Gel'fand, S. G. Gindikin and M. I. Graev, Izbrannye zadachi integral'noĭ geometrii (in Russian) [Selected problems in integral geometry] (Dobrosvet, 2000).
31. A. V. Borovik, I. M. Gel'fand and N. White, Combinatorial flag varieties, in memory of Gian-Carlo Rota, *J. Combin. Theory Ser. A* **91** (2000), No. 1–2, 111–136.
32. I. M. Gel'fand amd M. I. Graev, Louck polynomials and their relation to general hypergeometric functions and *GG*-functions (in Russian), *Dokl. Akad. Nauk* **372** (2000), No. 2, 151–154.
33. R. F. Booth, A. V. Borovik, I. M. Gel'fand and D. A. Stone, Lagrangian matroids and cohomology, *Ann. Comb.* **4** (2000), No. 2, 171–182.
34. V. S. Buslaev, A. M. Vershik and I. M. Gel'fand *et al.*, Mikhail Shlëmovich Birman (on the occasion of his seventieth birthday) (in Russian), *Usp. Mat. Nauk* **55** (2000), No. 1(331), 204–207; translation in *Russ. Math. Surv.* **55** (2000), No. 1, 201–205.
35. T. V. Alekseyevskaya, A. V. Borovik, I. M. Gel'fand and N. White, Matroid homology, in *The Gel'fand Mathematical Seminars*, 1996–1999, pp. 1–13 (Birkhäuser, 2000).
36. *The Gelfand Mathematical Seminars*, 1996–1999, edited by I. M. Gel'fand and V. S. Retakh (Birkhäuser, 2000).
37. I. M. Gel'fand and M. I. Graev, *GG*-functions and their relations to general hypergeometric functions, *Lett. Math. Phys.* **50** (1999), No. 1, 1–27.
38. V. I. Arnold, R. M. Borisyuk and I. M. Gel'fand *et al.*, Èmmanuil Èl'evich Shnol' (on the occasion of his seventieth birthday) (in Russian), *Usp. Mat. Nauk* **54** (1999), No. 3(327), 199–204; translation in *Russ. Math. Surv.* **54** (1999), No. 3, 677–683.
39. I. M. Gel'fand and M. I. Graev, Special functions associated with complex Lie groups (in Russian), *Dokl. Akad. Nauk* **364** (1999), No. 2, 151–154.
40. J. Barros-Neto and I. M. Gel'fand, Fundamental solutions for the Tricomi operator, *Duke Math. J.* **98** (1999), No. 3, 465–483.
41. I. M. Gel'fand and M. I. Graev, The Fourier transform of rapidly increasing functions and the calculation of integrals that arise in the theory of *GG*-functions (in Russian), *Dokl. Akad. Nauk* **360** (1998), No. 5, 598–602.

42. A. V. Borovik, I. M. Gel'fand and N. White, Symplectic matroids, *J. Algebraic Combin.* **8** (1998), No. 3, 235–252.
43. I. M. Gel'fand, V. Retakh and M. Shubin, Fedosov manifolds, *Adv. Math.* **136** (1998), No. 1, 104–140.
44. I. M. Gel'fand, M. I. Graev and V. S. Retakh, General gamma functions, exponentials and hypergeometric functions (in Russian), *Usp. Mat. Nauk* **53** (1998), No. 1(319), 3–60; translation in *Russ. Math. Surv.* **53** (1998), No. 1, 1–55.
45. P. Etingof, I. M. Gel'fand and V. Retakh, Nonabelian integrable systems, quasideterminants, and Marchenko lemma, *Math. Res. Lett.* **5** (1998), No. 1–2, 1–12.
46. T. V. Alekseyevskaya and I. M. Gel'fand, Incidence matrices, geometrical bases, combinatorial prebases and matroids, Proceedings of the 7th Conference on Formal Power Series and Algebraic Combinatorics (Noisy-le-Grand, 1995), *Disc. Math.* **180** (1998), No. 1–3, 23–44.
47. A. V. Borovik, I. M. Gel'fand and N. White, On exchange properties for Coxeter matroids and oriented matroids, *Disc. Math.* **179** (1998), No. 1–3, 59–72.
48. A. V. Borovik, I. M. Gel'fand and N. White, Coxeter matroid polytopes, *Ann. Comb.* **1** (1997), No. 2, 123–134.
49. I. M. Gel'fand and V. Retakh, Quasideterminants. I, *Selecta Math.* (*N.S.*) **3** (1997), No. 4, 517–546.
50. I. M. Gel'fand and M. I. Graev, *GG*-functions and their connection with general hypergeometric functions (in Russian), *Usp. Mat. Nauk* **52** (1997), No. 4(316), 3–48; translation in *Russ. Math. Surv.* **52** (1997), No. 4, 639–684.
51. A. V. Borovik, I. M. Gel'fand, A. Vince and N. White, The lattice of flats and its underlying flag matroid polytope, *Ann. Comb.* **1** (1997), No. 1, 17–26.
52. I. M. Gel'fand and M. I. Graev, *GG*-functions of several variables and their connection with general hypergeometric functions (in Russian), *Dokl. Akad. Nauk* **354** (1997), No. 5, 583–586.
53. I. M. Gel'fand and M. I. Graev, *GG*-functions of one variable (in Russian), *Dokl. Akad. Nauk* **354** (1997), No. 2, 155–158.
54. P. Etingof, I. M. Gel'fand and V. Retakh, Factorization of differential operators, quasideterminants, and nonabelian Toda field equations, *Math. Res. Lett.* **4** (1997), No. 2–3, 413–425.
55. I. M. Gel'fand, M. I. Graev and M. Zyskin, A problem of integral geometry on $K3$ connected with harmonic analysis on the group $SL(2, K)$, where $K$

is an arbitrary continuous locally compact field (in Russian), *Dokl. Akad. Nauk* **352** (1997), No. 1, 15–17.

56. I. M. Gel'fand, M. I. Graev and A. Postnikov, Combinatorics of hypergeometric functions associated with positive roots, in *The Arnold–Gel'fand Mathematical Seminars*, pp. 205–221 (Birkhäuser, 1997).
57. A. S. Fokas, I. M. Gel'fand and M. V. Zyskin, Nonlinear integrable equations and nonlinear Fourier transform, in *The Arnold–Gel'fand Mathematical Seminars*, pp. 139–170 (Birkhäuser, 1997).
58. J. C. Alvarez, I. M. Gel'fand and M. Smirnov, Crofton densities, symplectic geometry and Hilbert's fourth problem, in *The Arnold–Gel'fand Mathematical Seminars*, pp. 77–92 (Birkhäuser, 1997).
59. *The Arnold–Gel'fand Mathematical Seminars*. Geometry and singularity theory. Edited by V. I. Arnold, I. M. Gel'fand, V. S. Retakh and M. Smirnov (Birkhäuser, 1997).
60. *Algebraic Aspects of Integrable Systems. In Memory of Irene Dorfman*. Edited by A. S. Fokas and I. M. Gel'fand. Progress in Nonlinear Differential Equations and Their Applications, Vol. 26 (Birkhäuser, 1997).
61. L. V. Kantorovich, Selected works. Part I. Descriptive theory of sets and functions. Functional analysis in semi-ordered spaces. Translated from the Russian by A. B. Sossinskii [A. B. Sosinskiĭ]. With an introduction by I. M. Gel'fand. Edited and with a preface by S. S. Kutateladze. Classics of Soviet Mathematics, Vol. 3 (Gordon and Breach, 1996).
62. I. M. Gel'fand, M. I. Graev and S. A. Spirin, Hypergeometric functions and the Newton polytope associated with the action of the torus ($C_*$) on $\Lambda_k C_n$ (in Russian), *Dokl. Akad. Nauk* **348** (1996), No. 2, 155–158.
63. I. M. Gel'fand and M. M. Smirnov, Chern–Simons classes and cocycles on the Lie algebra of the gauge group, in *The Gel'fand Mathematical Seminars*, 1993–1995, pp. 101–122 (Birkhäuser, 1996).
64. I. M. Gel'fand and V. Retakh, Noncommutative Vieta theorem and symmetric functions, in *The Gel'fand Mathematical Seminars*, 1993–1995, pp. 93–100 (Birkhäuser, 1996).
65. A. S. Fokas and I. M. Gel'fand, A unified method for solving linear and nonlinear evolution equations and an application to integrable surfaces, in *The Gel'fand Mathematical Seminars*, 1993–1995, pp. 75–92 (Birkhäuser, 1996).
66. *The Gel'fand Mathematical Seminars*, 1993–1995. Edited by I. M. Gel'fand, J. Lepowsky and M. M. Smirnov (Birkhäuser, 1996).
67. A. V. Borovik, I. M. Gel'fand and N. White, Boundaries of Coxeter matroids, *Adv. Math.* **120** (1996), No. 2, 258–264.

68. A. S. Fokas, I. M. Gel'fand and Y. Kurylev, Inversion method for magnetoencephalography, *Inverse Probl.* **12** (1996), No. 3, L9–L11.
69. A. S. Fokas and I. M. Gel'fand, Surfaces on Lie groups, on Lie algebras, and their integrability. With an appendix by Juan Carlos Alvarez Paiva, *Comm. Math. Phys.* **177** (1996), No. 1, 203–220.
70. I. M. Gel'fand and M. I. Graev, Hypergeometric functions on a space of $p$-dimensional matrices (in Russian), *Dokl. Akad. Nauk* **346** (1996), No. 1, 7–10.
71. I. M. Gel'fand, G. L. Rybnikov and D. A. Stone, Projective orientations of matroids, *Adv. Math.* **113** (1995), No. 1, 118–150.
72. I. M. Gel'fand, D. Krob, A. Lascoux, B. Leclerc, V. S. Retakh and J.-Y. Thibon, Noncommutative symmetric functions, *Adv. Math.* **112** (1995), No. 2, 218–348.
73. I. M. Gel'fand and M. M. Smirnov, The algebra of Chern–Simons classes, the Poisson bracket on it, and the action of the gauge group, in *Lie Theory and Geometry*, pp. 261–288, Progr. Math., Vol. 123 (Birkhäuser, 1994).
74. I. M. Gel'fand and M. I. Graev, Hypergeometric functions on flag spaces (in Russian), *Dokl. Akad. Nauk* **338** (1994), No. 2, 154–157; translation in *Russ. Acad. Sci. Dokl. Math.* **50** (1995), No. 2, 204–209.
75. I. M. Gel'fand and M. M. Smirnov, Lagrangians satisfying Crofton formulas, Radon transforms, and nonlocal differentials, *Adv. Math.* **109** (1994), No. 2, 188–227.
76. I. M. Gel'fand and M. I. Graev, Projective nonunitary representations of current groups (in Russian), *Dokl. Akad. Nauk* **338** (1994), No. 3, 298–301; translation in *Russ. Acad. Sci. Dokl. Math.* **50** (1995), No. 2, 234–239.
77. L. J. Corwin, I. M. Gel'fand and R. Goodman, Quadratic algebras and skew-fields, in *Representation Theory and Analysis on Homogeneous Spaces* (New Brunswick, NJ, 1993), pp. 217–225, Contemp. Math., Vol. 177 (Amer. Math. Soc., 1994).
78. A. S. Fokas and I. M. Gel'fand, Integrability of linear and nonlinear evolution equations and the associated nonlinear Fourier transforms, *Lett. Math. Phys.* **32** (1994), No. 3, 189–210.
79. I. M. Gel'fand and M. Neumark, On the imbedding of normed rings into the ring of operators in Hilbert space. Corrected reprint of the 1943 original. Contemp. Math., Vol. 167, $C_*$-algebras: 1943–1993 (San Antonio, TX, 1993), pp. 2–19 (Amer. Math. Soc., 1994).
80. I. M. Gel'fand, M. I. Graev, V. S. Retakh and S. A. Spirin, $(r, s)$-exponentials (in Russian), *Dokl. Akad. Nauk* **336** (1994), No. 6, 730–732; translation in *Russ. Acad. Sci. Dokl. Math.* **49** (1994), No. 3, 577–581.

81. V. I. Arnol'd, M. Sh. Birman and I. M. Gel'fand *et al.*, Anatoliĭ Moiseevich Vershik (on the occasion of his sixtieth birthday) (in Russian), *Usp. Mat. Nauk* **49** (1994), No. 3(297), 195–204; translation in *Russ. Math. Surv.* **49** (1994), No. 3, 207–221.
82. I. M. Gel'fand and I. Zakharevich, The spectral theory for a pencil of skew symmetrical differential operators of the third order, *Comm. Pure Appl. Math.* **47** (1994), No. 8, 1031–1041.
83. A. S. Fokas and I. M. Gel'fand, Quadratic Poisson algebras and their infinite-dimensional extensions, *J. Math. Phys.* **35** (1994), No. 6, 3117–3131.
84. A. V. Borovik and I. M. Gel'fand, *WP*-matroids and thin Schubert cells on Tits systems, *Adv. Math.* **103** (1994), No. 2, 162–179.
85. I. M. Gel'fand, M. M. Kapranov and A. V. Zelevinsky, Discriminants, resultants, and multidimensional determinants, *Mathematics: Theory & Applications* (Birkhäuser, 1994).
86. A. S. Fokas and I. M. Gel'fand, Bi-Hamiltonian structures and integrability, in *Important Developments in Soliton Theory*, pp. 259–282, Springer Ser. Nonlinear Dynam. (Springer, 1993).
87. I. M. Gel'fand, A. V. Zelevinskiĭ and M. M. Kapranov, Correction to the paper: "Hypergeometric functions and toric varieties" [*Functional. Anal. i Prilozhen.* **23** (1989), No. 2, 12–26; (in Russian)], *Functional. Anal. i Prilozhen.* **27** (1993), No. 4, 91; translation in *Funct. Anal. Appl.* **27** (1993), No. 4, 295 (1994).
88. I. M. Gel'fand and M. I. Graev, Projective representations of the current group SU(1, 1)X (in Russian), *Functional. Anal. i Prilozhen.* **27** (1993), No. 4, 65–68; translation in *Funct. Anal. Appl.* **27** (1993), No. 4, 275–277.
89. I. M. Gel'fand, M. I. Graev and V. S. Retakh, $(R, S)$-hypergeometric functions of one variable (in Russian), *Dokl. Akad. Nauk* **333** (1993), No. 5, 567–570; translation in *Russ. Acad. Sci. Dokl. Math.* **48** (1994), No. 3, 591–596.
90. I. M. Gel'fand, M. I. Graev and V. S. Retakh, Hypergeometric functions on the external power $\Lambda_k C_n$ and on the Grassmannian $G_{k,n}$, their relationships and integral representations, *Russ. J. Math. Phys.* **1** (1993), No. 3, 289–311.
91. I. M. Gel'fand and M. I. Graev, Special representations of the group SU$(n, 1)$ and projective unitary representations of the current group SU$(n, 1)$X (in Russian), *Dokl. Akad. Nauk* **332** (1993), No. 3, 280–282; translation in *Russ. Acad. Sci. Dokl. Math.* **48** (1994), No. 2, 291–295.
92. I. M. Gel'fand and I. Zakharevich, On the local geometry of a bi-Hamiltonian structure, in *The Gel'fand Mathematical Seminars*, 1990–1992, pp. 51–112 (Birkhäuser, 1993).

93. I. M. Gel'fand and M. M. Smirnov, Crofton densities and nonlocal differentials, in *The Gel'fand Mathematical Seminars*, 1990–1992, pp. 35–50 (Birkhäuser, 1993).
94. I. M. Gel'fand and M. M. Kapranov, On the dimension and degree of the projective dual variety: A $q$-analog of the Katz–Kleiman formula, in *The Gel'fand Mathematical Seminars*, 1990–1992, pp. 27–33 (Birkhäuser, 1993).
95. L. Corwin and I. M. Gel'fand, Hopf algebra structures for the Heisenberg algebra. I, in *The Gel'fand Mathematical Seminars*, 1990–1992, pp. 11–17 (Birkhäuser, 1993).
96. *The Gel'fand Mathematical Seminars*, 1990–1992. Edited by L. Corwin, I. M. Gel'fand and J. Lepowsky (Birkhäuser, 1993).
97. I. M. Gel'fand, M. I. Graev and V. S. Retakh, The $q$-hypergeometric Gauss equation and the description of its solutions in the form of series and integrals (in Russian), *Dokl. Akad. Nauk* **331** (1993), No. 2, 140–143; translation in *Russ. Acad. Sci. Dokl. Math.* **48** (1994), No. 1, 40–45.
98. I. M. Gel'fand, M. I. Graev and V. S. Retakh, Reduction formulas for hypergeometric functions associated with the Grassmannian $G_{k,n}$ and description of these functions on strata of small codimension in $G_{k,n}$, *Russ. J. Math. Phys.* **1** (1993), No. 1, 19–56.
99. I. M. Gel'fand and M. I. Graev, *GG*-functions (in Russian), *Dokl. Akad. Nauk* **328** (1993), No. 6, 645–648; translation in *Russ. Acad. Sci. Dokl. Math.* **47** (1993), No. 1, 134–138.
100. L. J. Billera, I. M. Gel'fand and B. Sturmfels, Duality and minors of secondary polyhedra, *J. Combin. Theory Ser. B* **57** (1993), No. 2, 258–268.
101. I. M. Gel'fand and V. S. Retakh, Theory of noncommutative determinants, and characteristic functions of graphs (in Russian), *Functional. Anal. i Prilozhen.* **26** (1992), No. 4, 1–20, 96; translation in *Funct. Anal. Appl.* **26** (1992), No. 4, 231–246.
102. I. M. Gel'fand, M. I. Graev and V. S. Retakh, General hypergeometric systems of equations and series of hypergeometric type (in Russian), *Usp. Mat. Nauk* **47** (1992), No. 4(286), 3–82; 235; translation in *Russ. Math. Surv.* **47** (1992), No. 4, 1–88.
103. I. M. Gel'fand, M. I. Graev and V. S. Retakh, Difference analogues and $q$-analogues of general hypergeometric systems of differential equations (in Russian), *Dokl. Akad. Nauk* **325** (1992), No. 2, 215–220; translation in *Russ. Acad. Sci. Dokl. Math.* **46** (1993), No. 1, 30–35.
104. I. M. Gel'fand, M. M. Kapranov and A. V. Zelevinsky, Hyperdeterminants, *Adv. Math.* **96** (1992), No. 2, 226–263.

105. I. M. Gel'fand, M. I. Graev and V. S. Retakh, Generalized hypergeometric functions connected with an arbitrary finite or a locally compact continuous field (in Russian), *Dokl. Akad. Nauk* **323** (1992), No. 3, 394–397; translation in *Russ. Acad. Sci. Dokl. Math.* **45** (1992), No. 2, 343–347.
106. I. M. Gel'fand and O. Mathieu, On the cohomology of the Lie algebra of Hamiltonian vector fields, *J. Funct. Anal.* **108** (1992), No. 2, 347–360.
107. I. M. Gel'fand and B. L. Tsygan, On the localization of topological invariants, *Comm. Math. Phys.* **146** (1992), No. 1, 73–90.
108. I. M. Gel'fand and R. D. MacPherson, A combinatorial formula for the Pontrjagin classes, *Bull. Amer. Math. Soc. (N.S.)* **26** (1992), No. 2, 304–309.
109. I. M. Gel'fand, The works of L. V. Kantorovich (in Russian), On the eightieth anniversary of the birth of Academician L. V. Kantorovich (1912–1986), Part I. *Optimizatsiya* No. **50**(67) (1991), 131–134.
110. I. M. Gel'fand, Two archetypes in the psychology of Man, *Nonlinear Sci. Today* **1** (1991), No. 4, 11–16.
111. I. M. Gel'fand, M. M. Kapranov and A. V. Zelevinsky, Hypergeometric functions, toric varieties and Newton polyhedra (Okayama, 1990), pp. 104–121, *ICM-90 Satell. Conf. Proc.* (Springer, 1991).
112. I. M. Gel'fand, M. I. Graev and V. S. Retakh, Hypergeometric functions on the *k*th exterior power of the space $C_n$ and the Grassmannian $G_{k,n}$ and the connection between them (in Russian), *Dokl. Akad. Nauk SSSR* **320** (1991), No. 1, 20–24; translation in *Sov. Math. Dokl.* **44** (1992), No. 2, 375–380.
113. S. Gindikin and I. M. Gel'fand, *Geometry and Physics*, pp. ix–xii (Pitagora, 1991).
114. I. M. Gel'fand and V. S. Retakh, Determinants of matrices over non-commutative rings (in Russian), *Functional. Anal. i Prilozhen.* **25** (1991), No. 2, 13–25, 96; translation in *Funct. Anal. Appl.* **25** (1991), No. 2, 91–102.
115. I. M. Gel'fand, M. I. Graev and V. S. Retakh, Reduction formulas for hypergeometric functions on the Grassmannian $G_{k,n}$ and the description of hypergeometric functions on strata of small codimensions (in Russian), *Dokl. Akad. Nauk SSSR* **318** (1991), No. 4, 793–797; translation in *Sov. Math. Dokl.* **43** (1991), No. 3, 757–761.
116. K. V. Brushlinskiĭ, M. M. Gavrilov and I. M. Gel'fand *et al.*, Oleg Vyacheslavovich Lokutsievskiĭ (in Russian), *Usp. Mat. Nauk* **46** (1991), No. 2(278), 203–204; translation in *Russ. Math. Surv.* **46** (1991), No. 2, 233–235.
117. I. M. Gel'fand and I. W. Zakharevich, Veronese curves, and bi-Hamiltonian systems, *J. Funct. Anal.* **99** (1991), No. 1, 150–178.

118. I. M. Gel'fand and M. I. Graev, The Crofton function and inversion formulas in real integral geometry (in Russian), *Functional. Anal. i Prilozhen.* **25** (1991), No. 1, 1–6; translation in *Funct. Anal. Appl.* **25** (1991), No. 1, 1–5.
119. I. M. Gel'fand and D. B. Fairlie, The algebra of Weyl symmetrised polynomials and its quantum extension, *Comm. Math. Phys.* **136** (1991), No. 3, 487–499.
120. I. M. Gel'fand, M. I. Graev and V. S. Retakh, Hypergeometric functions on strata of small codimensions in $G_{k,n}$ (in Russian), Akad. Nauk SSSR Inst. Prikl. Mat. Preprint 1990, No. 138.
121. I. M. Gel'fand and S. G. Gindikin, Integral geometry and tomography (in Russian), Voprosy Kibernet. (Moscow) No. 157 (1990).
122. I. M. Gel'fand and M. I. Graev, Principal representations of the group $U(\infty)$, in *Representation of Lie Groups and Related Topics*, pp. 119–153, Adv. Stud. Contemp. Math., Vol. 7 (Gordon and Breach, 1990).
123. *Mathematical Problems of Tomography*. Edited by I. M. Gel'fand and S. G. Gindikin. Translated from Russian by S. Gelfand. Translation edited by A. Sossinsky [A. B. Sosinskiĭ]. Translations of Mathematical Monographs, Vol. 81 (Amer. Math. Soc., 1990).
124. I. M. Gel'fand, M. I. Graev and V. S. Retakh, Γ-series and general hypergeometric functions on the manifold of $k \times n$ matrices (in Russian), Akad. Nauk SSSR Inst. Prikl. Mat. Preprint 1990, No. 64.
125. I. M. Gel'fand, M. M. Kapranov and A. V. Zelevinsky, Generalized Euler integrals and $A$-hypergeometric functions, *Adv. Math.* **84** (1990), No. 2, 255–271.
126. I. M. Gel'fand, M. M. Kapranov and A. V. Zelevinsky, Newton polytopes of the classical resultant and discriminant, *Adv. Math.* **84** (1990), No. 2, 237–254.
127. I. M. Gel'fand, A. V. Zelevinskiĭ and M. M. Kapranov, Discriminants of polynomials in several variables and triangulations of Newton polyhedra (in Russian), *Algebra i Analiz* **2** (1990), No. 3, 1–62; translation in *Leningrad Math. J.* **2** (1991), No. 3, 449–505.
128. I. M. Gel'fand, M. I. Graev and I. I. Pyatezki-Shapiro, Representation theory and automorphic functions. Translated from the Russian by K. A. Hirsch. Reprint of the 1969 edition. *Generalized Functions*, Vol. 6 (Academic Press, 1990).
129. I. M. Gel'fand, A. V. Zelevinskiĭ and M. M. Kapranov, Discriminants of polynomials in several variables (in Russian), *Functional. Anal. i Prilozhen.* **24** (1990), No. 1, 1–4; translation in *Funct. Anal. Appl.* **24** (1990), No. 1, 1–4.

130. I. M. Gel'fand, Yu. L. Daletskiĭ and B. L. Tsygan, On a variant of noncommutative differential geometry (in Russian), *Dokl. Akad. Nauk SSSR* **308** (1989), No. 6, 1293–1297; translation in *Sov. Math. Dokl.* **40** (1990), No. 2, 422–426.
131. I. M. Gel'fand, *Lectures on Linear Algebra*, with the collaboration of Z. Ya. Shapiro. Translated from the second Russian edition by A. Shenitzer. Reprint of the 1961 translation. Dover Books on Advanced Mathematics (Dover, 1989).
132. I. M. Gel'fand, A. V. Zelevinskiĭ and M. M. Kapranov, Newton polyhedra of principal $A$-determinants (in Russian), *Dokl. Akad. Nauk SSSR* **308** (1989), No. 1, 20–23; translation in *Sov. Math. Dokl.* **40** (1990), No. 2, 278–281.
133. I. M. Gel'fand, A. V. Zelevinskiĭ and M. M. Kapranov, $A$-discriminants and Cayley–Koszul complexes (in Russian), *Dokl. Akad. Nauk SSSR* **307** (1989), No. 6, 1307–1311; translation in *Sov. Math. Dokl.* **40** (1990), No. 1, 239–243.
134. I. M. Gel'fand and G. L. Rybnikov, Algebraic and topological invariants of oriented matroids (in Russian), *Dokl. Akad. Nauk SSSR* **307** (1989), No. 4, 791–795; translation in *Sov. Math. Dokl.* **40** (1990), No. 1, 148–152.
135. I. M. Gel'fand, A. V. Zelevinskiĭ and M. M. Kapranov, Hypergeometric functions and toric varieties (in Russian), *Functional. Anal. i Prilozhen.* **23** (1989), No. 2, 12–26; translation in *Funct. Anal. Appl.* **23** (1989), No. 2, 94–106.
136. I. M. Gel'fand and I. S. Zakharevich, Spectral theory of a pencil of third-order skew-symmetric differential operators on $S$ (in Russian), *Functional. Anal. i Prilozhen.* **23** (1989), No. 2, 1–11; translation in *Funct. Anal. Appl.* **23** (1989), No. 2, 85–93.
137. I. M. Gel'fand, A. V. Zelevinskiĭ and M. M. Kapranov, Projective-dual varieties and hyperdeterminants (in Russian), *Dokl. Akad. Nauk SSSR* **305** (1989), No. 6, 1294–1298; translation in *Sov. Math. Dokl.* **39** (1989), No. 2, 385–389.
138. I. M. Gel'fand, *Collected Papers*. Vol. III. Edited by S. G. Gindikin, V. W. Guillemin, A. A. Kirillov, B. Kostant and S. Sternberg. With a foreword by Gindikin. With a contribution by Kostant (Springer-Verlag, 1989).
139. I. M. Gel'fand, A. V. Zelevinskiĭ and V. V. Serganova, General hypergeometric functions that are connected with a pair of homogeneous spaces (in Russian), *Dokl. Akad. Nauk SSSR* **304** (1989), No. 5, 1044–1049; translation in *Sov. Math. Dokl.* **39** (1989), No. 1, 182–187.

140. I. M. Gel'fand and M. I. Graev, Hypergeometric functions related to the Grassmannian $G_{3,6}$ (in Russian), *Mat. Sb.* **180** (1989), No. 1, 3–38; 142; translation in *Math. USSR-Sb.* **66** (1990), No. 1, 1–40.
141. I. M. Gel'fand, *Collected Papers*. Vol. II. Edited by S. G. Gindikin, V. W. Guillemin, A. A. Kirillov, B. Kostant and S. Sternberg. With a foreword by Kirillov. With contributions by G. Segal and C. M. Ringel (Springer-Verlag, 1988).
142. V. S. Buslaev, I. M. Gel'fand, O. A. Ladyzhenskaya, B. S. Pavlov, M. Z. Solomyak, L. D. Faddeev and D. R. Yafaev, Mikhail Shlëmovich Birman (on the occasion of his sixtieth birthday) (in Russian), *Usp. Mat. Nauk* **43** (1988), No. 3(261), 201–202; translation in *Russ. Math. Surv.* **43** (1988), No. 3, 233–235.
143. I. M. Gel'fand, A. V. Zelevinskiĭ and M. M. Kapranov, Equations of hypergeometric type and Newton polyhedra (in Russian), *Dokl. Akad. Nauk SSSR* **300** (1988), No. 3, 529–534; translation in *Sov. Math. Dokl.* **37** (1988), No. 3, 678–682.
144. A. L. Afendikov, L. R. Volevich, G. P. Voskresenskiĭ, I. M. Gel'fand, A. V. Zabrodin, O. V. Lokutsievskiĭ, O. A Oleĭnik, V. M. Tikhomirov and N. N. Chentsov, Konstantin Ivanovich Babenko (in Russian), *Usp. Mat. Nauk* **43** (1988), No. 2(260), 115–123; translation in *Russ. Math. Surv.* **43** (1988), No. 2, 139–151.
145. I. M. Gel'fand, V. S. Retakh and V. V. Serganova, Generalized Airey functions, Schubert cells and Jordan groups (in Russian), *Dokl. Akad. Nauk SSSR* **298** (1988), No. 1, 17–21; translation in *Sov. Math. Dokl.* **37** (1988), No. 1, 8–12.
146. I. M. Gel'fand and M. I. Graev, Strata in $G_{3,6}$ and the hypergeometric functions connected with them (in Russian), Akad. Nauk SSSR Inst. Prikl. Mat. Preprint 1987, No. 127.
147. I. M. Gel'fand, *Collected Papers*. Vol. I. With remarks by V. W. Guillemin and S. Sternberg (Springer, 1987).
148. I. M. Gel'fand and M. I. Graev, General hypergeometric functions on the Grassmannian $G_{3,6}$ (in Russian), Akad. Nauk SSSR Inst. Prikl. Mat. Preprint 1987, No. 123.
149. T. V. Alekseevskaya, I. M. Gel'fand and A. V. Zelevinskiĭ, Distribution of real hyperplanes and the partition function connected with it (in Russian), *Dokl. Akad. Nauk SSSR* **297** (1987), No. 6, 1289–1293; translation in *Sov. Math. Dokl.* **36** (1988), No. 3, 589–593.
150. A. N. Varchenko and I. M. Gel'fand, Heaviside functions of a configuration of hyperplanes (in Russian), *Functional. Anal. i Prilozhen.* **21** (1987), No. 4, 1–18, 96.

151. Yu. M. Berezanskiĭ, I. M. Gel'fand, M. G. Kreĭn, S. G. Kreĭn, Yu. A. Mitropol'skiĭ and A. V. Skorokhod, Yuriĭ L'vovich Daletskiĭ (on the occasion of his sixtieth birthday) (in Russian), *Usp. Mat. Nauk* **42** (1987), No. 4(256), 213–214.
152. I. M. Gel'fand, M. G. Kreĭn, V. A. Marchenko, N. K. Nikol'skiĭ and I. V. Ostrovskiĭ, Boris Yakovlevich Levin (on the occasion of his eightieth birthday) (in Russian), *Usp. Mat. Nauk* **42** (1987), No. 4(256), 207–210.
153. I. M. Gel'fand, M. I. Graev and A. V. Zelevinskiĭ, Holonomic systems of equations and series of hypergeometric type (in Russian), *Dokl. Akad. Nauk SSSR* **295** (1987), No. 1, 14–19; translation in *Sov. Math. Dokl.* **36** (1988), No. 1, 5–10.
154. I. M. Gel'fand and V. V. Serganova, Combinatorial geometries and the strata of a torus on homogeneous compact manifolds (in Russian), *Usp. Mat. Nauk* **42** (1987), No. 2(254), 107–134; 287.
155. V. A. Vasilev, I. M. Gel'fand and A. V. Zelevinskiĭ, General hypergeometric functions on complex Grassmannians (in Russian), *Functional. Anal. i Prilozhen.* **21** (1987), No. 1, 23–38.
156. I. M. Gel'fand, B. V. Lidskiĭ and V. A. Ponomarev, Preprojective reduction of the free modular lattice (in Russian), *Dokl. Akad. Nauk SSSR* **293** (1987), No. 3, 524–528.
157. I. M. Gel'fand and M. I. Graev, Hypergeometric functions connected with the Grassmannian $G_{3,6}$ (in Russian), *Dokl. Akad. Nauk SSSR* **293** (1987), No. 2, 288–293.
158. I. M. Gel'fand, R. M. Goresky, R. D. MacPherson and V. V. Serganova, Combinatorial geometries, convex polyhedra, and Schubert cells, *Adv. Math.* **63** (1987), No. 3, 301–316.
159. I. M. Gel'fand and V. V. Serganova, Strata of the maximal torus in a compact homogeneous space (in Russian), *Dokl. Akad. Nauk SSSR* **292** (1987), No. 3, 524–528.
160. I. M. Gel'fand and V. V. Serganova, On the general definition of a matroid and a greedoid (in Russian), *Dokl. Akad. Nauk SSSR* **292** (1987), No. 1, 15–20.
161. I. M. Gel'fand and A. V. Zelevinskiĭ, A canonical basis in irreducible representations of $\mathrm{gl}_3$ and its applications (in Russian), in *Group-Theoretic Methods in Physics*, Vol. 2 (in Russian) (Jūrmala, 1985), pp. 31–45 (Nauka, 1986).
162. I. M. Gel'fand and A. V. Zelevinskiĭ, Multiplicities and regular bases for $\mathrm{gl}_n$ (in Russian), *Group-Theoretic Methods in Physics*, Vol. 2 (in Russian) (Jūrmala, 1985), pp. 22–31 (Nauka, 1986).

163. I. M. Gel'fand and A. Zelevinsky, Multiplicities and proper bases for $gl_n$, *Group Theoretical Methods in Physics*, Vol. II (Yurmala, 1985), pp. 147–159 (VNU Sci. Press, 1986).
164. I. M. Gel'fand and A. Zelevinsky, Canonical basis in irreducible representations of $gl_3$ and its applications, in *Group Theoretical Methods in Physics*, Vol. II (Yurmala, 1985), pp. 127–146 (VNU Sci. Press, 1986).
165. I. M. Gel'fand and M. I. Graev, Description of all inversion formulas in a problem of integral geometry connected with the Grassmannian $G_{k,n}$ (in Russian), Akad. Nauk SSSR Inst. Prikl. Mat. Preprint 1986, No. 102.
166. I. M. Gel'fand, V. V. Minakhin and V. N. Shander, Integration in supermanifolds and the Radon supertransform (in Russian), *Functional. Anal. i Prilozhen.* **20** (1986), No. 4, 67–69.
167. I. G. Petrovskiĭ, Izbrannye trudy (in Russian) [Selected works] Sistemy uravneniĭ s chastnymi proizvodnymi. Algebraicheskaya geometriya. [*Systems of Partial Differential Equations. Algebraic Geometry*] Edited and with a preface by V. I. Arnol'd, N. N. Bogolyubov, A. N. Kolmogorov, O. A. Oleĭnik, S. L. Sobolev and A. N. Tikhonov. Compiled by Oleĭnik. With commentaries by A. N. Kolmogorov, L. R. Volevich, V. Ya. Ivriĭ, I. M. Gel'fand, G. E. Shilov, O. V. P. Palamodov, A. M. Gabrièlov and V. M. Kharlamov (Nauka, 1986).
168. I. M. Gel'fand and A. V. Zelevinskiĭ, Algebraic and combinatorial aspects of the general theory of hypergeometric functions (in Russian), *Functional. Anal. i Prilozhen.* **20** (1986), No. 3, 17–34; 96.
169. I. M. Gel'fand and A. B. Goncharov, Reconstruction of a compactly supported function from its integrals on lines intersecting a set of points in a space (in Russian), *Dokl. Akad. Nauk SSSR* **290** (1986), No. 5, 1037–1040.
170. V. A. Vasilev, I. M. Gel'fand and A. V. Zelevinskiĭ, The behavior of general hypergeometric functions in a complex domain (in Russian), *Dokl. Akad. Nauk SSSR* **290** (1986), No. 2, 277–281.
171. I. M. Gel'fand and A. B. Goncharov, A characterization of Grassmann manifolds (in Russian), *Dokl. Akad. Nauk SSSR* **289** (1986), No. 5, 1047–1052.
172. I. M. Gel'fand and M. I. Graev, A duality theorem for general hypergeometric functions (in Russian), *Dokl. Akad. Nauk SSSR* **289** (1986), No. 1, 19–23.
173. I. M. Gel'fand and S. I. Gelfand, Generalized hypergeometric equations (in Russian), *Dokl. Akad. Nauk SSSR* **288** (1986), No. 2, 279–283.
174. I. M. Gel'fand, General theory of hypergeometric functions (in Russian), *Dokl. Akad. Nauk SSSR* **288** (1986), No. 1, 14–18.

175. I. M. Gel'fand, M. I. Graev and A. M. Vershik, The commutative model of the principal representation of the current group SL(2, $R$)$X$ with respect to a unipotent subgroup, in *Group Theoretical Methods in Physics*, Vols. 1–3 (Zvenigorod, 1982), pp. 3–22 (Harwood Academic Publ., 1985).
176. I. M. Gel'fand and A. V. Zelevinsky, Representation models for classical groups and their higher symmetries, in *The Mathematical Heritage of Élie Cartan* (Lyon, 1984). Astérisque 1985, Numero Hors Serie, pp. 117–128.
177. I. M. Gel'fand and M. I. Graev, Some families of irreducible unitary representations of the group $U(\infty)$ (in Russian), Akad. Nauk SSSR Inst. Prikl. Mat. Preprint 1985, No. 51.
178. I. M. Gel'fand, M. I. Graev and A. M. Vershik, Models of representations of current groups, in *Representations of Lie Groups and Lie Algebras* (Budapest, 1971), pp. 121–179 (Akad. Kiadó, 1985).
179. Sh. A. Guberman, I. M. Gel'fand, G. I. Dzyuba and L. V. Kuznetsov, Use of the gestalt approach in distinguishing objects in an image in a problem of automated analysis of chest X-rays (in Russian), Medical diagnostic and prognostic problems from the mathematician's viewpoint. Voprosy Kibernet. (Moscow) No. 112 (1985), 148–172.
180. I. M. Gel'fand, Sh. A. Guberman, S. G. Gindikin, M. L. Izvekova, È. I. Kandel', M. Yu. Melikova, B. I. Rozenfel'd, M. N. Starkova, A. L. Syrkin and N. M. Chebotareva, Some problems of classification and prognosis from different fields of medicine (in Russian), Medical diagnostic and prognostic problems from the mathematician's viewpoint, Voprosy Kibernet. (Moscow) No. 112 (1985), 65–127.
181. I. M. Gel'fand, B. I. Rozenfeld and M. A. Shifrin, Structural organization of data in problems of medical diagnosis and prognosis (in Russian), Medical diagnostic and prognostic problems from the mathematician's viewpoint, Voprosy Kibernet. (Moscow) No. 112 (1985), 5–64.
182. Medical diagnostic and prognostic problems from the mathematician's viewpoint (in Russian), Edited by I. M. Gel'fand. Voprosy Kibernet. (Moscow) No. 112 (1985). Akad. Nauk SSSR, Nauchn. Sovet po Kompleksnoĭ Probleme "Kibernetika", Moscow, 1985, pp. 1–195.
183. V. I. Arnold, M. I. Vishik, I. M. Gel'fand, Yu. V. Egorov, A. S. Kalashnikov, A. N. Kolmogorov, S. P. Novikov and S. L. Sobolev, Olga Arsenevna Oleĭnik (on the occasion of her sixtieth birthday) (in Russian), *Usp. Mat. Nauk* **40** (1985), No. 5(245), 279–293.
184. I. M. Gel'fand and A. V. Zelevinskiĭ, Polyhedra in a space of diagrams and the canonical basis in irreducible representations of $gl_3$ (in Russian), *Functional. Anal. i Prilozhen.* **19** (1985), No. 2, 72–75.

185. R. G. Aĭrapetyan, I. M. Gel'fand, M. I. Graev and G. R. Oganesyan, The Plancherel theorem for an integral transform connected with a pair of Grassmann manifolds (in Russian), *Izv. Akad. Nauk Armyan. SSR Ser. Mat.* **19** (1984), No. 6, 467–483; 489.
186. I. M. Gel'fand and Yu. L. Daletskiĭ, Some formal differential structures connected with Lie superalgebras (in Russian), Akad. Nauk SSSR Inst. Prikl. Mat. Preprint 1984, No. 85.
187. I. M. Gel'fand and A. V. Zelevinskiĭ, Models of representations of classical groups and their hidden symmetries (in Russian), Akad. Nauk SSSR Inst. Prikl. Mat. Preprint 1984, No. 71.
188. I. M. Gel'fand, B. Ya. Levin, V. A. Marchenko, A. V. Pogorelov and S. L. Sobolev, Mikhail Iosifovich Kadets (on the occasion of his sixtieth birthday) (in Russian), *Usp. Mat. Nauk* **39** (1984), No. 6(240), 249–250.
189. I. M. Gel'fand, M. I. Graev and R. Roşu, The problem of integral geometry and intertwining operators for a pair of real Grassmannian manifolds, *J. Operator Theory* **12** (1984), No. 2, 359–383.
190. I. M. Gel'fand and A. V. Zelevinskiĭ, Models of representations of classical groups and their hidden symmetries (in Russian), *Functional. Anal. i Prilozhen.* **18** (1984), No. 3, 14–31.
191. I. M. Gel'fand, M. I. Graev and R. Roşu, The problem of integral geometry for $p$-dimensional planes in real projective space (the nonlocal variant), in *Operator Algebras and Group Representation*s, Vol. I (Neptun, 1980), pp. 192–207, Monogr. Stud. Math., Vol. 17 (Pitman, 1984).
192. R. G. Aĭrapetyan, I. M. Gel'fand, M. I. Graev and G. R. Oganesyan, The Plancherel theorem for an integral transformation related to a complex of $p$-dimensional planes in $CP_n$ and $C_n$ (in Russian), *Izv. Akad. Nauk Armyan. SSR Ser. Mat.* **18** (1983), No. 4, 271–282.
193. A. M. Vershik, I. M. Gel'fand and M. I. Graev, Commutative model of the representation of the group of flows connected with a unipotent subgroup (in Russian), *Functional. Anal. i Prilozhen.* **17** (1983), No. 2, 70–72.
194. I. M. Gel'fand and G. S. Smelëv, Geometric structures of double fibrations and their connection with certain problems of integral geometry (in Russian), *Functional. Anal. i Prilozhen.* **17** (1983), No. 2, 7–22.
195. I. M. Gel'fand and I. V. Cherednik, Abstract Hamiltonian formalism for classical Yang–Baxter sheaves (in Russian), *Usp. Mat. Nauk* **38** (1983), No. 3(231), 3–21.
196. R. G. Aĭrapetyan, I. M. Gel'fand, M. I. Graev and G. R. Oganesyan, Plancherel's theorem for the integral transformation connected with a complex of $p$-dimensional planes in $CP_n$ (in Russian), *Dokl. Akad. Nauk SSSR* **268** (1983), No. 2, 265–268.

197. I. M. Gel'fand and M. I. Graev, Integral transformations connected with two remarkable complexes in a projective space (in Russian), Akad. Nauk SSSR Inst. Prikl. Mat. Preprint 1982, No. 93.
198. I. M. Gel'fand and I. Ya. Dorfman, Integrable equations of KdV-H. Dym type (in Russian), *Current Problems in Mathematical Physics and Numerical Mathematics*, pp. 102–112; 336 (Nauka, 1982).
199. I. M. Gel'fand and I. V. Cherednik, Abstract Hamiltonian formalism for classical Yang–Baxter bundles (in Russian), Akad. Nauk SSSR Inst. Prikl. Mat. Preprint 1982, No. 208.
200. A. M. Vershik, I. M. Gel'fand and M. I. Graev, Commutative model of basic representation of the group with respect to a unipotent subgroup (in Russian), Akad. Nauk SSSR Inst. Prikl. Mat. Preprint 1982, No. 169.
201. I. M. Gel'fand and I. Ya. Dorfman, Hamiltonian operators and the classical Yang–Baxter equation (in Russian), *Functional. Anal. i Prilozhen.* **16** (1982), No. 4, 1–9; 96.
202. I. M. Gel'fand, M. I. Graev and R. Roshu, Nonlocal inversion formulas in a problem of integral geometry, connected with $p$-dimensional planes in real projective space (in Russian), *Functional. Anal. i Prilozhen.* **16** (1982), No. 3, 49–51.
203. G. Segal, An introduction to the paper: "Schubert cells, and the cohomology of the spaces" [*Usp. Mat. Nauk* **28** (1973), No. 3(171), 3–26] by J. N. Bernstein, I. M. Gel'fand and S. I. Gelfand, in *Representation Theory*, pp. 111–114, London Math. Soc. Lecture Note Ser., Vol. 69 (Cambridge Univ. Press, 1982).
204. M. S. Agranovich, I. M. Gel'fand, Yu. A. Dubinskiĭ, O. A. Oleĭnik, S. L. Sobolev and M. A. Shubin, Mark Iosifovich Vishik (on the occasion of his sixtieth birthday) (in Russian), *Usp. Mat. Nauk* **37** (1982), No. 4(226), 213–220.
205. I. M. Gel'fand and R. D. MacPherson, Geometry in Grassmannians and a generalization of the dilogarithm, *Adv. Math.* **44** (1982), No. 3, 279–312.
206. J. M. Vasiliev and I. M. Gel'fand, Possible common mechanism of morphological and growth-related alterations accompanying neoplastic transformation, *Proc. Nat. Acad. Sci. U.S.A.* **79** (1982), No. 8, 2594–2597.
207. R. G. Aĭrapetyan, I. M. Gel'fand, M. I. Graev and G. R. Oganesyan, Plancherel's formula for an integral transformation connected with a complex of straight lines intersecting an algebraic curve in $C_3$ and $CP_3$ (in Russian), *Akad. Nauk Armyan. SSR Dokl.* **75** (1982), No. 1, 9–15.
208. I. M. Gel'fand and V. A. Ponomarev, Gabriel's theorem is also valid for representations of equipped graphs by relations (in Russian), *Functional. Anal. i Prilozhen.* **15** (1981), No. 2, 71–72.

209. N. N. Bogolyubov, I. M. Gel'fand, R. L. Dobrushin, A. A. Kirillov, M. G. Kreĭn, D. A. Leĭtes, R. A. Minlos, Ya. G. Sinai and M. A. Shubin, Feliks Aleksandrovich Berezin. Obituary (in Russian), *Usp. Mat. Nauk* **36** (1981), No. 4(220), 185–190.
210. I. M. Gel'fand and I. Ya. Dorfman, Hamiltonian operators and infinite-dimensional Lie algebras (in Russian), *Functional. Anal. i Prilozhen.* **15** (1981), No. 3, 23–40.
211. I. M. Gel'fand, M. I. Graev and A. M. Veršik, Representations of the group of functions taking values in a compact Lie group, *Compositio Math.* **42** (1980/81), No. 2, 217–243.
212. I. M. Gel'fand and V. A. Ponomarev, Representations of graphs. Perfect subrepresentations (in Russian), *Functional. Anal. i Prilozhen.* **14** (1980), No. 3, 14–31; 96.
213. I. M. Gel'fand, S. G. Gindikin and M. I. Graev, Integral geometry in affine and projective spaces (in Russian), *Current Problems in Mathematics*, Vol. 16 (Russian), pp. 53–226; 228 (Akad. Nauk SSSR, 1980).
214. I. M. Gel'fand, S. G. Gindikin and M. I. Graev, Integral geometry for certain one-dimensional fibrations over projective space (in Russian), Akad. Nauk SSSR Inst. Prikl. Mat. Preprint 1980, No. 24.
215. I. M. Gel'fand, S. G. Gindikin and M. I. Graev, Integral geometry for one-dimensional fibrations of a general form over $RP_n$ (in Russian), Akad. Nauk SSSR Inst. Prikl. Mat. Preprint 1980, No. 60.
216. I. M. Gel'fand and M. I. Graev, Admissible $n$-dimensional complexes of curves in $R_n$ (in Russian), *Functional. Anal. i Prilozhen.* **14** (1980), No. 4, 36–44; 95.
217. I. M. Gel'fand and I. Ya. Dorfman, Schouten bracket and Hamiltonian operators (in Russian), *Functional. Anal. i Prilozhen.* **14** (1980), No. 3, 71–74.
218. V. Dlab, Structure des treillis linéaires libres (d'après I. M. Gel'fand et V. A. Ponomarev) (in French), Séminaire d'Algèbre Paul Dubreil et Marie-Paule Malliavin, 32ème année (Paris, 1979), pp. 10–34, Lecture Notes in Math., Vol. 795 (Springer, 1980).
219. I. M. Gel'fand, S. G. Gindikin and M. I. Graev, Problems of integral geometry in an affine space connected with integration of functions along planes (in Russian), Akad. Nauk SSSR Inst. Prikl. Mat. Preprint No. 152 (1979).
220. I. M. Gel'fand, S. G. Gindikin and M. I. Graev, A problem of integral geometry in $RP_n$, connected with integration of differential forms (in Russian), *Functional. Anal. i Prilozhen.* **13** (1979), No. 4, 64–66.

221. I. M. Gel'fand, S. G. Gindikin and M. I. Graev, Some questions connected with the integration of differential forms along planes in $RP_n$ and $CP_n$ (in Russian), Akad. Nauk SSSR Inst. Prikl. Mat. Preprint No. 126 (1979).
222. I. M. Gel'fand, S, G, Gindikin and M. I. Graev, Double fibrations and problems of integral geometry connected with the integration of differential forms over lines (in Russian), Akad. Nauk SSSR Inst. Prikl. Mat. Preprint No. 60 (1979).
223. A. M. Veršik, I. M. Gel'fand and M. I. Graev, Remarks on the representations of a group of functions with values in a compact Lie group (in Russian), Akad. Nauk SSSR Inst. Prikl. Mat. Preprint No. 17 (1979).
224. I. M. Gel'fand, S. G. Gindikin and M. I. Graev, Problems of integral geometry in projective space that are connected with the integration of differential forms (in Russian), Institute of Applied Mathematics Akad. Nauk SSSR. Preprint No. 79 (1979).
225. I. M. Gel'fand, S. G. Gindikin and M. I. Graev, Problems of integral geometry connected with integration of differential forms over lines in a three-dimensional projective space (in Russian), Institute of Applied Mathematics Akad. Nauk SSSR. Preprint No. 41 (1979).
226. I. M. Gel'fand, S. G. Gindikin and M. I. Graev, Problems of integral geometry connected with the integration of differential forms over straight lines in $R_3$ and $C_3$ (in Russian), Institute of Applied Mathematics Akad. Nauk SSSR. Preprint No. 24 (1979).
227. I. M. Gel'fand and V. A. Ponomarev, Model algebras and representations of graphs (in Russian), *Functional. Anal. i Prilozhen.* **13** (1979), No. 3, 1–12.
228. I. M. Gel'fand and I. Ya. Dorfman, Hamiltonian operators and algebraic structures associated with them (in Russian), *Functional. Anal. i Prilozhen.* **13** (1979), No. 4, 13–30.
229. I. M. Gel'fand and I. Ya. Dorfman, Hamiltonian operators and related algebraic structures (in Russian), Institute of Applied Mathematics Akad. Nauk SSSR. Preprint No. 84 (1979).
230. I. M. Gel'fand, S. G. Gindikin and Z. Ya. Shapiro, A local problem of integral geometry in a space of curves (in Russian), *Functional. Anal. i Prilozhen.* **13** (1979), No. 2, 11–31.
231. I. M. Gel'fand and L. A. Dikiĭ, Integrable nonlinear equations and the Liouville theorem (in Russian), *Functional. Anal. i Prilozhen.* **13** (1979), No. 1, 8–20; 96.
232. I. M. Gel'fand and L. A. Dikiĭ, A family of Hamiltonian structures connected with integrable nonlinear differential equations (in Russian), Institute of Applied Mathematics Akad. Nauk SSSR. Preprint No. 136 (1978).

233. J. Bernstein, I. M. Gel'fand and S. I. Gelfand, Structure locale de la catégorie des modules de Harish-Chandra (in French), *C. R. Acad. Sci. Paris Sér. A-B* **286** (1978), No. 11, A495–A497.
234. I. M. Gel'fand, B. L. Feĭgin and D. B. Fuks, Cohomology of infinite-dimensional Lie algebras and Laplace operators (in Russian), *Functional. Anal. i Prilozhen.* **12** (1978), No. 4, 1–5.
235. I. N. Bernšteĭn, I. M. Gel'fand and S. I. Gelfand, Algebraic vector bundles on and problems of linear algebra (in Russian), *Functional. Anal. i Prilozhen.* **12** (1978), No. 3, 66–67.
236. I. M. Gel'fand and L. A. Dikiĭ, The calculus of jets and nonlinear Hamiltonian systems (in Russian), *Functional. Anal. i Prilozhen.* **12** (1978), No. 2, 8–23; 95.
237. J. Bernstein, I. M. Gel'fand and S. I. Gelfand, Structure locale de la catégorie des modules de Harish-Chandra (in French), *C. R. Acad. Sci. Paris Sér. A-B* **286** (1978), No. 10, A435–A437.
238. I. M. Gel'fand, M. I. Graev and A. M. Veršik, Representations of the group of smooth mappings of a manifold into a compact Lie group, *Compositio Math.* **35** (1977), No. 3, 299–334.
239. I. M. Gel'fand and V. A. Ponomarev, Lattices, representations, and their related algebras. II (in Russian), *Usp. Mat. Nauk* **32** (1977), No. 1(193), 85–106; 271.
240. I. M. Gel'fand and S. G. Gindikin, Complex manifolds whose spanning trees are real semisimple Lie groups, and analytic discrete series of representations (in Russian), *Functional. Anal. i Prilozhen.* **11** (1977), No. 4, 19–27; 96.
241. I. M. Gel'fand and S. G. Gindikin, Nonlocal inversion formulas in real integral geometry (in Russian), *Functional. Anal. i Prilozhen.* **11** (1977), No. 3, 12–19; 96.
242. I. M. Gel'fand and L. A. Dikiĭ, The resolvent, and Hamiltonian systems (in Russian), *Functional. Anal. i Prilozhen.* **11** (1977), No. 2, 11–27; 95.
243. A. M. Veršik, I. M. Gel'fand and M. I. Graev, Representations of the group of smooth mappings of a manifold into a compact Lie group (in Russian), *Dokl. Akad. Nauk SSSR* **232** (1977), No. 4, 745–748.
244. I. M. Gel'fand and V. A. Ponomarev, Lattices, representations, and their related algebras. I (in Russian), *Usp. Mat. Nauk* **31** (1976), No. 5(191), 71–88.
245. I. M. Gel'fand and L. A. Dikiĭ, A Lie algebra structure in the formal calculus of variations (in Russian), *Functional. Anal. i Prilozhen.* **10** (1976), No. 1, 18–25.

246. I. N. Bernšteĭn, I. M. Gel'fand and S. I. Gelfand, Models of representations of Lie groups (in Russian), *Trudy Sem. Petrovsk. Vyp.* **2** (1976), 3–21.
247. I. M. Gel'fand, Ju. I. Manin and M. A. Šubin, Poisson brackets and the kernel of the variational derivative in the formal calculus of variations (in Russian), *Functional. Anal. i Prilozhen.* **10** (1976), No. 4, 30–34.
248. I. M. Gel'fand and L. A. Dikiĭ, Fractional powers of operators, and Hamiltonian systems (in Russian), *Functional. Anal. i Prilozhen.* **10** (1976), No. 4, 13–29.
249. D. B. Fuchs, A. M. Gabrièlov and I. M. Gel'fand, The Gauss–Bonnet theorem and the Atiyah–Patodi–Singer functionals for the characteristic classes of foliations, *Topology* **15** (1976), No. 2, 165–188.
250. A. M. Gabrièlov, I. M. Gel'fand and M. V. Losik, Atiyah–Patodi–Singer functionals for the characteristic classes of a tangent bundle (in Russian), *Functional. Anal. i Prilozhen.* **10** (1976), No. 2, 13–28.
251. P. S. Aleksandrov, I. M. Gel'fand, A. N. Kolmogorov, E. V. Maĭkov, V. P. Maslov, O. A. Oleĭnik, Ya. G. Sinai, O. G. Smoljanov and V. M. Tihomirov, In memory of Sergeĭ Vasilevič Fomin (in Russian), *Usp. Mat. Nauk* **31** (1976), No. 4(190), 199–212.
252. A. M. Gabrièlov, I. M. Gel'fand and M. V. Losik, A local combinatorial formula for the first Pontrjagin class (in Russian), *Functional. Anal. i Prilozhen.* **10** (1976), No. 1, 14–17.
253. P. S. Aleksandrov, I. M. Gel'fand, E. A. Gorin, V. V. Grušin, A. N. Kolmogorov, O. A. Oleĭnik, V. P. Palamodov and S. V. Fomīn, Georgiĭ Evgenevič Shilov. Obituary (in Russian), *Usp. Mat. Nauk* **31** (1976), No. 1(187), 217–228.
254. I. N. Bernšteĭn, I. M. Gel'fand and S. I. Gelfand, A certain category of *g*-modules (in Russian), *Functional. Anal. i Prilozhen.* **10** (1976), No. 2, 1–8.
255. I. N. Bernšteĭn, I. M. Gel'fand and S. I. Gelfand, Differential operators on the base affine space and a study of *g*-modules, in *Lie Groups and Their Representations* (Proc. Summer School, Bolyai János Math. Soc., Budapest, 1971), pp. 21–64 (Halsted Press, 1975).
256. I. M. Gel'fand and L. A. Dikiĭ, Asymptotic properties of the resolvent of Sturm–Liouville equations, and the algebra of Korteweg–de Vries equations (in Russian), *Usp. Mat. Nauk* **30** (1975), No. 5(185), 67–100.
257. I. N. Bernšteĭn, I. M. Gel'fand and S. I. Gelfand, Models of representations of compact Lie groups (in Russian), *Functional. Anal. i Prilozhen.* **9** (1975), No. 4, 61–62.

258. A. M. Gabrièlov, I. M. Gel'fand and M. V. Losik, Combinatorial computation of characteristic classes. I, II (in Russian), *Functional. Anal. i Prilozhen.* **9** (1975), No. 2, 12–28; *ibid.* **9** (1975), No. 3, 5–26.
259. I. M. Gel'fand and D. A. Kajdan, Representations of the group where is a local field, in *Lie Groups and Their Representations* (Proc. Summer School, Bolyai János Math. Soc., Budapest, 1971), pp. 95–118 (Halsted Press, 1975).
260. A. M. Veršik, I. M. Gel'fand and M. I. Graev, Representations of the group of diffeomorphisms (in Russian), *Usp. Mat. Nauk* **30** (1975), No. 6(186), 1–50.
261. Ju. M. Berezanskiĭ, I. M. Gel'fand, B. Ya. Levin, V. A. Marčenko and K. V. Maslov, Aleksandr Jakovlevič Povzner (on the occasion of his sixtieth birthday) (in Russian), *Usp. Mat. Nauk* **30** (1975), No. 5(185), 221–226 (1 plate).
262. I. M. Gel'fand and M. I. Graev, Square roots of a quasiregular representation of the group (in Russian), *Functional. Anal. i Prilozhen.* **9** (1975), No. 2, 64–66.
263. *Lie Groups and Their Representations*, Proceedings of the Summer School on Group Representations of the Bolyai János Mathematical Society (Budapest, August 16–September 3, 1971). Edited by I. M. Gel'fand (Halsted Press, 1975).
264. I. M. Gel'fand and D. B. Fuks, Foliations. II (in Russian), *Functional. Anal. i Prilozhen.* **8** (1974), No. 3, 7–11.
265. I. N. Bernšteĭn, I. M. Gel'fand and S. I. Gelfand, A new model for the representations of finite semisimple algebraic groups (in Russian), *Usp. Mat. Nauk* **29** (1974), No. 3 (177), 185–186.
266. I. M. Gel'fand and V. A. Ponomarev, Free modular lattices, and their representations (in Russian), Collection of articles dedicated to the memory of Ivan Georgievič Petrovskiĭ (1901–1973), IV. *Usp. Mat. Nauk* **29** (1974), No. 6(180), 3–58.
267. A. M. Gabrièlov, I. M. Gel'fand and M. V. Losik, The combinatorial computation of characteristic classes (in Russian), *Functional. Anal. i Prilozhen.* **9** (1974), No. 1, 54–55.
268. I. M. Gel'fand, B. L. Feĭgin and D. B. Fuks, Cohomology of the Lie algebra of formal vector fields with coefficients in its dual space and variations of characteristic classes of foliations (in Russian), *Functional. Anal. i Prilozhen.* **8** (1974), No. 2, 13–29.
269. A. M. Veršik, I. M. Gel'fand and M. I. Graev, Irreducible representations of the group and cohomology (in Russian), *Functional. Anal. i Prilozhen.* **8** (1974), No. 2, 67–69.

270. I. N. Bernšteĭn, I. M. Gel'fand and S. I. Gelfand, Schubert cells, and the cohomology of the spaces *G/P* (in Russian), *Usp. Mat. Nauk* **28** (1973), No. 3(171), 3–26.

271. A. M. Veršik, I. M. Gel'fand and M. I. Graev, Representations of the group SL(2, *R*), where *R* is a ring of functions (in Russian), *Usp. Mat. Nauk* **28** (1973), No. 5(173), 83–128.

272. I. N. Bernšteĭn, I. M. Gel'fand and V. A. Ponomarev, Coxeter functors, and Gabriel's theorem (in Russian), *Usp. Mat. Nauk* **28** (1973), No. 2(170), 19–33.

273. I. M. Gel'fand and D. B. Fuks, Foliations (in Russian), *Functional. Anal. i Prilozhen.* **7** (1973), No. 4, 29–37.

274. S. N. Mergeljan, I. M. Gel'fand and O. A. Oleĭnik, Rafaèl Aramovič Aleksandrjan (on the occasion of his fiftieth birthday) (in Russian), *Differencialnye Uravnenija* **9** (1973).

275. I. N. Bernšteĭn, I. M. Gel'fand and S. I. Gelfand, Schubert cells and the cohomology of a flag space (in Russian), *Functional. Anal. i Prilozhen.* **7** (1973), No. 1, 64–65.

276. I. N. Bernšteĭn, I. M. Gel'fand and S. I. Gelfand, Differential operators on a cubic cone (in Russian), *Usp. Mat. Nauk* **27** (1972), No. 1(163), 185–190.

277. I. M. Gel'fand and V. A. Ponomarev, Problems of linear algebra and classification of quadruples of subspaces in a finite-dimensional vector space, in *Hilbert Space Operators and Operator Algebras* (Proc. Internat. Conf., Tihany, 1970), pp. 163–237. Colloq. Math. Soc. Janos Bolyai, Vol. 5 (North-Holland, 1972).

278. I. M. Gel'fand and D. A. Každan, Representations of the group GL(*n*, *K*) where *K* is a local field (in Russian), *Functional. Anal. i Prilozhen.* **6** (1972), No. 4, 73–74.

279. I. M. Gel'fand, D. I. Kalinin and D. B. Fuks, The cohomology of the Lie algebra of Hamiltonian formal vector fields (in Russian), *Functional. Anal. i Prilozhen.* **6** (1972), No. 3, 25–29.

280. I. M. Gel'fand, D. A. Každan and D. B. Fuks, The actions of infinite-dimensional Lie algebras (in Russian), *Functional. Anal. i Prilozhen.* **6** (1972), No. 1, 10–15.

281. I. M. Gel'fand, The cohomology of infinite dimensional Lie algebras; some questions of integral geometry, in *Actes du Congrès International des Mathématiciens* (Nice, 1970), Tome 1, pp. 95–111 (Gauthier-Villars, 1971).

282. I. M. Gel'fand, Lektsii po lineĭnoĭ algebre (in Russian) [*Lectures on Linear Algebra*] Fourth edition, augmented. Izdat. (Nauka, 1971).

283. I. N. Bernšteĭn, I. M. Gel'fand and S. I. Gelfand, Structure of representations that are generated by vectors of higher weight (in Russian), *Functional. Anal. i Prilozhen.* **5** (1971), No. 1, 1–9.
284. I. M. Gel'fand, A. G. Kostjučenko, B. M. Levitan, Ya. B. Lopatinskiĭ, S. L. Sobolev and G. E. Shilov, Zaid Ismaĭlovič Halilov (on his sixtieth birthday) (in Russian), *Usp. Mat. Nauk* **26** (1971), No. 3(159), 221–224.
285. V. I. Arnold, I. M. Gel'fand, Yu. I. Manin, B. G. Moĭšezon, S. P. Novikov and I. R. Šafarevič, Galina Nikolaevna Tjurina. Obituary (in Russian), *Usp. Mat. Nauk* **26** (1971), No. 1, 207–211.
286. P. S. Aleksandrov, V. I. Arnold, I. M. Gel'fand, A. N. Kolmogorov, S. P. Novikov and O. A. Oleĭnik, Ivan Georgievič Petrovskiĭ (on his seventieth birthday) (in Russian), *Usp. Mat. Nauk* **26** (1971), No. 2, 3–24.
287. I. M. Gel'fand and D. A. Každan, Certain questions of differential geometry and the computation of the cohomologies of the Lie algebras of vector fields (in Russian), *Dokl. Akad. Nauk SSSR* **200** (1971), 269–272.
288. I. M. Gel'fand and V. A. Ponomarev, Quadruples of subspaces of finite-dimensional vector space (in Russian), *Dokl. Akad. Nauk SSSR* **197** (1971),762–765.
289. I. M. Gel'fand and D. B. Fuks, Cohomolgy of Lie algebras of vector fields with nontrivial coefficients (in Russian), *Functional. Anal. i Prilozhen.* **4** (1970), No. 3, 10–25.
290. I. M. Gel'fand and D. B. Fuks, Cycles that represent cohomology classes of the Lie algebra of formal vector fields (in Russian), *Usp. Mat. Nauk* **25** (1970), No. 5(155), 239–240.
291. I. M. Gel'fand and D. B. Fuks, Upper bounds for the cohomology of infinite-dimensional Lie algebras (in Russian), *Functional. Anal. i Prilozhen.* **4** (1970), No. 4, 70–71.
292. I. M. Gel'fand and D. B. Fuks, Cohomologies of the Lie algebra of tangent vector fields of a smooth manifold. II (in Russian), *Functional. Anal. i Prilozhen.* **4** (1970), No. 2, 23–31.
293. I. M. Gel'fand and D. B. Fuks, Cohomologies of the Lie algebra of smooth vector fields (in Russian), *Dokl. Akad. Nauk SSSR* **190** (1970) 1267–1270.
294. I. M. Gel'fand, M. I. Graev and Z. Ya. Shapiro, Integral geometry in a projective space (in Russian), *Functional. Anal. i Prilozhen.* **4** (1970), No. 1, 14–32.
295. I. N. Bernšteĭn, I. M. Gel'fand and S. I. Gelfand, Differential operators on the fundamental affine space (in Russian), *Dokl. Akad. Nauk SSSR* **195** (1970), 1255–1258.

296. I. M. Gel'fand, M. I. Graev and V. A. Ponomarev, A classification of the linear representations of the group SL(2, *C*) (in Russian), *Dokl. Akad. Nauk SSSR* **194** (1970), 1002–1005.
297. I. M. Gel'fand, M. I. Graev and Z. Ya. Shapiro, An integral geometry problem connected with a pair of Grassmann manifolds (in Russian), *Dokl. Akad. Nauk SSSR* **193** (1970), 259–262.
298. I. M. Gel'fand and D. B. Fuks, Cohomologies of the Lie algebra of formal vector fields (in Russian), *Izv. Akad. Nauk SSSR Ser. Mat.* **34** (1970), 322–337.
299. I. M. Guelfand, M. I. Graev and N. Ya. Vilenkin, Les distributions. Tome 5: Géométrie intégrale et théorie des représentations (in French), Traduit du russe par S. Vasilach et G. Gussi, Monographies Universitaires de Mathématiques, Vol. 34 (Dunod, 1970).
300. I. M. Gel'fand and A. S. Miščenko, A. S. Quadratic forms over commutative group rings, and *K*-theory (in Russian), *Functional. Anal. i Prilozhen.* **3** (1969), No. 4, 28–33.
301. I. M. Gel'fand and D. B. Fuks, Cohomologies of the Lie algebra of tangent vector fields of a smooth manifold (in Russian), *Functional. Anal. i Prilozhen.* **3** (1969), No. 3, 32–52.
302. I. M. Gel'fand and V. A. Ponomarev, Remarks on the classification of a pair of commuting linear transformations in a finite-dimensional space (in Russian), *Functional. Anal. i Prilozhen.* **3** (1969), No. 4, 81–82.
303. I. M. Gel'fand and G. E. Shilov, Verallgemeinerte Funktionen (Distributionen). II: Lineare topologische Räume. Räume von Grundfunktionen und verallgemeinerten Funktionen. (German) Übersetzung aus dem Russischen von Gerhard Tesch. Zweite Auflage. Hochschulbücher für Mathematik, Band 48 (VEB Deutscher Verlag der Wissenschaften, 1969).
304. I. M. Gel'fand and D. B. Fuks, Cohomologies of the Lie algebra of vector fields on a manifold (in Russian), *Functional. Anal. i Prilozhen.* **3** (1969), No. 2, 87.
305. I. M. Gel'fand, M. I. Graev and Z. Ya. Shapiro, Differential forms and integral geometry (in Russian), *Functional. Anal. i Prilozhen.* **3** (1969), No. 2, 24–40.
306. I. M. Gel'fand and A. A. Kirillov, Structure of the Lie field connected with a semisimple decomposable Lie algebra (in Russian), *Functional. Anal. i Prilozhen.* **3** (1969), No. 1, 7–26.
307. I. M. Gel'fand, M. I. Graev and I. I. Pyatezki-Shapiro, Representation theory and automorphic functions. Translated from the Russian by K. A. Hirsch (W. B. Saunders Co., Philadelphia, 1969).

308. I. M. Gel'fand and G. E. Shilov, *Generalized Functions*, Vol. 2. Spaces of fundamental and generalized functions. Translated from the Russian by Morris D. Friedman, Amiel Feinstein and Christian P. Peltzer (Academic Press, 1968).
309. I. M. Gel'fand and D. B. Fuks, Cohomologies of the Lie algebra of vector fields on the circle (in Russian), *Functional. Anal. i Prilozhen.* **2** (1968), No. 4, 92–93.
310. I. M. Gel'fand and M. I. Graev, Line complexes in the space $C_n$ (in Russian), *Functional. Anal. i Prilozhen.* **2** (1968), No. 3, 39–52.
311. I. M. Gel'fand and M. I. Graev, Representations of quaternion groups over locally compact fields and function fields (in Russian), *Functional. Anal. i Prilozhen.* **2** (1968), No. 1, 20–35.
312. I. M. Gel'fand and D. B. Fuks, Classifying spaces for principal bundles with Hausdorff bases (in Russian), *Dokl. Akad. Nauk SSSR* **181** (1968). 515–518.
313. I. M. Gel'fand and G. E. Shilov, *Generalized Functions*, Vol. 2. Spaces of fundamental and generalized functions. Translated from the Russian by M. D. Friedman, A. Feinstein and C. P. Peltzer (Academic Press, 1968).
314. I. M. Gel'fand and V. A. Ponomarev, Indecomposable representations of the Lorentz group (in Russian), *Usp. Mat. Nauk* **23** (1968), No. 2(140), 3–60.
315. I. M. Gel'fand and A. A. Kirillov, The structure of the quotient field of the enveloping algebra of a semisimple Lie algebra (in Russian), *Dokl. Akad. Nauk SSSR* **180** (1968), 775–777.
316. I. M. Gel'fand and M. I. Graev, Complexes of $k$-dimensional planes in the space $C_n$ and Plancherel's formula for the group $GL(n, C)$ (in Russian), *Dokl. Akad. Nauk SSSR* **179** (1968), 522–525.
317. I. M. Gel'fand and G. E. Shilov, *Generalized Functions*, Vol. 3. Theory of differential equations. Translated from the Russian by M. E. Mayer (Academic Press, 1967).
318. I. M. Gel'fand and M. I. Graev, Irreducible representations of the Lie algebra of the group $U(p, q)$: Description of the discrete series of representations (in Russian), 1967 High Energy Phys. Theory Elem. Particles (Internat. School Theoret. Phys., Yalta, 1966) (Russian), pp. 216–226.
319. I. M. Gel'fand and D. B. Fuks, The topology of noncompact Lie groups (in Russian), *Functional. Anal. i Prilozhen.* **1** (1967), No. 4, 33–45.
320. I. M. Gel'fand and D. B. Fuks, Topological invariants of noncompact Lie groups connected with infinite dimensional representations (in Russian), *Dokl. Akad. Nauk SSSR* **177** (1967), 763–766.

321. I. M. Gel'fand and D. B. Fuks, Cohomology of Lie groups with real coefficients (in Russian), *Dokl. Akad. Nauk SSSR* **176** (1967), 24–27.
322. I. M. Gel'fand and V. A. Ponomarev, The category of Harish-Chandra modules over the Lie algebra of the Lorentz group (in Russian), *Dokl. Akad. Nauk SSSR* **176** (1967), 243–246.
323. I. M. Gel'fand and M. I. Graev, Representations of the group of quaternions over a disconnected locally compact continuous field (in Russian), *Dokl. Akad. Nauk SSSR* **177** (1967), 17–20.
324. I. M. Gel'fand and V. A. Ponomarev, A classification of the indecomposable infinitesimal representations of the Lorentz group (in Russian), *Dokl. Akad. Nauk SSSR* **176** (1967), 502–505.
325. I. M. Gel'fand and G. E. Shilov, *Generalized Functions*, Vol. 3: Theory of differential equations. Translated from the Russian by M. E. Mayer (Academic Press, 1967).
326. I. M. Gel'fand and N. Y. Vilenkin, *Les Distributions*, Tome 4: Applications de l'analyse harmonique (in French), Traduit du russe par G. Rideau. Collection Universitaire de Mathématiques, No. 23 (Dunod, 1967).
327. I. M. Gel'fand, M. I. Graev and Z. Ya. Shapiro, Integral geometry on *k*-dimensional planes (in Russian), *Functional. Anal. i Prilozhen.* **1** (1967), 15–31.
328. I. M. Gel'fand, M. I. Graev and N. Ya. Vilenkin, *Generalized Functions*, Vol. 5: Integral geometry and representation theory. Translated from the Russian by Eugene Saletan (Academic Press, 1966).
329. I. M. Gel'fand, M. I. Graev and I. I. Pyatezki-Shapiro, Teoriya predstavleniĭ i avtomorfnye funktsii (in Russian) [Theory of representations and automorphic functions] Generalized functions, No. 6 Izdat (Nauka, 1966).
330. I. M. Gel'fand and A. A. Kirillov, Sur les corps liés aux algèbres enveloppantes des algèbres de Lie (in French), Inst. Hautes Études Sci. Publ. Math. No. 31 (1966), 5–19.
331. I. M. Gel'fand, M. I. Graev and N. Ya. Vilenkin, *Generalized Functions*, Vol. 5: Integral geometry and representation theory. Translated from the Russian by Eugene Saletan (Academic Press, 1966).
332. I. M. Gel'fand, Lektsii po lineĭnoĭ algebre (in Russian) [Lectures on linear algebra] Third revised and augmented edition Izdat (Nauka, 1966).
333. I. M. Gel'fand, M. I. Graev and Z. Ya. Shapiro, Integral geometry on manifolds of *k*-dimensional planes (in Russian), *Dokl. Akad. Nauk SSSR* **168** (1966), 1236–1238.
334. I. M. Gel'fand and A. A. Kirillov, On fields connected with the enveloping algebras of Lie algebras (in Russian), *Dokl. Akad. Nauk SSSR* **167** (1966), 503–505.

335. I. M. Gel'fand and M. I. Graev, Finite-dimensional irreducible representations of the unitary and complete linear group and special functions associated with them (in Russian), *Izv. Akad. Nauk SSSR Ser. Mat.* **29** (1965), 1329–1356.
336. I. M. Gel'fand and G. E. Shilov, *Les Distributions*, Tome 3: Théorie des équations différentielles (in French), Traduit par Serge Vasilach (Dunod, 1965).
337. I. M. Gel'fand and N. Ya. Vilenkin, *Generalized Functions*, Vol. 4: Applications of harmonic analysis. Translated from the Russian by Amiel Feinstein (Academic Press, 1964).
338. I. M. Gel'fand and G. E. Shilov, *Generalized Functions*, Vol. 1: Properties and operations. Translated from the Russian by Eugene Saletan (Academic Press, 1964).
339. I. M. Gel'fand, D. Raikov and G. E. Shilov, *Commutative Normed Rings*. Translated from the Russian, with a supplementary chapter (Chelsea, 1964).
340. I. M. Gel'fand and N. J. Wilenkin, *Verallgemeinerte Funktionen* (Distributionen). IV: Einige Anwendungen der harmonischen Analyse. Gel'fandsche Raumtripel. (German) Hochschulbücher für Mathematik, Band 50 (VEB Deutscher Verlag der Wissenschaften, 1964).
341. I. M. Gel'fand and N. Ya. Vilenkin, *Generalized Functions*, Vol. 4: Applications of harmonic analysis. Translated by Amiel Feinstein (Academic Press, 1964).
342. V. I. Bryzgalov, I. M. Gel'fand, I. I. Pyatezki-Shapiro and M. L. Zetlin, Homogeneous automata games and their simulation on digital computers (in Russian), *Avtomat. i Telemeh.* **25** (1964), 1572–1580.
343. I. M. Gel'fand, D. A. Raikow and G. E. Shilov, Kommutative normierte Algebren. (German) Übersetzung und wissenschaftliche Redaktion von Helmut Boseck. Mathematische Forschungsberichte, XIII (VEB Deutscher Verlag der Wissenschaften, 1964).
344. I. M. Gel'fand and G. E. Shilov, *Generalized Functions*, Vol. I: Properties and operations. Translated by Eugene Saletan (Academic Press, 1964).
345. I. M. Gel'fand and G. E. Shilov, *Verallgemeinerte Functionen* (Distributionen). III: Einige Fragen zur Theorie der Differentialgleichungen. (German) Hochschulbücher für Mathematik, Band 49 (VEB Deutscher Verlag der Wissenschaften, 1964).
346. I. M. Gel'fand, M. I. Graev and I. I. Pyatezki-Shapiro, Representations of adèle groups (in Russian), *Dokl. Akad. Nauk SSSR* **156** (1964), 487–490.
347. I. M. Gel'fand and G. E. Shilov, *Les Distributions*, Tome 2: Espaces fondamentaux (in French), Traduit par Serge Vasilach. Collection Universitaire de Mathématiques, XV (Dunod, 1964).

348. I. M. Gel'fand and I. I. Pyatezki-Shapiro, The Laplace operator on the Riemann surfaces and the theory of representation. 1963 Outlines Joint Sympos. Partial Differential Equations (Novosibirsk, 1963), pp. 87–88 Acad. Sci. USSR Siberian Branch, Moscow.
349. I. M. Gel'fand and M. I. Graev, The structure of the ring of finite functions on the group of second-order unimodular matrices with elements belonging to a disconnected locally compact field, *Dokl. Akad. Nauk SSSR* **153** 512–515 (in Russian); translated as *Sov. Math. Dokl.* **4** (1963), 1697–1700.
350. I. M. Gel'fand, I. I. Pyatezki-Shapiro and Yu. G. Fedorov, Determination of crystal structure by non-local search (in Russian), *Dokl. Akad. Nauk SSSR* **152** (1963), 1045–1048.
351. I. M. Gel'fand, Automorphic functions and the theory of representations, in *1963 Proc. Internat. Congr. Mathematicians* (Stockholm, 1962), pp. 74–85 (Inst. Mittag-Leffler, 1963).
352. I. M. Gel'fand and S. V. Fomin, *Calculus of Variations*. Revised English edition translated and edited by Richard A. Silverman (Prentice-Hall, 1963).
353. I. M. Gel'fand and I. I. Pyatezki-Shapiro, Automorphic functions and the theory of representations (in Russian), *Trudy Moskov. Mat. Obšč.* **12** (1963), 389–412.
354. I. M. Gel'fand and G. E. Shilov, Categories of finite-dimensional spaces (in Russian), *Vestnik Moskov. Univ. Ser. I Mat. Meh.* **1963** (1963), No. 4, 27–48.
355. I. M. Gel'fand, I. I. Pyatezki-Shapiro and M. L. Zetlin, On certain classes of games and robot games (in Russian), *Dokl. Akad. Nauk SSSR* **152** (1963), 845–848.
356. I. M. Gel'fand and M. I. Graev, Representations of the group of second-order matrices with elements in a locally compact field and special functions on locally compact fields (in Russian), *Usp. Mat. Nauk* **18** (1963), No. 4(112), 29–99.
357. I. M. Gel'fand, Some problems in the theory of quasilinear equations, *Amer. Math. Soc. Transl.* (2) **29** (1963), 295–381.
358. I. M. Gel'fand and M. I. Graev, Irreducible unitary representations of the group of second-order unimodular matrices with elements in a locally compact field (in Russian), *Dokl. Akad. Nauk SSSR* **149** (1963), 499–501.
359. I. M. Gel'fand and I. I. Pyatezki-Shapiro, Theory of representations and theory of automorphic functions, *Amer. Math. Soc. Transl.* (2) **26** (1963), 173–200.
360. I. M. Gel'fand, Some questions of analysis and differential equations, *Amer. Math. Soc. Transl.* (2) **26** (1963), 201–219.

361. I. M. Gel'fand and M. I. Graev, The Plancherel formula for the group of unimodular matrices of second order with elements in a locally compact field (in Russian), *Dokl. Akad. Nauk SSSR* **151** (1963), 262–264.
362. I. M. Gel'fand, M. I. Graev and N. Ya. Vilenkin, Obobshchennye functsii, Vyp. 5. Integralnaya geometriya i svyazannye s neĭ voprosy teorii predstavleniĭ (in Russian) [Generalized functions, No. 5. Integral geometry and related problems in the theory of representations] (Gosudarstv. Izdat. Fiz.-Mat. Lit., 1962).
363. I. M. Gel'fand and I. I. Pyatezki-Shapiro, Unitary representations in the space $G/\Gamma$, where $G$ is a group of real matrices of order $n$ and $\Gamma$ is a subgroup of integral matrices (in Russian), *Dokl. Akad. Nauk SSSR* **147** (1962), 275–278.
364. F. A. Berezin and I. M. Gel'fand, Some remarks on the theory of spherical functions on symmetric Riemannian manifolds, *Amer. Math. Soc. Transl.* (2) **21** (1962), 193–238.
365. I. M. Gel'fand and G. E. Shilov, *Verallgemeinerte Funktionen* (Distributionen). II: Lineare topologische Räume. Räume von Grundfunktionen und verallgemeinerten Funktionen. (German) Hochschulbücher für Mathematik, Bd. 48 (VEB Deutscher Verlag der Wissenschaften, 1962).
366. I. M. Gel'fand and M. I. Graev, An application of the horisphere method to the spectral analysis of functions in real and imaginary Lobatchevsky spaces (in Russian), *Trudy Moskov. Mat. Obšč.* **11** (1962), 243–308.
367. I. M. Gel'fand and M. I. Graev, Construction of irreducible representations of simple algebraic groups over a finite field (in Russian), *Dokl. Akad. Nauk SSSR* **147** (1962), 529–532.
368. I. M. Gel'fand and I. I. Pyatezki-Shapiro, Unitary representations in uniform spaces with discrete stationary group (in Russian), *Dokl. Akad. Nauk SSSR* **147** (1962), 17–20.
369. I. M. Gel'fand and M. I. Graev, Categories of group representations and the classification problem of irreducible representations (in Russian), *Dokl. Akad. Nauk SSSR* **146** (1962), 757–760.
370. I. M. Gel'fand and M. L. Zetlin, Some control methods for complex systems (in Russian), *Usp. Mat. Nauk* **17** (1962), No. 1, 3–25.
371. I. M. Gel'fand and G. E. Shilov, *Les Distributions* (in French), Traduit par G. Rideau. Collection Universitaire de Mathématiques, VIII (Dunod, 1962).
372. I. M. Gel'fand and S. V. Fomin, *Variatsionnoe Ischislenie* (in Russian) [Variational calculus] (Gosudarstv. Izdat. Fiz.-Mat. Lit., 1961).

373. I. M. Gel'fand and N. Ya. Vilenkin, *Obobshchennye Functsii*, Vyp. 4. Nekotorye primeneniya garmonicheskogo analiza. Osnashchennye gilbertovy prostranstva (in Russian) [Generalized functions, No. 4. Some applications of harmonic analysis. Equipped Hilbert spaces] (Gosudarstv. Izdat. Fiz.-Mat. Lit., 1961).
374. I. M. Gel'fand and M. I. Graev, Integral transformations connected with line complexes in a complex affine space (in Russian), *Dokl. Akad. Nauk SSSR* **138** (1961), 1266–1269.
375. I. M. Gel'fand, Integral geometry and its relation to the theory of representations (in Romanian), *Acad. R. P. Romîne An. Romîno-Soviet. Ser. Mat.-Fiz.* (3) **15** (1961), No. 2(37), 82–91.
376. I. M. Gel'fand, On elliptic equations (in Romanian), *Acad. R. P. Romîne An. Romîno-Soviet. Ser. Mat.-Fiz.* (3) **15** (1961), No. 1(36), 3–15.
377. I. M. Gel'fand and M. L. Zetlin, The principle of nonlocal search in automatic optimization systems, *Dokl. Akad. Nauk SSSR* **137** (1961), 295–298 (in Russian); translated as *Sov. Phys. Dokl.* **6** (1961), 192–194.
378. I. M. Gel'fand, *Lectures on Linear Algebra*. Translated from the revised second Russian edition by A. Shenitzer Interscience Tracts in Pure and Applied Mathematics, No. 9 (Interscience, 1961).
379. I. M. Gel'fand, Integral geometry and its relation to the theory of representations (in Russian), *Usp. Mat. Nauk* **15** No. 2(92), 155–164; translated as *Russ. Math. Surv.* **15** (1960), No. 2, 143–151.
380. I. M. Gel'fand and M. I. Graev, Integrals over hyperplanes of basic and generalized functions, *Dokl. Akad. Nauk SSSR* **135** (1960), 1307–1310 (in Russian); translated as *Sov. Math. Dokl.* **1** (1960), 1369–1372.
381. I. M. Gel'fand, D. A. Raĭkov and G. E. Shilov, *Kommutativnye Normirovannye Koltsa* (in Russian) [*Commutative Normed Rings*] Sovremennye Problemy Matematiki (Gosudarstv. Izdat. Fiz.-Mat. Lit., 1960).
382. I. M. Gel'fand, On elliptic equations, *Usp. Mat. Nauk* **15**, No. 3(93), 121–132 (in Russian); translated as *Russ. Math. Surv.* **15** (1960), No. 3, 113–123.
383. I. M. Gel'fand, Some questions of analysis and differential equations (in Romanian), *Acad. R. P. Romîne. An. Romîno-Soviet. Ser. Mat.-Fiz.* (3) **14** (1960), No. 1(32), 3–19.
384. I. M. Gel'fand and M. I. Graev, Fourier transforms of rapidly decreasing functions on complex semisimple groups, *Dokl. Akad. Nauk SSSR* **131** (1960), 496–499 (in Russian); translated as *Sov. Math. Dokl.* **1** (1960), 276–279.

385. I. M. Gel'fand and I. I. Pyatezki-Shapiro, Theory of representations of groups and theory of automorphic functions (in Romanian), *Acad. R. P. Romîne An. Romîno-Soviet. Ser. Mat.-Fiz.* (3) **14** (1960), No. 2(33), 60–84.
386. I. M. Gel'fand and M. L. Zetlin, Continuous models of control systems, *Dokl. Akad. Nauk SSSR* **131** (1960), 1242–1245 (in Russian); translated as *Sov. Math. Dokl.* **1** (1960), 409–412.
387. I. M. Gel'fand, On the subrings of a ring of continuous functions, *Amer. Math. Soc. Transl.* (2) **16** (1960), 477–479.
388. F. A. Berezin, I. M. Gel'fand, M. I. Graev and M. A. Naĭmark, Group representations, *Amer. Math. Soc. Transl.* (2) **16** (1960), 325–353.
389. I. M. Gel'fand, On some problems of functional analysis, *Amer. Math. Soc. Transl.* (2) **16** (1960), 315–324.
390. I. M. Gel'fand and G. E. Shilov, *Verallgemeinerte Funktionen* (Distributionen). I: Verallgemeinerte Funktionen und das Rechnen mit ihnen (in German), Hochschulbücher für Mathematik, Bd. 47 (VEB Deutscher Verlag der Wissenschaften, 1960).
391. I. M. Gel'fand and A. M. Jaglom, Integration in functional spaces and its applications in quantum physics, *J. Math. Phys.* **1** (1960), 48–69.
392. I. M. Gel'fand and D.-Š. Sya, On positive definite distributions (in Russian), *Usp. Mat. Nauk* **15** (1960), No. 1(91), 185–190.
393. I. M. Gel'fand and M. I. Graev, Geometry of homogeneous spaces, representations of groups in homogeneous spaces and related questions of integral geometry. I (in Russian), *Trudy Moskov. Mat. Obšč.* **8** (1959), 321–390.
394. I. M. Gel'fand and M. I. Graev, Expansion, in irreducible representations, of representations of the Lorentz group in spaces of functions given on symmetric spaces (in Russian), *Dokl. Akad. Nauk SSSR* **127** (1959), 250–253.
395. I. M. Gel'fand, Some questions of analysis and differential equations (in Russian), *Usp. Mat. Nauk* **14** (1959), No. 3(87), 3–19.
396. I. M. Gel'fand and A. M. Yaglom, Calculation of the amount of information about a random function contained in another such function, *Amer. Math. Soc. Transl.* (2) **12** (1959), 199–246.
397. I. M. Gel'fand, The structure of a ring of rapidly decreasing functions of a Lie group (in Russian), *Dokl. Akad. Nauk SSSR* **124** (1959), 19–21.
398. I. M. Gel'fand and I. I. Pyatezki-Shapiro, Theory of representations and theory of automorphic functions (in Russian), *Usp. Mat. Nauk* **14** (1959), No. 2(86), 171–194.

399. I. M. Gel'fand, Some problems of the theory of quasilinear equations (in Romanian), *Acad. R. P. Romíne. An. Romíno-Soviet. Ser. Mat.-Fiz.* (3) **13** (1959), No. 4(31), 5–83.
400. I. M. Gel'fand, Some problems in the theory of quasi-linear equations (in Russian), *Usp. Mat. Nauk* **14** (1959), No. 2(86), 87–158.
401. I. M. Gel'fand and I. I. Pyatezki-Shapiro, A theorem of Poincaré (in Russian), *Dokl. Akad. Nauk SSSR* **127** (1959), 490–493.
402. I. M. Gel'fand, A. S. Frolov and N. N. Čencov, The computation of continuous integrals by the Monte Carlo method (in Russian), *Izv. Vysš. Učebn. Zaved. Matematika* **1958** (1958), No. 5(6), 32–45.
403. I. M. Gel'fand, R. A. Minlos and Z. Ya. Shapiro, *Predstavleniy Agruppy Vrashcheni i gruppy Lorentsa, ikh primeneniya* (in Russian) [Representations of the rotation group and of the Lorentz group, and their applications] (Gosudarstv. Izdat. Fiz.-Mat. Lit., 1958).
404. I. M. Gel'fand and G. E. Shilov, *Nekotorye Voprosy Teorii Differentsial'nykh Uravneniĭ* (in Russian) [Some questions in the theory of differential equations] Obobščennye funkcii, Vypusk 3 (Gosudarstv. Izdat. Fiz.-Mat. Lit., 1958).
405. I. M. Gel'fand and G. E. Shilov, *Prostranstva Osnovnykh i Obobshchennykh Funktsiĭ* (in Russian) [Spaces of fundamental and generalized functions] Obobščennye funkcii, Vypusk 2 (Gosudarstv. Izdat. Fiz.-Mat. Lit., 1958).
406. I. M. Gel'fand and A. M. Jaglom, Über die Berechnung der Menge an Information über eine zufällige Funktion, die in einer anderen zufälligen Funktion enthalten ist (in German), 1958 Arbeiten zur Informationstheorie, II, pp. 7–56. Mathematische Forschungsberichte. VI (VEB Deutscher Verlag der Wissenschaften, 1958).
407. I. M. Gel'fand, A. N. Kolmogorov and A. M. Yaglom, Zur allgemeinen Definition der Information (in German), 1958 Arbeiten zur Informationstheorie, II, pp. 57–60. Mathematische Forschungsberichte. VI (VEB Deutscher Verlag der Wissenschaften, 1958).
408. I. M. Gel'fand and G. E. Shilov, Obobshchennye Funksii i Deĭstviya nad Nimi (in Russian) [Generalized functions and operations on them], Obobščennye funkcii, Vypusk 1. Generalized functions, part 1 (Gosudarstv. Izdat. Fiz.-Mat. Lit., 1958).
409. I. M. Gel'fand and Z. Ya. Shapiro, Homogeneous functions and their extensions, *Amer. Math. Soc. Transl.* (2) **8** (1958), 21–85.
410. I. M. Gel'fand and M. I. Graev, *Analogue of the Plancherel Formula for the Classical Groups*, American Mathematical Society Translations, Ser. 2, Vol. 9, pp. 123–154 (Amer. Math. Soc., 1958).

411. I. M. Gel'fand and V. B. Lidskiĭ, On the structure of the regions of stability of linear canonical systems of differential equations with periodic coefficients, *Amer. Math. Soc. Transl.* (2) **8** (1958), 143–181.
412. I. M. Gel'fand, On some problems of functional analysis (in Romanian), *Acad. R. P. Romîne. An. Romîno-Soviet. Ser. Mat.-Fiz.* (3) **11** (1957), No. 3(22), 58–6.
413. I. M. Gel'fand, *Some Aspects of Functional Analysis and Algebra*, 1957 Proceedings of the International Congress of Mathematicians, Amsterdam, 1954, Vol. 1, pp. 253–276 (Erven P. Noordhoff, North-Holland, 1957).
414. I. M. Gel'fand and M. A. Neumark, *Unitäre Darstellungen der klassischen Gruppen* (in German) (Akademie, 1957).
415. I. M. Gel'fand and A. M. Yaglom, Computation of the amount of information about a stochastic function contained in another such function (in Russian), *Usp. Mat. Nauk (N.S.)* **12** (1957), No. 1(73), 3–52.
416. I. M. Gel'fand, On subrings of the ring of continuous functions (in Russian), *Usp. Mat. Nauk (N.S.)* **12** (1957), No. 1(73), 249–251.
417. I. M. Gel'fand and G. E. Shilov, Fourier transforms of rapidly increasing functions and questions of the uniqueness of the solution of Cauchy's problem, *Amer. Math. Soc. Transl.* (2) **5** (1957), 221–274.
418. I. M. Gel'fand, D. A. Raĭkov and G. E. Shilov, Commutative normed rings, *Amer. Math. Soc. Transl.* (2) **5** (1957), 115–220.
419. F. A. Berezin, I. M. Gel'fand, M. I. Graev and M. A. Naĭmark, Representation of groups (in Russian), *Usp. Mat. Nauk (N.S.)* **11** (1956), No. 6(72), 13–40.
420. I. M. Gel'fand, On some problems of functional analysis (in Russian), *Usp. Mat. Nauk (N.S.)* **11** (1956), No. 6(72), 3–12.
421. F. A. Berezin and I. M. Gel'fand, Some remarks on the theory of spherical functions on symmetric Riemannian manifolds (in Russian), *Trudy Moskov. Mat. Obšč.* **5** (1956), 311–351.
422. I. M. Gel'fand, A. N. Kolmogorov and A. M. Yaglom, On the general definition of the amount of information (in Russian), *Dokl. Akad. Nauk SSSR (N.S.)* **111** (1956), 745–748.
423. I. M. Gel'fand and G. E. Shilov, Quelques applications de la théorie des fonctions généralisées (in French), *J. Math. Pures Appl.* (9) **35** (1956), 383–413.
424. I. M. Gel'fand, On identities for eigenvalues of a differential operator of second order (in Russian), *Usp. Mat. Nauk (N.S.)* **11** (1956), No. 1(67), 191–198.

425. I. M. Gel'fand and A. M. Yaglom, Integration in function spaces and its application to quantum physics (in Russian), *Usp. Mat. Nauk (N.S.)* **11** (1956), No. 1(67), 77–114.
426. I. M. Gel'fand and A. M. Yaglom, Methods of the theory of random processes in quantum physics (in Russian), *Vestnik Leningrad. Univ.* **11** (1956), No. 1, 33–34.
427. I. M. Gel'fand and M. I. Graev, Unitary representations of the real unimodular group (principal nondegenerate series), *Amer. Math. Soc. Transl.* (2) **2** (1956), 147–205.
428. I. M. Gel'fand and Z. Ya. Shapiro, Representations of the group of rotations of 3-dimensional space and their applications, *Amer. Math. Soc. Transl.* (2) **2** (1956), 207–316.
429. I. M. Gel'fand and S. V. Fomin, Geodesic flows on manifolds of constant negative curvature, *Amer. Math. Soc. Transl.* (2) **1** (1955), 49–65.
430. I. M. Gel'fand and B. M. Levitan, On the determination of a differential equation from its spectral function, *Amer. Math. Soc. Transl.* (2) **1** (1955), 253–304.
431. I. M. Gel'fand and V. B. Lidskiĭ, On the structure of the domains of stability of linear canonical systems of differential equations with periodic coefficients (in Russian), *Usp. Mat. Nauk (N.S.)* **10** (1955), No. 1(63), 3–40.
432. I. M. Gel'fand and A. G. Kostyučenko, Expansion in eigenfunctions of differential and other operators (in Russian), *Dokl. Akad. Nauk SSSR (N.S.)* **103** (1955), 349–352.
433. I. M. Gel'fand and Z. Ya. Shapiro, Homogeneous functions and their extensions (in Russian), *Usp. Mat. Nauk (N.S.)* **10** (1955), No. 3(65), 3–70.
434. I. M. Gel'fand and G. E. Shilov, On a new method in uniqueness theorems for solution of Cauchy's problem for systems linear partial differential equations (in Russian), *Dokl. Akad. Nauk SSSR (N.S.)* **102** (1955), 1065–1068.
435. I. M. Gel'fand and M. I. Graev, Analogue of the Plancherel formula for the classical groups (in Russian), *Trudy Moskov. Mat. Obšč.* **4** (1955), 375–404.
436. I. M. Gel'fand, Generalized random processes (in Russian), *Dokl. Akad. Nauk SSSR (N.S.)* **100** (1955). 853–856.
437. I. M. Gel'fand and M. I. Graev, Traces of unitary representations of the real unimodular group (in Russian), *Dokl. Akad. Nauk SSSR (N.S.)* **100** (1955), 1037–1040.
438. I. M. Gel'fand and R. A. Minlos, *Solution of The Quantum Field Equations* (Morris D. Friedman, 1954).
439. I. M. Gel'fand and R. A. Minlos, Solution of the quantum field equations (in Russian), *Dokl. Akad. Nauk SSSR (N.S.)* **97** (1954), 209–212.

440. I. M. Gel'fand and G. E. Shilov, Fourier transforms of rapidly increasing functions and questions of uniqueness of the solution of Cauchy's problem (in Russian), *Usp. Mat. Nauk* (*N.S.*) **8** (1953), No. 6(58), 3–54.
441. I. M. Gel'fand and M. I. Graev, Analogue of the Plancherel formula for real semisimple Lie groups (in Russian), *Dokl. Akad. Nauk SSSR* (*N.S.*) **92** (1953), 461–464.
442. I. M. Gel'fand and M. I. Graev, On a general method of decomposition of the regular representation of a Lie group into irreducible representations (in Russian), *Dokl. Akad. Nauk SSSR* (*N.S.*) **92** (1953), 221–224.
443. I. M. Gel'fand and M. I. Graev, Unitary representations of the real unimodular group (principal nondegenerate series) (in Russian), *Izvestiya Akad. Nauk SSSR. Ser. Mat.* **17** (1953), 189–248.
444. I. M. Gel'fand and B. M. Levitan, On a simple identity for the characteristic values of a differential operator of the second order (in Russian), *Dokl. Akad. Nauk SSSR* (*N.S.*) **88** (1953), 593–596.
445. I. M. Gel'fand, On the spectrum of non-self-adjoint differential operators (in Russian), *Usp. Mat. Nauk* (*N.S.*) **7** (1952), No. 6(52), 183–184.
446. I. M. Gel'fand and S. V. Fomin, Geodesic flows on manifolds of constant negative curvature (in Russian), *Usp. Mat. Nauk* (*N.S.*) **7** (1952), No. 1(47), 118–137.
447. I. M. Gel'fand and M. I. Graev, Unitary representations of real simple Lie groups (in Russian), *Dokl. Akad. Nauk SSSR* (*N.S.*) **86** (1952), 461–463.
448. I. M. Gel'fand and M. A. Naĭmark, Unitary representations of a unimodular group containing an identity representation of the unitary subgroup (in Russian), *Trudy Moskov. Mat. Obšč.* **1** (1952), 423–475.
449. I. M. Gel'fand and Z. Ya. Shapiro, Representations of the group of rotations in three-dimensional space and their applications (in Russian), *Usp. Mat. Nauk* (*N.S.*) **7** (1952), No. 1(47), 3–117.
450. I. M. Gel'fand and M. A. Naĭmark, Unitary representations of semisimple Lie groups. I. *Amer. Math. Soc. Transl.* **1952** (1952), No. 64, 42 pp.
451. I. M. Gel'fand, Remark on the work of N. K. Bari, Biorthogonal systems and bases in Hilbert space (in Russian), *Moskov. Gos. Univ. Učenye Zapiski Mat.* **148**(4) (1951), 224–225.
452. I. M. Gel'fand and B. M. Levitan, On the determination of a differential equation from its spectral function (in Russian), *Izvestiya Akad. Nauk SSSR. Ser. Mat.* **15** (1951), 309–360.
453. I. M. Gel'fand and S. V. Fomin, Unitary representations of Lie groups and geodesic flows on surfaces of constant negative curvature (in Russian), *Dokl. Akad. Nauk SSSR* (*N.S.*) **76** (1951), 771–774.

454. I. M. Gel'fand and B. M. Levitan, On the determination of a differential equation by its spectral function (in Russian), *Dokl. Akad. Nauk SSSR (N.S.)* **77** (1951), 557–560.
455. I. M. Gel'fand, Lekcii po lineĭnoĭ algebre (in Russian) [Lectures on Linear Algebra, 2nd edn.] (Gosudarstv. Izdat. Tehn.-Teor. Lit., 1951).
456. I. M. Gel'fand and M. A. Naĭmark, Unitarnye predstavleniya klassičeskih grupp (in Russian) [Unitary representations of the classical groups.], Trudy Mat. Inst. Steklov., Vol. 36 (Izdat. Nauk SSSR, 1950).
457. I. M. Gel'fand, Expansion in characteristic functions of an equation with periodic coefficients (in Russian), *Dokl. Akad. Nauk SSSR (N.S.)* **73** (1950), 1117–1120.
458. I. M. Gel'fand and M. L. Zetlin, Finite-dimensional representations of the group of unimodular matrices (in Russian), *Dokl. Akad. Nauk SSSR (N.S.)* **71** (1950), 825–828.
459. I. M. Gel'fand and M. A. Naĭmark, The relation between the unitary representations of the complex unimodular group and its unitary subgroup (in Russian), *Izvestiya Akad. Nauk SSSR. Ser. Mat.* **14** (1950), 239–260.
460. I. M. Gel'fand and M. L. Zetlin, Finite-dimensional representations of groups of orthogonal matrices (in Russian), *Dokl Akad. Nauk SSSR (N.S.)* **71** (1950), 1017–1020.
461. I. M. Gel'fand, Spherical functions in symmetric Riemann spaces (in Russian), *Dokl. Akad. Nauk SSSR (N.S.)* **70** (1950), 5–8.
462. I. M. Gel'fand, The center of an infinitesimal group ring (in Russian), *Mat. Sbornik N.S.* **26**(68) (1950), 103–112.
463. I. M. Gel'fand and M. A. Naĭmark, The trace in fundamental and supplementary series of representations of the complex unimodular group (in Russian), *Dokl. Akad. Nauk SSSR (N.S.)* **61** (1948), 9–11.
464. I. M. Gel'fand and A. M. Yaglom, Charge conjugation for general Lorentz invariant equations (in Russian), *Akad. Nauk SSSR. Zhurnal Eksper. Teoret. Fiz.* **18** (1948), 1105–1111.
465. I. M. Gel'fand and A. M. Yaglom, Pauli's theorem for general Lorentz invariant equations (in Russian), *Akad. Nauk SSSR. Zhurnal Eksper. Teoret. Fiz.* **18** (1948), 1096–1104.
466. I. M. Gel'fand and A. M. Yaglom, On Lorentz invariant equations to which correspond a definite charge and a definite energy (in Russian), *Dokl. Akad. Nauk SSSR (N.S.)* **63** (1948), 371–374.
467. I. M. Gel'fand and A. M. Yaglom, General Lorentz invariant equations and infinite-dimensional representations of the Lorentz group (in Russian), *Akad. Nauk SSSR. Zhurnal Eksper. Teoret. Fiz.* **18** (1948), 703–733.

468. I. M. Gel'fand and M. A. Naĭmark, The analogue of Plancherel's formula for the complex unimodular group (in Russian), *Dokl. Akad. Nauk SSSR (N.S.)* **63** (1948), 609–612.
469. I. M. Gel'fand and M. A. Naĭmark, On the connection between the representations of a complex semi-simple Lie group and those of its maximal compact subgroups (in Russian), *Dokl. Akad. Nauk SSSR (N.S.)* **63** (1948), 225–228.
470. I. M. Gel'fand and M. A. Naĭmark, Normed rings with involutions and their representations (in Russian), *Izvestiya Akad. Nauk SSSR. Ser. Mat.* **12** (1948), 445–480.
471. I. M. Gel'fand and A. M. Yaglom, General Lorentz invariant equations and infinite representations of the Lorentz group (in Russian), *Dokl. Akad. Nauk SSSR (N.S.)* **59** (1948), 655–658.
472. I. M. Gel'fand and M. A. Naĭmark, Unitary representations of the Lorentz group (in Russian), *Izvestiya Akad. Nauk SSSR. Ser. Mat.* **11** (1947), 411–504.
473. I. M. Gel'fand and M. A. Naĭmark, Supplementary and degenerate series of representations of the complex unimodular group (in Russian), *Dokl. Akad. Nauk SSSR (N.S.)* **58** (1947), 1577–1580.
474. I. M. Gel'fand and M. A. Naĭmark, Unitary representations of semisimple Lie groups. I (in Russian), *Mat. Sbornik N.S.* **21**(63) (1947), 405–434.
475. I. Gel'fand and M. A. Naĭmark, The principal series of irreducible representations of the complex unimodular group, *C. R. (Dokl.) Acad. Sci. URSS (N.S.)* **56** (1947), 3–4.
476. I. Gel'fand and M. A. Naĭmark, Unitary representations of the group of linear transformations of the straight line, *C. R. (Dokl.) Acad. Sci URSS (N.S.)* **55** (1947), 567–570.
477. I. M. Gel'fand, D. A. Raĭkov and G. E. Shilov, Commutative normed rings (in Russian), *Usp. Mat. Nauk (N. S.)* **1** (1946), No. 2(12), 48–146.
478. I. M. Gel'fand and M. A. Naĭmark, On unitary representations of the complex unimodular group, *C. R. (Dokl.) Acad. Sci. URSS (N.S.)* **54** (1946), 195–198.
479. I. Gel'fand and M. A. Naĭmark, Unitary representations of the Lorentz group, *Acad. Sci. USSR. J. Phys.* **10** (1946), 93–94.
480. I. Gel'fand and D. Raĭkov, Irreducible unitary representations of locally bicompact groups, *C. R. (Dokl.) Acad. Sci. URSS (N.S.)* **42** (1944), 199–201.
481. I. Gel'fand and D. Raĭkov, Irreducible unitary representations of locally bicompact groups, *Rec. Math. [Mat. Sbornik] N.S.* **13**(55) (1943), 301–316.

482. I. Gel'fand and M. A. Naĭmark, On the imbedding of normed rings into the ring of operators in Hilbert space, *Rec. Math.* [*Mat. Sbornik*] *N.S.* **12**(54) (1943), 197–213.
483. I. Gel'fand and G. Shilov, Über verschiedene Methoden der Einführung der Topologie in die Menge der maximalen Ideale eines normierten Ringes (in German), *Rec. Math.* [*Mat. Sbornik*] *N.S.* **9**(51) (1941), 25–39.
484. I. Gel'fand, Ideale und primäre Ideale in normierten Ringen (in German), *Rec. Math.* [*Mat. Sbornik*] *N.S.* **9**(51) (1941), 41–48.
485. I. Gel'fand, Über absolut konvergente trigonometrische Reihen und Integrale (in German), *Rec. Math.* [*Mat. Sbornik*] *N.S.* **9**(51) (1941), 51–66.
486. I. Gel'fand, Normierte Ringe (in German), *Rec. Math.* [*Mat. Sbornik*] *N.S.* **9**(51) (1941), 3–24.
487. I. Gel'fand, Zur Theorie der Charaktere der Abelschen topologischen Gruppen (in German) *Rec. Math.* [*Mat. Sbornik*] *N.S.* **9**(51) (1941), 49–50.
488. I. Gel'fand, Die direkte und umgekehrte Aufgabe der Bestimmung des Anziehungspotentials eines homogenen Kreissegments (in Russian), *Bull. Acad. Sci. URSS. Sér. Géograph. Géophys.* [Izvestia Akad. Nauk SSSR] 1941.
489. I. Gel'fand and D. Raĭkov, On the theory of characters of commutative topological groups, *C. R.* (*Dokl.*) *Acad. Sci. URSS* (*N.S.*) **28** (1940), 195–198.
490. I. Gel'fand, On one-parametrical groups of operators in a normed space, *C. R.* (*Dokl.*) *Acad. Sci. URSS* (*N.S.*) **25** (1939), 713–718.
491. I. Gel'fand, To the theory of normed rings. III. On the ring of almost periodic functions, *C. R.* (*Dokl.*) *Acad. Sci. URSS* (*N.S.*) **25** (1939), 573–574.
492. I. Gel'fand, To the theory of normed rings. II. On absolutely convergent trigonometrical series and integrals, *C. R.* (*Dokl.*) *Acad. Sci. URSS* (*N.S.*) **25** (1939), 570–572.

## Worked Related to I. M. Gel'fand

Vladimir Israel Moiseevich Gel'fand, *Eur. Math. Soc. Newsl.* No. 76 (2010), 25–27.

Israel Moiseevich Gel'fand (September 2, 1913–October 5, 2009) (in Russian), *Functional. Anal. i. Prilozhen.* **44** (2010), No. 1, 1–3.

Guichardet, Alain Quelques souvenirs sur I. M. Gel'fand (1913–2009) (in French) [Remembering I. M. Gel'fand (1913–2009)] *Gaz. Math.* No. 123 (2010), 95–100.

A. L. Afendikov and K. V. Brushlinskiĭ, I. M. Gel'fand and applied mathematics (in Russian), *Usp. Mat. Nauk* **64** (2009), No. 6(390), 181–186; translation in *Russ. Math. Surv.* **64** (2009), No. 6, 1149–1156.

Obituary [Gel'fand, Israel Moiseevich]. *Selecta Math.* (*N.S.*) **15** (2009), No. 4, 519–520.

*The Unity of Mathematics*, In Honor of *The Ninetieth Birthday of I. M. Gel'fand*. Papers from the conference held in Cambridge, MA, August 31–September 4, 2003. Edited by P. Etingof, V. Retakh and I. M. Singer. Progress in Mathematics, Vol. 244 (Birkhäuser, 2006).

A. N. Shiryaev and A. N. Kolmogorov, The teacher discusses his student. Four reviews by A. N. Kolmogorov on the works of I. M. Gel'fand (on the occasion of his 90th birthday) (in Russian), *Functional. Anal. i Prilozhen.* **37** (2003), No. 4, 3–12; translation in *Funct. Anal. Appl.* **37** (2003), No. 4, 243–250.

Special issue dedicated to I. M. Gel'fand on the occasion of his 90th birthday (in Russian), *Functional. Anal. i Prilozhen.* **37** (2003), No. 4.

I. M. Singer, Tribute to I. M. Gel'fand for his 80th birthday celebration, in *Functional Analysis on the Eve of the 21st Century*, Vol. II (New Brunswick, NJ, 1993), pp. xix–xxii, Progr. Math., Vol. 132 (Birkhäuser, 1996).

Mathematical publications of I. M. Gel'fand 1987–1995, in *Functional Analysis on the Eve of the 21st Century*, Vol. II (New Brunswick, NJ, 1993), pp. xiii–xviii, Progr. Math., Vol. 132 (Birkhäuser, 1996).

*Functional Analysis on the Eve of the 21st Century*. Vol. II, In honor of the eightieth birthday of I. M. Gel'fand. Papers from the conference held at Rutgers University, New Brunswick, New Jersey, October 24–27, 1993. Edited by S. Gindikin, J. Lepowsky and R. L. Wilson. Progress in Mathematics, Vol. 132 (Birkhäuser, 1996).

I. M. Singer, Tribute to I. M. Gel'fand for his 80th birthday celebration, in *Functional Analysis on the Eve of the 21st Century*, Vol. 1 (New Brunswick, NJ, 1993), pp. xix–xxii, Progr. Math., Vol. 131 (Birkhäuser, 1995).

Mathematical publications of I. M. Gel'fand 1987–1995, in *Functional Analysis on the Eve of the 21st Century*, Vol. 1 (New Brunswick, NJ, 1993), pp. xiii–xviii, Progr. Math., Vol. 131 (Birkhäuser, 1995).

I. M. Gel'fand (curriculum vitae), in *Functional Analysis on the Eve of the 21st Century*, Vol. 1 (New Brunswick, NJ, 1993), p. xii, Progr. Math., Vol. 131 (Birkhäuser, 1995).

*Functional Analysis on the Eve of the 21st Century*. Vol. I, In honor of the eightieth birthday of I. M. Gel'fand. Papers from the conference held at

Rutgers University, New Brunswick, New Jersey, October 24–27, 1993. Edited by S. Gindikin, J. Lepowsky and R. L. Wilson. Progr. Math., Vol. 131 (Birkhäuser, 1995).

Gel'fand receives MacArthur Fellowship. *Notices Amer. Math. Soc.* **41** (1994), No. 7, 797–798.

Interview with I. M. Gel'fand. Selected Lectures in Mathematics (Amer. Math. Soc., 1993). 1 videocassette (NTSC; 1/2 inch; VHS) (60 min.).

*I. M. Gel'fand Seminar.* Edited by S. Gelfand and S. Gindikin, Advances in Soviet Mathematics, 16, Part 2 (Amer. Math. Soc., 1993).

*I. M. Gel'fand Seminar.* Edited by S. Gelfand and S. Gindikin, Advances in Soviet Mathematics, 16, Part 1 (Amer. Math. Soc., 1993).

*Geometry and Physics.* Essays in honour of I. M. Gel'fand. Edited by S. Gindikin and I. M. Singer. Reprint of *J. Geom. Phys.* **5** (1988) (Pitagora Editrice and North-Holland, 1991).

N. N. Bogolyubov, S. G. Gindikin, A. A. Kirillov, A. N. Kolmogorov, S. P. Novikov and L. D. Faddeev, Israel Moiseevich Gel'fand (on the occasion of his seventieth birthday) (in Russian), *Usp. Mat. Nauk* **38** (1983), No. 6(234), 137–152.

List of the published works of I. M. Gel'fand, *Usp. Mat. Nauk* **29** (1974), No. 1(175), 241–246.

M. B. Berkinblit, Yu. M. Vasil'ev and M. L. Šik, The work of I. M. Gel'fand in biology (in Russian), *Usp. Mat. Nauk* **29** (1974), No. 1(175), 232–240.

O. V. Lokucievskiĭ and N. N. Zencov, The work of I. M. Gel'fand in applied and numerical mathematics (in Russian), *Usp. Mat. Nauk* **29** (1974), No. 1(175), 224–232.

S. G. Gindikin, A. A. Kirillov and D. B. Fuks, Work of I. M. Gel'fand in functional analysis, algebra and topology (in Russian), *Usp. Mat. Nauk* **29** (1974), No. 1(175), 195–223.

Israel Moiseevich Gel'fand (on the occasion of his sixtieth birthday) (in Russian), *Usp. Mat. Nauk* **29** (1974), No. 1(175), 193–195.

M. I. Višik, A. N. Kolmogorov, S. V. Fomin and G. E. Shilov, Israel Moiseevich Gel'fand. (On his fiftieth birthday) (in Russian), *Usp. Mat. Nauk* **19** (1964), No. 3(117), 187–205.

I. Singer, On a theorem of I. M. Gel'fand (in Russian), *Usp. Mat. Nauk* **17** (1962), No. 1(103), 169–176.

# ON THE DETERMINATION OF A DIFFERENTIAL EQUATION FROM ITS SPECTRAL FUNCTION

I. M. GEL'FAND AND B. M. LEVITAN

In this work there is given a method of reconstructing a differential equation of the second order from its spectral function $\rho(\lambda)$. This problem is reduced here to a certain linear integral equation. It is also explained how monotonic functions $\rho(\lambda)$ may serve as spectral functions of a differential equation of second order.

## Introduction

We shall consider a differential equation of the second order given on the interval $(0, \infty)$,

$$y'' + (\lambda - q(x))\, y = 0 \tag{1}$$

with boundary conditions

$$y(0) = 1, \quad y'(0) = h. \tag{1'}$$

The function $q(x)$ is assumed to be continuous on any finite interval. It is known that an arbitrary function with integrable square may be expressed as a Fourier integral in the eigenfunctions of this equation for a given $h$. More precisely, there exists a monotonic function $\rho(\lambda)$, bounded on each finite interval, such that for any function $f(x)$ with integrable square the following equation holds:

$$\int_0^{+\infty} f^2(x)\, dx = \int_{-\infty}^{+\infty} E^2(\lambda)\, d\rho(\lambda). \tag{2}$$

Here $E(\lambda)$ is the Fourier transform of the function $f(x)$, i.e.,

$$E(\lambda) = \int_0^{\infty} f(x)\varphi(x, \lambda)\, dx, \ * \tag{3}$$

where $\phi(x, \lambda)$ is the solution of equation (1) with the initial conditions (1').

We shall call the function $\rho(\lambda)$ the *spectral function* of equation (1) with conditions (1'). The meaning of the spectral function $\rho(\lambda)$ is particularly clear in the case of a discrete spectrum. In this case $\rho(\lambda)$ is the jump function with jumps at the spectral points $\lambda_1, \lambda_2, \cdots, \lambda_n, \cdots$. Moreover, the amount of the jumps equals

$$\frac{1}{\int_0^{\infty} \varphi^2(x, \lambda_n)\, dx}$$

and thus equation (2) becomes Parseval's equation for the Fourier series of the function $f(x)$.

Consequently, one may say that giving $\rho(\lambda)$ determines both the spectrum and the normalization of the eigenfunctions.

* For the sense in which this integral converges see §1.

This work is devoted to the solution of the following problem. Let a function $\rho(\lambda)$ be given. It is required to determine whether there exists an equation of the form (1) having the given $\rho(\lambda)$, and to give a method of actual computation of $q(x)$. The first problems of this type were treated by Ambarcumyan ([1]). The uniqueness of solution of the problem in this formulation was established by V. A. Marčenko ([2]). M. G. Kreĭn ([3]) then took up the question of the solvability of the inverse problem. By making use of his theory of the extension of positive definite functions, M. G. Kreĭn reduced to it the problem considered here and obtained a series of very interesting and important results. In particular, he completely solved by these methods the question of necessary and sufficient conditions for the existence of a cord with a distribution of mass having given spectra for two boundary conditions ([3a]).

In the present work, the above question is attacked with elementary means, namely, by reducing it to the solution of a certain integral equation. The idea of this work is very simple. Similarly to the way in which polynomials, orthogonal with respect to a given weight function, are constructed by orthogonalizing the powers of $x$, we construct from the spectral function $\rho(\lambda)$ the eigenfunctions $\phi(x, \lambda)$ by orthogonalizing the functions $\cos\sqrt{\lambda t}$ with respect to $\rho(\lambda)$.

We explain this in somewhat more detail. As is known, for the equation $y'' + \lambda y = 0$ and the boundary condition $y(0) = 1$, $y'(0) = 0$ we have $\rho(\lambda) = (2/\pi)\sqrt{\lambda}$ for $\lambda > 0$ and $\rho(\lambda) = 0$ for $\lambda < 0$.

In order to make the exposition stand out more we shall assume for a time that the function $\rho(\lambda)$ has the form

$$\rho(\lambda) = \begin{cases} \frac{2}{\pi}\sqrt{\lambda} + \sigma(\lambda) & \text{for} \quad \lambda > 0, \\ \sigma(\lambda) & \text{for} \quad \lambda < 0, \end{cases}$$

where the function $\sigma(\lambda)$ behaves itself sufficiently well at infinity; namely, we shall suppose that

$$\int_{-\infty}^{+\infty} |\lambda| \cdot |d\sigma(\lambda)| < +\infty.$$

We shall look for the eigenfunction $\phi(x, \lambda)$ in the form*

$$\varphi(x, \lambda) = \cos\sqrt{\lambda}\, x + \int_0^x K(x, t) \cos\sqrt{\lambda}\, t\, dt. \tag{4}$$

Here we introduce nonrigorous considerations which are justified for a wide class of functions $\rho(\lambda)$ in this work. We require that $\phi(x, \lambda)$ be an "orthonormal

* The fact that the eigenfunctions may be so represented in proved in ([4]) and ([5]). For a simpler proof of this see §1 of the present work

system'' with weight $\rho(\lambda)$, i.e., that

$$\int_{-\infty}^{+\infty} \varphi(x, \lambda)\varphi(y, \lambda)\, d\rho(\lambda) = \delta(x-y),$$

where $\delta(x)$ is Dirac's $\delta$-function. Since, because of (4), for $y < x$, $\phi(y,\lambda)$ is a combination of $\cos\sqrt{\lambda t}$ for $t \leq y$, then, inverting equation (4), we may convince ourselves that conversely $\cos\sqrt{\lambda y}$ is a combination of $\phi(t,\lambda)$ for $t \leq y$. Thus, $\cos\sqrt{\lambda y}$ is orthogonal to $\phi(x,\lambda)$ if $y < x$:

$$\int_{-\infty}^{+\infty} \varphi(x, \lambda) \cos\sqrt{\lambda}y\, d\rho(\lambda) = 0, \quad \text{if} \quad y < x.$$

Let us substitute here in place of $\phi(x,\lambda)$ its value from equation (4). We obtain

$$\int_{-\infty}^{+\infty} \cos\sqrt{\lambda}x \cos\sqrt{\lambda}y\, d\rho(\lambda) + \int_{-\infty}^{+\infty} \cos\sqrt{\lambda}y \int_{0}^{x} K(x, t)\cos\sqrt{\lambda}t\, dt\, d\rho(\lambda) = 0. \quad (5)$$

In this form the integral we have written down diverges. However, the equation may be given a precise meaning; namely, we represent $\rho(\lambda)$ in the form

$$\rho(\lambda) = \frac{2}{\pi}\sqrt{\lambda} + \sigma(\lambda),$$

where $\lambda > 0$ and note that the following symbolic equation holds:

$$\frac{2}{\pi}\int_{0}^{+\infty} \cos\sqrt{\lambda}t \cos\sqrt{\lambda}y \cdot d\sqrt{\lambda} = \delta(y-t),$$

$$y > 0, \quad t > 0.$$

Then, changing the order of integration in (5) and making use of the symbolic equation

$$\int_{0}^{+\infty} \delta(y-t) f(t)\, dt = f(y),$$

we obtain (noting that $x \neq y$ and, consequently, that $\delta(x-y) = 0$)

$$\int_{-\infty}^{+\infty} \cos\sqrt{\lambda}x \cos\sqrt{\lambda}y\, d\sigma(\lambda) + K(x, y) +$$

$$+ \int_{0}^{x} K(x, t)\left[\int_{-\infty}^{+\infty} \cos\sqrt{\lambda}t \cos\sqrt{\lambda}y\, d\sigma(\lambda)\right] dt = 0. \quad (6)$$

Since the function $\sigma(\lambda)$ is given, the function

$$f(x, y) = \int_{-\infty}^{+\infty} \cos\sqrt{\lambda}x \cos\sqrt{\lambda}y\, d\sigma(\lambda)$$

is also known.

Equation (6) for fixed $x$ is a linear integral equation for the unknown function $K(x, y)$ and has the form*

$$f(x, y) + \int_0^x f(y, t) K(x, t)\, dt + K(x, y) = 0. \tag{7}$$

We show that equation (7) is solvable. In finding $K(x, t)$ we may, from formula (4), also find the eigenfunctions $\phi(x, \lambda)$ and consequently reconstruct equation (1).

In order to show that the constructed functions $\phi(x, \lambda)$ satisfy some differential equation, we proceed in the following manner.

First we prove (§5) that the constructed functions $\phi(x, \lambda)$ satisfy the Parseval equation defined by formulas (2) and (3) above. Making use of this and the integral equation (7) we then obtain in §7 a functional equation for the functions $\phi(x, \lambda)$:

$$\varphi(x, \lambda) \cos \sqrt{\lambda} t = \frac{1}{2} \left[ \varphi(x + t, \lambda) + \varphi(x - t, \lambda) \right] + \int_{|x-t|}^{x+t} W(x, t, s) \varphi(s, \lambda)\, ds, \tag{8}$$

where the function $W(x, t, s)$ is constructed from the function $K(x, t)$. In order to obtain a differentiable equation we subtract $\phi(x, \lambda)$ from both sides of equation (8) and divide by $t^2$. We obtain

$$\varphi(x, \lambda) \frac{\cos \sqrt{\lambda} t - 1}{t^2} = \frac{\varphi(x + t, \lambda) - 2\varphi(x, \lambda) + \varphi(x - t, \lambda)}{2t^2} +$$
$$+ \frac{1}{t^2} \int_{|x-t|}^{x+t} W(x, t, s) \varphi(s, \lambda)\, ds.$$

Passing to the limit as $t \to 0$, we obtain

$$-\frac{1}{2} \lambda \varphi(x, \lambda) = \frac{1}{2} \varphi''(x, \lambda) - \frac{1}{2} q(x) \varphi(x, \lambda),$$

where we have written

$$-\frac{1}{2} q(x) = \lim_{t \to 0} \frac{1}{t^2} \int_{x-t}^{x+t} W(x, t, s)\, ds.$$

The existence of all limits is proved in §7.

We remark that the function $q(x)$ in equation (1) may be immediately constructed from $K(x, t)$ by means of the following formula deduced in §1:

$$\frac{1}{2} q(x) = \frac{dK(x, x)}{dx}.$$

For an approximate solution of the stated problem, one may use the nonlinear Volterra integral equation derived in §2.

* A rigorous deduction of this equation and precise conditions on the function $\sigma(\lambda)$ are given in §§3, 4.

We now formulate the basic result of this work. In the work there are given, in a known sense, necessary and sufficient conditions for the existence of an equation with a given function $\rho(\lambda)$. In fact, these conditions are sufficient for the existence of an equation with a continuous function $q(x)$, and are necessary if $q(x)$ has a continuous derivative. The following theorem is proved.

**Theorem.** *Let a monotonic function $\rho(\lambda)$ be given on the whole axis. Let us represent it in the form*

$$\rho(\lambda) = \begin{cases} \frac{2}{\pi}\sqrt{\lambda} + \sigma(\lambda) & \text{for} \quad \lambda \geqslant 0, \\ \sigma(\lambda) & \text{for} \quad \lambda \leqslant 0. \end{cases}$$

*Let us suppose that $\rho(\lambda)$ satisfies the following conditions.*

1. *For each $x > 0$ there exists the integral*

$$\int_{-\infty}^{0} e^{\sqrt{|\lambda|}\,x}\, d\rho(\lambda).$$

2. *The function*

$$a(x) = \int_{1}^{\infty} \frac{\cos\sqrt{\lambda}\,x}{\lambda}\, d\sigma(\lambda)$$

*has a continuous fourth derivative.*

*If the function $\rho(\lambda)$ satisfies conditions* 1 *and* 2 *then there exists a continuous function $q(x)$ and an $h$ such that $\rho(\lambda)$ is the spectral function of equation* (1) *with boundary conditions* (1′). *Conversely, if the function $q(x)$ has a continuous derivative then the spectral function $\rho(\lambda)$ corresponding to it satisfies conditions* 1 *and* 2.

From the function $\sigma(\lambda)$ it is possible to determine the degree of smoothness of $q(x)$, namely, if the function $a(x)$ has derivatives of order $n+4$, then $q(x)$ has $n$ derivatives. Conversely, one may show that if $q(x)$ has an $(n+1)$th continuous derivative, then $a(x)$ has continuous derivatives of order $n+4$.

Since all the conditions imposed upon the spectral function $\rho(\lambda)$ refer only to its behavior at $+\infty$, one may make the following interesting deduction. *The spectral function $\rho(\lambda)$ may be an arbitrary monotonic function on a finite interval.* Moreover, if on a finite interval one takes $\rho(\lambda)$ as an arbitrary monotonic function and extends it so that, beginning with some $\lambda > 0$, it has the form $(2/\pi)\sqrt{\lambda}$, then the function $f(x, y)$, and thus also $K(x, y)$ and consequently $q(x)$, will be analytic functions. In this manner one may construct an equation with analytic coefficients and with a spectral function $\rho(\lambda)$ as badly behaved as one wishes upon a finite interval.

The proofs carried out in this article are sometimes cumbersone, but are simple in concept. All proofs may be significantly simplified if it is assumed that for $\lambda > 0$

$$\rho(\lambda) = \frac{2}{\pi} \sqrt{\lambda} + \sigma(\lambda),$$

where

$$\int_1^\infty \lambda \, | d\sigma(\lambda) | < +\infty,$$

for the case of a continuous spectrum (and the analogous inequality for the case of the pure point spectrum).

In §8 we examine the case when the spectral function is an orthogonal spectral function. This is analogous to the so-called determined case of the moment problem. We note that for the indeterminate case we have found all (and not only the orthogonal) spectral functions. We also note that for the case when the whole spectrum lies on the semi-axis $(a, \infty)$ one always has the determined case (see §8).

In the last paragraph we consider the classical problem of eigenvalues for the equation

$$y'' + (\lambda - q(x)) y = 0$$

on the finite interval $(a, b)$ and with boundary conditions at $a$ and $b$. If we designate by $\phi_n(x)$ the eigenfunctions satisfying the conditions

$$\varphi_n(a) = 1, \qquad \varphi_n'(a) = 0,$$

and if $\lambda_n$ are the eigenvalues and

$$\rho_n = \int_a^b \varphi_n^2(x) \, dx,$$

then the result obtained may be formulated as follows: *For every sequence of numbers* $\lambda_n$ *and* $\rho_n$ *satisfying the usual asymptotic equalities one may construct a function* $q(x)$. For more details see the corresponding paragraph.

## §1. Some facts about differential equations of the second order

1. Let $q(x) (0 \le x \le \infty)$ be continuous in each finite interval. Let us consider the differential equation

$$y'' + (\lambda - q(x)) y = 0, \tag{1.1}$$

where $\lambda$ is a real number. Let us designate by $\phi_1(x, \lambda)$ the solution of equation (1.1) satisfying the initial conditions

$$\varphi_1(0, \lambda) = 1, \qquad \varphi_1'(0, \lambda) = 0,$$

and by $\phi_2(x, \lambda)$ the solution of the same equation satisfying the initial conditions

$$\varphi_2(0, \lambda) = 0, \qquad \varphi_2'(0, \lambda) = 1.$$

Further, let us designate by $h$ an arbitrary real number. Then the function

$$\varphi(x, \lambda) = \varphi_1(x, \lambda) + h\varphi_2(x, \lambda)$$

is also a solution of equation (1.1) and satisfies the boundary condition

$$\varphi'(0, \lambda) - h\varphi(0, \lambda) = 0. \tag{1.2}$$

If $h = 0$, then $\phi(x, \lambda) = \phi_1(x, \lambda)$. For $h = \infty$, we assume by definition that $\phi(x, \lambda) = \phi_2(x, \lambda)$.

It is known that for each function $q(x)$ and for each fixed $h$ there exists at least one nondecreasing function $\rho(\lambda)$ $(-\infty \leq \lambda \leq +\infty)$ such that, for each function $f(x) \in L_2(0, \infty)$, the functions

$$E_n(\lambda) = \int_0^n f(x)\,\varphi(x, \lambda)\,dx$$

converge in the mean square with the measure $\rho(\lambda)$ to some function $E(\lambda)$, i.e.,

$$\lim_{n\to\infty} \int_{-\infty}^{+\infty} [E(\lambda) - E_n(\lambda)]^2\,d\rho(\lambda) = 0.$$

In addition, Parseval's equality holds:

$$\int_{-\infty}^{+\infty} E^2(\lambda)\,d\rho(\lambda) = \int_0^{+\infty} f^2(x)\,dx. \tag{1.3}$$

The function $\rho(\lambda)$ is called the *spectral function* of equation (1.1) for conditions (1.2).

2. In this paragraph we derive formulas expressing the eigenfunctions $\phi(x, \lambda)$ of equation (1.1) in terms of $\cos\sqrt{\lambda}\,x$, and analogous formulas expressing $\cos\sqrt{\lambda}x$ in terms of $\phi(x, \lambda)$. These formulas were considered earlier in ([4]) and ([5]). A new derivation of these formulas is given here.

We shall suppose that there exists the function $K(x, t)$ $(t \leq x)$ having continuous partial derivatives of first and second order and such that

$$\varphi(x, \lambda) = \cos\sqrt{\lambda}\,x + \int_0^x K(x, t)\cos\sqrt{\lambda}\,t\,dt. \tag{1.4}$$

We shall explain what conditions the function $K(x, t)$ must satisfy in order that $\phi(x, \lambda)$ be a solution of equation (1.1). Differentiating (1.4) twice with respect to $x$, we obtain:

$$\varphi''(x, \lambda) = -\lambda\cos\sqrt{\lambda}x + \frac{dK(x, x)}{dx}\cos\sqrt{\lambda}x - \sqrt{\lambda}K(x, x)\sin\sqrt{\lambda}x + \\ + \left.\frac{\partial K(x, t)}{\partial x}\right|_{t=x}\cos\sqrt{\lambda}x + \int_0^x \frac{\partial^2 K(x, t)}{\partial x^2}\cos\sqrt{\lambda}t\,dt. \tag{1.5}$$

Integrating the expression

$$\lambda\int_0^x K(x, t)\cos\sqrt{\lambda}\,t\cdot dt$$

twice by parts, we obtain:

$$\lambda \int_0^x K(x, t) \cos \sqrt{\lambda} t \, dt = \sqrt{\lambda} \sin \sqrt{\lambda} x \; K(x, x) + \left. \frac{\partial K(x, t)}{\partial t} \right|_{x=t} \cos \sqrt{\lambda} x -$$

$$- \left. \frac{\partial K(x, t)}{\partial t} \right|_{t=0} - \int_0^x \frac{\partial^2 K(x, t)}{\partial t^2} \cos \sqrt{\lambda} t \; dt.$$

Substituting this expression and (1.5) in equation (1.1), we obtain after reduction

$$\frac{dK(x, x)}{dx} \cdot \cos \sqrt{\lambda} x + \left( \frac{\partial K(x, t)}{\partial t} + \frac{\partial K(x, t)}{\partial x} \right)_{x=t} \cos \sqrt{\lambda} x + \left. \frac{\partial K(x, t)}{\partial t} \right|_{t=0} -$$

$$- q(x) \cos \sqrt{\lambda} x + \int_0^x \left[ \frac{\partial^2 K}{\partial x^2} - \frac{\partial^2 K}{\partial t^2} - q(x) K(x, t) \right] \cos \sqrt{\lambda} t \, dt = 0.$$

In view of the uniqueness of the representation of a function as a Fourier-Stieltjes integral, there follows from the last equation the following partial differential equation:

$$\frac{\partial^2 K(x, t)}{\partial x^2} - q(x) K(x, t) = \frac{\partial^2 K(x, t)}{\partial t^2} \tag{1.6}$$

and the boundary conditions:

$$\left. \frac{\partial K(x, t)}{\partial t} \right|_{t=0} = 0, \tag{1.7}$$

$$\frac{dK(x, x)}{dx} = \frac{1}{2} q(x). \tag{1.8}$$

In order to find the function $K(x, x)$ from (1.8) one must know $K(0, 0)$. If $\phi(x, \lambda)$ is represented in the form (1.4), then it is obvious that

$$\varphi(0, \lambda) = 1, \quad \varphi'(0, \lambda) = K(0, 0). \tag{1.8 }$$

Thus, in order that $\phi(x, \lambda)$ satisfy the boundary condition (1.2) one must set $K(0, 0) = h$. Thus, from (1.8) follows

$$K(x, x) = h + \frac{1}{2} \int_0^x q(t) \, dt. \tag{1.9}$$

If the function $q(x)$ has a continuous derivative then, as is known, there exists a unique solution of equation (1.6) satisfying conditions (1.7) and (1.9). Hence there exists a function $K(x, t)$ satisfying (1.4). Solving equation (1.4) as a Volterra equation with unknown function $\cos \sqrt{\lambda t}$, we obtain

$$\cos \sqrt{\lambda} x = \varphi(x, \lambda) - \int_0^x K_1(x, t) \varphi(t, \lambda) \, dt. \tag{1.10}$$

Analogously we may show that $K_1(x, t)$ is the solution of the equation

$$\frac{\partial^2 K_1(x, t)}{\partial x^2} = \frac{\partial^2 K_1(x, t)}{\partial t^2} - q(t) K_1(x, t),$$

satisfying the conditions

$$\left(\frac{\partial K_1}{\partial t} - hK_1\right)_{t=0} = 0,$$

$$K_1(x, x) = h + \frac{1}{2}\int_0^x q(x)\,dx.$$

It is clear that formulas (1.4) and (1.10) also hold for $\lambda < 0$.

Furthermore, it is evident from formula (1.4) that, knowing $K(x, t)$, we also the eigenfunctions $\phi(x, \lambda)$ of equation (1.1), and then the equation itself since

$$\lambda - q(x) = -\frac{\varphi''(x, \lambda)}{\varphi(x, \lambda)}.$$

3. Let us now consider separately the case $h = \infty$. We set

$$\psi(x, \lambda) = \frac{\sin\sqrt{\lambda}x}{\sqrt{\lambda}} + \int_0^x L(x, t)\frac{\sin\sqrt{\lambda}t}{\sqrt{\lambda}}\,dt. \qquad (1.4')$$

Reasoning as above, we obtain for $L(x, t)$ the equation

$$\frac{\partial^2 L(x, t)}{\partial x^2} - q(x)L(x, t) = \frac{\partial^2 L(x, t)}{\partial t^2}$$

and the initial conditions

$$L(x, 0) = 0,$$

$$L(x, x) = \frac{1}{2}\int_0^x q(x)\,dx.$$

Solving equation (1.4′) for the unknown function $\frac{\sin\sqrt{\lambda t}}{\sqrt{\lambda}}$, we obtain

$$\frac{\sin\sqrt{\lambda}x}{\sqrt{\lambda}} = \psi(x, \lambda) - \int_0^x L_1(x, t)\psi(t, \lambda)\,dt. \qquad (1.10')$$

Thus, we have obtained formulas analogous to formulas (1.4) and (1.10). Also, the functions $L(x, t)$ and $L_1(x, t)$ have the same properties as the functions $K(x, t)$ and $K_1(x, t)$.

We assumed in paragraphs 2 and 3 that the function $q(x)$ had a continuous derivative. In this case, the functions $K(x, t)$ and $K_1(x, t)$ from paragraph 2 and the analogous functions from paragraph 3 have continuous second derivatives. If one assumes that $q(x)$ is simply continuous, one may proceed in the following manner. We approximate $q(x)$ by a sequence of functions $q_n(x)$ having a continuous derivative. The corresponding functions $K_n(x, t)$ have continuous second derivatives and converge to the function $K(x, t)$. The function $K(x, t)$ itself is known to have first derivatives but does not have second derivatives. However, the functions $\phi(x, \lambda)$ constructed from $K(x, t)$ have second derivatives and satisfy equation (1.1). Actually,

$$\varphi_n''(x, \lambda) + (\lambda - q_n(x))\varphi_n(x, \lambda) = 0.$$

Since $K_n \to K$, the functions $\phi_n$ converge uniformly in $x$ to $\phi(x,\lambda)$, which means that the $\phi_n''(x,\lambda)$ converge uniformly also, i.e., $\phi(x,\lambda)$ has a continuous second derivative and satisfies equation (1.1).

## §2. Derivation of the nonlinear integral equation

1. Let us consider first the case $h = \infty$. Integrating both sides of formula (1.10) from $0$ to $x$ and interchanging the order of integration, we obtain

$$\frac{\sin\sqrt{\lambda}\,x}{\sqrt{\lambda}} = \int_0^x \varphi(t,\lambda)\,dt - \int_0^x \varphi(t,\lambda)\,dt \int_t^x K_1(u,t)\,du =$$
$$= \int_0^x \varphi(t,\lambda)\left[1 - \int_t^x K_1(u,t)\,du\right] dt.$$

Let us fix the value of $x$ in this formula. Then the above equation shows that* $\frac{\sin\sqrt{\lambda}x}{\sqrt{\lambda}}$ is a Fourier transform in the eigenfunctions of equation (1.1) of the following function of $t$

$$\Phi(x,t) = \begin{cases} 1 - \int_t^x K_1(u,t)\,du, & t \leqslant x, \\ 0, & t > x. \end{cases}$$

From Parseval's equation (1.3) follows

$$\int_{-\infty}^{+\infty} \frac{\sin\sqrt{\lambda}\,x \cdot \sin\sqrt{\lambda}\,y}{\lambda}\,d\rho(\lambda) = \int_0^y \Phi(x,t)\,\Phi(y,t)\,dt =$$
$$= \int_0^y dt - \int_0^y dt \int_t^x K_1(u,t)\,du - \int_0^y dt \int_t^y K_1(u,t)\,dt +$$
$$+ \int_0^y dt \int_t^x K_1(u,t)\,du \int_t^y K_1(v,t)\,dv.$$

If in the last formula one places $x = y$, one obtains (since the right side has meaning) that for each $x > 0$ there exists the integral

$$\int_{-\infty}^{0} \frac{\mathrm{sh}^2\sqrt{|\lambda|}\,x}{|\lambda|}\,d\rho(\lambda),$$

that is to say, the integral

$$\int_{-\infty}^{0} \mathrm{ch}\sqrt{|\lambda|}\,x\,d\rho(\lambda).$$

This fact has been mentioned earlier by V. A. Marčenko (2).

* For $\lambda < 0$ one understands $\frac{\sin\sqrt{\lambda}x}{\sqrt{\lambda}}$ to be $\frac{\mathrm{sh}\sqrt{|\lambda|}x}{\sqrt{|\lambda|}}$.

Let us now proceed to the derivation of the integral equation for $K_1(x, y)$. Since

$$\frac{\sin \sqrt{\lambda}\, x}{\sqrt{\lambda}} = \int_0^x \cos \sqrt{\lambda}\, t\, dt,$$

it follows from the Parseval formula for ordinary Fourier integrals ($y \leq x$) that

$$\int_0^y dt = \frac{2}{\pi} \int_0^\infty \frac{\sin \sqrt{\lambda}\, x \sin \sqrt{\lambda}\, y}{\lambda}\, d(\sqrt{\lambda}).$$

Consequently ($y \leq x$)

$$F(x, y) = \int_{-\infty}^{+\infty} \frac{\sin \sqrt{\lambda}\, x \cdot \sin \sqrt{\lambda}\, y}{\lambda}\, d\sigma(\lambda) = -\int_0^y dt \int_t^x K_1(u, t)\, du -$$

$$-\int_0^y dt \int_t^y K_1(u, t)\, du + \int_0^y dt \int_t^x K_1(u, t)\, du \int_t^y K_1(v, t)\, dv, \tag{2.1}$$

where

$$\sigma(\lambda) = \begin{cases} \rho(\lambda) - \frac{2}{\pi}\sqrt{\lambda}, & \lambda \geqslant 0, \\ \rho(\lambda), & \lambda < 0. \end{cases}$$

Since the right-hand side of the equation has a mixed derivative, $\partial^2 F/\partial x \partial y$ exists and, differentiating both sides of equation (2.1), we obtain for $K_1(x, y)$ the non-linear integral equation ($y \leq x$):

$$\frac{\partial^2 F}{\partial x\, \partial y} = f(x, y) = -K_1(x, y) + \int_0^y K_1(x, t)\, K_1(y, t)\, dt. \tag{I}$$

By $F(x, y)$ we understand

$$F(x, y) = \int_{-\infty}^{+\infty} \frac{\sin \sqrt{\lambda}\, x \cdot \sin \sqrt{\lambda}\, y}{\lambda}\, d\sigma(\lambda). \tag{2.2}$$

In case $\sigma(\lambda)$ behaves sufficiently well at $+\infty$ (for example, $\operatorname{var}[\sigma(\lambda)] < +\infty$), $f(x, y)$ may be given immediately by the formula

$$f(x, y) = \frac{\partial^2 F}{\partial x \partial y} = \int_{-\infty}^{+\infty} \cos \sqrt{\lambda}\, x . \cos \sqrt{\lambda}\, y\, d\sigma(\lambda). \tag{2.2'}$$

Since $K_1(x, x-0)$ is a continuous function and $f(x, y) = f(y, x)$, it follows from equation (I) that the function $f(x, y)$ is continuous for all values of the arguments.

Let $\epsilon$ be an arbitrary positive number. It follows from formula (2.1) that the following limit exists uniformly in each finite region:

$$\lim_{\varepsilon\to 0}\frac{F(x+\varepsilon,\ y+\varepsilon)-F(x-\varepsilon,\ y+\varepsilon)-F(x+\varepsilon,\ y-\varepsilon)+F(x-\varepsilon,\ y-\varepsilon)}{4\varepsilon^2}=$$

$$=\frac{\partial^2 F}{\partial x\,\partial y}=\lim_{\varepsilon\to 0}\int_{-\infty}^{+\infty}\left(\frac{\sin\varepsilon\sqrt{\lambda}}{\varepsilon\sqrt{\lambda}}\right)^2\cos\sqrt{\lambda}\,x\cos\sqrt{\lambda}\,y\,d\sigma(\lambda).$$

In particular, setting $x = y$ we obtain uniformly in each finite interval for $\epsilon \to 0$

$$\int_{-\infty}^{+\infty}\frac{\sin^2\varepsilon\sqrt{\lambda}}{\lambda}\cos^2\sqrt{\lambda}\,x\,d\sigma(\lambda)=o(\varepsilon).$$

For $x = 0$ we obtain

$$\int_{-\infty}^{+\infty}\frac{\sin^2\varepsilon\sqrt{\lambda}}{\lambda}\,d\sigma(\lambda)=o(\varepsilon).$$

From the last two equations it follows that uniformly in each finite interval

$$\int_{-\infty}^{+\infty}\frac{\sin^2\varepsilon\sqrt{\lambda}}{\lambda}\cos\sqrt{\lambda}\,x\,d\sigma(\lambda)=o(\varepsilon). \tag{2.3}$$

Condition (2.3) will play an important role hereafter.

2. Analogously one derives an integral equation for $h = \infty$. Integrating both sides of equation (1.9′) with the limits $0$ and $x$, we obtain

$$\frac{1-\cos\sqrt{\lambda}\,x}{\lambda}=\int_0^x\psi(t,\lambda)\,\Phi(x,t)\,dt,$$

where

$$\Phi(x,t)=\begin{cases}1-\int_t^x L_1(u,t)\,du, & t\leqslant x,\\ 0, & t>x.\end{cases}$$

Consequently, it follows from Parseval's equation $(y \leq x)$:

$$\int_{-\infty}^{+\infty}\frac{(1-\cos\sqrt{\lambda}\,x)\,(1-\cos\sqrt{\lambda}\,y)}{\lambda^2}\,d\rho(\lambda)=\int_0^y dt-\int_0^y dt\int_t^x L_1(u,t)\,du-$$

$$-\int_0^y dt\int_t^y L_1(u,t)\,dt+\int_0^y dt\int_t^x L_1(u,t)\,du\int_t^y L_1(v,t)\,dv.$$

Since

$$\int_0^x\frac{\sin\sqrt{\lambda}\,t}{\sqrt{\lambda}}\,dt=\frac{1-\cos\sqrt{\lambda}\,x}{\lambda},$$

it follows from Parseval's formula for ordinary Fourier integrals that for $y \leq x$

$$\int_0^y dt=\frac{2}{3\pi}\int_0^\infty\frac{(1-\cos\sqrt{\lambda}\,x)\,(1-\cos\sqrt{\lambda}\,y)}{\lambda^2}\,d\left(\lambda^{3/2}\right).$$

Consequently,

$$F(x, y) = \int_{-\infty}^{+\infty} \frac{(1-\cos\sqrt{\lambda}\,x)\,(1-\cos\sqrt{\lambda}\,y)}{\lambda^2}\,d\sigma(\lambda) = -\int_0^y dt \int_t^x L_1(u, t)\,du -$$

$$-\int_0^y dt \int_t^y L_1(u, t)\,du + \int_0^y dt \int_t^x L_1(u, t)\,du \int_t^y L_1(v, t)\,dv,$$

where

$$\sigma(\lambda) = \begin{cases} \rho(\lambda) - \frac{2}{3\pi}\lambda^{3/2}, & \lambda \geqslant 0 \\ \rho(\lambda), & \lambda < 0. \end{cases}$$

Reasoning as in paragraph 1, we obtain for $L_1(x, y)$ the nonlinear integral equation $(y \leq x)$:

$$\frac{\partial^2 F(x, y)}{\partial x\,\partial y} = f(x, y) = -L_1(x, y) + \int_0^y L_1(x, t)\,L_1(y, t)\,dt.$$

In place of condition (2.3) we obtain the condition that for $\epsilon \longrightarrow 0$

$$\int_{-\infty}^{+\infty} \frac{\sin^2 \varepsilon\sqrt{\lambda}}{\lambda} \cdot \frac{\sin^2 \sqrt{\lambda}\,x}{\lambda}\,d\sigma(\lambda) = o(\varepsilon) \qquad (2.3')$$

uniformly in each finite interval.

## §3. Derivation of the linear integral equation

For each function $K(x, t)$ carrying $\cos\sqrt{\lambda x}$ into $\phi(x, \lambda)$ [formula (1.8)] we introduce a linear integral equation which is basic for the following work. It is easy to obtain it from the nonlinear equation. However, we shall derive it directly since further on (in the construction of $K(x, y)$ from $\rho(\lambda)$) we shall make use of the reasoning introduced below.

This equation has the following form:

$$f(x, y) + \int_0^x K(x, s)\,f(s, y)\,ds + K(x, y) = 0 \quad (y \leqslant x), \qquad \text{(II)}$$

where the function $f(x, y)$ is defined in the preceding section.

Let us proceed to the derivation of this equation. We show first that for $b < y < a < x$ the functions

$$\int_a^x \varphi(t, \lambda)\,dt \quad \text{and} \quad \int_0^y \cos\sqrt{\lambda}\,t\,dt$$

are orthogonal with the weight $\rho(\lambda)$, i.e., that

$$I = \int_{-\infty}^{+\infty} \left[\int_a^x \varphi(t, \lambda)\,dt\right]\left[\int_b^y \cos\sqrt{\lambda}\,t\,dt\right] d\rho(\lambda) = 0. \qquad (3.1)$$

In order to derive (3.1) we express $\cos\sqrt{\lambda t}$ in terms of $\phi(t,\lambda)$ according to formula (1.10):

$$h(b,y) = \int_b^y \cos\sqrt{\lambda}\, t\, dt =$$

$$= \int_b^y \varphi(t,\lambda)\,dt - \int_b^y dt \int_0^t K_1(t,s)\,\varphi(s,\lambda)\,ds =$$

$$= \int_b^y \varphi(t,\lambda)\,dt - \int_0^b \varphi(s,\lambda)\,ds \int_b^y K_1(t,s)\,dt - \int_b^y \varphi(s,\lambda)\,ds \int_s^y K_1(t,s)\,dt.$$

Consequently, the function $h(b,y)$ is a Fourier transform (in the functions $\phi(t,\lambda)$) of a function equal to $0$ outside of the interval $(b,y)$. Since the intervals $(b,y)$ and $(a,x)$ do not intersect, one has from Parseval's equation (1.3) $I = 0$.

In order to derive the integral equation (II) we now express in equation (3.1) $\phi(t,\lambda)$ in terms of $\cos\sqrt{\lambda t}$ [according to formula (1.4)]

$$\int_a^x \varphi(t,\lambda)\,dt = \int_a^x \cos\sqrt{\lambda}\, t\, dt + \int_0^a \cos\sqrt{\lambda}\, s\, ds \int_a^x K(t,s)\,dt +$$

$$+ \int_a^x \cos\sqrt{\lambda}\, s\, ds \int_s^x K(t,s)\,dt.$$

Thus,

$$I = \int_{-\infty}^{+\infty} \left[\int_a^x \cos\sqrt{\lambda}\, s\, ds\right]\left[\int_b^y \cos\sqrt{\lambda}\, s\, ds\right] d\rho(\lambda) +$$

$$+ \int_{-\infty}^{+\infty} \left[\int_0^a \cos\sqrt{\lambda}\, s . ds \int_a^x K(t,s)\,dt +\right.$$

$$\left. + \int_a^x \cos\sqrt{\lambda}\, s\, ds \int_s^x K(t,s)\,dt\right]\left[\int_b^y \cos\sqrt{\lambda}\, s\, ds\right] d\rho(\lambda) = 0. \qquad (3.1')$$

From Parseval's formula for ordinary Fourier integrals we have

$$\frac{2}{\pi}\int_0^\infty \left[\int_a^x \cos\sqrt{\lambda}\, s . ds\right]\left[\int_b^y \cos\sqrt{\lambda}\, s\, ds\right] d(\sqrt{\lambda}) = 0. \qquad (3.2)$$

Subtracting equation (3.2) from equation (3.1′), we obtain

$$\int_{-\infty}^{+\infty}\left[\int_a^x \cos\sqrt{\lambda}\,s\,ds\right]\left[\int_b^y \cos\sqrt{\lambda}\,s\,ds\right]d\sigma(\lambda)+\int_{-\infty}^{+\infty}\left[\int_0^a \cos\sqrt{\lambda}\,s\,ds\int_a^x K(t,s)\,dt+\right.$$

$$\left.+\int_a^x \cos\sqrt{\lambda}\,s\,ds\int_s^x K(t,s)\,dt\right]\left[\int_b^y \cos\sqrt{\lambda}\,s\,ds\right]d\sigma(\lambda)+$$

$$+\frac{2}{\pi}\int_0^\infty\left[\int_0^a \cos\sqrt{\lambda}\,s\,ds\int_a^x K(t,s)\,dt+\int_a^x \cos\sqrt{\lambda}\,s\,ds\int_s^x K(t,s)\,dt\right]\cdot$$

$$\cdot\left[\int_b^y \cos\sqrt{\lambda}\,s\,ds\right]d(\lambda^{1/2})=\int_{-\infty}^{+\infty}\frac{(\sin\sqrt{\lambda}\,x-\sin\sqrt{\lambda}\,a)\,(\sin\sqrt{\lambda}\,y-\sin\sqrt{\lambda}\,b)}{\lambda}\,d\sigma(\lambda)+$$

$$+\int_{-\infty}^{+\infty}\left[\int_0^a \cos\sqrt{\lambda}\,s\,ds\int_a^x K(t,s)\,dt+\int_a^x \cos\sqrt{\lambda}\,s\,ds\int_s^x K(t,s)\,dt\right]\cdot$$

$$\cdot\left[\int_b^y \cos\sqrt{\lambda}\,s\,ds\right]d\sigma(\lambda)+\int_b^y ds\int_a^x K(t,s)\,dt=0. \tag{3.3}$$

For the definition of $\sigma(\lambda)$ see the prededing section. The summand

$$\int_b^y ds\int_a^x K(t,s)dt$$

is obtained from the last summand of the left side by applying Parseval's equation for ordinary Fourier integrals.

As before, we set

$$F(x,y)=\int_{-\infty}^{+\infty}\frac{\sin\sqrt{\lambda}\,x\,\sin\sqrt{\lambda}\,y}{\lambda}\,d\sigma(\lambda).$$

If the expression for $F(x,y)$ could be differentiated under the integral sign, then the integral equation (II) could be obtained directly from (3.3) by differentiation. In order not to impose superfluous conditions upon $\sigma(\lambda)$ we proceed in the following manner. Let us set

$$H(x,s)=\begin{cases}\int_a^x K(t,s)\,dt, & 0\leqslant s\leqslant a,\\ \int_s^x K(t,s)\,dt, & a\leqslant s\leqslant x,\\ 0, & s>x.\end{cases}$$

Equation (3.3) may be written in the form

$$F(x,y)-F(x,b)-F(a,y)+F(a,b)+$$
$$+\int_{-\infty}^{+\infty}\left[\int_0^x H(x,s)\cos\sqrt{\lambda}\,s\,ds\right]\left[\int_b^y \cos\sqrt{\lambda}\,s\,ds\right]d\sigma(\lambda)+\int_b^y ds\left[\int_a^x K(t,s)\,dt\right]=0. \tag{3.4}$$

Since $H(x,x)=0$ and $\partial H/\partial s$ is bounded, then, integrating by parts, we obtain

$$F(x,y)-F(x,b)-F(a,y)+F(a,b)-$$
$$-\int_{-\infty}^{+\infty}\left[\int_0^x \frac{\partial H}{\partial s}\cdot\frac{\sin\sqrt{\lambda}s}{\sqrt{\lambda}}\,ds\right]\left[\frac{\sin\sqrt{\lambda}y-\sin\sqrt{\lambda}b}{\sqrt{\lambda}}\right]d\sigma(\lambda)+\int_b^y ds\int_a^x K(t,s)\,dt=0.$$

Changing the order of integration, we find

$$F(x,y)-F(x,b)-F(a,y)+F(a,b)-$$
$$-\int_0^x \frac{\partial H}{\partial s}\cdot[F(s,y)-F(s,b)]\,ds+\int_b^y ds\int_a^x K(t,s)\,dt=0.$$

Integrating by parts, we obtain the equation

$$\int_0^x \frac{\partial H}{\partial s}[F(s,y)-F(s,b)]\,ds=-\int_0^x H(x,s)\left[\frac{\partial F(s,y)}{\partial s}-\frac{\partial F(s,b)}{\partial s}\right]ds.$$

Consequently, the following equation holds:

$$F(x,y)-F(x,b)-F(a,y)+F(a,b)+$$
$$+\int_0^x H(x,s)\left[\frac{\partial F(s,y)}{\partial s}-\frac{\partial F(s,b)}{\partial s}\right]ds+\int_b^y ds\int_a^x K(t,s)\,dt=0. \tag{3.5}$$

Changing the order of integration, it is easy to establish that

$$\int_0^x H(x,s)\left[\frac{\partial F(s,y)}{\partial s}-\frac{\partial F(s,b)}{\partial s}\right]ds=\int_a^x dt\int_0^t K(t,s)\left[\frac{\partial F(s,y)}{\partial s}-\frac{\partial F(s,b)}{\partial s}\right]ds.$$

Thus it follows from equation (3.5) that

$$F(x,y)-F(x,b)-F(a,y)+F(a,b)+$$
$$+\int_a^x dt\int_0^t K(t,s)\left[\frac{\partial F(s,y)}{\partial s}-\frac{\partial F(s,b)}{\partial s}\right]ds+\int_b^y ds\int_a^x K(t,s)\,dt=0.$$

Differentiating the last equation with respect to $x$ and $y$, we obtain

$$\frac{\partial^2 F}{\partial x\,\partial y}+\int_0^x K(x,s)\frac{\partial^2 F(s,y)}{\partial s\cdot\partial y}\,ds+K(x,y)=0.$$

Setting $\partial^2 F/\partial x\,\partial y=f(x,y)$, we arrive at the integral equation:

$$f(x,y)+\int_0^x K(x,s)\,f(s,y)\,ds+K(x,y)=0\quad(y\leqslant x). \tag{II}$$

For each fixed $x$ equation (II) is a linear Fredholm equation with the symmetric kernel $f(x,y)$ and the unknown function $K(x,y)$ ($x$ fixed).

## §4. Investigation of the linear integral equation

1. Beginning with this section we shall take up the inverse problem, namely, the construction of the differential equation

$$y'' + (\lambda - q(x))\, y = 0$$

from the spectral function $\rho(\lambda)$. We shall obtain conditions on the function $\rho(\lambda)$ which are sufficient for there to exist a differential equation with a continuous function $q(x)$. These conditions are also necessary if one requires $q(x)$ to have a continuous derivative.

These conditions are as follows:

1. *For each* $x$ *there exists the integral*

$$\int_{-\infty}^{0} e^{\sqrt{|\lambda|}\, x}\, d\rho(\lambda)\,.$$

2. *Let us place* $\rho(\lambda) = \frac{2}{\pi}\sqrt{\lambda} + \sigma(\lambda)$ *for* $\lambda > 0$ *and* $\rho(\lambda) = \sigma(\lambda)$ *for* $\lambda < 0$. *We require that the function*

$$a(x) = \int_{1}^{\infty} \frac{\cos\sqrt{\lambda}\, x}{\lambda}\, d\sigma(\lambda) \tag{4.1}$$

*have a continuous fourth derivative.*

It follows easily from conditions 1 and 2 that the function

$$F(x, y) = \int_{-\infty}^{+\infty} \frac{\sin\sqrt{\lambda}\, x \cdot \sin\sqrt{\lambda}\, y}{\lambda}\, d\sigma(\lambda) \tag{4.2}$$

has continuous fourth derivatives.

We note that, conversely, from the differentiability of (4.2) follows the differentiability of (4.1).

The necessity of condition 1 was proved earlier. If $q(x)$ has a continuous derivative, then $K(x, t)$ has a continuous second derivative. It follows from the nonlinear integral equation that $f(x, y)$ has a continuous second derivative, which means that the function $F(x, y)$ defined by the formula

$$F(x, y) = \int_{-\infty}^{+\infty} \frac{\sin\sqrt{\lambda}\, x \cdot \sin\sqrt{\lambda}\, y}{\lambda}\, d\sigma(\lambda)\,,$$

has continuous fourth derivatives.

In view of condition 1,

$$\int_{1}^{\infty} \frac{\sin\sqrt{\lambda}\, x \cdot \sin\sqrt{\lambda}\, y}{\lambda} \cdot d\sigma(\lambda)$$

has also fourth derivatives. It follows from this that also $a(x)$ has fourth derivatives.

Conversely, it is clear that it follows from conditions 1 and 2 that $F(x, y)$ has continuous fourth derivatives.

For the time being, we exclude from consideration the case when the spectrum is purely discrete, tending to $+\infty$. This case will be treated separately in §9.

Since $f(x, y)$ is a continuous function by hypothesis,

$$f(x, y) = \lim_{\varepsilon\to 0} \frac{F(x+\varepsilon, y+\varepsilon) - F(x-\varepsilon, y+\varepsilon) - F(x+\varepsilon, y-\varepsilon) + F(x-\varepsilon, y-\varepsilon)}{4\varepsilon^2} =$$

$$= \lim_{\varepsilon\to 0} \int_{-\infty}^{+\infty} \left(\frac{\sin\sqrt{\lambda}\varepsilon}{\varepsilon\sqrt{\lambda}}\right)^2 \cos\sqrt{\lambda}\,x \cos\sqrt{\lambda}\,y \, d\sigma(\lambda). \tag{4.3}$$

As was shown in §2, it follows from this that as $\epsilon \to 0$ one has uniformly in each finite interval of variation of $x$

$$\int_{-\infty}^{+\infty} \frac{\sin^2\sqrt{\lambda}\varepsilon}{\lambda} \cdot \cos\sqrt{\lambda}\,x \, d\sigma(\lambda) = o(\varepsilon). \tag{4.3'}$$

From the mean-value theorem, we have for $\lambda < 0$

$$\frac{\sin\sqrt{\lambda}\,\varepsilon}{\sqrt{\lambda}\,\varepsilon} = \mathrm{ch}\sqrt{|\lambda|}\,\theta\varepsilon \quad (0<\theta<1).$$

Thus

$$\int_{-\infty}^{0} \left(\frac{\sin\sqrt{\lambda}\,\varepsilon}{\varepsilon\sqrt{\lambda}}\right)^2 \cos\sqrt{\lambda}x \, d\rho(\lambda) = \int_{-\infty}^{0} \mathrm{ch}^2\sqrt{\lambda}\,\theta\varepsilon \cdot \mathrm{ch}\sqrt{|\lambda|}\,x \, d\rho(\lambda) < A < \infty$$

uniformly in each finite interval. From this we have uniformly in each finite interval as $\epsilon \to 0$

$$\int_{-\infty}^{0} \frac{\sin^2\sqrt{\lambda}\,\varepsilon}{\lambda} \cos\sqrt{\lambda}\,x \, d\rho(\lambda) = o(\varepsilon)$$

which means, in view of (4.3′), that as $\epsilon \to 0$ we have

$$\int_{0}^{\infty} \frac{\sin^2\sqrt{\lambda}\,\varepsilon}{\lambda} \cos\sqrt{\lambda}\,x \, d\sigma(\lambda) = o(\varepsilon) \tag{4.4}$$

uniformly in each finite interval. Condition (4.4) plays an essential role in the following.

We need in addition the following inequality

$$\int_{a}^{\infty} \frac{d\rho(\lambda)}{\lambda} < +\infty, \quad a > 0.$$

Let us prove it. Setting $x = y = 0$ in (4.3) we have:

$$\int_{0}^{\infty} \left(\frac{\sin\sqrt{\lambda}\,\varepsilon}{\sqrt{\lambda}\,\varepsilon}\right)^2 d\sigma(\lambda) = O(1).$$

From the ordinary Parseval equation we have

$$\frac{2}{\pi}\int_0^\infty \left(\frac{\sin\sqrt{\lambda}\,\varepsilon}{\sqrt{\lambda}\varepsilon}\right)^2 d\sqrt{\lambda} = \varepsilon^{-1}.$$

We thus obtain for $\rho(\lambda)$

$$\int_0^\infty \left(\frac{\sin\sqrt{\lambda}\varepsilon}{\sqrt{\lambda}\varepsilon}\right)^2 d\rho(\lambda) = \frac{1}{\varepsilon} + O(1).$$

Since $\frac{\sin \epsilon\sqrt{\lambda}}{\epsilon\sqrt{\lambda}} \geq \frac{2}{\pi}$ for $\epsilon\sqrt{\lambda} \leq \frac{\pi}{2}$ we have

$$\int_0^\infty \left(\frac{\sin\sqrt{\lambda}\varepsilon}{\sqrt{\lambda}\varepsilon}\right)^2 d\rho(\lambda) \geqslant \int_0^{\frac{1}{\varepsilon^2}} \left(\frac{\sin\sqrt{\lambda}\varepsilon}{\sqrt{\lambda}\varepsilon}\right)^2 d\rho(\lambda) \geqslant$$

$$\geqslant \frac{4}{\pi^2}\int_0^{1/\varepsilon^2} d\rho(\lambda) = \frac{4}{\pi^2}\left(\rho\left(\frac{1}{\varepsilon^2}\right) - \rho(0)\right).$$

Thus

$$\rho\left(\frac{1}{\varepsilon^2}\right) - \rho(0) \leqslant \frac{\pi^2}{4}\left(\frac{1}{\varepsilon} + O(1)\right)$$

which means

$$\rho(\lambda) \leqslant c\sqrt{\lambda} + O(1).$$

It follows easily from this that $\int_a^\infty \lambda^{-1} d\rho(\lambda)$ converges.

2. Let there be given a function $\rho(\lambda)$ satisfying conditions 1 and 2 and having an infinite number of points of increase on some finite interval. Let us construct with it the function

$$F(x, y) = \int_{-\infty}^{+\infty} \frac{\sin\sqrt{\lambda}x \cdot \sin\sqrt{\lambda}y}{\lambda}\, d\sigma(\lambda) \text{ and } f(x, y) = \frac{\partial^2 F}{\partial x\, \partial y}.$$

We then have the following theorem.

**Theorem.** *For fixed x the integral equation*

$$f(x, y) + \int_0^x f(y, s) K(x, s)\, ds + K(x, y) = 0 \tag{4.5}$$

*for the unknown function K(x, y) has a solution which is unique.*

**Proof.** As is known, it is sufficient for the solvability of equation (4.5) that the homogeneous equation ($x$ fixed)

$$\int_0^x f(s, y) h(s)\, ds + h(y) = 0 \tag{4.6}$$

does not have a solution other than the trivial one. For this we show that the quadratic form

$$I(g) = \int_0^x\int_0^x f(s, y) g(s) g(y)\, ds\, dy + \int_0^x g^2(y)\, dy > 0 \tag{4.7}$$

for an arbitrary function $g(y)$ is not $0$.

Let us suppose to start with that $g(y)$ $(0 \leq y \leq x)$ is such that

1) $g(x) = 0$,

2) $g'(y)$ is continuous for $0 \leq y \leq x$.

Since $f(y, s) = \partial^2 F/\partial y\, \partial s$, we obtain, integrating the first summand in (4.7) by parts,

$$I(g) = \int_0^x g^2(y)\, dy + \int_0^x \int_0^x F(y, s)\, g'(y)\, g'(s)\, dy\, ds =$$

$$= \int_0^x g^2(y)\, dy + \int_0^x \int_0^x g'(y)\, g'(s)\, dy\, ds \int_{-\infty}^{+\infty} \frac{\sin\sqrt{\lambda} y}{\sqrt{\lambda}} \cdot \frac{\sin\sqrt{\lambda} s}{\sqrt{\lambda}}\, d\rho(\lambda) -$$

$$- \frac{2}{\pi} \int_0^x \int_0^x g'(y)\, g'(s)\, dy\, ds \int_0^\infty \frac{\sin\sqrt{\lambda} y}{\sqrt{\lambda}} \cdot \frac{\sin\sqrt{\lambda} s}{\sqrt{\lambda}}\, d(\sqrt{\lambda}) =$$

$$= \int_0^x g^2(y)\, dy + \int_{-\infty}^{+\infty} G^2(\lambda)\, d\rho(\lambda) - \frac{2}{\pi} \int_0^{+\infty} G^2(\lambda)\, d(\sqrt{\lambda}),$$

where

$$G(\lambda) = \int_0^x \frac{\sin\sqrt{\lambda}\, t}{\sqrt{\lambda}}\, g'(t)\, dt = \int_0^x \cos\sqrt{\lambda}\, t \cdot g(t)\, dt .$$

From Parseval's equation for ordinary Fourier integrals we have

$$\frac{2}{\pi} \int_0^\infty G^2(\lambda)\, d(\sqrt{\lambda}) = \int_0^x g^2(y)\, dy .$$

Thus

$$I(g) = \int_{-\infty}^{+\infty} G^2(\lambda) \cdot d\rho(\lambda) . \tag{4.8}$$

We shall show that $g(t) = 0$ follows from $I(g) = 0$. Since

$$G(\lambda) = \int_0^x \cos\sqrt{\lambda} t \cdot g(t)\, dt ,$$

$G(\lambda)$ is an entire analytic function of exponential growth in $\lambda$. Consequently, $G(\lambda)$ may have only isolated zeros with a limit point of $\infty$. If $\rho(\lambda)$ has an infinite number of points of increase on some finite interval, then it follows from the equation $I(g) = 0$, because of (4.8), that $G(\lambda) = 0$, i.e.,

$$\int_0^x \cos\sqrt{\lambda}\, t\, g(t)\, dt = 0$$

which means $g(t) = 0$.

Now let $g(y)$ be a continuous bounded function, not satisfying conditions 1) and 2). Let $g_n(y)$ converge in the mean square as $n \to \infty$ to $g(y)$ and satisfy conditions 1) and 2). It is easy to see that

$$\lim_{n\to\infty} I(g_n) = I(g) .$$

Consequently,

$$I(g) = \lim_{n\to\infty} \int_{-\infty}^{+\infty} G_n^2(\lambda)\, d\rho(\lambda),$$

where

$$G_n(\lambda) = \int_0^x g_n(y) \cos \sqrt{\lambda} y \, dy.$$

Since $g_n(y)$ converges in the mean to $g(y)$, $G_n(\lambda)$ converges uniformly to $G(\lambda)$. If $I(g) = 0$, i.e.,

$$\int_{-\infty}^{+\infty} G_n^2(\lambda)\, d\rho(\lambda) \to 0 ,$$

then, since the function under the integral is nonnegative and approaches a limit, this is possible only for $G(\lambda) \equiv 0$. From this we obtain as before that $g(t) = 0$.

Let us now show that the homogeneous equation (4.6) has only the trivial solution. Let $g(t)$ be a solution of equation (4.6). Multiply it by $g(y)$ and integrate from $0$ to $x$. We obtain $I(g) = 0$, which means $g(y) \equiv 0$. Thus, we have shown that the linear equation (4.5) is solvable.

Since, by hypothesis, $f(x, y)$ is a continuous function, it follows from equation (4.5) that for each fixed $x$ the function $K(x, y)$ is continuous in the variable $y$ for $y \leq x$. Furthermore, it follows from the same equation that if there exist the continuous derivatives $\partial^r f/\partial y^r$ $(r = 1, 2, \cdots, n)$, then there exist also the continuous derivatives $\partial^r K(x, y)/\partial y^r$.

3. The investigation of the behavior of the function $K(x, y)$ with respect to its first argument is somewhat more complicated. For this we make use of the following lemma:

**Lemma.** *Let there be given the integral equation*

$$g(x, a) = h(x, a) + \int_0^1 H(x, y; a)\, h(y, a)\, dy , \tag{4.9}$$

*in which the kernel and the free term $g(x, a)$ are continuous functions of the parameter $a$ and the independent variables. Then, if the homogeneous equation has only the trivial solution for $a = a_0$, the solution $h(x, a)$ is a continuous function of $x$ and $a$ in some neighborhood of the point $a = a_0$. If $H$ and $g$ have $n$ continuous derivatives with respect to $a$, then the same holds for $h(x, a)$.*

**Proof.** Let us place

$$H(x, y, a) = H(x, y, a_0) + H_1(x, y) = H_0 + H_1,$$

where $|H_1(x, y)| < \epsilon$ in case $a$ is in some sufficiently small neighborhood of $a_0$. Equation (4.9) may be written symbolically in the form

$$g = h + Hh = h + H_0 h + H_1 h.$$

Applying the operator $(E + H_0)^{-1}$ to both sides of the last equation, we obtain

$$(E + H_0)^{-1} g = h + (E + H_0)^{-1} H_1 h. \tag{4.10}$$

Since the norm of the operator $(E + H_0)^{-1} H_1$ may be made arbitrarily small, equation (4.10) may be solved by the method of successive approximations. The lemma is proved.

Let us now apply the lemma to equation (II). We shall investigate the neighborhood of some point $x_0$. In equation (II) change $s$ to $sx$ and $y$ to $yx$. We obtain the equation

$$f(y, yx) + x\int_0^1 K(x, sx) f(sx, yx)\, ds + K(y, yx) = 0,$$

i.e., the integral equation with the kernel $xf(sx, yx)$ and the free term $f(x, yx)$; here $x$ is a parameter. Since $f(x, y)$ is a continuous function by hypothesis, the continuity of the function $K(x, y)$ in the whole set of variables follows from the lemma. Furthermore, it follows from the same lemma that $K(x, y)$ has continuous derivatives with respect to $x$ of the same order as $f(x, y)$.

4. In order to prove the solubility of the equation for $h = \infty$ we set

$$F(x, y) = \int_{-\infty}^{+\infty} \frac{(1 - \cos\sqrt{\lambda}\, x)(1 - \cos\sqrt{\lambda}\, y)}{\lambda^2}\, d\sigma(\lambda),$$

where

$$\sigma(\lambda) = \begin{cases} \rho(\lambda) - \dfrac{3}{2\pi}\lambda^{3/2}, & \text{if} \quad \lambda \geqslant 0. \\ \rho(\lambda), & \text{if} \quad \lambda < 0 \end{cases}$$

and

$$f(x, y) = \frac{\partial^2 F}{\partial x \partial y}.$$

$L(x, y)$ satisfies the same equation as $K(x, y)$. In place of equation (4.4) we have

$$\int_0^\infty \frac{\sin^2 \varepsilon\sqrt{\lambda}}{\lambda} \cdot \frac{\sin^2 \varepsilon\sqrt{\lambda}\, x}{\lambda}\, d\sigma(\lambda) = o(\varepsilon) \tag{4.4'}$$

uniformly in each finite interval of variation of $x$. In place of $G(\lambda)$ we introduce

$$H(\lambda) = \int_0^x g(y) \frac{\sin\sqrt{\lambda}\, y}{\sqrt{\lambda}}\, dy.$$

The remainder of the proof goes over without change.

## §5. Derivation of Parseval's equation

1. In the preceding section we showed that for a given monotonic function $\rho(\lambda)$ satisfying conditions 1 and 2 there exists a solution $K(x, t)$ of the linear integral equation (4.5). More precisely, for the solvability of this equation it is sufficient to replace condition 2 by a weaker one: the existence for the function $F(x, y)$ of a continuous derivative $f(x, y) = \partial^2 F/\partial x \partial y$. By means of the function $K(x, t)$ it is possible to construct the functions $\phi(x, \lambda)$ by the formula

$$\varphi(x, \lambda) = \cos \sqrt{\lambda}\, x + \int_0^x K(x, t) \cos \sqrt{\lambda}\, t\, dt.$$

In this paragraph we shall show that the functions $\phi(x, \lambda)$ are "orthonormal" functions with the weight $\rho(\lambda)$, i.e., more precisely, Parseval's equation holds for them in the form in which it was given in §1, paragraph 1. We shall again not make complete use of condition 2, but shall assume that $f(x, y) = \partial^2 F/\partial x \partial y$ has continuous derivatives of the first order.

Let us proceed to the proof of Parseval's equation.

We shall show first that the functions

$$h_1(\lambda) = \int_a^x \varphi(t, \lambda)\, dt \quad \text{and} \quad h_2(\lambda) = \int_b^y \varphi(t, \lambda)\, dt$$

in the case when the intervals $(a, x)$ and $(b, y)$ do not intersect are orthogonal with the weight $\rho(\lambda)$, i.e., we shall prove the equation

$$\int_{-\infty}^{+\infty} h_1(\lambda)\, h_2(\lambda)\, d\rho(\lambda) = 0. \tag{5.1}$$

Without restricting the generality, we shall assume that $b < y \leq a < x$. Reversing the order of the reasoning used in §3 for the linear integral equation (II), we obtain

$$\int_{-\infty}^{+\infty} \left[\int_a^x \varphi(t, \lambda)\, dt\right] \left[\int_b^y \cos \sqrt{\lambda}\, t . dt\right] d\rho(\lambda) = 0, \quad b < y \leqslant a < x, \tag{5.2}$$

where

$$\varphi(t, \lambda) = \cos \sqrt{\lambda}\, t + \int_0^t K(t, s) \cos \sqrt{\lambda}\, s\, ds. \tag{5.3}$$

We shall show that from formula (5.2) one may obtain the following equation of which we have need:

$$\int_{-\infty}^{+\infty} \left[\int_a^x \varphi(t, \lambda)\, dt\right] \left[\int_b^y \varphi(t, \lambda)\, dt\right] d\rho(\lambda) = 0. \tag{5.4}$$

Indeed, it follows from formula (5.3) that

$$\int_b^y \varphi(t,\lambda)\,dt = \int_b^y \cos\sqrt{\lambda}\,t\,dt + \int_0^b \cos\sqrt{\lambda}\,s\,ds\int_b^y K(t,s)\,dt +$$

$$+\int_b^y \cos\sqrt{\lambda}\,s\,ds\int_s^y K(t,s)\,dt = \int_b^y \cos\sqrt{\lambda}\,t\,dt + \int_0^y H(s,y)\cos\sqrt{\lambda}\,s\,ds,$$

where the function

$$H(s,y) = \begin{cases} \int_b^y K(t,s)\,dt, & 0\leqslant s\leqslant b,\\ \int_s^y K(t,s)\,dt, & b\leqslant s\leqslant y \end{cases}$$

is continuous and has a derivative $\partial H/\partial s$.

Substituting the expression for $\int_b^y \phi(t,\lambda)dt$ in the left side of equation (5.4), and making use of equation (5.2), we obtain

$$\int_{-\infty}^{+\infty}\Big[\int_a^x \varphi(t,\lambda)\,dt\Big]\Big[\int_b^y \varphi(t,\lambda)\,dt\Big]\,d\rho(\lambda)=$$

$$=\int_{-\infty}^{+\infty}\Big[\int_a^x \varphi(t,\lambda)\,dt\Big]\Big[\int_0^y H(t,y)\cos\sqrt{\lambda}\,t\,dt\Big]\,d\rho(\lambda).$$

We shall now show that the last integral is equal to zero for each fixed $y < a$. For this we again make use of equation (5.2).

Let $P(s,y)$ $(s \leq y)$ be an arbitrary function. Replace $b$ by $s$ in formula (5.2), multiply both sides by $P(s,y)$, and integrate with respect to $s$ between the limits $0$ and $y$. We obtain (changing the order of integration)

$$\int_{-\infty}^{+\infty}\Big[\int_a^x \varphi(t,\lambda)\,dt\Big]\Big[\int_0^y P(s,y)\,ds\int_s^y \cos\sqrt{\lambda}\,t\,dt\Big]\,d\rho(\lambda) =$$

$$=\int_{-\infty}^{+\infty}\Big[\int_a^x \varphi(t,\lambda)\,dt\Big]\Big[\int_0^y \cos\sqrt{\lambda}\,t\,dt\int_0^t P(s,y)\,ds\Big]\,d\rho(\lambda) = 0.$$

Placing

$$H(t,y) = \int_0^t P(s,y)\,ds,$$

i.e.,

$$\frac{\partial H}{\partial t} = P(t,y),$$

we obtain (5.4).

Condition (5.4) is the orthogonality condition for the Fourier transform of the characteristic functions of nonintersecting intervals. In order to obtain Parseval's equation for step functions, one must still show that for an arbitrary interval $(a, b)$ $(a < b)$ the following equation holds:

$$\int_{-\infty}^{+\infty}\Big[\int_a^b \varphi(t,\lambda)\,dt\Big]^2 d\rho(\lambda) = b-a, \tag{5.5}$$

which is Parseval's equation for the characteristic function of the interval $(a, b)$.

Let us first show that a weaker condition holds: if $b-a \to 0$, then the equation

$$\int_{-\infty}^{+\infty}\Big[\int_a^b \varphi(t,\lambda)\,dt\Big]^2 d\rho(\lambda) = b-a+o(b-a) \tag{5.6}$$

is true; it is natural to call this the normalization condition. Let us place $b=\alpha+\epsilon$, $a=\alpha-\epsilon$. We shall first show that

$$\int_{-\infty}^{0}\Big[\int_{\alpha-\varepsilon}^{\alpha+\varepsilon} \varphi(t,\lambda)\,dt\Big]^2 d\rho(\lambda) = o(\varepsilon). \tag{5.7}$$

Since, by hypothesis, for $x \geq 0$

$$\int_{-\infty}^{0} \operatorname{ch}\sqrt{|\lambda|}\,x\,d\rho(\lambda) < \infty,$$

then because of formula (5.3) we have for all $t$

$$\int_{-\infty}^{0} \varphi^2(t,\lambda)\,d\rho(\lambda) < +\infty.$$

Thus, because of the Cauchy-Bunyakovskiĭ inequality,

$$\int_{-\infty}^{0}\Big[\int_{\alpha-\varepsilon}^{\alpha+\varepsilon} \varphi(t,\lambda)\,dt\Big]^2 d\rho(\lambda) \leqslant 2\varepsilon \int_{\alpha-\varepsilon}^{\alpha+\varepsilon}\Big(\int_{-\infty}^{0} \varphi^2(t,\lambda)\,d\rho(\lambda)\Big)dt = O(\varepsilon^2)$$

which holds uniformly in $\alpha$ for each finite interval. Thus, for the proof of (5.6) one must show that

$$\int_{0}^{\infty}\Big[\int_{\alpha-\varepsilon}^{\alpha+\varepsilon} \varphi(t,\lambda)\,dt\Big]^2 d\rho(\lambda) = 2\varepsilon + o(\varepsilon).$$

For $\lambda > 0$ we have

$$\int_{\alpha-\varepsilon}^{\alpha+\varepsilon} \varphi(t,\lambda)\,dt = \int_{\alpha-\varepsilon}^{\alpha+\varepsilon} \cos\sqrt{\lambda}\,t\,dt + \int_{\alpha-\varepsilon}^{\alpha+\varepsilon} dt \int_0^t K(t,s)\cos\sqrt{\lambda}\,s\,ds =$$

$$= \int_{\alpha-\varepsilon}^{\alpha+\varepsilon} \cos\sqrt{\lambda}\,t\,dt + \int_{\alpha-\varepsilon}^{\alpha+\varepsilon}\Big[\frac{\sin\sqrt{\lambda}\,t}{\sqrt{\lambda}}K(t,t) - \int_0^t \frac{\sin\sqrt{\lambda}\,s}{\sqrt{\lambda}}\frac{\partial K(t,s)}{\partial s}ds\Big]dt =$$

$$= \int_{\alpha-\varepsilon}^{\alpha+\varepsilon} \cos\sqrt{\lambda}\,t\,dt + \frac{O(\varepsilon)}{\sqrt{\lambda}}.$$

Then for $\lambda > 0$

$$\Big[\int_{\alpha-\varepsilon}^{\alpha+\varepsilon} \varphi(t,\lambda)\,dt\Big]^2 = \Big[\int_{\alpha-\varepsilon}^{\alpha+\varepsilon} \cos\sqrt{\lambda}t.dt\Big]^2 + 4\,\frac{\sin\sqrt{\lambda}\,\varepsilon}{\sqrt{\lambda}}\cdot\cos\sqrt{\lambda}\,\alpha\,\frac{O(\varepsilon)}{\sqrt{\lambda}} + \frac{O(\varepsilon^2)}{\lambda}.$$

Because of the boundedness of $\phi(t, \lambda)$ for arbitrary $\lambda > 0$,

$$\int_0^a \Big[\int_{\alpha-\varepsilon}^{\alpha+\varepsilon} \varphi(t, \lambda)\, dt\Big]^2 d\rho(\lambda) = O(\varepsilon^2). \tag{5.8}$$

Furthermore, we have

$$\int_a^\infty \Big[\int_{\alpha-\varepsilon}^{\alpha+\varepsilon} \varphi(t, \lambda)\, dt\Big]^2 d\rho(\lambda) = \int_a^\infty \Big[\int_{\alpha-\varepsilon}^{\alpha+\varepsilon} \cos \sqrt{\lambda}\, t\, dt\Big]^2 d\rho(\lambda) +$$

$$+ O(\varepsilon)\cdot \int_a^\infty \sin\sqrt{\lambda}\,\varepsilon\cdot\cos\sqrt{\lambda}\,\alpha \frac{d\rho(\lambda)}{\lambda} + O(\varepsilon^2)\int_a^\infty \frac{d\rho(\lambda)}{\lambda}. \tag{5.9}$$

Since (see §4)

$$\int_a^\infty \frac{d\rho(\lambda)}{\lambda} < \infty,$$

we have for $\epsilon \to 0$

$$\int_a^\infty \sin\sqrt{\lambda}\,\varepsilon\cdot\cos\sqrt{\lambda}\,\alpha\cdot \frac{d\rho(\lambda)}{\lambda} = o(1)$$

uniformly in $a$. From (5.8) and (5.9) and the estimate (4.4) follows

$$\int_0^\infty \Big[\int_{\alpha-\varepsilon}^{\alpha+\varepsilon} \varphi(t, \lambda)\, dt\Big]^2 d\rho(\lambda) = \int_a^\infty \Big[\int_{\alpha-\varepsilon}^{\alpha+\varepsilon} \cos\sqrt{\lambda}\, t\cdot dt\Big]^2 d\rho(\lambda) + o(\varepsilon) =$$

$$= \frac{2}{\pi}\int_0^\infty \Big[\int_{\alpha-\varepsilon}^{\alpha+\varepsilon} \cos\sqrt{\lambda}\, t\cdot dt\Big]^2 d\sqrt{\lambda} + 4\int_0^\infty \Big(\frac{\sin\sqrt{\lambda}\,\varepsilon}{\sqrt{\lambda}}\Big)^2 \cos^2\sqrt{\lambda}\alpha\, d\sigma(\lambda) + o(\varepsilon) =$$

$$= 2\varepsilon + o(\varepsilon),$$

i.e., condition (5.6) is proved.

We shall now show that from the orthogonality condition (5.4) and equation (5.6) follows equation (5.5), i.e., Parseval's equation for the characteristic function of the interval $(a, b)$..

Let $(a, b)$ be an arbitrary fixed finite interval. Set

$$a = a_0 < a_1 < a_2 < \cdots < a_{n-1} < a_n = b,$$

$$\max_{1 \leqslant i \leqslant n} (a_{i+1} - a_i) \to 0 \quad \text{as} \quad n \to \infty.$$

Because of (5.4) and (5.6),

$$\int_{-\infty}^{+\infty} \Big[\int_a^b \varphi(t, \lambda)\, dt\Big]^2 d\rho(\lambda) = \int_{-\infty}^{+\infty} \Big[\sum_{i=1}^n \int_{a_{i-1}}^{a_i} \varphi(t, \lambda)\, dt\Big]^2 d\rho(\lambda) =$$

$$= \sum_{i=1}^n \int_{-\infty}^{+\infty} \Big[\int_{a_{i-1}}^{a_i} \varphi(t, \lambda)\, dt\Big]^2 d\rho(\lambda) = \sum_{i=1}^n [(a_i - a_{i-1}) + o(a_i - a_{i-1})] \to b - a.$$

Since the left side in the last equation does not depend upon $n$, we obtain (5.5) by passing to the limit.

It is now easy to show Parseval's equation for an arbitrary step function. Let us consider a step function $f(x)$ of the form:

$$f(x) = d_i \text{ for } a_{i-1} \leqslant x < a_i \quad (i = 1, \ldots, n;\ a_0 = a,\ a_n = b),$$
$$f(x) = 0 \quad \text{outside the interval } (a, b).$$

From (5.4) and (5.6) follows

$$\int_{-\infty}^{+\infty} \left[\int_a^b f(x)\,\varphi(x,\lambda)\,dx\right]^2 d\rho(\lambda) = \sum_{i=1}^{n} d_i^2 (a_i - a_{i-1}) = \int_a^b f^2(x)\,dx.$$

We have thus obtained Parseval's equation for step functions. Since the set of step functions is dense in $L_2(0, \infty)$, Parseval's equation holds for all functions from $L_2(0, \infty)$.

We have proved the following theorem.

**Theorem.** *Let the function $\rho(\lambda)$ satisfy conditions 1 and 2 of §4* and let $K(x, y)$ be the solution of the integral equation* (II). *Let us place*

$$\varphi(x,\lambda) = \cos\sqrt{\lambda}\,x + \int_0^x K(x,t)\cos\sqrt{\lambda}\,t\,dt.$$

*Then for each function $f(x) \in L_2(0, \infty)$ the functions*

$$F_n(\lambda) = \int_0^n f(x)\,\varphi(x,\lambda)\,dx$$

*as $n \to \infty$ converge in the mean square (with weight $\rho(\lambda)$) to some function $F(\lambda)$ (the generalized Fourier transform of the function $f(x)$), i.e.,*

$$\lim_{n\to\infty} \int_{-\infty}^{+\infty} [F(\lambda) - F_n(\lambda)]^2\,d\rho(\lambda) = 0.$$

*In addition, Parseval's formula holds:*

$$\int_{-\infty}^{+\infty} F^2(\lambda)\,d\rho(\lambda) = \int_0^{\infty} f^2(x)\,dx.$$

If $g(x) \in L_2(0, \infty)$ is another function and $G(\lambda)$ is its Fourier transform, then

$$\int_{-\infty}^{+\infty} F(\lambda)\,G(\lambda)\,d\rho(\lambda) = \int_0^{\infty} f(x)\,g(x)\,dx.$$

Analogously, one may obtain Parseval's equation in the case $h = \infty$.

2. Parseval's equation is equivalent to mean-square convergence of the Fourier integral. In many cases it is necessary to have criteria of uniform convergence of a Fourier integral. We shall prove the following theorem:

**Theorem.** *Let $f(x)$ be a continuous function vanishing outside some finite interval $(0, a)$. Let us set*

* It is sufficient that the function $a(x)$ in condition 2 have three derivatives.

$$E(\lambda) = \int_0^a f(x)\,\varphi(x, \lambda)\,dx$$

*and assume that for all* $x \geq 0$

$$\int_{-\infty}^{+\infty} E(\lambda)\,\varphi(x, \lambda)\,d\rho(\lambda)$$

*converges absolutely and uniformly in each finite interval, and consequently represents a continuous function. Under these hypotheses*

$$f(x) = \int_{-\infty}^{+\infty} E(\lambda)\,\varphi(x, \lambda)\,d\rho(\lambda).$$

**Proof.** Let us set

$$h_n(x) = \int_{-n}^{n} E(\lambda)\,\varphi(x, \lambda)\,d\rho(\lambda).$$

Because of Parseval's equation,

$$\lim_{n \to \infty} \int_0^{\infty} [f(x) - h_n(x)]^2\,dx \to 0.$$

Consequently, and a fortiori for $N \geq a$,

$$\lim_{n \to \infty} \int_0^N [f(x) - h_n(x)]^2\,dx = 0. \tag{5.10}$$

By hypothesis, as $n \to \infty$, $h_n(x)$ converges uniformly in each finite interval to some function $h(x)$. Consequently, in (5.10) one may pass to the limit under the integral sign and obtain

$$\int_0^N [f(x) - h(x)]^2\,dx = 0.$$

Since $f(x)$ and $h(x)$ are continuous functions, there follows from the last equation

$$f(x) = h(x),$$

which was to be proved.

## §6. Derivation of the differential equation for $\phi(x, \lambda)$

1. In the preceding two sections we have constructed for functions $\rho(\lambda)$, satisfying conditions 1 and 2 of §4, a system of functions $\phi(x, \lambda)$ for which Parseval's equation was proved. These functions were constructed from the kernel $K(x, t)$ by means of the formula

$$\varphi(x, \lambda) = \cos\sqrt{\lambda}\,x + \int_0^x K(x, s)\cos\sqrt{\lambda}\,s\,ds. \tag{6.1}$$

Along with equation (6.1) we have also need of its inversion:

$$\cos\sqrt{\lambda}\,x = \varphi(x,\lambda) - \int_0^x K_1(x,s)\,\varphi(s,\lambda)\,ds. \tag{6.2}$$

We have shown that when condition 2 is fulfilled the function $K(x, t)$, and consequently also $K_1(x, t)$, has continuous second derivatives. In this paragraph we shall show that the functions $\phi(x, \lambda)$ are eigenfunctions of a differential equation of the form

$$y'' + (\lambda - q(x))\,y = 0, \tag{6.3}$$

where the function $q(x)$ is continuous and satisfies the given boundary conditions at $0$. For this we have first obtained a certain functional equation for the function $\phi(x, \lambda)$ from which we can later obtain a differential equation (6.3) for $\phi(x, \lambda)$. Let us multiply both sides of equation (6.1) by $\cos\sqrt{\lambda}\,t$. We obtain

$$\varphi(x,\lambda)\cos\sqrt{\lambda}\,t = \frac{1}{2}[\cos\sqrt{\lambda}(x+t) + \cos\sqrt{\lambda}(x-t)] +$$

$$+ \frac{1}{2}\int_0^x K(x,s)[\cos\sqrt{\lambda}(s+t) + \cos\sqrt{\lambda}(s-t)]\,ds.$$

Substituting every place on the right-hand side in place of $\cos\sqrt{\lambda}\,x$ its value from formula (6.2), we obtain

$$\varphi(x,\lambda)\cos\sqrt{\lambda}\,t = \frac{1}{2}[\varphi(x+t,\lambda) + \varphi(x-t,\lambda)] + \int_0^{x+t} W(x,t,s)\,\varphi(s,\lambda)\,ds, \tag{6.4}$$

where $W(x, t, s)$ is a certain function constructed from the functions $K(x, t)$ and $K_1(x, t)$, of whose exact form we shall have no need. We only mention that, since $K(x, t)$ and $K_1(x, t)$ are continuous for $t \leq x$, $W(x, t, s)$ is continuous for $s \leq x+t$.

It is quite essential that in (6.4) the integration should actually be carried out not from $0$ but from $|x-t|$. A rigorous proof of this fact is somewhat bulky. Consequently, we shall give some nonrigorous considerations from which the essence of the matter is clear. Let us multiply both sides of equation (6.4) by $\phi(s_0, \lambda)$, where $s_0 < x + t$, $s_0 \neq x - t$, and let us integrate both sides of the equation with $d\rho(\lambda)$. Since $\phi(x, \lambda)$ and $\phi(x_2, \lambda)$ for $x_1 \neq x_2$ are "orthogonal" with respect to $d\rho(\lambda)$, we obtain

$$W(x,t,s_0) = \int_{-\infty}^{+\infty} \varphi(x,\lambda)\cos\sqrt{\lambda}\,t\,\varphi(s_0,\lambda)\,d\rho(\lambda). \tag{6.5}$$

From this it is evident that $W(x, t, s_0)$ is symmetric in the variables $x$ and $s_0$. Since $W(x, t, s) = 0$ for $s_0 > x + t$, it follows from this symmetry that $W(x, t, s_0) = 0$ also in case $x > s_0 + t$, i.e., $s_0 < x - t$. This means that the integral (6.4) may be taken between the limits $x - t$ to $x + t$. We have reasoned unrigorously since the integral (6.5) diverges. The rigouous proof carried out below is based upon the same idea which is carried out above unrigorously.

Let us integrate both sides of equation (6.4) from $x$ to $t (t \leq x)$. We obtain

$$\int_0^x \varphi(u,\lambda)\,du \int_0^y \cos\sqrt{\lambda}\,v\,dv = h(x,y;\lambda) = \frac{1}{2}\int_0^x\int_0^y \varphi(u+v,\lambda)\,du\,dv +$$

$$+\frac{1}{2}\int_0^x\int_0^y \varphi(u-v,\lambda)\,du\,dv + \frac{1}{2}\int_0^x\int_0^y du\,dv \int_0^{u+v} W(u,v,t)\,\varphi(t,\lambda)\,dt.$$

Changing the order of integration, we find

$$h(x,y;\lambda) = \int_0^{x+y} Z(x,y,t)\,\varphi(t,\lambda)\,dt, \tag{6.6}$$

where $Z(x, y, t)$ is a continuous function in its variables. Formula (6.6) shows that, for arbitrary $x$ and $y$, $h(x, y; \lambda)$ is the Fourier transform of the continuous function $Z(x, y, t)$ vanishing outside the interval $(0, x+y)$. We shall show that for fixed $x$, $y$ and $t$ the integral

$$\int_{-\infty}^{+\infty} h(x,y;\lambda)\,\varphi(t,\lambda)\,d\rho(\lambda) \tag{6.7}$$

converges absolutely and uniformly with respect to $t$ in each finite interval.

We first estimate the function $\phi(t, \lambda)$. For $\lambda < 0$ we have

$$|\varphi(t,\lambda)| \leqslant 1 + \int_0^t |K(t,s)|\,dt < a(t), \tag{6.8}$$

where $a(t)$ is bounded in each finite interval. Thus for $0 \leq \lambda < +\infty$ the function $\phi(t, \lambda)$ is uniformly bounded in $t$ in each finite interval. For $\lambda < 0$ we have

$$|\varphi(t,\lambda)| \leqslant \operatorname{ch}\sqrt{|\lambda|}\,t\left(1+\int_0^t |K(t,s)|\,ds\right) < a(t)\operatorname{ch}\sqrt{|\lambda|}\,t. \tag{6.9}$$

To obtain an estimate of the integral (6.7) we split it into the sum of two integrals: one with limits $-\infty$ to $0$ and the other with the limits $0$ to $\infty$. Let us first consider the integral between the limits $0$ and $\infty$. Since the function

$$\psi(y,\lambda) = \int_0^y \cos\sqrt{\lambda}\,v\,dv = \int_0^y \varphi(v,\lambda)\,dv + \int_0^y \varphi(v,\lambda)\,dv \int_v^y K_1(t,v)\,dt$$

is the Fourier transform of the function

$$\Phi(y,v) = \begin{cases} 1 + \int_v^y K_1(t,v)\,dt, & v \leqslant y, \\ 0, & v > y, \end{cases}$$

and

$$\chi(x,\lambda) = \int_0^x \varphi(u,\lambda)\,du$$

is the Fourier transform of the function

$$e_x = \begin{cases} 1, & \text{if } u \leqslant x, \\ 0, & \text{if } u > x, \end{cases}$$

we have, by virtue of the Cauchy-Bunyakovskiĭ inequality, Parseval's equation, and the inequalities (6.8) (we recall that $h(x, y; \lambda) = \int_0^x \phi(u, \lambda) du \int_0^y \cos\sqrt{\lambda}\, v dv$)

$$\int_0^\infty |h(x, y; \lambda)|\, |\varphi(t, \lambda)|\, d\rho(\lambda) \leqslant a(t) \int_0^\infty |h(x, y, \lambda)|\, d\rho(\lambda) \leqslant$$

$$\leqslant a(t) \int_{-\infty}^{+\infty} |h|\, d\rho(\lambda) = a(t) \int_{-\infty}^{+\infty} |\chi(x, \lambda) \psi(y, \lambda)|\, d\rho(\lambda) \leqslant$$

$$\leqslant a(t) \sqrt{\int_{-\infty}^{+\infty} \chi^2(x, \lambda)\, d\rho(\lambda)} . \sqrt{\int_{-\infty}^{+\infty} \psi^2(y, \lambda)\, d\rho(\lambda)} = a(t) \sqrt{\int_0^y \Phi^2(u, v)\, dv} . \sqrt{x},$$

from which follows the absolute uniform convergence of the integral

$$\int_0^\infty h(x, y; \lambda)\, \varphi(t, \lambda)\, d\rho(\lambda).$$

We shall now consider the integral between the limits $-\infty$ and $0$. From the the estimate (6.9) follows

$$|h(x, y; \lambda)| \leqslant A(x, y) \operatorname{ch} \sqrt{|\lambda|}\, (x + y),$$

where the function $A(x, y)$ does not depend upon $\lambda$. Thus

$$\int_{-\infty}^{0} |h(x, y; \lambda)|\, |\varphi(t, \lambda)|\, d\rho(\lambda) \leqslant$$

$$\leqslant A(x, y)\, a(t) \int_{-\infty}^{0} \operatorname{ch} \sqrt{|\lambda|}\, (x + y) \operatorname{ch} \sqrt{|\lambda|}\, t . d\rho(\lambda).$$

Because of the conditions laid down upon $\rho(\lambda)$ the last integral converges. Thus (since because of (6.6) $h(x, y; \lambda)$ is the Fourier transform of the function $Z$) it follows from the theorem of paragraph 2, §5 that

$$Z(x, y, t = \begin{cases} \int_{-\infty}^{+\infty} h(x, y; \lambda)\, \varphi(t, \lambda)\, d\rho(\lambda), & t \leqslant x + y, \\ 0, & t > x + y. \end{cases}$$

Integrating $Z(x, y, s)$ with respect to $s$ between the limits $0$ and $t$, we obtain

$$Z_1(x, y, t) = \int_0^t Z(x, y, s)\, ds =$$

$$= \int_{-\infty}^{+\infty} \left( \int_0^x \varphi(u, \lambda)\, du \right) \left( \int_0^y \cos \sqrt{\lambda}\, v\, dv \right) \left( \int_0^t \varphi(s, \lambda)\, ds \right) d\rho(\lambda).$$

From the foregoing we obtain for $t > x + y$

$$\frac{\partial Z_1}{\partial t} = Z(x, y, t) = 0.$$

Since the function $Z_1(x, y, t)$ is symmetric in the variables $x$ and $t$,

$$\frac{\partial Z_1}{\partial x} = 0$$

for $x > t + y$, i.e., for $t < x - y$. Consequently, if $t$ is outside the interval $(x - y, x + y)$, then

$$\frac{\partial Z}{\partial x} = \frac{\partial}{\partial x}\left(\frac{\partial Z_1}{\partial t}\right) = \frac{\partial}{\partial t}\left(\frac{\partial Z_1}{\partial x}\right) = 0.$$

Differentiating the equation (6.6) with respect to $x$, we obtain

$$\varphi(x, \lambda) \int_0^t \cos\sqrt{\lambda}\, v\, dv = \int_{x-t}^{x+t} \frac{\partial Z}{\partial x} \cdot \varphi(s, \lambda)\, ds.$$

Differentiating the last equation with respect to $t$, we find

$$\varphi(x, \lambda) \cos\sqrt{\lambda}\, t = \frac{1}{2}[\varphi(x + t, \lambda) + \varphi(x - t, \lambda)] + \int_{x-t}^{x+t} W(x,t,s)\, \varphi(s,\lambda)\, ds. \quad (6.10)$$

Formula (6.10) allows one to obtain without particular difficulty the differential equation for $\phi(x, \lambda)$. Subtract $\phi(x, \lambda)$ from both sides of formula (6.10) and divide by $t^2$. We obtain

$$\varphi(x, \lambda) \frac{\cos\sqrt{\lambda}\, t - 1}{t^2} = \frac{\varphi(x+t,\lambda) - 2\varphi(x,\lambda) + \varphi(x - t,\lambda)}{2t^2} + \frac{1}{t^2}\int_{x-t}^{x+t} W(x, t, s)\, \varphi(s, \lambda)\, ds.$$

As $t \to 0$ the left side of this equation approaches $\frac{\lambda}{2}\phi(x, \lambda)$, the first term of the right side approaches $\frac{1}{2}\phi''(x, \lambda)$ (the differentiability of $\phi(x, \lambda)$ follows from the differentiability of $K$); consequently, the limit

$$\lim_{t\to 0} \frac{1}{t^2} \int_{x-t}^{x+t} W(x, t, s)\, \varphi(s, \lambda)\, ds$$

exists. Let us compute this limit.

Expanding the function $\phi(x, \lambda)$ in a Taylor series, we obtain

$$\varphi(s, \lambda) = \varphi(x, \lambda) + (s - x)\varphi'(x, \lambda) + O(t^2).$$

Thus

$$\int_{x-t}^{x+t} W(x, t, s)\, \varphi(s, \lambda)\, ds =$$

$$= \varphi(x,\lambda) \int_{x-t}^{x+t} W(x, t, s)\, ds + \varphi'(x, \lambda) \int_{x-t}^{x+t} (s - x)\, W(x, t, s)\, ds + O(t^3).$$

Since the function $W(x, t, s)$ is continuous for $x - t \leq s \leq x + t$, then for $t \to 0$

$$W(x, t, s) = W(x, t, x) + o(1).$$

Thus *

* We note that $$\int_{x-t}^{x+t} (s - x)\, W(x, t, x)\, ds = 0.$$

$$\int_{x-t}^{x+t} W(x, t, s)\,\varphi(s,\lambda)\,ds = \varphi(x,\lambda)\int_{x-t}^{x+t} W(x,t,s)\,ds +$$

$$+\varphi'(x,\lambda)\int_{x-t}^{x+t}(s-x)\big[W(x,t,x)+o(1)\big]\,ds + o(t^2) =$$

$$= \varphi(x,\lambda)\int_{x-t}^{x+t} W(x,t,s)\,ds + o(t^2).$$

Then,

$$\frac{1}{t^2}\int_{x-t}^{x+t} W(x,t,s)\,\varphi(s,\lambda)\,ds = \varphi(x,\lambda)\cdot\frac{1}{t^2}\int_{x-t}^{x+t} W(x,t,s)\,ds + o(1).$$

Since, for fixed $x$, $\phi(x,\lambda)$ cannot vanish identically in $\lambda$ and since in the last equation the limit on the left exists, the following limit exists

$$\lim_{t\to 0}\frac{1}{t^2}\int_{x-t}^{x+t} W(x,t,s)\,ds = -\frac{1}{2}q(x).$$

Consequently, $\phi(x,\lambda)$ satisfies the differential equation

$$-\lambda\varphi(x,\lambda) = \varphi''(x,\lambda) - q(x)\,\varphi(x,\lambda)$$

or

$$\varphi''(x,\lambda) + (\lambda - q(x))\,\varphi(x,\lambda) = 0.$$

From the last equation follows

$$q(x) - \lambda = -\frac{\varphi''(x,\lambda)}{\varphi(x,\lambda)}.$$

Since $\phi(x,\lambda)$ cannot vanish identically for fixed $x$, $q(x)$ is a continuous function. From formula (6.1) follows

$$\varphi(0,\lambda) = 1, \quad \varphi_x'(0,\lambda) = K(0,0) = h.$$

Thus $\rho(\lambda)$ determines not only the differential equation for $\phi(x,\lambda)$, but also the initial conditions.

2. Let us now consider the case $h=\infty$. Let

$$\psi(x,\lambda) = \frac{\sin\sqrt{\lambda}x}{\sqrt{\lambda}} + \int_0^x L(x,t)\,\frac{\sin\sqrt{\lambda}t}{\sqrt{\lambda}}\,dt, \tag{6.11}$$

$$\frac{\sin\sqrt{\lambda}x}{\sqrt{\lambda}} = \psi(x,\lambda) - \int_0^x L_1(x,t)\,\psi(t,\lambda)\,dt. \tag{6.12}$$

It follows from (6.11) for $y < x$ that

$$\psi(x,\lambda)\frac{\sin\sqrt{\lambda}y}{\sqrt{\lambda}} = \frac{\sin\sqrt{\lambda}x}{\sqrt{\lambda}}\cdot\frac{\sin\sqrt{\lambda}y}{\sqrt{\lambda}} + \int_0^x L(x,t)\,\frac{\sin\sqrt{\lambda}t}{\sqrt{\lambda}}\cdot\frac{\sin\sqrt{\lambda}y}{\sqrt{\lambda}}\,dt =$$

$$= \frac{1}{2}\int_{x-y}^{x+y}\frac{\sin\sqrt{\lambda}t}{\sqrt{\lambda}}\,dt + \int_0^x L(x,t)\Big[\int_{|t-y|}^{t+y}\frac{\sin\sqrt{\lambda}u}{\sqrt{\lambda}}\,du\Big]\,dt.$$

Reasoning exactly as in the preceding case, we obtain

$$\psi(x,\lambda)\frac{\sin\sqrt{\lambda}y}{\sqrt{\lambda}}=\frac{1}{2}\int_{x-y}^{x+y}\psi(t,\lambda)\,dt+\int_{x-y}^{x+y}W_1(x,y,t)\,\psi(t,\lambda)\,dt, \qquad (6.13)$$

where the function $W_1(x, y, t)$ is continuous for all of its arguments and, in particular, $W_1(x, y, t \pm x) = 0$ and $\partial W_1/\partial y$ is continuous for $x - y \leq t \leq x + y$. Differentiating (6.13) with respect to $y$, we obtain

$$\psi(x,\lambda)\cos\sqrt{\lambda}y=\frac{1}{2}[\psi(x+y,\lambda)+\psi(x-y,\lambda)]+\int_{x-y}^{x+y}\frac{\partial W_1(x,y,t)}{\partial y}\psi(t,\lambda)\,dt.$$

This equation is analogous to (6.4) and thus the following differential equation for $\psi$ holds:

$$\psi''(x,\lambda)+(\lambda-q(x))\,\psi(x,\lambda)=0.$$

## §7. Examples

1. Let us suppose that the function $\rho(\lambda)=\frac{2}{\pi}\sqrt{\lambda}+\sigma(\lambda)$ for $\lambda\geq 0$ and $\rho(\lambda)=0$ for $\lambda<0$ where the function $\sigma(\lambda)$ satisfies the following conditions:

1) $\rho(\lambda)$ is nondecreasing;

2) $\int_0^{\infty}\lambda^k|d\sigma(\lambda)|<+\infty, \quad k=0,1;$

3) $\rho(\lambda)$ has on some finite interval an infinite number of points of increase.

In the case under consideration

$$F(x,y)=\int_0^{\infty}\frac{\sin\sqrt{\lambda}x\cdot\sin\sqrt{\lambda}y}{\lambda}\,d\sigma(\lambda),$$

$$f(x,y)=\int_0^{\infty}\cos\sqrt{\lambda}x\cos\sqrt{\lambda}y.d\sigma(\lambda),$$

$$\frac{\partial f(x,y)}{\partial x}=-\int_0^{\infty}\sqrt{\lambda}\cos\sqrt{\lambda}x\cos\sqrt{\lambda}y.\,d\sigma(\lambda),$$

$$\frac{\partial^2 f}{\partial x^2}=\int_0^{\infty}\lambda\cos\sqrt{\lambda}x\cos\sqrt{\lambda}y\,d\sigma(\lambda).$$

In view of condition 2) all these functions are continuous. Thus all of our results are applicable and, consequently, there exists a differential operator of the second order $L(y)=y''-q(x)y$ with a continuous function $q(x)$ for which Parseval's equation can be written with the given function $\rho(\lambda)$. In particular, there follows from this the following interesting result.

*There exists an equation of the form*

$$y''+(\lambda-q(x))\,y=0, \quad 0\leqslant x<\infty,$$

*in which the function $\rho(\lambda)$ will be on a finite interval an arbitrary monotonic bounded function given in advance.*

2. In a series of particular cases we can actually find the function $q(x)$ for a spectral function $\rho(\lambda)$. Let us take one very simple example. Let

$$\rho(\lambda) = \frac{2}{\pi}\sqrt{\lambda} + \alpha\cdot e(\lambda - \lambda_0) \quad \text{for } \lambda > 0,$$

$$\rho(\lambda) = \alpha e(\lambda - \lambda_0) \quad \text{for } \lambda < 0,$$

where $e(\lambda - \lambda_0) = 0$ if $\lambda < \lambda_0$, and $e(\lambda - \lambda_0) = 1$ if $\lambda > \lambda_0$, and $a$ is a positive number.

The differential equation constructed with this $\rho(\lambda)$ must have a continuous spectrum for $\lambda > 0$ and a point of the discrete spectrum for $\lambda = \lambda_0$. In this case the function $\sigma(\lambda)$ has the form

$$\sigma(\lambda) = \alpha e(\lambda - \lambda_0)$$

and consequently

$$f(x, y) = \int_{-\infty}^{+\infty} \cos\sqrt{\lambda}\, x \cdot \cos\sqrt{\lambda}\, y \, d\sigma(\lambda) = \alpha \cos\sqrt{\lambda_0}\, x \cdot \cos\sqrt{\lambda_0}\, y.$$

The integral equation (II) will be an equation with the degenerate kernel $f(x, y)$ and has the form

$$\alpha \cos\sqrt{\lambda_0}\, x \cos\sqrt{\lambda_0}\, y + \alpha \int_0^x K(x, s) \cos\sqrt{\lambda_0}\, s, \cos\sqrt{\lambda_0}\, y \, ds + K(x, y) = 0.$$

From this we have

$$K(x, y) = \frac{-\alpha \cos\sqrt{\lambda_0}\, x . \cos\sqrt{\lambda_0}\, y}{1 + \alpha \int_0^x \cos^2\sqrt{\lambda_0}\, s \, ds}.$$

Let us now find the equation and the boundary conditions. From formula (1.8′) we have

$$h = K(0, 0) = -\alpha.$$

Furthermore, from formula (1.8) we have

$$\frac{1}{2} q(x) = \frac{dK(x, x)}{dx} = \frac{d}{dx}\left(\frac{-\alpha \cos^2\sqrt{\lambda_0}\, x}{1 + \alpha \int_0^x \cos^2\sqrt{\lambda_0}\, s \, ds}\right).$$

The corresponding eigenfunctions are computed by the formula

$$\varphi(x, \lambda) = \cos\sqrt{\lambda} x + \int_0^x K(x, t) \cos\sqrt{\lambda}\, t \, dt,$$

i.e.,

$$\varphi(x, \lambda) = \cos\sqrt{\lambda}\, x - \frac{\alpha \cos\sqrt{\lambda_0}\, x}{1 + \alpha \int_0^x \cos^2\sqrt{\lambda_0}\, s ds} \cdot \int_0^x \cos\sqrt{\lambda_0}\, t \cos\sqrt{\lambda}\, t dt.$$

As is evident from the definition of the function $\rho(\lambda)$, the spectrum consists of the numbers $\lambda \geq 0$ corresponding to the continuous spectrum and the point $\lambda_0$ of the discrete spectrum. The eigenfunction $\phi(x, \lambda_0)$ of the discrete spectrum has in this case the form

$$\varphi(x, \lambda_0) = \frac{\cos \sqrt{\lambda_0}\, x}{1 + \alpha \int_0^x \cos^2 \sqrt{\lambda_0}\, s\, ds}.$$

If $\lambda_0 < 0$, then instead of $\cos \sqrt{\lambda_0}\, x$ one must write $\cosh \sqrt{|\lambda|}\, x$.

Analogously one may find an equation and eigenfunctions for the case

$$\rho(\lambda) = \begin{cases} \frac{2}{\pi}\sqrt{\lambda} + \sigma(\lambda) & \text{for } \lambda \geqslant 0, \\ \sigma(\lambda) & \text{for } \lambda < 0, \end{cases}$$

where $\sigma(\lambda)$ is a step function with a finite number of jumps. In this case the integral equation also turns out to be degenerate.

In the beginning of the paper, we compared an arbitrary equation with the simple equation $y'' + \lambda y = 0$. Indeed, for this reason, we express the eigenfunctions by means of $\cos \sqrt{\lambda}\, t$. Without difficulty we may take some other equation instead of the "selected" equation. In particular, one may write formulas for the eigenfunctions of an equation obtained from the given one if to its spectral function $\rho(\lambda)$ one adds a step function with a finite number of jumps.

## §8. The condition of orthogonality of an expansion in a Fourier integral

Let $\phi(x, \lambda)$ be the solution of the equation

$$y'' + [\lambda - q(x)]\, y = 0,$$

satisfying boundary conditions

$$y(0) = 1, \quad y'(0) = h,$$

and let $\rho(\lambda)$ be a spectral function. The latter means, as we have said, that for each function $f(x) \in L_2(0, \infty)$ there exists the Fourier transform

$$E(\lambda) = \underset{n \to \infty}{\text{l.i.m.}} \int_0^n f(x)\, \varphi(x, \lambda)\, dx$$

and the Parseval equation holds:

$$\int_0^\infty f^2(x)\, dx = \int_{-\infty}^{+\infty} E^2(\lambda)\, d\rho(\lambda).$$

Let $\Delta = (\lambda, \lambda + \Delta)$ be an arbitrary finite interval. Let us set

$$E_\Delta(x) = \int_\Delta \varphi(x, \lambda)\, d\rho(\lambda).$$

The spectral function $\rho(\lambda)$ is called *orthogonal* if, for an arbitrary interval $\Delta$, $E_\Delta(x) \in L_2(0, \infty)$ and for any two finite intervals $\Delta$ and $\Delta'$

$$\int_0^\infty E_\Delta(x) E_{\Delta'}(x)\,dx = \int_{\Delta\Delta'} d\rho(\lambda). \tag{8.1}$$

In the general case the differential equation may have nonorthogonal spectral functions. In the present paragraph there is given a sufficient condition in order that the function $\rho(\lambda)$ satisfying the conditions of the preceding paragraph be an orthogonal spectral function. In particular, from this condition follows that $\rho(\lambda)$ is known to be orthogonal if it is concentrated on the interval $(c, +\infty)$, $c > -\infty$.

Let us denote by $L_2\{\rho\}$ the real Hilbert space of functions $g(\lambda)$ satisfying the condition

$$\int_{-\infty}^{+\infty} g^2(\lambda)\,d\rho(\lambda) < +\infty.$$

**Lemma.** *In order that the spectral function $\rho(\lambda)$ be orthogonal it it sufficient that the set of functions of the form*

$$F_a(\lambda) = \int_0^a f(x)\cos\sqrt{\lambda}\,x\,dx,$$

*be dense in $L_2\{\rho\}$, where $a$ is an arbitrary positive number and $f(x)$ is a continuous function having a continuous first derivative.*

**Proof.** First of all let us show that the functions $F_a(\lambda)$ belong to $L_2\{\rho\}$. The existence of the integral

$$\int_{-\infty}^{0} F_a^2(\lambda)\,d\rho(\lambda)$$

follows immediately from condition 1, §4. Furthermore, integrating by parts, we obtain

$$|F_a(\lambda)| = O\left(\frac{1}{\sqrt{\lambda}}\right).$$

Consequently,

$$|F_a(\lambda)|^2 = O\left(\frac{1}{\lambda}\right).$$

Thus it follows from the result of §4 that

$$\int_0^\infty |F_a(\lambda)|^2\,d\rho(\lambda) < +\infty.$$

Therefore, $F_a(\lambda) \in L_2\{\rho\}$.

Using the formula

$$\cos\sqrt{\lambda}\,x = \varphi(x,\lambda) - \int_0^x K_1(x,t)\,\varphi(t,\lambda)\,dt,$$

we obtain

$$F_a(\lambda) = \int_0^a f(x)\left[\varphi(x,\lambda) - \int_0^x K_1(x,t)\,\varphi(t,\lambda)\,dt\right]dx =$$

$$= \int_0^a \left[f(x) - \int_x^a K_1(t,x) f(t)\,dt\right]\varphi(x,\lambda)\,dx = \int_0^a g(x)\,\varphi(x,\lambda)\,dx,$$

where

$$g(x) = f(x) - \int_x^a K_1(t, x) f(t)\, dt. \tag{8.2}$$

Equation (8.2) is a Volterra equation. Thus $g(x)$ may be considered as an arbitrary function with a continuous derivative. It then remains to show that if the set of functions

$$\int_0^a g(x) \varphi(x, \lambda)\, dx$$

is dense in $L_2\{\rho\}$, then the spectral function $\rho(\lambda)$ is orthogonal. Let us denote by $F(\lambda)$ an arbitrary function from $L_2\{\rho\}$ and let the functions

$$F_n(\lambda) = \int_0^n f_n(x) \varphi(x, \lambda)\, dx$$

converge in the mean to $F(\lambda)$ as $n \to \infty$. It follows from Parseval's equation $(n > m)$ that

$$\int_0^n |f_n(x) - f_m(x)|^2\, dx = \int_{-\infty}^{+\infty} |F_n(\lambda) - F_m(\lambda)|^2\, d\rho(\lambda).$$

Thus the functions $f_n(x)$ converge in the mean square to some limit $f(x)$, and $F(\lambda)$ is the Fourier transform for $f(x)$, i.e., we have shown that an arbitrary function $F(\lambda) \in L_2\{\rho\}$ is a Fourier transform of some function $f(x)$.

Let

$$F_\Delta(\lambda) = \begin{cases} 1, & \text{if } \lambda \in \Delta, \\ 0, & \text{if } \lambda \overline{\in} \Delta. \end{cases}$$

The function $F_\Delta(\lambda) \in L_2\{\rho\}$. Thus, from the above, there exists a function $f_\Delta(x) \in L_2(0, \infty)$ whose Fourier transform is $F_\Delta(\lambda)$. We shall show that

$$f_\Delta(x) = \int_\Delta \varphi(x, \lambda)\, d\rho(\lambda). \tag{8.3}$$

From the last equation and Parseval's equation it follows that

$$\int_\Delta^\infty f_\Delta(x) f_{\Delta'}(x)\, dx = \int_{\Delta\Delta'} d\rho(\lambda),$$

i.e., the orthogonality of the spectral function $\rho(\lambda)$.

There now remains to prove equation (8.3). Let us denote by $h(x)$ $(x \geq 0)$ an arbitrary continuous function vanishing outside of a finite interval and let

$$H(\lambda) = \int_0^\infty h(x) \varphi(x, \lambda)\, dx.$$

Because of Parseval's equation

$$\int_0^\infty h(x) f_\Delta(x)\, dx = \int_{-\infty}^\infty H(\lambda) \cdot F_\Delta(\lambda)\, d\rho(\lambda) = \int_\Delta H(\lambda)\, d\rho(\lambda) =$$

$$= \int_\Delta d\rho(\lambda) \int_0^\infty h(x) \varphi(x, \lambda)\, dx = \int_0^\infty h(x) \int_\Delta \varphi(x, \lambda)\, d\rho(\lambda).$$

Since $h(x)$ is an arbitrary function, then it follows from the last equation that almost everywhere

$$f_\Delta(x) = \int_\Delta \varphi(x, \lambda)\, d\rho(\lambda).$$

The last equation, because of the continuity of the right-hand side, holds everywhere.

Let us point out a criterion of completeness for the set of functions $F_a(\lambda)$ in $L_2\{\rho\}$.

**Theorem.** *If there exists a number* $a < 2$ *such that for sufficiently large* $x > 0$

$$\int_{-\infty}^{0} e^{\sqrt{|\lambda|}x}\, d\rho(\lambda) < e^{x^\alpha}, \tag{8.4}$$

*then the set of functions* $F_a(\lambda)$ *is dense in* $L_2\{\rho\}$.

**Proof.** The set of functions $F_a(\lambda)$ generates in $L_2\{\rho\}$ a linear subspace. After taking the closure, we obtain the closed subspace $\mathfrak{M}\{\rho\}$. One must now show that if condition (8.4) is satisfied then $\mathfrak{M}\{\rho\} = L_2\{\rho\}$. Let us assume the contrary. Then in $L_2\{\rho\}$ one may find an element $g(\lambda)$ orthogonal to $\mathfrak{M}\{\rho\}$. In particular, for an arbitrary function

$$F_a(\lambda) = \int_0^a f(t) \cos\sqrt{\lambda}\, t\, dt$$

we have

$$\int_{-\infty}^{\infty} g(\lambda) F_a(\lambda)\, d\rho(\lambda) = 0. \tag{8.5}$$

Let us assume that for fixed $x > 0$ ($f(t) = 0$ for $t > a$)

$$f_x(t) = \frac{1}{2}\{f(x+t) + f(x-t)\}.$$

Let us introduce the function

$$F_a(\lambda;\ x) = \int_0^\infty f_x(t) \cos\sqrt{\lambda}\, t\, dt.$$

It is easy to see that

$$F_a(\lambda;\ x) = \cos\sqrt{\lambda}\, x F_a(\lambda).$$

Then, since $F_a(\lambda; x)$ belongs to $\mathfrak{M}\{\rho\}$, the following equation holds for each $x$:

$$\int_{-\infty}^{\infty} \cos\sqrt{\lambda}\, x\, g(\lambda) F_a(\lambda)\, d\rho(\lambda) = 0. \tag{8.6}$$

For $\lambda < 0$

$$|F_a(\lambda)| \leqslant C \cdot e^{\sqrt{|\lambda|}\, a}. \tag{8.7}$$

For sufficiently large $x$ the following estimate follows from the estimates (8.4) and (8.7) and the Cauchy-Bunyakovskiĭ inequality:

$$\left| \int_{-\infty}^{0} \cos \sqrt{\lambda} x g(\lambda) F_a(\lambda) \, d\rho(\lambda) \right| \leqslant C \int_{-\infty}^{0} \cos \sqrt{\lambda} x e^{\sqrt{|\lambda|}\, a} |g(\lambda)| d\rho(\lambda) \leqslant$$

$$\leqslant C \left( \int_{-\infty}^{0} \operatorname{ch}^2 \sqrt{|\lambda|} x e^{2\sqrt{|\lambda|}\, a} \, d\rho(\lambda) \right)^{1/2} \left( \int_{-\infty}^{0} |g(\lambda)|^2 \, d\rho(\lambda) \right)^{1/2} \leqslant C_1 \exp(x^\alpha). \quad (8.7')$$

The following theorem (7) holds:

*For each $x > 0$ let*

$$\int_{-\infty}^{\infty} \cos \sqrt{\lambda} x \, d\sigma(\lambda) = 0,$$

*where this integral converges absolutely for each $x$. If there exists $a < 2$ such that for all sufficiently large $x$*

$$\int_{-\infty}^{0} \cos \sqrt{\lambda} x |d\sigma(\lambda)| < e^{x^\alpha},$$

*then* $\sigma(\lambda) \equiv$ const.

From (8.6), (8.7′) and this result, it follows that for an arbitrary finite interval $\Delta$

$$\int_{\Delta} g(\lambda) F_a(\lambda) \, d\rho(\lambda) = 0. \quad (8.8)$$

For $\lambda > 0$ the set of functions $F_a(\lambda)$ is dense among all continuous functions. Thus it follows from (8.8) that for each interval $\Delta$ with positive end-points

$$\int_{\Delta} g(\lambda) \, d\rho(\lambda) = 0. \quad (8.9)$$

Let $N$ be an arbitrary positive number. From (8.9) follows:

$$\int_0^N g^2(\lambda) \, d\rho(\lambda) = \int_0^N g(\lambda) \, d \int_0^\lambda g(\mu) \, d\rho(\mu) = 0.$$

Letting $N \to \infty$, we obtain

$$\int_0^\infty g^2(\lambda) \, d\rho(\lambda) = 0.$$

Furthermore, from (8.9) and (8.5) follows

$$\int_{-\infty}^{0} g(\lambda) F_a(\lambda) \, d\rho(\lambda) = 0. \quad (8.10)$$

From condition 1, §4 it follows that

$$\int_{-\infty}^{0} d\rho(\lambda) < \infty.$$

Then, applying the Cauchy-Bunyakovskiĭ inequality, we obtain

$$\int_{-\infty}^{0} |g(\lambda)| \, d\rho(\lambda) \leqslant \left( \int_{-\infty}^{0} g^2(\lambda) \, d\rho(\lambda) \right)^{1/2} \left( \int_{-\infty}^{0} d\rho(\lambda) \right)^{1/2} < \infty.$$

Changing the order of integration in (8.10), we obtain

$$\int_0^a f(x)\,dx \int_{-\infty}^0 g(\lambda)\cos\sqrt{\lambda}\,x\,d\rho(\lambda) = 0.$$

Because of the arbitrariness of the functions $f(x)$ it follows from the last equation that

$$\int_{-\infty}^0 g(\lambda)\cos\sqrt{\lambda}\,x\,d\rho(\lambda) = 0.$$

Applying again the result indicated above we obtain

$$\int_\Delta g(\lambda)\,d\rho(\lambda) = 0.$$

Thus

$$\int_{-\infty}^0 g^2(\lambda)\,d\rho(\lambda) = 0,$$

which together with the equation obtained earlier gives

$$\int_{-\infty}^\infty g^2(\lambda)\,d\rho(\lambda) = 0,$$

and the theorem is proved.

Remark I. If $h = \infty$, then one may place

$$F_a(\lambda) = \int_0^a f(x)\frac{\sin\sqrt{\lambda}\,x}{\sqrt{\lambda}}\,dx.$$

In place of condition (8.4) one may require that for all sufficiently large $x$ the following estimate is satisfied:

$$\int_{-\infty}^0 \frac{\sin\sqrt{\lambda}\,x}{\sqrt{\lambda}}\,d\rho(\lambda) < \exp(x^\alpha),$$

where $a < 2$. The rest of the proof is analogous.

Remark II. If $\rho(\lambda) = \text{const.}$ for $\lambda < c$, $c > -\infty$, then for sufficiently large $x$

$$\int_{-\infty}^0 \cos\sqrt{\lambda}\,x\,d\rho(\lambda) = \int_c^0 \cos\sqrt{\lambda}\,x\,d\rho(\lambda) \leqslant Ae^{\sqrt{|c|}\,x}.$$

Thus the estimate (8.4) is satisfied and consequently the spectral measure is orthogonal.

## §9. The case of a single limit point at infinity of the spectrum

1. Up to now we have excluded the case when the spectrum has a single limit point at infinity. To this group of boundary problems belongs the classical Sturm-Liouville problem. In the present section we shall consider the case of a discrete spectrum on a half-line. In the following two sections we shall consider the clas-

sical Sturm-Liouville problem. We first introduce a concept. Let $E$ be some set of real numbers. Let us designate by $N(E)$ the number of numbers of the set $E$ included in the interval $(-N, N)$. Furthermore, let us set

$$\mu = \overline{\lim_{N\to\infty}} \frac{N(E)}{N}.$$

The number $\mu$ will hereafter be called the density of the set $E$.

Let us consider the Sturm-Liouville problem on the finite interval $(0, l)$:

$$y'' + \{\lambda - q(x)\} y = 0, \tag{9.1}$$

$$y'(0) - hy(0) = 0, \tag{9.2}$$

$$y'(l) + Hy(l) = 0, \tag{9.3}$$

where $h$ and $H$ are real numbers. Let us designate by $\lambda_1(l), \lambda_2(l), \cdots, \lambda_n(l), \cdots$ the eigenvalues of the boundary problem (9.1)–(9.3). As is known,

$$\sqrt{\lambda_n(l)} = \frac{\pi}{l} n + O\left(\frac{1}{n}\right). \tag{9.4}$$

Let $E(l)$ be the set of numbers $\sqrt{|\lambda_n(l)|}$ $(n = 1, 2, \cdots)$. From formula (9.4) follows that in this case

$$\mu = \frac{l}{\pi}.$$

Now let $g(x)$ be continuous in each finite interval. Let us consider equation (9.1) together with the boundary condition (9.2) on the interval $(0, \infty)$. It is known that if $q(x) \geq 0$ and tends monotonically to $\infty$, the spectrum is discrete, positive and has a unique limit point at infinity. In addition, the $n$th eigenfunction has $n$ zeros. Let us denote the points in the spectrum in this last case by $\lambda_1, \lambda_2, \cdots, \lambda_n, \cdots$, and let $E$ be the set of points $\sqrt{\lambda_n}$ $(n = 1, 2, \cdots)$. We shall show that the density of the set $E$ equals infinity.

Let us consider the boundary problem (9.1)–(9.3). Since the density of the set $E(l)$ equals $l/\pi$, one may select a number $N$ sufficiently large that

$$\frac{N[E(l)]}{N} > \frac{l}{2\pi}.$$

We shall denote by $\phi(x, \lambda)$ the solution of equation (9.1) satisfying the first boundary condition (9.2). It is clear that the number of zeros of the function $\phi(x, \lambda)$ in the interval $(0, \infty)$ is not less than the number of zeros of the same function in the interval $(0, l)$. Thus it follows from Sturm's theorem that the number of eigenvalues of the boundary problem (9.1)–(9.3) in the interval $(0, N)$ is not less than the number of eigenvalues of the boundary problem (9.1)–(9.3) on the infinite interval. Thus,

$$\frac{N(E)}{N} \geqslant \frac{N[E(l)]}{N} > \frac{l}{2\pi}.$$

Since $l$ is chosen arbitrarily, then

$$\overline{\lim_{N\to\infty}} \frac{N(E)}{N} = \infty, \tag{9.5}$$

which is what we wish to prove.

2. Let us now consider the inverse problem. Let the function $\rho(\lambda)$ satisfy conditions 1 and 2 of §4, and be a pure step function with a single limit point at infinity of the jumps. Let us denote the points of increase of the function $\rho(\lambda)$ by $\lambda_1, \lambda_2, \cdots, \lambda_n, \cdots$. Furthermore, let us denote by $E$ the set of points

$$\sqrt{|\lambda_n|} \quad (n = 1, 2, \cdots).$$

We shall assume that the density of the set $E$ is infinite. We shall now show that under these conditions the inverse problem is solvable.

The only place in which we have made use of the presence of limit points of increase of the function $\rho(\lambda)$ within a finite distance was in the solvability of equation (II):

$$f(x, y) + \int_0^x f(y, s) K(x, s)\, ds + K(x, y) = 0. \tag{II}$$

Thus we must show that if the density of the set $E$ is infinite then equation (II) is solvable. In §4 we showed that equation (II) is known to be solvable if it follows from equation

$$\int_{-\infty}^{\infty} G^2(\lambda)\, d\rho(\lambda) = 0, \tag{9.6}$$

where

$$G(\lambda) = \int_0^x g(t) \cos \sqrt{\lambda}\, t\, dt,$$

and $g(t)$ is a square-integrable function, that $G(\lambda) = 0$ and consequently that $g(t)$ vanishes almost everywhere.

Equation (9.6) can hold only if the jumps of the function $\rho(\lambda)$ coincide with the zeros of $G(\lambda)$. We shall show that this is impossible. Let us denote by $\lambda_{1,x}$, $\lambda_{2,x}, \cdots, \lambda_{n,x}, \cdots$ the set of all real zeros of the function $G(\lambda)$ and by $E_x$ the set of real numbers $\sqrt{|\lambda_{n,x}|}$ $(n = 1, 2, \cdots)$. Furthermore, let us designate by $\mu_x$ the density of the set $E_x$. It is known (8) that

$$\mu_x \leqslant \frac{x}{\pi}.$$

Since by hypothesis $\mu = \infty$, it follows from equation (9.6) that, for $x < \infty$, $G(\lambda) \equiv 0$; thus equation (II) is solvable.

If the points of increase of the function $\rho(\lambda)$ are bounded from below, then, in view of the results of the preceding paragraph, the eigenfunctions are orthogonal.

## §10. The case of a finite interval

Let us consider the equation

$$y'' + \{\lambda - q(x)\} y = 0 \tag{10.1}$$

on the interval $(0, l)$; we shall suppose that the function $q(x)$ is continuous in each interval $(0, l')$ where $l' < l < \infty$. Together with equation (10.1) we shall consider the boundary condition at the point $x = 0$:

$$y(0) = 1, \quad y'(0) = h. \tag{10.2}$$

Let $\phi(x, \lambda)$ be the solution of equation (10.1) satisfying conditions (10.2). Just as in the first section, one may show that if the function $K(x, t)$ is a solution of the partial differential equation

$$\frac{\partial^2 K}{\partial x^2} - q(x) K = \frac{\partial^2 K}{\partial t^2}$$

and satisfies the conditions

$$K(x, x) = h + \int_0^x q(t)\, dt \quad (x < l), \quad \left.\frac{\partial K}{\partial t}\right|_{t=0} = 0,$$

then

$$\varphi(x,\lambda) = \cos\sqrt{\lambda}\, x + \int_0^x K(x, t) \cos\sqrt{\lambda}\, t\, dt \quad (x < l). \tag{10.3}$$

Transforming the last equation, we obtain

$$\cos\sqrt{\lambda}\, x = \varphi(x, \lambda) - \int_0^x K_1(x, t)\, \varphi(t, \lambda)\, dt \quad (x < l).$$

Let $\rho(\lambda)$ be a spectral function for equation (10.1), i.e., for any function $f(x) \in L_2(0, l)$ there exists the Fourier transform

$$E(\lambda) = \underset{l' \to l}{\mathrm{l.\,i.\,m.}} \int_0^{l'} f(x)\, \varphi(x, \lambda)\, d\lambda$$

and Parseval's equation holds:

$$\int_0^l f^2(x)\, dx = \int_{-\infty}^{\infty} E^2(\lambda)\, d\rho(\lambda).$$

If $\rho(\lambda)$ is an orthogonal spectral functi^n, then the set of functions of the form

$$\int_0^{l'} f^2(x)\, \varphi(x, \lambda)\, dx \quad (l' < l),$$

where $f(x)$ is an arbitrary continuous function, is dense in $L_2\{\rho\}$. Let us set

$$\sigma(\lambda) = \begin{cases} \rho(\lambda) - \frac{2}{\pi}\sqrt{\lambda}, & \lambda \geqslant 0, \\ \rho(\lambda), & \lambda < 0, \end{cases} \tag{10.4}$$

$$F(x, y) = \int_{-\infty}^{\infty} \frac{\sin\sqrt{\lambda}\, x \cdot \sin\sqrt{\lambda}\, y}{\lambda}\, d\sigma(\lambda) \quad (x, y < l). \tag{10.5}$$

Just as in §§2 and 3 one may show that there exist continuous derivatives $(x, y < l)$

$$\frac{\partial^2 F}{\partial x \partial y} = f(x, y), \quad \frac{\partial f}{\partial x}, \quad \frac{\partial f}{\partial y}$$

and the following integral equations hold:

$$f(x, y) = K_1(x, y) - \int_0^y K_1(x, t) K_1(y, t)\, dt \quad (y \leqslant x,\ x < l), \tag{I}$$

$$f(x, y) + \int_0^x f(y, t) K(x, t)\, dt + K(x, y) = 0 \quad (y \leqslant x,\ x < l). \tag{II}$$

**2.** Let us consider the inverse problem. Let $\rho(\lambda)$ be a monotonic function bounded in each finite interval and satisfying the following two conditions:

1. *For each $x < 2l$ there exists the integral*

$$\int_{-\infty}^{\infty} \operatorname{ch} \sqrt{|\lambda|}\, x\, d\rho(\lambda).$$

2. *The function*

$$a(x) = \int_1^{\infty} \frac{\cos \sqrt{\lambda} x}{\lambda}\, d\sigma(\lambda)$$

*has a continuous fourth derivative if $0 \leq x < 2l$.* In this equation $\sigma(\lambda)$ is determined from (10.4).

It follows from condition 2 that the function $F(x, y)$, determined by equation (10.5), has the continuous partial derivatives

$$\frac{\partial^2 F}{\partial x \partial y} = f(x, y), \quad \frac{\partial^2 f}{\partial x^2}, \quad \frac{\partial^2 f}{\partial y^2}.$$

Furthermore, just as in §4, one may show that if $\rho(\lambda)$ has an infinite set of points of increase in any finite interval, then equation (II) is solvable for $x < l$. Equation (II) is also solvable in the case which was considered in the preceding section. Repeting the reasoning of §§5, 6, and 7, one may prove Parseval's equation and reconstruct the equation.

Thus, conditions 1 and 2 are sufficient in order that the function $\rho(\lambda)$ be a spectral function of equation (1) for the interval $(0, l)$.

Let us now see under what supplementary condition the function $\rho(\lambda)$ becomes an orthogonal spectral function. Just as in §8 one may show that in order that the function $\rho(\lambda)$ be an orthogonal spectral function, it is sufficient that (in addition to conditions 1 and 2) the set of functions

$$F_{l'}(\lambda) = \int_0^{l'} f(x) \cos \sqrt{\lambda} x\, dx \quad (l' < l),$$

be dense in $L_2\{\rho\}$, where $f(x)$ is an arbitrary continuous differentiable function. The last situation occurs in that case when it follows from the orthogonality of the function $g(\lambda) \in L_2\{\rho\}$ to all the functions $F_{l'}(\lambda)$, i.e., from the equation

$$\int_{-\infty}^{\infty} g(\lambda) F_{l'}(\lambda)\, d\rho(\lambda) = 0$$

that

$$\int_{-\infty}^{\infty} g^2(\lambda)\, d\rho(\lambda) = 0.$$

Let us denote by $N$ an arbitrary positive number. Changing the order of integration, we obtain

$$\int_{-\infty}^{N} g(\lambda) F_{l'}(\lambda)\, d\rho(\lambda) = \int_{0}^{l'} f(x) \int_{-\infty}^{N} g(\lambda) \cos \sqrt{\lambda}\, x\, d\rho(\lambda). \tag{10.6}$$

Let us suppose that as $N \to \infty$ the functions

$$\int_{-\infty}^{N} g(\lambda) \cos \sqrt{\lambda}\, x\, d\rho(\lambda)$$

converge in the mean square on the interval $(0, l)$ to some function* $G(x)$. Passing to the limit in equation (10.6), we obtain

$$\int_{-\infty}^{\infty} g(\lambda) F_{l'}(\lambda)\, d\rho(\lambda) = \int_{0}^{l'} f(x) G(x)\, dx = 0.$$

Since $f(x)$ is an arbitrary function, then it follows from the last equation that $G(x)$ vanishes almost everywhere on the interval $(0, l)$. Thus we have obtained the following theorem.

**Theorem.** *Let $g(\lambda)$ be an arbitrary function belonging to $L_2\{\rho\}$. If it follows that*

$$\int_{-\infty}^{\infty} g^2(\lambda)\, d\rho(\lambda) = 0,$$

*whenever, as $N \to \infty$, the functions*

$$\int_{-\infty}^{N} g(\lambda) \cos \sqrt{\lambda}\, x\, d\rho(\lambda)$$

*converge to 0 in the mean square on the interval $(0, l)$ then $\rho(\lambda)$ is an orthogonal spectral function.*

**Remark.** The case $h = \infty$ may be treated analogously.

## §11. The classical Sturm-Liouville problem

1. The results of the preceding sections allow one to characterize relatively easily the spectral functions corresponding to the classical Sturm-Liouville problem. Let the function $q(x)$ be continuous in the closed interval $(0, l)$. Let us denote by $\phi(x, \lambda)$ a solution of the equation

$$y'' + \{\lambda - q(x)\}\, y = 0, \tag{11.1}$$

satisfying the conditions

$$y(0) = 1, \quad y'(0) = h, \tag{11.2}$$

where $h$ is an arbitrary real number. Let $H$ be another real number. We are going to consider at the point $x = l$ the second boundary condition

$$(y' + Hy)_{x=l} = 0. \tag{11.3}$$

* For orthogonal spectral functions $\rho(\lambda)$ this always holds.

Let $\lambda_1, \lambda_2, \cdots, \lambda_n, \cdots$ be the eigenvalues of the boundary-value problem (11.1)–(11.3). As is known, the spectrum of the problem under consideration is bounded below. Let $\lambda_0$ be the smallest eigenvalue. Equation (11.1) may be written in the form

$$y'' + \{(\lambda - \lambda_0) - (q(x) - \lambda_0)\} y = 0.$$

Setting $\lambda - \lambda_0 = \mu$, we obtain a boundary-value problem with a non-negative spectrum. Thus, without destroying the generality of our reasoning, we may hereafter suppose that the spectrum is non-negative. The eigenfunctions $\phi(x, \lambda_n)$ corresponding to the different eigenvalues are mutually orthogonal, i.e.,

$$\int_0^l \varphi(x, \lambda_n) \varphi(x, \lambda_m)\, dx = 0 \quad (n \neq m).$$

Further, let us set

$$\rho_n = \int_0^l \varphi^2(x, \lambda_n)\, dx$$

(we recall that the eigenfunctions are selected so that they satisfy conditions (11.2)). For each function $f(x) \in L_2(0, l)$ Parseval's equation holds:

$$\int_0^l f^2(x)\, dx = \sum_{n=0}^{\infty} \frac{1}{\rho_n} \left\{ \int_0^l f(x) \varphi(x, \lambda_n)\, dx \right\}^2 .$$

If $H \neq \infty$, then the following asymptotic formula holds for the eigenvalues $\lambda_n$:

$$\sqrt{\lambda_n} = \frac{\pi}{l} n + O\left(\frac{1}{n}\right). \tag{11.4}$$

For the eigenfunctions one has the asymptotic formula:

$$\varphi(x, \lambda_n) = \cos n \frac{\pi}{l} x + O\left(\frac{1}{n}\right).$$

Consequently,

$$\rho_n = \frac{l}{2} + O\left(\frac{1}{n}\right). \tag{11.5}$$

In the case under consideration, the functions

$$K(x, t), \quad K_1(x, t), \quad f(x, t), \quad \frac{\partial f}{\partial x}, \quad \frac{\partial f}{\partial y} \quad (t \leqslant x)$$

are continuous up to the point $x = l$. Thus the integral equations (I) and (II) hold up to the points $y = l$ and $x = l$.

2. Let us now pass to the solution of the inverse problem. Let $\lambda_0, \lambda_1, \cdots, \lambda_n, \cdots$ be different non-negative numbers satisfying the asymptotic formula (11.4), and at the points $\sqrt{\lambda_n}$ let there be given positive jumps $\rho_n$ satisfying the asymptotic formula (11.5). We shall prove the following lemma.

**Lemma I.** *The system of functions* $\cos \sqrt{\lambda_n}\, x$ *is complete on the interval* $(0, l)$ *in the space* $L_2$.

**Proof.** Changing the numbers $\lambda_n$ to $\mu_n = l^2 \lambda_n / \pi^2$, we reduce the interval $(0, l)$ to the interval $(0, \pi)$. Let us denote $\sqrt{\mu_n}$ by $\tau_n$. Furthermore, using the following theorem of Levinson* [(9), p. 6]:

Let us denote by $\Lambda(u)$ the number of numbers $\tau_n$ which are $\leq u$. If

$$\int_1^v \frac{\Lambda(u)}{u}\,du > v - \frac{1}{8}\ln v - c, \tag{11.6}$$

where $c$ is a constant, then it follows from the equations

$$\int_0^\pi \cos\sqrt{\lambda_n}\,x\cdot f(x)\,dx = 0 \quad (n = 0, 1, \ldots, \quad f(x) \in L_2)$$

that $f(x) = 0$ almost everywhere, i.e., the system $\cos\sqrt{\mu_n}x$ is complete in $L_2$.

In our case $\tau_n = \sqrt{\mu_n} = n + O(1/n)$. Thus, for sufficiently large $u$, $\Lambda(u) = [u] + 1$ which means $\Lambda(u)/u \geq 1$. Thus, equation (11.6) is satisfied. Changing $x$ to $\pi x/l$ we see that the system $\cos\sqrt{\lambda_n}x$ is complete in $L_2(0, l)$.

**Lemma II.** *If the series* $\sum_{n=0}^{\infty} c_n \cos\sqrt{\lambda_n}x$ *converges to* $0$ *in the mean square on the interval* $(0, l)$, *then for each* $n$ *we have* $c_n = 0$.

**Proof.** In the proof of the preceding lemma we have already mentioned that we may restrict ourselves to the interval $(0, \pi)$. Let us define linear operator $A$ by the formula

$$A\cos nx = \cos\sqrt{\lambda_n}\,x.$$

We shall show that the operator $A$ may be represented in the form $A = E + P$, where $P$ is a completely continuous operator. In fact, let us select as a basis the functions $\cos nx$. Let the operator $A$ correspond to the matrix $a_{mn}$. Consequently, by definition,

$$a_{mn} = \frac{1}{\pi}\int_0^\pi \cos\sqrt{\lambda_n}\,x\cos mx\,dx =$$

$$= \frac{1}{\pi}\int_0^\pi \left(\cos nx + \frac{\alpha_n(x)}{n}\right)\cos mx\,dx = \delta_{mn} + \frac{1}{n}\frac{1}{\pi}\int_0^\pi \alpha_n(x)\cos mx\,dx = \delta_{mn} + p_{mn}.$$

Furthermore, we have as a result of the ordinary Parseval equation

$$\sum_{mn} p_{mn}^2 = \sum_n \frac{1}{n^2}\alpha_n{}^2(x) < +\infty.$$

Consequently, the operator $P$ with the matrix $p_{mn}$ is completely continuous.

Because of Lemma I, the operator $A$ takes the whole space $L_2(0, \pi)$ into the whole space $L_2(0, \pi)$. Consequently, because of the Fredholm theory, there exists a bounded inverse operator $B = A^{-1}$, where $B = E + Q$, and $Q$ is a completely continuous operator. Consequently,

$$B\cos\sqrt{\lambda_n}\,x = \cos nx.$$

Let us suppose that the series $\sum_{n=0}^{\infty} c_n \cos\sqrt{\lambda_n}x$ converges to $0$ in the mean

* Levinson considers the interval $(-\pi, \pi)$ and the functions $e^{i\tau_n x}$, $-\infty < \tau_n < +\infty$. Assuming $\tau_{-n} = -\tau_n$ and the function $f(x)$ to be even, we obtain the result used.

square. Applying the operator $B$, we find that the series $\Sigma_{n=0}^{\infty} c_n \cos nx$ converges to $0$ in the mean square, which means, in view of the orthogonality of the functions $\cos nx$, that $c_n = 0$ for $n = 0, 1, 2, \cdots$, which completes the proof.

On the basis of Lemmas I and II, we shall prove the following theorem:

**Theorem I.** *Let the positive numbers* $\lambda_0, \lambda_1, \cdots, \lambda_n, \cdots$ *obey the asymptotic formula* (11.4) *and let the positive numbers* $\rho_0, \rho_1, \cdots, \rho_n, \cdots$ *obey the asymptotic formula* (11.5). *Let*

$$\rho(\lambda) = \sum_{\lambda_n \leqslant \lambda} \frac{1}{\rho_n}.$$

*If the function* $\rho(\lambda)$ *satisfies condition 2 of the preceding section, then there exists a function* $K(x, t)$ *having continuous partial derivatives of the first and second orders for* $0 \le t \le x \le l$; *here the functions*

$$\varphi(x, \lambda_n) = \cos \sqrt{\lambda_n}\, x + \int_0^x K(x, t) \cos \sqrt{\lambda_n}\, t\, dt \tag{11.7}$$

*are solutions of the equation*

$$y'' + (\lambda_n - q(x)\, y) = 0, \tag{11.8}$$

*where*

$$q(x) = \frac{1}{2} \frac{dK(x, x)}{dx}, \quad (0 \leqslant x \leqslant l).$$

*The functions* $\phi(x, \lambda_n)$ *form a complete orthogonal system on the interval* $(0, l)$.

**Proof.** First we shall prove that the linear integral equation (II) is solvable for $x \le l$. On the basis of the result of §4 it is sufficient to prove that it follows from the equation

$$\int_0^{\infty} G^2(\lambda)\, d\rho(\lambda) = 0, \tag{11.9}$$

where $G(\lambda) = \int_0^x g(t) \cos \sqrt{\lambda}\, t\, dt$ $(0 \le x \le l)$ that $g(t) = 0$ almost everywhere. In view of the definition of the function $\rho(\lambda)$ equation (11.9) is possible only when for all $n = 0, 1, 2, \cdots$

$$G(\lambda_n) = \int_0^x g(t) \cos \sqrt{\lambda_n}\, t\, dt = 0 \quad (0 \leqslant x \leqslant l).$$

Because of Lemma I the last equation is possible only if $g(t) = 0$ almost everywhere. Thus equation (II) is solvable for all $x$ up to $x = l$. Repeating the reasoning of §§5 and 6, we prove the completeness of the system of functions (11.7) and obtain for them the differential equation (11.8).

Let us show that the functions $\phi(x, \lambda_n)$ are orthogonal. On the basis of the theorem proved in the preceding section it is sufficient to show that from the convergence to $0$ of the series $\Sigma_0^{\infty} \rho_n^{-1} a_n \cos \sqrt{\lambda_n} x$ follows that $a_n = 0$ for all $n$. But this is the import of Lemma II. Thus the theorem is completely proved.

3. Let us now show how to reconstruct the boundary condition at the end $x = l$. It follows from formula (11.7) that

$$\varphi(0, \lambda_n) = 1, \quad \varphi_x'(0, \lambda_n) = K(0, 0) = h. \tag{11.10}$$

Furthermore, we have

$$\varphi''(x, \lambda_n) + (\lambda_n - q(x))\,\varphi(x, \lambda_n) = 0,$$
$$\varphi''(x, \lambda_m) + (\lambda_m - q(x))\,\varphi(x, \lambda_m) = 0.$$

Multiply the first equation by $\phi(x, \lambda_m)$ and the second by $\phi(x, \lambda_n)$ and subtract one from the other. We obtain the identity

$$\varphi''(x, \lambda_n)\,\varphi(x, \lambda_m) - \varphi''(x, \lambda_m)\,\varphi(x, \lambda_n) = (\lambda_m - \lambda_n)\,\varphi(x, \lambda_n)\,\varphi(x, \lambda_m).$$

Noting that the first part of this identity is

$$[\varphi'(x, \lambda_n)\,\varphi(x, \lambda_m) - \varphi'(x, \lambda_m)\,\varphi(x, \lambda_n)]',$$

we obtain, integrating between the limits $0$ and $l$ and using orthogonality and the conditions (11.10), $\phi'(l, \lambda_n)\,\phi(l, \lambda_m) - \phi'(l, \lambda_m)\,\phi(l, \lambda_n) = 0$.

From the last equation follows

$$\frac{\varphi'(l, \lambda_n)}{\varphi(l, \lambda_n)} = \frac{\varphi'(l, \lambda_m)}{\varphi(l, \lambda_m)}.$$

Thus the ratio $\phi'(l, \lambda_n)/\phi(l, \lambda_n)$ is constant, which gives the second boundary condition.

**4.** If in the derivation of the integral equations (see §3) the function

$$f_x(t) = \begin{cases} 1 & \text{for} \quad t \leqslant x \\ 0 & \text{for} \quad t > x \end{cases}$$

is not expanded in a Fourier integral but in a Fourier series in $\cos n\frac{\pi}{l}x$, then for the function $F(x, y)$ one obtains the formula

$$F(x, y) = \sum_{n=1}^{\infty}\left[\frac{1}{\rho_n}\frac{\sin\sqrt{\lambda_n}\,x\cdot\sin\sqrt{\lambda_m}\,y}{\lambda_n} - \frac{2l}{\pi^2}\frac{\sin n\frac{\pi}{l}x\cdot\sin n\frac{\pi}{l}y}{n^2}\right] +$$
$$+ \frac{1}{\rho_0}\frac{\sin\sqrt{\lambda_0}\,x\,\sin\sqrt{\lambda_0}\,y}{\lambda_0} - \frac{xy}{l}.$$

Thus condition 2 of the preceding section may be reformulated as follows:

*The function*

$$a(x) = \sum_{n=1}^{\infty}\left[\frac{1}{\rho_n}\frac{\cos\sqrt{\lambda_n}x}{\rho_n} - \frac{2l}{\pi^2}\frac{\cos n\frac{\pi}{l}x}{n}\right]$$

*has a continuous fourth derivative for $0 \le x \le 2l$* (in the point $x = 2l$ it has a left-hand derivative and in the point $x = 0$ a right-hand one).

The following theorem holds.

**Theorem 2.** *If*

$$\sqrt{\lambda_n} = \frac{\pi}{l}n + \frac{b_1}{n} + \frac{b_3}{n^3} + O\left(\frac{1}{n^4}\right), \tag{11.11}$$

$$\rho_n = \frac{2}{l} + \frac{a_1}{n^2} + O\left(\frac{1}{n^4}\right), \tag{11.12}$$

*where $a_1$, $b_1$, $b_3$ are constants, then there exists a function $q(x)$ with given $\lambda_n$*

*and* $\rho_n$.

**Proof.** It is sufficient to show, in view of Theorem 1, that the function $a(x)$ has a continuous fourth derivative for $0 \leq x \leq 2l$. If the asymptotic formulas (11.11) and (11.12) are substituted in the expression for $a(x)$, then, on the one hand one may single out series which may be differentiated four times such that the four-times differentiated series converges uniformly. On the other hand, we may single out series which may be obtained by formal integration of the series $\Sigma \frac{\sin n\pi x/l}{n\pi/l}$. Since

$$\sum_{n=1}^{\infty} \frac{\sin n \frac{\pi}{l} x}{n \frac{\pi}{l}} = \frac{l-x}{2} \qquad (0 < x < 2l),$$

then for $0 \leq x \leq 2l$ to the formal derivatives of this series correspond continuous functions; and consequently, the theorem is proved.

**Remark 1.** If $h$ or $H = \infty$, then for $\lambda_n$ another asymptotic formula holds:

$$\sqrt{\lambda_n} = \frac{\pi}{l}\left(n + \frac{1}{2}\right) + O\left(\frac{1}{n}\right).$$

If also $h$ and $H = \infty$, then

$$\sqrt{\lambda_n} = \frac{\pi}{l}(n+1) + O\left(\frac{1}{n}\right).$$

In both cases $\rho_n$ obeys the asymptotic formula

$$\rho_n = \frac{1}{n^2 \frac{\pi^2}{l^2}}\left[1 + O\left(\frac{1}{n}\right)\right].$$

In the derivation of the integral equations the function $f_x(t)$ is then expanded in a series in $\Sigma \frac{\sin n\pi x/l}{n\pi/l}$. Then

$$F(x, y) = \sum_{n=0}^{\infty}\left[\frac{1}{\rho_n} \frac{(1-\cos\sqrt{\lambda_n}x)(1-\cos\sqrt{\lambda_n}y)}{\lambda_n^2} - \right.$$
$$\left. - \frac{l}{2}\left(1 - \cos n\frac{\pi}{l}x\right)\left(1 - \cos n\frac{\pi}{l}x\right)\right].$$

**Remark 2.** The differentiability of $q(x)$ is connected with the number of terms of the asymptotic expansion for $\rho_n$ and $\lambda_n$. For example, for infinite differentiability of $q(x)$ it is necessary and sufficient that the classical infinite asymptotic expansions hold for $\rho_n$ and $\lambda_n$.

## BIBLIOGRAPHY

[1] V. A. Ambarzumian. *Über eine Frage der Eigenwerttheorie.* Z. Physik 53, 690–695 (1929).

[2] V. A. Marčenko. *Certain questions of the theory of a second-order differential operator.* Dokl. Akad. Nauk SSSR (N.S.) 72, 457–460 (1950).

[3] M. G. Kreĭn. *Solution of the inverse Sturm-Liouville problem.* Dokl. Akad. Nauk SSSR (N.S.) 76, 21–24 (1951).

[3a] M. G. Kreĭn. *Determination of the density of a nonhomogeneous symmetric cord from its frequency spectrum.* Dokl. Akad. Nauk SSSR (N.S.) 76, 345–348 (1951).

[4] B. M. Levitan. *Application of generalized displacement operators to second-order linear differential equations.* Uspehi Mat. Nauk (N.S.) 4, no. 1 (29), 3–112 (1949); Amer. Math. Soc. Transl. no. 59 (1951).

[5] A. Ya. Povzner. *On differential equations of Sturm-Liouville type on a half-axis.* Mat. Sb. N.S. 23(65), 3–52 (1948); Amer Math. Soc. Transl. no. 5 (1950).

[6] G. Borg. *Eine Umkehrung der Sturm-Liouvilleschen Eigenwertaufgabe. Bestimmung der Differentialgleichung durch die Eigenwerte.* Acta Math. 78, 1–96 (1946).

[7] B. M. Levitan. *On a uniqueness theorem.* Dokl. Akad. Nauk SSSR (N.S.) 76, 485–488 (1951).

[8] E. C. Titchmarsh. *The zeros of certain integral functions.* Proc. London Math. Soc. (2) 25, 283–302 (1926).

[9] N. Levinson. *Gap and density theorems.* Colloq. Publ. vol. 26. Amer. Math. Soc., New York, 1940.

[10] L. A. Čudov. *The inverse Sturm-Liouville problem.* Mat. Sb. N.S. 25 (67), 451–456 (1949).

# AUTOMORPHIC FUNCTIONS AND THE THEORY OF REPRESENTATIONS

By I. M. GEL'FAND[1]

## 1. Introduction

According to F. Klein the investigation of geometrical objects may be reduced to the study of properties invariant under a group of transformations, i.e. to the study of homogeneous spaces.

By a homogeneous space $X$ we mean any manifold acted upon by a group of transformations which is a Lie group. We can therefore study in a unified manner the whole class of the most symmetrical objects, aesthetically perfect, such as sphere, Lobachevsky space, Grassmann's manifold, the space of positively-definite matrices, etc.

It is rather remarkable that we can reach the same aesthetic perfection in the study of the set of functions $f(x)$ defined on the homogeneous space $X$. A transformation $x \rightarrow xg$ in $X$ gives rise to a linear operator $T_g f(x) = f(xg)$ in the function space. So we are led to a representation of the group G, since the product of the transformations $T_{g_1}$ and $T_{g_2}$ corresponds to the product of the elements $g_1$ and $g_2$ of $G$.

Roughly speaking the problem is to decompose the function space into minimal invariant subspaces, or, which is the same, to decompose the representation into irreducible representations and to study invariant families involved.

The compact case of this problem (i.e. compact $X$ and $G$) was investigated by H. Weyl and E. Cartan who took as model the rotation group. The invariant families for the rotation group consist of spherical functions. The least invariant systems of functions, arising in the general case, we shall also call the spherical functions on $X$. But only after one rejects the compactness condition and passes over to infinite-dimensional representations can one fully appreciate the importance and interest of this problem.

In this report I should like to tell about some results obtained by me in collaboration with my friends I. I. Pyatezki-Shapiro and M. I. Graev. Some of the results were inspired by important works of Selberg and Godement. It should be said that only through systematic employment of the theory of infinite-dimensional representations can one obtain complete understanding of these results.

Each homogeneous space $X$ is associated with a group $G$ and its subgroup $\Gamma$ (stationary subgroup) consisting of those elements of $G$ which leave immobile some fixed point $x_0 \in X$. The first part of the report is devoted mainly to the consideration of the case in which this subgroup $\Gamma$ is a discrete group and the space $X$ has finite volume. The functions relating to this case we shall call automorphic functions. Thus automorphic functions are spheri-

(1) The author dedicates this report to his friends, L. Schwartz, M. Morse, J. Leray, O. A. Oleinik, J. G. Petrovskiỳ, L. Bers, R. Courant, P. Lax, whose kind assistance has greatly helped him.

cal functions, associated with a discrete stationary subgroup. Automorphic functions in this sense were introduced in 1950 in a paper by S. V. Fomin and myself [3]. The ordinary automorphic functions are included among these if we take the group of real $2\times 2$-matrices.

Permit me to cite another interesting and important example. Let us be given an arbitrary Riemann surface with a Riemann metric on it. Consider the manifold $X$ the points of which are taken to be linear elements (i.e. points with directions) of the initial surface. Suppose that the universal covering space of our Riemann surface is the upper halfplane. Then the group $G$ acting on $X$ will be the group of all linear fractional transformations with real coefficients and the stationary subgroup $\Gamma$ will be one of its discrete subgroups. If the original Riemann manifold is compact, i.e. if it is associated with an algebraic function, then the resulting representation of $G$ in $L_2(X)$ can be decomposed into a countable direct sum of representations. Every irreducible representation of the group $G$ under consideration can be characterised by a set of indices which we shall call the number of the representation. The numbers of the representations involved in the decomposition are invariants of the Riemann surface. They appear to form a complete system of invariants, and so it may happen that this question is connected with the classical problem of moduli in which such remarkable success was attained recently by Bers and Ahlfors. We can prove a weaker result: every continuous deformation of Riemann surface, leaving unchanged these invariants, is the identity.

So in the first half of the report we consider some general questions about the decomposition of a representation in a function space $L_2(X)$ on a homogeneous space $X$ with discrete stationary subgroup. The characterization of the resulting representations and the method of decomposition lead us to very interesting functions which we call Zeta-functions of the given homogeneous space. This system of functions is closely connected with the classical Zeta-function. In the case of the group of real $2\times 2$-matrices we can reduce these functions, using the duality theorems, to the functions, which were introduced by Selberg from particular considerations.

Now we wish to say a word about the methods. M. I. Graev and I have proposed [1] the so-called method of horyspheres for the investigation of representations in $L_2(X)$ in the case in which $\Gamma$ is continuous, but not discrete as in the case discussed above. This method can be applied with suitable modifications also to the case considered here. This method leads us to a series of theorems concerning the structure of the representations and to the Zeta-function of a homogeneous space. But the general idea of the horysphere method is not sufficient. It is necessary to study the operators $\int F(g)T_g dg$ in detail, where $F(g)$ is a finite function. This part of our work reminds one strongly of the theory of the $S$-matrix in quantum mechanics. The analogy is intrinsic. For example the Zeta-functions of a homogeneous space are quite analogous to the Heisenberg $S$-matrix.

In the second part of the report we consider analogous questions concerning the structure of representations in $X=G/\Gamma$ in the case of a semisimple Lie group over a finite field (in Chevalley-Dickson sense). So it is necessary to describe the representations of Lie groups over finite fields. These questions could serve as the theme of a separate report and we are forced to give only a short account of them. We obtain here a interesting system of Zeta-functions for a homogeneous space above a finite field.

The author has chosen as the theme of his report developments that are in the very earliest stages.

We hope that the many interesting problems and relationships which arise will compensate for the unavoidable lack of a description clear and comprehensible in every detail.

We hope that a reasonable bourbakisation of all facts given in this report and of other facts and problems of the theory of representations will lead to the creation of a domain of algebraic functional analysis in which these facts will be the main examples.

## 2. The case of a compact homogeneous space

If $G$ is a semisimple Lie group, $\Gamma$ a discrete subgroup and $X$ the space $G/\Gamma$ of left cosets, then evidently to each element $g \in G$ there corresponds a movement in $X$ transforming $x$ to $xg$. Denote by $L_2(X)$ the set of all functions on $X$ with integrable square. To each element $g \in G$ we make correspond the unitary shift operator $T_g$ in $L_2(X): T_g f(x) = f(xg)$. The operators $T_g$ form a representation of the group $G$.

The main problem is to decompose this representation into irreducible ones.

The operators $T_\varphi = \int_G \varphi(g) T_g dg$ where $\varphi(g)$ is a finite function on $G$ play an important role in the theory of infinite-dimensional representations. In our situation the use of operators $T_\varphi$ is based upon the fact that for every continuous finite function $\varphi(g)$ the operator $T_\varphi$ is an integral operator in $X$ with the kernel

$$K(x_1, x_2) = \sum_{\gamma \in \Gamma} \varphi(g_1^{-1} \gamma g_2), \tag{1}$$

where $g_1$ and $g_2$ are representatives of cosets $x_1$ and $x_2 \in X$. It is easy to verify that the series for the function $K(x_1, x_2)$ converges uniformly in every compact domain. Thus if $X$ is compact the operator $T_\varphi$ is completely continuous for every finite continuous function $\varphi(g)$.

It is easy to prove the following general proposition.

*If the unitary representation $g \to T_g$ of Lie group $G$ in a space $H$ is such that operator $T_\varphi$ is completely continuous for every finite continuous function $\varphi(g)$, then $H$ can be decomposed into a countable sum of irreducible representations (unitary) of the group $G$, where the multiplicity of each irreducible representation is finite.*

It follows at once from this proposition that when $X$ is compact the representation in $L_2(X)$ is decomposable into a countable sum of irreducible unitary representations of the group $G$. This fact does not exhaust all information about the irreducible representations contained in $L_2(X)$ which can be obtained by the use of the operators $T_\varphi$. There is a formula which gives in a reasonable sense a complete description of all irreducible unitary representations contained in $L_2(X)$.

Let $H_1, H_2, \ldots$ be irreducible non-equivalent representations contained in $L_2(X)$; $N_1, N_2, \ldots$, their multiplicities. Let $\pi_k(g)$ be the character of the irreducible representation $H_k$. The existence of such characters was proved by Godement and Harish Chandra. Then the following important formula

$$\sum_{m=1}^{\infty} \int_{F_m} \varphi(g^{-1}\gamma_m g)\, dg = \sum_{k=1}^{\infty} N_k \int \varphi(g)\, \pi_k(g)\, dg \tag{2}$$

is valid, where $\gamma_1, \gamma_2, \ldots$ is a sequence of mutually non conjugate elements of the group $\Gamma$ and $F_m$ is a fundamental domain for the centralizer $\Gamma_m$ of the element $\gamma_m$. This formula makes it often feasible to find out what representations enter in $L_2(X)$, and with what multiplicities. A particular case of this formula (namely the case in which the functions $\varphi(g)$ are invariant under $U$—the maximal compact subgroup of the semisimple group) can be essentially reduced by the use of the so-called duality theorem [2] to the previously known formula of Selberg [4].

Formula (2) may be used to investigate the asymptotic distribution of "numbers" of the representations, occurring in $L_2(X)$. The leading term of the asymptotic expansion will be obtained if we choose the function $\varphi(g)$ to be concentrated in a decreasing sequence of neighbourhoods of the unit of the group. Now we give following example.

Consider the asymptotic behaviour not of all representations but only of the so-called representations of class I. By definition these representations are the representations entering in the decomposition of $L_2(G/U)$ where $U$ is a maximal compact subgroup of $G$.

These representations are given in the following manner. Let $\mathfrak{A}$ be Cartan subalgebra of the symmetric space $G/U$ and let $\mathfrak{A}^+$ be the cone of the positive vectors in $\mathfrak{A}$ (i.e. vectors for which $(\alpha, \alpha) \geqslant 0$ for all positive roots $\alpha$). Then every representation of class I is uniquely determined by some vector in $\mathfrak{A}^+$. If now $X$ is a compact homogeneous space $G/\Gamma$ then the representations of class I entering into $L_2(X)$ form a countable set of points in $\mathfrak{A}^+$. Their asymptotic distribution is given by the following formula. Denote by $\alpha$ the positive roots of the symmetric space $G/U$ and by $v_\alpha$ the multiplicities of the roots of $G/U$. Then the formula

$$N(B_n) \sim C_G C_\Gamma \int_{B_n} \prod_{\alpha>0} (\varrho, \alpha)^{v_\alpha} d\varrho \tag{3}$$

is valid, where $B_n$ runs through an increasing sequence of subregions in $\mathfrak{A}^+$; $C_G$ is a constant depending only on $G$; $C_\Gamma$ is the volume of space $X = G/\Gamma$ and $N(B_n)$ is the number of irreducible representations entering in $L_2(X)$ indices of which belong to $B_n$. The proof is based on some results of F. I. Karpelevich and S. G. Gindikin.

Apparently a similar formula is valid for the other types of irreducible representations.

Since for the group of $2 \times 2$-matrices as was said, the "numbers" of the representations are invariants of the Riemann surface, the terms of the asymptotic expansion are also such. It would be very interesting if a few leading terms of this expansion were to play the role of moduli. The next terms of asymptotics demand apparently more averaging than the mere evaluation of $N(B_n)$.

Classical automorphic forms enter here in natural way. Among the representations of the group $G$ there are isolated ones, i.e. those occurring in isolated form already in $L_2(G)$. Among the isolated ones there are representations which are realizable by analytic functions [11, 12, 13]. These isolated representations, if they do occur in the decomposition of $L_2(X)$, are connected with automorphic forms in natural way. It is interesting to observe that there are still other isolated representations (Graev [6]). For every isolated representation one can deduce from the spur formula an explicit formula for the multiplicity $N_k$. In the case of classical automorphic forms such a formula was obtained by Hirzebruch [14] who used the Riemann–Roch theorem and by Selberg [4] who used a method close to ours.

## 3. Horyspheres and regular subgroups

The method of horyspheres, which has been elaborated elsewhere, mainly in the paper [1], is effective when $X$ is not compact. In general the method is as follows. Let $X$ be a homogeneous space acted upon by a group $G$. With the space $X$ one can associate a space $\Omega$ whose elements are taken to be surfaces in $X$ which we shall call horyspheres (the definition of horysphere will be given later). To each function defined on $X$ we make correspond its integrals over horyspheres, then the representation in the space of functions on $X$ maps homomorphically onto a representation in the space of functions on $\Omega$. As a consequence the problem of decomposing the representation in the space of functions on $X$ is reduced to the following two problems: (1) to find the kernel of the homomorphism; (2) to decompose the obtained set of functions on $\Omega$ into irreducible representations. In some important cases, discussed in [1], the kernel is zero and induced representation on $\Omega$ can be given a simple description. But in our case the kernel of homomorphism is not zero and the description is not trivial. Nevertheless the method of horyspheres is very productive in this situation.

In particular it makes it possible to separate quite effectively the series of representations occurring continuously in $L_2(X)$ from the series occurring discretely, or in other words to separate the discrete spectrum from the continuous one.

Let $X$ be an arbitrary homogeneous space; then the set of horyspheres in $X$ is not a homogeneous space itself and has not in general a Hausdorff structure. The structure of the space the points of which are transitive components in the space of horyspheres (i.e. the sets of horyspheres which can be carried one into another by the movements of the group) plays a fundamental role in the description of the spectrum of irreducible representations in $L_2(X)$.

Now we proceed to give the exact definitions. Let $G$ be a real semisimple Lie group and $g(t)$ a one-parameter subgroup. The set $Z$ consisting of all $z \in G$ such that

$$\lim_{t\to\infty} g(-t)\, z g(t) = 1$$

is called the horyspherical subgroup associated with the subgroup $g(t)$. For example if $G$ is the group of real $2\times 2$-matrices then every horyspheric subgroup is conjugate to the subgroup of all matrices of the form $\begin{pmatrix} 1 & \zeta \\ 0 & 1 \end{pmatrix}$.

If $G$ is the group of real matrices of order $\mathbf{n}$, then there exist as many nonconjugate horyspherical subgroups as there are different partitions of the the number $\mathbf{n}:\mathbf{n}=k_1+\ldots+k_s$ with positive integers $k$. (Partitions that are distinguished by the order are to be considered as the different ones.)

Now we proceed to define the horyspheres. Let $X$ be a homogeneous space of a group $G$.

*The orbits of horyspherical groups will be called horyspheres in X. The horysphere will be called compact if the set of points of which it consists, is compact.*

Now we proceed to describe the class of discrete subgroups $\Gamma$ which we shall be concerned with. First of all here belong these for which the volume of $G/\Gamma$ is finite. It is quite possible that all such subgroups satisfy the assumptions formulated below. For the groups of second order this follows from the results of Siegel. It would be very interesting to prove the assertion in the general case.

Now we introduce the following definition. A set $Y \subset X$ will be called a cylindrical set, if it can be split into mutually non-intersecting compact horyspheres which can be carried one into another by movements. Thus in a cylindrical set all the horyspheres are of the form $x_0 Zg$ where $Z$ is some fixed horyspherical subgroup.

Now we formulate the following fundamental definition.

*A discrete subgroup $\Gamma$ of the semisimple group $G$ will be called regular if the factorspace $G/\Gamma$ has a finite covering by regular bounded cylindrical sets and the intersection of each pair of them is compact* (the definition of regular bounded cylindrical sets is given below).

It is easy to verify that the factor-space $G/\Gamma$ for each regular discrete subgroup $\Gamma$ of semisimple group $G$ has finite volume.

We shall observe finally that except for the group of real $2\times 2$-matrices all presently known examples of irreducible discrete subgroups of semisimple Lie groups such that the factor space has finite volume are arithmetical groups that are constructed by the well-known construction of Borel.

We shall call linear algebraic group $G$ every group consisting of all complex $n\times n$-matrices the elements of which satisfy given polynomial relations. In the following we shall assume that the coefficients of these polynomials are rational numbers. We denote by $G_z$ the set of all matrices in $G$ whose elements are integers and whose determinant is equal to 1. In a similar way $G_R$ is the set of matrices in $G$ whose elements are real numbers. It is easy to see that $G_z$ is a discrete subgroup of the group $G_R$.

A. Borel and Harish Chandra [8] have proved that the volume of the factor-space $G_R/G_z$ is finite, if $G_R$ is semisimple group.

Apparently it can be proved by their methods that $G_z$ is a discrete regular subgroup of $G_R$.

Now we proceed to define the regular bounded cylindrical sets. It is not difficult to see that for every cylindrical set $Y$ there exists a horyspherical subgroup $Z$ and a set $S\subset G$, whose image in $G/\Gamma$ is $Y$, with the following properties:

(1) For every $g\in S$ and $z\in Z$ there exists $\delta\in\Delta=\Gamma\cap Z$ such that $\delta zg\in S$.

(2) If $g_1, g_2\in S$ and $g_1g_2^{-1}\in\Gamma$ then $g_1g_2^{-1}\in\Delta$.

The first condition means that $Y$ consists of horyspheres. The second that these horyspheres do not intersect.

If the set $Y$ is compact then one can choose $S$ so that it will have also the following properties:

(3) There exists a neighbourhood of unity $U_2$ such that if $g_1^{-1}\gamma g_2\in U$, where $g_1$, $g_2\in S$, then $\gamma\in\Delta$.

(This condition is more strong than condition 2.)

(4) For every $z\in Z$ there exists a compact neighbourhood of unity $U_z$ such that $g^{-1}zg\in U_z$ for any $g\in S$.

Condition (4) plays in this theory the fundamental role.

We shall agree to call bounded any cylindrical set $Y$ for which there exist sets $S$ and $Z$ with the properties (1)–(4). Bounded cylindrical sets are in general non-compact and this seems to be the geometrical reason for the existence in semisimple groups of discrete subgroups such that the factor-space $G/\Gamma$ has finite volume and at the same time is non-compact.

Let $Y$ be a bounded cylindrical set. It can be shown that all elements $g \in S$ are representable in the form $g = zat$, where $a$ belongs to a certain accompanying subgroup $A$ of $Z$,(1) and $t$ belongs to a certain compact set $T$ in $G$. We shall agree to call a normal subgroup $\tilde{Z}$ of $Z$ allowable, or simply allowable, if the intersection of the Lie algebra of the group $\tilde{Z}$ with every root subspace is either vacuous or contains the root subspace.

A bounded cylindrical set will be called regular if

(5) For each allowable subgroup $\tilde{Z}$ of $Z$ (including the group $Z$ itself) the factor-space $\tilde{Z}/(\tilde{Z} \cap \Gamma)$ is compact.

Thus finally the regular bounded cylindrical sets are characterized by the existence of the sets $S, Z, A$ for which the conditions (1)–(5) hold.

## 4. The separation of the continuous spectrum of representation from the discrete one

We shall suppose as formerly that we have a homogeneous space $X = G/\Gamma$, where the discrete group $\Gamma$ is regular and $G$ is a real semisimple group.

By the use of the horysphere method we can decompose the space $L_2(X)$ into a direct orthogonal sum of two spaces. One of them, which we denote $L_2^0(X)$, is decomposable into a discrete direct sum of irreducible representations. The other is decomposable into representations of continuous spectra. However there may be representations which enter into the second space discretely, the so-called representations of the complementary series. They get into the second space because they are, so to speak, the analytical continuation of the continuous spectrum involved. One more justification of this fact is that they hit exactly at the singular points of Zeta-functions corresponding to the given continuous series. It is very much like the complementary discrete spectrum of quantum mechanics that hits exactly at the zeros of the $S$-matrix.

The decomposition into the direct sum is carried out in the following way. To every function $f(x) \in L_2(X)$ we make correspond its integrals over *compact* horyspheres and we denote by $L_2^0(X)$ the space of functions which have their integral over any compact horysphere equal to zero; $L_2'(X)$ is the orthogonal complement of $L_2^0(X)$.

Then *the space $L_2^0(X)$ can be decomposed into a countable number of irreducible representations*. The proof is based on a study of asymptotic properties of the kernel of the integral operator $\int \varphi(g) T_g dg$ at the regular cylindrical sets.

---

(1) A commutative subgroup $A$ is called accompanying if each element of $A$ is semisimple and if $A$ is generated by oneparametrical subgroups $g_1(t), \ldots, g_n(t)$ to each of it $Z$ is associated.

## 5. Functions in the space of horyspheres

We divide the set of compact horyspheres into transitive families $\Omega_i$ (i.e. sets of horyspheres which can be carried one into another by movements). When $\Gamma$ is regular there will be only a finite number of such families. The families $\Omega_i$ are partially ordered in a natural way, namely $\Omega_i < \Omega_j$ if each horysphere of $\Omega_i$ can be split into horyspheres of $\Omega_j$.

Now we take one of the transitive horysphere families $\Omega_i$ and make correspond to each function $f(x) \in L_2(X)$ its integrals over horyspheres of $\Omega_i$. We obtain a function $\check{f}(\omega)$ defined on $\Omega_i$. The inner product defined in $L_2(X)$ is carried in a natural way into the set of these functions. Thus in $L_2(\Omega_i)$ there is defined a quadratic form. To this form there corresponds an operator $M$ defined by

$$\int_{\Omega_i} f\bar{\check{\varphi}}\, d\omega = [\check{f},\ M\varphi],$$

where $[\check{f}_1, \check{f}_2]$ is the inner product inherited from $L_2(X)$. This operator plays a fundamental role. It is permutable with the movements in $\Omega_i$. As a consequence this operator, by the decomposition of representation in $L_2(\Omega_i)$ into irreducible ones, will be, in every system of representations equivalent to the given one, a matrix whose order coincides with the number of equivalent representations. In the general case this number is less than or equal to the order of Weyl group. Thus after the decomposition there will arise a matrix of order equal to the order of Weyl group. (For a group of $\mathbf{n} \times \mathbf{n}$-matrices this order is equal to $\mathbf{n}!$). This matrix depending on the "number" of the representation is by definition the Zeta-function of the space $X$. These functions, associated with the "number" of representation, are closely connected with such functions as the Riemann Zeta-function and its generalizations. It is not excluded that a deeper development of this theory will throw light on the blank spaces in the theory of the classical Zeta-function. The results just formulated give us a chance to describe the representations involved. It could be done in detail but lack of time prevents us from doing so. We merely point out that the spectrum of the representation of the component, associated with any $\Omega_i$, has Lebesgue type and multiplicity 1. The indices of the representations fill up several linear subspaces of Cartan algebra. The description of the allowable subspaces is carried out by induction. The dimension of these spaces is easy to calculate. It is equal to the dimension of a group which is associated with $\Omega_i$ and which consists of all the homeomorphisms permutable with the movements of $G$ in $\Omega_i$.

## 6. Example: the matrix group $G$ and the subgroup $\Gamma$ of matrices with integer elements

As an example we take the subgroup $\Gamma$ of integer elements of the real $\mathbf{n} \times \mathbf{n}$-matrix group. Each transitive family of compact horyspheres is defined by a partition $\mathbf{n} = \mathbf{n}_1 + \ldots + \mathbf{n}_k$. The maximal family $\Omega_1$ is defined by the partition $\mathbf{n} = 1 + \ldots + 1$ and consists of orbits of the triangular group. Every irreducible representation of $G$ can be defined by an index which is a vector in a space of dimension $n-1$. The set of indices of all the irreducible representations involved in the decomposition of $L_2(X)$, where $X=G/\Gamma$, is called the spectrum of $L_2(X)$. To each $\Omega_i$ there corresponds a spectrum consisting of not more than a countable number of linear subspaces. To $\Omega_1$ itself there corresponds a spectrum of multiplicity 1 filling up the whole space. The spectrum of $L_2(X)$ is the union of the subspaces mentioned above and of a countable number of points.

Now we shall exhibit the Zeta-function corresponding to the homogeneous space $G/\Gamma$. We shall restrict ourselves to the Zeta-function of highest dimension. The Weyl group in this case is the symmetric group of degree $n$. Thus the Zeta-function is a matrix of order $n!$. We shall write down the first row because the rank of the matrix is equal to 1 and all other rows are easily expressed in terms of the first. (Each representation associated with $\Omega_1$ enters in $L_2(X)$ only one time!) So we have: if $\sigma$ is a permutation of the symmetric group and $x=(x_1,\ldots,x_n)$ is the "number" of a representation $(x_1+\ldots+x_n=0)$, then

$$\zeta_\sigma(x)=\prod_{\substack{i>j\\ \sigma(i)<\sigma(j)}} B\left(\frac{1}{2},\frac{x_i-x_j}{2}\right)\frac{\zeta(x_i-x_j+1)}{\zeta(x_i-x_j)}, \tag{5}$$

where $\sigma=(\sigma_1,\ldots,\sigma_n)$.

It is not difficult to write this formula in standard root notation. Written in this form it can be generalized.

The functional equation for the Zeta-function which is a consequence of the general theory of representations, is of the form

$$\zeta_{\sigma_1\sigma_2}(x)=\zeta_{\sigma_2}(x)\,\zeta_{\sigma_1}(\sigma_2 x). \tag{6}$$

## 7. The representations of finite Chevalley groups

The success of the theory of representations depends after all on a happy construction of irreducible representations. In the case of the complex semisimple Lie groups a simple construction of irreducible representations was given (in 1947–1950 by M. A. Naimark and the author [9] (later we shall speak about the construction). However, if we go to other classes of semisimple groups—the real semisimple groups, the matrix groups over finite fields etc.—the construction of Naimark and author does not yield all the representations but only a small part of them.

Now we shall discuss the classification of the irreducible representations of semisimple matrix groups over a finite field. We shall deal with the groups which were considered in the well-known papers by Dickson [15] and Chevalley [16]. Included here for example are the unimodular matrix groups over finite fields, the matrix groups leaving invariant some quadratic form etc.

Already in the last century in one of the first papers on the theory of representations Frobenius found the characters of the unimodular group of $2\times 2$-matrices over a finite field. Later (in 1928) Hecke gave the construction for half of the irreducible representations of this group. The construction for the other groups was absent.

Now we proceed to construct the irreducible representations. In the case of a complex semisimple Lie group $G$ the construction of the principal series of irreducible representations is as follows. We consider the homogeneous space $G/Z$ of the right cosets of $G$ with respect to a maximal nilpotent subgroup $Z$. We call this space the principal affine space associated with $G$. The set of homeomorphisms of $G/Z$ permutable with the movements form a commutative group which we shall call the homothety group. A function defined on $G/\Gamma$ will be called homogeneous if it is multiplied by a constant under a homothety.

The irreducible representations of the principal series are realized in spaces of homogeneous functions on $G/Z$; the operator of representation is defined as the shift operator. We have already noted that this construction when transferred to the simple Dickson–Chevalley groups over a finite field, will give us only a small part of all irreducible representations. Now we shall explain how to construct *all* irreducible representations. Again we consider a representation in the space of functions on $G/Z$ where $Z$ is a maximal nilpotent subgroup. An operator of the representation will be defined not as mere shift but as a shift operator multiplied by a fixed function depending on a point of the space and on an element of the group

$$T_\chi(g)f(z)=f(zg)\,\alpha(z,g).$$

It is easy to show that the function $\alpha(z,g)$ is defined essentially by a one-dimensional representation $\chi(z)$ of the group $Z$. For brevity we shall call the $\chi(z)$ simply characters. It can be shown that by the decomposition of these representations we shall obtain all the irreducible representations. This follows from the fact that in every representation of $G$ there exists a vector which is an eigenvector with respect to the element of $Z$. Now we introduce a partial ordering into the set of characters: let $\chi_1<\chi_2$ if from $\chi_2(s)=1$ it follows $\chi_1(s)=1$ for each $s\in Z$ such that $s=e^{E_\alpha t}$ where $\alpha$ is a root of $G$; the maximal characters $\chi(s)$ will be called the characters of general position. It turns out that the representations $T^\chi(g)$ corresponding to the characters $\chi$ of general position do contain each irreducible representation of $G$ not more than one time. The irreducible representations entering into these $T_\chi(g)$ we shall call the principal representations of $G$, all others will be called degenerate. The number of all irreducible representations of $G$ is a polynomial of $k$, where $k$ is the order of field under consideration. It turns out that the number of degenerate representation is a polynomial of lower degree. In this sense we can say that the principal representations are almost all irreducible representations of $G$. It can be shown that the dimensions of the principal representations of $G$ can be expressed as polynomials in $k$ of degree $N$, where $N$ is the dimension of group $Z$, and that the dimensions of the representations of the degenerate series are expressed as polynomial of lower degree. As an illustration we give a formula for the dimensions of the different principal representations of the unimodular matrix group of the $\mathbf{n}$th order. The principal representations are split into several series. Each series is defined by a partition $\mathbf{n}=\mathbf{n}_1+\ldots+\mathbf{n}_l$ of $\mathbf{n}$ into sum of positive integers. The dimensions of the representations of this series are equal to(1)

(1) (Added in proof.): There is a very interesting paper of I. A. Green. *Trans. Amer. Math. Soc.* (1955).

$$\frac{(k-1)(k^2-1)\dots(k^l-1)}{(k^{n_1}-1)\dots(k^{n_e}-1)}, \tag{7}$$

where $k$ is order of the field under consideration. There are exceptions to this formula, the so-called singular representations. The number of these divided by the total number of representations goes to zero with increasing $k$.

Now we shall explain how to split into series the principal representations of the arbitrary group $G$. We combine into one series the principal representations contained in the same $T_\chi(g)$ and with the same multiplicities. It seems that the representations belonging to same series are "contrived in the same manner"; their dimensions coincide; formulae for the characters are written in the same way; when realized they give rise to the same special functions.

Since the principal representations enter in $T_\chi(g)$ of general position with the multiplicity 1, the realization of the representation is comparatively easy. It is sufficient to consider the ring of operators, permutable with the representation $T_\chi(g)$, where $\chi$ is a character of general position. This ring is commutative (!), and so it decomposes into a direct sum of complex fields. Each summand defines one of the principal representations of $G$. It should be noted that the procedure leads to interesting classes of special functions on $G$. Thus in the case of the unimodular group of second order the one-dimensional components of the ring of operators permutable with $T_\chi(g)$ can be expressed as certain sums known as the Kloosterman sums. It is remarkable that the summation is carried sometimes over a certain "contour" in a quadratic extension of the field. Depending on the set of summation we obtain representations of one or another principal series.

## 8. Zeta-functions associated with homogeneous space over a finite field

The representations of the group $G$ over a finite field being constructed we have the possibility to define the Zeta-function. Here we restrict ourselves to the Zeta-function associated with principal representations with $\chi \equiv 1$; we shall not carry out the "analytic continuation" to the other series. Let $G$ be a Dickson-Chevalley group over a finite field, let $\Gamma$ be any of its subgroups and let $Z$ be a maximal nilpotent subgroup. In order to construct the Zeta-function we consider the functions $\varphi(g)$ constant on the cosets of $G$ with respect to $Z$, i.e. the functions satisfying the equation $\varphi(\zeta g) = \varphi(g)\chi(\zeta)$. The operator $M$ acting in the space of these functions is given by

$$M\varphi = \psi(g) = \sum_{\substack{\gamma \in \Gamma \\ \zeta \in Z}} \varphi(\gamma\zeta g)\chi'(\zeta). \tag{8}$$

Then $\psi(\zeta g) = \psi(g)\chi(\zeta)$. The operator $M$ is permutable with the movement $T_{g_0}$ of the group: $T_{g_0}\varphi(g) = \varphi(gg_0)$. So after the decomposition of the representation into irreducible ones it will be given by a matrix which depends on the "number" of the representation and has order equal to the order of Weyl group. By the definition the matrix is the Zeta-function of group $G$ respect to the subgroup $\Gamma$. It depends on the index $\pi$ of representation which is a multiplicative character of the Cartan subgroup. We shall give the explicit form of this function. For this purpose we construct the function $k(g)$ equal for each $g$ to the number of ways in which $g$ can be written in the form $g = \zeta_1\gamma\zeta_2$, where $\gamma \in \Gamma$ and $\zeta_1$ and $\zeta_2$ are in $Z$. Each $g$ is of the form

$g=\zeta'\delta\, s\,\zeta''$, where $\delta$ is in the Cartan subgroup and $s$ is in the Weyl group; therefore $k(g)$ depends only on $\delta$ and $s$: $k(g)=k(\delta,s)$. The Fourier transform of $k(\delta,s)$ with respect to $\delta$ is now the Zeta-function:

$$\zeta_s(\pi)=\sum_{\delta} k(\delta,s)\,\pi(\delta).$$

We shall not write down the form of this function for the representations of other principal series.

## 9. Reprsentations associated with groups over other fields

We shall not consider the case of groups associated with the field of $p$-adic numbers which were studied by Mautner and Bruhat. We hope that the indicated constructions are applicable to this case and make it possible to describe all representations.

Very interesting is the problem of studying the representations of the Chevalley groups over the field of algebraic functions over the field of complex numbers. Here it is not evident how to define a representation. Apparently the following definition is the most natural. We require $T_g$ to be an operator defined every where except a manifold of lower dimension. We require $T_g$ to depend continuously on $g$ and to satisfy the equation $T_{g_1}T_{g_2}=T_{g_1g_2}$ everywhere except a manifold of lower dimension. These groups are examples of infinite-dimensional groups for which the problem of constructing their irreducible representations is quite real and interesting. Also in this case the constructions indicated above give many important types of representations. It is possible, however, to exhibit the examples showing that the construction does not exhaust all degenerate representations.

### REFERENCES

[1]. Гельфанд, И. М. & Граев, И. М. И., *Труды Моск. Матем. общ.*, 8 (1959), 323–390.

[2]. Гельфанд, И. М. & Пятецкий-Шациро, И. М. И., *Успехи матем. наук*, XIV 2 (1959), 172–194.

[3]. Гельфанд, И. М. & Фомин, И. С. В., *Успехи матем. наук*, 7:1 (47) (1952), 118–137.

[4]. Selberg, A., *Tata Inst. of Fund. Res., Bombay*, 1960, 147–164.

[5]. Selberg, A., *J. Ind. Math. Soc.*, 20 (1956), 47–87.

[6]. Граев, М. И. *Докл. А. Н. С.С.С.Р.*, 113 (1957), 966–969.

[7]. Selberg, A., *Analytic Functions*, vol. 2, 1960.

[8]. Borel, A. & Harish-Chandra, *Ann. Math.*, 75 3 (1962).

[9]. Гельфанд, И. М. & Наймарк, И. М. А., *Труды матем. ин-та А.Н. С.С.С.Р.*, 36 (1950), 1–288).

[10]. Harish-Chandra, *Proc. Nat. Acad. Sci. USA.*, 45 (1959), 570–573.

[11]. Bargman, *Ann. Math.*, 48 (1947), 568–640.

[12]. Гельфанд, И. М. & Граев, И. М. И., *Изв. А.Н. С.С.С.Р, сер. матем.*, 17 (1953), 189–238.

[13]. Граев, М. И., *ТрудыМоск. матем общ.*, VII (1959).

[14]. Hirzebruch, *Analytic Functions*, vol. 2, 1960.

[15]. Dickson, L., *Linear Groups*.

[16]. Chevalley, C., *Tôhoku Math. J.*, 7 (1955), 14–66.

*Actes, Congrès intern. math.*, 1970. Tome 1, p. 95 à 111.

# THE COHOMOLOGY OF INFINITE DIMENSIONAL LIE ALGEBRAS; SOME QUESTIONS OF INTEGRAL GEOMETRY

by I. M. GEL'FAND

This report is concerned with certain results and problems arising in the theory of the representation of groups. In the last twenty years much has been achieved in this field and—most important—its almost boundless possibilities have become apparent.

Indeed, its problems, touching on the interests of algebraic geometry, on many questions of the algebraic number theory, analysis, quantum field theory and geometry, as well as its inner symmetry and beauty have resulted in the growing popularity of the theory of representations.

It is impossible to list even briefly its main achievements, and this is not the aim of this communication. Nevertheless, one cannot omit mentioning the outstanding papers by Harish-Chandra, Selberg, Langlands, Kostant, A. Weil, which considerably advanced the development of the theory of representations and opned up new relationships; and, since we do not go into these questions, we will not be able to touch upon many of the deep notions and results of the theory of representations.

We feel that the methods which have arisen in the theory of representation of groups may be used in a considerably more general non-homogeneous situation. We will give some examples:

1. The proof of the fact that the spectrum of a flow on symmetric spaces of constant negative curvature is a Lebesgue spectrum [1] was based on methods of the theory of representations, namely the decomposition of representations into irreducible ones. One of the most useful methods of decomposing representations into irreducible representations is the orisphere method [5]. In the works of Sinai, Anosov, Margulis [2], [3], [4], only the orispheres are considered and groups symmetries are left out. This rendered possible the study of the spectrum of dynamic systems in a considerably more general situation.

2. The theorem of Plancherel and the method of orispheres gives rise to the consideration of more general problems of integral geometry, taking place in a non-homogeneous situation [5], [6], [7], [8], [9], [10], [11], [12], [13], [14], [15], [16], [17].

3. If we have a manifold and its mapping, the study of distributions " constant on the inverse image of each point " of this mapping is an extremely interesting problem, special examples of which were studied in the homogeneous situation (functions in

four-dimensional space, invariant relative to the Lorenz group, functions constant on classes of conjugate elements of a semi-simple Lie group [18], etc.). There are various aspects of this problem which are considerably more interesting and important than may seem at first glance. Of course, the main interest of the problem is the study of these distributions at singularies of the mapping. To be more precise, suppose $X$ is a manifold ($C^\infty$-analytical, algebraic) and $\mathscr{G}$ is some (perhaps infinitely dimensional) Lie algebra of smooth vector fields. One wishes to describe the space of unvariant distributions.

A more natural statement of the problem is obtained by replacing the distributions by generalised sections of a vector bundle which vary according to a given finite dimensional representation. Unfortunately consideration of length prevent me from giving a series of existing examples. Those examples are particularly interesting when $X$ has only a finite number of orbits relative to $\mathscr{G}$. For interesting example in the non-homogeneous situation see [34].

4. The theory of representation of groups makes the consideration of interesting examples possible and shows the importance of studying the ring of all the regular differential operators on those algebraic manifolds which are homogeneous spaces. It is quite natural to wish to describe the structure of the ring $R$ of regular differential operators on any algebraic manifold. Perhaps, as in [19], [20], it would be helpful to consider the quotient skew-field of the ring $R$. Another interesting problem is the description of the involutions of this ring $R$.

In this report I would like to tell about certain problems which were studied by my friends and myself while thinking about questions connected with representation theory.

## I. Representations of semisimple Lie algebras.

0. Suppose $\mathscr{G}$-is a semisimple Lie algebra. The study of representations is essentially the study of a category of $\mathscr{G}$-modules. The choice of the particular category of $\mathscr{G}$-modules considered in the algebraic problems of the theory of representations is essential. Suppose $\mathscr{k}$ is a fixed subalgebra of $\mathscr{G}$. $\mathscr{G}$-the module will be called $(\mathscr{g}, \mathscr{k})$ finite iff 1° it is a finitely generated $\mathscr{U}(\mathscr{G})$-module and 2° as an $\mathscr{U}(\mathscr{G})$-module it is the algebraic direct sum of finite dimensional irreducible representations of $\mathscr{k}$ and in this decomposition each of the irreducible representation appears only a finite number of times.

The following two cases are very interesting:

1° $\mathscr{G}$ is a real semisimple algebra, $\mathscr{k}$ is the subalgebra corresponding to the maximum compact subgroup. The corresponding $(\mathscr{g}, \mathscr{k})$-modules were considered by V. A. Ponamaryov and the author and were called by them " Harish-Chandra modules ".

2° $\mathscr{G}$ is a real Lie algebra, $\mathscr{k}$ is a Cartan subalgebra or, more generally, the semisimple part of the parabolic subalgebra.

1. Let us consider in more detail the category of Harish-Chandra modules in the case when $\mathscr{G}$ is the algebra of a complex semisimple Lie group.

Further, each module is the direct sum of modules on each of which the Laplace operators have only one eigen-value.

Consider an example. Suppose $G$ is a simply connected Lie group over the algebra $\mathscr{G}$, $B$ – its Borel subgroup, $\mathscr{N}$ is a unipotent radical of $B$, $H$ – a Cartan subgroup. Consider the indecomposable finite dimensional representation $\rho$ of the group $H$. Note that since $H = C^* \times C^* \times \ldots \times C^*$ the question of the finite dimensional representations of $H$ is reduced to the determination of a finite number of pairwise commutative matrices. Let us extend this representation $\rho$ of the group $H$ to a representation of the group $B$ and consider, further, the representation of the group $G$ induced by this representation $B$. The representation thus obtained will be called a Jordan representation. In the case when $\rho$ is of dimension one, we obtain the well-known representation of the principal series. Thus we have constructed, using the representation of the group $H$, a representation of the group $G$. Note that the description of the canonical form of the representation of $H$ is in some sense an unsolvable problem if the rang of $H$ is greater than 1 [21].

If we consider the representation of the algebra $\mathscr{G}$ thus constructed only on the space of vectors which vary over the finite dimensional representation of the maximum compact subgroup, we will obtain Harish-Chandra modules. Apparently the following hypothesis holds: at the points of general position all the indecomposable Harish-Chandra modules are all Jordan representations (*).

For $SL(2, \mathbb{C})$ this statement follows from work of Zhelobenko. The most interesting is the study of Harish-Chandra modules at singular points. Of course, the problem of listing all the Jordan modules is already a badly stated (unsolvable) problem, since it is based on the classification of systems of pairwise commutative matrices. However, it is not clear whether it is possible to solve this problem at a singular point, considering the Jordan modules as given. If such a solution were possible, it would have exceptional interest.

The problem of describing Harish-Chandra modules was completely solved by V. A. Ponomaryov and the author for the Lie algebra of the group $SL(2, \mathbb{C})$ [22], [23], [24]. Then these representations were constructed as a group representation (and not only as an algebra representation) by M. I. Graev and the authors cited above [25].

The classification of indecomposable Harish-Chandra modules is carried out in two stages.

1. The problem is reduced to a problem in linear algebra.
2. The linear algebra problem obtained for $SL(2, \mathbb{C})$ generalises the problem of describing the canonical form of pairs of matrices $A$, $B$ such that $AB = BA = 0$. To solve this problem we apply the Maclane relation theory, which allows us to use the relations $A^\#$ and $B^\#$, inverse to the degenerate operators $A$ and $B$, as well as the monomials $A^{\# k_1} B^{\# k_2} A^{\# k_3} \ldots$.

The Harish-Chandra modules at a singular point may be divided into two classes.

(*) To be more precise, each Harish-Chandra module is decomposed into direct sum of submodules on which the Laplace operators have precisely one eigen-value. The set of eigen-values thus obtained is called singular if the representation of the fundamental series with the same eigen-values of the Laplace operator are reducible. The points of general position will be exactly the non-singular points.

The modules of first class are uniquely defined by any set of natural numbers, the modules of the second class are determined by any set of natural numbers together with one complex number $\lambda$. It is thus interesting to note that at singular points the module space is not discrete. The most convenient canonical form of Harish-Chandra modules are given in [25].

In the case of $SL(2, \mathbb{R})$ the problem of classifying Harish-Chandra modules is easily reduced to a problem in linear algebra; explicitely the category of Harish-Chandra modules at a given singular point is isomorphic to the following category of diagrams in the category of finite dimensional linear spaces:

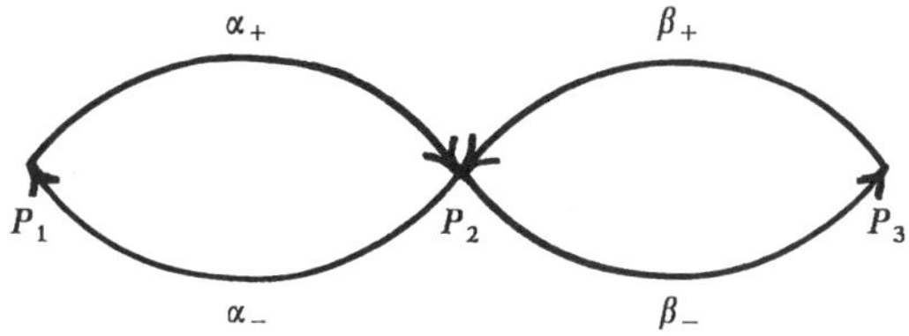


with the condition $\alpha_+\alpha_- = \beta_+\beta_- = \gamma$, where $\gamma$ is nilpotent. The question of the classification of the objects of this category is aparantly solvable but leads to considerable difficulties.

CONJECTURE. — The category of Harish-Chandra modules for any semisimple group with given eigen-values of Laplace operators is equivalent to a certain category of diagrams in the category of finite dimensional linear spaces.

2. This and the following section of the report summarise some results of I. N. Bernstein, S. I. Gel'fand and the author.

Suppose $\mathscr{G}$ is a semisimple Lie algebra over $\mathbb{C}$, $b$ is its Borel subalgebra, $\mathfrak{u}$ is a radical and $\mathscr{h}$ is a Cartan subalgebra. Consider the following category $\mathcal{O}$. Its objects are $(\mathscr{G}, \mathscr{h})$ – finite modules $M$, satisfying the following condition: for every vector $\xi \in M$ the space $\mathscr{U}(u)\xi$ is finite dimensional. This category is most useful for the application of the theory of highest weights. In this category, let us chose a class of objects which will be called elementary. All the others will be constructed from them and their factor modules by step by step extensions.

Suppose $\chi$ is a linear functional over $\mathscr{h}$. Denote by $M_\chi \mathscr{U}(\mathscr{G})$-module, generated by $f_\chi$, with the relations $nf_\chi = 0$ and $hf_\chi = (\chi - \rho, h).f_\chi$ for all $h \in \mathfrak{h}$ and $n \in \mathfrak{u}$. Here $\rho$ denotes the half-sum of the positive roots. By studying the modules $M_\chi$ we get extensive information on the representation of the algebras $\mathscr{G}$, including finite dimensional ones. We now state a few theorems on $M_\chi$ modules and their morphisms.

THEOREM 1 (Verma). — Let the modules $M_{\chi_1}$ and $M_{\chi_2}$ be given. Two cases are possible:

1° $$\mathrm{Hom}\,(M_{\chi_1}, M_{\chi_2}) = 0;$$

and

2° $$\mathrm{Hom}\,(M_{\chi_1}, M_{\chi_2}) \approx \mathbb{C},$$

then any non-trivial homomorphism $M_{\chi_1}$ into $M_{\chi_2}$ is an embedding.

To state the next theorem we must introduce a partial ordering in the Weyl group $W$. Suppose $s_1, s_2 \in W$. We shall say that $s_1 > s_2$ iff there exist reflexions $\sigma_1, \ldots, \sigma_r$ in $W$ such that $s_1 = \sigma_1 \ldots \sigma_r s_2$ and $l(\sigma_{i+1} \ldots \sigma_r s_2) = l(\sigma_{i-1} \ldots \sigma_r s_2) + 1, i = 1, \ldots, r$, where $l(s)$ is the length of the element $s \in W$.

THEOREM 2. — Let $M_{\chi_1}$ and $M_{\chi_2}$ be given. $M_{\chi_1}$ imbeds into $M_{\chi_2}$ if and only if,

1. There exists such an $\chi$ that Re $\chi$ lies in the positive Weyl chamber and such a pair of elements $s_1, s_2 \in W$, $s_1 > s_2$ that $\chi_1 = s_1\chi$, $\chi_2 = s_2\chi$.

2. $\chi_1 - \chi_2 = \Sigma n_i\alpha_i$, where $n_i$ are integers, $\alpha_i$ are simple roots.

The module $M_{\chi_0}$ is richest in submodules for integer values of $\chi_0$ from the positive Weyl chamber. It follows from theorem 2 that $M_{\chi_0}$ contains a submodule $M_{s\chi_0}$ for all $s \in W$. In this case the embedding of $M_{s\chi_0}$ into $M_{\chi_0}$ is determined in the following way. Suppose $s_{\alpha_i}$ is the reflection with respect to the simple roots $\alpha_i$, $s = s_{\alpha_i} \ldots s_{\alpha_k}$ is the decomposition of minimum length. Let

$$\chi_i = s_{\alpha_i} s_{\alpha_{i+1}} \ldots s_{\alpha_k} \chi_0 .$$

Then

$$f_{s\chi_0} = af_{\chi_0},$$

where

$$a = E_{-\alpha_1}^{\frac{(\chi_2 - \chi_1, \alpha_1)}{(\alpha_1, \alpha_1)}} . E_{-\alpha_2}^{\frac{(\chi_3 - \chi_2, \alpha_2)}{(\alpha_2, \alpha_2)}} \ldots E_{-\alpha_k}^{\frac{(\chi_0 - \chi_k, \alpha_k)}{(\alpha_k, \alpha_k)}}$$

Since the minimum representation $s$ in the form of the product of $s_{\alpha_i}$ is not unique, whereas the injection $M_{s\chi_0}$ into $M_{\chi_0}$ is uniquely determined, the theorem gives relations between " chains " of the type described. In the general case the embedding is more complicated.

The relations between $M_{s\chi_0}$ may easily be shown by the following commutative diagram. The vertices of the diagram are numbered by the elements $s$ of the Weyl group and correspond to the modules $M_{s\chi_0}$. If $s_1 < s_2$, then an arrow going from $s_2$ to $s_1$ is drawn. The mapping is defined by the embedding of $M_{s_2\chi_0}$ into $M_{s_1\chi_0}$. We obtain a commutative diagram. It is not difficult, using this diagram, to get in particular, a resolution of the finite dimensional representation by free $\mathcal{U}(u)$-modules.

The finite dimensional representation with highest weight $\chi_0 - \rho$ is of the form

$$M = M_{\chi_0} / \sum_{s \neq l} M_{s\chi_0} .$$

The theorems stated above and this diagram contain, in this case, the formulas of Kostant, Weyl's formulas for characters, the Borel-Weil theorem and the Harish-Chandra theorem concerning the left ideals of enveloping algebras.

3. *The ring of differential operators on the principal affine space and the generalisation of the Segal-Bargman representation to any compact group.*

Suppose $G$ is a complex semisimple Lie group, $\mathcal{N}$ is the maximum unipotent subgroup, $H$ — a Cartan subgroup. The manifold $A = \mathcal{N} \backslash G$ is called the principal affine space of the group $G$. It is an algebraic quasi affine manifold. It is interesting to consider the ring $\mathscr{D}$ of regular differential operators on $A$. Suppose $f(g)$ ranges over all the regular algebraic functions (polynomials) on the group $G$. We will give a method allowing to construct for any such function a differential operator on $A$. Since $H$ normalises $\mathscr{N}$, the transformation $g \to hg$ may be carried over to $A$ (left translations [5]). Using these left translations we can assign to every element of the Lie algebra $\not{h}$ of the group $H$ a differential operator on $A$. The commutative ring of differential operators on $A$ generated by these operators will be denoted, following [20], by $W_u$. Suppose $\pi$ is the natural map of $G$ into $A$. Denote by $\pi^*$ extension of the functions over $A$ to functions over $G$ induced by $\pi$. The operation $\pi_*$, mapping the functions on $G$ into functions on $A$ is less obvious and supplements, in our case, the operation of averaging the function over the subgroup. The construction of $\pi_*$ is carried out in the following way.

Suppose $f(g)$ is a regular algebraic function on $G$. Consider it as the linear combination of matrix elements of finite dimensional irreducable representations in the basis of weight vectors $H$. Threw out all the elements of this sum except the summands corresponding to those matrix elements whose first index is the highest weight of the corresponding representations. Denote by $\pi_* f$ the function thus obtrained.

Suppose $f$ is a fixed function on $G$. Define the operator $f$ in the functions by the formula

$$\bar{f}(\varphi) = \pi_*(f\pi_*(\varphi))$$

THEOREM 1. — There exists an element $w \in W_u$ such that $w_0 f$ is a regular differential operator on $A$. Conversely, every regular differential operator on $A$ may be represented in the form $\Sigma w_i \,.\, \bar{f_i}$, $w_i \in W_u$ where $f_i$ are functions on $G$.

Suppose $\mathscr{K}$ is the quotient field of the $W_u$ ring, $\mathscr{F}(G)$ is the ring of regular algebraic functions on $G$. The map constructed in theorem 1 may be expanded to the map

$$i : \mathscr{D} \bigotimes_{W_u} \mathscr{K} \to \mathscr{F}(G) \bigotimes_{\mathbb{C}} \mathscr{K}$$

THEOREM 2. — $i$ is a linear space isomorphism over $\mathscr{K}$, compatible with the right translations by elements of $G$.

Note that the fact of the existence of an isomorphism of the spaces above was obtained earlier in a joint paper of A. A. Kirillov and the author [20].

For the group $SU(2)$ there exists an extremely useful realisation of the whole series of representations of this group due to Segal and Bargmann. This realisation is in the Hilbert space of analytic functions of two complex variables, square, integrable with weight $e^{-|z_1|^2-|z_2|^2}$. We will point out a generalisation of this construction for any compact Lie group.

Suppose $K$ is a simply connected compact Lie group of rang $r$, $G$ — its complexification, $A$ — the principal affine space of the group $G$. Introduce the weight function $e^{-H(a)}$, $a \in A$. Suppose $\rho_i$ is the $i'$-th fundamental representation of $G$, let $\xi_i$ denote the vector of highest weight in $\rho_i$. Put

$$H_i(g) = (\rho_i(g)\xi_i\,,\, \rho_i(g)\xi_i),$$

where ( , ) is the scalar product in the space of the representation $\rho_i$ invariant relative to $K$. It is clear that $H_i(g)$ is a function on $A$ and we can then put

$$H(a) = \sum_{i=1}^{r} H_i(a).$$

Now consider the analytic functions on $A$ which are square integrable with weight $e^{-H(a)}$. Call the Hilbert space of all these functions a " generalised Segal-Bargmann space ". The group $K$ thus obtained acts on it in a natural way and the unitary representation thus obtained contains every irreducible one exactly once. Let us call any operator with polynomial regular algebraic coefficients a " differential operator on $A$ ".

CONJECTURE. — The operator conjugate (in the generalised Segal-Bargmann space) with a regular differential operator is again a regular differential operator.

The involutions which arise in the ring of regular differential operators are far from trivial. Thus, for the case of $SU(n)$ the operator, say, conjugate with multiplication by a simple first order function, is a differential operator of the $(n - 1)$-st order. The techniques developed in the previous section apparently will turn out to be very useful in the study of the ring of differential operators on $A$, in particular, for the proof of the conjecture stated aboce. The fact of the matter is that the construction of the involution itself is most conveniently carried out in the terms developed there. Using this method the conjecture was checked for $SU(3)$.

We state another problem. Let the real form of the group $G$ be given. Its unitary representation naturally gives rise to an involution in the enveloping algebra $\mathcal{U}(\mathcal{G})$. We must find all the possible extensions of this involution from $\mathcal{U}(\mathcal{G})$ to the ring of all the regular differential operators on $A$. In the simpliest examples these extended involutions correspond to series of unitary representations (of real groups) contained in the regular one. It would be interesting to list the involutions in the ring of regular differential operators on any quasiaffine algebraic manifold.

It would also be interesting to consider the factor space of the group $G$, not only over the maximal unipotent group, but also over any orispherical subgroup.

## II. **Integral geometry.**

In this paragraph I will only consider one elementary example [17]. The derivation of the Plancherel formula for $G = GL(n, \mathbb{C})$ is based on the following problem in integral geometry. Denote by $\mathcal{N} \in G$ the set of all the upper triangular matrices with units on the diagonal. Suppose the function $f(x)$, $x \in G$ is given. Let

$$\varphi(x_1, x_2) = \int_{\mathcal{N}} f(x_1^{-1} z x_2) dz,$$

where $x_1$ and $x_2$ are any matrices. The problem is: given $\varphi(x_1, x_2)$ find $f(x)$. It suffices to solve the problem when $x = e$ is the unit matrix. We can assume that the fonction $f$ is given on $\mathbb{C}^{n^2}$ and the equation $y = x_1^{-1} z x_2$ for fixed $x_1$ and $x_2$ defines in $\mathbb{C}^{n^2}$ a plane of dimension $\dfrac{n(n-1)}{2}$.

Now replace our problem with the following, at first glance meaningless, problem. Consider the space $H_{n^2,k}\left(k = \dfrac{n(n-1)}{2}\right)$ of all the $k$-dimensional planes in $\mathbb{C}^{n^2}$. For all $h \in H_{n^2,k}$ consider the function

$$\varphi(h) = \int_h f(x)dx.$$

We must now recover $f(x)$. In the paper [10] this problem is solved in the following manner. Using the function $\varphi$ and its derivatives construct a differential $(k, h)$ form $\mathscr{H}\varphi$ on the Grassman manifold $G_{n^2,k}$ of $k$-dimensional planes containing the point $x$. This form $\mathscr{H}\varphi$ is closed and the value of $f(x)$ is equal to $\int_{\gamma_0} \mathscr{H}\varphi$, where $\gamma_0$ is any cycle homologic to the set of all $k$-dimensional planes containing the point $x$ and lying in a fixed $k+1$-dimensional plane passing through the point (Euler's cycle (*)). As to the integral over the other $k$-dimensional cycles in the basis of Schubert cells in $G_{n^2,k}$, it is equal to zero.

In our case the function $\varphi(x_1, x_2)$ is known not on the whole manifold $H_{n^2,k}$ but only on a certain submanifold. The submanifold of $H_{n^2,k}$ will from now on be called the " complex of $k$-dimensional planes ". The complex is called permissible if the form $\mathscr{H}\varphi$ on this complex is determined by the values of the function $\varphi$ on this complex only. In the case when $\varphi$ is given on a permissible complex we can recover $f(x)$ by using the formula

$$f(x) = C_\gamma \int_\gamma \mathscr{H}\varphi,$$

where $\gamma$ is a cycle lying in the complex; thus to find $C_\gamma$ it suffices to decompose the cycle $\gamma$ over the Schubert cell basis. In our case the complex will consist of planes of the form $h_{x_1,x_2} = \{\, y/y = x_1^{-1}zx_2 \,\}$ and has dimension $n^2$. It turns out to be permissible. The set of these planes of this complex which contain the point $e$ has the necessary dimension $k$ and forms a cycle. The coefficient of the Euler cycle is equal to $n!$ Considering the form $\mathscr{H}\varphi$ only on the complex, we will obtain the classical inversion formula

$$f(e) = [(2i)^k \pi^{2k} n!]^{-1} \int \prod_{q<p} \left(\frac{\partial}{\partial \delta_p} - \frac{\partial}{\partial \delta_q}\right)\left(\frac{\partial}{\partial \bar{\delta}_p} - \frac{\partial}{\partial \bar{\delta}_q}\right) \times \varphi(\mathscr{C}^{-1}\delta\mathscr{C})\,|_{\delta=e} \bigwedge_{q<p} d\mathscr{C}_{qp} \bigwedge_{q<p} d\bar{\mathscr{C}}qp.$$

Apparently one can obtain the Paley-Wiener theorem for $GL(n, \mathbb{C})$, in a similar manner; in other words, obtain conditions on $\varphi$, which imply the decrease of $f$ at infinity. To do this we embed $GL(n, \mathbb{C})$ not into $\mathbb{C}^{n^2}$ but into $\mathbb{CP}^{n^2}$ and consider the problem as a projective problem of integral geometry (see [15]). Since in this case we can recover $f(x)$ in the points at infinity as well, the Paley-Wiener conditions will consist in the

(*) Note that other problems of integral geometry give rise to integration over other cycles in $G_{n^2,k}$; see, for example [16].

following: the function $f'$ and its derivatives (recovered by using $\varphi$) must be equal to zero at all the points of infinity.

## III. **Cohomology of infinite algebras.**

0. *This part of the report contains results obtained jointly by D. B. Fuks and the author.*

We know how difficult it is to describe any reasonable category of representations. On the other hand, the problem of determining cohomology groups is a sumpler one. Here we list results about the cohomology of Lie algebras of vector spaces, which show that these cohomologies are reasonable, are not equal to zero and are not infinite dimensional.

Recall that the cohomology $H^*(\mathscr{G}; M) = \sum_q H^q(\mathscr{G}; M)$ of the topological algebra $\mathscr{G}$ with coefficients in the $\mathscr{G}$-module is defined as the cohomology of the complex $C(\mathscr{G}; M) = \{ c^q(\mathscr{G}; M), d^a(\mathscr{G}; M) \}$ where $c^a(\mathscr{G}; M)$ is the space of continuous skew-symetric $q$-linear functionals on $\mathscr{G}$ ranging over $M$, and the differential $d^a = d^r(\mathscr{G}; M)$ is defined by the formula

$$(d^qL)(\xi_1,\ldots,\xi_{q+1}) = \sum_{1\leqslant s<t\leqslant q+1} (-1)^{s+t-1}L([\xi_s, \xi_t], \xi_1,\ldots,\hat{\xi}_s,\ldots,\hat{\xi}_t,\ldots,\xi_{q+1}) - \sum_{1\leqslant s\leqslant q} (-1)^s\xi_sL(\xi_1,\ldots,\hat{\xi}_s,\ldots,\xi_{q+1}).$$

If $M$ is a ring, and the operators on $\mathscr{G}$ are its differentials, then the complex $C(\mathscr{G}; M)$ has a natural multiplicative structure.

1. *Problems and examples.*

The main example of an infinitely dimensional Lie algebra will be the algebra of smooth vector fields on a smooth manifold.

Suppose $M$ is a closed orientable connected smooth (*) manifold. Denote by $\mathfrak{A}(M)$ the Lie algebra of smooth tangent vector fields on $M$ with Poisson brackets for commuting. The first of the problems considered is a follows. Define the cohomology ring $\mathfrak{H}^*(M) = H^*(\mathfrak{A}(M); \mathbb{R})$ of the algebra $\mathfrak{A}(M)$ with coefficients in the unit representation, i. e., in the field $\mathbb{R}$ of real numbers with a trivial $\mathfrak{A}(M)$-module structure. This ring obviously is a differential invariant of the manifold $M$. Looking ahead we shall say that the space $H^q(\mathfrak{A}(M); \mathbb{R})$ will turn out to be finite dimensional for any $q$ (see [28]). The problem of computing the ring $\mathfrak{H}^*(M)$ is not as of yet completely solved.

We would like to point out the difference between the method of constructing invarients of manifolds by using objects of differential geometry (the Lie algebra of vector fields) and the usual method of constructing differential invariants. Whereas usually the differential form representing a Pontryagin of Chern class on the manifold $X$ is built up from the individual object (by using the metric) on the manifold,

(*) By smooth we always mean of class $C^\infty$.

in our case the invariants are constructed using the infinite dimensional set of all smooth vector fields on the manifold.

As an example consider the case when $M$ is the circle $S^1$. We can show that the ring $H^*(S^1)$ is generated by a two-dimensional generator $a$ and a three-dimensional generator, the two being related only by the skewsymetry condition.

Further the generators $a \in \mathfrak{H}^2(S^1)$, $\mathscr{C} \in \mathfrak{H}^3(S^1)$ are represented by cocycles $A \in C^2(\mathfrak{A}(S^1); \mathbb{R})$, $B \in C^3(\mathfrak{A}(S^1); \mathbb{R})$ given by the formulas

$$A(f, g) = \int_{S^1} \begin{vmatrix} f'(x) & f''(x) \\ g'(x) & g''(x) \end{vmatrix} dx$$

$$B(f, g, h) = \int_{S^a} \begin{vmatrix} f(x) & f'(x) & f''(x) \\ g(x) & g'(x) & g''(x) \\ h(x) & h'(x) & h''(x) \end{vmatrix} dx$$

When the dimension of the manifold $M$ increases the ring $\mathfrak{H}^*(M)$ becomes considerably richer; thus the ring $\mathfrak{H}^*(S^2)$ has 10 generators, and the ring $\mathfrak{H}^*(S^1 \times S^2)$, 20 generators (see [29]).

The cohomology of the Lie algebra of smooth vector fields is intimately connected with the cohomology of Lie algebras of formal vector fields. By a formal vector field at the point $O$ of the space $R^n$ we mean a linear combination of the form $\Sigma p_i(x_1, \ldots, x_n)e_i$ where $e_1, \ldots, e_n$ are the standard basis vectors of the space $R^n$ and $p_i(x_1, \ldots, x_n)$, the formal power series with real coefficients in the coordinates $x_1, \ldots, x_n$ of the space. The set of formal vector fields is, in an obvious sense, a linear topological space, and a natural commutation operation transforms it into a topological Lie algebra. This algebra is denoted by $W_n$.

2. *The algebra of formal vector fields. The cohomology of the algebra $W_n$ with coefficients in,R.*

In order to state the final result it is necessary to describe a certain auxilliary topological space $X_n$ $(n = 1, 2, \ldots)$. Suppose $\mathscr{N} \geqslant 2n$ and let $p_i E(N, n) \to G(N, n)$ be the canonical $U(n)$ bundle over the (complex) Grasman manifold $G(\mathscr{N}, n)$. The usual ($W$-complex of the manifold $G(\mathscr{N}, n)$ has the following property: the $2n$-th skeleton $[G(\mathscr{N}, n)]_{2n}$ does not depend on $\mathscr{N}$ when $\mathscr{N} \geqslant 2n$. The inverse image of the set $[G(\mathscr{N}, n)]$ under the map $p$ will be denoted by $X_n$.

The space $X_1$ is a three-dimensional sphere, the other spaces do not have such a simply visualised description. We have the following.

THEOREM 2.1. — For all $q$, $n$ there is an isomorphism

$$H^q(W_n ; R) = H^q(X_n ; R).$$

Multiplication in the ring $H^*(W_n ; R)$ (as well as in the ring $H^*(X_n ; R)$) is trivial, i. e., the product of any two elements of positive dimension is equal to zero.

The cohomology of the space $X_n$ may be computed by using standard topological methods. For example, it is trivial for $0 < q \leq 2n$ and for $q > n(n + 2)$.

Theorem 2.1 is the central result of the article [30]. Its proof uses a somewhat modified version of the Serre-Hoschild spectral sequence [31] corresponding to the subalgebra of the algebra $W_n$, generated by the elements $x_i e_j$ (*); this subalgebra is isomorphic to $\mathcal{G}l(n, R)$. Beginning with the second member, this spectral sequence turns out to be isomorphic to the Leray-Serre spectral sequence of the bundle $X_n \to [G(\mathcal{N}, n)]_{2n}$ with fibre $U(n)$.

It turns out also that each element $\alpha \in H^q(W_n ; \mathbb{R})$ is represented by such a cocycle $A \in C^q(W_n ; \mathbb{R})$, that $A(\xi_1, \ldots, \xi_q)$ depends only on the 2-jets of formal vector fields $\xi_1, \ldots, \xi_q$ (see [30]).

To study the cohomology of $W_n$ with coefficients in other modules (and to describe those modules) it is important to know the structure of the subalgebras

$$\ldots \subset L_k \subset \ldots \subset L_0 \subset W_n$$

where $L_k$ consists of vector spaces whose components are series without terms of power less than or equal to $K$.

*The relation between the cohomology of the algebras $W_n$ and $L_0$.* The following general fact is easily generalised to the case of the cohomology of infinite dimensional Lie algebras.

Suppose $B$ is an subalgebra of Lie algebra $A$; $M$ — some $B$-module; $\hat{M}$ — an induced $A$-module (i. e. $\hat{M} = \mathrm{Hom}_{[B]}(M, [A])$ where $[A]$, $[B]$ are enveloping algebras for $A$, $B$). Then

$$H^*(A ; \hat{M}) = H^*(B ; M).$$

We will apply this statement in the case when $M$ is a tensor representation of the algebra $L_0$ (i. e. a finite dimensional representation obtained from the representation of the algebra $\mathcal{G}l(n; R)$ by means of the projection $L_0 \to L_0/L_1 = \mathcal{G}l(n; R)$). At the same time the induced representation $\hat{M}$ of the algebras $W_n$ is none other than the space of the corresponding formal tensor fields. For example, if $M = R$ is the unit representation of the algebra $L_0$, then $\hat{M}$ is the space $F(R^n)$ of formal power series in $n$ variables with the natural action of the algebra $W_n$ ; if $M$ is the space $\Lambda^r(R^n)'$ of skewsymetric $r$-linear forms in $R^n$, then $\hat{M}$ is the space $\Omega^r$ of formal exterior differential forms of $r$ order in $R^n$.

*The cohomology of the algebra $W_n$ with coefficients in the spaces of formal exterior differential forms.* The space

$$H^*(W_n, \Omega^*) = \sum_{q,\mathfrak{A}} H^q(W_n ; \Omega^r)$$

is obviously a bigraduated algebra (over $R$), isomorphic, as we just found out, to $H^*(L_0 ; \Lambda^*(R^n)')$.

THEOREM 2.2. — The bigraduated ring $H^*(W_n ; \Omega^*) = H^*(L_0; \Lambda^*(R^n)')$ is multiplicatively generated by $2n$ generators

$$\rho_i \in H^{2i-1}(L_0; \Lambda^0(R^n)') \qquad (i = 1, \ldots, n)$$
$$\tau_i \in H^i \quad (L_0; \Lambda^i(R_n)') \qquad (i = 1, \ldots, n)$$

These generators are connected only by the following relations $\rho_i\rho_k = -\rho_k\rho_i$ ; $\rho_k\tau_i = \tau_i\rho_k$ ; $\tau_i\tau_k = \tau_k\tau_i$ ; $\tau_1^{i_1}\tau_2^{i_2} \ldots \tau_n^{i_n} = 0$ if $i_1 + 2i_2 + \ldots + ni_n > n$.

(*) $i, j = 1, \ldots, n$.

In particular, the ring $H^*(L_0; R) = H^*(L_0; \Lambda^0(R^n)') = H^*(W_n; F(R^n))$ is an exterior algebra in generators of dimension 1, 3, 5, ..., $2n-1$. i. e.

$$H^*(W_n; F(R^n)) = H^*(gl(n, R); R).$$

Moreover,

$$H^q(L_0; \Lambda^r(R^n)') = \begin{cases} 0 & \text{where} \quad q < r \\ H^r(L_0; \Lambda^r(R^n)') \otimes H^{q-r}(gl(n, R); R) & \text{where} \quad q \geqslant r \end{cases}$$

while the dimension of the space $H^r(L_0, \Lambda^r(R^n)')$ is equal to the number of ways in which the number $r$ may be represented as the sum of natural numbers.

The computation of the cohomology of $L_0$ with coefficients in the tensor representation reduces to the computation of the cohomology of the algebra $L_1$ with coefficients in $\mathbb{R}$. In a similar way for jets, to the cohomology of $L_k$ with coefficients in $\mathbb{R}$.

Apparently the following statement holds.

CONJECTURE. — For any $n$ the spaces $H^q(L_k; R)$ are finite dimensional.

For $n=1$ the dimension of the space $H^q(L_k; R)$ equals $C_q^{k-1} + C_{q+1}^{k-1}$, $q, k = 0, 1, \ldots$).

Using previously mentioned results to compute the cohomology of the algebra $L_0$ with tensor coefficients we can deduce that the classes of cohomology of the algebra $W_n$ (even $W_1$) with coefficients in tensor fields is not always representable by cocycles depending only on 2-jets of their arguments (in contrast with the cases of constant and skewsymetric coefficients).

We have been unsuccessful, so far, in computing the cohomology $H^*(A, \mathbb{R})$ for other Cartan algebras. Note that all these cohomologies are connected with very important standard complexes. For this complex consists of the polynomials $P(\alpha_1, \ldots, \alpha_q); (\beta_1, \ldots, \beta_q)$, $\alpha_i \in \mathbb{R}^n$, $\beta_i \in (\mathbb{R}^n)'$; the polynomial $P$ is skewsymetric under the simultaneous interchange of $\alpha_i, \beta_i$ with $\alpha_j, \beta_j$. The differential is given by the formula

$$\begin{aligned} dP(\alpha_1, \ldots, \alpha_{q+1}; \beta_1, \ldots, \beta_{q+1}) \\ = \Sigma(-1)^{s+t}(\alpha_s, \beta_t) - (\alpha_t, \beta_s))P(\alpha_f + \alpha_t, \alpha_f, \ldots, \hat{\alpha}_f, \ldots, \hat{\alpha}_t, \ldots; \\ \beta_f + \beta_t, \beta_1, \ldots, \hat{\beta}_f, \ldots, \hat{\beta}_t, \ldots, \beta_{q+1}). \end{aligned}$$

Usually, the infinite dimensional Lie algebras which arise in the formal theory are factor subcomplexes of this complex.

3. *The algebra of smooth vector fields. Cohomology with coefficients in R.*

Suppose $M$ is a compact connected orientable smooth $n$-dimensional manifold without boundary, $\mathfrak{A}(M)$ – the Lie algebra of smooth tengent yest fields on $M$. In the standard complex $C(M) = \{ C^a(M) = C(\mathfrak{A}(M); R)d^a \}$ we introduce a filtration $0 = C_0(M) \subset C_1(M) \subset \ldots \subset C(M)$ where $C_k(M) = \{ C_k^a(M) \}$ is a subcomplex of the complex $C(M)$, defined in the following way. A cochain $L \in C^a(M)$ belongs to $C_k^q(M)$ if it equals zero on any $C^a$ the vector fields $\xi_1, \ldots, \xi_q$ such that for any $k$ points of the manifold $M$ one of the fields $\xi_1, \xi_2, \ldots, \xi_q$ equals zero in the neighbourhood of each of these points. For example, $C_0^a(M) = 0$; $C_1^q(M)$ consists of such cochains $L$ that $L(\xi_1, \ldots, \xi_q) = 0$ when the supports of the fields $\xi_1, \ldots, \xi_q$ are pairwise nonintersecting; $C^q_{q-1}(M)$ consists of such cochains $L$ that $L(\xi_1,\ldots,\xi_q)=0$ when the supports of the fields $\xi_1,\ldots,\xi_q$ have no common intersection to all of them; $C^q_k(M)=C^q(M)$ when $k \geqslant q$. It is clear that $C_k(M)$ for all $k$ is a subcomplex of the complex $C(M)$ and that $C^q_k(M)C^r_l(M) \subset C^{q+r}_{k+l}(M)$.

To compute the cohomology of the factor complex $C_k(M)/C_{k-1}(M)$ we have defined a spectral sequence, the first term of which may be expressed by using the cohomology of the manifold $M$ and the algebra $W_n$. A special role is played by the complex $C_1(M)$. This complex we shall call a diagonal complex.

CONJECTURE. — The image of the cohomology of the diagonal complex $C_1(M)$ in $\mathfrak{H}^*(M)$ under the embedding $C_1(M) \to C(M)$ multiplicatively generates all of the ring $\mathfrak{H}^*(M)$. In particular the ring $\mathfrak{H}^*(M)$ is always finitely generated.

*Remark.* — This is true for the second term of the spectral sequence.

Let us describe a spectral sequence which converges to the cohomology of the diagonal complex. It arises in connection with two different filtrations of the diagonal complex of the manifold. In order to describe the first filtration, note that the $q$-cochains of the diagonal complex $C_1(M)$ are determined by distributions (more precisely, by the generalized sections of a certain fibre bundle) on $M^q$ which are supported by the diagonal. The $m$-th term $C^q_{1,m}$ of the first filtration consists of those distributions which have an order (relative to $\Delta$) less than or equal to $m$.

To define the second filtration fix a triangulation of the manifold

$$M = M_n \supset M_{n-1} \supset \ldots \supset M_0$$

where $M_i$ is the $i$-dimensional skeleton, and the $m$-th term $C_{1,m}$ of the filtration consider those $q$-cochains which are realised by distributions whose support is $M_m \subset \Delta$.

Knowing the cohomology of $W_n$ can construct a spectral sequence which allows us to compute the cohomology of the diagonal complex.

THEOREM 3.1. — There exists a spectral sequence $\mathscr{E} = \{E^{p,q}_r, d^{p,q}_r\}$ which converges to the cohomology of the diagonal complex $\mathfrak{H}^*(M)$ such that

$$E^{p,q}_r = H^{p+n}(M) \otimes H^q(W_n\,;\,R);$$

$E^{p,q}_r$, in particular, can be different from zero only when $-n \leqslant p \leqslant 0$.

Let us clarify the operation of " globalizing " the formal cohomology: construct a mapping of the space $E^{-r,q+r}_r = H^{n-r}(M) \otimes H^{q+r}(W_n, R)$ into $C^q_1(M)$. This mapping is not uniqual determined: it depends on the choice of the system of local coordinates on $M$. Suppose $\Gamma = \{U_1,\ldots,U_{\mathcal{N}}\}$ is a coordinate covering of $M$ with coordinates $y_{k_1},\ldots,y_{k_n}$ on $U_i$ and $\{\rho_i\}$ is a decomposition of unity consistent with this covering. In order to construct the element $\mathscr{I}(a\otimes\Psi)(a \in H^{n+r}(W_n, R), \Psi \in H^{n-v}(M))$ find a cochain $\alpha \in C^{n+v}(W_n; R)$ representing the closed form $\omega$ from the class $\Psi$. Set

$$\mathscr{I}(a_n \otimes \Psi)(\xi_1,\ldots,\xi_q) = \int_M \omega \Lambda [\sum_{k=1}^{N} \rho_k \varphi(\alpha, U_k\,;\,\xi_1,\ldots,\xi_q)]$$

where $\varphi(\alpha, U_i; \xi_1, \ldots, \xi_q)$ is a form on $U_k$, which equals

$$\sum_{1 \leqslant i < \ldots < i_r \leqslant n} \alpha(\xi_1(u, U_k), \ldots, \xi_q(u, U_i), e_{k,i_1 \ldots} e_{k,i_r}) \times dy_i \wedge \ldots \wedge dy_{i_v}$$

at the point $u \in U_i$, where the $\xi_i$ are considered as a formal field in the neighbourhood of the point $u$ under the coordinates $y_{k_i}$. The theorem is proved in [29] (statement 1.4).

*The cohomology with coef cients in the spaces of smooth sections of smooth vector bundles.* Suppose $A$ is a finite dimensional $GL(n, R)$ module and suppose $M$ is a smooth connected manifold (we do not assume $M$ either orientable, or compact, or without boundary). Denote by $\alpha$ the vector bundle over $M$ with fiber isomorphic to $A$, induced by the tangent bundle and by means of the representation of the group $CL(n, R)$ in $A$. By $\mathscr{A}$ denote the space of smooth sections of the fiber bundle $\alpha$. The space $\mathscr{A}$ has an obvious $\mathfrak{A}(M)$ module structure. Our goal is the study of the cohomology of the algebra $\mathfrak{A}(M)$ with coefficients in the $\mathfrak{A}(M)$ module $\mathscr{A}$.

In the complex $C(M; A) = \{ C^a(\mathfrak{A}(M); \mathscr{A}); d^a \}$ we will introduce a filtration similar to the one considered above for $C(M)$. We shall say that the cocycle $L \in C^q(\mathfrak{A}(M); \mathscr{A})$ has filtration no greater than $k$ if the section $L(\xi_1, \ldots, \xi_q)$ of the bundle $\alpha$ is equal to zero for any point $x \in M$ with the following property: for any points $x_1, \ldots, x_k \in M$ one of the vector fields $\xi_1, \ldots, \xi_q$ equals zero in the neighbourhood of each of the points $x_1, \ldots, x_k, x$.

The space of $q$-dimensional cocycles which have filtration no greater than $k$ is denoted by $C^q_k(\mathfrak{A}(M); \mathscr{A})$. It is clear that $C_k(M; \mathscr{A}) = \{ C^q_k(\mathfrak{A}(M); \mathscr{A}) \}$ is a subcomplex of the complex $C(M; \mathscr{A})$.

The subcomplex $C_0(M; \mathscr{A})$ is called " diagonal ". We denote it by $C_\Delta(M; \mathscr{A})$.

THEOREM 3.5. — We have the following spectral sequence $\{ E_r^{p,q}, d_r^{p,q} \}$ which converges to $\mathfrak{H}^*_\Delta(M; \mathscr{A})$ and is such that $E_r^{p,q} = H^p(M; R) \otimes H^q(L_0; A)$. In the multiplicative case the spectral sequence is a multiplicative one and the isomorphism considered above is an isomorphism of rings.

CONJECTURE. — $H^*_\Delta(M, \mathscr{A}) = H(T_M, R) \otimes \mathrm{Hom}_{CL(n)}(A, H^*(L, R))$ where $T$ is the principal $U(n)$ bundle over $M$ induced by the complexification of the tangent bundle.

This conjecture has been proved in the case when $A = \Lambda^q$ is the exterior power of the standard representation. The case $q = 0$ was independently studied by Locik [33].

In the end of this part of the report I would like to introduce a general concept of formal differential geometry. It arises when one formalises and generalises the methods of construction of Pontryagin and Chern classes (by means of metrics and connections); also in the expression of the index of a differential operator in terms of the symbol and the metric of the manifold.

Suppose we have an algebra $W_n$ of formal vector fields. Consider the jet space and, in it, a invariant algebraic submanifold $X$. Examples of such manifolds are the space of all symmetric tensors of rang 2, the set of all affine connections.

Let us define the complex $\Omega(X)$. Any rational map of $X$ into the complex of formal differential forms will be called a chain of $\Omega(X)$, the differential will be obtained by differentiation in the image. Set $\Omega(X) = \mathrm{Hom}(X, \Omega)$, where $\Omega$ is the complex of formal differential forms, and call the maps of the rational cohomology of $\Omega(X)$-generalised Chern classes. It can be shown, in the case when $X$ is the manifold of symmetric tensors of rang 2, that they coincide with Pontryagin classes ($q < n$).

## REFERENCES

[1] I. M. Gel'fand, S. V. Fomin. — Geodesic flows on surfaces of constant negative curvature (Геодезические потоки на поверхностях постоянной отрицательной кривизны). Успехи математически ких наук, т. XУI, 1 (1952), pp. 118-137.

[2] Y. G. Sinai. — Classical dynamic systems with countable Lebesgue spectrum (2) (Классические динамические системы со счетнократным лебеговым спектром). Изв. АН СССР, сер. матем. 30, 1 (1966), pp. 15-68.

[3] D. V. Anosov. — Geodesic flows on closed Riemann manifolds of negative curvature (Геодезические потоки на замкнутых римановых многообразиях отрицательной кривизны). Труды Матем. ин-та АН СССР, 90 (1967), pp. 1-210.

[4] G. A. Margulis. — On some applications of ergodic theory to the study of manifolds of negative curvature (О некоторых применениях эргодической теории к изучению многообразий отрицательной кривизны). Функц. анализ, 3, вып. 4 (1969), pp. 89-90.

[5] I. M. Gel'fand, M. I. Graev. — The geometry of homogeneous spaces, representation theory in homogeneous spaces and related questions of integral geometry (Геометрия однородных пространств, представления групп в однородных пространствах и связанные с ними вопросы интегралвной геометрии). Труды Моск. матем. об-ва, 8 (1959), pp. 321-390.

[6] I. M. Gel'fand. — Automorphic functions and the theory of representations (Автоморфные функции и теория представлений). *Proceedings of the International Congress of Mathematicians*, Stockholm (1962), pp. 74-85.

[7] I. M. Gel'fand, M. I. Graev, I. I. Pyatetskii-Shapiro. — Representation theory and automorphic functions (Теория представлений и автоморфные функции). Изд. « Наука » (1966), pp. 1-512.

[8] Harish-Chandra. — *Eisenstein series over finite fields*. Stone Jubilee volume, Springer.

[9] Helgason. — *Seminare Bourbaki* (1965).

[10] I. M. Gel'fand, M. I. Graev, Z. Y. Shapiro. — Integral geometry on $\kappa$-dimensional planes (Интегральная геометрия на $\kappa$-мерных плоскостях). Функц. анал., т. 1, вып. 1 (1967), pp. 15-31.

[11] I. M. Gel'fand, M. I. Graev. — Permissible complexes of planes in $C^n$ (Допустимые комилексы прямых в пространстве $C^n$). Функц. анализ, т. 2, вып. 3 (1968).

[12] —, —, Z. Y. Shapiro. — Differential forms and integral geometry (Дифференциальные формы и интегральн ная геометрия). Функц. анализ, т. 3, вып. 2 (1969).

[13] Langlands. — *On the functional equations satisfied by Eisenstein series* (1967).

[14] I. M. Gel'fand, M. I. Graev, N. Y. Vilenkin. — Integral geometry and related questions of representation theory (Интегральная геометрия и связанные с ней вопросы теории представлений). Гос. изд. физ.-мат. лит. (1962), pp. 1-656.

[15] —, —, Z. Y. — SHAPIRO. Integral geometry in projective space (Интегральная геометрия в проективном пространстве). Функц. анализ, т. 4, вып. 1 (1970), pp. 14-32,

[16] —, —, —. — Problems of integral geometry related with a pair of Grassman manifolds (Задачи интегральной геометрии, связанные с парой грассмановых многообразий). ДАН СССР, 193, N° 2 (1970).

[17] —, —. — Complexes of $\kappa$-dimensional planes and the Plancherel formula for GL($n$, $\mathbb{C}$) (Комплексы $\kappa$-мерных плоскостей и формула Планшереля для GL($n$, $\mathbb{C}$)). ДАН СССР, 179 N° 3 (1968).

[18] HARISH-CHANDRA. — Invariant distributions on Lie algebras. *Am. J. Math.*, LXXXVI, 2 (1964), pp. 272-309.

[19] I. M. GEL'FAND, A. A. KIRILLOV. — Sur les corps liés aux algèbres enveloppantes des algèbres de Lie (О телах, связанных с обертывающими алгебрами алгебр Ли). Institut des hautes études scientifiques, Publications Mathématiques, (1966).

[20] —, —. — Structure of the Lie skew-field, related to a semisimple split Lie algebra (Структура тела Ли, связанного с полупростой расщепимой алге брой Ли). Функц. анализ, 3, вып. 1 (1969).

[21] I. M. GEL'FAND, V. A. PONOMAREV. A remark on the classification of pairs of commuting linear operators in finite dimensional space (Замечания о классификации пары коммутирующих линейных преобразований в конечномерном пространстве). Функц. анализ, 3, вып. 4 (1969), pp. 81-82.

[22] —, —. — Undecomposable representations of the Lorenz group (Неразложимые представления группы Лоренца). Успехи матем. наук, XXIII, вып. 2 (140), 1968 г.

[23] —, —. — The category of Harish-Chandra modules over the Lie algebra of the Lorenz group (Категория модулей Хариш-Чандры над алгеброй Ли группы Лоренца). ДАН СССР, N° 2, 176 (1967), pp. 243-246.

[24] —, —. — The classification of undecomposable infinitesimal representations of the Lorenz group. (Классификация неразложимых инфинитезимальных представлений группы Лоренца). ДАН СССР, N° 3, 176 (1967).

[25] —, —, M. I. GRAEV. — The classification of linear representations of the group SL (2, C) (Классификация линейных представлений группы SL (2, C)). ДАН СССР, N° 5, 194 (1970).

[26] I. N. BERNSTEIN, I. M. GEL'FAND, S. I. GEL'FAND. — Differential operators on the principal affine space (Дифференциальные операторы на основном пространстве). ДАН СССР, N° 6, 195 (1970).

[27] V. BARGMANN. — On a Hilbert space of analytic functions and an associated integral transform. I, II. *Com. Pure and Appl. Math.*, XIV, 3 (1961), pp. 187-214 and XX, 1 (1967), pp. 1-101.

[28] I. M. GEL'FAND, D. B. FUKS. — The cohomology of the Lie algebra of tangent vector fields of a smooth manifolds, I (Когомологии алгебры Ли касательных векторных полей гладкого многообразия). Функц. анализ, 3, вып. 3, pp. 32-52.

[29] —, —. — The cohomology of the Lie algebra of tangent vector fields of a smooth manifolds, II (Когомологии алгебры Ли касательных векторных полей гладкого многообразия). Функц. анализ, вып. 2, 4 (1970), pp. 23-32.

[30] —, —. — The cohomology of the Lie algebra of formal vector fields (Когомологии алгебры Ли формальных векторных полей). Изв. АН СССР, сер. мат., 34 (1970), pp. 322-337.

[31] I. P. SERRE, C. HOCHSHILD. — Cohomology of group extensions. *Trans. Am. Math. Soc.*, 74 (1953), pp. 110-134.

[32] I. M. GEL'FAND, D. B. FUKS. — The cohomology of the Lie algebra of vector fields with non-trivial coefficients (Когомологии алгебры Ли векторных полей с нетривиальными коэффициентами). Функц. анализ, вып. 3, 4 (1970).
[33] M. V. LOCIK. — Concerning the cohomology of infinite dimensional Lie algebras of vector fields (О когомологиях бесконечномерных алгебр Ли векторных полей). Функц. анал., вып. 2, 4 (1970), pp. 43-53.
[34] I. N. BERNSTEIN, S. I GEL'FAND. — The function $\rho^\lambda$ meromorphic (Мероморфность функции $\rho^\lambda$). Функц. анал., вып. 1, 3 (1969).

University of Moscow,
Department of Mathematics,
Moscow V 234 (U. R. S. S.)

# HYPERGEOMETRIC FUNCTIONS, TORIC VARIETIES AND NEWTON POLYHEDRA

I.M.Gelfand[1], M.M.Kapranov[2], A.V.Zelevinsky[2]

## Introduction

In this talk we give a survey of our recent results on multidimensional hypergeometric functions [GZK 1,2,7]. Before developing the general theory we briefly discuss main features of the classical Gauss function $F(x)={}_2F_1(a,b;c;x)$. By definition, $F(x)$ is the solution of the hypergeometric equation

$$x(1-x)\frac{d^2F}{dx^2}+[c-(a+b+1)x]\frac{dF}{dx}-abF=0 \tag{1}$$

regular at $x=0$ and normalized by $F(0)=1$. Here $a,b$ and $c$ are complex parameters.

It is well-known [BE] that $F(x)$ can be expanded as the power series

$$F(x)=\sum_{n\geq 0}\frac{(a)_n(b)_n}{(c)_n}\frac{x^n}{n!}, \tag{2}$$

where $(a)_n=\Gamma(a+n)/\Gamma(a)=a(a+1)\ldots(a+n-1)$. The analytic continuation of this series can be given by the Euler integral

$$F(x)=\frac{\Gamma(c)}{\Gamma(b)\Gamma(c-b)}\int_0^1\frac{t^{b-1}(1-t)^{c-b-1}}{(1-tx)^a}\,dt, \tag{3}$$

converging for $Re(c)>Re(b)>0$.

Several classical attempts were made to develop multi-dimensional generalizations of the Gauss function (cf. e.g. [AK]). Their starting point is consideration of power series in several variables similar to (2). The most general series of this type were introduced by J.Horn [H]. Important results on these series were obtained by M.Sato in his study of prehomogeneous vector spaces [S].

[1]Laboratory of Biorganic Chemistry, Moscow State University, Corpus A, Moscow I 7-34, USSR

[2]Scientific Council for Cybernetics, 40, Vavilova St., 117333 Moscow, USSR

105

We develop an unified approach to special functions represented by Horn series. This includes the construction and study of a holonomic system of linear differential equations satisfied by these series as well as their integral representations generalizing (3). Our approach is closely connected with geometry of toric varieties and combinatorics of convex polytopes. We associate a system of generalized hypergeometric functions to every finite subset A of an integer lattice $Z^{n-1}$, and call them A-*hypergeometric functions*. In the corresponding Horn series the summation is over some part of the lattice of affine relations among elements of A.

Note that hypergeometric functions on the Grassmannian $Gr_p(C^{p+q})$ studied in [A], [G], [VGZ] are a particular kind of A-hypergeometric functions. The corresponding set A is the set of vertices of the product $\Delta^{p-1}\times\Delta^{q-1}$ of two simplices.

In §1 for any finite $A\subset Z^{n-1}$ we introduce a system of linear differential equations satisfied by A-hypergeometric functions. We call it the A-*hypergeometric system*. This system first appeared in [GGZ], where it was proven to be holonomic. We calculate its characteristic cycle in terms of certain Newton polytopes. In particular, this gives the number of linearly independent solutions at a generic point.

In §2 we prove that each hypergeometric series in the sense of Horn satisfies a certain A-hypergeometric system. Moreover, we construct a class of bases in the space of A-hypergeometric functions consisting of series of this type. It is amazing that these bases correspond to certain triangulations of the convex hull of A.

In §3 we study integral representations of A-hypergeometric functions. The corresponding integrals are of the form

$$\int_\sigma \prod P_i(x_1, \ldots, x_k)^{\alpha_i} x_1^{\beta_1}\ldots x_k^{\beta_k}\, dx_1 \ldots dx_k \qquad (4)$$

for some Laurent polynomials $P_i$; the integrals are considered as functions of the coefficients of $P_i$. Here $\sigma$ is some k-cycle; the precise meaning of the integral will be explained in §3. It is natural to call the integrals of type (4) *generalized Euler integrals*. They include as special cases the classical Euler integral (3) as well as more general integrals given by Appell

[AK]. For the hypergeometric functions on Grassmannians this construction gives us the integrals of products of powers of linear functions studied in [A], [VGZ].

We shall prove that any integral of type (4) satisfies a certain A-hypergeometric system. Conversely, we show that for any A the integrals of type (4) form the complete system of solutions of the A-hypergeometric equations.

## §1. A-hypergeometric system

1.1. Notation and assumptions. Let A be a finite subset of an integral lattice $Z^{n-1}$. Denote by $L=L(A)\subset Z^A$ the lattice of affine relations among elements of A, i.e., the set of integer vectors $(a_\omega)$, $\omega\in A$, such that $\sum_{\omega\in A} a_\omega \omega=0$, $\sum_{\omega\in A} a_\omega=0$. All the constructions below will depend only on the affine geometry of A, i.e., on the lattice $L(A)\subset Z^A$. Therefore we can and will assume that A lies in fact in $Z^n$ and satisfies the following two conditions:

(a) A generates $Z^n$ as an Abelian group.

(b) There is a group homomorphism $h:Z^n\longrightarrow Z$ such that $h(\omega)=1$ for any $\omega\in A$.

Let $V=C^A$ be the space of vectors $(v_\omega)$, $\omega\in A$, where $v_\omega\in C$. For any $a\in L$ we define the differential operator $\square_a$ on V by

$$\square_a=\prod_{\omega:a_\omega>0}(\partial/\partial v_\omega)^{a_\omega}-\prod_{\omega:a_\omega<0}(\partial/\partial v_\omega)^{-a_\omega}.$$

Note that (b) implies $\sum_{\omega\in A} a_\omega=0$ for any $a\in L$, so $\square_a$ is homogeneous.

Define also differential operators

$$Z_i=\sum_{\omega\in A}\omega_i v_\omega(\partial/\partial v_\omega),\quad i=1,\ldots,n$$

on V, where $\omega_i$ is the i-th coordinate of $\omega\in A\subset Z^n$.

1.2. Definition. *Let $\gamma=(\gamma_1,\ldots,\gamma_n)$ be a complex vector. The A-hypergeometric system with parameters $\gamma$ is the following system of linear differential equations on a function $\Phi(v)$, $v\in V$:*

$$\square_a\Phi=0\ (a\in L),\qquad Z_i\Phi=\gamma_i\Phi\qquad (i=1,\ldots,n). \tag{5}$$

Holomorphic solutions of (5) will be called *A-hypergeometric functions*.

This system was introduced and studied in [GGZ], [GZK 1,2]. The second group of equations in (5) means that $\Phi(v)$ is homogeneous under the action of the complex torus $(C^*)^n$ on V given by $(\lambda v)_\omega=\lambda^\omega v_\omega$, $\omega\in A$, where $\lambda=(\lambda_1,\ldots,\lambda_n)$, $\lambda^\omega=\prod \lambda_i^{\omega_i}$. Therefore, $\Phi$ can be written as $v^\theta F(x)$, where $v^\theta$ is a monomial and $x\in V/(C^*)^n$ is the image of v. Thus A-hypergeometric functions can be reduced to functions in $|A|-n$ variables.

1.3. Theorem. ([GGZ]) *The system* (5) *is holonomic, and so the number of its linearly independent solutions at a generic point is finite.*

We shall denote by $Hyp(\gamma)$ the sheaf on V whose sections are A-hypergeometric functions. By general theory of holonomic systems, $Hyp(\gamma)$ is a constructible sheaf. By $Hyp(\gamma)_v$ we denote the space of local holomorphic solutions near v, i.e., the stalk of $Hyp(\gamma)$ at v.

1.4. The number of solutions. Let $Q\subset R^n$ be the convex hull of A. This is a convex polytope of dimension n-1 lying in the hyperplane $h(u)=1$. We introduce the volume form $Vol_{n-1}$ on this hyperplane by setting the volume of an elementary simplex on the lattice $\{u\in Z^n: h(u)=1\}$ to be equal to 1.

1.5. Theorem. ([GZK 1,2]). *The number of linearly independent holomorphic solutions of the hypergeometric system* (5) *at a generic point of* V *is equal to* $Vol_{n-1}(Q)$.

1.6. The characteristic variety. To make precise the notion of "generic points" we shall introduce the D-module corresponding to the system (5), and its characteristic variety. Let $D_V$ be the sheaf of rings of differential operators on V with holomorphic voefficients. Define a left ideal $J_\gamma\subset D_V$ by the formula $J_\gamma=\sum D_V\Box_a+\sum D_V(Z_i-\gamma_i)$. Its elements are "differential consequences" of the equations (5), i.e., they vanish on all A-hypergeometric functions. As usual, we associate to the system (5) the $D_V$-module $M_\gamma=D_V/J_\gamma$, see [K], [Bj]. By definition, the *characteristic variety* (or *singular support*) $SS(M_\gamma)$ is the subvariety in the cotangent bundle $T^*V$ defined by principal symbols of all differential operators from $J_\gamma$. The characteristic variety of a holonomic D-module governs the singularities of the corresponding system: one of the irreducible components of $SS(M_\gamma)$ is V (the zero section of the cotangent bundle), and the set of singularities of (5) is the union of projections of all other components. We define the *generic stratum* $V_{gen} \subset V$ as the complement of this union in V.

Let $V^*$ be the dual vector space to V, and $(\xi_\omega)$, $\omega \in A$ be the coordinates dual to $(v_\omega)$. For each $a \in L$ consider the polynomial $\hat{\Box}_a(\xi) = \prod_{\omega: a_\omega > 0} \xi_\omega^{a_\omega} - \prod_{\omega: a_\omega < 0} \xi_\omega^{-a_\omega}$ which is the symbol of the differential operator $\Box_a$. Let S be the subvariety in $V^*$ defined by equations $\hat{\Box}_a(\xi) = 0$ for all $a \in L$. As shown in [GZK 2], S is a toric variety (not necessarily normal). More precisely, consider the action of the torus $(C^*)^n$ on $V^*$ given by $(\lambda\xi)_\omega = \lambda^{-\omega}\xi_\omega$, $\omega \in A$. For any face $\Gamma \subset Q$, including Q and $\emptyset$ (where Q is the convex hull of A) consider the subvariety $S(\Gamma) \subset S$ defined by equations $\xi_\omega = 0$ for $\omega \notin \Gamma$.

1.7 Proposition. ([GZK 1,2]). a) *S is the closure of the orbit of the point* $(1,\ldots,1)$.

b) *There is a* 1-1 *correspondence* $\Gamma \longrightarrow S_0(\Gamma)$ *between faces of Q and orbits of* $(C^*)^n$ *on S such that the closure of* $S_0(\Gamma)$ *is* $S(\Gamma)$. *In particular,* $S_0(Q)$ *is the orbit of* $(1,\ldots,1)$.

We shall identify each of the cotangent bundles $T^*V$ and $T^*V^*$ with $V \times V^*$ and hence with each other. Let $\varphi: T^*V^* \longrightarrow T^*V$ be the corresponding isomorphism. For any subvariety $Z \subset V^*$ denote $T^*_Z(V^*)$ its conormal bundle (if Z is singular then by $T^*_Z(V^*)$ we mean the closure of the conormal bundle to the set of smooth points of Z). This subvariety is Lagrangian and conic, i.e., invariant under the action of $C^*$ on the fibers of the cotangent bundle (in our case each of these fibers is identified with V). It is well known [Gi] that each conic Lagrangian subvariety in $T^*V^*$ has the form $T^*_Z(V^*)$.

1.8. Theorem. *The irreducible components of* $SS(M_\gamma)$ *are precisely the* $\varphi(T^*_{S(\Gamma)}V^*)$ *for all faces* $\Gamma \subset Q$ *(including* $G = \emptyset, Q$*).*

Since each $S(\Gamma)$ is invariant under the action of $C^*$ on $V^*$, the variety $\varphi(T^*_{S(\Gamma)}V^*)$ has the form $T^*_{V(\Gamma)}V$ for some $V(\Gamma) \subset V$. In fact, $V(\Gamma)$ is the projectively dual variety to $S(\Gamma)$. In particular, $V(\emptyset) = V$.

1.9. Corollary. *We have* $V_{gen} = V - \bigcup_\Gamma V(\Gamma)$, *where* $\Gamma$ *runs over all non-empty faces of Q.*

It is easy to see that $V(\Gamma)$ is an irreducible variety defined over Z. When codim$(\Gamma)=1$ we call the irreducible polynomial on V vanishing on $V(\Gamma)$ the $A\cap\Gamma$-discriminant. These discriminants are studied in [GZK 3,4,5,6].

1.10. Example. Let $\Delta^{p-1}\subset R^p$ be the standard simplex, i.e., the convex hull of the standard basis vectors. Let $A\subset Z^{p+q}$ be the set of vertices of the product $\Delta^{p-1}\times\Delta^{q-1}=Q$. (Here the number n from n.1.1. is equal to p+q-1). Then V is naturally identified with the space of all p×q complex matrices. As usual we regard V as an affine chart on the Grassmannian $Gr_p(C^{p+q})$. The corresponding A-hypergeometric system is the system from [GGe] written in the local coordinates, see also [GGZ]. Faces $\Gamma\subset Q$ correspond to pairs of subsets $I\subset\{1,...,p\}$, $J\subset\{1,...,q\}$. The corresponding subvariety $V(\Gamma)$ consists of matrices whose (I,J)-submatrix is of non-maximal rank. Therefore, the generic stratum $V_{gen}$ is the set of matrices with all minors non-zero. In terms of the Grassmannian this means that all Plucker coordinates are non-zero. The volume of Q in our normalisation is equal to $\binom{p+q-2}{p-1}$. In the case p=q=2 each A-hypergeometric function $\Phi(v)$ can be factorized as $\Phi(v)=v^{\theta}F(x)$, where $x=v_{11}v_{22}/v_{12}v_{21}$ (cf.remark after definition 1.2). It is easy to verify that the equations (5) on $\Phi$ reduce to the Gauss equation (1) for F. This observation was the starting point of the whole theory.

1.11. The characteristic cycle. Let $\underline{SS}(M_\gamma)$ be the characteristic cycle of the D-module $M_\gamma$ (cf.[K], [Gi]). This is the formal linear combination of irreducible components $T^*_{V(\Gamma)}V$ of $SS(M_\gamma)$ with some integral positive multiplicities $m_\Gamma$. These multiplicities are important numerical invariants of $M_\gamma$. In particular, $m_\emptyset$ (the multiplicity of the zero section) equals the number of linearly independent solutions of (5) at a generic point. In general, $m_\Gamma$ is defined as the multiplicity in the sense of commutative algebra of the $O_{T^*V}$-module $gr(M_\gamma)$ along the subvariety $T^*_{V(\Gamma)}V$, where $gr(M_\gamma)$ is the associated graded module to $M_\gamma$ with respect to some good filtration, see [Gi],[K]. We shall describe these multiplicities in terms of volumes of certain polytopes similar to Q.

Let $\Gamma$ be a face of Q. Denote by $R\Gamma\subset R^n$ the real vector space generated by $\Gamma$, and by Lin $(\Gamma\cap A)$ the subgroup in $R^n$ generated by

$\Gamma\cap A$. Put $Z\Gamma = Z^n \cap R\Gamma$ and $i(\Gamma,A) = [Z\Gamma : \mathrm{Lin}_Z(\Gamma\cap A)]$. Consider the lattice $Z^n/Z\Gamma$ in the vector space $R^n/R\Gamma$. This lattice induces the volume form Vol on $R^n/R\Gamma$ as in n.1.4. Consider two polytopes in $R^n/R\Gamma$: $P(\Gamma)$, which is the convex hull of images of all $\omega\in A$, and $Q(\Gamma)$, the convex hull of images of all $\omega\in A-\Gamma$.

<u>1.12. Theorem.</u> *For any face $\Gamma\subset Q$ the multiplicity $m_\Gamma$ is equal to* $i(\Gamma,A)(\mathrm{Vol}P(\Gamma)-\mathrm{Vol}Q(\Gamma))$.

<u>Remarks.</u> a) In [GZK 1,2] the factor $i(\Gamma,A)$ was lost; the correct formulation appeared in [GZK 6].

b) For $\Gamma=\emptyset$ we have $R^n/R\Gamma=R^n$, $Q(\Gamma)=Q$, the convex hull of A, and $P(\Gamma)$ is the pyramid with vertex 0 and base Q. Our normalisation implies $\mathrm{Vol}(P(\Gamma))=\mathrm{Vol}_{n-1}(Q)$, $\mathrm{Vol}(Q(\Gamma))=0$, and we arrive at theorem 1.5.

## §2. Multidimensional hypergeometric series

We shall consider Laurent series in r variables of the form $\sum c(m_1,\dots,m_r)x_1^{m_1}\dots x_r^{m_r}$ or simply $\sum c(m)x^m$, where m runs over $Z^r$.

<u>2.1.</u> <u>Definition</u> (J.Horn [H]). *A formal Laurent series* $F(x)=\sum c(m)x^m$ *is hypergeometric if there exist non-zero rational functions* $b_i(m)$ *such that* $c(m+e_i)=b_i(m)c(m)$, $i=1,\dots,r$, *where* $e_i\in Z^r$ *are the standard basis vectors.*

The functions $b_i$ must satisfy the conditions

$$b_i(m+e_j)b_j(m)=b_j(m+e_i)b_i(m). \tag{6}$$

This means that the $b_i$ form a 1-cocycle of the group $Z^r$ with coefficients in the multiplicative group $C(m)^*$ of rational functions in r variables. Here $Z^r$ acts on $C(m)^*$ by the formula $(m'b)(m)=b(m+m')$. Such a cocycle was called by M.Sato [S] a *b-function*.

If two cocycles are cohomologous then the corresponding series are related by a (pseudo-)differential operator. Namely, suppose $b'_i(m)=a(m+e_i)a(m)^{-1}b_i(m)$ for some rational function $a(m)$. Write $a(m)=p(m)/q(m)$, where p,q are polynomials. Then the series $F(x)$ and $F'(x)=\sum c'(m)x^m$, where $c'(m+e_i)=b'_i(m)c'(m)$ are connected by the equality

$$\mathrm{const}.p(x_1\partial/\partial x_1,\dots,x_r\partial/\partial x_r)F(x)=q(x_1\partial/\partial x_1,\dots,x_r\partial/\partial x_r)F'(x).$$

This means that if F(x) converges somewhere then so does F′ and the analytic behaviour of F′ is determined by that of F.

2.2 Description of cocycles. Let $D=(d_{jp})$ be an integral $N\times r$ matrix for some N, and let $\mu=(\mu_1,\dots,\mu_r)\in(C^*)^r$, $\theta=(\theta_1,\dots,\theta_N)\in C^N$. Put

$$b_i(m)=b_i(D,\mu,\theta;m)=\mu_i \prod_{j=1}^{N}\left(\theta_j+1+\sum_{p=1}^{r} d_{jp}m_p\right)^{-1}_{d_{ji}}$$

(where $(x)_d=\Gamma(x+d)/\Gamma(x)$, cf. Introduction).

The $b_i(m)$ form a cocycle. The cocycle condition (6) follows from the equality $b_i(m)=c(m+e_i)/c(m)$, where

$$c(m)=c(D,\mu,\theta;m)=\mu^m/\prod_{j=1}^{N}\Gamma\left(\theta_j+1+\sum_{p=1}^{r} d_{jp}m_p\right). \tag{7}$$

Denote by $L=L(D)\subset Z^N$ the lattice generated by the columns $d_p=(d_{1p},\dots,d_{Np})$ of D.

2.3. Proposition. *If $\theta,\theta'\in C^N$ are congruent modulo the lattice* L(D) *then the cocycles* $\{b_i(D,\mu,\theta;m)\}$ *and* $\{b_i(D,\mu,\theta';m)\}$ *are cohomologous.*

Proof. Suppose that $\theta'=\theta+\sum_{p=1}^{r}k_pd_p$, $k_p\in Z$. Then by definition $b_i(D,\mu,\theta';m)=b_i(D,\mu,\theta;m+k)$, where $k=(k_1,\dots,k_r)$. It suffices to consider the case when $k=e_p$, the standard basis vector of $\mathbb{Z}^r$. By the cocycle condition (6) for $b_i(m)=b_i(D,\mu,\theta;m)$ we have $b_i(m+e_p)/b_i(m)=b_p(m+e_i)/b_p(m)$, and the RHS is the coboundary of the 0-cochain $b_p$.

2.4. Theorem. *Each cocycle in $H^1(Z^r,C(m)^*)$ is cohomologous to* $\{b_i(D,\mu,\theta;m)\}$ *for some* $D,\mu,\theta$ *as above.*

It seems that the first proof of this theorem was given by O.Ore [O]. It was also rediscovered by M.Sato [S].

2.5.Assumptions on $D,\mu,\theta$. For $D,\mu,\theta$ as above let $F(D,\mu,\theta;x)=\sum_{m\in Z^r}c(D,\mu,\theta;m)x^m$, where the coefficients $c(D,\mu,\theta;m)$ are given by (7). In our study of the behaviour of $F(D,\mu,\theta;x)$ we shall restrict ourselves to the case when:

(1) $\mu=1=(1,\dots,1)$.

(2) $N>r$ and $\sum_{j=1}^{N} d_{jp}=0$ for all p.

(3) $rk(D)=r$, and the lattice $L(D)$ is primitive, i.e., the g.c.d. of all $r\times r$-minors of D is equal to 1.

(4) There is an r-element subset $J\subset\{1,\dots,N\}$ such that $\det(d_{jp})_{j\in J}$ is non-zero and $\theta_j\in Z$ for $j\in J$.

This can be justified as follows.

For (1): by definition, $F(D,\mu,\theta;x)=F(D,1,\theta;\mu x)$.

For (2): this will guarantee the existence of a holonomic system satisfied by $F(D,\mu,\theta;x)$ with regular singularities everywhere including the infinite points. The series not satisfying (2) should be considered as "confluent". They are either divergent everywhere (as the series $\sum n!x^n$) or have too large domain of convergence (as the series $\sum (n!)^{-1}x^n$).

For (3): if $rk(D)<r$ then after a monomial change of variables $x_1=y^{\omega_i}$, $\det(\omega_1,\dots,\omega_r)=\pm 1$, we shall have $F=(\sum_{k=-\infty}^{+\infty} y_1^k)G(y_2,\dots,y_r)$; so it does not converge. If L is not primitive then by a similar change of variables with $\det(\omega_1,\dots,\omega_r)=$g.c.d.(all $r\times r$-minors of D) we can obtain a series in y for which the corresponding lattice will be primitive.

For (4): this condition assures us that coefficients of our series vanish outside some affine simplicial cone in $R^r$. We shall see that this provides a non-trivial domain of convergence for F.

2.6. A-hypergeometric system for Horn series. We shall show that each series of the form $F(D,\theta;x)=F(D,1,\theta;x)$ satisfying conditions (1)-(4) of n.2.5, can be extended to a solution of certain A-hypergeometric system. Moreover, series of this type will allow us to construct a complete system of A-hypergeometric functions.

Consider the lattice $Z^N/L(D)$ of rank $n=N-r$. We shall denote this lattice simply $Z^n$. Denote the projection $Z^N\longrightarrow Z^n$ as well as the corresponding projection of complexifications $C^N\longrightarrow C^n$ by pr. Let $\omega_j=pr(e_j)\in Z^n$ be the image of the j-th standard basis vector in $Z^N$, $j=1,\dots,N$. Set $A=\{\omega_1,\dots,\omega_N\}$. Conditions (2) and (3) from n.2.5 imply that A satisfies the conditions a) and b) from n.1.1. Furthermore, L(D) coincides with the lattice L(A) introduced in n.1.1; we shall denote it simply by L.

Let $V=C^A$ be the vector space with coordinates $v_\omega, \omega\in A$, or simply $v_1,\dots,v_N$. Denote $x_p = \prod_{j=1}^{N} v_j^{d_{jp}}$, $p=1,\dots,r$. Thus $F(D,\theta;x)$ becomes a function (or, rather, a formal series) on V.

2.7. Proposition. *The series* $v^\theta F(D,\theta;x)$ *formally satisfies the* A-*hypergeometric system* (5) *with vector of parameters* $pr(\theta)$.

Proof. We can write the series $v^\theta F(D,\theta;x)$ in the form

$$\Phi_\theta(v) = \sum_{a\in L} v^{\theta+a} / \prod_{j=1}^{N} \Gamma(\theta_j+a_j+1) \tag{8}$$

Now the statement becomes Proposition 1 of [GZK 2], §1.

2.8. Convergence of Horn series. We say that a subset $I\subset\{1,\dots N\}$ is *a base* if the $\omega_j$, $j\in I$, form a basis of $R^n$. (This defines a matroid structure on $\{1,\dots,N\}$). It is easy to see that I is a base if and only if the r×r-minor $\det(d_{jp})$, $j\in\{1,\dots,N\}-I$, is non-zero. If $\theta_j\in Z$ for $j\in\{1,\dots,N\}-I$ we shall say that $\Phi_\theta(v)$ and $F(D,\theta;x)$ are *associated with* I.

Introduce the "logarithmic" space $R^N$ with coordinates $u_1,\dots,u_N$. For each base $I\subset\{1,\dots,N\}$ and a point $u\in R^N$ define a linear form $\varphi_{I,u}$ on $R^n$ which takes the value $u_j$ at $\omega_j$ for $j\in I$. Let $C(I)\subset R^N$ be the cone consisting of such u that $\varphi_{I,u}(\omega_j)\le u_j$ for $j\in\{1,\dots,N\}-I$. Clearly, C(I) is isomorphic to $(R_+)^{N-n}\times R^n$.

2.9. Proposition. *If* $I\subset\{1,\dots,N\}$ *is a base then any series* $\Phi_\theta(v)$ *associated with* I *has non-empty domain of convergence. Moreover, there exists a vector* $k\in C(I)$ *such that* $\Phi_\theta(v)$ *converges whenever* $(-\log|v_1|,\dots,-\log|v_N|)\in C(I)+k$.

Proof. See [GZK 2], Proposition 1.2.

Remark. Precise description of the domain of convergence for Horn series amounts to the study of A-discriminants which was undertaken in [GZK 3-6].

2.10 Regular triangulations of the Newton polytope and bases in the space of solutions. Now we fix $\gamma\in C^n$ and construct a complete system of solutions to the A-hypergeometric system with parameters $\gamma$. Let $\Pi(\gamma)\subset C^N$ be the inverse image $pr^{-1}(\gamma)$, see n.2.6.

This is an affine subspace of dimension $r=N-n$ parallel to the complexification of the lattice L.

Let $I\subset\{1,\dots,N\}$ be a base. Denote by $\Delta(I)\subset\mathbb{R}^n$ the simplex with vertices $\omega_i$, $i\in I$. It has dimension n-1 and lies in the affine span of A, i.e., in the hyperplane $\{h(u)=1\}$, see §1.

2.11. Proposition. *The number of distinct series* $\Phi_\theta(v)$, $\theta\in\Pi(\gamma)$, *which are associated with* I *is equal to* $\mathrm{Vol}_{n-1}\Delta(I)$ *where* $\mathrm{Vol}_{n-1}$ *is the volume form introduced in* §1.

Proof. Let $\Pi_Z(\gamma,I)=\{\theta\in\Pi(\gamma):\ \theta_j\in Z \text{ for } j\in\{1,\dots,N\}-I\}$. Clearly $\Pi_Z(\gamma,I)$ is an integral lattice in $\Pi(\gamma)$ which is invariant under translations by elements of L. By (8), for each $a\in L$, $\theta\in\Pi_Z(\gamma,I)$ we have $\Phi_{\theta+a}(v)=\Phi_\theta(v)$. Thus the number of distinct series equals $|\Pi_Z(\gamma,I)/L|$ which is easily seen to coincide with $\mathrm{Vol}_{n-1}\Delta(I)$.■

2.12. Definition. *A set of bases* T *is a triangulation of* (Q,A) *if* $\bigcup_{I\in T}\Delta(I)=Q$, *and for any* $I_1,I_2\in T$ *the simplices* $\Delta(I_1)$ *and* $\Delta(I_2)$ *meet at a common face (possibly empty).*

For a triangulation T we set $C(T)=\bigcap_{I\in T}C(I)$ (see n.2.8); this is a cone in the "logarithmic" space. Let $\mathrm{Vert}(T)\subset A$ denote the set of vertices of T, i.e., $\mathrm{Vert}(T)=\bigcup_{I\in T}I$. We can describe C(T) as follows.

For any $u\in\mathbb{R}^N$ define a continuous piecewise linear function $\varphi_{T,u}:Q\longrightarrow\mathbb{R}$ whose restriction to each $\Delta(I)$ coincides with $\varphi_{I,u}$ introduced in n.2.8. The following statement is obvious.

2.13. Proposition. *The cone* C(T) *consists of those* $u\in\mathbb{R}^N$ *for which* $\varphi_{T,u}$ *is convex, and* $\varphi_{T,u}(\omega_j)\le u_j$ *for* $j\in\{1,\dots,N\}-\mathrm{Vert}(T)$

2.14. Definition. *A triangulation* T *of* (Q,A) *is regular if* C(T) *has nonempty interior.*

In other words, T is said to be regular if it admits a *strictly* convex continuous T-piecewise linear function.

2.15. Proposition. *For each* A *as above there exists a regular triangulation of* (Q,A).

For the proof see [GZK 2], §1, Proposition 4.

2.16. Definition. *A vector of parameters* $\gamma\in\mathbb{C}^n$ *is* T-*nonresonant if the sets* $\Pi_Z(\gamma,I)$ *for* $I\in T$ *are mutually disjoint.*

It suffices, e.g., that the components $\gamma_i$ of $\gamma$ and 1 be linearly independent over Q.

Set $\Pi_Z(\gamma,T)=\bigcup_{I\in T}\Pi_Z(\gamma,I)$. By Proposition 2.11 for a T-nonresonant $\gamma$ we have $\mathrm{Card}(\Pi_Z(\gamma,T))=\mathrm{Vol}_{n-1}Q$.

2.17. Theorem. *Let* T *be a regular triangulation of* (Q,A) *and* $\gamma$ *be a* T*-nonresonant vector of parameters. Then the series* $\Phi_\theta(v)$, $\theta\in\Pi_Z(\gamma,T)$, *have a common domain of convergence and form a basis of the space of* A*-hypergeometric functions in this domain.*

By proposition 2.9., the common domain of convergence contains the set $U(T,k)=\{v\in V:(-\log|v_1|,\dots,-\log|v_N|)\in C(T)+k\}$ for some $k\in C(T)$. This explains the significance of the regularity condition for T.

2.18. Remark. By theorem 2.17, the set U(T,k) as above does not meet the singularities of A-hypergeometric system, i.e., the union $\bigcup_\Gamma V(\Gamma)$ from corollary 1.9. This explains us why triangulations are so relevant in the study of discriminants, cf. [GZK 3,4,5,6].

Some applications of theorem 2.17 to classical multivariate hypergeometric functions (of Horn, Appell, Lauricella etc.) can be found in [GZK 2], §3 and [GG 2].

## §3. Generalized Euler integrals

Proofs of all the results of this section will appear in [GZK 7].

3.1. Notation. Let $A_1,\dots,A_m\subset Z^k$ be arbitrary finite subsets. To each element $\omega=(\omega_1,\dots,\omega_k)\in Z^k$ we associate the Laurent monomial $x^\omega=x_1^{\omega_1}\dots x_k^{\omega_k}$ in k variables $x_1,\dots,x_k$. We shall regard the vector space $C^{A_i}$ as the space of Laurent polynomials of the form $P_i(x)=\sum v_\omega x^\omega$, $\omega\in A_i$.

Let $\alpha=(\alpha_1,\dots,\alpha_m)\in C^m$, $\beta=(\beta_1,\dots,\beta_k)\in C^k$ be complex vectors. We shall study the integral

$$F_\sigma(\alpha,\beta;P)=\int_\sigma \prod P_i(x_1,\dots,x_k)^{\alpha_i}\, x_1^{\beta_1}\dots x_k^{\beta_k}dx_1\dots dx_k, \qquad (9)$$

where $P=(P_1,\ldots,P_m)\in\prod C^{A_i}$, as a multivalued function of P. Since the integrand is also multivalued we have to explain the meaning of this integral and the domain of integration.

3.2. Precise definition of the integral. Denote the region $(C^*)^k-\cup\{P_i=0\}$ by $U(P)=U(P_1,\ldots,P_m)$. Consider the one-dimensional local system (i.e., locally constant sheaf of C-vector spaces) $L(\alpha,\beta,P)$ on U(P) defined by monodromy exponents $\alpha_i$ around $\{P_i=0\}$ and $\beta_j$ around $\{x_j=0\}$. We shall sometimes abbreviate $L(\alpha,\beta,P)$ as L(P) or simply L. A section of L over a simply connected region $U\subset U(P)$ can be viewed as a function $f:U\longrightarrow C$ such that f is a scalar multiple of some branch of $P^\alpha x^\beta=\prod P_i(x_1,\ldots,x_k)^{\alpha_i}x_1^{\beta_1}\ldots x_k^{\beta_k}$. A (singular) p-chain with coefficients in L is a finite formal sum $\sum(\delta,f_\delta)$, where each $\delta:\Delta^p\longrightarrow U(P)$ is a singular p-simplex in U(P), and $f_\delta$ is a section of $\delta^*(L)$ over $\Delta^p$. For each k-chain $\sigma$ we define the integral (9) as $\sum_\delta\int_{\Delta^k}f_\delta dx_1\ldots dx_k$, where the $x_j$'s are viewed as functions $x_j(\delta(t))$ on $\Delta^k$.

Denote by $C_p(U(P),L)$ the space of p-chains defined above. The boundary operator $d:C_p(U(P),L)\longrightarrow C_{p-1}(U(P),L)$ is defined in a standard way (see, e.g., [St]). Let $H_p(U(P),L)$ be the homology of this chain complex.

We shall consider the integrals $F_\sigma(\alpha,\beta;P)$ only when $\sigma$ is a k-cycle, i.e., $d\sigma=0$. Then $F_\sigma(\alpha,\beta;P)$ depends only on the homology class of $\sigma$. We fix $\alpha,\beta$ and consider $F_\sigma$ as a multivalued analytic function of the coefficients of all the $P_i$, i.e., on the space $\prod C^{A_i}$. More precisely, choose some initial $P=(P_1,\ldots,P_m)$ and a k-cycle $\sigma=\sum(\delta,f_\delta)$ in U(P). Then for P′ sufficiently close to P all simplices $\delta$ occuring in $\sigma$ will lie in U(P′). There is a unique k-cycle $\sigma'=\sum(\delta,f'_\delta)$ with coefficients in $L(\alpha,\beta,P')$ such that the $f'_\delta$ are obtained from $f_\delta$ by analytic continuation.

We define the germ of our multivalued function by

$$F_\sigma(\alpha,\beta;P')=\int_{\sigma'}\prod P'_i(x_1,\ldots,x_k)^{\alpha_i}x_1^{\beta_1}\ldots x_k^{\beta_k}dx_1\ldots dx_k$$

3.3. Remark. The correspondence $\sigma\longrightarrow\sigma'$ defines a mapping $G_{P,P'}:H_k(U(P),L(P))\longrightarrow H_k(U(P'),L(P'))$ for P′ sufficiently close to P. Intuitively, P can be "more singular" than P′. Then $G_{P,P'}$ is not an isomorphism. When P and P′ are generic, $G_{P,P'}$ is the isomorphism of parallel transport with respect to the Gauss-Manin connection.

3.4. Lemma. For any $\varphi_1,\ldots,\varphi_m \in Z^k$ we have the equality

$$F_\sigma(\alpha,\beta;x^{\varphi_1}P_1,\ldots,x^{\varphi_m}P_m) = F_\sigma(\alpha,\beta+\sum \alpha_i\varphi_i;P_1,\ldots,P_m).$$

This follows at once from the definitions.■

According to this lemma we can and will assume that each $A_i$ contains 0. We shall assume also that the union of $A_i$ generates the Abelian group $Z^k$ (otherwise the integral either can be reduced to this case or is equal to 0).

3.5. From $(A_1,\ldots,A_m)$ to one set of monomials (the Cayley trick). We shall construct an A-hypergeometric system having $F_\sigma$ as its solution. Consider the lattice $Z^m$ with the basis $e_1,\ldots,e_m$ and let $A=\cup(\{e_i\}\times A_i)\subset Z^m\times Z^k=Z^{m+k}$. Clearly this set lies in the hyperplane $h(u)=1$, where $h:Z^{m+k}\longrightarrow Z$ is the sum of first m coordinates. On the level of Laurent polynomials this amounts to associating to a collection $(P_1,\ldots,P_m)$ of polynomials in $x=(x_1,\ldots,x_k)$ a new polynomial $P(y,x)=\sum y_iP_i(x)$, where $y=(y_1,\ldots,y_m)$. This substitution was used by Cayley in elimination theory.

The proof of the next lemma is straigtforward.

3.5. Lemma. *If the union of $A_i$ generates $Z^k$ as an Abelian group, and each $A_i$ contains 0 then A generates $Z^{m+k}$.*■

We shall use the natural identification $C^A=\prod C^{A_i}$ and denote a sequence $(P_1,\ldots,P_m)$ and the corresponding polynomial P(y,x) by the same letter P. Under this identification $C^A_{gen}$ corresponds to the set of $(P_1,\ldots,P_m)$ such that all hypersurfaces $(P_i=0)\subset(C^*)^k$ are smooth, intersect each other transversely, and the same conditions hold "at infinity", i.e., on a suitable compactification.

3.6. Theorem. *For any $P=(P_1,\ldots,P_m)\in C^A$ and any $\sigma\in H_k(U(P),L(\alpha,\beta,P))$ the function $P'\longrightarrow F_\sigma(\alpha,\beta;P')$ in a neighborhood of P satisfies the A-hypergeometric system (2) with* *parameters $\gamma$, where $(\gamma_1,\ldots,\gamma_n)=(\alpha_1,\ldots,\alpha_m,-\beta_1-1,\ldots,-\beta_k-1)$.*

3.7. Remark. The integrals we are considering are proper, i.e., taken over compact cycles. Therefore we do not have problems of convergence. It is more common in the theory of special functions to integrate over some non-compact regions naturally connected with the integrand. For example, if all $P_i$ have real coefficients one can consider the integral of type (9) over some connected component of $R^k\cap U(P)$. If such an integral has good properties of convergence Theorem 3.6 is also true for it since it is proven in a purely formal way.

3.8. The non-resonance condition. Let $A=\cup(\{e_i\}\times A_i)\subset Z^m\times Z^k=Z^{m+k}$ be as above. It is clear that any set $A\subset Z^n$ satisfying the conditions a) and b) from n.1.1 can be transformed to such a form by an automorphism of $Z^n$, at least for $m=1$, $k=n-1$. Denote $K\subset R^n$ the convex cone generated by $A\subset Z^n$. Clearly, each face of K is a cone over some face of Q, the convex hull of A. For each face $\Gamma\subset K$ of codimension 1 let $\text{Lin}(\Gamma)\subset C^n$ be the C-linear span of $\Gamma$.

We say that a vector of parameters $\gamma=(\gamma_1,\ldots,\gamma_n)\in C^n$ is *non-resonant* (for A) if for each face $\Gamma\subset K$ of codimension 1 we have $\gamma\notin Z^n+\text{Lin}(\Gamma)$.

Now we can formulate the converse statement to Theorem 3.6, namely that generalized Euler integrals form a complete set of solutions for any A-hypergeometric system. Denote by

$$E=E(\alpha,\beta,P):H_k(U(P),L(P))\longrightarrow \text{Hyp}(\gamma)_P$$

the mapping $\sigma\longrightarrow F_\sigma(\alpha,\beta;P)$.

3.9. Theorem. *Suppose that $\alpha$ and $\beta$ in Theorem 3.6 are such that $\gamma$ is non-resonant for A. Then for each $P=(P_1,\ldots,P_m)\in\prod C^{A_i}$ the mapping $E(\alpha,\beta,P)$ is an isomorphism.*

Note that the non-resonance of $\gamma$ in Theorem 3.9 implies that all $\alpha_i\neq 0$.

Our proof of Theorem 3.9 is based on the following

3.10. Theorem. *If $P\in C^A_{gen}$ (see 1.9) and $\gamma$ is non-resonant then the monodromy representation of $\pi_1(C^A_{gen},P)$ on $\text{Hyp}(\gamma)_P$ is irreducible.*

3.11. Remark. Let $z_1,\dots,z_n$ be the coordinates on the open orbit $S_0$ of the torus in S (see 1.6). The non-resonance condition means that the multivalued function $z^\gamma$ on $S_0$ has non-trivial monodromy along each orbit $S(\Gamma)$ of codimension 1. It seems probable that a weaker condition suffices for Theorem 3.9 (but not for Theorem 3.10), viz., that $z^\gamma$ has a singularity along each $S(\Gamma)$ for $\operatorname{codim}(\Gamma)=1$. (Thus it can either ramify, or have a pole). This means that $\gamma$ does not lie in $(Z^n\cap K)+\operatorname{Lin}(\Gamma)$. Among these "semi-non-resonant" $\gamma$ there are some integer points, namely $\gamma\in Z^n$ such that $(-\gamma)$ lies strictly within K. The study of Euler integrals and hypergeometric functions for such $\gamma$ is very important since they are connected with polylogarithms.

3.12. The sheaf $H(\alpha,\beta)$. Now we give sheaf-theoretic versions of Theorems 3.9 and 3.10. Let us introduce a constructible sheaf $H(\alpha,\beta)$ on $\prod C^{A_i}$ whose stalk at any point P is $H_k(U(P),L(\alpha,\beta,P))$. Let $U=\{(P,x)\in(\prod C^{A_i})\times(C^*)^k:\ x\in U(P)\}$ be the disjoint union of all $U(P)$. There is the local system $L(\alpha,\beta)$ on U whose restriction on each $U(P)$ is $L(\alpha,\beta,P)$. Its sections are branches of functions of the form $(P,x)\longrightarrow \text{const}.P^\alpha x^\beta$. Let $\pi:U\longrightarrow\prod C^{A_i}$ be the projection. Define $H(\alpha,\beta)=R^k\pi_!L(\alpha,\beta)$. Here $R^k\pi_!$ is the k-th direct image with proper supports, see [Bo]. By definition, the stalk $H(\alpha,\beta)_P$ is $H^k_c(U(P),L(\alpha,\beta,P))$, the *cohomology* with compact supports. By Poincaré duality, one has isomorphisms

$$H^k_c(U(P),L(\alpha,\beta,P))=H^{2k-k}(U(P),L(\alpha,\beta,P)^*)^*=H_k(U(P),L(\alpha,\beta,P)).$$

It is clear that the mapping $E(\alpha,\beta,P)$ from Theorem 3.9 is induced by a morphism of sheaves $E(\alpha,\beta):H(\alpha,\beta)\longrightarrow \operatorname{Hyp}(\gamma)$ by taking stalks. Note that the mapping $G_{P,P'}$ (see remark 3.3) is a special case of "transport" map defined for any constructible sheaf.

3.13. Theorem. *In the conditions of Theorem 3.9 the morphism $E(\alpha,\beta)$ is an isomorphism of sheaves.*

Consider also the complex $R\pi_*L(\alpha,\beta)$, the full direct image.

3.14. Theorem. *$R\pi_*L(\alpha,\beta)$ is an irreducible perverse sheaf, and the canonical morphism $R\pi_!L(\alpha,\beta)\longrightarrow R\pi_*L(\alpha,\beta)$ is an isomorphism.*

3.15. Remark. We have a natural restriction morphism $res: R^k\pi_* L(\alpha,\beta)_P \longrightarrow H^k(U(P), L(\alpha,\beta,P))$. In general, this morphism is not an isomorphism (in contrast with the case of $R^k\pi_!$ and $H^k_c$). Its image consists of cohomology classes of cocycles which can be extended to $H^k(U(P'), L(\alpha,\beta,P'))$ for all $P'$ close to P. By Poincaré duality, $H^k(U(P), L(\alpha,\beta,P))$ is isomorphic to $H^{lf}_k(U(P), L(\alpha,\beta,P))$, the homology defined by means of locally finite chains, see [VGZ]. For an "extendable" cycle $\sigma \in H^{lf}_k(U(P), L(\alpha,\beta,P))$ we could define the Euler integral as in [VGZ]. But Theorem 3.14 means that the space of extendable cycles is just the image of the canonical map $H_k(U(P), L(\alpha,\beta,P)) \longrightarrow H^{lf}_k(U(P), L(\alpha,\beta,P))$, and various "extensions" of a cycle $\sigma$ correspond to various compact cycles homologous to $\sigma$.

3.16. Example: hypergeometric functions on the Grassmannian $Gr_p(C^{p+q})$ (cf. Example 1.10). These functions were originally defined in [G] in terms of generalized Euler integrals of type (9), where all $P_i$ are (inhomogeneous) linear functions. In this case all $A_i$ $(i=1,\dots,q)$ are equal to $\{0, e_1, \dots, e_{p-1}\} \subset Z^{p-1}$, where the $e_j$ are standard basis vectors. The construction in 3.5 leads to the realization of $C^A$ as the space of polynomials $P(y,x) = \sum_{i=1}^{q} v_{i0} y_i + \sum_{i=1}^{q} \sum_{j=1}^{p-1} v_{ij} y_i x_j$. The polytope Q is the product $\Delta^{p-1} \times \Delta^{q-1}$ of two simplices. It has p+q faces of codimension 1. The corresponding non-resonance conditions have the form $\alpha_i \notin Z$, $i=1,\dots,q$; $\beta_j \notin Z$, $j=1,\dots,p-1$; $(\sum \alpha_i) + (\sum \beta_j) \notin Z$, cf. [VGZ].

The set A can be transformed to a form $\cup(\{e'_i\} \times A'_i)$, $i=1,\dots,m$, $A'_i \subset Z^k$, in three different ways. The first was just described (here $m=q$, $k=p-1$). The second has $m=p$, $k=q-1$, and the transformation is given by transposition of the matrix $(v_{ij})$. The third has $m=1$, $k=p+q-2$. So we obtain three kinds of Euler integrals for the same hypergeometric system which are taken over cycles of different dimension. The relation between integrals of first two kinds is closely connected with the duality studied in [GG 1]. Both these integrals can be obtained from the integrals of the third kind by an appropriate iterated integration.

## REFERENCES


[A] *Aomoto K.* On the structure of integrals of power products of linear functions.-Sci.Papers Coll.Gen. Educ.Univ. Tokyo, 1977, v.27, no.2, 49-61.

[AK] *Appell P., Kampé de Fêriet J.* Fonctions hypergéométriques et hypersphériques; polynomes d'Hermite.-Paris, Gauthier-Villars, 1926.

121

[BE] *Bateman H., Erdelyi A.* Higher transcendental functions.- vol.1, McGraw-Hill, 1953.

[Bj] *Bjork J.E.* Rings of differential operators.- North-Holland, 1979.

[Bo] *Borel A.* (ed.). Seminar on intersection homology.-Birkhäuser, Boston, 1984.

[G] *Gelfand I.M.* General theory of hypergeometric functions.-Doklady AN SSSR, 1986, v.288, no.1, 14-18 [Sov. Math. Dokl., 1987, v.33, 9-13].

[GGe] *Gelfand I.M, Gelfand S.I.* Generalized hypergeometric equations.-Doklady AN SSSR, 1986, v.288, no.2, 279-283 [Sov. Math. Dokl., 1987, v.33, 643-646].

[GG1] *Gelfand I.M., Graev M.I.* A duality theorem for general hypergeometric functions.-Doklady AN SSSR, 1986, v.289, no.1, 19-23 [Sov. Math. Dokl., 1987, v.34, 9-13].

[GG2] *Gelfand I.M., Graev M.I.* Hypergeometric functions associated with the Grassmannian $G_{3,6}$.-Mat.Sbornik, 1989, v.180, no.1, 3-38 [Math. USSR Sb., 1990, v.66, no.1, 1-40].

[GGZ] *Gelfand I.M., Graev M.I., Zelevinsky A.V.* Holonomic systems of equations and series of hypergeometric type.-Doklady AN SSSR, 1987, v.295, no.1, 14-19 [Sov. Math. Dokl., 1988, v.36, 5-10].

[GZK 1] *Gelfand I.M., Zelevinsky A.V., Kapranov M.M.* Equations of hypergeometric type and Newton polytopes.-Doklady AN SSSR, 1988, v.300, no.3, 529-534 [Sov. Math. Dokl., 1988, v.37, 678-683].

[GZK 2] *Gelfand I.M., Zelevinsky A.V., Kapranov M.M.* Hypergeometric functions and toric varieties.-Funkc.Anal., 1989, v.23, no.2, 12-26.

[GZK 3] *Gelfand I.M., Zelevinsky A.V., Kapranov M.M.* A-discriminants and Cayley-Koszul complexes.-Doklady AN SSSR, 1989, v.307, no.6, 1307-1310 [Sov. Math. Dokl., 1990, v.40, no.1, 239-243].

[GZK 4] *Gelfand I.M., Zelevinsky A.V., Kapranov M.M.* Newton polytopes of principal A-determinants.-Doklady AN SSSR, 1989, v.308, no.1, 20-23.

[GZK 5] *Gelfand I.M., Zelevinsky A.V., Kapranov M.M.* On discriminants of polynomials in several variables.-Funkc.Anal., 1990, v.24, no.1, 1-4.

[GZK 6] *Gelfand I.M., Zelevinsky A.V., Kapranov M.M.* Discriminants of polynomials in several variables and triangulations of Newton polytopes.- Algebra i analiz, 1990, v.2,no.3, 1-62.

[GZK 7] *Gelfand I.M., Kapranov M.M., Zelevinsky A.V.* Generalized Euler integrals and A-hypergeometric functions.-to appear in Adv.Math.

[Gi] *Ginsburg V.A.* Characteristic varieties and vanishing cycles.-Invent. Math, 1986, v.84, 327-402.

[H] *Horn J.* Ueber die Konvergenz der hypergeometrischen Reihen zweier und dreier Veranderlichen.-Math.Ann,1889,Bd.34, S.544-600.

[K] *Kashiwara M.* Systems of micro-differential equations. Birkhäuser, Boston, 1983.

[O] *Ore O.* Sur la forme de fonctions hypergeometriques de plusieurs variables.-J.Math.Pures et Appl., 1930, v.9, no.4, 311-327.

[S] *Sato M.* Singular orbits of a prehomogeneous vector space and hypergeometric finctions.-to appear in Nagoya Math.Journal.

[St] *Steenrod N.* Homology with local coefficients.-Ann. Math., 1945, v.44, 610-627.

[VGZ] *Vasiliev V.A., Gelfand I.M., Zelevinsky A.V.* General hypergeometric functions on complex Grassmannians.-Funkc.Anal., 1987, v.21, no.1, 23-38.

# THE ACTIONS OF INFINITE-DIMENSIONAL LIE ALGEBRAS

I. M. Gel'fand, D. A. Kazhdan,
and D. B. Fuks

The actions of Lie algebras on smooth manifolds are the subject of a classical and far advanced theory. It is not surprising that the main concepts of this theory may be extended to the infinite-dimensional case; however, in the latter, the theory acquires in an unexpected way what appears to us to be a new and very interesting content. We discuss this infinite-dimensional theory here, and at the end we indicate the connection with problems of calculating the cohomologies of infinite-dimensional Lie algebras.

## §1. Definition of the Action. Principal Homogeneous Spaces

1. First we give the main definitions in the most general form. Let A be a nuclear commutative associative **C**-algebra with a unit element, and let S denote the set of closed maximal ideals of A, that is, the set of continuous ring homomorphisms $A\to \mathbf{C}$. By a tangent vector to S at the point $s\in S$ we mean a continuous linear mapping d: $A\to \mathbf{C}$, such that $d(a_1a_2)=s(a_1)\,d(a_2)+s(a_2)\,d(a_1)$. The set $\tau_sS$ of these tangent vectors is, in an obvious sense, a linear topological space over **C**; it is called the tangent space to S at s. By a vector field on S we mean a derivation of A, that is, a continuous linear mapping D: $A\to A$, such that $D(a_1a_2)=a_1D(a_2)+a_2D(a_1)$. The set $\mathfrak{A}(S)$ of these vector fields is, in an obvious sense, a topological Lie algebra and a topological A-module. For every $s\in S$ the composition $D\mapsto s\circ D$ defines a homomorphism of the A-module $\mathfrak{A}(S)$ into the **C**-module $\tau_sS$, compatible with the homomorphism s: $A\to \mathbf{C}$. The image of a vector field under this homomorphism v(s): $\mathfrak{A}(S)\to\tau_sS$ is called the value of the field at the point $s\in S$.

Let $\mathfrak{g}$ be a nuclear Lie algebra. We say that $\mathfrak{g}$ acts on S, or that S is a $\mathfrak{g}$-space, if there is a continuous homomorphism of $\mathfrak{g}$ into the Lie algebra $\mathfrak{A}(S)$. If the action $\varphi$: $\mathfrak{g}\mapsto\mathfrak{A}(S)$ has the property that the mapping of the A-module $A\otimes_{\mathbf{C}}\mathfrak{g}$ into the A-module $\mathfrak{A}(S)$ defined by $a\otimes g\mapsto a\varphi(g)$ is an isomorphism, then S is called a principal homogeneous $\mathfrak{g}$-space.

2. Next we consider the significance of these definitions in a geometrical situation. The simplest geometrical situation is the case when S is a smooth manifold (finite- or infinite-dimensional) and A is the algebra of complex-valued smooth functions on S. In this case the tangent vectors, the vector fields, and the action of the Lie algebra obviously have the usual meaning, and we need interpret only the concept of a principal homogeneous $\mathfrak{g}$-space.

PROPOSITION 1.1. Let $\varphi$: $\mathfrak{g}\to\mathfrak{A}(S)$ denote the action of the Lie algebra $\mathfrak{g}$ on the smooth manifold S. For S to be a principal homogeneous space relative to this action, it is necessary and sufficient that, for each $s\in S$ the through homomorphism

$$\mathfrak{g}\xrightarrow{\varphi}\mathfrak{A}(S)\xrightarrow{v(s)}\tau_sS$$

is an isomorphism.

We do not use the necessity of the condition in what follows; therefore we leave the proof of it to the reader; we prove the sufficiency of the condition.

Moscow State University. Translated from Fuktsional'nyi Analiz i Ego Prilozheniya, Vol. 6, No. 1, pp. 10-15, January-March, 1972. Original article submitted October 1, 1971.

We must prove that the formula $a \otimes g \mapsto a\varphi(g)$ determines an isomorphism $A \otimes_C \mathfrak{g} \to \mathfrak{A}(S)$. It is obvious that this is a monomorphism, since if $\sum a_i \otimes g_i \mapsto 0$, then for every $s \in S$, under the homomorphism $v(s)\circ\varphi$, the element $\sum s(a_i)\, g_i \in \mathfrak{g}$ goes into the zero element, and this contradicts our assumption. To prove that the mapping is an epimorphism it is sufficient to note that $A \otimes \mathfrak{g}$ is the space of smooth functions on S with values in $\mathfrak{g}$. The element of $A \otimes \mathfrak{g}$ that goes into a given element $\alpha \in \mathfrak{A}(S)$ is the function $s \mapsto (\varphi \circ v(s))^{-1} v(s)\, \xi$.

Comment. Proposition 1.1 reveals the extreme poorness of the contact of a principal homogeneous space in the finite-dimensional case. For, if $\mathfrak{g}$ is the Lie algebra of a finite-dimensional simply connected Lie group G, then the connected components of the principal homogeneous $\mathfrak{g}$-space S must be factor spaces of G with respect to discrete subgroups. On the contrary, in the infinite-dimensional case this concept is meaningful, as is confirmed by the following example.

3. A basic example. Let M be a smooth n-dimensional manifold. Let S(M) denote the space of $\infty$-jets of the mappings $\mathbf{R}^n \to M$ which are regular at the point $0 \in \mathbf{R}^n$; more precisely, S(M) is the inverse limit of the sequence

$$M = S_0(M) \leftarrow S_1(M) \leftarrow S_2(M) \leftarrow \ldots,$$

where $S_k(M)$ denotes the space of k-jets of the mappings $\mathbf{R}^n \to M$ which are regular at $0 \in \mathbf{R}^n$, and the arrows denote the obvious submersion. It is clear that S(M) is an infinite-dimensional smooth manifold, and that the algebra A of smooth complex-valued functions on S(M) is the direct limit of the sequence of algebras $A_k$ of smooth functions on $S_k(M)$.

We make the Lie algebra $W_n$ of formal vector fields at $0 \in \mathbf{R}^n$ act on S(M); the elements of $W_n$ are interpreted as $\infty$-jets of one-parameter families of diffeomorphisms $\mathbf{R}^n \to \mathbf{R}^n$, that is, tangent vectors to $S(\mathbf{R}^n)$ at the point corresponding to the identity mapping $\mathbf{R}^n \to \mathbf{R}^n$. Moreover, this interpretation enables us to determine an isomorphism of the space $W_n$ onto this tangent space. Since a point of S(M) is an $\infty$-jet of the mapping $\mathbf{R}^n \to M$, it also determines a mapping $S(\mathbf{R}^n) \to S(M)$ which induces an isomorphism of the tangent spaces. Thus, all tangent spaces of the manifold S(M) may be canonically identified with the tangent space to $S(\mathbf{R}^n)$, that is, with $W_n$. This also determines a homomorphism $W_n \to \mathfrak{A}(S(M))$ (its permutability with commutation may be verified trivially), that is, the action of $W_n$ on S(M).

By Proposition 1.1, S(M) is a principal homogeneous $W_n$-space.

4. We return to the general theory developed in Sec. 1. Let $A$, $S$, $\mathfrak{A}(S)$ have the same meaning as in Sec. 1, and let B be a nuclear space. By an exterior differential form of degree q on S with values in B we mean a continuous A-linear mapping $\Lambda^q_A \mathfrak{A}(S) \to A \otimes_C B$ (here $\Lambda^q_A \mathfrak{A}(S)$ is the skew-symmetrized q-th power of the A-module $\mathfrak{A}(S)$ relative to tensor multiplication over A). Let $\Omega^q = \Omega^q(S, B)$ denote the space of these forms and let $\Omega^* = \bigoplus_q \Omega^q$. Exterior differentiation d: $\Omega^q \to \Omega^{q+1}$ is defined by the usual formula:

$$(d\alpha)(D_1, \ldots, D_{q+1}) = \sum_{1 \leqslant s < t \leqslant q+1} (-1)^{s+t-1} \alpha([D_s, D_t], D_1, \ldots \hat{D}_s \ldots \hat{D}_t \ldots, D_{q+1}) + \sum_{1 \leqslant s \leqslant q+1} (-1)^s D_s \alpha(D_1, \ldots \hat{D}_s, \ldots, D_{q+1}).$$

It is clear from the A-linearity of the differential form $\Lambda^q_A \mathfrak{A}(S) \to A \otimes_C B$, that, for every $s \in S$ it induces a linear (over **C**) mapping $\Lambda^q_C T_s S \to B$ (this accounts for the name).

The complex $\{\Omega^*(S, B), d\}$ is called the de Rham complex of the manifold S with coefficients in B.

We assume now that S is a principal homogeneous $\mathfrak{g}$-space. Then the mapping $\omega : \mathfrak{A}(S) \to A \otimes \mathfrak{g}$, inverse to $\mathrm{id} \otimes \varphi$ (see Sec. 1) is a 1-form on S with values in $\mathfrak{g}$.

PROPOSITION 1.2. (The Maurer-Cartan formula).

$$d\omega = -\frac{1}{2}[\omega, \omega],$$

that is, $d\,\omega(D_1, D_2) = -\frac{1}{2}[\omega(D_1), \omega(D_2)]$ for any $D_1, D_2 \in \mathfrak{A}(S)$. (The commutator on the right-hand side of the last equality is taken in $A \otimes \mathfrak{g}$; it is defined by the formula $[a_1 \otimes g_1, a_2 \otimes g_2] = a_1 a_2 \otimes [g_1, g_2]$.)

To prove the proposition it is sufficient to verify the equality $\omega([D_1, D_2]) - D_1\omega(D_2) + D_2\omega(D_1) = [\omega(D_1), \omega(D_2)]$ for elements $D_i$ of the form $a_i\varphi(g_i)$, where $a_i \in A$, and $g_i \in \mathfrak{g}$. By taking into account that $\omega = (\mathrm{id} \otimes \varphi)^{-1}$ and that $[a_1D_1, a_2D_2] = D_1(a_2)D_2 - D_2(a_1)D_1 + a_1a_2[D_1, D_2]$ for any $D_1, D_2 \in \mathfrak{A}(S)$, we obtain that $\omega([D_1, D_2]) = a_1a_2 \otimes [g_1, g_2] + D_1(a_2) \otimes g_2 - D_2(a_1) \otimes g_1$, whence follows the desired equality.

## § 2. Geometric $\mathfrak{g}$-Complexes

1. The concept of a geometric $\mathfrak{g}$-complex, defined below, is not essentially new (see, for example, [6]). We are interested in the interaction of this concept with those introduced in Sec. 1.

Let $\mathfrak{g}$ be a Lie algebra and let $\{K = \bigoplus_q K^q, d\}$ be a complex for which the differential d has degree +1. We say that the geometric action of $\mathfrak{g}$ in K is defined (or K is a geometric $\mathfrak{g}$-complex, or simply, a $\mathfrak{g}$-complex) if, for each $g \in \mathfrak{g}$ there is a linear mapping $\bar{g}: K \to K$ of degree $-1$ such that the following equalities hold for any $g_1, g_2 \in \mathfrak{g}$

1°. $\overline{(\alpha g_1 + \beta g_2)} = \alpha\bar{g}_1 + \beta\bar{g}_2 \quad (\alpha, \beta \in \mathbf{C})$,

2°. $\bar{g}_1\bar{g}_2 = -\bar{g}_2\bar{g}_1$,

3°. $\overline{[g_1, g_2]} + \bar{g}_1 d\bar{g}_2 - \bar{g}_2 d\bar{g}_1 + d\bar{g}_1\bar{g}_2 - \bar{g}_2\bar{g}_1 d = 0$.

When we put $a(g) = \bar{g}d + d\bar{g}$, we obtain a representation of $\mathfrak{g}$ in the endomorphisms of the complex K. It is easy to appreciate, however, that not every such representation is obtained in this way from a geometric action (for example, the operators $a(g)$ arising from the geometric action induce trivial homological endomorphisms).

If K is a geometric $\mathfrak{g}$-complex, we denote by $K^{\mathfrak{g}}$ the subcomplex of K consisting of those $k \in K$ such that $\bar{g}k = a(g)k = 0$ (or, $\bar{g}k = \bar{g}\,dk = 0$) for all $g \in \mathfrak{g}$.

2. The fundamental examples. A. The de Rham complex, with arbitrary coefficients, of the $\mathfrak{g}$-space S (see Sec. 1.1) has the natural structure of a geometric $\mathfrak{g}$-complex. The geometric action is defined by the formula $(g\alpha)(D_1, \ldots, D_{q-1}) = \alpha(\varphi(g), D_1, \ldots, D_{q-1})$, where $g \in \mathfrak{g}$; $\alpha \in \Omega^{q-1}$ (S; the space of coefficients), $D_1, \ldots, D_{q-1} \in \mathfrak{A}(S)$ and $\varphi$ denotes the action $\mathfrak{g} \to \mathfrak{A}(S)$ of $\mathfrak{g}$ in S.

B. Let $\mathfrak{g}$ be a subalgebra of the Lie algebra $\mathfrak{h}$. The standard complex of cochains of $\mathfrak{h}$ (with arbitrary coefficients) has the natural structure of a geometric $\mathfrak{g}$-complex (here it would be appropriate to recall the definition of a standard complex, but we delay this slightly since in Sec. 3 we shall require more extensive information about cochains and cohomlogies of Lie algebras; if necessary, the reader should now consult the beginning of Sec. 3). The geometric action is given by the formula $(gL)(h_1, \ldots, h_{q-1}) = L(g, h_1, \ldots, h_{q-1})$, where $g \in \mathfrak{g}$, $L \in C^{q-1}$ ($\mathfrak{g}$; the module of coefficients), and $h_1, \ldots, h_{q-1} \in \mathfrak{h}$.

3. We clarify Example A in the finite-dimensional case. Let S be a finite-dimensional manifold, and suppose that a finite-dimensional Lie group G, whose Lie algebra is $\mathfrak{g}$, acts on S smoothly and without fixed points. In this case it is obvious that the operator $a(g): \Omega^q(S) \to \Omega^{q-1}(S)$ (coefficients in $\mathbf{C}$), determined by an element $g \in \mathfrak{g}$ coincides with the orginary Lie differentiation along the vector field corresponding to g under the action $\mathfrak{g} \to \mathfrak{A}(S)$.

PROPOSITION 2.1. The subcomplex $(\Omega^{\cdot}(S))^{\mathfrak{g}}$ of the complex $\Omega^{\cdot}(S)$ coincides with the image of the monomorphism $p^{\cdot}: \Omega^{\cdot}(S/G) \to \Omega^{\cdot}(S)$, where $S/G$ denotes the factor manifold, and p is a projection.

Proof. If the form $\alpha \in \Omega^{\cdot}(S)$ belongs to $(\Omega^{\cdot}(S))^{\mathfrak{g}}$ it means that $a(g)\alpha = 0$ and that $\bar{g}\alpha = 0$ for $g \in \mathfrak{g}$; the first condition is equivalent to the condition that $\alpha$ is a G-invariant form, while the second condition is equivalent to the condition that $\alpha$ is annihilated by the substitution of a vector tangential to the fiber. Together these conditions mean that $\alpha$ is a form lifted from $S/G$.

Remark. A similar result holds for infinite-dimensional manifolds if the Lie algebra $\mathfrak{g}$ corresponds to a Lie group G, and if S and $S/G$, are smooth manifolds.

4. If $K_1 = \bigoplus_q K_1^q$, $K_2 = \bigoplus_q K_2^q$ are two $\mathfrak{g}$-complexes, we may canonically form the $\mathfrak{g}$-complex $\mathrm{Hom}(K_1, K_2) = \bigoplus_{-\infty<q<\infty} \mathrm{Hom}^q(K_1, K_2)$, where $\mathrm{Hom}^q(K_1, K_2) = \prod_r \mathrm{Hom}(K_1^r, K_2^{r+q})$, the boundary operator is given by the formula $(d\rho)(k) = d(\rho(k)) - \rho(dk)$, and the geometric action of the algebra $\mathfrak{g}$ is defined by the formula $(\bar{g}\rho)(k) = \bar{g}(\rho(k)) - \rho(\bar{g}k)$.

## §3. Application: The Homologies of the Diagonal Complex

1. In Sec. 3 we shall be concerned with things connected with cohomologies of infinite-dimensional Lie algebras. We recall certain relevant definitions and results (for the details, see [1-3]).

By the standard complex of cochains of the topological Lie algebra $\mathfrak{g}$ (over **C**) with coefficients in the $\mathfrak{g}$-module B we mean the complex $\{C^q, d^q\}$, where $C^q$ denotes the space of continuous skew-symmetric q-linear functionals over $C^q$ with values in B, and the differential $d^q : C^q \to C^{q+1}$ is defined by the formula

$$(d^qL)(g_1, \ldots, g_{q+1}) = \sum_{1 \leqslant s < t \leqslant q+1} (-1)^{s+t-1} L([g_s, g_t], g_1, \ldots \hat{g}_s \ldots \hat{g}_t \ldots, g_{q+1}) + \sum_{1 \leqslant s \leqslant q+1} (-1)^s g_s L(g_1, \ldots \hat{g}_s \ldots, g_{q+1}).$$

The homologies of this standard complex are called the cohomologies of the algebra $\mathfrak{g}$ with coefficients in B, and we use the notation $H^*(\mathfrak{g}; B) = \bigoplus_q H^q(\mathfrak{g}; B)$.

When B is a ring, in particular, when B = **C**, the standard complex is multiplicative, and $H^*(\mathfrak{g}; B)$ is a graded ring.

The cohomologies of $W_n$ with coefficients in **C** were found in [3]† namely, the authors constructed the isomorphism $H^*(W_n; \mathbf{C}) \to H^*(X_n; \mathbf{C})$, where $X_n$ is the inverse-image of the 2n-skeleton of the base BU(n) of the universal U(n)-fibering in the total space of this fibering (it is assumed that the space BU(n) is split into cells of even dimensions).

It was shown in [1] that, for a compact orientable manifold M, the cohomologies of the Lie algebra of the smooth vector fields on M with coefficients in **C** are finite-dimensional in each dimension.‡ In a certain sense the calculation of these cohomologies is reduced to the calculation of the homologies of the "diagonal complex," that is, the subcomplex of $\{C^q(\mathfrak{A}(M); \mathbf{C}); d^q\}$ defined in the following way. The chain $L \in C^q(\mathfrak{A}(M); \mathbf{C})$ is said to be diagonal if, for any vector fields $D_1, \ldots, D_q \in \mathfrak{A}(M)$ whose supports are nonintersecting in aggregate, the value $L(D_1, \ldots, D_q)$ is zero. The diagonal chains form a subcomplex of the diagonal complex; we call this subcomplex diagonal also and denote it by $C_1^*$.

To calculate the homologies of the diagonal complex the authors of [1] constructed a spectral sequence, the second term of which is calculated from the formula $E_2^{s,t} = H_{-s}(M) \otimes H^t(W_n; \mathbf{C})$.

2. The subsequent reasoning is based on a study of the geometric $W_n$-complex $K = \mathrm{Hom}(C^*(W_n; \mathbf{C}), \Omega^*(S(M)))$ (see Secs. 2.2 and 2.4).

PROPOSITION 3.1. The canonical isomorphism $[(K^{-q})^{W_n}]' = C_1^q(\mathfrak{A}(M))$ holds for all q (the prime denotes conjugation). This isomorphism permutes with the boundary operators d': $[(K^{-q})^{W_n}]' \to [(K^{-q-1})^{W_n}]'$ and $d: C_1^q(\mathfrak{A}(M)) \to C_1^{q+1}(\mathfrak{A}(M))$.

Proof. We confine ourselves to the construction of the mapping $[(K^{-q})^{W_n}]' \to C_1^q(\mathfrak{A}(M))$. By using the fact that S(M) is a principal homogeneous $W_n$-space, in particular, by using the Maurer-Cartan formula, one may verify automatically that this mapping is an isomorphism and that it permutes with the differentials.

We assume that there is a functional on the space $(K^{-q})^{W_n} = \bigoplus_r \mathrm{Hom}(C^r(W_n; \mathbf{C}), \Omega^{r-q}(S(M)))$ and, we construct, in terms of it, a chain of the diagonal complex. We take q vector fields $D_1, \ldots, D_q \in \mathfrak{A}(M)$. For each r they determine a homomorphism $C^r(W_n; \mathbf{C}) \to \Omega^{r-q}(S(M))$, since to the chain $L \in C^r(W_n; \mathbf{C})$ corresponds the form which relates to the vectors $a_1, \ldots, a_{r-q} \in \tau_s\mathfrak{A}(M)$ tangential to $\mathfrak{A}(M)$ at some point, the number $L(D_1, \ldots, D_q, a_1, \ldots, a_{r-q})$ (we may regard the fields $D_1, \ldots, D_q$ as elements of $W_n$, because the fixing of a point of S(M) corresponds to the fixing of a local coordinate system of M; we may regard the vectors $a_1, \ldots, a_{r-q}$ as elements of $W_n$, since $\tau_s S(M)$ is identified with $W_n$). It is obvious that the element of $\mathrm{Hom}(C^r(W_n; \mathbf{C}), \Omega^{r-q}(S(M)))$ we have obtained is contained in $K^{-q}$, so that our functional has a certain value

†Gel'fand and Fuks [3] found the cohomologies of the real algebra $W_n$ with real coefficients; however, the result is not affected by replacing **R** by **C**.

‡An attentive reader of [1] will observe that this result holds without the assumptions of compactness and orientability.

at this element. We take it to be $L(D_1, \ldots, D_q)$. It is obvious that the chain $L_1$ lies in the diagonal complex. The required homomorphism is given by the juxtaposition $L \mapsto L_1$.

3. The complex K is a geometric $L_0$-complex, where $L_0$ is the algebra of formal vector fields vanishing at $0 \in R^n$. We study the complex $K^{L_0}$.

PROPOSITION 3.2. The inclusion $K^{W_n} \to K^{L_0}$ induces a homological isomorphism.

Proof. This is obvious when $M = \mathbf{R}^n$, since the homologies of both complexes are isomorphic (with a change of enumeration) to the cohomologies of the algebra $W_n$. To go from here to the general case it is sufficient to fix a triangulation of M and use it to define a filtration in the complex K (it is defined with the help of the usual filtration in the de Rham complex of S(M), fibered over M), and hence in the complexes $K^{L_0}$ and $K^{W_n}$. This filtration determines spectral sequences, which are convergent to homologies of the complexes $K^{L_0}$ and $K^{W_n}$, and the established result for the case $M = \mathbf{R}^n$ means that, for these spectral sequences the second terms are mapped isomorphically under the homomorphism induced by the inclusion $K^{W_n} \to K^{L_0}$.

4. By using Propositions 3.1 and 3.2 the problem of calculating the homologies of the diagonal complex is reduced to the problem of calculating the homologies of $K^{L_0}$. It is clear from here that the homologies of the diagonal complex depend only on topological properties of the tangent, fibering of the manifold M. For, let $S^0(M)$ denote the space of $\infty$-jets of the regular mappings of $\mathbf{R}^n$ into the tangent spaces of M that map 0 into 0. It is clear that $S^0(M)$ is the total space of the fibering over M induced by the tangent space; it is also clear that $S^0(M)$ is an $L_0$-space (but not a $W_n$-space), and as an $L_0$-space it is defined by a tangent fibering. At the same time, by using any Riemannian metric on M we may construct an $L_0$-diffeomorphism $S^0(M) \to S(M)$, so that $\mathrm{Hom}(C^r(W_n; \mathbf{C}), \Lambda^{r-q}(S(M)))^{L_0} = \mathrm{Hom}(C^r(W_n; \mathbf{C}), \Omega^{r-q}(S^0(M)))^{L_0}$.

We may also use Propositions 3.1 and 3.2 when calculating the cohomologies of the diagonal complex. For example, the following theorem holds.

THEOREM. If the manifold M is parallelizable, then

$$H_1^q(\mathfrak{A}(M)) = \sum_{t-s=q} H_s(M) \otimes H_c^t(W_n; \mathbf{C}).$$

Indeed, in this case the fibering $S(M) \to M$ has a cross section. Hence, there is a mapping $\Omega^*(S(M)) \to \Omega^*(M)$, which induces a mapping $K \to \mathrm{Hom}(C^*(W_n; \mathbf{C}), \Omega^*(M))$ under which the complex $K^{L_0}$ is mapped isomorphically onto $\mathrm{Hom}(C^*(W_n; \mathbf{C}), \Omega^*(M))$. It remains to apply Propositions 3.1 and 3.2, and the Kunneth formula.

## CONCLUSION

The short paper [4] preceded the present article. M. V. Losik [5] recently obtained new results about the homologies of the diagonal complex; the results in Sec. 3.4 follow from his work.

## LITERATURE CITED


1. I. M. Gel'fand and D. B. Fuks, "Cohomologies of the Lie algebra of tangent vector fields of a smooth manifold," Funktsional. Analiz i Ego Prilozhen., 3, No. 3, 32-52 (1969).
2. I. M. Gel'fand and D. B. Fuks, "Cohomologies of the Lie algebra of tangent vector fields of a smooth manifold. II," Funktsional. Analiz. i Ego Prilozhen., 4, No. 2, 23-32 (1970).
3. I. M. Gel'fand and D. B. Fuks, "Cohomologies of the Lie algebra of formal vector fields," Izv. Akad. Nauk SSSR. Ser. Matem., 34, 322-337 (1970).
4. I. M. Gel'fand and D. A. Kazhdan, "Some questions on differential geometry and the calculation of the cohomologies of the Lie algebras of vector fields," Dokl. Akad. Nauk SSSR, 200, No. 2, 269-272 (1971).
5. M. V. Losik, "On the cohomologies of the Lie algebra of vector fields with coefficients in the trivial identity representation," Funksional. Analiz i Ego Prilozhen., 6, No. 1, 24-36 (1972).
6. H. Cartan, Notions d'algèbre differentielle; application aux groups de Lie et aux varietés où opere un group de Lie, Coll. de Toplogie, Bruxelles, 1950, Georges Thone, Liege, Masson et Cie., Paris (1951), 15-27.

**Curriculum Vitae**
Phillip A. Griffiths

Date of Birth: October 18, 1938, North Carolina, USA

## Education:

| | |
|---|---|
| 1962 | Ph.D., Princeton University |
| 1959 | B.S., Wake Forest University |

## Employment History:

*Current Positions*

| | |
|---|---|
| July 2009– | Professor Emeritus of Mathematics, Institute for Advanced Study |
| 1999– | Chair, Science Initiative Group, Institute for Advanced Study |

*Advisory Positions*

| | |
|---|---|
| 2007– | Carnegie Fellow, Carnegie Corporation of New York |
| 2002– | Distinguished Presidential Fellow for International Affairs, National Academy of Sciences |
| 2001– | Senior Advisor, The Andrew W. Mellon Foundation |

*Past Positions*

| | |
|---|---|
| 2004–2009 | Professor of Mathematics, Institute for Advanced Study |
| 1991–2003 | Director, Institute for Advanced Study |
| 1983–1991 | Provost and James B. Duke Professor of Mathematics, Duke University |
| 1983 | Dwight Parker Robinson Professor of Mathematics, Harvard University |
| 1975–1976 | Miller Fellow, University of California at Berkeley |
| 1972–1983 | Professor of Mathematics, Harvard University |
| 1968–1970 | Member, School of Mathematics, Institute for Advanced Study |
| 1968–1972 | Professor of Mathematics, Princeton University |
| 1967–1968 | Visiting Professor, Princeton University |
| 1964–1967 | Faculty Member, University of California at Berkeley |
| 1962–1964 | Miller Fellow, University of California at Berkeley |

## Awards and Honors:

| | |
|---|---|
| 2008 | Brouwer Prize, Royal Dutch Mathematical Society |

| | |
|---|---|
| 2008 | Wolf Foundation Prize in Mathematics |
| 2004 | Honorary Doctor of Science degree, Duke University |
| 2002 | Ordem Nacional do Mérito Científico, Ministry of Science and Technology, Brazil |
| 2002 | Doctores Honoris Causa, University of Oslo |
| 1983 | Honorary degree, University of Peking |
| 1980–1982 | Guggenheim Fellow |
| 1979 | Docteur Honoris-Causa from Angers University in France |
| 1979 | Dannie Heineman Prize, Academy of Sciences at Göttingen |
| 1973 | Honorary degree, Wake Forest University |
| 1971 | LeRoy P. Steele Prize given by the American Mathematical Society |

**Member of:**

*Boards*

| | |
|---|---|
| 2008–2013 | Board of Trustees, Mathematical Sciences Research Institute (MSRI); Chair, 2010–2013 |
| 2006– | Zurich Natural Catastrophe Advisory Board |
| 2004– | Board of Directors, Math for America |
| 2004–2005 | Board of Trustees, Italian Institute of Technology |
| 2004–2005 | Board of Directors, InventQjaya, Malaysia |
| 2001– | Board of Directors, GSI Group |
| 1999– | Board of Directors, Oppenheimer Funds |
| 1994– | Oversight Board, IAS/Park City Mathematics Institute |
| 1994–1999 | Board of Directors, Bankers Trust New York Corporation |
| 1983– | Board of Trustees, Woodward Academy |
| 1983–1991 | Board of Directors, Research Triangle Institute |
| 1983–1991 | Board of Trustees, North Carolina School of Science & Mathematics |

*Committees:*

| | |
|---|---|
| 2010– | Member, National Research Council Committee to Develop Science Education Standards |
| 2007–2008 | Chair, Visiting Committee, National Mathematical Centre, Abuja, Nigeria |
| 2005–2006 | Chair, Mathematical Sciences Education Board, The National Academies |
| 2004– | Member, Board on African Science Academy Development |
| 2004–2008 | Chair, Independent Scientific Advisory Board, African Institute of Science & Technology |

| | |
|---|---|
| 2004–2007 | Chair, Temporary Nominating Group for Africa, National Academy of Sciences |
| 2003–2007 | Member, Honorary Presidential Advisory Council on Science and Technology, Nigeria |
| 1999–2006 | Secretary, International Mathematical Union |
| 1997–1999 | Member, Advisory Board, Center for Science, Mathematics & Engineering Education, NRC |
| 1995–1998 | Chair, Program Committee, International Congress of Mathematicians |
| 1993 | Chair, Defense Conversion and Community Assistance Program, State of New Jersey |
| 1992–1999 | Chair, Committee on Science, Engineering and Public Policy, NAS/NAE/IOM |
| 1992 | Chair, Commission on Physical Sciences, Mathematics, and Applications, NRC |
| 1986–1991 | Chair, Board on Mathematical Sciences, National Research Council (NRC) |

*Memberships*

| | |
|---|---|
| 2003– | Indian Academy of Sciences (Foreign Associate) |
| 2002– | Council on Foreign Relations |
| 2001– | TWAS, The Academy of Sciences for the Developing World (Foreign Associate) |
| 2001– | Accademia Nazionale dei Lincei (Foreign Associate) |
| 1996– | American Philosophical Society |
| 1995– | American Academy of Arts and Sciences |
| 1991–1996 | National Science Board |
| 1979– | National Academy of Sciences |

*Editorial Boards*

| | |
|---|---|
| 2008– | MSRI and Nankai Book Series on Mathematics |
| 2001– | Advances in Function Theory |
| 2001– | Annals of Math Studies |
| 1997–2007 | Annals of Mathematics |
| 1994– | Selecta Mathematica |
| 1983– | Duke Mathematical Journal |
| 1980–1992 | Compositio Mathematica |
| 1980–1990 | Journal of Differential Geometry |

## List of Publications


1. P. Griffiths, Mumford–Tate groups, to appear in *Mumford–Tate Goups and Domains: Their Geometry and Arithmetic*, eds. M. Green, P. A. Griffiths and M. Kerr (Princeton Univ. Press, 2012).
2. M. Green, P. Griffiths and M. Kerr, Mumford–Tate domains, *Boll. Unione Mat. Ital.* (9) **3** (2010), No. 2, 281–307.
3. M. Green, P. Griffiths and M. Kerr, Néron models and limits of Abel–Jacobi mappings, *Compos. Math.* **146** (2010), No. 2, 288–366.
4. J. Carlson, M. Green and P. Griffiths, Variations of Hodge structure considered as an exterior differential system: Old and new results, *SIGMA Symmetry Integrability Geom. Methods Appl.* **5** (2009), Paper 087, 40 pp.
5. M. Green, P. Griffiths and M. Kerr, Some enumerative global properties of variations of Hodge structures, *Mosc. Math. J.* **9** (2009), No. 3, 469–530.
6. M. Green, P. Griffiths and M. Kerr, Néron models and boundary components for degenerations of Hodge structure of mirror quintic type, in *Curves and Abelian Varieties*, pp. 71–145, Contemp. Math., Vol. 465 (Amer. Math. Soc., 2008).
7. J. Carlson and P. Griffiths, What is ... a period domain?, *Notices Amer. Math. Soc.* **55** (2008), No. 11, 1418–1419.
8. M. Green and P. Griffiths, Algebraic cycles and singularities of normal functions, in *Algebraic Cycles and Motives*, Vol. 1, pp. 206–263, London Math. Soc. Lecture Note Ser., Vol. 343 (Cambridge Univ. Press, 2007).
9. M. Green and P. Griffiths, Algebraic cycles and singularities of normal functions. II, in *Inspired by S. S. Chern*, pp. 179–268, Nankai Tracts Math., Vol. 11 (World Scientific, 2006).
10. M. Green and P. Griffiths, On the tangent space to the space of algebraic cycles on a smooth algebraic variety, Annals of Mathematics Studies, Vol. 157 (Princeton Univ. Press, 2005).
11. P. Griffiths, Hodge theory and geometry, *Bull. London Math. Soc.* **36** (2004), No. 6, 721–757.
12. M. Green and P. Griffiths, Formal deformation of Chow groups, in *The Legacy of Niels Henrik Abel*, pp. 467–509 (Springer, 2004).
13. P. Griffiths, The legacy of Abel in algebraic geometry, in *The Legacy of Niels Henrik Abel*, pp. 179–205 (Springer, 2004).
14. M. Green, P. Griffiths and K. H. Paranjape, Cycles over fields of transcendence degree 1, *Michigan Math. J.* **52** (2004), No. 1, 181–187.

15. J. J. Kohn, P. Griffiths, H. Goldschmidt, E. Bombieri, B. Cenkl, P. Garabedian and L. Nirenberg, Donald C. Spencer (1912–2001), *Notices Amer. Math. Soc.* **51** (2004), No. 1, 17–29.
16. P. Griffiths, *Selected Works of Phillip A. Griffiths with Commentary*, Part 4. *Differential Systems*, edited by R. L. Bryant and D. R. Morrison (Amer. Math. Soc. and Int. Press, 2003).
17. P. Griffiths, *Selected Works of Phillip A. Griffiths with Commentary*, Part 3. *Variations of Hodge Structures*, edited by C. H. Clemens and D. R. Morrison (Amer. Math. Soc. and Int. Press, 2003).
18. P. Griffiths, *Selected Works of Phillip A. Griffiths with Commentary*, Part 2. *Algebraic Geometry*, edited by E. Arbarello and J. Harris (Amer. Math. Soc. and Int. Press, 2003).
19. P. Griffiths, *Selected Works of Phillip A. Griffiths with Commentary*, Part 1. *Analytic Geometry*, edited by M. Cornalba, M. L. Green and W. Schmid (Amer. Math. Soc. and Int. Press, 2003).
20. M. Green and P. Griffiths, An interesting 0-cycle, *Duke Math. J.* **119** (2003), No. 2, 261–313.
21. R. Bryant, P. Griffiths and D. Grossman, Exterior differential systems and Euler–Lagrange partial differential equations, in *Chicago Lectures in Mathematics* (Univ. of Chicago Press, 2003).
22. M. Green and P. Griffiths, Hodge-theoretic invariants for algebraic cycles, in *Int. Math. Res. Not.* (2003), No. 9, 477–510.
23. M. Green and P. Griffiths, The regulator map for a general curve, in *Symposium in Honor of C. H. Clemens* (Salt Lake City, UT, 2000), pp. 117–127, Contemp. Math., Vol. 312 (Amer. Math. Soc., 2002).
24. M. Green and P. Griffiths, Abel's differential equations, Special issue for S. S. Chern, *Houston J. Math.* **28** (2002), No. 2, 329–351.
25. P. Griffiths, A classical complex analyst encounters a post-modern mathematical object, *Bol. Asoc. Mat. Venez.* **8** (2001), No. 2, 183–195.
26. P. Griffiths, Trends for science and mathematics in the 21st century, in *Mathematics and the 21st century* (Cairo, 2000), pp. 3–11 (World Scientific, 2001).
27. P. Griffiths, Mathematics and the sciences: Is interdisciplinary research possible?, in *Mathematics Towards the Third Millennium* (Rome, 1999) (Atti Accad. Naz. Lincei Cl. Sci. Fis. Mat. Natur. Rend. Lincei (9) 2000), Special Issue, pp. 65–76.
28. P. Griffiths, Mathematics at the turn of the millennium (in Spanish), translated from the English, *Gac. R. Soc. Mat. Esp.* **3** (2000), No. 1, 23–41.
29. P. Griffiths, Mathematics at the turn of the millennium, *Amer. Math. Monthly* **107** (2000), No. 1, 1–14.

30. P. Griffiths, Mathematics at the turn of the millennium, *Current Sci.* **77** (1999), No. 6, 750–758.
31. R. Bryant, P. Griffiths and L. Hsu, Toward a geometry of differential equations, in *Geometry, Topology, & Physics*, pp. 1–76, Conf. Proc. Lecture Notes Geom. Topology, Vol. IV (Int. Press, 1995).
32. R. Bryant, P. Griffiths and L. Hsu, Hyperbolic exterior differential systems and their conservation laws. II, *Selecta Math.* (*N.S.*) **1** (1995), No. 2, 265–323.
33. R. Bryant and P. Griffiths, Characteristic cohomology of differential systems. II, Conservation laws for a class of parabolic equations, *Duke Math. J.* **78** (1995), No. 3, 531–676.
34. R. Bryant, P. Griffiths and L. Hsu, Hyperbolic exterior differential systems and their conservation laws. I, *Selecta Math.* (*N.S.*) **1** (1995), No. 1, 21–112.
35. R. Bryant and P. Griffiths, Characteristic cohomology of differential systems. I, General theory, *J. Amer. Math. Soc.* **8** (1995), No. 3, 507–596.
36. P. Griffiths and J. Harris, Principles of algebraic geometry, reprint of the 1978 original. Wiley Classics Library (John Wiley & Sons, 1994).
37. P. Griffiths and S. S. Chern, Always changing, always the same, in *Chern — A Great Geometer of the Twentieth Century*, pp. 117–120 (Int. Press, 1992).
38. P. Griffiths, Some aspects of exterior differential systems, in *Complex Geometry and Lie Theory* (Sundance, UT, 1989), pp. 151–173, Proc. Sympos. Pure Math., Vol. 53 (Amer. Math. Soc., 1991).
39. R. Bryant, S. S. Chern, R. B. Gardner, H. L. Goldschmidt, and P. Griffiths, Exterior differential systems, in Mathematical Sciences Research Institute Publications, Vol. 18 (Springer, 1991).
40. P. Griffiths and D. V. Morgan, Ratsional'naya teoriya gomotopiĭ i differentsial'nye formy (in Russian) [Rational homotopy theory and differential forms], translated from the English and with an appendix by A. V. Pazhitnov (Nauka, 1990).
41. P. Griffiths, Introduction to algebraic curves, translated from the Chinese by Kuniko Weltin. Translations of Mathematical Monographs, Vol. 76 (Amer. Math. Soc., 1989).
42. P. Griffiths and G. R. Jensen, Differential systems and isometric embeddings, in Annals of Mathematics Studies, Vol. 114. The William H. Roever Lectures in Geometry (Princeton Univ. Press, 1987).
43. P. Griffiths, Vneshnie differentsial'nye sistemy i variatsionnoe ischislenie (in Russian) [Exterior differential systems and the calculus of variations], translated from the English by S. K. Lando. Translation edited and with

a preface by V. I. Arnol'd and an appendix by A. M. Vershik and V. Ya. Gershkovich (Mir, 1986).

44. R. Bryant and P. Griffiths, Reduction for constrained variational problems and $\int \frac{\kappa^2}{2} ds$, *Amer. J. Math.* **108** (1986), No. 3, 525–570.
45. P. Griffiths, Mathematical sciences: A unifying and dynamical resource, *Notices Amer. Math. Soc.* **33** (1986), No. 3, 463–479.
46. P. Griffiths, Linearizing flows and a cohomological interpretation of Lax equations, *Amer. J. Math.* **107** (1985), No. 6, 1445–1484.
47. P. Griffiths and J. Harris, On the Noether–Lefschetz theorem and some remarks on codimension-two cycles, *Math. Ann.* **271** (1985), No. 1, 31–51.
48. E. Arbarello, M. Cornalba, P. Griffiths and J. Harris, *Geometry of Algebraic Curves*, Vol. I. Grundlehren der Mathematischen Wissenschaften [Fundamental Principles of Mathematical Sciences], Vol. 267 (Springer, 1985).
49. P. Griffiths, Linearizing flows and a cohomology interpretation of Lax equations, in *Seminar on Nonlinear Partial Differential Equations* (Berkeley, Calif., 1983), pp. 37–46, Math. Sci. Res. Inst. Publ., Vol. 2 (Springer, 1984).
50. P. Griffiths, Infinitesimal invariant of normal functions, in *Topics in Transcendental Algebraic Geometry* (Princeton, N.J., 1981/1982), pp. 305–316, Ann. of Math. Stud., Vol. 106 (Princeton Univ. Press, 1984).
51. P. Griffiths and L. Tu, Infinitesimal variation of Hodge structure and the generic global Torelli theorem, in *Topics in Transcendental Algebraic Geometry* (Princeton, N.J., 1981/1982), pp. 227–237, Ann. of Math. Stud., Vol. 106 (Princeton Univ. Press, 1984).
52. P. Griffiths and L. Tu, Asymptotic behavior of a variation of Hodge structure, in *Topics in Transcendental Algebraic Geometry* (Princeton, N.J., 1981/1982), pp. 63–74, Ann. of Math. Stud., Vol. 106 (Princeton Univ. Press, 1984).
53. P. Griffiths and L. Tu, Infinitesimal variation of Hodge structure, in *Topics in Transcendental Algebraic Geometry* (Princeton, N.J., 1981/1982), pp. 51–61, Ann. of Math. Stud., Vol. 106 (Princeton Univ. Press, 1984).
54. P. Griffiths and L. Tu, Curvature properties of the Hodge bundles, in *Topics in Transcendental Algebraic Geometry* (Princeton, N.J., 1981/1982), pp. 29–49, Ann. of Math. Stud., Vol. 106 (Princeton Univ. Press, 1984).
55. P. Griffiths and L. Tu, Variation of Hodge structure, in *Topics in Transcendental Algebraic Geometry* (Princeton, N.J., 1981/1982), pp. 3–28, Ann. of Math. Stud., Vol. 106 (Princeton Univ. Press, 1984).

56. R. Bryant, P. Griffiths and D. Yang, Characteristics and existence of isometric embeddings, *Duke Math. J.* **50** (1983), No. 4, 893–994.
57. P. Griffiths, Exterior differential systems and the calculus of variations, in Progress in Mathematics, Vol. 25 (Birkhäuser, 1983).
58. P. Griffiths, Infinitesimal variations of Hodge structure. III, Determinantal varieties and the infinitesimal invariant of normal functions, *Compositio Math.* **50** (1983), No. 2–3, 267–324.
59. P. Griffiths and J. Harris, Infinitesimal variations of Hodge structure. II, An infinitesimal invariant of Hodge classes, *Compositio Math.* **50** (1983), No. 2–3, 207–265.
60. J. Carlson, M. Green, P. Griffiths and J. Harris, Infinitesimal variations of Hodge structure. I, *Compositio Math.* **50** (1983), No. 2–3, 109–205.
61. R. Bryant and P. Griffiths, Some observations on the infinitesimal period relations for regular threefolds with trivial canonical bundle, in *Arithmetic and Geometry*, Vol. II, pp, 77–102, Progr. Math., Vol. 36 (Birkhäuser, 1983).
62. E. Berger, R. Bryant and P. Griffiths, The Gauss equations and rigidity of isometric embeddings, *Duke Math. J.* **50** (1983), No. 3, 803–892.
63. P. Griffiths, Poincaré and algebraic geometry, *Bull. Amer. Math. Soc.* (*N.S.*) **6** (1982), No. 2, 147–159.
64. R. Bryant, S. S. Chern and P. Griffiths, Exterior differential systems, in *Proc. of the 1980 Beijing Symposium on Differential Geometry and Differential Equations*, Vols. 1–3 (Beijing, 1980), pp. 219–338 (Science Press, 1982).
65. P. Griffiths and D. Kharris, Printsipy algebraicheskoĭ geometrii. Tomes 1, 2 (in Russian) [Principles of algebraic geometry. Vols. 1, 2], translated from the English by V. I. Danilov, G. B. Shabat and V. V. Shokurov, translation edited and with a preface by V. A. Iskovskikh (Mir, 1982).
66. P. Griffiths, An observation on normal functions, in Symposia Mathematica, XXIV (Sympos., INDAM, Rome, 1979), pp. 347–353 (Academic Press, 1981).
67. P. Griffiths and J. W. Morgan, Rational homotopy theory and differential forms, in Progress in Mathematics, Vol. 16 (Birkhäuser, 1981).
68. S. S. Chern and P. Griffiths, Corrections and addenda to our paper: "Abel's theorem and webs" [Jahresber. Deutsch. Math.-Verein. **80** (1978), No. 1–2, 13–110], *Jahresber. Deutsch. Math.-Verein.* **83** (1981), No. 2, 78–83.
69. E. Berger, R. Bryant and P. Griffiths, Some isometric embedding and rigidity results for Riemannian manifolds, *Proc. Nat. Acad. Sci. U.S.A.* **78** (1981), No. 8, part 1, 4657–4660.

70. P. Griffiths, An introduction to the theory of special divisors on algebraic curves, in CBMS Regional Conference Series in Mathematics, Vol. 44 (Amer. Math. Soc., 1980).
71. M. Green and P. Griffiths, Two applications of algebraic geometry to entire holomorphic mappings, in *The Chern Symposium 1979* (Proc. Internat. Sympos., Berkeley, Calif., 1979), pp. 41–74 (Springer, 1980).
72. J. Carlson and P. Griffiths, Infinitesimal variations of Hodge structure and the global Torelli problem, in *Journées de Géometrie Algébrique d'Angers, Juillet 1979/Algebraic Geometry*, Angers, 1979, pp. 51–76 (Sijthoff & Noordhoff, 1980).
73. P. Griffiths and J. Harris, On the variety of special linear systems on a general algebraic curve, *Duke Math. J.* **47** (1980), No. 1, 233–272.
74. P. Griffiths, Some problems in complex analytic geometry with growth conditions, in *Proceedings of the International Congress of Mathematicians* (Helsinki, 1978), pp. 645–651 (Acad. Sci. Fennica, 1980).
75. P. Griffiths amd J. Harris, Algebraic geometry and local differential geometry, *Ann. Sci. École Norm. Sup.* (4) **12** (1979), No. 3, 355–452.
76. P. Griffiths, A theorem concerning the differential equations satisfied by normal functions associated to algebraic cycles, *Amer. J. Math.* **101** (1979), No. 1, 94–131.
77. P. Griffiths, Complex analysis and algebraic geometry, *Bull. Amer. Math. Soc. (N.S.)* **1** (1979), No. 4, 595–626.
78. S. S. Chern and P. Griffiths, Linearization of webs of codimension one and maximum rank, in *Proceedings of the International Symposium on Algebraic Geometry* (Kyoto Univ., Kyoto, 1977), pp. 85–91 (Kinokuniya Book Store, 1978).
79. P. Griffiths, Complex differential and integral geometry and curvature integrals associated to singularities of complex analytic varieties, *Duke Math. J.* **45** (1978), No. 3, 427–512.
80. P. Griffiths and J. Harris, On Cayley's explicit solution to Poncelet's porism, *Enseign. Math.* (2) **24** (1978), No. 1–2, 31–40.
81. P. Griffiths amd J. Harris, Residues and zero-cycles on algebraic varieties, *Ann. of Math.* (2) **108** (1978), No. 3, 461–505.
82. P. Griffiths and J. Harris, Principles of algebraic geometry, in *Pure and Applied Mathematics* (Wiley-Interscience [John Wiley & Sons], 1978).
83. S. S. Chern and P. Griffiths, An inequality for the rank of a web and webs of maximum rank, *Ann. Scuola Norm. Sup. Pisa Cl. Sci.* (4) **5** (1978), No. 3, 539–557.
84. S. S. Chern and P. Griffiths, Abel's theorem and webs. Jahresber, *Deutsch. Math.-Verein.* **80** (1978), No. 1–2, 13–110.

85. P. Griffiths and J. Harris, Two proofs of a theorem concerning algebraic space curves, in *Proceedings of the Eighth National Mathematics Conference* (Arya-Mehr Univ. Tech., Tehran, 1977), pp. 350–370 (Arya-Mehr Univ. Tech., 1977).
86. P. Griffiths and J. Harris, A Poncelet theorem in space, *Comment. Math. Helv.* **52** (1977), No. 2, 145–160.
87. P. Griffiths, On Abel's differential equations, in *Algebraic Geometry* (J. J. Sylvester Sympos., Johns Hopkins Univ., Baltimore, Md., 1976), pp. 26–51 (Johns Hopkins Univ. Press, 1977).
88. P. Griffiths, P. Deligne, J. Morgan and D. Sullivan, Real homotopy theory of Kähler manifolds (in Russian), translated from the English by Ju. I. Manin. *Usp. Mat. Nauk* **32** (1977), No. 3(195), 119–152, 247.
89. P. Griffiths and J. King, Teoriya Nevanlinny i golomorfnye otobrazheniya algebraicheskikh mnogoobraziĭ (in Russian) [Nevanlinna theory and holomorphic mappings between algebraic varieties], translated from the English (Mir; distributed by Imported Publications, Chicago, 1976).
90. M. Cowen and P. Griffiths, Holomorphic curves and metrics of negative curvature, *J. Anal. Math.* **29** (1976), 93–153.
91. P. Griffiths, Entire holomorphic mappings in one and several complex variables, The fifth set of Hermann Weyl Lectures, given at the Institute for Advanced Study, Princeton, N. J., October and November 1974. Annals of Mathematics Studies, No. 85 (Princeton Univ. Press; University of Tokyo Press, Tokyo, 1976).
92. P. Griffiths, Variations on a theorem of Abel, *Invent. Math.* **35** (1976), 321–390.
93. P. Griffiths and W. Schmid, Recent developments in Hodge theory: A discussion of techniques and results, in *Discrete Subgroups of Lie Groups and Applicatons to Moduli* (Internat. Colloq., Bombay, 1973), pp. 31–127 (Oxford Univ. Press, 1975).
94. M. Cornalba and P. Griffiths, Some transcendental aspects of algebraic geometry, in *Algebraic Geometry* (Proc. Sympos. Pure Math., Vol. 29, Humboldt State Univ., Arcata, Calif., 1974), pp. 3–110 (Amer. Math. Soc., 1975).
95. P. Griffiths, Differential geometry and complex analysis, in *Differential Geometry* (Proc. Sympos. Pure Math., XXVII, Part 2, Stanford Univ., 1973), pp. 43–64 (Amer. Math. Soc., 1975).
96. P. Deligne, P. Griffiths, J. Morgan and D. Sullivan, Real homotopy theory of Kähler manifolds, *Invent. Math.* **29** (1975), No. 3, 245–274.
97. M. Cornalba and P. Griffiths, Analytic cycles and vector bundles on non-compact algebraic varieties, *Invent. Math.* **28** (1975), 1–106.

98. P. Griffiths, On Cartan's method of Lie groups and moving frames as applied to uniqueness and existence questions in differential geometry, *Duke Math. J.* **41** (1974), 775–814.
99. P. Griffiths, Two results in the global theory of holomorphic mappings, in *Contributions to Analysis* (*A Collection of Papers Dedicated to Lipman Bers*), pp. 169–183 (Academic Press, 1974).
100. P. Griffiths, On the Bezout problem for entire analytic sets, *Ann. of Math.* (2) **100** (1974), 533–552.
101. J. Carlson and P. Griffiths, The order functions for entire holomorphic mappings, in *Value Distribution Theory*, Part A (Proc. Tulane Univ. Program, 1972–1973), pp. 225–248 (Dekker, 1974).
102. P. Griffiths, Topics in algebraic and analytic geometry, written and revised by John Adams. Mathematical Notes, No. 13 (Princeton Univ. Press; University of Tokyo Press, Tokyo, 1974).
103. P. Griffiths, Some remarks on Nevanlinna theory, in *Value Distribution Theory* (Proc. Tulane Univ. Program, Tulane Univ., New Orleans, LA, 1972–1973), Part A, pp. 1–11 (Dekker, 1974).
104. P. Griffiths and J. King, Nevanlinna theory and holomorphic mappings between algebraic varieties, *Acta Math.* **130** (1973), 145–220.
105. P. Griffiths, A Schottky–Landau theorem for holomorphic mappings in several complex variables, Symposia Mathematica, X (Convegno di Geometria Differenziale, INDAM, Rome, 1971), pp. 229–243 (Academic Press, 1972).
106. P. Griffiths, Function theory of finite order on algebraic varieties. I(B), *J. Differential Geom.* **7** (1972), 45–66.
107. J. Carlson and P. Griffiths, A defect relation for equidimensional holomorphic mappings between algebraic varieties, *Ann. of Math.* (2) **95** (1972), 557–584.
108. C. H. Clemens and P. Griffiths, The intermediate Jacobian of the cubic threefold, *Ann. of Math.* (2) **95** (1972), 281–356.
109. P. Griffiths, Holomorphic mappings: Survey of some results and discussion of open problems, *Bull. Amer. Math. Soc.* **78** (1972), 374–382.
110. P. Griffiths, A transcendental method in algebraic geometry, in *Actes du Congrès International des Mathématiciens* (Nice, 1970), Tome 1, pp. 113–119 (Gauthier-Villars, 1971).
111. P. Griffiths and V. Šmid, Locally homogeneous complex manifolds (in Russian), translated from the English (*Acta Math.* **123** (1969), 253–302) by B. L. Žitnickiĭ. *Usp. Mat. Nauk* **26** (1971), No. 5(161), 117–168.
112. P. Griffiths, Complex-analytic properties of certain Zariski open sets on algebraic varieties, *Ann. of Math.* (2) **94** (1971), 21–51.

113. P. Griffiths, Some transcendental methods in the study of algebraic cycles, in *Several Complex Variables*, II (Proc. Internat. Conf., Univ. Maryland, College Park, Md., 1970), pp. 1–46. Lecture Notes in Math., Vol. 185 (Springer, 1971).
114. P. Griffiths, Function theory of finite order on algebraic varieties. I(A), *J. Differential Geom.* **6** (1971/72), 285–306.
115. P. Griffiths, Two theorems on extensions of holomorphic mappings, *Invent. Math.* **14** (1971), 27–62.
116. P. Griffiths, Holomorphic mapping into canonical algebraic varieties, *Ann. of Math.* (2) **93** (1971), 439–458.
117. P. Griffiths, Periods of integrals on algebraic manifolds. III, Some global differential-geometric properties of the period mapping, *Inst. Hautes Études Sci. Publ. Math.* No. **38** (1970), 125–180.
118. P. Griffiths, A theorem on periods of integrals of algebraic manifolds, *Rice Univ. Studies* **56** (1970), No. 2, 143–152.
119. P. Griffiths, Deformations of complex structure, in *1970 Global Analysis* (Proc. Sympos. Pure Math., XV, Berkeley, Calif., 1968), pp. 251–273 (Amer. Math. Soc., 1970).
120. P. Griffiths, Periods of integrals on algebraic manifolds: Summary of main results and discussion of open problems, *Bull. Amer. Math. Soc.* **76** (1970), 228–296.
121. P. Griffiths, On the periods of certain rational integrals. I, II, *Ann. of Math.* (2) **90** (1969), 460–495; *ibid.* (2) **90** (1969), 496–541.
122. P. Griffiths and W. Schmid, Locally homogeneous complex manifolds, *Acta Math.* **123** (1969), 253–302.
123. P. Griffiths, Hermitian differential geometry, Chern classes, and positive vector bundles, in *1969 Global Analysis* (*Papers in Honor of K. Kodaira*), pp. 185–251 (Univ. Tokyo Press, 1969).
124. P. Griffiths, Some results on algebraic cycles on algebraic manifolds, in *1969 Algebraic Geometry* (Internat. Colloq., Tata Inst. Fund. Res., Bombay, 1968), pp. 93–191 (Oxford Univ. Press, 1969).
125. P. Griffiths, On the periods of integrals on algebraic manifolds, *Rice Univ. Studies* **54** (1968), No. 4, 21–38.
126. P. Griffiths, Periods of integrals on algebraic manifolds. II, Local study of the period mapping, *Amer. J. Math.* **90** (1968), 805–865.
127. P. Griffiths, Periods of integrals on algebraic manifolds. I, Construction and properties of the modular varieties, *Amer. J. Math.* **90** (1968), 568–626.
128. P. Griffiths, Some remarks and examples on continuous systems and moduli, *J. Math. Mech.* **16** (1967), 789–802.

129. P. Griffiths, The residue calculus and some transcendental results in algebraic geometry. I, II, *Proc. Nat. Acad. Sci. U.S.A.* **55** (1966), 1303–1309; 1392–1395.
130. P. Griffiths, Some results on locally homogeneous complex manifolds, *Proc. Nat. Acad. Sci. U.S.A.* **56** (1966), 413–416.
131. P. Griffiths, The extension problem in complex analysis. II, Embeddings with positive normal bundle, *Amer. J. Math.* **88** (1966), 366–446.
132. P. Griffiths, On the existence of a locally complete germ of deformation of certain $G$-structures, *Math. Ann.* **159** (1965), 151–171.
133. P. Griffiths, The extension problem for compact submanifolds of complex manifolds. I, The case of a trivial normal bundle, in *1965 Proc. Conf. Complex Analysis* (Minneapolis, 1964), pp. 113–142 (Springer, 1965).
134. P. Griffiths, Deformations of $G$-structures. Part B: Deformations of geometric $G$-structures, *Math. Ann.* **158** (1965), 326–351.
135. P. Griffiths, Hermitian differential geometry and the theory of positive and ample holomorphic vector bundles, *J. Math. Mech.* **14** (1965), 117–140.
136. P. Griffiths, On the theory of variation of structures defined by transitive, continuous pseudogroups, *Osaka J. Math.* **1** (1964), 175–199.
137. P. Griffiths, Deformations of $G$-structures. Part A: General theory of deformations, *Math. Ann.* **155** (1964), 292–315.
138. P. Griffiths, Deformations of holomorphic mappings, *Illinois J. Math.* **8** (1964), 139–151.
139. P. Griffiths, Some remarks of automorphisms, analytic bundles, and embeddings of complex algebraic varieties, *Proc. Nat. Acad. Sci. U.S.A.* **49** (1963), 817–820.
140. P. Griffiths, On the differential geometry of homogeneous vector bundles, *Trans. Amer. Math. Soc.* **109** (1963), 1–34.
141. P. Griffiths, Some geometric and analytic properties of homogeneous complex manifolds. II, Deformation and bundle theory, *Acta Math.* **110** (1963), 157–208.
142. J. A. Wolf and P. Griffiths, Complete maps and differentiable coverings, *Michigan Math. J.* **10** (1963), 253–255.
143. P. Griffiths, Some geometric and analytic properties of homogeneous complex manifolds. I, Sheaves and cohomology, *Acta Math.* **110** (1963), 115–155.
144. P. Griffiths, On a theorem of Chern, *Illinois J. Math.* **6** (1962), 468–479.
145. P. Griffiths, On certain homogeneous complex manifolds, *Proc. Nat. Acad. Sci. U.S.A.* **48** (1962) 780–783.

# Introductory Comments to Part 1

## Vector Bundles

The origin of the paper "*The Extension Problem for Compact Submanifolds of Complex Manifolds, I. The Case of a Trivial Normal Bundle*" was an attempt by me to understand how sheaf cohomology behaves in a family of compact complex manifolds. In particular, I wanted to understand why, although the ranks of cohomology groups may jump, the Euler characteristic remains constant for a family of holomorphic vector bundles over a family of compact complex manifolds. This result had initially been proved by Kodaira and Spencer, using harmonic forms, and more generally by Serre and Grothendieck in an algebraic context. What I found was that, intuitively at least, there are two reasons that a cohomology group may jump. The first is that a cocycle representing the cohomology class in question does not extend as a cocycle. The second is that a cocycle representing the cohomology class extends as a cocycle, but for $t \neq 0$ becomes a coboundary. The interesting fact is that these two phenomena are exactly paired in adjacent dimensions. This immediately gives the invariance of the Euler characteristic and also gives some insight into the geometric consequences of the jumping of cohomology classes. For example, the phenomenon of jumping of complex structure (which means there is a family of compact complex manifolds lying over the disc, all of which are biholomorphic for $t \neq 0$ but which is not a trivial family) by the above mechanism may be easily seen to exactly correspond to the presence of infinitesimal automorphisms over the central fiber which disappear for $t \neq 0$.

"*The Extension Problem in Complex Analysis II. Embeddings with Positive Normal Bundle*" has an interesting story. The thesis problem that Don suggested to me was an extremely attractive one that he and Louis Nirenberg had worked on. They had observed that if you have a compact complex manifold of dimension greater than or equal to two that is embedded with the same positive normal bundle into two different complex manifolds, then if the two embeddings agree up to a certain finite order that can be calculated from the normal bundle, there is a unique formal isomorphism between neighborhoods of the embedded compact complex manifold. What this meant in local coordinates was that once the two embeddings agree up to a certain finite order, there are then *uniquely* determined formal power series that establish a formal isomorphism between the two germs of embeddings. The problem was to prove that this formal power series converges. Hans Grauert had considered the same equivalence-of-embeddings problem under the assumption that the normal bundle was negative; in this case, although there is no uniqueness, he had proved the equivalence of two germs of embeddings that agreed to sufficiently high order. Hence Grauert's paper "*Über Modifikationen und exzeptionelle analytische Mengen,*" was very influential for me at that time.

Don and Louis had devised some extremely ingenious and deep successive estimates on the coefficients in the power series; these, however, were insufficient to establish convergence. I thought about this problem on and off for a couple of years and at some point realized that perhaps it should be approached geometrically. That is, if one could embed both germs of embeddings in a complex projective space in such a way that the equivalence up to a finite order corresponded to geometric osculation of the two embeddings along the compact complex manifolds, then the statement that there was a unique formal power series establishing an isomorphism between the two germs of embeddings would have the geometric analog that the two submanifolds of complex projective space actually coincided. In this way, the Nirenberg-Spencer problem could be reduced to essentially a linear problem dealing with projective embeddings of germs of neighborhoods of a compact complex manifold with positive normal bundle. The essential geometric point here is that a tubular neighborhood of a compact complex submanifold with positive normal bundle has both pseudo-convexity in the fiber direction and pseudo-concavity in the manifold direction. Thus, one is in effect faced with $\bar{\partial}$-Neumann-type problems, except that the boundary has nondegenerate Levi form but of indefinite signature.

It was therefore necessary to extend the methods of Joe Kohn to cover this situation. I was able to make good use of the fact that Joe had given the marvelous seminar on his work and that a geometric understanding of the basic estimate in the $\bar{\partial}$-Neumann problem could be carried over in its essential aspects to the mixed pseudo-convex, pseudo-concave case. A similar observation had been made independently by Andreotti and Grauert, who worked somewhat differently using complete metrics and the $L^2$ methods of Hörmander rather than the boundary value methods of Joe. In any case, I was in this way able to settle the Nirenberg-Spencer problem. This project also got me fairly deeply into Hermitian differential geometry because it was necessary to put the requisite estimates in intrinsic geometric form. The goal here was to show that the geometry of the neighborhood, whose principal invariant is the Levi form, lines up with the geometry of the submanifold as determined by the curvature of the normal and related vector bundles.

"*Hermitian Differential Geometry, Chern Classes, and Positive Vector Bundles*" is largely expository. One of the motivations for writing it was to clarify two broad notions of positivity that were floating around at the time. One, due primarily to Nakano, was the notion of positivity that arose in his extension of the Kodaira vanishing theorem to vector bundles. This notion was used extensively by Calabi and Vesentini in their paper, "*On Compact, Locally Symmetric Kähler Manifolds,*" which had a large influence on me as a student. The other notion of positivity was the existence of a Hermitian metric under which the associated Levi form of a tubular neighborhood in the dual bundle would be positive definite. This definition was the differential geometric analog of the notion of positivity used by Grauert. In effect, Grauert's notion of positivity was that there exists something like a Finsler metric on the dual bundle that would render the Levi form of the unit tubular neighborhood positive definite. Grauert positivity is essentially the same as ampleness, in the same sense as Hartshorne. The two notions of positivity are related but not the same. Essentially, the Nakano notion requires that a quadratic form on tensors in the tensor product of two vector spaces be positive definite, whereas the Grauert notion only requires the positive definiteness of the quadratic form on decomposable tensors.

In order to understand this situation, it was necessary to develop some foundational notions in Hermitian differential geometry. In hindsight, the notion that was most fruitful was the introduction of the second fundamental form of a holomorphic sub-bundle of a complex vector bundle with a Hermitian metric. A consequence is the important curvature-decreasing property of holomorphic sub-bundles of holomorphic vector bundles. This property reflects the general fact that complex differential geometry is in a way simpler, or one might say richer, than ordinary Riemannian geometry in that curvatures tend to have signs. This curvature-decreasing property of sub-bundles has far reaching implications in Hermitian differential geometry, both for vector bundles and for complex submanifolds, and subsequently it was widely used by Kobayashi and others. Yet another question addressed in this paper was whether or not positive vector bundles had positive Chern classes. This issue, together with that of *numerical* ampleness criteria for vector bundles, has been largely settled through the work of David Gieseker, Spencer Bloch, Bill Fulton and others.

### Period Spaces and Hodge Theory

The paper on "*Locally Homogeneous Complex Manifolds*" which I wrote with Wilfried Schmid, has a somewhat complex background. As noted earlier, my thesis had been a study of compact homogeneous complex manifolds. These were the C-spaces of H. C. Wang, including the non-Kählerian spaces. The Kählerian spaces had already been extensively studied by Borel, Weil and Bott.

From an entirely different direction, I was strongly influenced during the mid-sixties by three papers on algebraic curves. One paper was "Riemann surfaces and the theta function" by Joseph Lewittes, who studied the Riemann theta divisor and related the singularity structure of that divisor to special divisors on the algebraic curves. The other paper was "On a theorem of Torelli" by Aldo Andreotti, who gave an extraordinarily beautiful geometric proof of the Torelli theorem for algebraic curves. The third paper, whose philosophy was very influential even though my work did not directly derive from it, was "On period relations for abelian integrals on algebraic curves" by Aldo Andreotti and Alan Mayer, who applied the heat equation to the study of moduli of algebraic curves.

Each of these papers had focused attention on the period matrix of algebraic curves, particularly on the variational properties of that period matrix when the curve moves in moduli. This, coupled with my exposure to general deformation theory from Kodaira and Spencer, led me quite naturally to consider what happens to the period matrix or Hodge decomposition of a complex Kähler manifold when it varies in moduli. The normalized period matrix of a smooth algebraic curve is a point in the Siegel upper-half-plane, which is a bounded symmetric domain. Analogously, one of the principal ingredients in studying general variation of Hodge structure would be to construct a space of all possible period matrices—now called polarized Hodge structures—that might arise in a family of algebraic varieties.

These classifying spaces for Hodge structures were constructed in another paper in Part 3, and they turn out to have surprising properties. In particular, they are not Hermitian symmetric spaces or even Kählerian in any natural way. Rather, they have convexity/concavity properties reminiscent of those I encountered in the "Extension Problem in Complex Analysis II" discussed above.

Wilfried Schmid was a graduate student at Berkeley during this time, and in his thesis he developed a general theory of the cohomology of these non-compact complex manifolds. This theory showed, among other things, that the Levi form had the convexity/concavity properties mentioned above. He found that the cohomology of the natural homogeneous line bundle was concentrated in one particular dimension, and the structure of that cohomology as a topological vector space was consistent with what had been predicted by Bob Langlands for the general discrete series representations of the Lie groups that acted on these classifying spaces for Hodge structures.

In this paper, Wilfried and I discussed the general differential geometric and function theoretic properties of these classifying spaces for Hodge structures. One of the most important results was the calculation that these spaces had a canonical Hermitian metric whose holomorphic sectional curvatures were negative in the so-called horizontal or transverse directions. This means that the period matrix space fibers over the corresponding symmetric space with compact complex submanifolds as fibers. The fibering, however, is not holomorphic. The variation of Hodge structure in a family of algebraic varieties turns out to be, in terms of the natural metric, orthogonal to the tangent spaces to these compact subvarieties.

A consequence of negative curvature in the horizontal directions is that holomorphic mappings into the classifying space for Hodge structures behave exactly like holomorphic mappings into bounded symmetric domains—provided we impose the horizontality condition. In other words, one should think of the study of holomorphic maps into classifying spaces for Hodge structures as a relative one. This result turned out to have widespread implications for the study of variations of Hodge structure.

"*Recent Developments in Hodge Theory: A Discussion of Techniques and Results.*" Parts of this paper anticipate the introductory comments to "Periods of Integrals on Algebraic Manifolds, III." The mid- to late 1960s were wonderful years for transcendental algebraic geometry. At IHES, Deligne was developing his theory of mixed Hodge structures, which taught us, among other things, that one could treat the cohomology of singular, non-compact and relative algebraic varieties in much the same fashion as one treats the cohomology of the classical smooth projective varieties, at least in terms of their Hodge structures. In Berkeley, Herb Clemens and Alan Landman were proving what is now called the monodromy theorem, the basic result concerning the action on homology when one turns around a singularity.

The theorem on regular singular points also emerged during this time. Although I never wrote up the result, there is an interesting story here. During the mid-sixties in Berkeley, we ran several seminars on the classics in transcendental algebraic geometry. In particular, we studied the works of Picard and Picard-Simart, Poincaré and Lefschetz. Picard makes extensive use of the so-called Picard-Fuchs equation, which is now called the Gauss-Manin connection. Picard tacitly assumed that these equations have regular singular points. In fact, when one looks at the original papers by Fuchs, they are concerned mainly with the calculation of examples that always exhibit the regular singular point behavior for the differential equations satisfied by the periods. I think that to these classically trained mathematicians the fact that the Picard-Fuchs equations have regular singular points is obvious. During the seminars in Berkeley, I had formulated the same result in a general way.

Here again is an interesting sub-story. Volume II of Picard and Simart contains a proof of what is now called the algebraic deRham theorem for algebraic surfaces. In the seminar at Berkeley, I showed how Picard's proof could be extended to the general case. Once one has the algebraic deRham theorem, the theorem on regular singular points has meaning. Regularity was stated and proved by means of growth estimates on the periods. In fact, given Lefschetz' explicit description of cycles representing a homology basis, the result is, to a complex analyst, evident. The proof was never published, however, because to me it was unsatisfactory that this result of purely algebraic geometry required analytic methods to prove. In particular, the result is substantively different from the properties of Hodge structure that are derived analytically from the theory of Kähler manifolds, for the theorem on regular singular points can be formulated purely algebraically and involves no use of the differentiable structure. It seemed to me, therefore, that the result should be proved purely algebraically. Such a proof was subsequently given by Nick Katz and appears in more or less definitive form in the marvelous book by Pierre Deligne, *Equations Différentielles à Points Singuliers Réguliers.*

Finally, as mentioned above, in Schmid's thesis can be seen the beginnings of his later successful attempt on the Langlands conjecture concerning the explicit realization of a discrete series representation by $L^2$ cohomology on homogeneous line bundles defined over the classifying spaces for Hodge structures. In addition, Wilfried had provided a deep analysis of how the Hodge structure of an algebraic variety behaves as the variety acquires a singularity. This analysis of the asymptotic behavior of period matrices, which for obvious reasons connects very nicely with Deligne's theory of mixed Hodge structures, still stands as the seminal work in this area. (Significant extensions of Wilfried's results to the case of several variables were done later by Cattani and Kaplan and by Cattani, Kaplan and Schmid. Extensions to the case of variation of mixed Hodge structure were done by Joseph Steenbrink and Steve Zucker, with valuable input by Deligne.) It seemed to Wilfried and me that it would be worthwhile to try to pull together all these developments in some systematic form. In particular, we wanted to bring the work of Deligne to the attention of those of us who worked in more traditional transcendental algebraic geometry and to suggest how one might prove his results from a more analytic or differential geometric perspective. This was the motivation behind this paper with Wilfried.

The paper with Pierre Deligne, John Morgan and Dennis Sullivan on "*Real Homotopy Theory of Kähler Manifolds*" proves the result that the rational homotopy type of a simply connected smooth algebraic variety is a formal consequence of its cohomology. Roughly speaking, and ignoring questions of torsion and the fundamental group, this means that all the information in the homotopy groups of an algebraic variety is already contained in the cohomology ring. The proof of this result was based on Dennis Sullivan's rational homotopy theory via differential forms. Dennis had conceived this wonderful theory in the early seventies, when he described it in a seminar at the Institute for Advanced Study that John Morgan and I attended. This led John, Eric Friedlander and me to put together a seminar in the summer of 1972 in Florence, Italy, where we attempted to present Sullivan's work *de novo*, along with necessary background materials and complete proofs of the main results. These Florence notes were later converted into a book by John and me, *Rational Homotopy Theory and Differential Forms.*

Missing at the time, however, was the application to compact Kähler manifolds. From the very beginning Dennis had said he was convinced that the presence of a Kähler metric should have implications on the rational homotopy type via the use of differential forms. In hindsight, given Sullivan's theory, the result is rather easy to prove, but the proof took some experimentation with special cases before the result in this paper was actually formulated. I think this delay might have been due to our difficulty in imagining that such a strong result would hold.

**Analytic Geometry**

"*Holomorphic Mappings into Canonical Algebraic Varieties.*" During my period at Berkeley, Chern was particularly interested in the relation between differential geometry and complex function theory, which he approached via the study of holomorphic mappings between complex manifolds. Perhaps the key result in this area was the so-called Ahlfors lemma, which states that holomorphic mappings from the unit disc with the Poincaré metric into a negatively curved complex manifold are distance decreasing. This was Lars' intrinsic formulation of the classical Schwartz-Pick lemma.

Chern and his colleagues and collaborators generalized this to volume forms. Coming from an algebro-geometric perspective, volume forms are closely related to the so-called Bergman metric of algebraic varieties which have very ample canonical bundles. Finally, going back to the classic work of Picard-Simart, one finds the theorem that a holomorphic mapping from the complex plane into a compact Riemann surface of genus greater than or equal to two is necessarily constant.

Although Picard derived this from the uniformization theorem, it also follows from the Ahlfors lemma. As a metric on the compact Riemann surface one may either use the constant negative curvature metric coming from uniformization or, in the non-hyperelliptic case, the Bergman metric coming from the holomorphic one-forms.

Putting all these threads together, one is led naturally to the result that if one has an n-dimensional algebraic variety with ample canonical bundle, then a mapping of $\mathbb{C}^n$ into it must have identically vanishing Jacobian. In fact, the proof has been more or less sketched above. This result represented a confluence of function theory and algebraic geometry and perhaps was one of the first of its type. Subsequent to it came a great deal of activity in the area by Kobayashi, Kodaira and others.

"*Defect Relation for Equidimensional Holomorphic Mappings between Algebraic Varieties.*" As mentioned above, during the mid-sixties, Chern was very interested in holomorphic mappings between complex manifolds. The quantitative version of the classical Picard theorem is the fantastic defect relations that were discovered by Rolf Nevanlinna. To me the defect relations remain one of the most beautiful statements in mathematics.

As Chern presented this theory in seminars from his geometric perspective, it had the flavor of non-compact algebraic geometry. A natural question was whether there was an analog of the Nevanlinna defect relations in the n-dimensional case for equidimensional holomorphic mappings into general algebraic varieties. Although Rolf Nevanlinna's proof of his defect relations had been achieved by methods of classical function theory, especially the use of Jensen's theorem and its consequence—the so-called "lemma on the logarithmic derivative"—a more differential geometric proof of the defect relations had been given by his brother Frithiof Nevanlinna. This was very much in the spirit of a quantitative version of the Ahlfors lemma referred to above.

Jim Carlson, in his thesis for Princeton, had observed that algebraic varieties represented as cyclic coverings of sufficiently high degree of complex projective space branched along a smooth hypersurface had very ample canonical bundle. Since the Bergman metric of the covering manifold is invariant under the covering group, this meant that there is induced on the complement of the branch locus a negatively curved volume form. This suggested to us that in general, the complement of a divisor larger than the canonical divisor should have some sort of negatively curved volume form. Putting together the threads coming from Frithiof Nevanlinna's proof of the defect relation with those suggested in Jim Carlson's thesis, we were able to construct a negatively curved volume form on an arbitrary smooth algebraic variety minus a divisor with normal crossings and with Chern class bigger than that of the canonical bundle. Along the divisor, the volume form behaves asymptotically like the Poincaré volume form on a product of punctured discs. This then leads quite naturally to an extension of the Nevanlinna defect relations to the equidimensional case.

"*Complex Differential and Integral Geometry and Curvature Integrals Associated to Singularities of Complex Analytic Varieties.*" Before discussing the material in this paper, I would like to offer some background. When I was at Berkeley, Chern gave a marvelous course on integral geometry. The subject had particular appeal for me because, as I look at it, the usual statements that one has only in complex algebraic geometry can usually be applied more generally. For example, recall that in the complex projective plane, the degree of an algebraic curve is, by the Wirtinger theorem, equal to the (suitably normalized) area of the curve. Therefore, the Bézout theorem states that the number of intersection points of two algebraic curves in the projective plane is equal to the product of the areas. In particular, the number of intersection points of an algebraic curve with a line is equal to the area of the curve. The Crofton formula in integral geometry states the same result in the real Euclidean plane: namely, the length of a smooth curve in the plane is equal to the average number of intersections of that curve with a line. Here, "average" means that one imagines the set of lines in the plane acted on transitively by the group of Euclidean motions. As a consequence, one can simply integrate over the group the function given on the space of lines by counting the number of intersection points with the curve in question.

The above paper grew out of the Gergen lectures I gave at Duke University in 1978. The first part is concerned with the extension to the complex case of the classical formula given by Hermann Weyl for the volume of a tube around a submanifold in Euclidean space. In the complex case, this formula is especially simple because the coefficients involve only the combinations of the curvature that appear in the Chern forms.

Following the derivation of the tube formula is an extensive discussion of integral geometry in the complex case, in which the analog of the so-called kinematic formula is given. This is a rather interesting result, which takes the classic kinematic formula in the real case and derives the complex analog, where again we see that the curvature appears only in combinations given by the Chern forms.

The last part of the Gergen lectures concerned curvature integrals around singularities of complex analytic varieties. Linda Ness, a student of David Mumford, had proved in her Harvard thesis that when one considers a family of algebraic curves acquiring a node, even though the curvature function of the induced metric blows up, the total curvature remains finite as one approaches the singular point. This idea was extended by Rémi Langevin in *Courbure et singularités complexes*. Studying the work of Ness and Langevin it struck me that when a general algebraic variety acquires singularities, although its Chern forms would blow up, their integrals should remain finite. It was particularly interesting that, loosely speaking, the limit of the integral was not equal to the integral of the limit. I called their difference a *Plücker defect*, since one could in fact trace this difference to the terms of the classical Plücker formulas. The end result was a formula expressing the Milnor number for a singularity as a curvature integral.

"*Analytic Cycles and Vector Bundles over Non-compact Varieties.*" One of the deepest and most striking theorems to emerge from the 1950s was Grauert's non-abelian version of the Oka principle. Grauert's result implies, among other things, that on a Stein manifold the category of holomorphic vector bundles maps isomorphically to the category of topological vector bundles. Using standard algebraic topology, this implies that all the even-dimensional rational cohomology of a Stein manifold is represented by analytic cycles.

In the compact case in algebraic geometry are the famous Hodge $(p, p)$ conditions that a homology class be represented by an algebraic cycle. However, if one starts with a compact algebraic variety and deletes the hyperplane at infinity, one obtains an affine algebraic variety which is in particular a Stein manifold. Thus, all the rational, even-dimensional homology is represented by analytic cycles. (To be precise, one is capturing only the primitive homology. From the work of Lefschetz, however, one knows that the primitive homology forms the building blocks for all the homology, so nothing essential is lost.)

A natural question concerns the growth behavior of the analytic cycles representing homology classes which could not be algebraic. Maurizio Cornalba and I found an interesting relationship. The deviation from being of Hodge-type $(p, p)$ puts a lower bound on the growth of any analytic cycle representing a homology class coming from the compact variety. The projective area of the analytic cycle has to grow at least as $r^{2k}$ where $k$ is the deviation of the cycle from being of pure Hodge type. This then suggested to us that the filtration by order of growth on analytic cycles representing homology classes coming from the compact variety should be related to the Hodge filtration.

In order to prove this, one would need to prove Grauert's theorems with precise growth conditions. In this paper, we attempted to do this, and although we were able to give a new proof of Grauert's theorem with some growth conditions, they are far from optimal, and therefore this program to determine the minimal orders of growth of analytic cycles representing homology classes on affine algebraic varieties remains unfinished. In fact, the whole question of extending linear function theory with growth conditions to the non-abelian case remains to me a very interesting unfinished area of complex analysis.

"*Two Applications of Algebraic Geometry to Entire Holomorphic Mappings.*" Returning to the earlier work on value distribution theory, one of the conjectures derived from the theory was that an algebraic variety is measure hyperbolic in the sense of Kobayashi if and only if it is of general type. The paper "Holomorphic Mappings into Canonical Algebraic Varieties" contains a first result along these lines. The conjecture is obviously true for algebraic curves, and in the paper with Mark Green we were able to prove it for algebraic surfaces, modulo what turned out to be a nontrivial technical detail. The proof relied on the classification of algebraic surfaces and reduced the question to finding a nondegenerate holomorphic mapping from polycylinder with at least one unbounded radius into an algebraic $K3$ surface.

Also in this paper we picked up on a theme of A. Bloch that extended the qualitative version of Picard's theorem to holomorphic curves in abelian varieties. Bloch's highly original proof was based on the use of jet differentials, and it was this aspect that Mark and I focused on. We were able to give a new proof of Bloch's theorem along the lines of the original Bloch proof; a related proof was given independently by Ochiai.

The arithmetic side of this subject was developed by Paul Vojta, Serge Lang and others. Their work led to conjectures by Lang, whose interpretation in the analytic setting is that an entire holomorphic mapping from the complex line into an algebraic variety of general type is necessarily degenerate in the sense that the image lies in a proper algebraic subvariety. For abelian varieties, this result essentially reduces to the theorem of Bloch. To prove such a theorem it seemed to Mark and myself that it would be necessary to interpolate between something of dimension one and something of dimension n; that is, in some way to thicken a holomorphic curve in the algebraic variety. We felt that the use of jet differentials was a promising approach to this. The basis for this feeling is that the general type condition translates into an ample supply of jet differentials, as shown in our paper. However, there are major technical problems in this approach. Recently these difficulties have been overcome in a spectacular fashion for generic surfaces of high degree in projective 3-space in M. McQuillan, "*Holomorphic curves on hyperplane sections of 3-folds*" and Demailly & El Goul, "*Hyperbolicity of generic surfaces of high degree in projective 3-space.*"

# Introductory Comments to Part 2

I have always been interested in explicit constructions in algebraic geometry. For example, the Kodaira-Spencer-Nirenberg construction of deformations using power series methods left a deep impression on me. Of course, the most famous construction issue in algebraic geometry is that of algebraic cycles. Here the famous Hodge conjecture stated first by Hodge some seventy years ago remains the central unsolved problem. The problem is not only unsolved, but other than the Lefschetz $(1,1)$ theorem of the 1920s and its implications, I do not know of a single instance of the *construction* of an algebraic cycle in a given Hodge class.

When I studied Riemann surfaces, I particularly enjoyed Riemann's explicit construction of Jacobi inversion via the Riemann theta divisor. Of course, one can prove the Jacobi inversion theorem by an elementary complex function theoretic argument using Abel's theorem and the implicit function theorem, but the richness in the explicit construction of Jacobi inversion using the theta divisor had special appeal.

In the same sense, Lefschetz' proof of the $(1,1)$ theorem referred to above was again extremely explicit. Following his analysis via the Lefschetz pencil method, one first develops a deep understanding of the topology of an algebraic surface. Then through using the normal functions introduced by Poincaré, one moves to a deep understanding of the Hodge theory of the algebraic surface. These two steps are then combined to give a proof of the Lefschetz $(1,1)$ theorem, which appears very naturally as a consequence of the understanding of the topology and the Hodge theory of the surface.

Even though later Kodaira and Spencer gave a much slicker sheaf theoretic proof of this result, I always admired the highly constructive global geometric character of the Lefschetz argument. So a theme in many of the papers in this volume is that of constructing something on an algebraic variety via analytic techniques. Because of my background in differential geometry, I am drawn to the use of infinitesimal methods whenever these apply - for example, in iteration schemes such as the one used by Kodaira, Nirenberg and Spencer in the work described above.

## Cycles and Deformation Theory

"*Some Results on Algebraic Cycles on Algebraic Manifolds*" was, in a way, an essay on algebraic cycles and intermediate Jacobians. Perhaps the most interesting result of the paper was the computation of the differential of the map that assigns to the Chern classes of a holomorphic vector bundle, viewed as algebraic cycles, their Abel-Jacobi images in the intermediate Jacobian. Although the conclusion was simple to state, the proof itself was surprisingly difficult. It led me to the construction of singular differential forms and explicit residue computations very much in the spirit of what came later in arithmetic algebraic geometry with the construction of Green's currents. These methods were also related to the paper by Bott and Chern, which was done at about the same time.

"*Complex Analytic Properties of Certain Zariski Open Sets on Algebraic Varieties.*" One of the most striking results of the theory of complex Riemann surfaces is the uniformization theorem. I was particularly interested in Poincaré's attempts to prove this theorem by what are essentially algebro-geometric techniques. Namely, he viewed the inverse of the uniformization map as given by the ratio of solutions to a second order algebraic ordinary differential equation of a very special type on the algebraic curve. The number of complex parameters in this equation is $3g-3$, and among the families of all such equations there should be a unique one that gives the uniformization of the compact Riemann surface by the upper half plane.

Poincaré was able to prove quite easily the uniqueness of a uniformization map, but the existence from his approach was much more difficult. Essentially what he showed by a parameter count was that the differential of the map from algebraic curves to uniformizing maps is locally one to one. As noted earlier, it is injective, which means that he had to prove that the map is onto. Poincaré then considered the moduli space of Riemann surfaces and argued that as the algebraic curve acquires a singularity, the uniformizing map blows up; i.e. it acquires a parabolic transformation. This was, of course, strongly related to what became known as the Picard-Lefschetz transformation or, from a differential topological point of view, a Dehn twist.

To me, Poincaré's proof was convincing but not complete. Nevertheless, the idea of trying to construct a uniformization by solutions to algebraic differential equations was highly appealing, and I wondered whether similar methods might work in higher dimensions. In this regard, Lefschetz had proved that a small enough Zariski neighborhood of any point on a smooth algebraic variety is a $K(\pi, 1)$. This makes it at least plausible that the universal cover might be realized as a bounded domain in some complex Euclidean space.

I was able to prove that this is indeed the case, making use of the ideas of Poincaré mentioned above, together with results of Lipman Bers on the simultaneous uniformization of Riemann surfaces by quasi-Fuchsian groups. This result in my paper suffers, however, because, as far as I know, there is no uniqueness of the universal cover realized as a bounded domain. That is, it is possible that two uniformizing maps might be constructed that are not related via a global holomorphic bitransformation of the ambient complex Euclidean space.

"*The Intermediate Jacobian of the Cubic Threefold.*" In the mid-sixties, Francesco Gherardelli of Florence, Italy, was visiting at Berkeley where he was working on the intermediate Jacobian of the cubic threefold. The cubic threefold was in a certain sense the first unknown threefold, at least in the Hodge theoretic sense. Since the canonical bundle of the cubic threefold is negative, its intermediate Jacobian is a principally polarized abelian variety of dimension five. The generalized Hodge conjecture would imply the surjectivity of the map on algebraic one-cycles homologous to zero to the intermediate Jacobian. The cubic threefold was much studied classically by Fano, who had discovered a beautiful surface of lines on it. Gherardelli posed the question of whether or not the mapping from the

Albanese variety of the surface of lines to the intermediate Jacobian is an isogeny. What are the properties of this mapping? For example, what is the equivalence relation on algebraic one-cycles that it defines?

After becoming interested in Gherardelli's questions, Herb Clemens and I studied the cubic threefold for several years, mostly while I was a visiting Member at the Institute for Advanced Study from 1968-1970. Herb was a Member at the Institute then as well, and later took a faculty position at Columbia University while keeping his residence in Princeton. We had written a preliminary version of the paper which attempted to prove that the mapping of the Albanese variety of the surface of lines to the intermediate Jacobian was an isogeny of degree nine. After writing up the preprint, Herb went to Chile for a year of follow-up to his time in the Peace Corps helping to establish a mathematical community in that country. When he returned, we resumed work on the cubic threefold and discovered an error in our computation of the degree of the map from the Albanese to the intermediate Jacobian. We also came across a recent paper by Enrico Bombieri and Peter Swinnerton-Dyer, "On the Local Zeta Function of a Cubic Threefold," which stimulated us to try to understand in greater depth the geometry of the cubic threefold through the Fano surface.

After several false starts everything fell into place, and in fact we were able to prove analogs of all the main classical results on divisors on algebraic curves for the intermediate Jacobian of the cubic threefold. This included a proof that the Albanese of the surface of the Fano surface is isomorphic to the intermediate Jacobian. It also included properties of the theta divisor; the difference from the curve case is that whereas the singularities of the theta divisor for curves falls in dimension $g-4$ where $g$ is the genus, here they fell in dimension $g-5$. For that reason, the theta divisor, which is intrinsically associated to the Hodge structure of the cubic threefold, looks different from the theta divisor of an algebraic curve. Thus, the classical characterization of rational surfaces by Castelnuovo and Enriques does not generalize to higher dimensions. This allowed us to show that, although unirational, the cubic threefold is nonrational. It is interesting that, at about this time, the nonrationality of the cubic fourfold was proved by Manin and Iskovskikh, and independently Mike Artin and David Mumford gave an example of a threefold that was unirational but not rational.

The study of the cubic threefold and the subsequent writing work gave us a chance to write what was in effect an essay on aspects of algebraic geometry which, although rooted in the classical study of the subject, began to lead into less charted territory. In particular, this paper gave me, and I think Herb as well, enormous pleasure in being able to combine classical geometry with modern topology and Hodge theory to bring a definitive understanding to what remains to this day one of the most beautiful algebraic varieties. The relationship between the Hodge structure and the geometry of an algebraic variety continues to be a central interest of mine.

## Abel's Theorem

"*Variations on a Theorem of Abel.*" An aspect of classical algebraic geometry that I have found appealing can be described under the general heading of "functional equations" or "addition theorems." An obvious example that was central to classical function theory and the study of algebraic curves of genus one was the addition theorem for elliptic functions. Almost all these addition theorems seem to have their roots in a theorem of Abel, which might be said to express the addition theorems in infinitesimal form.

A particular result along these lines, derived by Sophus Lie, states the following: If you have, in the complex plane, pieces of complex analytic arcs, together with a holomorphic differential form on each arc, and if for an open set in the space of lines meeting these arcs the infinitesimal version of Abel's theorem is satisfied, then the pieces of analytic arc can be extended to an algebraic curve and the differentials are the restrictions of differentials of the first kind on the curve. Lie's interest in this result focused on the problem of determining which principally polarized abelian varieties were Jacobian varieties of algebraic curves. His idea of the necessary and sufficient condition was that the theta divisor should be of so-called translation type, and he sought infinitesimal conditions on the theta divisor that would ensure this to be the case. In principle, using theta functions, these infinitesimal conditions should then be expressible in terms of the period matrix of the curve, thereby giving the so-called Schottky relations. I learned all of this from Bernard Saint-Donat's "Variétés de translation et théorème de Torelli."

Since another of my interests had been the construction of objects in algebraic geometry, both functions and cycles, I found the theorem of Lie of particular interest. In this paper I discovered and proved an $n$-dimensional version of this result. Again, writing this paper provided me an opportunity to write what might be imagined as the mathematical analog of a historical novel. That is, although the ultimate rationale will be new results that are proved, the opportunity to tell the story of Abel's theorem in a historical context leading up to the result proved in the paper was something I very much enjoyed.

"*A Poncelet Theorem in Space.*" The Poncelet theorem may be seen as a major step in the development of algebraic geometry. The question posed by Poncelet is to find the conditions on a pair of conics so that there is a polygon inscribed in one and circumscribed about the other. On the face of it, this problem appears to arise from classical projective geometry, having somewhat the same flavor as Pascal's theorem. However, it is by no means an elementary result. By elementary result, for example, I would include the Bezout theorem, in that it is reasonably easy to convince oneself that it is true, even if providing a rigorous argument requires considerable background work. In the case of Poncelet's theorem, the statement is elementary but it masks something quite deep. In fact, since the geometric object given by the pairs consisting of a point on one of the conics together with a line through that point tangent to the other conic is an algebraic curve of genus one, it turns out that the construction underlying Poncelet's theorem is in effect another form of the addition law on an elliptic curve.

I wrote two papers with Joe Harris on the Poncelet theorem. One of these, included in this volume, gives an extension of the Poncelet theorem in the plane to a similar configuration in the space. The other one, whose origins can be traced to the works of Cayley, gives the explicit conditions on a pair of conics that there be a polygon inscribed in the first and circumscribed about the second. These two papers, even though they did not break new ground in modern algebraic geometry, were, I hope, useful for my colleagues and especially for graduate students in that they could serve as examples of how easily stated and interesting geometric questions may lead quickly into deep mathematics.

"*Residues and zero-cycles on algebraic varieties.*" This paper, also written with Joe Harris, again reflected our interest in the interface between complex analysis and algebraic geometry. In complex analysis, one of the most beautiful and useful techniques is the residue theorem. The residue theorem on a compact Riemann surface is a manifestation of duality. This is quite evident in, for example, the geometric proof of the Riemann-Roch theorem in the form of the solution to the Mittag-Leffler problem. This theorem gives the necessary and sufficient conditions on a set of principal parts (or Laurent tails) that there be a global meromorphic function having exactly those singularities. Explicitly, these conditions dictate that the global residue theorem should hold when paired against any holomorphic differential form.

In this paper, we wanted to explore the constraints imposed by duality on configurations of points on an $n$-dimensional algebraic variety. The general residue theorem had appeared in the book that Joe and I wrote and this paper was devoted to applications of the global residue theorem to global geometric problems. One such geometric problem has its roots in the classical result of Castelnuovo, which gives an upper bound on the genus of a nondegenerate curve in projective space in terms of the degree of the curve. In his thesis, Joe had studied analogs of the Castelnuovo bound in higher dimensions. In this paper, we used the global residue theorem to, among other things, derive a bound in terms of its degree on the geometric genus on a general $n$-dimensional projective variety.

The paper on "Abel's Theorem and Webs" was a summary of more extensive work that Chern and I did. A web consists of a set of foliations of the same codimension whose tangent spaces are in general position at each point. The linearization problem asks when can the foliations be simultaneously linearized; i.e., when is there a coordinate system in which the foliations are all given by linear equations. The classical example of a linearizable web in the plane arises by taking an algebraic plane curve and in the dual space taking through each point the set of tangent lines from that point to the curve. Via projective duality this is essentially the same as taking the points of intersection (which are lines in the dual space) with an algebraic curve.

In this situation, Abel's theorem has an implication which is called an abelian relation for the webs. The study of webs was carried out in Hamburg in the 1930s by Blaschke and especially by G. Bol, and Chern became acquainted with this when he was in Hamburg. It was known that the set of abelian equations forms a finite-dimensional complex vector space, and Chern was able to prove that in general the dimension of that vector space is less than or equal to the number referred to above that Castelnuovo had found when he gave a formula for the maximum genus of an algebraic curve of degree d in projective space of dimension r.

Chern retained his interest in the subject, and during the year I spent on sabbatical in Berkeley we fell to discussing how to generalize the classical theorem of Bol, which said that, with one exception, the webs of maximal rank, i.e., those that have the maximum of linearly independent abelian equations, had to come from the construction above. This is very much in the spirit of the converse to Abels theorem that I was discussing in the comments on the paper "Variations on a Theorem of Abel." In any case, Chern and I had a marvelous time working on webs. We did not solve completely the main problem we were interested in—a problem that remains open to this day and is still generating some mathematical interest. Chern and I both enjoyed this interplay between differential geometry and algebraic geometry.

## Algebraic and Differential Geometry

"*Complex Analysis and Algebraic Geometry.*" The relationship between algebraic geometry and differential geometry was also a theme of the paper on "Algebraic Geometry and Local Differential Geometry" that Joe Harris and I wrote. This paper had several origins. The underlying theme is that the local consequences of infinitesimal constraints imposed on an algebraic variety can, in many interesting cases, be "integrated," and from that integration, global conclusions drawn. We wanted to prove what was then a conjecture: that a projective algebraic variety of dimension n, whose space of tangents failed to fill out a variety of dimension 2n, must in fact be singular.

During the time we were working on the paper, this conjecture was given a very beautiful global proof by Fulton and Hansen. In our paper, Joe and I were able to prove the result in low dimensions, showing that as a consequence of the assumption, the algebraic variety had the very special property of being in some sense developable by linear subspaces and singularities necessarily developed along the cusp lines. Thus, in low dimensions we were able to prove both the conjectured result and, perhaps more interestingly, to give a list of the algebraic varieties whose tangent variety is degenerate.

The technique is rooted in the theory of exterior differential systems developed by Elie Cartan. Namely, the geometric assumption that there be some non-generic behavior of the infinitesimal structure of an algebraic variety is expressed by the vanishing of certain differential forms defined on a suitably adapted subbundle of the bundle of frames lying over the variety. The exterior differential system defined by the vanishing of these differential forms typically fails to be involutive, and one must then adjoin to the original equations their exterior derivatives. Continuing to iterate this method leads to successively greater constraints imposed on the algebraic variety, and in cases where the method works, one is able to integrate these constraints and understand the special structure of the variety. We were able to prove a number of global results for example, that a subvariety of an abelian variety whose top Chern class vanishes must itself be ruled by abelian subvarieties. Interesting rigidity results that are very much in the flavor of this paper have recently been given by Robert Bryant in a paper entitled "Rigidity and Quasi-rigidity of Extremal Cycles in Hermitian Symmetric Spaces."

## Loci of Divisors

*"On the Variety of Special Linear Systems on a General Algebraic Curve."* When one studies specific geometric problems on algebraic curves, the special linear systems on the curve almost always play an essential role in the analysis. The subject of special divisors has classical roots, beginning with Clifford's theorem and the Riemann singularity theorem in the last century and continuing through the works of the Italian school. For example, the theorem of Castelnuovo referred to above may be interpreted as giving an upper bound on just how special a very ample linear series may be and in determining those cases where this extremal behavior is achieved. These so-called Castelnuovo curves are very special among the $3g - 3$ dimensional family of all curves.

A natural question is: What is the most special linear series that exists on a general curve? An answer to this was proposed by Castelnuovo and Enriques, and Severi gave a suggested method of proof. Their answer was given in terms of the so-called Brill-Noether matrix, which expresses the differential of the map from the symmetric product of the algebraic curve to its Jacobian. The rank of the Brill-Noether matrix precisely reflects the index of speciality of the divisor in question. From the data consisting of the degree and genus of the curve, one writes down the so-called Brill-Noether number, and the conjectured result was that the Brill-Noether number should be non-negative for all linear series on a general curve.

The existence half, namely the result that on every curve there should exist linear series realizing each non-negative Brill-Noether number, had been proved by George Kempf in "Schubert methods with an application to algebraic curves" (Publ. Mathematisch Centrum (1972)) and by Steve Kleiman and Dan Laksov in "On the existence of special divisors" (Amer. J. Math. 94 (1972), 431-436). Their proof was done by interpreting the Brill-Noether matrix intrinsically as a mapping between vector bundles. They were then able to compute that the homology class of certain degeneracy loci must be non-zero and therefore to conclude that these degeneracy loci were non-empty.

In the late 1970's, Enrico Arbarello and Edoardo Sernesi had been clarifying and extending earlier work of K. Petri. They came to understand that a conjecture of Petri would, using deformation theoretic methods, imply a proof of the Brill-Noether conjecture. At this point Maurizio Cornalba, Joe Harris and I got interested in trying to prove Petri by relating the natural maps that arise to higher order obstructions to deforming curves together with a linear series on the curve. The formalism was extremely suggestive but we were not able to push the argument through. To this day, to me the acid test of any higher order deformation theoretic formalism is that it yields a proof of Petri. In any case, Enrico and Maurizio went on to use infinitesimal methods to prove the Brill-Noether conjecture for 2-dimensional linear series.

Now the underlying issue is that no one has ever "seen" a general curve, and aside from variational methods, there were no techniques for proving that some statement is true for a general curve. One way around this was to degenerate the curve into a special curve, together with certain configurations on that curve, which one might hope to directly analyze. In this way, Joe and I were able to prove the Brill-Noether statement in general, but only later was David Gieseker able to prove the Petri conjecture. The whole method of degeneration was developed systematically by Joe, David Eisenbud, David Gieseker and others during the 1980's and has proved to be an extremely powerful tool for analyzing linear series and other geometric properties of algebraic curves.

"*On the Noether-Lefschetz Theorem and some remarks on codimension-two cycles.*" In the late 19th century, Max Noether had asserted that the only curves on a general surface of degree at least four in projective three-space are complete intersections. Lefschetz had given a proof of this by using monodromy arguments; these are based on the fact that for degree at least four, a general surface has a non-zero holomorphic two-form (in modern terminology, its Hodge structure is of weight two).

The argument Lefschetz gave serves as a prototype for extensive use of monodromy arguments that I and others have used in recent years. These monodromy arguments, together with variational techniques, seem to be among the most effective methods in the explicit calculation of examples (understanding, of course, that the results are only valid for generic members of families). Joe and I were interested in alternative ways of looking at the Noether-Lefschetz theorem, and in particular of estimating the codimension of the locus on which smooth surfaces in three-space acquired non-complete intersection curves. This analysis was the object of this paper. Subsequent to it and the paper that Joe and I wrote on infinitesimal variation of Hodge structure, discussed in the introduction to Part 3, there has been extensive and very interesting work by Mark Green, Claire Voisin and others on the Noether-Lefschetz loci.

# Introductory Comments to Part 3

In the introduction to Part 1, I reminisced about the mathematical environment in the years when I was a Miller Fellow and then assistant professor at Berkeley. I commented that one of my main mathematical influences during this period was Andreotti's proof of the Torelli theorem for algebraic curves. Andreotti's beautiful argument associated to the Hodge structure of the curve a geometric object - in this case the theta divisor - and from that theta divisor showed how to reconstruct the algebraic curve. This to me was a model of what a result in transcendental algebraic geometry might be.

Also during this period, people were discussing the conjecture that the period matrix (it wasn't yet called the Hodge structure) for a polarized $K3$ surface uniquely determines the surface. I'm not entirely sure of its historical origins; perhaps it was André Weil who first formulated the conjecture that the Torelli theorem should hold for $K3$ surfaces (the $K3$ refers to Kummer, Kneser and Kodaira). In any case, this was to me a very appealing problem, and I spent some time unsuccessfully working on it. A few years later, a very beautiful proof was given by Piatetski-Shapiro and Shafarevich.

In any case, I began to wonder why it would not be a general rule that the Hodge structure of an algebraic variety should determine the variety. (Because of the Enriques surface it is not always true, although even this example works if one extends Hodge theory to allow twisted coefficients.) After all, to say that a continuous mapping between two algebraic varieties which induces an isomorphism on homology also induces an isomorphism on their Hodge structures is equivalent to saying that the graph of the homeomorphism is a Hodge class on the product of the two varieties. If the Hodge conjecture is true, for two varieties to have the same Hodge structure will imply that there is at least a virtual birational correspondence between them.

Thinking about the Torelli problem got me to thinking about the spaces of all possible Hodge structures and wondering whether one might not be able to prove Torelli-type theorems by showing that the map from moduli to the space of period matrices was one to one. Here one would of course have to first prove that the differential of the map was injective, and then through some other argument show that the mapping was globally one to one. Indeed, the proof of the global Torelli theorem for $K3$ surfaces given some years later by Piatetski–Shapiro and Shafarevich was of this type.

The proof that the mapping in effect had degree one was done by a careful geometric analysis of $K3$ surfaces containing special configurations of algebraic curves. So in any case it was really the Torelli theorem that got me to thinking about the space of period matrices. As an aside, I should mention that to me

Hodge theory, which is the central and deepest aspect of transcendental algebraic geometry, was always of interest - not in and of itself, but as a means of studying the geometry of algebraic varieties.

As I observed earlier, the period of the 1960's in Berkeley, and then the late 60's and early 70's when I was in Princeton, was an especially active time for transcendental algebraic geometry. In the Berkeley seminars with Herb Clemens, Alan Landman, Wilfried Schmid and others we studied in detail the book by Lefschetz, *L'Analysis situs et la géométrie algébrique*, as well as other classics in transcendental algebraic geometry that I mentioned earlier, the books by Picard and Simart, the papers by Poincaré, and the recently developed deformation theory of Kodaira and Spencer. All these sources pointed naturally toward considering how the Hodge structure of an algebraic variety varies when the variety moves in moduli. The Berkeley seminar was continued during the Princeton years under the title - which I had inherited from Don Spencer - of the "Nothing Seminar." Among those who attended this seminar were Herb Clemens, Spencer Bloch, Steve Zucker, and a superb group of graduate students.

## Periods of Integrals

"*Periods of Integrals on Algebraic Manifolds, I, II, and III.*" The papers "*Periods of Integrals on Algebraic Manifolds, I and II*" were written during the period 1963–65. I would observe that during this time there was very little pressure on young mathematicians to publish. It was generally felt we should fully develop ideas before putting them in print. So even though these papers were prepared during this period, it was some years before they appeared, in part because for me the objective was to find applications of the theory to geometric problems.

In "*Periods of Integrals on Algebraic Manifolds, I*" is a construction of what is now called the classifying space for polarized Hodge structures. As I mentioned in the comments in Part 1, the big surprise here was that the space of Hodge structures is no longer a bounded symmetric domain. It is a homogeneous space of a real semi-simple Lie group factored by the centralizer of a torus and therefore in a natural way carries an invariant complex structure. It has a curious property: when you fiber the classifying space over the associated symmetric space, the fibers are complex subvarieties, but the fibration itself is not holomorphic.

From this property I realized that non-standard phenomena were occurring. One of the main objectives of the paper was to calculate the curvature for the natural homogeneous line bundles over the classifying spaces. The curvature form turned out to be nonsingular, but only positive definite in the directions transverse to the compact subvarieties appearing as fibers over the corresponding symmetric space. Restricted to those fibers, the curvature form was negative definite.

This very much suggested that these classifying spaces had the character of a bounded symmetric domain in the horizontal directions and of a homogenous compact Kähler manifold in the vertical directions. The latter I had encountered in my thesis when I was seeking to generalize the Borel-Weil-Bott theorem. In any case, the curvature computation suggested that the cohomology of powers of the natural homogeneous line bundle would occur only in dimension equal to that of the compact subvarieties. In fact, if one passes to the quotient space of the classifying space by a co-compact Lie group, the then-existing vanishing theorems, together with the Riemann-Roch formula, imply that this is indeed the case. All of this led Bob Langlands to suggest that this theory might very naturally connect with his conjectures concerning the realization of the discrete representations of the semi-simple Lie groups in question. This connection was referred to in the introductory comments to Part 1.

In "*Periods of Integrals on Algebraic Manifolds, II*," is the calculation of how the Hodge structure of an algebraic variety varies infinitesimally when the variety itself moves in moduli. The essential results here are that the Hodge filtration varies holomorphically and, to first order, shifts only by one. The geometric meaning of this latter statement is that the mapping from moduli to the classifying space for Hodge structures is transverse to the fibering of the classifying space over the symmetric space, the so-called "horizontality" or "transversality" condition.

When I first arrived at this infinitesimal period relation, the immediate implication that struck me was that *not all polarized Hodge structures can arise from algebraic geometry.* That is, if one considers all families of algebraic varieties and associates to each variety its Hodge structure, except in a few exceptional cases the space of Hodge structures can never be filled out. (This is even true if one enlarges the Hodge structures to include so-called Hodge motives.)

This of course means that as far as Hodge theory goes, life in higher dimensions and co-dimensions is going to be radically different from the classical situation for algebraic curves and abelian varieties, and the situation of $K3$ surfaces which I referred to earlier. In this paper, the intrinsic form of the differential of the period mapping was given. It is obtained by composing the Kodaira-Spencer mapping with a cup product in cohomology. This enables one to calculate in examples just how the Hodge structure varies infinitesimally with moduli, and in this paper a number of calculations of this sort were given.

This paper also considered the intermediate Jacobian and the Abel-Jacobi mapping from cycles homologous to zero into the intermediate Jacobian. Intermediate Jacobians and the Abel-Jacobi map had been discussed by Hodge in the early 1950s and had been introduced systematically some years later by André Weil. At the time we were having the seminar on these topics in Berkeley, Weil's papers had disappeared from the library, so I had to try to reconstruct them from memory. When I tried to do this, I assumed that Weil's definition would have been such that the intermediate Jacobian would vary holomorphically with the parameters of the variety. It turned out later that in fact this was not the case; Weil's intermediate Jacobians have the property of being a polarized abelian variety but do not vary holomorphically. The intermediate Jacobians that were discussed in this paper have the property of varying holomorphically with moduli, but the curvature form of the natural line bundle has mixed signature. It is, however, positive on the image of algebraic families of cycles, a situation quite similar to the variation of Hodge structure of algebraic varieties. In general, mappings to Hodge structures or to intermediate Jacobian could never be surjective. But on the image there is a line bundle with positive polarization so that one has a *relative theory.* In other words, what one had come to expect classically would be present - but only on the image of families of algebraic varieties.

In 1968, a marvelous conference was held at the Tata Institute of Fundamental Research in Bombay. It was the first time I had been to a major meeting in algebraic geometry (it was also the first time I had been outside the USA). This one was particularly interesting because several Russian mathematicians, including Manin, were there, and I also met Grothendieck in person.

He and I talked at some length, and he explained to me Deligne's theory of mixed Hodge structures, which was just then being developed. I had actually been in correspondence with him about a year before concerning the monodromy theorem and the theorem on regular singular points. He had seen an announcement of these results and had written to me expressing some interest. Prompted, I believe, by the statement of the monodromy theorem, he had found an arithmetic proof using $\ell$-adic cohomology that the eigenvalues of the monodromy matrix were roots of unity, but he had not found a proof of the result about the index of unipotency.

In any case, in Bombay he mentioned to me a very interesting conjecture, which later became the theorem on the fixed part. Grothendieck's conjecture was that, given a global family of algebraic varieties whose general member was smooth, and given an invariant cohomology class, the Hodge $(p, q)$ components are also invariant. This would be a vast generalization of the classic theorem of the fixed part for the Jacobian variety of a family of curves on a surface, the latter stating that the invariant part of the cohomology corresponds exactly to the restriction of the Picard variety of the surface mapping to the Jacobian of the curve.

I was of course very interested in Grothendieck's statement, and while I was in Bombay I gave a proof for the case of the second cohomology of surfaces whose monodromy around each singular fiber is a finite group. The proof was an elementary argument based on the curvature properties of the Hodge bundle. Grothendieck was very interested in this, and we had at that point several discussions about variation of Hodge structure and the infinitesimal period relation. He even explained to me his early thinking on what was later to become his conjectural theory of motives.

To say the least, it was a most interesting meeting. When I went back to the States, I tried to give general proofs of the statement Grothendieck had made and was able to do these for variations of Hodge structure whose singular fibers each have a finite monodromy. These arguments, which are given in "*Periods III*," are based on rather subtle curvature properties of the Hodge bundle. Basically, the alternation of signs, together with the infinitesimal period relation, exactly make the argument work out right. These arguments were later extended by Schmid to an arbitrary of variation of Hodge structure using his results on the asymptotic behavior of a variation of Hodge structure around a singular point.

Actually, Spencer Bloch and I wrote a sequel to this paper, which exists only in preprint form. There we proved a theorem concerning the global behavior of normal functions. Namely, if one is given a smooth, even-dimensional algebraic variety and takes a Lefschetz pencil, then associated to each normal function is its fundamental class, which is a Hodge class that is an integral class of type $(n, n)$ on the ambient manifold. The result we proved was the converse; that is, given a Hodge class, there exists a normal function whose fundamental class is the given Hodge class. Moreover, the normal function is itself unique up to a global normal function induced from the intermediate Jacobian of the ambient manifold.

Of course, what we had in mind was to use some inductive procedure to produce algebraic cycles. The fact that the condition to be a Hodge class exactly allowed one to construct a global analytic object at first gave us some hope. The problem is, of course, that the Jacobian inversion theorem completely breaks down in higher codimension, so the classic argument used by Lefschetz was not applicable. The result Spencer and I proved was then greatly extended and put in much more satisfactory form by Steve Zucker. As far as I know, this is the only global analytic and, unfortunately, not yet algebraic object that has been associated to a Hodge class. Below is a further comment on this issue.

"*On the Periods of Certain Rational Integrals, I and II.*" This set of papers actually had two origins. One was the desire to give a proof of the following fact. When one realizes the primitive cohomology of a smooth hypersurface in projective space by residues, the filtration by order of pole of meromorphic differential forms of top degree with singularities along the hypersurface exactly corresponds to the Hodge filtration in the primitive cohomology of the hypersurface. It turns out that the cohomology of the complement can be described quite nicely using the Jacobian ring of the defining equation of the hypersurface. This gives an efficient means of proving things like the local Torelli theorem for hypersurfaces.

A second motivation for this paper was to study the global behavior of the normal function in a Lefschetz pencil of hyperplane sections of a smooth hypersurface in projective space. When the smooth hypersurface in projective space has nontrivial primitive algebraic cycles, i.e., algebraic cycles not induced by intersection with cycles in the ambient projective space, then for global topological reasons the corresponding normal function is nontrivial. On the other hand, by choosing the Lefschetz pencil generically, one can arrange by a Noether-Lefschetz-type argument that the odd cohomology of the generic member of the Lefschetz pencil has no nontrivial sub-Hodge structures. This then produces examples of algebraic cycles that are homologous but not algebraically equivalent to zero.

Later, Herb Clemens went on to study the situation in more depth and prove that not only did such cycles exist, but in fact the countable group of algebraic cycles homologous to zero but not algebraically equivalent to zero may not be even finitely generated. This result, together with Mumford's result on the infinite dimensionality of the Chow variety of surface with non-zero geometric genus, showed convincingly that the situation for algebraic cycles in higher codimension is radically different from the classical case of divisors. This remains an extremely active area of research through the present time.

One interesting note about this paper is that although results were proved only for hypersurfaces in projective space, it was obvious from the proof that similar results are valid for sufficiently ample divisors in an arbitrary smooth algebraic variety. However, the explicit polynomial calculations in the case of projective space were especially appealing to me, so I chose to simply write up the arguments in this case and indicate that similar results are true in greater generality. This brought howls of complaint from the referee, who wrote that I was practicing reactionary mathematics. Fortunately, the Annals editors took the referees comments with a grain of salt and published the paper in the form in which I had written it. To this day, I still feel that in many cases it is more instructive to give an argument in a particularly interesting special case and indicate that the general case follows by similar arguments, rather than always try to give the proof in the greatest generality.

"*Periods of Integrals on Algebraic Manifolds: Summary of main results and discussion of open problems.*" This was an expository paper written during a period of intense activity in transcendental algebraic geometry. It was also written just after I had spent a very enjoyable period at IHES, my first introduction to France. In the paper, I listed a number of open problems and, following the lead of the Michelin guides to restaurants, put one, two, or three stars by these problems to indicate accessibility by current techniques (1), possibly accessible by current techniques (2), or seemed to require some totally new idea (3). One of the three-star problems was the Torelli theorem for polarized K3 surfaces. This expository paper had just been written when the Piatetski-Shapiro-Shafarevich proof of this result appeared.

## Variations of Hodge Structures

"*Infinitesimal Variation of Hodge Structure (I), (II), and (III).*" As noted earlier, one of the most striking aspects of Hodge theory in higher dimensions is that a general Hodge structure is not realizable geometrically. Therefore, there are conditions on a Hodge structure that arise from algebraic geometry, conditions which, as far as I know, are still not described even by conjecture. This contrasts with the case of principally polarized abelian varieties; all of these arise in geometry, and there is a beautiful characterization of those that are Jacobians of curves in terms of the singularity structure of the theta divisor.

In any case, since I certainly have no idea what sort of invariant might be attached to a Hodge structure in order to tell if it arises from geometry, a reasonably natural alternative is to consider a first-order variation of Hodge structure. (In fact one could, and probably should, consider higher-order variations as well.)

In the case of algebraic curves, there already exists a rich structure on the infinitesimal variation of its Hodge structure: namely, this is essentially described by the quadrics through the canonical curve. Additionally, earlier works such as that in "*Periods of Certain Rational Integrals I,*" indicated that in higher dimensions there might similarly be some intimate link between geometry and the first order variation of the Hodge structure. Exploring what this should mean, and in particular determining systematically what the invariants of the first order variation of Hodge structure are and how they might be interpreted geometrically in special cases, is the theme of these three papers.

In an earlier paper not included in these volumes, Jim Carlson and I used infinitesimal variation of Hodge structure methods to prove a global Torelli theorem for a very special class of smooth hypersurfaces in projective space. Later, by an extremely ingenious argument, Ron Donagi extended this result to almost all hypersurfaces. I believe that Donagi's theorem and the subsequent work by Mark Green, Claire Voisin and others shows convincingly that the infinitesimal variation of Hodge structure may contain rich geometric information.

I would like to comment briefly on the third in this series of papers. As I noted earlier, the one global analytic invariant that is attached to a Hodge class is the essentially unique normal function on any family of hypersurface sections of a fixed degree. The problem in trying to use this to construct cycles is that we don't know that on each hypersurface section the point in the corresponding intermediate Jacobian is invertible. The question of the image of the Abel-Jacobi map is similar to, and probably as difficult as, that of determining which Hodge structures arise from geometry. This suggests that one might seek to determine what geometric information there is in the first order variation of the normal function.

In the third paper in the series, I attempted to define the first order invariant of the normal function and to illustrate by some fairly simple examples how it might be interpreted geometrically. Later, the correct definition of the infinitesimal invariant was given by Mark Green. Also, Mark, Claire Voisin, G. Ceresa, A. Collino and others were able to use that invariant very effectively in applications to various questions in geometry. A far-reaching extension of the type of results obtained by infinitesimal and variational methods has been given by M. Nori in his paper, "Algebraic cycles and Hodge-theoretic connectivity."

As for using the infinitesimal invariant of the normal function associated to a Hodge class, it always seemed to me that given a primitive algebraic cycle on an even-dimensional smooth algebraic variety, if we consider the corresponding normal function on the family of hypersurface sections, the equations of that cycle should in some way be related to the infinitesimal invariant of the normal function. Of course the cycle is not unique, so the relationship cannot be simple-minded. In any case, this line of investigation was something that remained elusive when I tried it in the late 70s and early 80s.

In closing, I would like to note an interesting analogous question: Does a global variation of Hodge structure, all of whose irreducible factors are nontrivial, necessarily arise from geometry?

The above reflections on the development of variation of Hodge structure essentially cover the period up to the latter half of the 1970s, a period that has been referred to as "neoclassical Hodge theory." A good account of developments since the 1970s is given in the paper "An Overview of Recent Advances in Hodge Theory" by Jean-Luc Brylinski and Steven Zucker, a systematic and detailed account of the vast extensions of the theory. In the introduction to the paper, they note that "the algebro-geometric aspects [of neoclassical Hodge theory] never achieved full success," referring to the fact that the major problems involving cycles were not accessible to the theory of variation of Hodge structures and mixed Hodge structures. As best I can tell, that situation remains true to this day. That is, although the formal aspects of Hodge theory have been quite satisfactorily extended to great generality, the fundamental issues of relating Hodge theory to geometry remain.

# Introductory Comments to Part 4

Since my days as a graduate student I have been interested in differential equations. What appealed to me was not so much the finer analytic aspects - although these are both interesting and necessary - as much as trying as far as possible to understand the use of differential equations to study geometric or physical phenomena. For this effort, Cartan's theory of exterior differential systems provides a natural and appealing framework for constructing the invariants associated to classes of differential equations in an intrinsic geometric fashion. In fact Cartan's *method of equivalence*, as it is called, applies to provide a "quasi-algorithm" for determining a complete set of invariants for any G-structure.

By "quasi-algorithm" I mean an algorithmic procedure which, at certain stages, requires choices. Thus, for example, one may be led to two different intrinsic Cartan connections by making different choices at some stage in the algorithm. This gives the process the flavor of an art as much as a science. For example, the method of equivalence, which seems to have been in the back of Chern's mind throughout his mathematical work, is one of the most powerful computational/geometric tools I know of.

## Moving Frames and Differential Geometry

"*On Cartan's Method of Lie Groups and Moving Frames as Applied to Uniqueness and Existence Questions in Differential Geometry.* "This paper is mainly expository. The method of moving frames of Cartan, which has its roots in the geometric works of the 19th century (especially those of Darboux), has always been of considerable appeal to me. I first encountered it in studying the papers of Chern; as is well known, throughout his mathematical life, he has systematically used moving frames with great effectiveness. One may view moving frames as one manifestation of the Cartan philosophy that whenever there are natural parameters in a geometric situation, these parameters should be included as part of the structure and their invariants should enter into its study.

My goal in this paper was to provide an exposition of some of the underlying principles of the method and to illustrate those principles in specific geometric examples. Some of the geometric examples were new. In particular, the application of the moving frame method to give an extension and new proof of Calabi's rigidity theorem (by understanding the relation between the rigidity result and the second main theorem as formulated by Hermann Weyl) was highly satisfying.

Subsequent to this paper, the issue of systematically determining the invariants for higher order contact of submanifolds of homogeneous spaces was taken up by Gary Jensen in his monograph, *Higher Order Contact of Submanifolds of*

*Homogeneous Spaces*, and by Mark Green in a paper entitled "The Moving Frame, Differential Invariants and Rigidity Theorems for Curves in Homogeneous Spaces." These works provide a fairly definitive set of results for the general theory.

## Differential Systems and Hodge Structures

"*Some Observations on the Infinitesimal Period Relations for Regular Threefolds with Trivial Canonical Bundle.*" As discussed in some of the introductory comments for Part 3, a variation of Hodge structure arising from geometry is subject to infinitesimal constraints. These constraints are naturally expressed as an exterior differential system and the system has been studied extensively by a number of people, especially Jim Carlson and Domingo Toledo.

In the case of the third cohomology of a smooth algebraic variety of dimension three whose Hodge number $h^{3,0}$ is one, these constraints have a very nice form from the perspective of the theory of exterior differential systems. For this special case, Robert Bryant and I worked out systematically how one could "deprolong" or "integrate" these infinitesimal period relations to obtain something like a local normal form for a variation of Hodge structure.

Much to our surprise, this paper turned out to hold some interest for physicists. In fact, the development at the interface between theoretical physics and complex algebraic geometry, including the role of variation of Hodge structure in some of the early considerations of mirror symmetry, has been one of the most interesting mathematical events I have observed. My Duke colleague Dave Morrison, with his strong background in complex algebraic geometry and his knowledge of the language and thinking of theoretical physicists, has played an especially important role bridging these two communities.

"*Reduction for Constrained Variational Problems and* $\int \frac{1}{2}\kappa^2 ds$." During the late 1970s and early 1980s, I took pleasure in teaching the undergraduate differential equations and mechanics courses at Harvard. The wonderful book *Mathematical Methods of Classical Mechanics*, by Arnold, had just came out in English translation and provided an excellent primary source.

I had become particularly interested in functionals defined on integral curves of an exterior differential system, which essentially amounts to the study of constrained variational problems. When I was in China, I gave a course on how one might approach the calculus of variations from the perspective of exterior differential systems and wrote a book on this subject. Of course the idea of using Cartan's method in the calculus of variations was not new; it went back to Cartan himself and had been used by others since. These works had, however, been concerned mainly with the foundations. What especially attracted me was not so much a desire to provide a conceptual framework but the challenge of constructing a computational framework for the derivation of the Euler-Lagrange systems, the computation of the various conservation laws, and the integration of the equations, when possible. The book referred to above is mainly a collection of exercises to illustrate the exterior differential system approach.

One of the functionals I studied in some detail is the integral of the square of the curvature along curves in the plane and the extension of this and related functionals to higher dimensions. I had come across the $\kappa^2$ functional in the works of Blaschke, and there, by a rather ingenious computation, he had shown that its Euler-Lagrange equations are in a real sense integrable.

Formulating the question in the context of exterior differential systems led to a derivation of Blaschke's result that was both conceptually and computationally simple and elegant. It later turned out that this functional dates back at least to Euler and had been the subject of investigation off and on since that time. In any case, Robert Bryant and I became interested in studying its global properties. This can be effectively done by utilizing the conservation laws arising from the symmetries in the problem, and in order to do this in the natural setting, one needed to extend the Marsden-Weinstein method of reduction of the Euler-Lagrange systems arising from variational problems with symmetries to the constrained case. This we did in the paper, and were also able to use this technique to globally determine the integral curves of the $\int \kappa^2$ functional.

## Integrability

"*Linearizing Flows and a Cohomological Interpretation of Lax Equations.*" The early 1980's was a period of intensive work generating solutions to various special differential equations through algebro-geometric methods. Although there had been work on the Korteweg de Vries equations in this country in the 1960's, the integration of this and related equations through algebro-geometric means was a development that came mainly from the Russian school in the 1970's. This extremely interesting subject was one that was periodically presented at the Tuesday Harvard algebraic geometry seminar, and two papers that I found particularly remarkable were "Completely Integrable Systems, Euclidean Lie Algebras, and Curves" and "Linearization of Hamiltonian Systems, Jacobi Varieties, and Representation Theory" by Adler and von Moebeke. In these papers, the authors used methods from the theory of compact Riemann surfaces to integrate essentially all the classical integrable equations. (Only later would they do the Kowaleska equation by these methods.)

The underlying mathematical object was a so-called Lax equation $A_t = [A, B]$. Here, A and B are in a Lie algebra; the fact that B is only determined up to the centralizer of A means there is some evident non-uniqueness in the Lax equations. To an algebraic geometer, this non-uniqueness suggests that a cohomology group may be lurking somewhere. In fact this turns out to be the case, and the work of Adler and von Moebeke could be extended and interpreted in intrinsic cohomological form. That was the objective of this paper.

## The Characteristic Variety and Its Geometry

"*Some Aspects of Exterior Differential Systems.*" Starting around 1980, Chern, Robert Bryant, Robby Gardner, Hubert Goldschmidt and I began working on a book about exterior differential systems. Our objective was to lay out the foundational aspects, including the cohomological methods of the subject dating back to the work of Don Spencer and others in the 1960's, and especially to illustrate these foundational aspects through examples. This project went on for many years and was only completed in the late 1980's. The current paper gave an overview of some of the basic aspects of exterior differential systems, especially those dealing with the rather elusive notion of involution and the central concept of the characteristic variety, which is in some sense the principal invariant of an exterior differential system.

"*Characteristic Cohomology of Differential Systems (I) and (II).*" During the 1980's, Robert Bryant and I were working on a number of aspects of exterior differential systems. One set of papers, not in these volumes, concerned the rigidity and existence of isometric imbeddings. Together with Deane Yang, we were able to settle a longstanding problem, namely the existence of local $C^\infty$ isometric imbeddings of a three-dimensional Riemannian manifold in six-dimensional Euclidean space. This was a well-known problem, which had been worked on by Nirenberg and Chern, among others. In the setting of exterior differential systems, one was naturally led to the geometry of this problem, especially to the characteristic variety referred to above.

What distinguishes the three-dimensional case from the classical embedding of surfaces in Euclidean three-space is that in the three-dimensional case, no matter what the sign of the curvature, there are always characteristics, which implies that the linearized equations are of principal type rather than simply elliptic or hyperbolic or parabolic. In fact, in the three-dimensional case, the characteristic variety is generically a non-singular real cubic curve, and in particular is always non-empty. In any case, understanding the geometry of this system allowed us to set up an iterative Nash-Moser scheme adapted to the geometry of the situation and led to the solution of the isometric embedding problem in this case.

In the late 1980's, Robert Bryant and I continued our collaboration in the two papers on characteristic cohomology. For us, their origin was our interest in the geometry of conservation laws and, in particular, some very interesting work by Vinogradov and his colleagues during the 1970's. Drawing in part on earlier works of Don Spencer, Vinogradov had introduced what he calls the $\mathcal{C}$ spectral sequence, and from it found several interesting results concerning differential equations, especially the so-called two-line theorem.

The conservation laws sit in the characteristic cohomology of an exterior differential system, and in the first of these two papers Robert and I set out to systematically establish the foundational aspects of the story. In particular, we were able to generalize Vinogradov's two-line theorem to the involutive overdetermined case, which gave the so-called $r$-line theorem, where $r$ measures the degree of overdeterminedness of the exterior differential system.

In the second paper, we used the general results to analyze in complete detail the conservation laws for second order parabolic equations for one unknown function in two independent variables. The structure here, which came out of a detailed use of the equivalence method, was perhaps the first case where the set of conservation laws of a class of differential equations was completely determined.

One interesting problem that surfaced in this study is the possible extension of a result of Thom. In the 1950's, Thom had proved that, on a contact manifold, the homology computed from chains consisting of integral manifolds of the contact system is the same, up to the middle dimension, as ordinary homology. In the paper "*Characteristic Cohomology of Differential Systems I,*" we came across a very natural possible generalization of this result of Thom which would state a similar result, but where the cohomology is now computed from chains consisting of integral manifolds of an involutive exterior differential system and, most interestingly, where the expected dimension of the isomorphism would be up to what is called the Cartan integer. To us, this is an attractive, albeit probably very difficult, problem in the general program of Gromov.

"*Hyperbolic exterior differential systems and their conservation laws I and II.*" These two papers, written with Robert Bryant and Lucas Hsu, study in detail the exterior differential system given on a four-dimensional manifold by a pair of differential forms of degree two. These systems turn out again to have an extremely rich structure. In the equivalence method, Cartan identified what might be called "flat models," which served to guide the equivalence method. The flat model appears as the end point of one branch of the equivalence method where all of the invariants are constant, so that the intrinsic pseudoconnection is the Maurer-Cartan form of a Lie group. In the setting of a pair of two-forms on a four-dimensional manifold, the equivalence method turns out to have several branches, and the end points of these various branches lead to several different "almost flat models" or, if one likes, EDSs with maximal symmetry. The determination of these and of the conservation laws of the different classes was the objective of this set of papers.

# MUMFORD-TATE GROUPS

PHILLIP GRIFFITHS



## I. Definitions and basic properties

Mumford-Tate groups are the basic symmetry groups of Hodge theory. They lie at the interface of Hodge theory with

- algebraic geometry (variations of Hodge structure)
- representation theory of real, non-compact semi-simple Lie groups (discrete series)
- arithmetic algebraic geometry (theory of Shimura varieties)

Much of the use of Mumford-Tate groups has been in the classical case of weight one polarized Hodge structures (theory of Shimura varieties). Since this topic will be covered elsewhere in the summer school, this set of lectures will emphasize the non-classical, higher weight case which is much less explored.

The first four of these lectures will be largely self-contained, making use of the lectures of Lê Dung Tráng, Cattani, Tu, El Zein, Voisin, and Carlson at this summer school. Lectures V and VI will use some of the basic structure theory of semi-simple complex and real Lie groups and Lie algebras, and will be largely without proofs. At the end we indicate, in very general terms, where some potentially interesting directions for further work might lie.

A. Notations
B. Polarized Hodge structures
C. Mumford-Tate groups
D. Basic properties

### A. Notations.

- $V$ is a vector space over $\mathbb{Q}$; $V_{\mathbb{R}} = V \otimes_{\mathbb{Q}} \mathbb{R}$ etc.;
- $\check{V}$ is the dual vector space;
- $Q : V \otimes V \to \mathbb{Q}$ is a non-degenerate bilinear form with $Q(v,w) = (-1)^n Q(w,v)$ where $n$ will be the weight of the Hodge structure;
- $\mathrm{GL}(V)$, $\mathrm{GL}(V_{\mathbb{R}})$ etc. are the general linear groups;
- $G = \mathrm{Aut}(V,Q)$ is the subgroup of $\mathrm{GL}(V)$ preserving $Q$.

Lectures based in large part on joint work with Mark Green and Matt Kerr.

**Definition.** A $\mathbb{Q}$-*algebraic group* will be a subgroup of a $\mathrm{GL}(V)$ defined by polynomial equations with $\mathbb{Q}$-coefficients.[1]

**Example.** $G$ is a $\mathbb{Q}$-algebraic group.

A connected, commutative $\mathbb{Q}$-algebraic group is called an *algebraic torus.*

**Examples.** $\mathbb{Q}^*$ is an algebraic torus.

$$T = \left\{ \begin{pmatrix} a & b \\ -b & a \end{pmatrix} : a^2 + b^2 = 1, \quad a, b \in \mathbb{Q} \right\}$$

is an algebraic torus.

We may also define $\mathbb{R}$-algebraic groups.

**Example.** $\mathrm{Res}_{\mathbb{C}/\mathbb{R}}\, \mathbb{C}^*$, the restriction of scalars from $\mathbb{C}$ to $\mathbb{R}$ of $\mathbb{C}^*$, is an $\mathbb{R}$-algebraic group. Throughout these lectures it will be denoted simply by $\mathbb{S}$. It is the semi-direct product of its subgroups $\mathbb{R}^*$ (*scaling*) and $S^1 = \{z \in \mathbb{S} : |z| = 1\}$.

For a $\mathbb{Q}$-algebraic group $M$, $M(\mathbb{Q})$, $M(\mathbb{R})$, $M(\mathbb{C})$ will denote the set of its $\mathbb{Q}$, $\mathbb{R}$, $\mathbb{C}$-valued points.

- For a subgroup $\Gamma \subset M$, $\overline{\Gamma}^{\mathbb{Q}}$ will denote its $\mathbb{Q}$-Zariski closure.

**Definition.** The *Lie algebra* $\mathfrak{g}$ of $G$ is the tangent space to $G$ at the identity.

**Examples.** $\mathrm{gl}(V) = \mathrm{End}(V) \cong V \otimes \check{V}$ with the bracket

$$[X, Y](v) = X(Y(v)) - Y(X(v)), \qquad X, Y \in \mathrm{gl}(V), v \in V.$$

For $G$ as above

$$\mathfrak{g} = \{X \in \mathrm{gl}(V) : Q(Xu, v) + Q(u, Xv) = 0, \quad u, v \in V\}.$$

B. **Polarized Hodge structures.**

**Definition** (i). A *Hodge structure of weight* $n$ is given by a *Hodge decomposition*

$$\begin{cases} V_{\mathbb{C}} = \bigoplus_{p+q=n} V^{p,q} \\ \overline{V}^{p,q} = V^{q,p}. \end{cases}$$

**Definition** (ii). A *Hodge structure of weight* $n$ is given by a *Hodge filtration*

$$\begin{cases} F^n \subset F^{n-1} \subset \cdots \subset F^0 = V_{\mathbb{C}} \\ F^p \oplus \overline{F}^{n-p+1} \xrightarrow{\sim} V_{\mathbb{C}}, \qquad p = n, \ldots, 1. \end{cases}$$

**Definition** (iii). A *Hodge structure of weight* $n$ is given by a non-constant homomorphism of $\mathbb{R}$-algebraic groups

$$\widetilde{\varphi} : \mathbb{S} \to \mathrm{GL}(V_{\mathbb{R}})$$

such that over $\mathbb{C}$ the eigenspace decomposition of $\widetilde{\varphi} : \mathbb{S} \to \mathrm{GL}(V_{\mathbb{C}})$ is

$$\begin{cases} V_{\mathbb{C}} = \bigoplus_{p+q=n} V^{p,q} \\ \widetilde{\varphi}(z)v = z^p \bar{z}^q v, \qquad v \in V^{p,q}. \end{cases}$$

[1]The basic reference is Borel, *Linear algebraic groups.*

*Equivalence of definitions.*

$$\text{(i)} \Rightarrow \text{(ii)} \qquad F^p = \bigoplus_{p' \geqq p} V^{p',q'}$$

$$\text{(ii)} \Rightarrow \text{(i)} \qquad V^{p,q} = F^p \cap \overline{F}^q$$

$$\text{(i)} \Leftrightarrow \text{(iii)} \qquad V^{p,q} = z^p \bar{z}^q \text{ eigenspace for } \widetilde{\varphi}(\mathbb{S}).$$

The *Weil operator* $C \in \mathrm{GL}(V_\mathbb{R})$ is defined by

$$C = \widetilde{\varphi}(i) \qquad (= i^{p-q} \text{ on } V^{p,q}).$$

We denote the Hodge structure by $(V, \widetilde{\varphi})$, or sometimes by just $V_{\widetilde{\varphi}}$.

**Definition.** A *polarized Hodge structure* $(V, Q, \varphi)$ of weight $n$ is given by

$$\varphi : S^1 \to G(\mathbb{R})$$

such that (i) the characters of $\varphi$ lie in $[-n, n]$, and (ii) the *Hodge-Riemann bilinear relations*

$$\begin{cases} Q(F^p, F^{n-p+1}) = 0 \\ Q(v, C\overline{v}) > 0, \qquad 0 \neq v \in V_\mathbb{C} \end{cases}$$

are satisfied.

These relations are equivalent to

$$\begin{cases} Q(V^{p,q}, V^{p',q'}) = 0, \qquad p' \neq n-p \\ i^{p-q} Q(v, \overline{v}) > 0, \qquad 0 \neq v \in V^{p,q}. \end{cases}$$

**Remark.** The action of $\varphi(S^1)$ on $V_\mathbb{C}$ decomposes into eigenspaces $V^{p,q} = \overline{V}^{q,p}$ where $\varphi(z) = z^{p-q}$ $(= z^p \bar{z}^q)$ on $V^{p,q}$. The character group

$$X(S^1) \cong \mathbb{Z}$$

where $m \in \mathbb{Z}$ gives the character $\chi_m(z) = z^m$.

**Note.** The difference between $\widetilde{\varphi}$ and $\varphi$ is that the scaling action of $\widetilde{\varphi}$ gives the weight

$$\widetilde{\varphi}(r) = r^n \, \mathrm{id}_V \, .$$

We may define a Hodge structure, without specifying the weight, to be given by

$$\widetilde{\varphi} : \mathbb{S} \to \mathrm{GL}(V_\mathbb{R})$$

with the following condition: On $\mathbb{Q}^* \subset \mathbb{S}$ the action of $\widetilde{\varphi}(\mathbb{Q}^*)$ decomposes over $\mathbb{Q}$ into weight spaces

$$V = \bigoplus_n V^{(n)}$$

and $(V^{(n)}, \widetilde{\varphi}\,|_{V^{(n)}})$ is a Hodge structure of weight $n$. Hodge structures of weight $n$ are sometimes referred to as *pure Hodge structures*. Polarized Hodge structures are always assumed to be pure.

**Basic example.** $V = \mathbb{Q}^2 =$ column vectors $v = \binom{v_1}{v_2}$,

$$\begin{cases} Q(u, v) = {}^t v Q u, \qquad Q = \begin{pmatrix} 0 & 1 \\ -1 & 0 \end{pmatrix} \\ V_\mathbb{C} = V^{1,0} \oplus V^{0,1} \\ V^{1,0} = \mathbb{C} v_{1,0}, \qquad v_{1,0} = \binom{v_1}{v_2}. \end{cases}$$

The condition

$$iQ(v_{1,0}, \overline{v}_{1,0}) > 0$$

is

$$(-i)(v_1\overline{v}_2 - \overline{v}_1 v_2) > 0\,.$$

This implies $v_2 \neq 0$ and when we scale to have

$$v_{1,0} = \begin{pmatrix} \tau \\ 1 \end{pmatrix}$$

the above condition becomes $\operatorname{Im}\tau > 0$. We set

$$v_\tau = \frac{1}{\sqrt{2\operatorname{Im}\tau}}\begin{pmatrix} \tau \\ 1 \end{pmatrix}$$
$$\Rightarrow iQ(v_\tau, \overline{v}_\tau) = 1\,.$$

Then $v_\tau, \overline{v}_\tau$ give a *Hodge basis.*

Denoting by $\varphi_\tau$ the circle giving the Hodge structure, in terms of the Hodge basis $\varphi_\tau$ has the matrix

$$\varphi_\tau(z) = \begin{pmatrix} z & 0 \\ 0 & z^{-1} \end{pmatrix}, \qquad z^{-1} = \bar{z}\,.$$

**Example.** The Tate structure $\mathbb{Q}(n) := 2\pi i\mathbb{Q} \subset \mathbb{C}$ with Hodge structure of pure type $(-n, -n)$.

Hodge structures and polarized Hodge structures admit the usual operations of linear algebra: $\oplus$, Hom, $\otimes$, $\wedge$ and Sym. A subspace $W \subset V$ is a *sub-Hodge structure* of $(V, \widetilde{\varphi})$, not assumed to be pure, if

$$\varphi(\mathbb{S})W_{\mathbb{R}} \subseteq W_{\mathbb{R}}\,.$$

In the polarized case $(V, Q, \varphi)$ we have that $Q\mid_W$ is non-singular and

$$V = W \oplus W^\perp$$

is a direct sum of polarized Hodge structures. Polarized Hodge structures form a semi-smiple abelian category.

Because of

**Hodge's Theorem.** *The cohomology group $H^n(X, \mathbb{Q})$ of a smooth, projective algebraic variety has a polarized Hodge structure of weight $n$.*

polarized Hodge structures are the basic objects of Hodge theory. The basic example is the polarized Hodge structure on $H^1(E_\tau, \mathbb{Q})$ where

$$E_\tau = \mathbb{C}/\mathbb{Z} + \mathbb{Z}\tau\,.$$

**Note.** In practice we will usually be given a lattice $V_{\mathbb{Z}}$ with $V = V_{\mathbb{Z}} \otimes_{\mathbb{Z}} \mathbb{Q}$. For the theory of Mumford-Tate groups it is better to work over $\mathbb{Q}$ than over $\mathbb{Z}$.

For a Hodge structure $(V, \widetilde{\varphi})$ of even weight $n = 2p$, we define the space of *Hodge vectors* by

$$\operatorname{Hg}_{\widetilde{\varphi}} = V \cap V^{p,q}\,.$$

These are the rational vectors of Hodge type $(p, p)$. The famous Hodge conjecture is that the Hodge vectors in $H^{2p}(X, \mathbb{Q})$ are represented by the fundamental classes of algebraic cycles.

C. **Mumford-Tate groups.**

**Definition.** The *Mumford-Tate group* $M_{\widetilde{\varphi}}$ associated to a Hodge structure $(V, \widetilde{\varphi})$ is the smallest $\mathbb{Q}$-algebraic subgroup of $\mathrm{GL}(V)$ such that

$$\widetilde{\varphi}(\mathbb{S}) \subset M_{\widetilde{\varphi}}(\mathbb{R}).$$

It is not required that $V_{\widetilde{\varphi}}$ be of pure weight. If $(V_{\widetilde{\varphi}}) = (V_{1,\widetilde{\varphi}_1}) \oplus (V_{2,\widetilde{\varphi}_2})$ is a direct sum, then $M_{\widetilde{\varphi}} \subseteq M_{\widetilde{\varphi}_1} \times M_{\widetilde{\varphi}_2}$ and the projection onto each factor is surjective but equality may not hold (e.g., if $V_{\widetilde{\varphi}_1} = V_{\widetilde{\varphi}_2}$ then $M_{\widetilde{\varphi}}$ is the diagonal subgroup of the product).

**Definition.** The *Mumford-Tate group* $M_{\varphi}$ associated to a polarized Hodge structure $(V, Q, \varphi)$ is the smallest $\mathbb{Q}$-algebraic subgroup of $G$ such that

$$\varphi(S^1) \subset M_{\varphi}(\mathbb{R}).$$

**Note.** In the literature $M_{\widetilde{\varphi}}$ is usually called the Mumford-Tate group and $M_{\varphi}$ the special Mumford-Tate group or Hodge group. Because of the centrality in these lectures of polarized Hodge structures, we shall refer to both as Mumford-Tate groups and let the subscripts $\varphi$ and $\widetilde{\varphi}$ specify to which we are referring. For pure Hodge structures we shall see that $M_{\widetilde{\varphi}}$ and $M_{\varphi}$ differ only by the scaling action; $M_{\widetilde{\varphi}}$ will be the semi-direct product of $M_{\varphi}$ and $\mathbb{G}_{m,\mathbb{Q}}$.

**Basic example (continued).** For the above example, we first have $G = \mathrm{SL}_2(\mathbb{Q})$. The change of basis from the Hodge basis to the $\mathbb{Q}$-basis is

$$A_{\tau} = \frac{1}{\sqrt{2 \operatorname{Im} \tau}} \begin{pmatrix} \tau & \overline{\tau} \\ 1 & -1 \end{pmatrix},$$

and so the matrix of $\varphi_{\tau}(z)$ in the $\mathbb{Q}$-basis is

$$\psi_{\tau}(z) := A_{\tau} \varphi_{\tau}(z) A_{\tau}^{-1}.$$

The entries of $\psi_{\tau}$ are not particularly illuminating quadratic expressions in $\tau$ and $\overline{\tau}$ and $z$ with $\mathbb{Q}$-coefficients. To satisfy an additional equation over $\mathbb{Q}$ beyond $\det \psi_{\tau} = 1$ suggests that $\tau$ must be an algebraic number. We shall see later that there are two cases:

(i) $\mathbb{Q}(\tau)$ is a purely imaginary quadratic extension of $\mathbb{Q}$, and $M_{\varphi_{\tau}}$ is the element of norm one in $\mathbb{Q}(\tau)^*$;

(ii) $M_{\varphi_{\tau}} = \mathrm{SL}_2(\mathbb{Q})$.

**Example.** For $\mathbb{Q}(n)$, the Mumford-Tate group is $\mathbb{G}_{m,\mathbb{Q}}$ for $n \neq 0$ and is trivial for $n = 0$.

We shall now formulate and prove the basic property for $M_{\varphi}$; subsequently, we shall do the same for $M_{\widetilde{\varphi}}$. In each case the basic property is an answer to the question:

*What are the defining equations for the $\mathbb{Q}$-algebraic group $M_{\varphi}$, respectively $M_{\widetilde{\varphi}}$?*

For this we consider a polarized Hodge structure $(V, Q, \varphi)$ and let

$$\begin{cases} T^{k,l} = V^{\otimes^k} \otimes \check{V}^{\otimes^l}, \\ T^{\bullet,\bullet} = \underset{k,l \geqq 0}{\oplus} T^{k,l} \end{cases}$$

be the tensor spaces and tensor algebra of $V$. We then have

$$\begin{cases} \mathrm{Hg}_{\varphi}^{k,l} \subset T^{k,l} \\ \mathrm{Hg}_{\varphi}^{\bullet,\bullet} = \oplus \mathrm{Hg}_{\varphi}^{k,l} \end{cases}$$

consisting of the Hodge tensors in $T^{k,l}$ and the algebra of Hodge tensors.

**Basic Property (I).** *$M_\varphi$ is the subgroup of $G$ fixing $\mathrm{Hg}_{\varphi}^{\bullet,\bullet}$.*

**Step one:** *If $t \in \mathrm{Hg}_{\varphi}^{k,l}$, then $M_\varphi$ fixes $t$.*

*Proof.* Since $t$ is rational, fixing it defines a $\mathbb{Q}$-algebraic subgroup $G(t)$ of $\mathrm{GL}(V)$. If $t$ is of Hodge type $(p,p)$ where $n(k-l) = 2p$, then $\varphi(z)t = z^{p-p}t = t$. If follows that

$$\varphi(S^1) \subseteq G(t)(\mathbb{R}),$$

and by the minimality of $M_\varphi$ we conclude that $M_\varphi \subset G(t)$. This gives

$$M_\varphi \subseteq \mathrm{Fix}\,\mathrm{Hg}_{\varphi}^{\bullet,\bullet}.$$

**Step two:** *If $M_\varphi$ stabilizes the line $\mathbb{Q}t$ spanned by $t \in T^{k,l}$, then $t \in \mathrm{Hg}_{\varphi}^{k,l}$ is a Hodge tensor.*

*Proof.* Since $\varphi(S^1) \subset M_\varphi(\mathbb{C})$, if $M_\varphi$ stabilizes $\mathbb{Q}t$ then $t$ will be an eigenvector for $\varphi(z)$ for a general $z \in S^1$. Hence $t$ is of pure Hodge type, and since it is rational it must be a Hodge tensor.

**Chevalley's theorem** (in our context). *Let $M$ be a closed $\mathbb{Q}$-algebraic subgroup of $\mathrm{GL}(V)$. Then $M$ is the stabilizer of a line in $\overset{m}{\underset{i=1}{\oplus}} T^{k_i,k_i}$.*

*Completion of the proof of the basic property (I).* There exists

$$\tau = (t_1, \dots, t_m) \in \overset{m}{\underset{i=1}{\oplus}} T^{k_i,k_i}$$

such that setting $L = \mathbb{Q} \cdot \tau$,

$$M_\varphi = \{g \in \mathrm{GL}(V) : g(L) \subseteq L\}.$$

Then $g \in M_\varphi$ fixes each line $\mathbb{Q}t_i \subset T^{k_i,k_i}$, and by step two $t_i$ is a Hodge tensor. Thus

$$\mathrm{Fix}\,\mathrm{Hg}_{\varphi}^{\bullet,\bullet} \subseteq \bigcap_i \mathrm{Fix}(t_i) = M_\varphi. \qquad \square$$

*Proof of Chevalley's theorem.* From the open embedding $\mathrm{GL}(V) \hookrightarrow \mathrm{End}(V)$ of algebraic varieties comes the injection of coordinate rings

$$\mathbb{Q}[\mathrm{GL}(V)] \hookleftarrow \mathbb{Q}[\mathrm{End}(V)] \cong \underset{k\geq 0}{\oplus} \mathrm{Sym}^k(\mathrm{End}(V)^\vee) \subseteq \underset{k\geq 0}{\oplus} T^{k,k}V.$$

The action of $\mathrm{GL}(V)$ on itself by conjugation extends to the adjoint action on $\mathrm{End}(V)$, and these induce compatible actions on the coordinate rings which, moreover, are compatible with the action of $\mathrm{GL}(V)$ on $\oplus T^{k,k}V$. Write $S := \mathbb{Q}[\mathrm{End}(V)]$. If we choose a basis for $V$, we may think of $S$ as polynomials $P(X_j^i)$ in the matrix entries of $X \in \mathrm{End}(V)$.

The stabilizer of the ideal

$$I(M) \subseteq \mathbb{Q}[\mathrm{GL}(V)]$$

of $M$, viewed as a subvariety of $\mathrm{GL}(V)$, is the largest algebraic subgroup contained in the Zariski closure of $M$. But since $M$ is Zariski closed and a subgroup, the stabilizer is just $M$. Note that $S$ and

$$I := I(M) \cap S$$

inherit nonnegative $\mathrm{GL}(V)$-invariant gradings from the above injection of coordinate rings, and $M$ is also the subgroup of $\mathrm{GL}(V)$ stabilizing $I$ in $S$. This follows from the above because $\mathbb{Q}[\mathrm{GL}(V)] = \mathbb{Q}[\mathrm{End}(V)][\frac{1}{\det}]$ and det is nonvanishing on $M$.

Since $M$ is an algebraic variety in $\mathrm{GL}(V)$, $I(M)$ is finitely generated; the same goes for $I$, the ideal of the Zariski closure of $M$ in $\mathrm{End}(V)$. Let $P_i \in I^{\leq k}$ be a generating set for $I$. Since $I^{\leq k}$ generates $I$ and the action is compatible with products, $M$ is the stabilizer of $I^{\leq k}$ in $S^{\leq k}$, and this is, by linear algebra, the same as the stabilizer of

$$L := \wedge^d I^{\leq k} \subseteq \wedge^d S^{\leq k}$$

where $\dim I^{\leq k} = d$ and $\wedge^d S^{\leq k} \subset \underset{\ell \geq 0}{\oplus} T^{\ell,\ell} V$. □

For a general Hodge structure $(V, \widetilde{\varphi})$, not necessarily of pure weight, using $\mathrm{Gr}_0^W \mathrm{Hg}_{\widetilde{\varphi}}^{\bullet,\bullet}$ to denote the weight zero part $\underset{i}{\oplus} \mathrm{Hg}_{\widetilde{\varphi}}^{k_i,k_i}$ of $\mathrm{Hg}_{\widetilde{\varphi}}^{\bullet,\bullet}$, essentially the same argument gives the

**Basic Property (II).** *$M_{\widetilde{\varphi}}$ is the subgroup of* $\mathrm{GL}(V)$ *fixing* $\mathrm{Gr}_0^W \mathrm{Hg}_{\widetilde{\varphi}}^{\bullet,\bullet}$.

We next have

*If $(V, Q, \varphi)$ is a polarized Hodge structure, then the following properties of a subspace $W \subset V$ are equivalent:*

(i) *$W$ is invariant under $M_\varphi$;*
(ii) *$W$ is a sub-Hodge structure.*

*Proof of* (i) $\Rightarrow$ (ii). $W_{\mathbb{R}}$ is invariant under $M_\varphi(\mathbb{R})$ which contains $\varphi(S^1)$; thus the circle acts on $W_{\mathbb{R}}$ to give a sub-Hodge structure, necessarily polarized, of $(V, Q, \varphi)$.

*Proof of* (ii) $\Rightarrow$ (i). As previously noted, since $W$ is a sub-Hodge structure, we have the $Q$-orthogonal direct sum decomposition

$$V = W \oplus W^{\perp}.$$

If $G(W) \subset G$ is the $\mathbb{Q}$-algebraic group of those $g \in G$ with $g(W) \subseteq W$, then we have $\varphi(S^1) \subset G(W)(\mathbb{R})$. By the minimality of $M_\varphi$, we then have $M_\varphi \subseteq G(W)$. □

**Corollary.** *$M_\varphi$ is a reductive algebraic group.*

The above properties of $M_\varphi$ are also valid for $M_{\widetilde{\varphi}}$. An additional property is: Let $\rho : \mathrm{GL}(V) \to \mathrm{GL}(V_\rho)$ be a representation. Then setting $\widetilde{\varphi}_\rho = \rho \circ \widetilde{\varphi}$, $(V_\rho, \widetilde{\varphi}_\rho)$ is a Hodge structure and

$$M_{\widetilde{\varphi}(\rho)} = \rho(M_{\widetilde{\varphi}}).$$

The proof is an exercise.

For properties of algebraic groups see Borel, loc. cit. A consequence of the corollary is that $M_\varphi$ is an almost direct product $M_1 \times \cdots \times M_k \times A$ of simple $\mathbb{Q}$-algebraic groups $M_i$ and an algebraic torus $A$.

Below we will see that $A$ is anisotropic; i.e., $A(\mathbb{R})$ is compact.

*Exercise.* Any non-trivial reductive subgroup of $\mathrm{SL}_2$ is a torus $A$. If $A$ is anisotropic, it is conjugate to the standard torus given earlier.

## II. CM Hodge structures

For reasons of exposition, in this lecture we shall assume that our Hodge structures $(V, \varphi)$ are of pure weight $n$, *simple*, and *polarizable*. Simple means that there are no non-trivial, proper sub-Hodge structures (and thus simple $\Rightarrow$ pure weight). Polarizable means that there exists a polarization — we do not specify what it is.[2] We also set $\widetilde{G} = \mathrm{GL}(V)$.

The internal symmetries of $(V, \varphi)$ are encoded in its *endomorphism algebra*

$$\mathcal{E}_\varphi = \{X \in \mathrm{End}_\varphi(V, V)\} .$$

The notation $\mathrm{End}_\varphi(V, V)$ means that as maps of $V_\mathbb{R}$ to itself, for all $z \in S^1$ we have

$$[X, \varphi(z)] = 0 .$$

Identifying $\mathrm{End}(V, V)$ with $V \otimes \check{V}$, using the basic property we have

$$\mathcal{E}_\varphi = \mathrm{Hg}^{1,1} ,$$

or equivalently

$$\mathcal{E}_\varphi = [\mathrm{End}(V, V)]^{M_\varphi}$$

are the $M_\varphi$-invariants in $\mathrm{End}(V, V)$.

Using the assumption that $(V, \varphi)$ is simple, it follows that $\mathcal{E}_\varphi$ is a division algebra and $\mathcal{E}_\varphi^* = \mathcal{E}_\varphi \backslash \{0\}$ is a subgroup of $\widetilde{G}(\mathbb{Q})$. In fact,

$$\mathcal{E}_\varphi^* = Z_{\widetilde{G}}(M_\varphi)$$

is the centralizer of $M_\varphi$ in $\widetilde{G}$.

The existence of a polarization $Q$ implies that $\mathcal{E}_\varphi$ is a special type of division algebra; namely, it has a positive (anti-) involution $\iota$: For $X \in \mathcal{E}_\varphi$ and $u, v \in V$, $\iota(X) = X^\dagger$ is defined by

$$Q(Xu, v) = Q(u, X^\dagger v) .$$

The involution property $(X^\dagger)^\dagger = (-1)^n X$ is immediate; the positivity is a consequence of the 2[nd] Hodge-Riemann bilinear relation (we do not need its explicit form). Division algebras with this property have been classified into four types. For our purposes the ones of type IV are of interest. Denoting by $L$ the center of $\mathcal{E}_\varphi$, division algebras of this type are described as follows:

(i) $L$ is a *CM-field*. That is, it is a totally imaginary extension of $\mathbb{Q}$ having a totally real subfield $L_0$ with $[L : L_0] = 2$;

(ii) $\mathcal{E}_\varphi$ is a division algebra of rank $d^2$ over $L$, where $d^2 = [\mathcal{E}_\varphi : L]$, and

(iii) $\mathcal{E}_\varphi \otimes_\mathbb{Q} \mathbb{R} \cong \underbrace{M_d(\mathbb{C}) \times \cdots \times M_d(\mathbb{C})}_{e}$ where $e = [L_0 : \mathbb{Q}]$ and where

$$\iota(A_1, \ldots, A_e) = ({}^t\bar{A}_1, \ldots, {}^t\bar{A}_e) .$$

**Definition.** $(V, \varphi)$ is a *CM-Hodge structure* if $M_\varphi$ is a torus.

**Proposition.** *If $(V, \varphi)$ is a CM-Hodge structure, then $\mathcal{E}_\varphi$ is a CM-field.*

**Step one:** $\mathcal{E}_\varphi^*$ is a $\mathbb{Q}$-algebraic group and contains a maximal torus $T$ defined over $\mathbb{Q}$ (cf. Borel). Since it is maximal and centralizes $M_\varphi$, we have $T \supset M_\varphi$.

Any sum of non-zero elements of $T(\mathbb{Q}) \subset \mathcal{E}_\varphi^*$ belongs to $\mathcal{E}_\varphi^* \subset \widetilde{G}$ and commutes with $T$, hence by maximality belongs to $T(\mathbb{Q})$. Thus there is a field $L$ with $\eta : L \hookrightarrow \mathcal{E}_\varphi$ and $\eta(L^*) = T(\mathbb{Q})$. For any $w \in V$, the subspace $W = \eta(L)w$ of $V$ is stabilized

[2] By the basic property I, $\varphi$ will preserve any polarization.

by $M_\varphi$; hence is a sub-Hodge structure. Thus $W = V$ and $V$ is a 1-dimensional vector space over $L$. Also $[L : \mathbb{Q}] = \dim V := r$, and we have

$$\mathcal{E}^*_\varphi = T(\mathbb{Q}) = Z_{\widetilde{G}}(M_\varphi)\,.$$

**Step two:** A *Hodge frame* is given by a basis $\omega = \{\omega_j \in V^{p_j,q_j}, \omega_{n-j} = \overline{\omega}_j\}$ for $V_\mathbb{C}$ such that

$$i^{p_j - q_j} Q(\omega_j, \omega_{n-k}) = \delta_{jk}$$

(think of $v_\tau$ and $\bar{v}_\tau$ in the standard example). We may choose $\omega$ to diagonalize the action of $M_\varphi$ and hence of $\varphi(\mathbb{S})$. This basis defines and diagonalizes a maximal torus of $\widetilde{G}$ centralizing $M_\varphi$ and *a priori* defined over $\mathbb{R}$. But since $Z_{\widetilde{G}}(M_\varphi) = T(\mathbb{Q})$, this torus is necessarily $T$ and consequently $\omega$ diagonalizes $\eta(L)$.

**Step three:** Let $\gamma \in L$ be a primitive element for the extension $L/\mathbb{Q}$ and

$$P(\lambda) = \prod_{j=1}^{r} (\lambda - \eta_j(\gamma))$$

its minimal polynomial over $\mathbb{Q}$, where $\eta_i : L \hookrightarrow \mathbb{C}$ are embeddings. This is also a minimal polynomial for $\eta(\gamma)$, and so up to reordering in the Hodge basis we have

$$\eta(l) = \operatorname{diag}\{\eta_1(l), \ldots, \eta_r(l)\}$$

for all $l \in L$. Since $\eta(L) \subset \widetilde{G}(\mathbb{R})$ the eigenvalues of $\eta(\gamma)$ on $\omega_j$ and $\omega_{r-j} = \overline{\omega}_j$ must be conjugate, which implies that $\eta_{r-j} = \overline{\eta}_j$. Except when $n$ is even and $r$ is odd, a case that can be ruled out although we shall not do so here, this shows that $L$ is totally imaginary.

**Step four:** The Rosati involution gives

$$\sigma := \eta^{-1} \circ \dagger \circ \eta \in \mathrm{Gal}(L/\mathbb{Q}),$$

and the defining properties of $\dagger$ and of the Hodge frame give

$$\eta_j \circ \sigma = \overline{\eta}_j$$

for all $j$. (This computation is a nice exercise.) The existence of such an involution is the defining property for a CM field. □

**Standard example (continued):** The above argument simplifies in this case. If $M_\tau$ is a torus, then it follows from $\mathcal{E}_\varphi = [\mathrm{End}_\varphi(V, V)]$ that $V_\tau$ has an extra endomorphism other than $\pm \mathrm{id}_V$. If $\Lambda_\tau = \mathbb{Z} + \mathbb{Z}\tau$ is the lattice with $E_\tau = \mathbb{C}/\Lambda_\tau$, then there is a complex number $\xi$ such that

$$\xi(\Lambda_\tau) \subseteq m\Lambda_\tau$$

for some $m \in \mathbb{Z}$. Then we have

$$\begin{cases} \xi &= a + b\tau \\ \xi\tau &= c + d\tau \end{cases}$$

where $a, b, c, d \in \mathbb{Z}$. This gives an equation

$$\tau^2 + \alpha\tau + \beta = 0$$

where $\alpha, \beta \in \mathbb{Q}$.

Conversely, suppose that $\tau$ satisfies such an equation. Then multiplication by $\tau$ gives an isogeny of $E_\tau$ to itself, which we may think of as a Hodge class $[\tau]$ in

$H^2(E_\tau \times E_\tau, \mathbb{Q})$. The induced action of $[\tau]$ as an element in $\mathrm{End}_\tau(H^1(E_\tau, \mathbb{Q})) = \mathrm{End}_\tau(V)$ diagonalizes with distinct eigenvalues with respect to some Hodge frame $v_\tau, \bar{v}_\tau$. Since the eigenvalues are distinct, the only matrices that commute with it are diagonal. It follows that $M_\tau$ is a torus with

$$\begin{cases} M_\tau(\mathbb{C}) = \mathrm{diag}(z, 1/z) \\ M_\tau(\mathbb{R}) = \mathrm{diag}(\exp(ix), \exp(-ix)) \,. \end{cases}$$

Writing out the elements of this form which are rational with respect to a rational basis of $V$ we have $z = \exp(ix)$ in $\mathbb{Q}(\tau)$. (This is another nice exercise.) From this it follows that $M_\tau$ is isomorphic to the elements of norm one in $\mathbb{Q}(\tau)^*$.

There is a converse to the above:

*Suppose that $\mathcal{E}_\varphi$ has an embedded field $\eta : L \hookrightarrow \mathcal{E}_\varphi$ of degree $r = \dim V$. Then $M_\varphi$ is abelian.*

*Proof.* Since $\eta(L) \subset \mathrm{Hg}_\varphi^{1,1}$, by the basic property $M_\varphi$ must centralize $\eta(L)$. Furthermore, $\eta(L^*)$ gives the $\mathbb{Q}$-points of a torus of dimension $[L : \mathbb{Q}]$. Since by assumption $[L : \mathbb{Q}] = \dim V$, this torus is maximal in $\widetilde{G}$, and hence it contains $M_\varphi$. □

Anticipating the discussion in the next lecture of the period domain $D = \{$set of polarized Hodge structures on $(V, Q)$ with given Hodge numbers$\}$ as a homogeneous complex manifold

$$D = G(\mathbb{R})/H_\varphi$$

where $H_\varphi = \{g \in G(\mathbb{R}) : [g, \varphi] = 0\}$ is the compact isotropy group preserving a reference polarized Hodge structure $(V, Q, \varphi)$, we have:

*If $M_\varphi(\mathbb{R})$ is contained in $H_\varphi$, then $M_\varphi$ is a torus and $\varphi$ is a CM Hodge structure.*

*Proof.* First, the Mumford-Tate group $M_\varphi$ commutes with $\mathcal{E}_\varphi$. Since the action of $M_\varphi(\mathbb{R})$ preserves the Hodge structure it follows that $M_\varphi(\mathbb{Q}) \subseteq \mathcal{E}_\varphi$. Thus $M_\varphi(\mathbb{Q})$ is commutative, and by the Zariski density of the $\mathbb{Q}$-points in a connected, linear $\mathbb{Q}$-algebraic group, $M_\varphi$ is commutative. Hence it is a torus and the previous result applies. □

It may be shown that the CM polarized Hodge structures are dense in the period domain. They play a central role in the theory of Shimura varieties. Their role in the study of higher weight Hodge structures is in its early stages.

## III. Mumford-Tate domains

Given $(V,Q)$ and a set of *Hodge numbers* $h^{p,q}=h^{q,p}$, $p+q=n$ and $\sum_{p+q=n} h^{p,q} = \dim V$, the *period domain* $D$ is the set of polarized Hodge structures $(V,Q,\varphi)$ with $\dim V^{p,q}=h^{p,q}$. Given $\varphi\in D$, we recall that a Hodge frame $\omega$ is $\{\omega_j\in V^{p_j,q_j}$, $\omega_{r-j}=\overline{\omega}_j$ and $i^{p_j-q_j}Q(\omega_j,\overline{\omega}_k)=\delta_{jk}\}$. Picking a reference point $\varphi$, the set of Hodge frames is identified with $G(\mathbb{R})$, and the subgroup $H_\varphi$ of $G(\mathbb{R})$ fixing $\varphi$ is compact. Hence,

$$D=G(\mathbb{R})/H_\varphi$$

is a homogeneous manifold.

In fact, $D$ is a *homogeneous complex manifold.* One way to see this is to consider the *compact dual* $\check{D}$, consisting of all filtrations $F^\bullet=\{F^n\subset F^{n-1}\subset\cdots\subset F^0=V_\mathbb{C}\}$ satisfying the first Hodge-Riemann bilinear relations

$$Q(F^p,F^{n-p+1})=0\,.$$

It may be shown that $\check{D}$ is acted on transitively by $G(\mathbb{C})$ with stability group of $F^\bullet\in\check{D}$ a parabolic subgroup $P$. Thus

$$\check{D}=G(\mathbb{C})/P$$

is a compact, complex manifold. It is in fact a rational, projective variety defined over $\mathbb{Q}$, as may be seen from the $G(\mathbb{C})$-equivariant embeddings

$$\check{D}\subset\prod_{p=n}^{[\eta/2]}\mathrm{Grass}(f^p,V_\mathbb{C})\subset\prod_{p=n}^{[\eta/2]}\mathbb{P}(\wedge^{f_p}V_\mathbb{C})$$

where $f^p=h^{n,0}+\cdots+h^{p,n-p}$, and where the second inclusion is the *Plücker embedding.* Then

$$D\subset\check{D}$$

is an open $G(\mathbb{R})$-orbit of a fixed point $\varphi\in D$, and as such has an induced complex structure.

**Basic example (continued).** In the case $n=1$, $h^{1,0}=1$ we have

$$\mathcal{H}\subset\check{\mathcal{H}}=\mathbb{P}^1\,.$$

Another way to see the complex structure on $D$ is to observe that $\mathrm{End}(V,V)\cong V\otimes\check{V}$ has a weight-zero polarized Hodge structure and $\mathfrak{g}\subset\mathrm{End}(V,V)$ is a sub-Hodge structure. In fact

$$\mathfrak{g}_\mathbb{C}=\bigoplus_i\mathfrak{g}^{-i,i}$$

where

$$\mathfrak{g}^{-i,i}=\left\{X\in\mathfrak{g}_\mathbb{C}:X(V^{p,q})\subset V^{p-i,q+i}\right\}$$

shifts the Hodge decomposition $i$-places to the right.[3] We have

$$\begin{cases}\overline{\mathfrak{g}}^{i,-i}=\mathfrak{g}^{-i,i}\\ \mathfrak{g}^{0,0}=\mathfrak{h}_{\varphi,\mathbb{C}}=\text{the complexified Lie algebra of } H_\varphi\\ [\mathfrak{g}^{-i,i},\mathfrak{g}^{-j,j}]\subseteq\mathfrak{g}^{-(i+j),i+j}.\end{cases}$$

[3]We shall drop the subscript "$\varphi$" on the $\mathfrak{g}^{-i,i}$ and on the $\mathfrak{m}^{-i,i}$ below.

Setting

$$\mathfrak{g}^{-} = \underset{i>0}{\oplus} \mathfrak{g}^{-i,i}, \quad \mathfrak{g}^{+} = \overline{\mathfrak{g}^{-}}$$

the complexified tangent space[4]

$$\begin{aligned} T_{\varphi} D_{\mathbb{R}} \otimes \mathbb{C} &\cong \mathfrak{g}_{\mathbb{C}} / \mathfrak{h}_{\varphi,\mathbb{C}} \\ &\cong \mathfrak{g}^{-} \oplus \overline{\mathfrak{g}^{-}}\,. \end{aligned}$$

Thus, $D$ has a $G(\mathbb{R})$-invariant almost complex structure whose $(1,0)$ part at the identity coset is $\mathfrak{g}^{-}$. By the bracket relation above this almost complex structure is integrable. Note that if $\varphi \in D \subset \check{D}$ is chosen as a reference point, then

$$\mathfrak{p} = \underset{i \leqq 0}{\oplus} \mathfrak{g}^{-i,i} = \mathfrak{h}_{\varphi,\mathbb{C}} \oplus \mathfrak{g}^{+}\,.$$

Note also that

$$\mathfrak{w} := \mathfrak{g}^{-1,1} \subset \mathfrak{g}^{-}$$

gives a $\mathbb{G}(\mathbb{C})$-invariant distribution

$$W \subset T\check{D}\,.$$

A holomorphic map

$$\Phi : S \to \check{D}\,,$$

where $S$ is a complex manifold and $\Phi(s) = F_s^{\bullet}$ is a holomorphically varying filtration, is said to satisfy the *infinitesimal period relation* (IPR) if

$$\Phi_* : TS \to W\,.$$

We may think of this as

$$dF_s^p \subset F_s^{p-q} \otimes \Omega^1_{S,s}\,.$$

A *local variation of Hodge structure* is given by $\Phi : S \to D$ as above that satisfies the IPR. If we set $W^{\perp} = I \subset T^*D$, then for local holomorphic sections $\theta$ of $I$ we have

$$\begin{cases} \Phi^*(\theta) = 0\,, \\ \Phi^*(d\theta) = d\Phi^*(\theta) = 0\,. \end{cases}$$

The second relation gives the *integrability conditions* associated to the differential constraint $dF_s^p \subset F_s^{p-1} \otimes \Omega^1_{S,s}$. They are equivalent to saying that

$$\Phi_*(T_s S) := E \subset \mathfrak{m}^{-1,1}$$

is an abelian subalgebra.

**Definition.** Let $\varphi \in D$ have a Mumford-Tate group $M_{\varphi}$. Then the associated *Mumford-Tate domain* is the $M_{\varphi}(\mathbb{R})$-orbit

$$D_{M_{\varphi}} = M_{\varphi}(\mathbb{R}) \cdot \varphi \subset D$$

of $\varphi$.

**Proposition.** *$D_{M_{\varphi}}$ is a homogeneous complex submanifold of the period domain $D$.*

*Proof.* The basic observation is that the Lie algebra of $M_{\varphi}$

$$\mathfrak{m}_{\varphi} \subset \mathfrak{g}$$

---

[4]Here only, we denote by $T_{\varphi} D_{\mathbb{R}}$ the real tangent space to $D$ considered as a real manifold. Elsewhere, $T_{\varphi} D$ will denote the usual holomorphic tangent space (thus $T_{\varphi} D \cong (T_{\varphi} D_{\mathbb{R}} \otimes \mathbb{C})^{1,0}$).

is a sub-Hodge structure of $\mathfrak{g}$. Then, as above

$$\mathfrak{m}_{\varphi,\mathbb{C}} = \bigoplus_i \mathfrak{m}^{-i,i}$$

where $\mathfrak{m}^{-i,i} = \mathfrak{m}_{\varphi,\mathbb{C}} \cap \mathfrak{g}^{-i,i}$, and where $\mathfrak{m}^{0,0} = \mathfrak{m}_{\varphi,\mathbb{C}} \cap \mathfrak{h}_{\varphi,\mathbb{C}}$ is the complexified Lie algebra of the isotropy group $H_{M_\varphi} = H_\varphi \cap M_\varphi(\mathbb{R})$: thus

$$D_{M_\varphi} = M_\varphi(\mathbb{R})/H_{M_\varphi}\,.$$

We have that $\mathfrak{m}_\varphi^- = \mathfrak{m}_\mathbb{C} \cap \mathfrak{g}^-$ gives the $(1,0)$ part of an $M_\varphi(\mathbb{R})$-invariant almost complex structure, which is integrable for the same reason as in the period domain case. □

A basic fact is given by the

**Proposition.** *$M_\varphi(\mathbb{R})$ contains a compact maximal torus.*

*Proof.* We have observed above that $M_\varphi(\mathbb{R})$ is reductive, and thus it is an almost direct product $M_1 \times \cdots \times M_l \times A$ of simple, real Lie groups $M_i$ and an abelian part $A$.[5] We shall assume that $M_\varphi(\mathbb{R})$ is simple, and comment that the argument may be extended to the general case. For notational simplicity, we shall drop the subscript $\varphi$ — thus $M_\varphi = M$, etc.

We first note that

$$H_M = Z_M(\varphi(S^1))$$

is the centralizer in $M(\mathbb{R})$ of the circle $\varphi(S^1)$. In fact, writing $z \in S^1$ as $z = e^{2\pi i\xi}$, we have that $\varphi(z) = e^{i(2p-n)\xi}$ on $V^{p,n-p}$. It follows that for $g \in M(\mathbb{R})$

$$g\varphi(z) = \varphi(z)g \Leftrightarrow g(V_\varphi^{p,q}) \subseteq V_\varphi^{p,q}\,.$$

Let $B$ be a maximal connected abelian subgroup of $M(\mathbb{R})$ containing $\varphi(S^1)$. Then

$$B \subseteq Z_M(\varphi(S^1)) = H_M\,,$$

and since $H_M$ is compact, so is $B$. Because $B$ is maximal abelian, rank $M(\mathbb{R}) = \dim B$. □

**Corollary.** $\mathrm{SL}_n$ *is not a Mumford-Tate group for $n \geqq 3$.*

The other extreme to $M(\mathbb{R})$ being simple is when $M = T$ is an algebraic torus. The non-trivial characters of $T$ acting on $V_\mathbb{C}$ occur in conjugate pairs, so that $T$ is anisotropic. In this case the component of the Mumford-Tate domain containing $\varphi$ is just the point $\varphi$, which is a CM-polarized Hodge structure.

**Basic example (continued):** $\varphi = \sqrt{-d} \in \mathcal{H}$. Then

$$M_\varphi = \left\{ \begin{pmatrix} \alpha & \sqrt{d}\beta \\ -\sqrt{d}\beta & \alpha \end{pmatrix} : \alpha, \beta \in \mathbb{Q} \text{ and } \alpha^2 + d\beta^2 = 1 \right\},$$

and

$$D_{M_\varphi} = \sqrt{-d}$$

is a point in $\mathcal{H}$.

Now we come to Noether-Lefschetz loci.

[5] $M_\varphi$ may be simple as a $\mathbb{Q}$-algebraic group but $M_\varphi(\mathbb{R})$ may only be semi-simple as a real Lie group.

**Definition.** Let $\varphi \in D$. Then the *Noether-Lefschetz locus*

$$\mathrm{NL}_\varphi = \left\{ \varphi' \in D : \mathrm{Hg}^{\bullet,\bullet}_{\varphi'} \supseteq \mathrm{Hg}^{\bullet,\bullet}_{\varphi} \right\}.$$

In other words, the Noether-Lefschetz locus is the set of polarized Hodge structures whose algebra of Hodge tensors contains the given algebra $\mathrm{Hg}^{\bullet,\bullet}_{\varphi}$. Classically, the Noether-Lefschetz loci were defined in the parameter space of a family of smooth projective varieties to be the subvariety where a Hodge class remains a Hodge class. For geometric reasons, the above would seem to be a more natural concept. Denoting by $Z^\circ$ the component containing $\varphi$ of a subvariety $Z$ of $D$ with $\varphi \in Z$, we have the

**Proposition.** $D_{M_\varphi}{}^\circ = \mathrm{NL}_\varphi{}^\circ$.

*Proof.* We clearly have $D_{M_\varphi}{}^\circ \subseteq \mathrm{NL}_\varphi{}^\circ$. Since $D_{M_\varphi}{}^\circ$ is smooth, it will suffice to show equality of the Zariski tangent spaces at $\varphi$. Identifying $T_\varphi D$ with $\mathfrak{g}^-$ and $T_\varphi D_{M_\varphi}$ with $\mathfrak{m}^-$, we have

$$T_\varphi \mathrm{NL}_\varphi = \left\{ X \in \mathfrak{g}^- : X(\zeta) = 0 \text{ for all } \zeta \in \mathrm{Hg}^{\bullet,\bullet}_\varphi \right\}.$$

Here, for $\zeta \in \mathrm{Hg}^{k,l}_\varphi$ of Hodge type $(p,p)$ where $n(k-l) = 2p$ and for $X \in \mathfrak{g}^{-i,i}$, $X(\zeta)$ is of Hodge type $(p-i, p+i)$. Thus the above statement in brackets expresses the condition that, to first order in the direction $X$, $\zeta$ remains a Hodge class.

On the other hand, by the basic property of Mumford-Tate groups, the $\mathfrak{m}^-$-part of the complexification of the tangent space $T_e M_\varphi \subset T_e G$ is

$$\{X \in \mathfrak{g}^- : X(\zeta) = 0 \text{ for all } \zeta \in \mathrm{Hg}^{\bullet,\bullet}_\varphi\},$$

which by the above gives

$$T_\varphi \mathrm{NL}_\varphi = T_\varphi D_{M_\varphi}.$$

□

For applications to algebraic geometry we consider a local variation of Hodge structure

$$\Phi : S \to D,$$

and for a point $s_0 \in S$ we set

$$\mathrm{NL}_{s_0}(S) = \Phi^{-1}(\mathrm{NL}_{\Phi(s_0)}).$$

A classical question is to estimate the codimension of $\mathrm{NL}_{s_0}(S)$ in $S$. More precisely, if $n = 2p$ and $\zeta \in V^{p,p}_{\varphi(s_0)} \cap V$ is a Hodge class, classically one has been interested in the locus $\mathrm{NL}_{s_0}(\zeta) \subset S$ where $\zeta$ remains a Hodge class. An obvious estimate is

$$\mathrm{codim}_S \, \mathrm{NL}_{s_0}(\zeta) \leqq h^{2p,0} + \cdots + h^{p+1,p-1}$$

since the RHS is the number of conditions to be a Hodge class. This estimate can be significantly sharpened and applied to the algebra of Hodge tensors as follows: Set

$$E = \Phi_*(T_{s_0} S) \subseteq \mathfrak{g}^{-1,1}.$$

There are maps

$$\mathrm{Sym}^{p-2} E \otimes (E \cap \mathfrak{m}^{-1,1}) : V^{2p,0}_{\Phi(s_0)} \to V^{p+1,p-1}_{\Phi(s_0)}.\,[6]$$

Denote by $\sigma_m$ the rank of this map.

[6]Here $p \geqq 2$; the classical case $p = 1$ is less interesting.

**"Proposition":** *We have*

$$\operatorname{codim}_S \mathrm{NL}_{s_0}(S) \leqq \dim\left(\mathfrak{g}^{-1,1}/\mathfrak{m}^{-1,1}\right) - \sum_m \sigma_m .$$

The quotation marks mean that there are some mild technical assumptions needed that do not effect the essential geometric content of the result. The term $\mathfrak{g}^{-1,1}/\mathfrak{m}^{-1,1}$ represents the part of the normal space to $D_{M_\varphi}$ in $D$ that satisfies the IPR. The terms $\sigma_m$ reflect the integrability conditions in the IPR. It may be shown by examples that the above estimate is sharp. In practice it says that for a given Hodge class $\zeta$ and when $p \geqq 2$, the codimension of $\mathrm{NL}_{s_0}(\zeta)$ is usually *much less* than the naive estimate $h^{2p,0} + \cdots + h^{p+1,p-1}$. If the Hodge conjecture is true it means that there are more algebraic cycles than predicted by naive dimension counts.

## IV. Variation of Hodge structure and Mumford-Tate domains

A *global variation of Hodge structure*

$$\Phi : S \to \Gamma\backslash D$$

is given by the following data:

- $S$ is a smooth, quasi-projective variety;
- $\Gamma \subset G(\mathbb{R})$ is a discrete group;[7]
- $\Phi$ is a locally liftable holomorphic mapping whose local lifts satisfy the IPR.

Since $\Gamma$ is discrete, the intersection of $\Gamma$ with the compact isotropy group $H_\varphi$ is a finite group. Being locally liftable means that around each point $s_0 \in S$ there is a neighborhood $U$ and a diagram

$$\begin{array}{ccc} & & D \\ & \overset{\widetilde{\Phi}}{\nearrow} & \downarrow \\ U & \xrightarrow{\Phi} & \Gamma\backslash D. \end{array}$$

Because of this there is an induced mapping

$$\Phi_* : \pi_1(S, s_0) \to \Gamma$$

and the image is called the *monodromy group*. It is *the* fundamental invariant of a global variation of Hodge structure.

Using the local liftings we may obtain a diagram

$$\begin{array}{ccc} \widetilde{S} & \xrightarrow{\widetilde{\Phi}} & D \\ {\scriptstyle\pi}\downarrow & & \downarrow \\ S & \xrightarrow{\Phi} & \Gamma\backslash D \end{array}$$

where $\widetilde{S} \to S$ is the universal covering. It is sometimes convenient to think of the properties of $\Phi$ as given by the $\Gamma$-invariant properties of $\widetilde{\Phi}$. For example, the Mumford-Tate group $M_\Phi$ of $\Phi$ may be defined as follows: Outside of a countable union $Z$ of proper subvarieties of $\widetilde{S}$, the Mumford-Tate groups $M_{\widetilde{\Phi}(\tilde{s})} \subset G$ are constant and serve to define $M_\Phi$. More precisely, choosing $s_0 \in S$ to be the image of $\tilde{s}_0 \in \widetilde{S}\backslash Z$, we may think of $V_\mathbb{C}$ as the fibre over $s_0 \in S$ of the flat vector bundle $\mathbf{V}_\mathbb{C} = \widetilde{S} \times_\Gamma V_\mathbb{C} \to S$, and then $\mathrm{Hg}^{\bullet,\bullet}_{s_0}$ gives a subspace of the tensor algebra of $V \subset V_\mathbb{C} = \mathbf{V}_{\mathbb{C},s_0}$ that, as a subspace, is invariant under parallel translation around all paths $\gamma \in \pi_1(S, s_0)$. Then $M_\Phi = \mathrm{Fix}\,\mathrm{Hg}^{\bullet,\bullet}_{s_0}$.

A basic property of the monodromy group is the following

**Theorem** (Schmid). *If $v \in V_\mathbb{C}^\Gamma$ is a monodromy invariant vector, then the Hodge $(p,q)$ components $v_s^{p,q}$ are constant and $\Gamma$-invariant.*

For applications of this theorem, we assume that there is a lattice $V_\mathbb{Z} \subset V$ with $\Gamma \subset G_\mathbb{Z}$. Then the intersection of $\Gamma$ with any compact subgroup of $\mathrm{GL}(V_\mathbb{R})$ is a finite group. We shall also work up to isogeny, i.e. up to finite coverings $S' \to S$ with the induced variation of Hodge structure $\Phi' : S' \to \Gamma'\backslash D$ where $\Gamma'$ is a subgroup of

[7]In practice, $\Gamma$ will usually be a subgroup of an arithmetic group, an arithmetic group being one that is commensurable with $G_\mathbb{Z}$.

finite index in $\Gamma$. In particular, referring to the above, we observe that $\Gamma$ acts on the space $\mathrm{Hg}_{s_0}^{\bullet,\bullet}$ by a finite group. This is because the polarizing form is definite on the spaces $\mathrm{Hg}_{s_0}^{k,l}$. Passing to a a finite covering and relabeling we may assume that $\mathrm{Hg}_{s_0}^{\bullet,\bullet}$ is *pointwise* fixed by $\Gamma$. This implies that

$$(*) \qquad \Gamma \subset M_\Phi\,.$$

A consequence of Schmid's theorem is the

**Theorem of the fixed part:** *If $U = V^\Gamma$ is the subspace of $\Gamma$-invariant vectors in $V$, then $Q\big|_U$ is non-singular, $V = U \oplus U^\perp$ and the variation of Hodge structure splits into a direct sum whose $U$-part is trivial.*

*Proof.* If $u \in U_\mathbb{C}$, then by Schmid's theorem the Hodge $(p,q)$ components $u(\widetilde{\Phi}(\tilde{s}))^{(p,q)}$ are constant and $\Gamma$-invariant. Thus the $U$-part of $\widetilde{\Phi}(\tilde{s})$ is a constant sub-Hodge structure of $\widetilde{\Phi}(\tilde{s})$. □

Another consequence is the

**Rigidity theorem:** *If we have two global variations of Hodge structure*

$$\begin{cases} \Phi : S \to \Gamma\backslash D \\ \Phi' : S \to \Gamma\backslash D \end{cases}$$

*where $\Phi(s_0) = \Phi'(s_0)$ and the induced monodromy representations coincide, then $\Phi = \Phi'$.*

This is because the identity map $\mathrm{id}_V \in \mathrm{Hom}(V,V) = V \otimes \check{V}$ is $\Gamma$-invariant and of Hodge type $(0,0)$ at $\tilde{s}_0$, hence it is everywhere of type $(0,0)$.

For a third application, we recall that $\overline{\Gamma}^{\mathbb{Q}}$ denotes the $\mathbb{Q}$-Zariski closure of $\Gamma \subset G$. It is the smallest $\mathbb{Q}$-algebraic subgroup of $G$ containing $\Gamma$. By $(*)$ above we have

$$\overline{\Gamma}^{\mathbb{Q}} \subseteq M_\Phi\,.$$

**Semi-simplicity of monodromy theorem:** *$\overline{\Gamma}^{\mathbb{Q}}$ is a semi-simple $\mathbb{Q}$-algebraic subgroup of* $\mathrm{Aut}(V)$.

In fact, the argument given in the proof of the structure theorem below will show that $\overline{\Gamma}^{\mathbb{Q}}$ is a normal subgroup of the derived group $DM_\Phi$.

When the global variation of Hodge structure *arises from geometry*, meaning that we have a family

$$\pi : \mathcal{X} \to S$$

of smooth projective varieties $X_s = \pi^{-1}(s)$ and

$$\Phi(s) = H^n(X_s, \mathbb{Q})_{\mathrm{prim}}\,,$$

the above results are due to Deligne as a consequence of his mixed Hodge theory.

Let now

$$M_\Phi = M_1 \times \cdots \times M_l \times A$$

be the almost product decomposition of the reductive $\mathbb{Q}$-algebraic group $M_\Phi$ into simple factors $M_i$ and an abelian part $A$. Denote by $D_i \subset D$ the $M_i(\mathbb{R})$-orbit of $\widetilde{\Phi}(\tilde{s}_0)$. Then we have the

**Structure theorem:** (i) *The $D_i$ are homogeneous complex submanifolds of $D$.* (ii) *After passing to a finite covering of $S$, the monodromy group splits as a direct product*

$$\Gamma = \Gamma_1 \times \cdots \times \Gamma_k, \qquad k \leqq l$$

*where* $\overline{\Gamma}_i^{\mathbb{Q}} = M_i$.

The proof will give the following:

(i) Setting $P = M_1 \times \cdots \times M_k$ and $R = M_{k+1} \times \cdots \times M_l \times A$, the Mumford-Tate group

$$M_\Phi = P \times R\,,$$

and we denote by $D_P$ and $D_R$ the corresponding sub-domains of $D$;

(ii) the global variation of Hodge structure is given by

$$\Phi : S \to \Gamma \backslash D_P \times D_R$$

and is constant in the $D_R$-factor;

(iii) the monodromy group $\Gamma$ and the $\mathbb{Q}$-algebraic group $P$ have the same tensor invariants.

In particular, although $\Gamma$ may not be a subgroup of finite index in the arithmetic group $P_{\mathbb{Z}}$, they at least have the same tensor invariants.

The main step in the proof of the structure theorem is this:

$$\overline{\Gamma}^{\mathbb{Q}} \text{ is a normal subgroup of the derived group } DM_\Phi\,.$$

*Proof.* For any tensor representation $\rho : V \to V_\rho$ we have a corresponding variation of Hodge structure $\Phi_\rho$ and a monodromy group $\Gamma_\rho \subset \mathrm{Aut}(V_\rho)$. By the theorem of the fixed part,

$$U_\rho := \ker(\overline{\Gamma}_\rho^{\mathbb{Q}} - I) \subset V_\rho$$

gives a constant sub-variation of Hodge structure. Recalling that the Mumford-Tate group

$$M_{\Phi_\rho} = \rho(M_\Phi)\,,$$

we infer that $\rho(M_\Phi)$ stabilizes $U_\rho \subset V_\rho$. Given $g \in M_\Phi(\mathbb{Q})$ and $\gamma \in \overline{\Gamma}_\rho^{\mathbb{Q}}$,

$$(**) \qquad \rho(g)\gamma\rho(g)^{-1} \text{ fixes } U_\rho \text{ pointwise.}$$

We now let $\Gamma' \subset \mathrm{Aut}(V)$ be the largest subgroup such that $\Gamma'$ fixes all tensors of weight zero that are fixed by $\Gamma$. Then $\overline{\Gamma}^{\mathbb{Q}} \subseteq \Gamma'$, and by the argument used in the proof of the basic property (Chevalley's theorem where $\rho$ are the respresentations on $T^{k,k}$), it follows that $\overline{\Gamma}^{\mathbb{Q}} = \Gamma'$ and hence by $(**)$, $\overline{\Gamma}^{\mathbb{Q}}$ is a normal subgroup of $M_\Phi$. Since the orbit of the abelian part $A(\mathbb{R})$ of $M(\mathbb{R})$ is a point, $A(\mathbb{R})$ is compact and $\Gamma \cap A$ is a finite group. After passing to a finite index subgroup, we may then assume that $\Gamma \cap A = \{e\}$. It then follows that $\overline{\Gamma}^{\mathbb{Q}}$ is a semi-simple subgroup of $DM_\Phi$. □

We may now infer that $\Gamma$ splits as in the statement of the structure theorem. Moreover, by the argument just given we have that

$$\overline{\Gamma}_i^{\mathbb{Q}} = M_i\,. \qquad \square$$

Because of the structure theorem we see that Mumford-Tate domains are the natural target spaces for period maps. In the next lectures we shall see that different

Mumford-Tate domains may be manifestations of a more basic object, namely a Hodge domain.

## V. Hodge representations

In this lecture we will discuss the questions: Given a semi simple $\mathbb{Q}$-algebraic group $M$

(a) When is $M$ the Mumford-Tate group of a polarized Hodge structure?
(b) In how many ways can such an $M$ be realized as a Mumford-Tate group?

Essentially complete answers are known to these question and we shall describe and illustrate them, largely without proofs. This will require some facts from the structure theory of Lie groups and representation theory, which we will briefly review as needed.

The basic concept is the following:

**Definition.** A *Hodge representation* $(M, \rho, \varphi)$ is given by a representation defined over $\mathbb{Q}$

$$\rho : M \to \mathrm{Aut}(V, Q)$$

together with a circle

$$\varphi : S^1 \to M(\mathbb{R})$$

such that, setting $\varphi_\rho = \rho \circ \varphi$, $(V, Q, \varphi_\rho)$ is a polarized Hodge structure.

Without essential loss of generality we can and do assume that $M(\mathbb{R})$ is simple and that $\rho : M(\mathbb{R}) \to \mathrm{Aut}(V_{\mathbb{R}}, Q)$ is irreducible. These $\rho$ are then classified using the root and weight structures on the Lie algebra $\mathfrak{m}_{\mathbb{R}}$. There are two issues:

(i) For which $\rho : M(\mathbb{R}) \to \mathrm{GL}(V_{\mathbb{R}})$ is there an invariant bilinear form $Q$?
(ii) Given $\rho : M(\mathbb{R}) \to \mathrm{Aut}(V_{\mathbb{R}}, Q)$, for which circles $\varphi_\rho = \rho \circ \varphi : S^1 \to M(\mathbb{R})$ is $(V, Q, \varphi_\rho)$ a polarized Hodge structure?

We have seen in lecture two that if a Hodge representation exists, $M(\mathbb{R})$ contains a compact maximal torus $T$ with $\varphi(S^1) \subset T$. The action of $T$ on $V_{\mathbb{C}}$ decomposes

$$V_{\mathbb{C}} = \oplus V_\omega$$

where the weight spaces $V_\omega$ on which $T = \mathfrak{t}/\Lambda$ acts by a character $\exp i \langle \omega, \bullet \rangle$ where $\omega \in \mathrm{Hom}(\Lambda, \mathbb{Z})$ occur in conjugate pairs: $V_{-\omega} = \overline{V}_\omega$. Restricting the action to $\varphi(S^1) \subset T$ gives a Hodge structure

$$V_{\mathbb{C}} = \bigoplus_{p+q=n} V^{p,q}_{\varphi_\rho}$$

where $V^{p,q}_{\varphi_\rho}$ is a sum of weight spaces.[8] Because the form $Q$ is preserved, the first Hodge-Riemann bilinear relation

$$Q(V^{p,q}_{\varphi_\rho}, V^{p',q'}_{\varphi_\rho}) = 0, \qquad p' \neq n - p$$

will be satisfied. The difficult one is the second bilinear relation

$$i^{p-q} Q(V^{p,q}_{\varphi_\rho}, \overline{V}^{p,q}_{\varphi_\rho}) > 0\,.$$

The answer will be expressed in terms of congruences "mod 2" reflecting on the parity of $Q \bmod 2$, and "mod 4" reflecting the fact that the second Hodge-Riemann bilinear relation depends on $p - q \bmod 4$.

The easiest case to analyze is the adjoint representation. Here we recall that if $(M, \rho, \varphi)$ is a Hodge representation, then so is $(M, \mathrm{Ad}, \varphi)$. Specifically, we have

$$\mathfrak{m} \subset \mathrm{End}(V) \cong V \otimes \check{V}$$

[8] There is a subtlety here in that since $\varphi(S^1)$ does not have a scaling action, the weight $n$ of the Hodge structure is not uniquely defined. We shall let $n$ be the integer such that $V^{n,0}_{\varphi_\rho} \neq 0$.

and the bilinear form $Q$ on $V$ induces the Cartan-Killing form $B : \mathfrak{m} \otimes \mathfrak{m} \to \mathbb{Q}$. Then $(\mathfrak{m}, B, \operatorname{Ad}\varphi)$ gives a polarized Hodge structure of even weight.

The conditions on $\varphi$ that $(\mathfrak{m}, B, \operatorname{Ad}\varphi)$ give a polarized Hodge structure will now be explained. Under the action of the Cartan sub-algebra $\mathfrak{t}_{\mathbb{C}}$ on $\mathfrak{m}_{\mathbb{C}}$, one has the root space decomposition

$$\mathfrak{m}_{\mathbb{C}} = \mathfrak{t}_{\mathbb{C}} \oplus \left( \bigoplus_{\alpha \in \mathfrak{r}} \mathfrak{m}_\alpha \right) .$$

Here, the set of *roots*

$$\mathfrak{r} \subset i\check{\mathfrak{t}}$$

are purely imaginary linear functions on the Lie algebra $\mathfrak{t}$ of $T$. The root spaces $\mathfrak{m}_\alpha$ are 1-dimensional and occur in conjugate pairs. For a maximal compact subgroup $K \subset M(\mathbb{R})$ we have $T \subset K$ and a Cartan decomposition

$$\mathfrak{m}_{\mathbb{R}} = \mathfrak{k} \oplus \mathfrak{p}$$

where

$$\begin{cases} [\mathfrak{k}, \mathfrak{p}] \subseteq \mathfrak{p} \\ [\mathfrak{p}, \mathfrak{p}] \subseteq \mathfrak{k} . \end{cases}$$

We denote by $\alpha_1, \dots, \alpha_d$ the *compact roots* which belong to $\mathfrak{k}$, meaning that the root space

$$\mathfrak{m}_{\alpha_j} \subset \mathfrak{k}_{\mathbb{C}} ,$$

and by $\beta_1, \dots, \beta_e$ the *non-compact roots* belonging to $\mathfrak{p}$.

Writing $T = \mathfrak{t}/\Lambda$, the circle $\varphi : S^1 \to T$ is given by the exponential of the line through a lattice point $\mathfrak{l}_\varphi \in \Lambda$. We then have:

$(\mathfrak{m}, B, \operatorname{Ad}\varphi)$ *is a polarized Hodge structure if, and only if,*

$$\begin{cases} \langle \alpha_j, \mathfrak{l}_\varphi \rangle \equiv 0 \pmod 4 \\ \langle \beta_k, \mathfrak{l}_\varphi \rangle \equiv 2 \pmod 4 . \end{cases}$$

*Notational caveat.* The roots are purely imaginary linear functions on $\Lambda \subset \mathfrak{t}$; we have omitted the "$i$" in the above congruences.

The proof of the above is based on the relation

$$\mathfrak{m}^{-k,k} = \bigoplus_{\left\{ \substack{\alpha \in \mathfrak{r} \\ \langle \alpha, \mathfrak{l}_\varphi \rangle = -2ki} \right.} \mathfrak{m}_\alpha ,$$

and the fact that

$$\begin{cases} B < 0 & \text{on } (\mathfrak{m}_{\alpha_j} \oplus \mathfrak{m}_{-\alpha_j})_{\mathbb{R}} \\ B > 0 & \text{on } (\mathfrak{m}_{\beta_k} \oplus \mathfrak{m}_{-\beta_k})_{\mathbb{R}} . \end{cases}$$

In practice the possible polarized Hodge structures can be read off from the above and standard root tables. If there is one such, then it turns out there are many.

**Example.** For $G_2$ with the standard root diagram

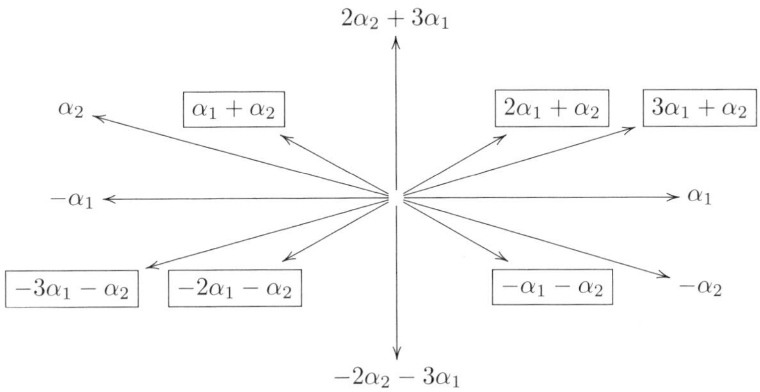


where the non-compact roots are framed, the minimal weight of a polarized Hodge structure is six with

$$\begin{aligned}
\mathfrak{g}_2^{0,0} &= \operatorname{span}\{e_{\alpha_2}\} \\
\mathfrak{g}_2^{1,5} &= \operatorname{span}\{e_{-3\alpha_1-\alpha_2}, e_{-\alpha_1}\} \\
\mathfrak{g}_2^{2,4} &= \operatorname{span}\{e_{\alpha_1+\alpha_2}, e_{-2\alpha_1-\alpha_2}\} \\
\mathfrak{g}_2^{3,3} &= \operatorname{span}\{\mathfrak{t}_{\mathbb{C}}, e_{2\alpha_2+3\alpha_1}, e_{-2\alpha_2-3\alpha_1}\}\,.
\end{aligned}$$

**Note.** In the classical case of weight one polarized Hodge structures, no exceptional group may occur as a Mumford-Tate group.

Among the main results that come out of the analysis are

(A) *The adjoint group of a simple $\mathbb{Q}$-algebraic group $M$ is a Mumford-Tate group if, and only if, $M$ contains an anisotropic maximal torus.*

To explain the next result, we define a map

$$\psi : \mathfrak{r} \to \mathbb{Z}/2\mathbb{Z}$$

by

$$\psi(\alpha) = \begin{cases} 0 & \text{if } \alpha \text{ is a compact root} \\ 1 & \text{if } \alpha \text{ is a non-compact root.} \end{cases}$$

Since the Cartan involution $\theta : \mathfrak{m}_{\mathbb{R}} \to \mathfrak{m}_{\mathbb{R}}$ given by $\theta = 1$ on $\mathfrak{k}$ and $\theta = -1$ on $\mathfrak{p}$ is a Lie algebra homomorphism, this map extends linearly to a homomorphism

$$\psi : R \to \mathbb{Z}/2\mathbb{Z}$$

where $R \subset i\check{\mathfrak{t}}$ is the $\mathbb{Z}$-span of the roots, and we define $\Psi : R \to \mathbb{Z}/4\mathbb{Z}$ by

$$\Psi(\alpha) = \begin{cases} 0 & \text{if } \psi(\alpha) = 0 \\ 2 & \text{if } \psi(\alpha) = 1\,. \end{cases}$$

Thinking of $\mathfrak{l}_\varphi \in \Lambda$ as defining an element of $\operatorname{Hom}(R, \mathbb{Z})$, the above condition that $(\mathfrak{m}, B, \operatorname{Ad}\varphi)$ give a polarized Hodge structure may be expressed as

$$\mathfrak{l}_\alpha \equiv \Psi \pmod 4\,;$$

i.e., $\mathfrak{l}_\alpha$ should give a lifting of $\Psi$ in the diagram

$$\begin{array}{ccc} & & \mathbb{Z} \\ & \overset{\mathfrak{l}_\varphi}{\nearrow} & \downarrow \\ R & \xrightarrow{\Psi} & \mathbb{Z}/4\mathbb{Z}\,. \end{array}$$

The list of the non-compact real forms of the simple Lie algebras for which this condition can be satisfied is the following:

$A_r$ su$(p,q)$, $p+q=r+1$, $0 \le p,q \le r+1$; sl$_2$

$B_r$ so$(2p,2q+1)$, $p+q=r$, $0 \le p,q \le r$

$C_r$ sp$(p,q)$, $p+q=r$, $0 \le p,q \le r$; sp$(r)$

$D_r$ so$(2p,2q)$, $p+q=r$, $0 \le p,q \le r$; so$^*(2r)$

$E_6$ EII, EIII

$E_7$ EV, EVI, EVII

$E_8$ EVIII, EIX

$F_4$ FI, FII

$G_2$ G

Those that do not have solutions to the above conditions are

$$\text{sl}(n,\mathbb{R}) \text{ for } n \ge 3$$

$$\text{sl}(n,\mathbb{H}) \text{ for } n \ge 2$$

EI

EIV

As noted, it is more difficult to have a Hodge representation of odd weight. The following is the list of those that do.

$$\text{su}(4k), \text{so}(4k+2) \quad \text{(compact case)}$$

$$\text{su}(2p, 4k-2p), \text{su}(2p+1, 2q+1)$$

$$\text{so}(4p+2, 2q+1), \text{so}(2p, 2q) \ \text{for}\ p+q \text{ odd}$$

$$\text{so}^*(4k)$$

$$\text{sp}(r)$$

$$\text{EV}, \text{EVII}$$

Finally, without getting into the details we will let $P \subset i\mathfrak{t}$ denote the weight lattice and let $\lambda \in P$ be a highest weight. Define

$$P' = \mathbb{Z}\lambda + R\,,$$

so that we have

$$P \supseteq P' \supseteq R\,.$$

We let $\delta \in \mathbb{Z}^+$ be the minimal positive integer such that $\delta\lambda \in R$. Then for $M$ with $M(\mathbb{R})$ having the maximal torus

$$T = \mathfrak{t}/\Lambda$$

where $\Lambda = \operatorname{Hom}(P', \mathbb{Z})$, we have

(B) *There exists $\mathfrak{l}_\varphi \in \Lambda$ and an invariant form $Q$ such that the representation $\rho_\lambda$ with highest weight $\lambda$ gives a Hodge representation $(M, \rho_\lambda, \varphi_{\rho_\lambda})$ if, and only if,*

$$\Psi(\delta\lambda) \equiv \delta m \pmod 4$$

*for some integer $m$.*

This result provides a practical test for determining the Hodge representations.

**Example** ($G_2$-continued). We consider the standard representation of $G_2$ in $\mathrm{SO}(3,4)$. We use block matrices

$$\begin{matrix} & \overbrace{\quad}^{3} \;\; \overbrace{\quad}^{4} \\ \begin{matrix} 3\{ \\ 4\{ \end{matrix} & \begin{pmatrix} A & B \\ {}^tB & C \end{pmatrix} \end{matrix}$$

where $A = -{}^tA$, $C = -{}^tC$ is an element of $\mathrm{so}(3,4)$. Then $\mathfrak{g}_{2,\mathbb{R}} \subset \mathrm{so}(3,4)$ is defined by seven independent linear equations (whose exact form is not necessary for our purposes). Setting $E = \left(\begin{smallmatrix} 0 & -1 \\ 1 & 0 \end{smallmatrix}\right)$ and

$$H_1 = \left(\begin{array}{cc|cc} E & 0 & 0 & 0 \\ 0 & 0 & 0 & 0 \\ \hline 0 & 0 & E & 0 \\ 0 & 0 & 0 & 0 \end{array}\right) \qquad H_2 = \left(\begin{array}{cc|cc} 0 & 0 & 0 & 0 \\ 0 & 0 & 0 & 0 \\ \hline 0 & 0 & E & 0 \\ 0 & 0 & 0 & E \end{array}\right),$$

$H_1$ and $H_2$ give a basis for $\mathfrak{t}_\mathbb{C}$; the exponentials of $2\pi i$ times $\mathbb{R}H_1 + \mathbb{R}H_2$ give a maximal torus $T \subset M(\mathbb{R})$, and $T \cong \mathbb{R}^2/\Lambda$ where $\Lambda = \mathbb{Z}H_1 + \mathbb{Z}H_2$. We have

*The co-character $\varphi$ whose differential is $\mathfrak{l}_\varphi = l_1H_1 + l_2H_2$ gives a Hodge representation for* **every** *representation of $G_2$ if, and only if, the conditions*

$$l_1 \equiv 0 \pmod 4$$
$$l_2 \equiv 2 \pmod 4$$

*are satisfied.*

For the standard representation on $\mathbb{Q}^7$ we obtain the following illustrative cases:

(a) $l_1 = 0$, $l_2 = 2$. Then the weight $n = 2$ and the Hodge numbers $h^{2,0} = 2$, $h^{1,1} = 3$.

(b) $l_1 = 4$, $l_2 = -2$. Then the weight $n = 4$ and $h^{4,0} = 1$, $h^{3,1} = 2$, $h^{2,2} = 1$.

(c) $l_1 = 4$, $l_2 = 6$. Then $n = 6$ and all $h^{p,q} = 1$.

The Mumford-Tate groups in cases (a) and (c) appeared (not as Mumford-Tate groups, of course) in the classic 1905 paper of Elie Cartan in which $G_2$ was first realized geometrically as a transformation group.

## VI. Hodge domains

Given the pair $(M, \varphi)$ consisting of a semi-simple $\mathbb{Q}$-algebraic group $M$ containing an anisotropic maximal torus and a non-trivial circle $\rho : S^1 \to M(\mathbb{R})$, there may be many different $\rho$'s giving Hodge representations $(M, \rho, \varphi)$. For each of these there is a Mumford-Tate domain $D_{(M,\rho,\varphi)}$ contained in the corresponding period domain. A natural question is:

*What is the relation among the $D_{(M,\rho,\varphi)}$'s for a given $(M, \varphi)$?*

With the notation $D_{\mathfrak{m},\varphi}$ for the Mumford-Tate domain for $(\mathfrak{m}, B, \operatorname{Ad}\varphi)$ we have the result:

**Theorem.** *As homogeneous complex manifolds together with the invariant distribution in the tangent bundle given by the IPR,*

$$D_{(M,\rho,\varphi)} \cong D_{\mathfrak{m},\varphi}\,.$$

*Proof.* Denoting by $H_\varphi = Z_{M(\mathbb{R})}(\varphi(S^1))$ the compact centralizer in $M(\mathbb{R})$ of the circle $\varphi(S^1)$, we have

$$D_{(M,\rho,\varphi)} \cong M(\mathbb{R})/H_\varphi$$

independently of $\rho$. Remark that the finite center $Z_M$ of $M(\mathbb{R})$ is contained in the compact maximal torus $T \subset H_\varphi$, so that the RHS above may be replaced by the images of $M(\mathbb{R})$ and $H_\varphi$ in the adjoint group $M(\mathbb{R})_a = M(\mathbb{R})/Z_M$.

If we identify the tangent space to $D_{\mathfrak{m},\varphi}$ at the identity coset with $\mathfrak{m}^-$, then the subbundle $W$ of the tangent bundle has fibre $\mathfrak{w} = \mathfrak{m}^{-1,1}$; thus

$$\mathfrak{w} = \left\{ \begin{array}{c} (-2i)\text{-eigenspace of } \varphi(S^1) \\ \text{acting on } \mathfrak{m}^-\,. \end{array} \right\}$$ □

**Definition.** A *Hodge domain* is given by $D_{\mathfrak{m},\varphi} = M(\mathbb{R})/H_\varphi$ where $H_\varphi = Z_{M(\mathbb{R})}(\varphi(S^1))$ and $(M, \varphi)$ is as above.

Thus, Hodge domains are the *universal parameter spaces* for families of polarized Hodge structures whose generic Mumford-Tate group is $M$ and whose stability group of a fixed polarized Hodge structure is $H_\varphi$.

**Example.** This first non-classical case is when

$$D_{\mathfrak{m},\varphi} = \mathrm{SU}(2,1)/T\,.$$

The root diagram is (the non-compact roots are framed)

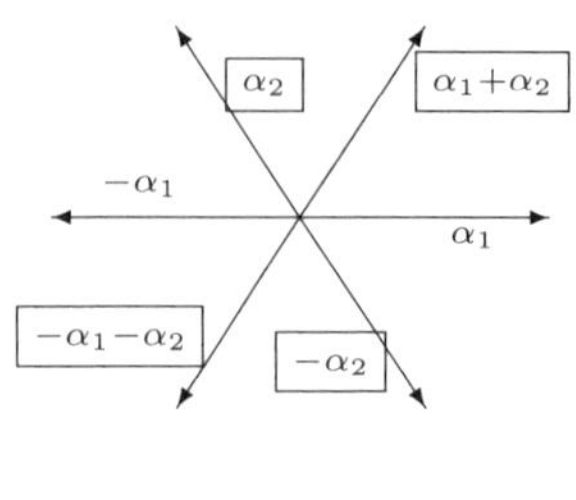


The co-characters that give Hodge domains are those $\mathfrak{l}_\varphi \in \Lambda \subset \mathfrak{t}$ such that

$$\begin{cases} \langle \alpha_1, \mathfrak{l}_\varphi \rangle = 4k_1 & k_1 \neq 0 \\ \langle \alpha_2, \mathfrak{l}_\varphi \rangle = 4k_2 + 2 & \end{cases}$$

where $k_1, k_2$ are integers. The root plane is divided into six Weyl chambers corresponding to the inequalities

| | Inequalities | Basis for $\mathfrak{m}^-$ |
|---|---|---|
| (i) | $k_1 > 0, k_2 \geq 0$ | $e_{\alpha_1}, e_{-\alpha_2}, e_{-\alpha_1-\alpha_2}$ |
| (ii) | $k_1 + k_2 \geq 0, k_2 < 0$ | $e_{-\alpha_1}, e_{\alpha_2}, e_{-\alpha_1-\alpha_2}$ |
| (iii) | $k_1 + k_2 \geq 0, k_1 < 0$ | $e_{\alpha_1}, e_{-\alpha_2}, e_{\alpha_1-\alpha_2}$ |
| (i)* | $k_1 < 0, k_2 < 0$ | $e_{\alpha_1}, e_{\alpha_2}, e_{\alpha_1+\alpha_2}$ |
| (ii)* | $k_1 + k_2 < 0, k_2 \geq 0$ | $e_{\alpha_1}, e_{-\alpha_2}, e_{\alpha_1+\alpha_2}$ |
| (iii)* | $k_1 + k_2 < 0, k_1 > 0$ | $e_{-\alpha_1}, e_{\alpha_2}, e_{\alpha_1+\alpha_2}$. |

Here (i)* is the conjugate complex structure to (i), etc.. Note that $\varphi$ induces a complex structure on $\mathrm{SU}(2,1)/T$, which is different in each of the six cases.

The complex structures (i)–(iii) may be visualized as follows (cf. [1] in the references): Taking $V = \mathbb{Q}^3$ and for $Q$ the form $\mathrm{diag}(-1,1,1)$, we have the unit ball $B = \{[z_0, z_1, z_2] : |z_1|^2 + |z_2|^2 < |z_0|^2\} \subset \mathbb{P}^2$. In $\mathbb{P}^2 \times \check{\mathbb{P}}^2$ there is the standard incidence correspondence

$$I = \{(P, L) : P \in L\},$$

which is the flag variety for the complex group $\mathrm{SL}(3,\mathbb{C}) \cong \mathrm{SU}(2,1)(\mathbb{C})$ acting on $\mathbb{P}^2$. The pictures of the three complex structures are

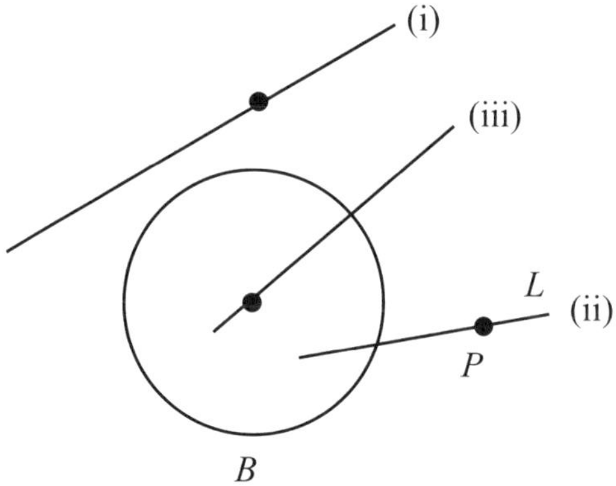


The complex structure (ii) is the non-classical one; i.e., it does *not* fibre holomorphically or anti-holomorphically over an Hermitian symmetric space. There is a unique choice of $\varphi$ for which the IPR is non-trivial and non-integrable; it is when $k_2 = -1$, $k_1 = 1$. Then

$$\mathfrak{m}^{-1,1} = \mathrm{span}\{e_{\alpha_2}, e_{-\alpha_1-\alpha_2}\}$$

and

$$[e_{\alpha_2}, e_{-\alpha_1-\alpha_2}] = ce_{-\alpha_2}, \qquad c \neq 0,$$

shows that the bracket of the 2-plane field given by the IPR is non-trivial. It is a contact structure; geometrically the 2-plane field is spanned by the two tangent vectors given by the conditions that to $1^{\text{st}}$ order either $P$ or $L$ does not move. The polarized Hodge structures on su(2, 1) have Hodge numbers

$$h^{4,0} = 1,\ h^{3,1} = 2,\ h^{2,2} = 2\,.$$

It is not known if there is a family of smooth, projective fourfolds having a non-trivial period map to a quotient of this $D_{\mathfrak{m},\varphi}$.[9]

We mention this example because it is indicative of the rich geometry of Mumford-Tate domains, and because it has been the subject of some very interesting work in representation theory and automorphic cohomology that goes beyond the classical case of Shimura varieties (cf. loc. cit.). In this regard we have:

- *The real forms $M(\mathbb{R})$ of non-compact Hodge groups are exactly those that have non-trivial discrete series representations in $L^2(M(\mathbb{R}))$;*
- *The $\mathbb{Q}$-algebraic Hodge groups whose real forms are non-compact are those for which one may hope to have cuspidal automorphic representations in $L^2(M(\mathbb{Q})\backslash M(\mathbb{A}))$.*

Aside from the work cited above, the rich confluence of

- algebraic geometry/Hodge theory
- representation theory (discrete series)
- arithmetic (automorphic forms)

present in the case when the Mumford-Tate domains parametrize the complex points of Shimura varieties seems to be largely unexplored in the higher weight case. In particular, one may ask what if any role the enhanced data $(M, \varphi)$, beyond just the $\mathbb{Q}$-algebraic group $M$, might play? For example, when there is a "$\varphi$" there is a well-defined notion of CM points in quotients $\Gamma\backslash D_{\mathfrak{m},\varphi}$ of Hodge domains by arithmetic groups. In the classical case these CM points are a central part of the story. In general, even though in the non-classical case $\Gamma\backslash D_{\mathfrak{m},\varphi}$ is not an algebraic variety, the CM points — and more generally the Noether-Lefschetz loci — have an arithmetic structure arising from the "$\varphi$" and that may play an important role.

## Selected references[10]

[1] H. Carayol, Cohomologie automorphe et compactifications partielles de certaines variétés de Griffiths-Schmid, *Compos. Math.* **141** (2005), 1081–1102.

[2] J. Carlson, S. Mueller-Stach, and C. Peters, *Period Mappings and Period Domains*, Cambridge Univ. Press, 2003, 559pp.

[3] M. Green, P. Griffiths, and M. Kerr, Mumford-Tate Groups and Domains, in preparation.

[4] B. Moonen, An introduction to Mumford-Tate groups, 2004, notes. Available at:
http://staff.science.uva.nl/~bmoonen/MyPapers/OwnWork.html

[5] V. Rotger, Shimura varieties and their canonical models, notes on the course, Centre de Recerca Matemàtica, Bellaterra, Spain, Sept. 20–23, 2005.

[7] W. Schmid, Discrete series, *Proc. Symp. Pure Math.* **61** (1997), 83–113.

[8] ———, Variation of Hodge structure: The singularities of the period mapping, *Invent. Math.* **22** (1973), 211–319.

[9] Since the IPR is non-integrable, they could not be Calabi-Yau's.

[10] These are a small sample; [2]–[5] contain extensive further references to the literature.

**Curriculum Vitae**

Mikhail Leonidovich Gromov

Date of Birth: December 23, 1943, Boksitogorsk, USSR

**Education:**

Masters degree (1965), Doctorate (1969) and Post-doctoral Thesis (1973) from Leningrad University

**Employment History:**

| | |
|---|---|
| 1982– | Permanent Professor, Institut des Hautes Études Scientifiques, Bures-sur-Yvette, France |
| 1996– | Jay Gould Professor of Mathematics at the Courant Institute of Mathematical Sciences, New York University |
| 1991–1996 | Professor of Mathematics, University of Maryland, College Park |
| 1981–1982 | Professor, Université de Paris VI |
| 1974–1981 | Professor, State University of New York at Stony Brook |
| 1967–1974 | Assistant Professor, Leningrad University, Soviet Union |

**Awards:**

| | |
|---|---|
| 1971 | Moscow Mathematical Society Prize |
| 1981 | Oswald Veblen Prize in Geometry, American Mathematical Society |
| 1984 | Élie Cartan Prize, French Academy of Sciences |
| 1989 | Prix de l'Union des Assurances de Paris |
| 1992 | Honorary Doctorate, University of Geneva |
| 1993 | Wolf Prize |
| 1997 | Leroy P. Steele Prize for a Seminal Contribution to Research, American Mathematical Society |
| 1997 | Lobatchewski Medal, Kazan State University |
| 1999 | Balzan Prize, International Balzan Prize Foundation |
| 2002 | Kyoto Prize in Basic Sciences, Inamori Foundation |
| 2004 | Frederic Esser Nemmers Prize in Mathematics, Northwestern University |
| 2005 | János Bolyai Prize, Hungarian Academy of Sciences |

**Membership Honorary Societies:**

| | |
|---|---|
| 1989 | Foreign Associate Member, US National Academy of Sciences |
| 1989 | Foreign Honorary Member of the American Academy of Arts and Sciences |

1989 Associé étranger de l'Académie des Sciences, France
1997 Membre de l'Institut de France, Académie des Sciences, France

## List of Publications

1. M. L. Gromov, Singularities, expanders and topology of maps, I. Homology versus volume in the spaces of cycles, *Geom. Funct. Anal.* **19** (2009), No. 3, 743–841.
2. M. L. Gromov, Erratum to: Isoperimetry of waists and concentration of maps GAFA 13:1 (2003), pp. 178–215, *Geom. Funct. Anal.* **18** (2009), No. 5, 1786.
3. M. L. Gromov, Mendelian dynamics and Sturtevant's paradigm, *Geometric and Probabilistic Structures in Dynamics*, pp. 227–242, Contemp. Math., Vol. 469 (Amer. Math. Soc., 2008).
4. T. Delzant and M. L. Gromov, Courbure mésoscopique et théorie de la toute petite simplification (in French) [Mesoscopic curvature and small cancellation theorem], *J. Topol.* **1** (2008), No. 4, 804–836.
5. M. L. Gromov, Entropy and isoperimetry for linear and non-linear group actions, *Groups Geom. Dyn.* **2** (2008), No. 4, 499–593.
6. M. L. Gromov, *Metric Structures for Riemannian and Non-Riemannian Spaces*, based on the 1981 French original. With appendices by M. Katz, P. Pansu and S. Semmes. Translated from the French by Sean Michael Bates. Reprint of the 2001 English edition. Modern Birkhäuser Classics (Birkhäuser, 2007).
7. T. Delzant and M. L. Gromov, Cuts in Kähler groups, *Infinite Groups: Geometric, Combinatorial and Dynamical Aspects*, pp. 31–55, Progr. Math., Vol. 248 (Birkhäuser, 2005).
8. M. Bertelson and M. L. Gromov, Dynamical Morse entropy, in *Modern Dynamical Systems and Applications*, pp. 27–44 (Cambridge Univ. Press, 2004).
9. M. L. Gromov, On the entropy of holomorphic maps, *Enseign. Math.* (2) **49** (2003), No. 3–4, 217–235.
10. A. Carbone and M. L. Gromov, Functional labels and syntactic entropy on DNA strings and proteins, in *Logic and Complexity in Computer Science* (Créteil, 2001). *Theoret. Comput. Sci.* **303** (2003), No. 1, 35–51.
11. M. L. Gromov, Isoperimetry of waists and concentration of maps, *Geom. Funct. Anal.* **13** (2003), No. 1, 178–215.
12. M. L. Gromov, Random walk in random groups, *Geom. Funct. Anal.* **13** (2003), No. 1, 73–146.
13. M. L. Gromov, Small cancellation, unfolded hyperbolicity, and transversal measures, in *Essays on Geometry and Related Topics*, Vols. 1, 2, pp. 371–399, Monogr. Enseign. Math., Vol. 38 (Enseignement Math., 2001).

14. M. L. Gromov, CAT($\kappa$)-spaces: Construction and concentration, *Zap. Nauchn. Sem. S.-Peterburg. Otdel. Mat. Inst. Steklov.* (*POMI*) **280** (2001), *Geom. i Topol.* 7; 100–140; 299–300; translation in *J. Math. Sci* (*N. Y.*) **119** (2004), No. 2, 178–200.
15. J.-P. Bourguignon, E. Calabi, J. Eells, O. Garcia-Prada and M. L. Gromov, Where does geometry go? A research and education perspective, in *Global Differential Geometry*: *The Mathematical Legacy of Alfred Gray* (Bilbao, 2000), pp. 442–457, Contemp. Math., Vol. 288 (Amer. Math. Soc., 2001).
16. M. L. Gromov, Mesoscopic curvature and hyperbolicity, in *Global Differential Geometry*: *The Mmathematical Legacy of Alfred Gray* (Bilbao, 2000), pp. 58–69, Contemp. Math., Vol. 288 (Amer. Math. Soc., 2001).
17. M. L. Gromov, Possible trends in mathematics in the coming decades (in Ukrainian), translated from *Notices Amer. Math. Soc.* **45** (1998), No. 7, 846–847 by Grigoriĭ Shurenkov and Ol'ga Yadrenko. *Svītī Mat.* **7** (2001), No. 1, 3–5.
18. M. L. Gromov, Possible trends in mathematics in the coming decades, *Mathematics Unlimited — 2001 and Beyond*, pp. 525–527 (Springer, 2001).
19. A. Carbone and M. L. Gromov, Mathematical slices of molecular biology. With an introduction in French by Éric Westhof, *Gaz. Math.* No. 88, suppl. (2001).
20. M. L. Gromov, Spaces and questions, GAFA 2000 (Tel Aviv, 1999), *Geom. Funct. Anal.* **2000**, Special Volume, Part I, pp. 118–161.
21. M. L. Gromov, Three remarks on geodesic dynamics and fundamental group, *Enseign. Math.* (2) **46** (2000), No. 3–4, 391–402.
22. M. L. Gromov, Topological invariants of dynamical systems and spaces of holomorphic maps, I. *Math. Phys. Anal. Geom.* **2** (1999), No. 4, 323–415.
23. M. L. Gromov, *Metric Structures for Riemannian and Non-Riemannian Spaces*. Based on the 1981 French original. With appendices by M. Katz, P. Pansu and S. Semmes. Translated from the French by Sean Michael Bates. Progress in Mathematics, Vol. 152 (Birkhäuser, 1999).
24. M. L. Gromov, Endomorphisms of symbolic algebraic varieties, *J. Eur. Math. Soc.* (*JEMS*) **1** (1999), No. 2, 109–197.
25. M. L. Gromov, Quantitative homotopy theory, in *Prospects in Mathematics* (Princeton, NJ, 1996), pp. 45–49 (Amer. Math. Soc., 1999).
26. M. L. Gromov, Possible trends in mathematics in the coming decades, *Notices Amer. Math. Soc.* **45** (1998), No. 7, 846–847.
27. Y. Eliashberg and M. L. Gromov, Lagrangian intersection theory: Finite-dimensional approach, in *Geometry of Differential Equations*, pp. 27–118, Amer. Math. Soc. Transl. Ser. 2, Vol. 186 (Amer. Math. Soc., 1998).

28. M. L. Gromov, G. Henkin and M. Shubin, Holomorphic $L_2$ functions on coverings of pseudoconvex manifolds, *Geom. Funct. Anal.* **8** (1998), No. 3, 552–585.
29. M. L. Gromov, G. Henkin and M. Shubin, $L_2$ holomorphic functions on pseudo-convex coverings, in *Operator Theory for Complex and Hypercomplex Analysis* (Mexico City, 1994), pp. 81–94, Contemp. Math., Vol. 212 (Amer. Math. Soc., 1998).
30. Y. Eliashberg and M. L. Gromov, Lagrangian intersections and the stable Morse theory, *Boll. Un. Mat. Ital. B* (7) **11** (1997), No. 2, suppl., 289–326.
31. M. L. Gromov, Systoles and intersystolic inequalities, in *Actes de la Table Ronde de Géométrie Différentielle* (Luminy, 1992), pp. 291–362, Sémin. Congr., Vol. 1 (Soc. Math. France, 1996).
32. M. L. Gromov, Carnot–Carathéodory spaces seen from within, in *Sub-Riemannian Geometry*, pp. 79–323, Progr. Math., Vol. 144 (Birkhäuser, 1996).
33. M. L. Gromov, Positive curvature, macroscopic dimension, spectral gaps and higher signatures, in *Functional Analysis on the Eve of the 21st Century*, Vol. II (New Brunswick, NJ, 1993), pp. 1–213, Progr. Math., Vol. 132 (Birkhäuser, 1996).
34. M. L. Gromov, Geometric reflections on the Novikov conjecture, in *Novikov Conjectures, Index Theorems and Rigidity*, Vol. 1 (Oberwolfach, 1993), pp. 164–173, London Math. Soc. Lecture Note Ser., Vol. 226 (Cambridge Univ. Press, 1995).
35. M. L. Gromov and M. Shubin, Erratum to: "von Neumann spectra near zero" [*Geom. Funct. Anal.* **1** (1991), No. 4, 375–404], *Geom. Funct. Anal.* **5** (1995), No. 4, 729.
36. M. L. Gromov and M. A. Shubin, The Riemann–Roch theorem for elliptic operators and solvability of elliptic equations with additional conditions on compact subsets, *Invent. Math.* **117** (1994), No. 1, 165–180.
37. M. L. Gromov, Metric invariants of Kähler manifolds, *Differential Geometry and Topology* (Alghero, 1992), pp. 90–116 (World Scientific, 1993).
38. M. L. Gromov, Asymptotic invariants of infinite groups, in *Geometric Group Theory*, Vol. 2 (Sussex, 1991), pp. 1–295, London Math. Soc. Lecture Note Ser., Vol. 182 (Cambridge Univ. Press, 1993).
39. M. L. Gromov and M. A. Shubin, The Riemann–Roch theorem for elliptic operators and solvability of elliptic equations with additional conditions on compact subsets, in *Journées "Équations aux Dérivées Partielles" (Saint-Jean-de-Monts, 1993)*, Exp. No. XVIII (École Polytech., 1993).

40. M. L. Gromov and M. A. Shubin, The Riemann–Roch theorem for elliptic operators, in *I. M. Gel'fand Seminar*, pp. 211–241, Adv. Soviet Math., Vol. 16, Part 1 (Amer. Math. Soc., 1993).
41. A. Connes, M. L. Gromov and H. Moscovici, Group cohomology with Lipschitz control and higher signatures, *Geom. Funct. Anal.* **3** (1993), No. 1, 1–78.
42. M. L. Gromov and M. A. Shubin, Near-cohomology of Hilbert complexes and topology of non-simply connected manifolds, in *Méthodes Semi-Classiques*, Vol. 2 (Nantes, 1991). *Astérisque*, No. 210 (1992), pp. 9–10; pp. 283–294.
43. M. L. Gromov and R. Schoen, Harmonic maps into singular spaces and $p$-adic superrigidity for lattices in groups of rank one, *Inst. Hautes Études Sci. Publ. Math.* No. 76 (1992), 165–246.
44. M. L. Gromov, Stability and pinching, in *Geometry Seminars. Sessions on Topology and Geometry of Manifolds* (in Italian) (Bologna, 1990), pp. 55–97 (Univ. Stud. Bologna, 1992).
45. Yu. Burago, M. L. Gromov and G. A. D. Perel'man, Aleksandrov spaces with curvatures bounded below (in Russian), *Usp. Mat. Nauk* **47** (1992), No. 2(284), 3–51; 222; translation in *Russ. Math. Surv.* **47** (1992), No. 2, 1–58.
46. Y. Eliashberg and M. L. Gromov, Embeddings of Stein manifolds of dimension $n$ into the affine space of dimension $3n/2+1$, *Ann. Math* (2) **136** (1992), No. 1, 123–135.
47. M. L. Gromov, Spectral geometry of semi-algebraic sets, *Ann. Inst. Fourier (Grenoble)* **42** (1992), No. 1–2, 249–274.
48. M. L. Gromov and M. A. Shubin, The Riemann–Roch theorem for general elliptic operators, *C. R. Acad. Sci. Paris Sér. I Math.* **314** (1992), No. 5, 363–367.
49. J. Cheeger, K. Fukaya and M. L. Gromov, Nilpotent structures and invariant metrics on collapsed manifolds, *J. Amer. Math. Soc.* **5** (1992), No. 2, 327–372.
50. M. L. Gromov, Sign and geometric meaning of curvature, *Rend. Sem. Mat. Fis. Milano* **61** (1991), 9–123 (1994).
51. J. Cheeger and M. L. Gromov, Chopping Riemannian manifolds, in *Differential Geometry*, pp. 85–94, Pitman Monogr. Surveys Pure Appl. Math., Vol. 52 (Longman, 1991).
52. M. L. Gromov and P. Pansu, Rigidity of lattices: An introduction, in *Geometric Topology: Recent Developments* (Montecatini Terme, 1990), pp. 39–137, Lecture Notes in Math., Vol. 1504 (Springer, 1991).

53. G. D'Ambra and M. L. Gromov, Lectures on transformation groups: Geometry and dynamics, in *Surveys in Differential Geometry* (Cambridge, MA, 1990), pp. 19–111 (Lehigh Univ., 1991).
54. M. L. Gromov and M. A. Shubin, von Neumann spectra near zero, *Geom. Funct. Anal.* **1** (1991), No. 4, 375–404.
55. Y. Eliashberg and M. L. Gromov, Convex symplectic manifolds, in *Several Complex Variables and Complex Geometry*, Part 2 (Santa Cruz, CA, 1989), pp. 135–162, Proc. Sympos. Pure Math., Vol. 52, Part 2 (Amer. Math. Soc., 1991).
56. M. L. Gromov, Foliated Plateau problem, II, Harmonic maps of foliations, *Geom. Funct. Anal.* **1** (1991), No. 3, 253–320.
57. M. L. Gromov, Foliated Plateau problem, I, Minimal varieties, *Geom. Funct. Anal.* **1** (1991), No. 1, 14–79.
58. M. L. Gromov, Kähler hyperbolicity and $L_2$-Hodge theory, *J. Differential Geom.* **33** (1991), No. 1, 263–292.
59. M. L. Gromov, Convex sets and Kähler manifolds, in *Advances in Differential Geometry and Topology*, pp. 1–38 (World Scientific, 1990).
60. M. L. Gromov, *Differentsial'nye sootnosheniya s chastnymi proizvodnymi* (in Russian) [Partial differential relations], translated from English and with a preface and an appendix by N. M. Mishachëv (Mir, 1990).
61. J. Cheeger and M. L. Gromov, Collapsing Riemannian manifolds while keeping their curvature bounded, II, *J. Differential Geom.* **32** (1990), No. 1, 269–298.
62. A. Connes, M. L. Gromov and H. Moscovici, Conjecture de Novikov et fibrés presque plats (in French) [The Novikov conjecture and almost flat vector bundles], *C. R. Acad. Sci. Paris Sér. I Math.* **310** (1990), No. 5, 273–277.
63. M. L. Gromov, Soft differential equations, in *IX International Congress on Mathematical Physics* (Swansea, 1988), pp. 374–376 (Hilger, 1989).
64. J. Bourgain and M. L. Gromov, Estimates of Bernstein widths of Sobolev spaces, in *Geometric Aspects of Functional Analysis* (1987–88), pp. 176–185, Lecture Notes in Math., Vol. 1376 (Springer, 1989).
65. M. L. Gromov, Oka's principle for holomorphic sections of elliptic bundles, *J. Amer. Math. Soc.* **2** (1989), No. 4, 851–897.
66. M. L. Gromov, Sur le groupe fondamental d'une variété kählérienne (in French) [On the fundamental group of a Kähler manifold], *C. R. Acad. Sci. Paris Sér. I Math.* **308** (1989), No. 3, 67–70.
67. M. L. Gromov, Soft and hard symplectic geometry, A plenary address presented at the *International Congress of Mathematicians* held in

Berkeley, California, August 1986. Introduced by Jeff Cheeger, ICM Series (Amer. Math. Soc., 1988).

68. M. L. Gromov, H. B. Lawson, Jr. and W. Thurston, Hyperbolic 4-manifolds and conformally flat 3-manifolds, *Inst. Hautes Études Sci. Publ. Math.* No. 68 (1988), 27–45 (1989).
69. M. L. Gromov, Width and related invariants of Riemannian manifolds, in *On the Geometry of Differentiable Manifolds* (Rome, 1986), *Astérisque*, No. 163–164 (1988), 6; 93–109; 282.
70. M. L. Gromov, Rigid transformations groups, *Géométrie Différentielle* (Paris, 1986), pp. 65–139, Travaux en Cours, Vol. 33 (Hermann, 1988).
71. M. L. Gromov, Dimension, nonlinear spectra and width, in *Geometric Aspects of Functional Analysis* (1986/87), pp. 132–184, Lecture Notes in Math., Vol. 1317 (Springer, 1988).
72. M. L. Gromov and I. Piatetski-Shapiro, Nonarithmetic groups in Lobachevsky spaces, *Inst. Hautes Études Sci. Publ. Math.* No. 66 (1988), 93–103.
73. M. L. Gromov, Soft and hard symplectic geometry, in *Proc. of the International Congress of Mathematicians*, Vols. 1, 2 (Berkeley, Calif., 1986), pp. 81–98 (Amer. Math. Soc., 1987).
74. M. L. Gromov, Cauchy–Riemann equation in Lagrange intersection theory, in *Periodic Solutions of Hamiltonian Systems and Related Topics* (Il Ciocco, 1986), pp. 175–176, NATO Adv. Sci. Inst. Ser. C Math. Phys. Sci., Vol. 209 (Reidel, 1987).
75. M. L. Gromov, Hyperbolic groups, in *Essays in Group Theory*, pp. 75–263, Math. Sci. Res. Inst. Publ., Vol. 8 (Springer, 1987).
76. M. L. Gromov, Monotonicity of the volume of intersection of balls, in *Geometrical Aspects of Functional Analysis* (1985/86), pp. 1–4, Lecture Notes in Math., Vol. 1267 (Springer, 1987).
77. M. L. Gromov and V. D. Milman, Generalization of the spherical isoperimetric inequality to uniformly convex Banach spaces, *Compositio Math.* **62** (1987), No. 3, 263–282.
78. M. L. Gromov and W. Thurston, Pinching constants for hyperbolic manifolds, *Invent. Math.* **89** (1987), No. 1, 1–12.
79. M. L. Gromov, Entropy, homology and semialgebraic geometry, in *Séminaire Bourbaki*, Vol. 1985/86. *Astérisque*, No. 145–146 (1987), 5; 225–240.
80. M. L. Gromov, Partial differential relations, in *Ergebnisse der Mathematik und ihrer Grenzgebiete* (3) [Results in Mathematics and Related Areas (3)], Vol. 9 (Springer-Verlag, 1986).

81. M. L. Gromov, Large Riemannian manifolds, in *Curvature and Topology of Riemannian Manifolds* (Katata, 1985), pp. 108–121, Lecture Notes in Math., Vol. 1201 (Springer, 1986).
82. J. Cheeger and M. L. Gromov, Collapsing Riemannian manifolds while keeping their curvature bounded, I, *J. Differential Geom.* **23** (1986), No. 3, 309–346.
83. J. Cheeger and M. L. Gromov, $L_2$-cohomology and group cohomology, *Topology* **25** (1986), No. 2, 189–215.
84. M. L. Gromov, Isometric immersions of Riemannian manifolds, in *The Mathematical Heritage of Élie Cartan* (Lyon, 1984). *Astérisque* **1985**, Numero Hors Serie, 129–133.
85. W. Ballmann, M. L. Gromov and V. Schröder, *Manifolds of Nonpositive Curvature*, Progress in Mathematics, Vol. 61 (Birkhäuser, 1985).
86. M. L. Gromov, Pseudo Holomorphic curves in symplectic manifolds, *Invent. Math.* **82** (1985), No. 2, 307–347.
87. J. Cheeger and M. L. Gromov, Bounds on the von Neumann dimension of $L_2$-cohomology and the Gauss–Bonnet theorem for open manifolds, *J. Differential Geom.* **21** (1985), No. 1, 1–34.
88. J. Cheeger and M. L. Gromov, On the characteristic numbers of complete manifolds of bounded curvature and finite volume, *Differential Geometry and Complex Analysis*, pp. 115–154 (Springer, 1985).
89. M. L. Gromov and V. D. Milman, Brunn theorem and a concentration of volume phenomena for symmetric convex bodies, in *Israel Seminar on Geometrical Aspects of Functional Analysis* (1983/84), V (Tel Aviv Univ., 1984).
90. M. L. Gromov, Infinite groups as geometric objects, in *Proc. of the International Congress of Mathematicians*, Vols. 1, 2 (Warsaw, 1983), pp. 385–392 (PWN, 1984).
91. M. L. Gromov, Asymptotic geometry of homogeneous spaces, in *Conference on Differential Geometry on Homogeneous Spaces* (Turin, 1983), *Rend. Sem. Mat. Univ. Politec. Torino* **1983**, Special Issue, pp. 59–60 (1984).
92. M. L. Gromov and V. D. Milman, A topological application of the isoperimetric inequality, *Amer. J. Math.* **105** (1983), No. 4, 843–854.
93. M. L. Gromov and H. B. Lawson, Jr., Positive scalar curvature and the Dirac operator on complete Riemannian manifolds, *Inst. Hautes Études Sci. Publ. Math.* No. 58 (1983), 83–196.
94. M. L. Gromov, Filling Riemannian manifolds, *J. Differential Geom.* **18** (1983), No. 1, 1–147.

95. M. L. Gromov, Volume and bounded cohomology, *Inst. Hautes Études Sci. Publ. Math.* No. **56** (1982), 5–99.
96. J. Cheeger, M. L. Gromov and M. Taylor, Finite propagation speed, kernel estimates for functions of the Laplace operator, and the geometry of complete Riemannian manifolds, *J. Differential Geom.* **17** (1982), No. 1, 15–53.
97. M. L. Gromov, Hyperbolic manifolds (according to Thurston and Jørgensen), in *Bourbaki Seminar*, Vol. 1979/80, pp. 40–53, Lecture Notes in Math., Vol. 842 (Springer, 1981).
98. M. L. Gromov, Structures métriques pour les variétés riemanniennes (in French) [Metric structures for Riemann manifolds], edited by J. Lafontaine and P. Pansu. Textes Mathématiques, Vol. 1 (CEDIC, 1981).
99. M. L. Gromov, Groups of polynomial growth and expanding maps, *Inst. Hautes Études Sci. Publ. Math.* No. 53 (1981), 53–73.
100. M. L. Gromov, Hyperbolic manifolds, groups and actions, in *Riemann Surfaces and Related Topics: Proc. of the 1978 Stony Brook Conference* (State Univ. New York, Stony Brook, N.Y., 1978), pp. 183–213, Ann. of Math. Stud., Vol. 97 (Princeton Univ. Press, 1981).
101. M. L. Gromov, Curvature, diameter and Betti numbers, *Comment. Math. Helv.* **56** (1981), No. 2, 179–195.
102. M. Brin and M. L. Gromov, On the ergodicity of frame flows, *Invent. Math.* **60** (1980), No. 1, 1–7.
103. M. L. Gromov and H. B. Lawson, Jr., The classification of simply connected manifolds of positive scalar curvature, *Ann. of Math.* (2) **111** (1980), No. 3, 423–434.
104. M. L. Gromov and H. B. Lawson, Jr., Spin and scalar curvature in the presence of a fundamental group, I, *Ann. of Math.* (2) **111** (1980), No. 2, 209–230.
105. M. L. Gromov, Synthetic geometry in Riemannian manifolds, *Proc. of the International Congress of Mathematicians* (Helsinki, 1978), pp. 415–419 (Acad. Sci. Fennica, 1980).
106. M. L. Gromov, Homotopical effects of dilatation, *J. Differential Geom.* **13** (1978), No. 3, 303–310.
107. M. L. Gromov, Almost flat manifolds, *J. Differential Geom.* **13** (1978), No. 2, 231–241.
108. M. L. Gromov, Manifolds of negative curvature, *J. Differential Geom.* **13** (1978), No. 2, 223–230.
109. M. L. Gromov, Convex integration of differential relations, I (in Russian), *Izv. Akad. Nauk SSSR Ser. Mat.* **37** (1973), 329–343.

110. M. L. Gromov and Ya. M. Eliashberg, Construction of a smooth mapping with a prescribed Jacobian, I (in Russian), *Functional. Anal. i Prilozhen.* **7** (1973), No. 1, 33–40.
111. M. L. Gromov, Degenerate smooth mappings (in Russian), *Mat. Zametki* **14** (1973), 509–516.
112. M. L. Gromov, Smoothing and inversion of differential operators (in Russian), *Mat. Sb. (N.S.)* **88**(130) (1972), 382–441.
113. M. L. Gromov, A topological technique for the construction of solutions of differential equations and inequalities, in *Actes du Congrès International des Mathématiciens* (Nice, 1970), Tome 2, pp. 221–225 (Gauthier-Villars, 1971).
114. M. L. Gromov and Ya. M. Eliashberg, Construction of nonsingular isoperimetric films (in Russian), Boundary value problems of mathematical physics, Vol. 7. *Trudy Mat. Inst. Steklov.* **116** (1971), 18–33; 235.
115. M. L. Gromov and Ya. M. Eliashberg, Elimination of singularities of smooth mappings (in Russian), *Izv. Akad. Nauk SSSR Ser. Mat.* **35** (1971), 600–626.
116. M. L. Gromov and Ya. M. Eliashberg, Nonsingular mappings of Stein manifolds (in Russian), *Functional. Anal. i Prilozhen.* **5** (1971), No. 2, 82–83.
117. M. L. Gromov and V. A. Rohlin, Imbeddings and immersions in Riemannian geometry (in Russian), *Usp. Mat. Nauk* **25** (1970), No. 5(155), 3–62.
118. M. L. Gromov, Isometric imbeddings and immersions (in Russian), *Dokl. Akad. Nauk SSSR* **192** (1970), 1206–1209.
119. M. L. Gromov, Stable mappings of foliations into manifolds (in Russian), *Izv. Akad. Nauk SSSR Ser. Mat.* **33** (1969), 707–734.
120. M. L. Gromov, Simplexes inscribed on a hypersurface (in Russian), *Mat. Zametki* **5** (1969), 81–89.
121. M. L. Gromov, Mappings of foliations into manifolds which are equipped with supplementary structure (in Russian), *Vestnik Leningrad. Univ.* **23** (1968), No. 19, 167.
122. M. L. Gromov, Transversal mappings of foliations (in Russian), *Dokl. Akad. Nauk SSSR* **182** (1968), 255–258.
123. M. L. Gromov, The number of simplexes in subdivisions of finite complexes (in Russian), *Mat. Zametki* **3** (1968), 511–522.
124. M. L. Gromov, On a geometric hypothesis of Banach (in Russian), *Izv. Akad. Nauk SSSR Ser. Mat.* **31** (1967), 1105–1114.

# Forum

## Possible Trends in Mathematics in the Coming Decades

*Mikhael Gromov*

Here are a few brief remarks on possible trends in mathematics for the coming decades.

1. Classical mathematics is a quest for structural harmony. It began with the realization by ancient Greek geometers that our 3-dimensional continuum possessed a remarkable (rotational and translational) symmetry (groups $O(3)$ and $\mathbb{R}^3$) which permeates the essential properties of the physical world. (We stay intellectually blind to this symmetry no matter how often we encounter and use it in everyday life while generating or experiencing mechanical motion, e.g., walking. This is partly due to noncommutability of $O(3)$, which is hard to grasp.) Then deeper (noncommutative) symmetries were discovered: Lorentz and Poincaré in relativity, gauge groups for elementary particles, Galois symmetry in the algebraic geometry and number theory, etc. And similar mathematics appears once again on a less fundamental level, e.g., in crystals and quasicrystals, in self-similarity for fractals, dynamical systems and statistical mechanics, in monodromies for differential equations, etc.

*Mikhael Gromov is professor of mathematics at the Institut des Hautes Études Scientifiques, France. His e-mail address is* gromov@ihes.fr.

*Article adapted from Appendix 3 of the "Report of the Senior Assessment Panel of the International Assessment of the U.S. Mathematical Sciences", National Science Foundation, March 1998.*

The search for symmetries and regularities in the structure of the world will stay at the core of the pure mathematics (and physics). Occasionally (and often unexpectedly) some symmetric patterns discovered by mathematicians will have practical as well as theoretical applications. We saw this happening many times in the past; for example, integral geometry lies at the base of the x-ray tomography (CAT scan), the arithmetic over prime numbers leads to generation of perfect codes, and infinite-dimensional representations of groups suggests a design of large economically efficient networks of a high connectivity.

2. As the body of mathematics grew, it became itself subject to a logical and mathematical analysis. This has led to the creation of mathematical logic and then of the theoretical computer science. The latter is now coming of age. It absorbs ideas from the classical mathematics and benefits from the technological progress in the computer hardware which leads to a practical implementation of theoretically devised algorithms. (Fast Fourier transform and fast multipole algorithm are striking examples of the impact of pure mathematics on numerical methods used every day by engineers.) And the logical computational ideas interact with other fields, such as the quantum computer project, DNA-based molecular design, pattern formation in biology, the dynamics of the brain, etc. One expects that in several decades computer science will develop ideas on even deeper mathematical levels which will be followed by radical progress in the industrial application of computers, e.g., a (long overdue) breakthrough in artificial intelligence and robotics.

3. There is a wide class of problems, typically coming from experimental science (biology, chemistry, geophysics, medical science, etc.), where one has to deal with huge amounts of loosely structured data. Traditional mathematics, probability theory, and mathematical statistics work pretty well when the structure in question is essentially absent. (Paradoxically, the lack of structural organization and of correlations on the local level lead to a high

degree of overall symmetry. Thus the Gauss law emerges in the sums of random variables.) But often we have to encounter structured data where classical probability does not apply. For example, mineralogical formations or microscopic images of living tissues harbor (unknown) correlations which have to be taken into account. (What we ordinarily "see" is not the "true image" but the result of the scattering of some wave: light, x-ray, ultrasound, seismic wave, etc.) More theoretical examples appear in percolation theory, in self-avoiding random walk (modelling long molecular chains in solvents), etc. Such problems, stretching between clean symmetry and pure chaos, await the emergence of a new brand of mathematics. To make progress one needs radical theoretical ideas, as well as new ways of doing mathematics with computers and closer collaboration with scientists in order to match mathematical theories with available experimental data. (The wavelet analysis of signals and images, context-dependent inverse scattering techniques, geometric scale analysis, and x-ray diffraction analysis of large molecules in crystallized form indicate certain possibilities.)

We shall need ... the creation of a new breed of mathematical professionals able to mediate between pure mathematics and applied science. The cross-fertilization of ideas is crucial for the health of the science and mathematics.

Both the theoretical and industrial impacts of this development will be enormous. For example, an efficient inverse scattering algorithm would revolutionize medical diagnostics, making ultrasonic devices at least as efficient as current x-ray analysis.

4. As the power of computers approaches the theoretical limit and as we turn to more realistic (and thus more complicated) problems, we face the "curse of dimension", which stands in the way of successful implementations of numerics in science and engineering. Here one needs a much higher level of mathematical sophistication in computer architecture as well as in computer programming, along with the ideas indicated above in (2) and (3). Successes here may provide theoretical means for performing computations with high power growing arrays of data.

5. We must do a better job of educating and communicating ideas. The volume, depth, and structural complexity of the present body of mathematics make it imperative to find new approaches for communicating mathematical discoveries from one domain to another and drastically improving the accessibility of mathematical ideas to nonmathematicians. As matters stand now, we mathematicians often have little idea of what is going on in science and engineering, while experimental scientists and engineers are in many cases unaware of opportunities offered by progress in pure mathematics. This dangerous imbalance must be restored by bringing more science to the education of mathematicians and by exposing future scientists and engineers to core mathematics. This will require new curricula and a great effort on the part of mathematicians to bring fundamental mathematical techniques and ideas (especially those developed in the last decades) to a broader audience. We shall need for this the creation of a new breed of mathematical professionals able to mediate between pure mathematics and applied science. The cross-fertilization of ideas is crucial for the health of the science and mathematics.

6. We must strengthen financing of mathematical research. As we use more computer power and tighten collaboration with science and industry, we need more resources to support the dynamic state of mathematics. Even so, we shall need significantly less than other branches of science, so that the ratio of return/investment remains highest for mathematics, especially if we make a significant effort to popularize and apply our ideas. So it is important for us to make society well aware of the full potential of mathematical research and of the crucial role of mathematics in near- and long-term industrial development.

Uspekhi Mat. Nauk 47:2 (1992), 3 – 51 Russian Math. Surveys 47:2 (1992), 1 – 58

# A.D. Alexandrov spaces with curvature bounded below[1]

**Yu. Burago, M. Gromov, and G. Perel'man**

CONTENTS



## §1. Introduction

**1.1.** In this paper we develop the theory of (basically finite-dimensional) metric spaces with curvature (in the sense of Alexandrov) bounded below [1], [2]. We are talking, roughly speaking, about spaces with an intrinsic metric, for which the conclusion of Toponogov's angle comparison theorem is true (although only in the small); for precise definitions see §2. These spaces are defined axiomatically by their local geometric properties, without the techniques of analysis. They may have metric and topological singularities, in particular, they may not be manifolds. The class considered includes all limit spaces of sequences of complete Riemannian manifolds with sectional curvature uniformly bounded below. Alexandrov spaces arise naturally if Riemannian manifolds are considered from the viewpoint of synthetic geometry and one avoids the excessive assumptions of smoothness connected

---

[1] We have used this spelling of Alexandrov, rather than Aleksandrov, because of the title of the paper (in English) mentioned on p. 4. (Ed.)

with the use of analytic technique. Spaces with singularities may appear as limits of sequences of ordinary Riemannian manifolds, and therefore the first are necessary for studying the latter.

**1.2.** We recall that the class of Riemannian manifolds which have dimension and diameter bounded above, and sectional curvatures bounded below, by fixed numbers, is precompact in the Hausdorff topology (see [17], [19], [20]), but the limit spaces may not be Riemannian.

There are two reasons for the appearance of singularities in passing to the limit. The first is that the curvature may not be bounded above; the second is the appearance of so-called collapsing, when the dimension decreases on passing to the limit. If the curvature is bounded on both sides and there is no collapsing, then the limit space is a Riemannian manifold (of class $C^{2-\varepsilon}$ for any $\varepsilon > 0$), see [19], [28], [29], [15]. (Note that the $C^{2-\varepsilon}$-smoothness of the limit manifolds, proved in [28], [29], [15], is also implied by earlier results of Nikolaev [26], see also [8].)

The local structure of the limit space has been thoroughly studied [13] when collapsing occurs with two-sided bounds for the curvature: the limit in this case is the quotient space of a Riemannian manifold (of class $C^{2-\varepsilon}$) with respect to the isometric action of a Lie group, whose identity component is nilpotent.

The simplest singularities that occur when the curvature is not bounded above can be seen on non-smooth (for example polyhedral) convex hypersurfaces in $\mathbb{R}^n$. When collapsing occurs the limit space may have much more complicated singularities.

The topology and geometry of spaces that are limits of sequences of Riemannian manifolds with curvature bounded only below were studied (not as the basic object) in [23] and [22].

We emphasize that the synthetic approach used in this paper enables us to study the structure of a space without depending on whether it is obtained as the result of collapsing, or is the limit of Riemannian manifolds of the same dimension, or has some other origin. Grove has, independently of the authors, also found it necessary to consider spaces with curvature bounded below irrespective of their origin. We thank Professor K. Grove for letting us see his manuscript containing the plan of research and some results about spaces with curvature bounded below.

**1.3.** Spaces with curvature bounded below, along with spaces with curvature bounded above, were introduced by Alexandrov [1], [2] as generalized Riemannian manifolds as long ago as the fifties. In his research Alexandrov paid much attention, in essence, to the axiomatic question of describing such spaces in terms of the excesses of triangles (the excess is the amount by which the sum of the angles of a triangle exceeds $\pi$). In subsequent papers by Alexandrov, Reshetnyak, Nikolaev and others (see the surveys [4], [8]) spaces with curvature bounded above were investigated, mostly by assuming, a priori, that the geodesic between any two points in the domain being considered was unique. We note that, without additional assumptions, the condition that the curvature is bounded above is not, generally speaking, inherited when passing to the limit. Berestovskii and Nikolaev proved that a space with a two-sided bound for the curvature (and without boundary) is a Riemannian manifold (of smoothness $C^{2-\varepsilon}$ for any $\varepsilon > 0$); for more details see [8].

**1.4.** We describe briefly the contents of the paper. In §2 we give the basic definitions and those known facts about the spaces being considered that we shall need later on. §4 contains a list of standard constructions leading to spaces with curvature bounded below; in particular, it extends the collection of examples of such spaces given in §2. Some are given separately in §3. There it is proved, roughly speaking, that for spaces with curvature bounded below Toponogov's comparison theorem remains true. In other words, it is proved that the condition that the curvature is bounded below, which in defining the class under consideration enters, a priori, as a local condition, must in fact be satisfied globally, that is, for "arbitrarily large" objects.

The presentation in §3 may seem somewhat more difficult than in the next three sections. However, the reader who is interested only in questions of local structure may omit §3 without loss of understanding. In the rest of the paper one can simply start from stronger axioms, assuming that the condition that the curvature is bounded below is satisfied, not in the small, but globally. However, the basis for certain examples will be made more difficult. Possibly the aesthetic feelings of the reader will suffer.

In §§5–7 we establish the foundations for the theory of finite-dimensional spaces with curvature bounded below. We bear in mind that we have to create the apparatus we need and the first stages for studying the local structure of such spaces $X$.

The starting point is the concept of an explosion of the space $X$ at a point $p$ that is characterized by the presence of a collection of geodesics going out from $p$, which to some extent calls to mind the axes of a Cartesian coordinate system. From this basic concept it is proved that the set $X_0$, which is to a certain extent the burst points of $X$, is open and everywhere dense in $X$. Later on it is established that for any $\varepsilon > 0$ the set $X_0$ can be chosen so that a certain neighbourhood of every point of $X_0$ is $\varepsilon$-almost isometric to a domain in $\mathbb{R}^m$, where $m$ is the Hausdorff dimension of $X$.

The tangent cone $C_p$ to the space $X$ at a point $p$ plays an important role in studying the structure of $X$ in the neighbourhood of a given point $p$. In the case of a Riemannian manifold $C_p$ is simply the tangent space at the point $p$, but for general convex hypersurfaces the cone $C_p$ coincides with the usual tangent cone. The tangent cone $C_p$ exists at any point $p \in X$ and to a first approximation it describes the geometry of $X$ near $p$. The cone $C_p$ has the same dimension as $X$ and is a cone over a compact space of curvature $\geqslant 1$.

Basically some specific local questions are solved in the next two sections. Thus in §8 we get the following natural generalization of the precompactness theorem mentioned in 1.2: for any numbers $n$, $k$, $D$ the metric space (with Hausdorff metric), consisting of all spaces of curvature $\geqslant k$ that have Hausdorff dimension not greater than $n$ and diameter not exceeding $D$, is compact.

The subject of §9 is those situations when spaces near to a given space, in the Hausdorff metric, turn out to be also near to the given space in the Lipschitz metric.

In §10 we study the Hausdorff measure on spaces with curvature bounded below. The basic results are an estimate of the Hausdorff dimension of the singularities and a theorem on the weak convergence of Hausdorff measures for non-collapsing, (Hausdorff) convergent sequences of Alexandrov spaces.

In the next two sections we develop the apparatus of functions with directional derivatives, which are a good substitute for smooth functions on the spaces under consideration. The basic result of §12 is that there are no singularities of codimension 1 other than the boundary.

Everywhere, except in the last section, assertions are given with complete proofs, but there are more details in the first half of the paper and not so many details in the rest. The exception is in the last section. It contains a resumé of subsequent results obtained recently by the authors and also some unsolved questions and conjectures. A full presentation of these results is contained in the preprint: G. Perel'man, "A.D. Alexandrov's spaces with curvatures bounded from below. II".

**1.5.** This paper is noticeably different from its preliminary version, whch was distributed as a preprint with the same name. Apart from editorial corrections we have made changes of a more fundamental nature. Thus §§11, 14, 15, 16 in the preliminary version were not included in this new edition; §10 about the Hausdorff measure has been added, and so also have the more general theorems on almost isometry in §9; we give a completely new proof of the "generalized theorem of Toponogov".

## §2. Basic concepts

In this section we present the known [5], [1], [2] facts on spaces with curvature bounded below in a form convenient for us.

**2.1.** The metric space $(M, |\cdot, \cdot|)$ is called an *intrinsic metric space* if for any $x, y \in M$, $\delta > 0$ there is a finite sequence of points $z_0 = x, z_1, \dots, z_k = y$ such that $|z_i z_{i+1}| < \delta$ $(0 \leqslant i \leqslant k-1)$ and $\sum_{i=0}^{k-1} |z_i z_{i+1}| < |xy| + \delta$. A subset in the space $M$ with intrinsic metric is said to be (metrically) *convex* if the metric induced on it from $M$ is intrinsic. A *geodesic* is a curve whose length is equal to the distance between its ends. In a locally compact complete space with intrinsic metric any two points can be joined by a geodesic, possibly not unique (without completeness this property is only satisfied locally). A limit of geodesics is again a geodesic and moreover with the limit length. A collection of three points $p, q, r \in M$ and three geodesics $pq, pr, qr$ is called a *triangle* in $M$ and is denoted by $\Delta pqr$.

**2.2.** Let us fix the real number $k$. A *k-plane* will be a two-dimensional complete simply-connected Riemannian manifold of curvature $k$, that is, a Euclidean plane, a sphere, or a Lobachevskii plane. With a triad of points $p, q, r$ in a space $M$ with intrinsic metric we associate a triangle $\tilde{\Delta} pqr$ on the $k$-plane with vertices $\tilde{p}, \tilde{q}, \tilde{r}$ and sides of lengths $|\tilde{p}\,\tilde{q}| = |pq|$, $|\tilde{p}\,\tilde{r}| = |pr|$, $|\tilde{q}\,\tilde{r}| = |qr|$. For $k \leqslant 0$ the triangle $\tilde{\Delta} pqr$ always exists and is unique up to a rigid motion, and for $k > 0$ it exists only with the additional assumption that the perimeter of $\Delta pqr$ is less than $2\pi/\sqrt{k}$. We let $\tilde{\measuredangle} pqr$ denote the angle at the vertex $\tilde{q}$ of the triangle $\tilde{\Delta} pqr$.

**2.3.** *Definition.* A locally complete space $M$ with intrinsic metric is called a *space with curvature* $\geqslant k$ if in some neighbourhood $U_x$ of each point $x \in M$ the following condition is satisfied:

(D) for any four (distinct) points $(a;\, b,\, c,\, d)$ in $U_x$ we have the inequality $\tilde{\measuredangle} bac + \tilde{\measuredangle} bad + \tilde{\measuredangle} cad \leqslant 2\pi$.

If the space $M$ is a one-dimensional manifold and $k > 0$, then we require in addition that diam $M$ does not exceed $\pi/\sqrt{k}$.

*Remark.* If the space $M$ satisfies all the conditions of the definition, apart from local completeness, then it may be converted into a space with curvature $\geqslant k$ by local completion. We use the a priori condition of local completeness only in §§5, 6.

**2.4.** Berestovskii [5] has shown that condition (D) in Definition 2.3 can be replaced by the following condition:

(E) any four points in $U_x$ admit an isometric embedding into some $k'$-plane, $k' \geqslant k$, where the number $k'$ depends on the four points chosen.

We shall not use this version of the definition.

**2.5.** For spaces in which, locally, any two points are joined by a geodesic, in particular for locally compact spaces, the condition (D) in Definition 2.3 can be replaced by the condition:

(A) for any triangle $\Delta pqr$ with vertices in $U_x$ and any point $s$ on the side $qr$ the inequality $|ps| \geqslant |\tilde{p}\tilde{s}|$ is satisfied, where $\tilde{s}$ is the point on the side $\tilde{q}\tilde{r}$ of the triangle $\tilde{\Delta} pqr$ corresponding to $s$, that is, such that $|qs| = |\tilde{q}\,\tilde{s}|$, $|rs| = |\tilde{r}\,\tilde{s}|$.

To prove the implication (D) $\Rightarrow$ (A) it is sufficient to substitute the four points $(s;\, p,\, q,\, r)$ in condition (D) and use the following elementary observation.

**Lemma.** *Let triangles pqs, prs be given on a k-plane, which are joined to each other in an exterior way along the side ps, and let there also be given a triangle bcd, where* $|bc| = |pq|$, $|bd| = |pr|$, $|cd| = |qs|+|sr|$, *and* $|bc|+|bd|+|cd| < 2\pi/\sqrt{k}$ *in the case* $k > 0$. *Then* $\measuredangle psq + \measuredangle psr \leqslant \pi$ $[\geqslant \pi]$ *if and only if* $\measuredangle pqs \geqslant \measuredangle bcd$ *and* $\measuredangle prs \geqslant \measuredangle bdc$ *[respectively* $\measuredangle pqs \leqslant \measuredangle bcd$ *and* $\measuredangle prs \geqslant \measuredangle bdc$*].*

We give below the proof of the converse implication (A) $\Rightarrow$ (D), see 2.8.2.

**2.6.** Let $\gamma$, $\sigma$ be geodesics with origin $p$, and let $q$, $r$ be points on these geodesics at distances $x = |pq|$, $y = |pr|$ from the point $p$. The angle $\tilde{\measuredangle} qpr$ is denoted by $\omega_k(x, y)$. Obvioously condition (A) may be replaced by the following condition:

(B) for any geodesics $\gamma$, $\sigma$ in $U_x$ the function $\omega_k(x, y)$ is non-increasing with respect to each of the variables $x$, $y$.

**2.7.** The limit $\lim_{x,y\to 0} \omega_k(x, y)$ is called the *angle between* $\gamma$, $\sigma$; clearly this limit does not depend on $k$. If (B) is satisfied, then the angle between the geodesics exists. It is easily verified that the angles between three geodesics with common origin satisfy the triangle inequality.

The following assertion is easily deduced from (A) and (B).

(C) For any triangle $\Delta pqr$ contained in $U_x$ none of its angles is less than the corresponding angle of the triangle $\tilde{\Delta} pqr$ on the $k$-plane.

($C_1$) The sum of adjacent angles is equal to $\pi$, that is, if $r$ is an interior point of the geodesic $pq$, then for any geodesic $rs$ we have $\measuredangle prs + \measuredangle qrs = \pi$.

Conversely (A) and (B) are easily deduced from (C) and ($C_1$).

A consequence of the condition ($C_1$) is that geodesics do not branch. This means that if two geodesics have two common points, then either they are continuations of a third geodesic, their intersection, beyond its distinct ends, or these common points are the ends of both geodesics. In particular, if a geodesic is extendable, then it is extendable in a unique way.

**2.8.** We list a few other known properties of spaces of curvature bounded below which follow easily from (C) and ($C_1$).

**2.8.1.** If the geodesics $p_i q_i$ converge to $pq$, while the geodesics $p_i r_i$ converge to $pr$, then $\measuredangle qpr \leqslant \liminf_{l\to\infty} \measuredangle q_i p_i r_i$.

This follows easily from (C).

**2.8.2.** If $pa$, $pb$, $pc$ are geodesics, then $\measuredangle apb + \measuredangle bpc + \measuredangle cpa \leqslant 2\pi$.

In fact, if $d \in ap$, then by property ($C_1$)

$$\measuredangle adb + \measuredangle adc + \measuredangle bdc \leqslant (\measuredangle adb + \measuredangle bdp) + (\measuredangle adc + \measuredangle cdp) = 2\pi.$$

All that remains is to use property 2.8.1 when $d \to p$ if the geodesics $pb$, $pc$ are unique. Otherwise the points $b$ and $c$ can be replaced by points in $pb$ and $pc$ and the argument repeated.

Now condition (D) follows immediately from the property just proved and (C).

**2.8.3.** ***Lemma on the limit angle.*** *If the points $a_i$ of a geodesic pa converge to p and $b \notin pa$, then* $\limsup_{i\to\infty} \measuredangle aa_ib \leqslant \measuredangle apb$ *for any geodesics pb, $a_ib$.*

This property, unlike the preceding ones, is of a non-local character. However, it is easily proved from the local property (C) in exactly the same way as in the case of Riemannian manifolds, see for example [16].

**2.9. Examples of spaces of curvature $\geqslant k$.**

(1) Riemannian manifolds without boundary or with locally convex boundary, whose sectional curvatures are not less than $k$.

(2) Convex subsets in such Riemannian manifolds.

(3) Convex hypersurfaces in such Riemannian manifolds, see [9].

(4) Hilbert space ($k = 0$).

(5) Quotient space $\mathbb{R}^3/\sim$, where $x \sim -x$, that is, the cone over $\mathbb{R}P^2$ is a space of curvature $\geqslant 0$. This example will be justified below.

(6) Locally finite simplicial spaces glued (along all the faces) from simplexes of a space of constant curvature $k$ and of fixed dimension, so that every face of codimension 1 belongs to precisely two simplexes and the sum of the dihedral angles for each face of codimension 2 does not exceed $2\pi$.

The last example will be justified in 4.5. There will also be described in §4 natural constructions which enable more complicated examples to be constructed.

## §3. Globalization theorem

**3.1.** Spaces with curvature bounded below were defned in §2 using local conditions. In general, the analogues of these conditions "in the large" may not be satisfied. It is sufficient to consider, for example, a plane from which a closed disc has been removed. However, for complete spaces the global conditions may be deduced from the corresponding local ones. In dimension 2 this was first discovered by Alexandrov. For Riemannian manifolds of arbitrary dimension the corresponding theorem was proved by Toponogov. The arguments in §2 proving that conditions (A), (B), (C, $C_1$) and (D) are equivalent for locally compact spaces are still completely valid if these conditions are replaced by the corresponding assertions "in the large". Therefore it is sufficient to deduce just one such global condition from the local ones.

We give the proof for condition (D), which is related to a more general situation.

**3.2.** ***Globalization theorem.*** *Let M be a complete space with curvature $\geqslant k$. Then for any quadruple of points (a; b, c, d) we have the inequality* $\tilde{\measuredangle} bac + \tilde{\measuredangle} bad + \tilde{\measuredangle} cad \leqslant 2\pi$.

**3.3.** *Remarks.* a) The theorem has to be formulated more precisely when $k > 0$. In this case we assume a priori that the perimeters of the triangles $bac$, $bad$, $cad$ are less than $2\pi/\sqrt{k}$. In 3.6–3.7 we shall show that in fact under the conditions of the theorem the perimeter of any triangle in $M$ does not exceed $2\pi/\sqrt{k}$. The proof of the theorem when $k > 0$ requires, in comparison with the case $k \leqslant 0$, some additional arguments which will be enclosed in square brackets.

b) Throughout the proof of the theorem and also in 3.7 we assume that there exist geodesics between any two points of $M$. This is done only to simplify the arguments, as it enables us to use phrases of the type "we choose a point $r$ on the geodesic $pq$ so that ..." . In the general case, it is necessary in such a situation to choose a point $r$ from a sequence constructed from the points $p$, $q$ and a sufficiently small $\delta > 0$ in accordance with Definition 2.1 of an intrinsic metric.

**3.4. Proof of the globalization theorem.**

**3.4.1.** The *size* of the quadruple $(a; b, c, d)$ is $S(a; b, c, d)$, where this denotes the greatest of the perimeters of the triangles $\Delta bac$, $\Delta bad$, $\Delta cad$, while the *excess* $E(a; b, c, d)$ is $\max\{\tilde{\measuredangle} bac + \tilde{\measuredangle} cad + \tilde{\measuredangle} bad - 2\pi, 0\}$.

The following is the key to the proof of the theorem.

***Lemma on the subadditiveness of the excess.*** *Let $p$, $q$, $r$, $s$ be points in a space with intrinsic metric, and let $t$ lie on the geodesic $pq$. Then*

(1) $S(p; q, r, s) \geqslant \max\{S(p; t, r, s), S(t; p, q, s), S(t; p, q, r)\}$.

(2) $E(p; q, r, s) \leqslant E(p; t, r, s) + E(t; p, q, s) + E(t; p, q, r)$.

[If $k > 0$, then additional assumptions are necessary for assertion (2) to be justified. It is sufficient to assume that

$$\max\{|pq|, |pr|, |ps|\} + 10|pt| < \pi/2\sqrt{k}.]$$

Assertion (1) is obviously implied by the triangle inequality. To prove assertion (2) we need a numerical refinement of Lemma 2.5.

***Technical lemma.*** *Let triangles $pqr$, $prx$, $qry$ be given on a $k$-plane such that $|rx| = |ry|$, $|pq| = |px| + |qy|$, $\measuredangle qyr + \measuredangle pxr \geqslant \pi$. Then $\measuredangle qpr - \measuredangle xpr \leqslant \measuredangle qyr + \measuredangle pxr - \pi$. [If $k > 0$, then in addition we assume that* $\max\{|pr|, |pq|\} + 10|px| < \pi/2\sqrt{k}$.]

*Proof.* We may suppose that the triangles $prx$, $qry$ lie in the triangle $pqr$ and do not overlap. Let $z$ be a point on the side $pq$ such that $|qz| = |qy|$, $|pz| = |px|$. We calculate the sum of the complete angles for $x$, $y$, $z$. We have $5\pi = (\measuredangle qyr + \measuredangle pxr) + (\pi + \delta(\Delta ryx) - \measuredangle yrx) + (\pi + \delta(\Delta xyz)) + (\pi + \delta(\Delta qyz) - \measuredangle yqz) + (\pi + \measuredangle xpr - \measuredangle qpr + \delta(\Delta xpz))$, where $\delta(\Delta)$ denotes the excess of the triangle $\Delta$, that is, the sum of its angles in excess of $\pi$.

Now in the case $k \leqslant 0$ the assertion of the technical lemma is obvious, since all excesses $\delta(\Delta)$ are non-posiitve.

[In the case $k > 0$ the condition $\max\{|pr|, |pq|\} + 10|px| < \pi/2\sqrt{k}$ guarantees that the total area of the triangles $\Delta ryx$, $\Delta qyz$, $\Delta xyz$, $\Delta xpz$ is less than half the total area of the spherical lunes with angles $\measuredangle yrx$ and $\measuredangle yqz$, that is, the sum of the excess is less than $\measuredangle yrx + \measuredangle yqz$, and this is just sufficient to make our deduction. This proves the technical lemma. ■]

We turn to the proof of (2). It is now sufficient for us to add together the inequalities arising in the technical lemma

$$E(t; q, r, p) \geqslant \tilde{\measuredangle}\, qpr - \tilde{\measuredangle}\, tpr,$$
$$E(t; q, s, p) \geqslant \tilde{\measuredangle}\, qps - \tilde{\measuredangle}\, tps$$

and use the definitions

$$E(p; t, r, s) = \tilde{\measuredangle}\, tpr + \tilde{\measuredangle}\, tps + \tilde{\measuredangle}\, rps - 2\pi,$$

$$\tilde{\measuredangle}\, qpr + \tilde{\measuredangle}\, qps + \tilde{\measuredangle}\, rps - 2\pi = E(p; q, r, s).$$

The lemma on subadditiveness is proved. ■

**3.4.2.** We now turn to the proof of Theorem 3.2 itself. Let us assume that the assertion of the theorem is false. Then there exist a point $p \in M$ and a number $l > 0$ [$l < 2\pi/\sqrt{k}$ if $k > 0$] such that:

(a) The excess of any quadruple of size $\leqslant (0.99)\,l$ lying in the ball $B_p(100l)$ is zero.

(b) There is a quadruple of size $\leqslant l$ lying in the ball $B_p(10\,l)$ and having a positive excess.

In fact, otherwise we could find a sequence of points $p_i$ and numbers $l_i$ satisfying condition (b) and such that $l_{i+1} \leqslant (0.99)\,l_i$, $|p_i p_{i+1}| \leqslant 1000 l_i$. But then in view of the fact that the space is complete, the sequence $p_i$ would have a limit point in an arbitrary small neighbourhood of which the condition (D) is violated.

**3.4.3.** The basic stage in the proof of the theorem is to check that a quadruple $(a; b, c, d)$ of size $\leqslant l$ lying in the ball $B_p(20\,l)$ has zero excess if

$$(1) \qquad |ab| \leqslant 0.01l, \quad |cd| \geqslant 0.1l + \max\{|ac|, |ad|\}.$$

[If $k > 0$, then we include, in addition, the condition $\max\{|ac|, |ad|\} < \pi/2\sqrt{k}$.] We assume that there is a quadruple $(a; b, c, d)$ for which this is not so and let $\delta = E(a; b, c, d) > 0$. We take the point $x$ on the geodesic $ac$ at a distance $\varepsilon \leqslant (0.01)\,l$ from $a$.

[In the case $k > 0$ in addition we assume that $10\varepsilon + \max\{|ac|, |ad|\} < \pi/2\sqrt{k}$.]

By applying the lemma on subadditiveness we get $\delta = E(a; b, c, d) \leqslant E(x; a, c, d)$, since the sizes of the quadruples $(x; a, b, c)$ and $(a; x, b, d)$ are less than $(0.99)\,l$, this follows from (1). We now take the point $y$ on the geodesic $xd$ at a distance $\varepsilon$ from $x$ and, again applying the lemma on subadditiveness, we get $E(y; x, c, d) \geqslant E(x; a, c, d) \geqslant \delta$. In addition,

$S(y; x, c, d) \leqslant l$, and this quadruple satisfies the inequalities analogous to (1). Finally

$$(2) \qquad |yc|+|yd| \leqslant |ac|+|ad|-\delta^2\varepsilon/2.$$

In fact, by adding the inequalities

$$\tilde{\measuredangle}\, xab \geqslant \tilde{\measuredangle}\, cab \qquad (\text{by Lemma 2.15, since } E(x; a, b, c) = 0),$$
$$2\pi \geqslant \tilde{\measuredangle}\, xab + \tilde{\measuredangle}\, dab + \tilde{\measuredangle}\, xad \quad (\text{since } E(a; x, d, b) = 0),$$
$$\tilde{\measuredangle}\, dac + \tilde{\measuredangle}\, dab + \tilde{\measuredangle}\, cab = 2\pi + \delta,$$
$$\pi \geqslant \tilde{\measuredangle}\, dac,$$

we get $\tilde{\measuredangle} xad \leqslant \pi - \delta$. Similarly $\tilde{\measuredangle} yxc \leqslant \pi - \delta$. Thus

$$|dx| \leqslant |da| + (1-\delta^2/4)\varepsilon,$$
$$|cx| = |ca| - \varepsilon,$$
$$|dy| = |dx| - \varepsilon,$$
$$|cy| \leqslant |cx| + (1-\delta^2/4)\varepsilon,$$

from which we get (2).

Thus from the given quadruple we have constructed another quadruple that satisfies the same sort of bounds but "essentially less" than those expressed by inequality (2). After not more than $2l/\delta^2\varepsilon$ such steps we get a contradiction.

**3.4.4.** We now reduce the general case to the particular case just considered. Let $(a; b, c, d)$ be a quadruple of size $\leqslant l$ in the ball $B_p(10\,l)$, with positive excess. We take the point $x$ on the geodesic $ab$ at a distance $\varepsilon \leqslant (0.001)\,l$ from $a$ [if $k > 0$, we assume further that $10\varepsilon < 2\pi/\sqrt{k} - l$] and consider the quadruples $(a; x, c, d)$, $(x; a, b, c)$, $(x; a, b, d)$. By the lemma on subadditiveness at least one of these quadruples has positive excess, and they all have size $\leqslant l$. So we have reduced the general case to the study of quadruples $(a; b, c, d)$ such that $|ab| = \varepsilon$.

Let $(a; b, c, d)$ be such a quadruple. Our immediate aim is to reduce everything to a consideration of a quadruple $(x; y, z, t)$ of size $\leqslant l$ for which $|xy| \leqslant 2\varepsilon$, $|xz| \leqslant 2\varepsilon$. We shall divide the geodesic $ac$ by the points $a_0 = a, a_1, \ldots, a_n = c$ into segments of length $\leqslant \varepsilon$. The lemma on subadditiveness implies that at least one of the quadruples $(a_0; a_1, b, d)$, $(a_1; a_0, b, c)$, $(a_1; a_0, c, d)$ has positive excess. If this is one of the first two quadruples, then our aim is achieved. Otherwise one of the quadruples $(a_1; a_2, a_0, d)$, $(a_2; a_1, a_0, c)$, $(a_2; a_1, c, d)$ has positive excess. At this stage also, if it is one of the first two, our aim is achieved, otherwise we continue "thinning out" the quadruples with positive excess until we get a quadruple of the desired kind.

Finally let $(a; b, c, d)$ be a quadruple of size $\leqslant l$ with positive excess for which $|ab| \leqslant 2\varepsilon$, $|ac| \leqslant 2\varepsilon$. We choose a point $x$ on the geodesic $ad$ such that $|cx| = |dx|$. Clearly in this case $|cx| \leqslant l/4+2\varepsilon$. The lemma on subadditiveness enables us to assert that at least one of the quadruples $(a; x, b, c)$, $(x; a, b, d)$, $(x; a, c, d)$ has positive excess. But the first of these has size $\leqslant (0.99)l$, consequently its excess is zero. Without loss of generality we may suppose that just the last quadruple has positive excess. We choose a point $y$ on the geodesic $ax$ at a distance $\varepsilon$ from $x$. By the lemma on subadditiveness one of the quadruples $(x; y, c, d)$, $(y; x, a, c)$, $(y; x, a, d)$ must have positive excess. But the last quadruple has excess 0, since $\tilde{\measuredangle} dyx = 0$, the second quadruple also has zero excess, since its size is small, and the first quadruple corresponds to the particular case that was considered in section 3.4.3. The theorem is proved. ■

**3.5.** *Remark.* If $M$ is not complete, then the proof given above enables us to assert that condition (D) is satisfied for those quadruples that are at a distance which considerably exceeds their size from the points of the completion of $M$.

**3.6. Theorem.** *Let $M$ be a complete space with curvature $\geqslant k$, $k > 0$. Then* diam $M \leqslant \pi/\sqrt{k}$.

**3.6.1.** *Beginning of the proof.* We first consider the exceptional case when the neighbourhood of some point $p \in M$ is a section of a geodesic. Let $q$, $r$ be points of this geodesic such that $|pq| = |pr| = |qr|/2 < \pi/2\sqrt{k}$. Then all the points $s \in M$ can be separated into two (perhaps intersecting) classes: those for which $\tilde{\measuredangle} spq = 0$, that is, $|qs| = ||ps|-|pq||$, and those for which $\tilde{\measuredangle} spr = 0$, that is, $|rs| = ||pr|-|ps||$. Moreover in each class there is at most one point $s$ with a given distance $|ps| < \pi/\sqrt{k}$, and the set of values $|ps|$, realizable by the points $s$ in a single class, is an interval—this follows from condition (D) and the fact that $M$ is complete. By isometrically moving the points $q$, $p$, $r$ and repeating the arguments it is not difficult to establish that $M$ contains an open convex subset which is a one-dimensional manifold. Hence clearly $M$ is also this manifold and according to our Definition 2.3 diam $M \leqslant \pi/\sqrt{k}$.

**3.6.2.** If the space $M$ is locally compact, then the proof can now be completed fairly simply. Let $p$ be the midpoint of the geodesic $qr$ for which $\pi/\sqrt{k} < |qr| < (1.1)\pi/\sqrt{k}$. We may suppose that there are points $s \in M$ near $p$ that do not lie on $qr$. Let $s'$ be the point of $qr$ nearest to $s$. A simple calculation, based on condition (C), shows that for $s$ sufficiently near to $p$, $|qs| < |qs'|$, $|rs| < |rs'|$, which contradicts the fact that $qr$ is a geodesic.

For completeness we give a formal proof without local compactness.

**3.6.3.** We consider the general case. We need the following observation. Let the points $a$, $b$ be such that $\pi/\sqrt{k} < |ab| < (1.01)\pi/\sqrt{k}$. We consider sequences of points $x_i, y_i \in M$ such that

$$|ay_i|+|by_i| \to |ab|, \quad |x_iy_i|+\max\{|ay_i|, |by_i|\} < 0.99 \cdot \pi\sqrt{k}.$$

Then $|ax_i|+|bx_i| \to |ab|$.

In fact we choose another sequence $z_i \in M$ such that $|y_i z_i| \to 0$, $\tilde{\measuredangle} y_i z_i b \to \pi$, $\tilde{\measuredangle} z_i y_i a \to \pi$. Then by applying condition (D) to the sets $(y_i;\, z_i,\, x_i,\, a)$ and $(z_i;\, y_i,\, x_i,\, b)$ we get $\overline{\lim}(\tilde{\measuredangle} a y_i x_i + \tilde{\measuredangle} z_i y_i x_i) \leqslant \pi$, $\overline{\lim}(\tilde{\measuredangle} b z_i x_i + \tilde{\measuredangle} y_i z_i x_i) \leqslant \pi$. On the other hand, $\lim(\tilde{\measuredangle} y_i z_i x_i + \tilde{\measuredangle} z_i y_i x_i) = \pi$ if $|x_i y_i|$ is not zero. Therefore $\overline{\lim}(\tilde{\measuredangle} a y_i x_i + \tilde{\measuredangle} b y_i x_i) \leqslant \pi$ and this means, in view of $|a y_i| + |b y_i| > \pi/\sqrt{k}$, that $|a x_i| + |b x_i| \to |ab|$.

**3.6.4.** We now show that, for any $l \in [(0.9)\pi/2\sqrt{k},\, (0.99)\pi/2\sqrt{k}]$, any sequence of points $s_i \in M$ such that $|a s_i| \to l$ and $|a s_i| + |b s_i| \to |ab|$ has limit points and moreover at most two. In this way we reduce the theorem to the exceptional case already considered. So, let $s_i$ be a sequence of points with the properties indicated. If our assertion is false, then we can choose three subsequences $s_i^1,\, s_i^2,\, s_i^3$ such that the distances $|s_i^1,\, s_i^2|$, $|s_i^1,\, s_i^3|$, $|s_i^2,\, s_i^3|$ are not zero. By applying condition (D) to the quadruples $(a;\, s_i^1,\, s_i^2,\, s_i^3)$ we may, without loss of generality, conclude that $\tilde{\measuredangle} s_i^1 a s_i^2 \leqslant 2\pi/3$ for all $i$. If now the $x_i$ lie half way between $s_i^1$ and $s_i^2$, then the sequences $x_i,\, y_i = s_i^1$ satisfy the conditions of 3.6.3 and consequently $|a x_i| + |b x_i| \to |ab|$. But on the other hand if we estimate $|a x_i|$ and $|b x_i|$, by considering the quadruples $(x_i;\, s_i^1,\, s_i^2,\, a)$ and $(x_i;\, s_i^1,\, s_i^2,\, b)$ respectively (in essence we are using condition (A) here), then it is not difficult to show that $|a x_i| + |b x_i|$ must be distinct from $|ab|$, which gives a contradiction. ■

**3.7. *Theorem.*** *Let $M$ be a complete space with curvature $\geqslant k$, $k > 0$. Then for any three points $a,\, b,\, c \in M$ we have $|ab| + |ac| + |bc| \leqslant 2\pi/\sqrt{k}$ and condition (D) is satisfied for all quadruples of size $2\pi/\sqrt{k}$.* (Here we suppose that, if $|ab| = \pi/\sqrt{k}$, $|ac| + |bc| = \pi/\sqrt{k}$, then $\tilde{\measuredangle} bac = \tilde{\measuredangle} abc = 0$, $\tilde{\measuredangle} acb = \pi$.)

*Proof.* We first check the second assertion of the theorem for a quadruple $(a;\, b,\, c,\, d)$ of size $2\pi/\sqrt{k}$. We may suppose that $\max\{|ab|,\, |ac|,\, |ad|\} < \pi/\sqrt{k}$. Let $b_\alpha$ denote a variable point of the geodesic $ab$. Then for $b_\alpha$ near to $a$ the perimeters of the triangles $a b_\alpha c$, $a b_\alpha d$ are less than $2\pi/\sqrt{k}$. In addition, if the perimeter of the triangle $abc$ is less than $2\pi/\sqrt{k}$, then by applying condition (D) to the quadruple $(b_\alpha;\, a,\, b,\, c)$ and using Lemma 2.5 we get $\tilde{\measuredangle} b_\alpha a c \geqslant \tilde{\measuredangle} bac$. Finally, if the perimeter of $a b_\alpha d$ is near to $2\pi/\sqrt{k}$, then $\tilde{\measuredangle} b_\alpha a d$ is near to $\pi$. Thus, if the excess of the quadruple $(a;\, b,\, c,\, d)$ were positive, we could construct a quadruple $(a;\, b_\alpha,\, c,\, d)$ with positive excess where the perimeters of the triangles $a b_\alpha c$, $a b_\alpha d$ will be less than $2\pi/\sqrt{k}$. Similarly one can get rid of the triangle $acd$ if its perimeter is equal to $2\pi/\sqrt{k}$.

We now return to the first assertion of the theorem and assume that the perimeter of the triangle $abc$ is greater than $2\pi/\sqrt{k}$. It may be supposed that $\max\{|ab|,\, |ac|,\, |bc|\} < \pi/\sqrt{k}$. We choose a point $x$ on the geodesic $bc$ such that $2\pi/\sqrt{k} < |ac| + |ax| + |cx| < |ab| + |ac| + |bc|$. Let $y$ be a point of the geodesic $xa$ such that the perimeter of the triangle $cxy$ is equal to $2\pi/\sqrt{k}$.

Clearly $|xb| < |xy|$. Since $|xc| < \pi/\sqrt{k}$ and $|xy| < \pi/\sqrt{k}$, then $\tilde{\measuredangle} cxy = \pi$. By now applying condition (D) to the quadruple $(x;\, c,\, y,\, b)$ and taking into account that $|xb| < |xy|$ we get $|xy| = |xb| + |by|$. But then $|xa| = |xb| + |ba|$, which contradicts the choice of $x$. The theorem is proved. ■

## §4. Natural constructions

**4.1. Direct product.**
***Proposition.*** *The direct product* (with the metric determined by Pythagoras' theorem) *of a finite number of spaces with curvature* $\geqslant 0$ *is a space with curvature* $\geqslant 0$.

This assertion will become almost obvious if we use the following (equivalent) modification of condition (D).

(D′) For any quadruple $(a;\, b,\, c,\, d)$ lying in $U_x$ it is possible [impossible] to construct a quadruple $(\bar{a};\, \bar{b},\, \bar{c},\, \bar{d})$ on the $k$-plane such that the segments $\bar{a}\bar{b}$, $\bar{a}\bar{c}$, $\bar{a}\bar{d}$ divide the complete angle at the point $\bar{a}$ into three angles each $\leqslant \pi$, where $|\bar{a}\bar{b}| = |ab|$, $|\bar{a}\bar{c}| = |ac|$, $|\bar{a}\bar{d}| = |ad|$ and $|\bar{b}\bar{c}| \geqslant |bc|$, $|\bar{b}\bar{d}| \geqslant |bd|$, $|\bar{c}\bar{d}| \geqslant |cd|$ [respectively $|\bar{b}\bar{c}| < |bc|$, $|\bar{c}\bar{d}| < |cd|$, $|\bar{b}\bar{d}| < |bd|$].

**4.2. The cone.**
**4.2.1.** Let $X$ be a metric space. The *cone over* $X$ with vertex $A$ is the quotient space $C_A(X) = X \times [0, \infty]/\sim$, where $(x_1, a_1) \sim (x_2, a_2) \sim A \Leftrightarrow$ $\Leftrightarrow a_1 = a_2 = 0$. Let $\Pi : C_A(X)\setminus A \to X$ be the natural projection. The metric of the cone is defined from the cosine formula $|\bar{x}_1\bar{x}_2|^2 =$ $= a_1^2 + a_2^2 - 2a_1a_2 \cos(\min\{|x_1x_2|, \pi\})$, where $\bar{x}_1 = (x_1, a_1)$, $\bar{x}_2 = (x_2, a_2)$.

**4.2.2.** It is easily seen that if the metric of $X$ is intrinsic, then the metric of $C_A(X)$ is also intrinsic. The converse assertion, generally speaking, is false but it becomes true if for any two points $x_1, x_2 \in X$ there are arbitrarily nearby points $y_1, y_2 \in X$ such that $|y_1y_2| < \pi$. If $\gamma$ is a geodesic in $X$ of length not greater than $\pi$, then $\Pi^{-1}(\gamma)$ is a convex subset in $C_A(X)$ isometric to the plane sector with angle equal to the length of $\gamma$. Conversely, if $\bar{\gamma}$ is a geodesic in $C_A(X)$ not passing through $A$, then $\Pi(\bar{\gamma})$ is a geodesic in $X$.

**4.2.3.** ***Proposition.*** *Let $X$ be a complete metric space. The following two conditions are equivalent*:

(a) *$X$ is a space with curvature* $\geqslant 1$.

(b) *$C_A(X)$ is not a straight line and is a space with curvature* $\geqslant 0$.

*Proof.* (a) $\Rightarrow$ (b). The fact that $C_A(X)$ is not a straight line is obvious.

We assume that there is a quadruple $(\bar{a};\, \bar{b},\, \bar{c},\, \bar{d})$ in $C_A(X)$ that violates condition (D). We may suppose that $A$ is not one of these points and consider their projections $a = \Pi(\bar{a}), \ldots, d = \Pi(\bar{d})$. By applying (D′) we can find a corresponding quadruple $(a_0;\, b_0,\, c_0,\, d_0)$ on the sphere $S^2$ for which $|a_0b_0| = |ab|$, $|a_0c_0| = |ac|$, $|a_0d_0| = |ad|$ but $|b_0c_0| \geqslant |bc|$, $|b_0d_0| \geqslant |bd|$, $|c_0d_0| \geqslant |cd|$.

If we think of Euclidean space $E^3$ as the cone $C_{A_0}(S^2)$ with projection $\Pi_0$, we can find a quadruple $(\bar{a}_0; \bar{b}_0, \bar{c}_0, \bar{d}_0)$ in $E^3$ such that $|A_0\bar{a}_0| = |A\bar{a}|, \dots, |A_0\bar{d}_0| = |A\bar{d}|$ and $\Pi_0(\bar{a}_0) = a_0, \dots, \Pi_0(\bar{d}_0) = d_0$. Now it is clear that $|\bar{a}_0\bar{b}_0| = |\bar{a}\,\bar{b}|$, $|\bar{a}_0\bar{c}_0| = |\bar{a}\,\bar{c}|$, $|\bar{a}_0\bar{d}_0| = |\bar{a}\,\bar{d}|$ and $|\bar{b}_0\bar{c}_0| \geqslant |\bar{b}\,\bar{c}|$, $|\bar{b}_0\bar{d}_0| \geqslant |\bar{b}\,\bar{d}|$, $|\bar{c}_0\bar{d}_0| \geqslant |\bar{c}\,\bar{d}|$, consequently the quadruple $(\bar{a}_0; \bar{b}_0, \bar{c}_0, \bar{d}_0)$ violates condition (D)—a contradiction.

(b) $\Rightarrow$ (a). We shall first show that the metric $X$ is intrinsic by verifying the condition formulated in 4.2.2. Let $x_1, x_2 \in X$, $\bar{x}_1 \in \Pi^{-1}(x_1)$, $\bar{x}_2 \in \Pi^{-1}(x_2)$. If $|\bar{x}_1\bar{x}_2| \neq |\bar{x}_1A|+|\bar{x}_2A|$, then $|x_1x_2| < \pi$ and our condition is satisfied. If however $|\bar{x}_1\bar{x}_2| = |\bar{x}_1A|+|\bar{x}_2A|$, then the segments $\bar{x}_1A$, $A\bar{x}_2$ form a geodesic. Since $C_A(X)$ is not a straight line and since geodesics do not branch, we can find a point $\bar{y}_1$, arbitrarily near to $\bar{x}_1$, for which $|\bar{y}_1\bar{x}_2| < |\bar{y}_1A|+|\bar{x}_2A|$. Thus one can find a point $y_1 = \Pi(\bar{y}_1)$, arbitrarily near to $x_1$, for which $|y_1x_2| < \pi$, which is what we required.

Now we suppose one can find a quadruple $(a; b, c, d)$ in $X$ that violates condition (D′). We may suppose, by using the globalization theorem, that the pairwise distances between the points of the quadruple are all less than $\pi/2$. Let $(a_0; b_0, c_0, d_0)$ be a quadruple on the unit sphere $S^2$ for which $|a_0b_0| = |ab|$, $|a_0c_0| = |ac|$, $|a_0d_0| = |ad|$, $|b_0c_0| < |bc|$, $|b_0d_0| < |bd|$, $|c_0d_0| < |cd|$, and the segments $a_0b_0$, $a_0c_0$, $a_0d_0$ divide the complete angle at the point $a_0$ into three angles each $\leqslant\pi$. In the Euclidean space $E^3 = C_{A_0}(S^2)$ we find a quadruple $(\bar{a}_0; \bar{b}_0, \bar{c}_0, \bar{d}_0)$ of points lying in the same plane for which $\Pi_0(\bar{a}_0) = a_0, \dots, \Pi_0(\bar{d}_0) = d_0$ and we construct the corresponding quadruple $(\bar{a}; \bar{b}, \bar{c}, \bar{d})$ in $C_A(X)$ so that $\Pi(\bar{a}) = a, \dots, \Pi(\bar{d}) = d$, $|A\bar{a}| = |A_0\bar{a}_0|, \dots, |A\bar{d}| = |A_0\bar{d}_0|$. We have $|\bar{a}_0\bar{b}_0| = |\bar{a}\,\bar{b}|$, $|\bar{a}_0\bar{c}_0| = |\bar{a}\,\bar{c}|$, $|\bar{a}_0\bar{d}_0| = |\bar{a}\,\bar{d}|$, $|\bar{b}\,\bar{c}| > |\bar{b}_0\bar{c}_0|$, $|\bar{b}\,\bar{d}| > |\bar{b}_0\bar{d}_0|$, $|\bar{c}\,\bar{d}| > |\bar{c}_0\bar{d}_0|$. But $\tilde{\measuredangle}\,\bar{b}_0\bar{a}_0\bar{c}_0+\tilde{\measuredangle}\,\bar{b}_0\bar{a}_0\bar{d}_0+ +\tilde{\measuredangle}\,\bar{c}_0\bar{a}_0\bar{d}_0 = 2\pi$, consequently $\tilde{\measuredangle}\,\bar{b}\,\bar{a}\,\bar{c}+\tilde{\measuredangle}\,\bar{c}\,\bar{a}\,\bar{d}+\tilde{\measuredangle}\,\bar{b}\,\bar{a}\,\bar{d} > 2\pi$, which contradicts the condition (D). ■

**4.3.** The construction of the cone can be modified by using the spherical or hyperbolic cosine formula instead of the Euclidean cosine formula.

**4.3.1. Spherical suspension.**
Let $X$ be a metric space of diameter $\leqslant\pi$. The *spherical suspension* is the quotient space $S(X) = X\times[0, \pi]/\sim$, where $(x_1, a_1) \sim (x_2, a_2) \Leftrightarrow a_1 = a_2 = 0$ or $a_1 = a_2 = \pi$ with the canonical metric $\cos|\bar{x}_1\bar{x}_2| = \cos a_1 \cos a_2 + \sin a_1 \sin a_2 \cos|x_1x_2|$, where $\bar{x}_1 = (x_1, a_1)$, $\bar{x}_2 = (x_2, a_2)$.

***Proposition.*** *Let $X$ be a complete metric space of diameter $\leqslant\pi$. Then the following two conditions are equivalent.*

(a) *$X$ is a space with curvature $\geqslant 1$.*

(b) *$S(X)$ is not a circle and is a space with curvature $\geqslant 1$.*

**4.3.2. Elliptic cone.**
Let $X$ be a metric space of diameter $\leqslant \pi$. The *elliptic cone over* $X$ is the quotient space $EC(X) = X \times [0, \infty]/\sim$, where $(x_1, a_1) \sim (x_2, a_2) \Leftrightarrow$ $\Leftrightarrow a_1 = a_2 = 0$ with the canonical metric $\cosh|\bar{x}_1\bar{x}_2| = \cosh a_1 \cosh a_2 - $ $- \sinh a_1 \sinh a_2 \cos|x_1x_2|$, where $\bar{x}_1 = (x_1, a_1)$, $\bar{x}_2 = (x_2, a_2)$.

***Proposition.*** *Let X be a complete space with diameter $\leqslant \pi$. Then the following two conditions are equivalent.*
(a) *$X$ is a space with curvature $\geqslant 1$.*
(b) *$EC(X)$ is not a straight line and is a space with curvature $\geqslant -1$.*

**4.3.3. Parabolic cone.**
Let $X$ be a metric space. The *parabolic cone over* $X$ is the product $PC(X) = X \times \mathbb{R}$ with a canonical metric corresponding to the metric of hyperbolic space represented as the twisted product of a horosphere by a straight line.

***Proposition.*** *Let X be a complete space. Then the following two conditions are equivalent.*
(a) *$X$ is a space with curvature $\geqslant 0$.*
(b) *$PC(X)$ is a space with curvature $\geqslant -1$.*

**4.3.4. Hyperbolic cone.**
Let $X$ be a metric space. The *hyperbolic cone over* $X$ is the product $HC(X) = X \times \mathbb{R}$ with a canonical metric corresponding to the metric of hyperbolic space represented as the twisted product of a hyperplane by a straight line.

***Proposition.*** *Let X be a complete space. Then the following two conditions are equivalent.*
(a) *$X$ is a space with curvature $\geqslant -1$.*
(b) *$HC(X)$ is a space with curvature $\geqslant -1$.*

Propositions 4.3.1 – 4.3.4 are proved in the same way as Propositon 4.2.3.

**4.4. Join.**
Let $X$, $Y$ be complete spaces with curvature $\geqslant 1$. Then the direct product of the cones $C_A(X) \times C_A(Y)$ is a space with curvature $\geqslant 0$ and at the same time is the cone over the join $X * Y = X \times Y \times [0, \pi/2]/\sim$, where $(x_1, y_1, a_1) \sim (x_2, y_2, a_2) \Leftrightarrow a_1 = a_2 = 0$ and $x_1 = x_2$ or $a_1 = a_2 = \pi/2$ and $y_1 = y_2$, with metric $\cos|(x_1, y_1, a_1), (x_2, y_2, a_2)| = \cos a_1 \cos a_2 \cos|x_1x_2| +$ $+ \sin a_1 \sin a_2 \cos|y_1y_2|$. Thus the join $X * Y$ is also a complete space with curvature $\geqslant 1$.

**4.5.** We can now justify Example 2.9 (6). In fact, a locally finite simplicial space made by gluing simplicial spaces of constant curvature $k$ can be represented locally as a domain in the space obtained from a lower-dimensional finite simplicial space with curvature $\geqslant 1$ by using the constructions 4.2, 4.1 in the case $k = 0$, using 4.3.1 in the case $k > 0$, and using 4.3.2, 4.3.4 in the case $k < 0$. Thus Example 2.9 (6) is justified by induction on the dimension if we take into account the globalization theorem.

**4.6. An equidistant fibred space** (see [6]).

Let $X$ be a metric space represented as a disjoint union of closed equidistant subsets $X_\mu$, $\mu \in M$. (We say that the subsets $X_{\mu_1}$ and $X_{\mu_2}$ are equidistant if the distances $|x_2X_{\mu_1}|$ and $|x_1X_{\mu_2}|$ do not depend on $x_i \in X_{\mu_i}$.) In this case there is a natural metric on the space $M$ defined by the formula $|x_1x_2| = \inf\{|\bar{x}_1\bar{x}_2|, \bar{x}_1 \in \Pi^{-1}(x_1), \bar{x}_2 \in \Pi^{-1}(x_2)\}$, where $\Pi : X \to M$ is a natural projection.

***Proposition.*** *If $X$ is a space with curvature $\geqslant k$, then $M$ is a space with curvature $\geqslant k$.*

*Proof.* The fact that $M$ is a locally complete space with intrinsic metric is obvious. We assume that condition (D) is violated for the quadruple $(a; b, c, d)$ in $M$. Let the quadruple $(\bar{a}; \bar{b}, \bar{c}, \bar{d})$ in $X$ be such that $\Pi(\bar{a}) = a, \ldots, \Pi(\bar{d}) = d$ and the distances $|\bar{a}\bar{b}|$, $|\bar{a}\bar{c}|$, $|\bar{a}\bar{d}|$ do not differ much from the corresponding distances $|ab|$, $|ac|$, $|ad|$. Then since $|\bar{b}\bar{c}| \geqslant |bc|$, $|\bar{b}\bar{d}| \geqslant |bd|$, $|\bar{c}\bar{d}| \geqslant |cd|$, the quadruple $(\bar{a}; \bar{b}, \bar{c}, \bar{d})$ also violates condition (D)—a contradiction. ■

***Corollary.*** *Let $X$ be a space with curvature $\geqslant k$ and let the group $G$ act isometrically on $X$. Then the quotient space $\overline{X/G}$, whose points correspond to the closure of the orbits of $G$, is a space with curvature $\geqslant k$.*

## §5. Burst points

**5.1. Content of the section.**

All the examples considered above of a space with curvature bounded below included "singular" points, which have arbitrarily small neighbourhoods differing greatly, either just metrically or metrically as well as topologically, from a small neighbourhood of a point in a Riemannian manifold. In all these examples such singular points formed a sparse set. We shall see below that this is the situation in any (finite-dimensional) space with curvature bounded below.

In this section a condition is introduced, expressed in terms of distances, which, if satisfied for a certain point, ensures that it is non-singular in the sense that such a point has neighbourhoods that are homeomorphic and even almost isometric to a ball in Euclidean space. The points satisfying this condition form, in a (finite-dimensional) space with curvature bounded below, an open everywhere dense subset which is a topological manifold.

Let us describe one of the principal ideas. If a maximal (depending on dimension) number of pairwise orthogonal (or at least "linearly independent") geodesics pass through a point, then this point is automatically non-singular.

Generally speaking it is not clear whether such points exist. However, this condition can be modified to make it stable: a point turns out to be non-singular if a maximal number of pairs of geodesics emanate from it so that the geodesics of a single pair have "almost opposite" direction, while the geodesics from distinct pairs are "almost perpendicular".

**5.1.1.** *Remarks.* 1) From this section on we are dealing mostly with the local properties of spaces with curvature bounded below. For local considerations we shall suppose, without especially saying so, that everything takes place in a domain $U_x$, where conditions (D), (A), (B), (C) are satisfied. In addition, the explicit constants in our estimates will be for the case $k = 0$, since this greatly simplifies the calculations.

2) In this and the next section, to keep the argument simple, we assume that geodesics exist between the points we consider. In the general case one has to make some trivial modifications.

**5.2.** *Definition.* Let $M$ be a space with curvature not less than $k$. A point $p \in M$ is said to be an $(n, \delta)$*-burst point* if there are $n$ pairs of points $a_i$, $b_i$ distinct from $p$ such that

$$(1) \qquad \tilde{\measuredangle}\, a_i p b_i > \pi - \delta, \quad \tilde{\measuredangle}\, a_i p a_j > \pi/2 - \delta, \\ \tilde{\measuredangle}\, a_i p b_j > \pi/2 - \delta, \quad \tilde{\measuredangle}\, b_i p b_j > \pi/2 - \delta,$$

where $i \neq j$. The collection of these points $a_i$, $b_i$ will be called an $(n, \delta)$*-explosion* (or simply an *explosion*) at the point $p$. Condition (1) also implies the upper bounds $\tilde{\measuredangle} a_i p a_j < \pi/2 + 2\delta$, $\tilde{\measuredangle} a_i p b_i < \pi/2 + 2\delta$, $\tilde{\measuredangle} b_i p b_j < \pi/2 + 2\delta$, where $i \neq j$.

Clearly the set of $(n, \delta)$-burst points is open. It is not difficult to see that for $(n, \delta)$-burst points $p$ the explosion $\{a_i, b_i\}$ can be chosen from points arbitrarily near to $p$. If $M$ is locally compact, then the fact that $p \in M$ is an $(n, \delta)$-burst point is equivalent to the existence of $2n$ geodesics $pa_i$, $pb_i$ between which the angles satisfy inequalities analogous to (1).

**5.3.** *Definition.* A map $\varphi$ from an open set $U$ in the metric space $M$ into the metric space $M_1$ is said to be $\varepsilon$*-open* if for any points $x \in U$ and $\bar{y} \in M_1$ such that $\{z \in M : |xz| \leqslant \varepsilon^{-1} |\varphi(x)\bar{y}|_1\} \subset U$ one can find a point $y \in U$ such that $|xy| \leqslant \varepsilon^{-1} |\varphi(x)\bar{y}|_1$ and $\varphi(y) = \bar{y}$. (Here $|\cdot, \cdot|$ and $|\cdot, \cdot|_1$ denote the distances in $M$ and $M_1$ respectively.)

Clearly an $\varepsilon$-open map is open, and a continuous one-to-one $\varepsilon$-open map $\varphi$ is a homeomorphism, where the map inverse to the restriction of $\varphi$ to a sufficiently small subset $U_1 \subset U$ is Lipschitz with coefficient $\varepsilon^{-1}$.

**5.4. Theorem.** *Let $p$ be a $(n, \delta)$-burst point with explosion $(a_i, b_i)$, where $\delta < 1/2n$. Then the formula $\varphi(q) = (|a_1 q|, |a_2 q|, \ldots, |a_n q|)$ gives a $(1 - 2n\delta)/\sqrt{n}$-open Lipschitz map from a certain neighbourhood of the point $p$ into $\mathbb{R}^n$. If, in addition, there are no $(n+1, 4\delta)$-burst points near $p$, then the* *map $\varphi$ is a bi-Lipschitz homeomorphism between a certain neighbourhood of the point $p$ and a domain in $\mathbb{R}^n$.*

**5.5.** *Remark.* Below (see 9.4) we shall show that under the conditions of the second half of the theorem, if $\delta > 0$ is sufficiently small, then $\varphi$ is an "almost isometric" map from a certain neighbourhood of the point $p$ onto a domain in $\mathbb{R}^n$. The map $\varphi$ is called the *explosion map*.

To prove the second assertion of Theorem 5.4 we need the following lemma which will be useful later on.

**5.6. Lemma.** *Let $p, q, r, s$ be points in a space with curvature $\geqslant k$. If $|qs| < \delta\cdot\min\{|pq|, |rq|\}$ and $\tilde{\measuredangle} pqr > \pi-\delta_1$, then*

$$|\tilde{\measuredangle} pqs+\tilde{\measuredangle} rqs-\pi|<10\delta+2\delta_1.$$

(In particular, if geodesics exist, then the angles $\tilde{\measuredangle} pqs$, $\tilde{\measuredangle} rqs$ are little different from the corresponding angles $\measuredangle pqs$, $\measuredangle rqs$.)

*Proof.* The inequality $\tilde{\measuredangle} psq+\tilde{\measuredangle} rqs-\pi < \delta_1$ follows directly from condition (D) for the quadruple $(q; p, s, r)$. To prove the second inequality we first note that $\tilde{\measuredangle} psr > \pi-4\delta-\delta_1$—this follows from a consideration of the plane triangles $\tilde{\Delta} prs$, $\tilde{\Delta} pqr$.

In addition, it is obvious that $\tilde{\measuredangle} pqs+\tilde{\measuredangle} psq \geqslant \pi-2\delta$, $\tilde{\measuredangle} rqs+\tilde{\measuredangle} rsq \geqslant \pi-2\delta$. Therefore the second assertion of the lemma follows from condition (D) for the quadruple $(s; p, q, r)$. ∎

**5.7. Corollary.** *Let $p, q, r, s, t$ be points in a space with curvature $\geqslant k$ such that $|qs| < \delta\cdot\min\{|pq|, |rq|\}$, $\tilde{\measuredangle} pqr > \pi-\delta$, and also $||pq|-|ps|| < \delta\cdot|qs|$ and $\tilde{\measuredangle} qts > \pi-\delta$. Then each of the angles $\tilde{\measuredangle} ptq$, $\tilde{\measuredangle} pts$, $\tilde{\measuredangle} rtq$, $\tilde{\measuredangle} rts$ satisfies the inequality*

$$|\tilde{\measuredangle}\ldots-\pi/2|<100\delta+2\delta_1.$$

The proof is not complicated and we leave it to the reader.

**5.8.** *Proof of Theorem* 5.4. To prove the first part it is sufficient to verify that $\varphi$ is $(1-2n\delta)$-open with respect to the distance in $\mathbb{R}^n$ as determined by the norm $||x|| = \sum_{i=1}^{n}|x^i|$, where $x = (x^1, \ldots, x^n)$. Let the point $q$ be near to $p$, let $\varphi(q) = \bar{q}$, and let $\bar{r}$ be close to $\bar{q}$. We find the point $r\in\varphi^{-1}(\bar{r})$ as the limit of a sequence constructed inductively by the rule: $q_1 = q$, and if $q_{j-1}$ is already constructed, then we take as $q_j$ the point on the two-link broken line $a_\alpha q_{j-1}b_\alpha$ which is at a distance $\bar{r}^\alpha$ from $a_\alpha$, where $\alpha$ is the index for which $||a_iq_{j-1}|-\bar{r}^i|$ is maximal.

For definiteness, let $q_j$ lie on the geodesic $q_{j-1}b_\alpha$. Since $\tilde{\measuredangle} a_\alpha q_{j-1}q_j \geqslant$ $\geqslant \tilde{\measuredangle} a_\alpha q_{j-1}b_\alpha > \pi-\delta$, then $||a_\alpha q_{j-1}|-\bar{r}^\alpha| > |q_{j-1}q_j|\cos\delta$. If we take into account that $\tilde{\measuredangle} a_iq_{j-1}q_j \geqslant \tilde{\measuredangle} a_iq_{j-1}b_\alpha > \pi/2-\delta$ when $i\neq\alpha$, and on the other hand $\tilde{\measuredangle} a_iq_{j-1}q_j \leqslant 2\pi-\tilde{\measuredangle} a_iq_{j-1}a_\alpha-\tilde{\measuredangle} a_\alpha q_{j-1}q_j < \pi/2+2\delta$, we get, when

$|q_{j-1}q_j|$ is sufficiently small and $i \neq \alpha$,

$$||a_iq_j|-\bar{r}^i| \leqslant ||a_iq_j|-|a_iq_{j-1}||+||a_iq_{j-1}|-\bar{r}^i| \leqslant$$
$$\leqslant 2\delta\cdot|q_jq_{j-1}|+||a_iq_{j-1}|-\bar{r}^i| \leqslant (2\delta/\cos\delta)\cdot||a_\alpha q_{j-1}|-\bar{r}^\alpha|+||a_iq_{j-1}|-\bar{r}^i|.$$

Hence $\Delta_j = \|\varphi(q_j)-\bar{r}\| = \sum_{i\neq\alpha}||a_iq_j|-\bar{r}^i| \leqslant (n-1)2\delta/\cos\delta\,||a_\alpha q_{j-1}|-\bar{r}^\alpha|+$ $+\Delta_{j-1}-||a_\alpha q_{j-1}|-\bar{r}^\alpha| = \Delta_{j-1}-(1-2(n-1)\delta/\cos\delta)\,||a_\alpha q_{j-1}|-\bar{r}^\alpha|$. Thus $\Delta_{j-1}-\Delta_j \geqslant (\cos\delta-2(n-1)\delta)\,|q_{j-1}q_j| > (1-2n\delta)\,|q_{j-1}q_j|$, and in addition

$$\leqslant \Delta_{j-1}(1-(1-2(n-1)\delta/\cos\delta)/n) < \Delta_{j-1}\left(1-\frac{1-2n\delta}{n}\right)$$ according to the

choice of $\alpha$, $\Delta_j$. Therefore, if $\delta < 1/2n$, the sequence $q_j$ converges to some point $r$, where $\varphi(r) = \bar{r}$ and $|qr| \leqslant (1-2n\delta)\,|\overline{q\,r}|$.

To prove the second part of the theorem all that remains is for us to verify that $\varphi$ is one-to-one in a sufficiently small neighbourhood $U$ of the point $p$. We choose the neighbourhood $U$ so small that the pairs $(a_i, b_i)$ form an $(n, \delta)$-explosion for every point $q \in U$ and $\operatorname{diam} U < (0.01)\delta \min_{1\leqslant i\leqslant n}\{|pa_i|, |pb_i|\}$. If we had $\varphi(x) = \varphi(y)$ for $x, y \in U$, then, according to Corollary 5.7, the pairs $(a_i, b_i)$, together with the pair $(x, y)$, would form an $(n+1, 4\delta)$-explosion for the midpoint of the geodesic $xy$, which contradicts the condition. ■

We need the following lemma in order to characterize the dimension of the space with the help of burst points.

**5.9. *Lemma.*** *Given an arbitrarily small $\delta' > 0$ there are points having $(n, \delta')$-explosions in any neighbourhood of a point having an $(n, \delta)$-explosion with $\delta < 1/8n$.*

*Proof.* Let the point $p$ have an $(n, \delta)$-explosion $(a_i, b_i)$ with $\delta < 1/8n$. We construct the desired $(n, \delta')$-burst point $p'$ as the last term of the sequence $p_0 = p, \ldots, p_n = p'$ in which each point $p_j$ has an $(n, 4\delta)$-explosion $(a_i^j, b_i^j)$ such that $\tilde{\measuredangle} a_i^j p_j b_i^j > \pi-\delta'/4$ for $i \leqslant j$ and $\tilde{\measuredangle} a_i^j p_j a_{i'}^j > \pi/2-\delta'$, $\tilde{\measuredangle} b_i^j p_j b_{i'}^j >$ $> \pi/2-\delta'$, $\tilde{\measuredangle} a_i^j p_j b_i^j > \pi/2-\delta'$ if $i \neq i'$ and $i, i' \leqslant j$.

The sequence is constructed inductively. Let $p_j$ be already constructed; then we take as $p_{j+1}$ the midpoint of the geodesic $p_jq_j$, where $q_j$ is a point very near to $p_j$ for which $|p_ja_i^j| = |q_ja_i^j|$ when $i \neq j+1$, but $|p_ja_{j+1}^j| \neq |q_ja_{j+1}^j|$; the construction of this point is ensured by the first part of the proof of Theorem 5.4. We take as points of the explosion $a_i^{j+1} = a_i^j$ and $b_i^{j+1} = b_i^j$ when $i \neq j+1$, $a_{j+1}^{j+1} = p_j$, $b_{j+1}^{j+1} = q_j$. To verify this it is obvious that we only need $\tilde{\measuredangle} p_jp_{j+1}a_i^j > \pi/2-\delta'$ for $i < j+1$ and analogous inequalities with $b_i^j$ instead of $a_i^j$ and/or $q_j$ instead of $p_j$ and/or $\delta$ instead of $\delta'$ for $i > j+1$. These inequalities are ensured by Corollary 5.7 and the induction assumption. ■

## §6. Dimension

**6.1.** In a space with curvature bounded below, the number $n$ is called the *burst index* near $p$ if, for an arbitrarily small $\delta > 0$, there are $(n, \delta)$-burst points in any neighbourhood of this point but the analogous condition with $n$ replaced by $n+1$ is not satisfied ($n$ is a natural number or 0). If there is no such $n$, then we suppose the burst index to be $\infty$. We show that in such a space the burst indices near the various points are equal and coincide with its Hausdorff dimension. In fact, in what follows (up to §10) the Hausdorff measure and Hausdorff dimension do not play an essential role; instead we use the rough volume and the rough dimension defined below.

**6.2.** *Definition. The rough a-dimensional volume* $Vr_a(X)$ of a bounded set $X$ in a metric space is $\limsup_{\varepsilon\to 0} \varepsilon^a \beta_X(\varepsilon)$, where $\beta_X(\varepsilon)$ is the largest possible number of points $a_i \in X$ that are at least $\varepsilon$ pairwise distant from each other. The value $\beta_X(\varepsilon) = \infty$ is not excluded a priori.

$\inf\{a : Vr_a(X) = 0\} = \sup\{a : Vr_a(X) = \infty\}$ is called the *rough dimension* $\dim_r X$. Obviously $\dim_r X \geqslant \dim_H X$, where $\dim_H$ is the Hausdorff dimension. We also note that if $f : X \to Y$ is a Lipschitz map, then $\dim_H f(X) \leqslant \dim_H X$ and $\dim_r f(X) \leqslant \dim_r X$; if $f$ is bi-Lipschitz, then $\dim_H f(X) = \dim_H X$, $\dim_r f(X) = \dim_r X$.

**6.3.** ***Lemma.*** *Let $u, v \in M$ be points in a space with curvature not less than $k$, and let $U$ and $V$ be their neighbourhoods. Then, if $U$ and $V$ are sufficiently small,* $\dim_r U = \dim_r V$.

*Proof.* Since $u$ and $v$ may be joined in $M$ by a sequence of near points, then we may suppose that $u$ and $v$ are so near to each other and $U$ and $V$ are so small that conditions (A), (B), (C), (D) are satisfied for triangles with vertices in $U \cup V$.

Let $\dim_r V > a \geqslant 0$. Then $\limsup_{\varepsilon\to 0} \varepsilon^a \beta_V(\varepsilon) = \infty$ and hence for some sequence $\varepsilon_i \to 0$ we have $\varepsilon_i^a \beta_V(\varepsilon_i) \geqslant c > 0$. For every $i$ there is a collection of points $c_1, \ldots, c_{N_i} \in V$ which are at least $\varepsilon_i$ pairwise distant from each other and such that $\varepsilon_i^a N_i \geqslant c$. We fix in $U$ a ball $B = B_u(R)$ and some geodesics $uc_j$ and we choose points $b_j$ on them so that $|ub_j| = \frac{R}{D}|uc_j|$, where $D = \sup\{|ux| : x \in V\}$. Clearly the points $b_j$ lie in $B$ and by condition (B) are not less than $\varepsilon_i' = \varepsilon R/D$ distant from each other. Now we have

$$(\varepsilon_i')^a \beta_U(\varepsilon_i') \geqslant (R/D)^a \varepsilon_i^a N_i \geqslant c(R/D)^a,$$

consequently $Vr_a(U) > 0$ and in this way $\dim_r U \geqslant \dim_r V$.

If we exchange $U$ and $V$ we get the assertion of the lemma. ■

**6.4.** ***Lemma.*** *Let $p \in M$ be a point in a space with curvature bounded below. Then for a sufficiently small neighbourhood $U \ni p$ the burst index of $M$ near $p$ is equal to* $\dim_r U$ *and equal to* $\dim_H U$.

*Proof.* We assume first that the burst index of $M$ near $p$ is equal to $\infty$, that is, for an arbitrarily large $n$ and an arbitrarily small $\delta > 0$ there are $(n, \delta)$-burst points in any neighbourhood $U \ni p$. By Theorem 5.4 there is some neighbourhood $U_1 \subset U$ of the $(n, \delta)$-burst point $p_1 \in U$ that admits a Lipschitz open map into $\mathbb{R}^n$, whence we have $\dim_r U \geqslant \dim_r U_1 \geqslant n$ and $\dim_H U \geqslant \dim_H U_1 \geqslant n$. Since this is true for any natural number $n$, then in this case $\dim_r U = \dim_H U = \infty$.

Now let the burst index of $M$ near $p$ be $n$ and let $n$ be a natural number (the case $n = 0$ is trivial—in this case $M$ is a point). Then by definition of the burst index and by Lemma 5.9 there are no $(n+1, 1/8(n+1))$-burst points in some neighbourhood $U \ni p$. Therefore Theorem 5.4 ensures that there is a bi-Lipschitz homeomorphism from some neighbourhood $U_1 \subset U$ of an $(n, 1/100n)$-burst point $p_1 \in U$ onto a domain in $\mathbb{R}^n$. This immediately implies that $\dim_r U_1 = n$, $\dim_H U_1 = n$. By Lemma 6.3 we get $\dim_r U = \dim_r U_1 = n$ and finally $\dim_H U = n$, since $\dim_H U_1 \leqslant \dim_H U \leqslant \dim_r U$. ■

**6.5. Corollary.** *The burst indices near the various points of a space with curvature bounded below are equal. They coincide with its Hausdorff dimension. If the burst indices are finite, then they coincide with the topological dimension of the space.*

*Proof.* The first assertion is implied by Lemmas 6.4 and 6.3, the second by Lemma 6.4 (it is easy to verify that a space is separable if the burst indices are finite). To prove the third assertion we note that on the one hand it is known that the topological dimension of any metric space does not exceed its Hausdorff dimension, and on the other hand, if the burst index of the space near a point is equal to $n$, then by Theorem 5.4 some neighbourhood of an $(n, 1/100n)$-burst point $p_1$ near $p$ is homeomorphic to $\mathbb{R}^n$ and hence has topological dimension $n$. ■

**6.6.** *Remark.* We have not been able to prove that a space with curvature bounded below which has finite topological dimension must have finite burst indices.

By a finite-dimensional space with curvature bounded below (FDSCBB) we shall mean a space with finite burst indices. By Corollary 6.5 one can talk about the dimension of a FDSCBB without danger of ambiguity. Later on we shall be dealing exclusively with FDSCBB.

**6.7. Corollary.** *If $M$ is an $n$-dimensional* SCBB, *then for any $\delta > 0$ the set of $(n, \delta)$-burst points is everywhere dense in $M$.*

A point in a FDSCBB is said to be an *$n$-burst point* if it is an $(n, \delta)$-burst point for a sufficiently small (depending on the context) $\delta > 0$.

**6.8. Corollary.** *In a complete* FDSCBB *any bounded set is precompact.*

In fact an argument analogous to the proof of Lemma 6.3, but using, instead of condition (B), the global version of it that comes from Theorem 3.2, shows that in a complete $n$-dimensional SCBB any bounded set has a finite $n$-dimensional rough volume and further has finite $\varepsilon$-nets for all $\varepsilon > 0$.

**6.9.** *Remark.* The results of §§5, 6 retain part of their strength if the inequality in condition (D) is replaced by a weaker inequality $\tilde{\measuredangle} bac + \tilde{\measuredangle} cad + \tilde{\measuredangle} bad \leqslant \leqslant 2\pi + \varepsilon_m$, where $\varepsilon_m$ is a positive number depending only on the upper bound $m$ of the Hausdorff dimensions of the spaces under consideration. (This inequality is, in a certain sense, an integral bound for the negative curvature.) For sufficiently small $\varepsilon_m > 0$ one can determine inductively, starting with the largest $n$, positive numbers $\delta_n$, $1 \leqslant n \leqslant m+1$, such that Theorem 5.4 retains its strength when $0 < \delta \leqslant \delta_n$ and in the second assertion $(n+1, 4\delta)$ is replacd by $(n+1, \delta_{n+1})$. If now in defining the burst index near a point we only take into account the $(n, \delta_n)$-burst points, then the sets $M_n$, $0 \leqslant n \leqslant m$, consisting of the points of $M$ near which the burst index is $n$ form a "stratification" of the space $M$, that is, the union of all $M_n$ with numbers less than $l$ is open in $M$ for any $l$, and in every $M_n$ the set of $(n, \delta_n)$-burst points (which is a topological $n$-manifold) is open and dense.

## §7. The tangent cone and the space of directions

**7.1. Content of the section.**

The tangent cone at a point of a FDSCBB, which we define in this section, generalizes the concept of the tangent space at a point of a Riemannian manifold and of the tangent cone at a point of a convex hypersurface. To a first approximation it characterizes the metric in a neighbourhood of the point. The tangent cone to a point of a FDSCBB is a cone (in the sense of 4.2) over a compact space with curvature $\geqslant 1$, whose dimension is 1 less than that of the original space; thus the tangent cone is (by Proposition 4.2.3) a space of non-negative curvature and its dimension is the dimension of the space. As in the case of a convex hypersurface the tangent cone depends semicontinuously on the point of the space. In addition we show that the tangent cone changes continuously when the point moves along the interior of a geodesic.(1)

One natural definition of a tangent cone would be: one has to take the Hausdorff limit of "blown up" small neighbourhoods of the chosen point. For technical reasons we begin with another definition, which is based on the concept of the space of directions.

(1) In fact the tangent cone here does not change at all. This was recently proved by A. Petrunin.

**7.2. Space of directions.**
Let $p \in M$ be a point in a FDSCBB. Geodesics with origin $p$ are considered to be equivalent if they form a zero angle with each other; in this case one of them is a part of the other. The set of equivalence classes of geodesics with origin $p$ is endowed with a metric in which the distance is the angle between the geodesics at the point $p$. This metric space is denoted by $\Sigma_p'$ and its metric completion is called the *space of directions* at the point $p$ and is denoted by $\Sigma_p$.

**7.2.1.** *Remark.* Usually (see [8]) the space of directions is defined somewhat differently: one considers those curves with origin $p \in M$ that form a zero upper angle with themselves at $p$ ("having direction"); two such curves are considered equivalent if they form a zero upper angle with each other at $p$. The set of equivalence classes with the natural angle metric is called the *space of directions* $\Sigma_p^*$.

It can be shown that in the case of a FDSCBB this definition gives the same space $\Sigma_p$ as in 7.2. We know a proof of this which is simple but is based on the following Theorem 7.3.

**7.3.** ***Theorem.*** *The space of directions at any point of a* FDSCBB *is compact.*

*Proof.* To show that $\Sigma_p$ is compact it is sufficient to find a finite $\varepsilon$-net in $\Sigma_p'$ for any $\varepsilon > 0$. Therefore the theorem follows from the particular case $m = 0$ of the following lemma.

**7.4.** ***Lemma.*** *For any integers $n \geqslant 1$, $0 \leqslant m \leqslant n$, and a real number $\varepsilon > 0$ there exist a real number $\delta = \delta(n, m, \varepsilon) > 0$ and a natural number $N = N(n, m, \varepsilon)$ with the following property. In an n-dimensional* SCBB *there cannot exist points $p$, $a_i$, $b_i$, $c_j$, $1 \leqslant i \leqslant m$, $1 \leqslant j \leqslant N$, such that*

$$(1)\quad \tilde{\measuredangle}\, c_j p a_i > \pi/2 - \delta, \quad \tilde{\measuredangle}\, c_j p b_i > \pi/2 - \delta, \quad \tilde{\measuredangle}\, c_j p c_{j'} > \varepsilon \text{ when } j \neq j'$$

*and the pairs $(a_i, b_i)$ form an $(m, \delta)$-explosion at $p$.*

To prove Lemma 7.4 we need the following technical assertion, which will be used later on.

**7.5.** ***Lemma.*** *Let $\{pa_i\}$ be a finite collection of geodesics in a* FDSCBB. *Then for any $\delta > 0$ there is a neighbourhood $U$ of the point $p$ (depending on the collection of geodesics and on $\delta$) such that the angles of all the triangles $\Delta pqr$ with vertices $q$, $r$ on the parts of the geodesics $pa_i$ which are in $U$ differ from the corresponding angles of the triangles $\tilde{\Delta} pqr$ by no more than $\delta$.*

*Proof.* It is sufficient to consider the case of two geodesics $pa$, $pb$. Let $R > 0$ be so small that if $a_1 \in pa$, $b_1 \in pb$ are such that $|pa_1| \leqslant R$, $|pb_1| \leqslant R$, then $\measuredangle a_1 p b_1 - \tilde{\measuredangle} a_1 p b_1 < \delta/2$. We now consider any $\Delta p a_1 b_1$ with $a_1 \in pa$, $b_1 \in pb$ and $|pa_1| < (0.1)\delta R$, $|pb_1| < (0.1)\delta R$. Let the point $b_2 \in pb$ be such that $|pb_2| = R$. Clearly, if we add the triangles $\tilde{\Delta} p a_1 b_1$ and $\tilde{\Delta} a_1 b_1 b_2$

on the plane externally along the side $\tilde{a}_1\tilde{b}_1$, then we obtain a convex quadrangle, and by comparing this with the triangle $\tilde{\Delta} a_1pb_2$ we get

$$\begin{aligned}\tilde\measuredangle a_1pb_1 + \tilde\measuredangle a_1b_2b_1 - \tilde\measuredangle a_1pb_2 - \tilde\measuredangle a_1b_2p &= \\ = (\tilde\measuredangle pa_1b_2 - \tilde\measuredangle pa_1b_1 - \tilde\measuredangle b_1a_1b_2) &+ (\pi - \tilde\measuredangle pb_1a_1 - \tilde\measuredangle b_2b_1a_1) \geqslant \\ &\geqslant \pi - \tilde\measuredangle pb_1a_1 - \tilde\measuredangle b_2b_1a_1.\end{aligned}$$

Therefore

$$\begin{aligned}\measuredangle pb_1a_1 - \tilde\measuredangle pb_1a_1 \leqslant (\measuredangle pb_1a_1 + \measuredangle b_2b_1a_1) &- (\tilde\measuredangle pb_1a_1 + \tilde\measuredangle b_2b_1a_1) \leqslant \\ \leqslant (\tilde\measuredangle a_1pb_1 - \tilde\measuredangle a_1pb_2) &+ \tilde\measuredangle a_1b_2b_1 - \tilde\measuredangle a_1b_2p < \delta,\end{aligned}$$

since the difference in the brackets does not exceed $\delta/2$ and $\tilde\measuredangle a_1b_2b_1 \leqslant \delta/2$. Similarly we get $\measuredangle pa_1b_1 - \tilde\measuredangle pa_1b_1 < \delta$. ■

**7.6.** *Proof of Lemma* 7.4. We use inverse induction with respect to $m$ for fixed $n$. In the case $m = n$ one can take $N = 1$, $\delta = 1/100n$ since, if the points $p$, $a_i$, $b_i$, $c$ satisfied conditions (1), then a point $p_1$ of the geodesic $pc$ that is sufficiently near $p$ would be a $(n+1, 2\delta)$-burst point, which is impossible.

Induction step. We take $\delta$ and $N$ such that

$$\delta(m, \varepsilon) < 0.001\varepsilon\delta(m+1, \varepsilon/2), \tag{2}$$

$$\frac{1}{2}\left(N(m, \varepsilon) - N\left(m+1, \frac{\varepsilon}{2}\right)\right) > \frac{1000}{\delta(m, \varepsilon)} N\left(m+1, \frac{\varepsilon}{2}\right). \tag{3}$$

Let the points $p$, $a_i$, $b_i$, $c_j$ satisfy conditions (1) for given $m$, $\varepsilon$ and chosen $N$, $\delta$. By making a small shift in $p$ along the geodesic $pc_N$ and renaming $a_{m+1} = p$, $b_{m+1} = c_N$ we may suppose that the points $p$, $a_i$, $b_i$, $c_j$, $1 \leqslant i \leqslant m+1$, $1 \leqslant j \leqslant N-1$, satisfy the conditions analogous to (1) in which $\delta$ is replaced by $2\delta$ and the inequalities $\tilde\measuredangle a_{m+1}pc_j > \frac{\pi}{2} - \delta$, $\tilde\measuredangle b_{m+1}pc_j > \frac{\pi}{2} - \delta$ are excluded. By making a shift in the points $a_i$, $b_i$, $c_j$ towards $p$ we may suppose that they are in a zone where Lemma 7.5 applies. By the inductive hypothesis there cannot exist more than $N\left(m+1, \frac{\varepsilon}{2}\right) - 1$ of the points $c_j$ for which $\tilde\measuredangle a_{m+1}pc_j > \frac{\pi}{2} - 3\delta$ and $\tilde\measuredangle b_{m+1}pc_j > \frac{\pi}{2} - 3\delta$. Therefore in view of (3) we may suppose that among the $c_j$ there are at least $1000\delta^{-1}N\left(m+1, \frac{\varepsilon}{2}\right)$ of them for which $\frac{\varepsilon}{2} \leqslant \tilde\measuredangle a_{m+1}pc_j \leqslant \measuredangle a_{m+1}pc_j \leqslant \leqslant \tilde\measuredangle a_{m+1}pc_j + \delta \leqslant \frac{\pi}{2} - 2\delta$. In turn one can find at least $N\left(m+1, \frac{\varepsilon}{2}\right)$ of these points such that for any two of them $|\varphi_j - \varphi_{j'}| < (0.01)\delta$, where $\varphi_j = \tilde\measuredangle a_{m+1}pc_j$. If we now take points $p_1$ on $pa_{m+1}$ and $\bar{c}_j$ on $pc_j$ such that $|p\bar{c}_j| = |pp_1|(\cos\varphi_j)^{-1}$, then the points $p_1$, $a_i$, $b_i$, $\bar{c}_j$ provide a contradiction to the inductive hypothesis. In fact the inequality $\tilde{\measuredangle}\bar{c}_j p\bar{c}_{j'} > \frac{\varepsilon}{2}$ follows easily from our construction and the condition $|\varphi_j - \varphi_{j'}| < (0.01)\delta$. The relation between $|pp_1|$ and $|p\bar{c}_j|$ and Lemma 7.5 implies that $\tilde{\measuredangle} a_{m+1} p_1 \bar{c}_j > \frac{\pi}{2} - 10\delta$, $\tilde{\measuredangle} b_{m+1} p_1 \bar{c}_j > \frac{\pi}{2} - 10\delta$. Finally to prove the inequalities $\tilde{\measuredangle} a_i p_1 \bar{c}_j >$ $> \frac{\pi}{2} - \delta\left(m+1, \frac{\varepsilon}{2}\right)$, $1 \leqslant i \leqslant m$ (and the analogous inequalities with $b_i$ instead of $a_i$) it is sufficient to show that $||a_i\bar{c}_j| - |a_i p_1||$ is bounded above by $|p_1\bar{c}_j|$. Since $\varphi_j \geqslant \frac{\varepsilon}{2}$, then $|p_1\bar{c}_j| > (0.1)\varepsilon(|p_1 p| + |p_1\bar{c}_j|)$. On the other hand $||a_i\bar{c}_j| - |a_i p_1|| \leqslant ||a_i p| - |a_i p_1|| + ||a_i\bar{c}_j| - |a_i p|| \leqslant 5\delta(|p_1 p| + |p\bar{c}_j|)$. This, in view of condition (2), implies the required inequality. ■

**7.7.** *Definition.* The *tangent cone* $C_p$ at the point $p$ of a FDSCBB $M$ is the cone over the space of directions $\Sigma_p$.

The map $\exp_p$ is defined in the standard way with values in $M$, and with domain of definition a certain subset $C_p' \subset C_p$; this subset, generally speaking, may not contain any neighbourhood of the vertex of the cone $C_p$. (The inverse map $\exp_p^{-1}$, considered as a multi-valued map, is defined on all $M$; this map, generally speaking, is not continuous. It becomes single-valued (but is still not continuous) if it is considered as a map from the set of geodesics with origin $p$ in $C_p$.)

**7.8.** Let $X$, $Y$ be a metric space, $\nu > 0$. The map $f: X \to Y$ is called a (*Hausforff*) *$\nu$-approximation* if $||f(x_1)f(x_2)| - |x_1x_2|| < \nu$ for any $x_1, x_2 \in X$ and $Y$ is contained in the $\nu$-neighbourhood $U_\nu(f(X))$. We say that the compact metric spaces $X_i$ converge in the sense of Hausdorff to a compact metric space $X$ if for arbitrarily small $\nu > 0$ and sufficiently large numbers $i$ there exist $\nu$-approximations $f_i: X_i \to X$. There are other equivalent definitions of Hausdorff convergence for compact spaces; Hausdorff convergence can also be defined for non-compact spaces with a base point (see [19]).

**7.8.1.** ***Theorem.*** *Let $(M, \rho)$ be a* FDSCBB *and let $p \in M$. Then the spaces with base point $(M, p, \lambda\rho)$ converge in the sense of Hausdorff as $\lambda \to \infty$ to the tangent cone $C_p$.*

*Proof.* Let $B^\lambda(R)$ denote the ball in $M$ in the metric $\lambda\rho$ with centre $p$ and radius $R$, and let $B(R)$ be the ball in $C_p$ of radius $R$ with centre at the vertex of the cone $C_p$. By definition ([19], 3.14) we need to verify that $B^\lambda(R+\lambda^{-1})$ converges to $B(R)$ in the Hausdorff metric as $\lambda \to \infty$. For this, according to [19], 3.5(b), it is sufficient to verify that for any $\varepsilon > 0$ there exists an $\varepsilon$-net in $B(R)$ which is the limit of $\varepsilon$-nets in $B^\lambda(R+\lambda^{-1})$.

As $\overline{B}(R)$ is compact, we can choose there a finite $\varepsilon/2$-net. With a suitable choice of this net and for $\lambda$ sufficiently large the map $\xi \to \exp_p \lambda^{-1}\xi$ takes this into an $\varepsilon$-net in $B^\lambda(R+\lambda^{-1})$ where the $(\lambda\rho)$-distances between the points are near to the corresponding distances between points of the original net. The reader will not find it difficult to convert this argument into a rigorous proof. ■

**7.9. Corollary.** *Under the conditions of Theorem* 7.8.1 *the tangent cone $C_p$ is a space of non-negative curvature.*

In fact the metric of $C_p$ is intrinsic, as a Hausdorff limit of the intrinsic metrics $\lambda\rho$ (see [19]). If $(M, \rho)$ is a space with curvature $\geqslant k$, then $(M, \lambda\rho)$ is a space with curvature not less than $k\lambda^{-2}$. It is immediately clear from Definition 2.3 that the limit space $C_p$ is a space with curvature not less than

$$0 = \lim_{\lambda\to\infty} k\lambda^{-2}. \blacksquare$$

**7.10. Corollary.** *If, under the conditions of Theorem* 7.8.1, $\dim M > 1$, *then the space of directions $\Sigma_p$ is a space with curvature not less than* 1.

This is implied directly by Corollary 7.9 and Proposition 4.2.3.

**7.11. Corollary.** *Under the conditions of Theorem* 7.8.1

$$\dim \Sigma_p = \dim M - 1$$

*or, what is equivalent,* $\dim C_p = \dim M$.

To prove that $\dim C_p \geqslant \dim M$ it is sufficient to compare the rough volume of the small ball $B_p(R) \subset M$ with the rough volume of the ball $B(R) \subset C_p$. The corresponding inequality is easily obtained by using the map $B_p(R) \to B(R)$ inverse to the exponential map (see 7.7).

To prove the opposite inequality $\dim \Sigma_p \leqslant \dim M - 1$ it is sufficient to note the following. Let the point $q' \in \Sigma_p'$ have an $n$-explosion $(a_i', b_i')$. It may be supposed that $a_i', b_i'$ are arbitrarily near to $q'$ $(a_i', b_i' \in \Sigma_p')$, and let $q'$ be represented by the geodesic $pz$. Let $q$ be an interior point of the geodesic $pz$ sufficiently near to $p$. Then $q$ has an $(n+1)$-explosion $(a_i, b_i)$, where $a_{n+1}$, $b_{n+1}$ are points on the geodesic $pz$ and, when $i \leqslant n$, $a_i$, $b_i$ are points representing $a_i', b_i'$ on the geodesics, such that $|pa_i| = |pb_i| = |pq|$. ■

**7.12. Corollary.** *A homogeneous finite-dimensional space with curvature bounded below is a Riemannian manifold.*

In fact, according to a paper of Berestovskii [7], for this it is sufficient to prove that the assertion of Theorem 7.8.1 is satisfied at least at one (and hence at every) point of the space and moreover $C_p$ is isometric to $\mathbb{R}^n$; the last fact is implied immediately by 9.5, 6.7.

**7.13.** Let $X$, $Y$ be compact metric spaces. By definition $X \leqslant Y$ if there exists a non-contracting (not necessarily continuous) map $f: X \to Y$, that is, such that $|f(x)f(y)| \geqslant |xy|$. It is easy to see that $X \leqslant Y$ if any finite subset of some everywhere dense subspace $\chi \subset X$ admits a non-contracting map into $Y$. It can also be verified that if $X \geqslant Y \geqslant X$, then $X$ and $Y$ are isometric. We shall write $\liminf_{i\to\infty} X_i \geqslant X$ if for any Hausdorff subsequence $X_{i_k} \xrightarrow{H} Y$ that converges in the metric we have $Y \geqslant X$. Similarly one can give a meaning to the inequality $\limsup_{i\to\infty} X_i \leqslant X$.

**7.14.** ***Theorem*** (on the semicontinuity of tangent cones). *If $q_i, p \in M$ are points in a* FDSCBB *and $q_i \to p$, then* $\liminf \Sigma_{q_i} \geqslant \Sigma_p$.

*Proof.* Let $\{p_j\}$ be a subsequence of $\{q_i\}$ such that $\Sigma_{p_j} \xrightarrow{H} \Sigma$. Let us verify that $\Sigma \geqslant \Sigma_p$. We take a finite $\varepsilon$-net $A_\varepsilon = \{a_1', \ldots, a_m'\}$ in $\Sigma_p'$ so that the geodesic $pa_i$ goes out in the direction $a_i'$. Let $a_{ji}' \in \Sigma_{p_j}$ be the direction of the geodesic $p_j a_i$. We have (see 2.8.1) $\liminf_{j\to\infty} \measuredangle a_i p_j a_{i'} \geqslant \measuredangle a_i p a_{i'}$, that is, $\liminf_{j\to\infty} |a_{ji}' a_{ji'}'| \geqslant |a_i a_{i'}|$. Some point $b_{ji} \in \Sigma$ will correspond to each point $a_{ji}'$ such that $|a_{ji}' a_{ji'}'| - |b_{ji} b_{ji'}| \to 0$, consequently $\liminf_{j\to\infty} |b_{ji} b_{ji'}| \geqslant |a_i a_{i'}|$.

Thus we can define a non-contracting map $A_\varepsilon \to \Sigma$ by considering one of the limit points of the sequence $b_{ji}$ as the image of the point $a_i$. ■

**7.15.** Our next aim is to prove that the tangent cone is continuous when a point moves along the interior of a geodesic. In fact we shall prove, after some preliminary considerations, a somewhat stronger assertion.

Let $p$ be an interior point of the geodesic $ab$ in the FDSCBB $M$. Then the points $a', b' \in \Sigma_p$ that are images of the geodesics $pa$ and $pb$ respectively are at a maximal possible distance $\pi$ in the space $\Sigma_p$ with curvature $\geqslant 1$. Hence it is not difficult to deduce from the angle comparison theorem in $\Sigma_p$ that $\Sigma_p$ is the spherical suspension (in the sense of 4.3.1) over a space of one less dimension with curvature $\geqslant 1$, which naturally identifies with

$$\Lambda_p = \left\{\xi \in \Sigma_p : |\xi a'| = \frac{\pi}{2}\right\} = \left\{\xi \in \Sigma_p : |\xi b'| = \frac{\pi}{2}\right\}.$$

**7.16.** ***Theorem.*** *Let $p$ be an interior point of the geodesic $ab$ in the* FDSCBB *$M$, and let the sequence $\{p_i\}$ of points in $ab$ converge to $p$. Then $\Lambda_{p_i}$ converges to $\Lambda_p$ in the Hausdorff metric.*

*Proof.* In order to show that $\liminf_{i\to\infty} \Lambda_{p_i} \geqslant \Lambda_p$ it is sufficient to repeat with small alterations the proof of Theorem 7.14, taking into account that if the point $x \in M$ is near to $p$ and $(|px| \ll \min\{|pa|, |pb|\})$, then for sufficiently large $i$ the angles $\measuredangle xpa$ and $\measuredangle xp_i a$ are near.

To prove the opposite inequality lim sup $\Lambda_{p_i} \leqslant \Lambda_p$ we consider the following construction. Let the points $r$, $p$, $q$ be situated on the geodesic $ab$ so that the order is: $a$, $r$, $p$, $q$, $b$. Let the points $r_1$, $r_2$ be near to $r$, and let the points $p_1$, $p_2$ lie on the geodesics $qr_1$, $qr_2$ so that $\frac{|qp_1|}{|qr_1|} = \frac{|qp_2|}{qr_2|} = \frac{|qp|}{|qr|}$. We assume that the angles $\measuredangle r_1rp$, $\measuredangle r_2rp$ are near to $\pi/2$ and $|r_1r| = |r_2r| \ll \ll |pr| \ll |pq| \leqslant \min\{|pa|, ||pb|\}$. Then, as we shall show below, the angles $\measuredangle p_1pq$, $\measuredangle p_2pq$ are near to $\pi/2$, while $|pp_1|$ and $|pp_2|$ are near to $|rr_1|\cdot\frac{|qp|}{|qr|}$. Therefore, since $|p_1p_2| \geqslant |r_1r_2|\cdot\frac{|qp|}{|qr|}$ by condition (B), then $\measuredangle p_1pp_2$ cannot be substantially less than $\measuredangle r_1rr_2$. Thus the construction described, taking $p_i$ as $r$, enables us to establish, using standard arguments, that $\limsup_{i\to\infty} \Lambda_{p_i} \leqslant \Lambda_p$ and thus to conclude the proof of the theorem. All that remains is for us to verify the assertions stated here without proof.

**7.17.** ***Lemma.*** *Let the points $a$, $r$, $p$, $q$, $b$, $r_1$, $p_1$ be as in the construction described above. Then there exists* $\delta > 0$ (not depending on the given points) *such that if* $|rr_1| < \delta^2|pr|$, $|\measuredangle r_1rp - \pi/2| < \delta$, $|qr| < \delta \min\{|ar|, |bq|\}$, *then*

a) $\measuredangle r_1qr < (1+2\delta)\dfrac{|rr_1|}{|qr|}$,

b) $\widetilde{\measuredangle} r_1rq > \dfrac{\pi}{2} - 2\delta$,

c) $\left|\dfrac{|pp_1|}{|rr_1|}\cdot\dfrac{|qr|}{|qp|} - 1\right| < 1 + 5\delta$,

d) $\left|\measuredangle p_1pq - \dfrac{\pi}{2}\right| < 3\delta$.

*Proof.* By applying condition (C) and the cosine formula we get

$$|ar_1| \leqslant (|ar|^2 + |rr_1|^2 + 2|ar||rr_1|\cos\measuredangle r_1rp)^{1/2} \leqslant$$
$$\leqslant |ar| + |rr_1|\cos\measuredangle r_1rp + \frac{|rr_1|^2}{2|ar|},$$
$$|br_1| \leqslant (|qr_1|^2 + |qb|^2 + 2|qr_1||qb|\cos\measuredangle r_1qr)^{1/2} \leqslant$$
$$\leqslant |qr_1| + |qb| - \frac{|qr_1||qb|}{|qr_1| + |qb|}(1 - \cos\measuredangle r_1qr),$$
$$|qr_1| \leqslant (|qr|^2 + |rr_1|^2 - 2|qr||rr_1|\cos\measuredangle r_1rp)^{1/2} \leqslant$$
$$\leqslant |qr| - |rr_1|\cos\measuredangle r_1rp + \frac{|rr_1|^2}{2|qr|}.$$

In addition

$$|ar| + |qr| + |qb| = |ab| < |ar_1| + |br_1|.$$

By adding these four inequalities we get

$$\frac{|qr_1||qb|}{|qr_1|+|qb|}(1-\cos \measuredangle\, r_1qr) < \frac{1}{2}|rr_1|^2\left(\frac{1}{|ar|}+\frac{1}{|qr|}\right)$$

or

$$1-\cos \measuredangle\, r_1qr < \frac{|rr_1|^2}{2|qr|^2}\cdot\left(1+\frac{|qr|}{|ar|}\right)\cdot\frac{|qr|}{|qr_1|}\cdot\left(1+\frac{|qr_1|}{|qb|}\right) <$$
$$< (1+3\delta)\frac{|rr_1|^2}{2|qr|^2},$$

hence $\measuredangle\, r_1qr < (1+2\delta)\dfrac{|rr_1|}{|qr|}$.

To prove (b) we note that if $\tilde{\measuredangle} r_1rq \leqslant \frac{\pi}{2}-2\delta$, then

$$|qr_1| = (|rq|^2+|rr_1|^2-2|rq||rr_1|\cos\tilde{\measuredangle}\, r_1rq)^{1/2} < |rq|-\frac{3}{2}\delta|rr_1|.$$

On the other hand, by condition (C), taking into account that $\measuredangle arr_1 < \frac{\pi}{2}+\delta$, we get

$$|ar_1| \leqslant (|ar|^2+|rr_1|^2-2|ar||rr_1|\cos\measuredangle\, arr_1)^{1/2} < |ar|+\frac{3}{2}\delta|rr_1|.$$

By combining these two inequalities with the triangle inequality $|ar|+|rq| = |aq| < |ar_1|+|qr_1|$ we get a contradiction.

Assertion c) follows from assertion b), conditions (B), (C), and assertion a):

$$(1-2\delta)\frac{|rr_1|}{|qr|} < \tilde{\measuredangle}\, rqr_1 \leqslant \tilde{\measuredangle}\, pqp_1 \leqslant \measuredangle\, r_1qr < (1+2\delta)\frac{|rr_1|}{|qr|}.$$

We note that c) and the conditions of the lemma imply that $|pp_1| < 2\delta^2|pq|$. The lower bound $\measuredangle p_1pq \geqslant \tilde{\measuredangle} p_1pq > \frac{\pi}{2}-3\delta$ follows in an elementary way from b) and c). To conclude the proof of the lemma it remains for us to verify that $\measuredangle rpp_1 > \frac{\pi}{2}-3\delta$. Let us assume that this is not so. Then

$$|p_1r| \leqslant (|pr|^2+|pp_1|^2-2|pr||pp_1|\cos\measuredangle\, rpp_1)^{1/2} < |pr|-2\delta|pp_1|.$$

On the other hand, condition (A) implies that $|p_1r| \geqslant |\tilde{p}_1\tilde{r}|$, where $\tilde{p}_1$ is a point on the side $\tilde{q}\tilde{r}_1$ of the triangle $\tilde{\Delta}qrr_1$, dividing it in the same ratio as the point $p_1$ divides the geodesic $qr$. Thus

$$|p_1r| \geqslant \left(|pr|^2+\frac{(|rr_1||pq|)^2}{|qr|^2}+2|pr|\frac{|rr_1||pq|}{|rq|}\cos\tilde{\measuredangle}\, qrr_1\right)^{1/2} >$$
$$> |pr|-\delta\frac{|rr_1||pq|}{|qr|}.$$

These inequalities imply that $2|pp_1| < \dfrac{|rr_1||pq|}{|qr|}$, which contradicts c). ■

**7.18.** *Remark.* A stronger assertion holds for a convex hypersurface in $\mathbb{R}^n$ and this follows immediately from the results of Milka [25]: the tangent cones at the interior points of geodesics on a convex hypersurface are mutually isometric (but of course not necessary congruent as subsets of $\mathbb{R}^n$). We do not know whether an analogous assertion is true for a FDSCBB (see p. 22).

**7.19.** We shall separate the points of a FDSCBB into two classes: interior points and boundary points. The boundary of a FDSCBB is defined by induction in the following way.

A one-dimensional SCBB $M$ is a manifold and the boundary of $M$ is by definition the boundary of this manifold. Now let $M$ be an $n$-dimensional SCBB, where $n > 1$. The point $p \in M$ is considered to be a *boundary point* if the space of directions $\Sigma_p$ is an $(n-1)$-dimensional SCBB with non-empty boundary. If however $\Sigma_p$ does not have a boundary, then $p$ is called an *interior* point of $M$.

We note that all $(n, \delta)$-burst points of $M$ for sufficiently small $\delta > 0$ are interior points (see 9.6). With regard to the boundary see also §12 and 13.3.

## Conventions and notation

In what follows we shall, without specifically saying so, use the following notation.

**1.** If $A$ is a subset of the FDSCBB $M$ and the point $p \in M$ does not belong to $A$, then $A' \subset \Sigma_p$, as a rule, denotes the set of directions of all possible geodesics $pA$, that is, geodesics $pa$ such that $a \in A$ and $|pa| = |pA|$. Moreover, within the limits of a single argument, $A'$ may denote subsets of the spaces of directions at various points, say $A' \subset \Sigma_p$ and $A' \subset \Sigma_q$; we remove ambiguity by indicating in every case the point at which we are dealing with the space of directions. In certain especially stipulated cases $A' \in \Sigma_p$ may denote the direction of some single geodesic $pA$.

**2.** We let $c$ denote positive "constants". This notation ignores whether such "constants" depend on the dimension of the FDSCBB (and also on the lower bound of the curvature). The "constants" depending on certain parameters $a$, $\Pi$, $\xi$ are denoted by $c(a, \Pi, \xi)$. If we wish to emphasize that the given "constant" does not depend on the parameter $a$ (but may depend on the other parameters), we write $c[\widehat{a}]$.

**3.** We let $\varkappa$ denote positive functions that are infinitesimally small at zero. Moreover the functions $\varkappa$ may also depend on parameters (which are not necessarily present in the notation), but for any fixed values of parameters the function $\varkappa$ is infinitesimally small in all of its variables. For example, the function $\varkappa[\widehat{a}](\delta, R)$ may depend on certain additional parameters but not on $a$, and for any fixed values of parameters $\varkappa[\widehat{a}](\delta, R) \to 0$ as $\delta, R \to 0$. As a rule the "constants" $c$ and the functions $\varkappa$ could be explicitly indicated.

Moreover the variables of $\varkappa$ are assumed to be so small that for all $\varkappa$ and $c$ taking part in the argument that have to be compared, we have $\varkappa < c$.

**4.** We let $\delta$ and $R$ (possibly with indices) denote positive real numbers which should be considered to be arbitrarily small. Here, as a rule, the "ceiling" of admissible values of $\delta$ could be explicitly indicated, depending on the dimension and the "degree of degeneracy" of the FDSCBB under consideration (see below 8.7 and 12.1), whereas the "ceiling" of admissible values of $R$ depends on random circumstances.

## §8. Estimates of rough volume and the compactness theorem

In this section we gather together several assertions which concern estimates of the rough volume of a FDSCBB. These estimates are qualitative in character and do not pretend to be exact. The theorem that the class of FDSCBB with bounded diameter is compact turns out to be a simple consequence of an estimate of the rough volume.

**8.1.** We introduce a function $\psi(k, D)$ defined as follows: $\psi(k, D) = \max\{|pr|(\measuredangle pqr)^{-1}$ for triangles $\Delta pqr$ on the $k$-plane such that $|pq| \leqslant D$, $|qr| \leqslant D$, $|pr| \geqslant 2||pq|-|qr||\}$.

It is easily seen that $\psi(k, D)$ is a finite positive number, $\psi(k, D) \underset{D\to 0}{=} O(D)$ for any fixed $k$, and for fixed $k > 0$ the function $\psi(k, D)$ is bounded. We shall often omit the variable $k$ in the function $\psi$.

**8.2. *Lemma.*** *Let $p \in M$ be a point in a complete n-dimensional* SCBB; *suppose that $A \subset M$, $\Gamma \subset \Sigma_p$ are such that for every point $a \in A$ other than $p$ there is a geodesic $pa$ such that $a' \in \Gamma$. Then*

$$Vr_n(A) \leqslant Vr_{n-1}(\Gamma)\cdot 2D_1\psi^{n-1}(D),$$

*where* $D = \operatorname{diam}(A \cup \{p\})$, $D_1 = \max_{a\in A}|ap| - \min_{a\in A}|ap|$.

*Proof.* If we take $\beta_A(\varepsilon)$ points in $A$ whose distance from each other is at least $\varepsilon$, there are at least $\beta_A(\varepsilon)\cdot\left(\frac{2D_1}{\varepsilon}+1\right)^{-1}$ of them such that their distances to $p$ differ pairwise by not more than $\varepsilon/2$. Thus by the angle comparison theorem we get $\beta_A(\varepsilon)\cdot\left(\frac{2D_1}{\varepsilon}+1\right)^{-1}$ points in $\Gamma$ at a pairwise distance of at least $\varepsilon/\psi(D)$. We obtain the inequality $\beta_\Gamma\left(\frac{\varepsilon}{\psi(D)}\right) \geqslant \beta_A(\varepsilon)\left(\frac{2D_1}{\varepsilon}+1\right)^{-1}$ from which we get the assertion of the lemma if we let $\varepsilon$ tend to 0. ■

**8.3. *Corollary.*** *For complete n-dimensional spaces with curvature $\geqslant 1$ we have the bound $Vr_n(M) \leqslant c(n)$. In addition $\beta_M(\varepsilon) \leqslant c(n)\varepsilon^{-n}$ for all $\varepsilon > 0$.*

The proof is by an obvious induction taking into account that diam $M \leqslant \pi$.

**8.4. Corollary.** *For complete n-dimensional spaces with curvature* $\geqslant k$ *and with* diam $M \leqslant D$ *we have the bound* $Vr_n(M) \leqslant c(n, k, D)$. *In addition,* $\beta_M(\varepsilon) \leqslant c(n, k, D)\varepsilon^{-n}$ *for all* $\varepsilon > 0$.

**8.5.** Let $\mathfrak{M}(n, k, D)$ denote the metric space (with Hausdorff metric) consisting of complete spaces with curvature not less than $k$, dimension not greater than $n$, and diameter not greater than $D$. We consider $\mathfrak{M}(n, k, D)$ as a subspace (which is complete in the Hausdorff metric) of the space of all compact metric spaces.

***Theorem.*** *The space* $\mathfrak{M} = \mathfrak{M}(n, k, D)$ *is compact.*

*Proof.* Let us check first that $\mathfrak{M}$ contains its limit points. In fact, if $M_i \in \mathfrak{M}$, where $M$ is a compact metric space and $M_i \to M$ in the sense of Hausdorff, then obviously diam $M \leqslant D$ and by [19], Proposition 3.8 the metric of $M$ is intrinsic. The fact that this metric has curvature not less than $k$ follows trivially from the definitions. Finally, the condition dim $M \leqslant n$ is satisfied, since the presence in $M$ of a (greater than $n$)-burst point would imply the presence of such points in $M_i$. We now check that $\mathfrak{M}$ is precompact. It was shown in [19] (see also [20]) that for this it is sufficient to show that for any $\varepsilon > 0$ there is a natural number $N(\varepsilon)$ such that any space $M \in \mathfrak{M}$ has an $\varepsilon$-net of $N(\varepsilon)$ points. But the existence of such an $N(\varepsilon)$ follows directly from Corollary 8.4. ■

The idea of the proof of the next lemma is due to Cheeger [10].

**8.6. *Lemma.*** *Let* $p_1, \ldots, p_m$ *be points in a compact n-dimensional* SCBB. *We let* $p'_{ij}$ *denote the set of directions of the geodesics* $p_i p_j$ *in* $\Sigma_{p_j}$. *We assume that for each number* $i$, $1 \leqslant i \leqslant m$, *there are directions* $\xi_i^- \in p'_{i-1,i}$ *and* $\xi_i^+ \in p'_{i+1,i}$ *such that* $|\xi_i^-, \xi_i^+| > \pi - \delta$, *where* $i+1 = 1$ *for* $i = m$, $i-1 = m$ *for* $i = 1$. (*We shall say that such points* $p_i$ *almost generate a closed geodesic.*) *Let* $D = $ diam $M$, $\delta_1 = D^{-1}\max|p_i p_{i+1}|$. *Then* $Vr_n(M) \leqslant \varkappa(\delta, \delta_1) D\psi^{n-1}(D)$, *where the function* $\varkappa$ *depends on* $m$ *as a parameter.*

*Proof.* Let $r \in M$ be such that $|p_i r| > \delta_1^{-1/2} \max\limits_j |p_j p_{j+1}|$ for all $i$. Then

$$(1) \qquad \tilde{\measuredangle} r p_{i+1} p_i + \tilde{\measuredangle} r p_i p_{i+1} > \pi - \varkappa(\delta_1).$$

On the other hand, for the direction $r'_i \in \Sigma_{p_i}$ of any geodesic $p_i r$ we have

$$(2) \qquad |r'_i \xi_i^-| + |r'_i \xi_i^+| < \pi + \delta,$$

whence

$$(3) \qquad \measuredangle r p_i p_{i-1} + \measuredangle r p_i p_{i+1} < \pi + \delta.$$

If we combine the inequalities (1) and (3) for all $i$, we may conclude that all the inequalities (1)–(3) are nearly equalities, in particular $|\tilde{\measuredangle} r p_i p_{i-1} + \tilde{\measuredangle} r p_i p_{i+1} - \pi| < \varkappa(\delta, \delta_1)$ and $|\tilde{\measuredangle} r p_i p_{i+1} + \tilde{\measuredangle} r p_{i+1} p_i - \pi| < \varkappa(\delta, \delta_1)$, whence

$|\tilde{\measuredangle} rp_ip_{i-1}+\tilde{\measuredangle} rp_{i+1}p_i| < \varkappa(\delta, \delta_1)$. If $i_0$ is that number for which $|p_ir|$ is minimal, then $\tilde{\measuredangle} rp_{i_0}p_{i_0+1} > \frac{\pi}{2}-\varkappa(\delta_1)$ and $\tilde{\measuredangle} rp_{i_0}p_{i_0-1} > \frac{\pi}{2}-\varkappa(\delta_1)$. Therefore for any $i$ we get $\left|\measuredangle rp_ip_{i+1}-\frac{\pi}{2}\right| < \varkappa(\delta, \delta_1)$. If we take into account that the inequalities (2) are nearly equalities, we can assert, in particular that $\left||\xi_1^+r_1'|-\frac{\pi}{2}\right| < \varkappa(\delta, \delta_1)$. If we apply Lemma 8.2 and Corollary 8.3 we get $Vr_{n-1}\left(\left\{r_1' \in \Sigma_{p_1} : \left||\xi_1^+r_1'| - \frac{\pi}{2}\right| < \varkappa(\delta, \delta_1)\right\}\right) < \varkappa(\delta, \delta_1)$, and again by Lemma 8.2 $Vr_n(\{r \in M : |p_ir| > (\delta_1^{-1/2} \max |p_jp_{j-1}|$ for all $i\}) < \varkappa(\delta, \delta_1)D\psi^{n-1}(D)$. To finish the proof it is sufficient to note that by Lemma 8.2

$$Vr_n(\{r\in M: |p_1r|<(\delta_1^{-1/2}+m)\max_j |p_jp_{j-1}|\})<\varkappa(\delta_1)\,D\psi^{n-1}(D). \blacksquare$$

**8.7.** *Remark.* $Vr_n(M)(\text{diam}\, M\cdot\psi^{n-1}(\text{diam}\, M))^{-1}$ can serve as a non-degeneracy characteristic of the dimension for an $n$-dimensional compact SCBB $M$. The results of this section imply that the existence of a (positive) lower bound for this ensures a uniform lower bound for $Vr_{n-1}(\Sigma_p)$ over points $p \in M$ and also a lower bound for the "lengths of almost closed geodesics". In the case of non-compact FDSCBB there are similar arguments which show that rough volumes $Vr_{n-1}(\Sigma_p)$ and the "lengths of almost closed geodesics" are uniformly separated from zero on compact subsets.

## §9. Theorems on almost isometry

**9.1.** We shall say that a complete finite-dimensional space $M$ with curvature $\geqslant 1$ has an $(m, \delta)$-explosion $(A_i, B_i)$, $1 \leqslant i \leqslant m$, if $A_i, B_i \subset M$ are compact subsets such that $|A_i, B_i| > \pi-\delta$, $|A_i, B_j| > \frac{\pi}{2}-\delta$, $|A_i, A_j| > \frac{\pi}{2}-\delta$, $|B_i, B_j| > \frac{\pi}{2}-\delta$ when $i \neq j$. Clearly a point $p$ in a FDSCBB has an $(m, \delta)$-explosion if and only if its space of directions $\Sigma_p$ has an $(m, \delta)$-explosion in the sense of the definition given above. We recall that for sufficiently small $\delta > 0$ an $(n, \delta)$-burst point in an $n$-dimensional space with curvature bounded below has a neighbourhood that admits a bi-Lipschitz homeomorphism onto a ball in $\mathbb{R}^n$. In this section we show that this homeomorphism is almost isometric for small $\delta > 0$ (that is, the distortion coefficient of the distance is near to 1). Correspondingly an $(n-1)$-dimensional space with curvature $\geqslant 1$ that has an $(n, \delta)$-explosion turns out to be an almost isometric unit sphere $S^{n-1}$ (cf. Otsu, Shiohama, and Yamaguchi [27]).

**9.2.** The following assertions, which are based directly on the angle comparison theorem and some elementary spherical geometry, are used constantly in what follows.

**Lemma.** *Let $M$ be a complete finite-dimensional space with curvature $\geqslant 1$.*

a) *Let the sets $A_i, B_i$ form an $(m, \delta)$-explosion in $M$; $p \in M$, $|pA_i| > \frac{\pi}{2}-\delta$, $|pB_i| > \frac{\pi}{2}-\delta$ for all $i$. Then the sets $A_i'$, $B_i'$ form an $(m, \varkappa(\delta))$-explosion in $\Sigma_p$.*

b) *Let $A$, $B$ form a $(1, \delta)$-explosion in $M$; $p, q \in M$, $|pA| > \frac{\pi}{2}-\delta$, $|pB| > \frac{\pi}{2}-\delta$. Let the points $\tilde{A}$, $\tilde{B}$, $\tilde{p}$, $\tilde{q}$ be given on the unit sphere $S^2$ such that $|\tilde{A}\tilde{B}| = \pi$, $|\tilde{A}\tilde{p}| = |\tilde{B}\tilde{p}| = \frac{\pi}{2}$, $||\tilde{p}\tilde{q}|-|pq|| < \delta$, and also*

$$(1) \qquad |\measuredangle \tilde{A}\tilde{p}\tilde{q}|-|A'q'||<\delta, \quad |\measuredangle \tilde{B}\tilde{p}\tilde{q}-|B'q'||<\delta,$$

*where $A'$, $B'$, $q' \subset \Sigma_p$, and $q'$ is the direction of one of the geodesics $pq$. Then $||Aq|-|\tilde{A}\tilde{q}|| < \varkappa(\delta)$ and $||Bq|-|\tilde{B}\tilde{q}|| < \varkappa(\delta)$.*

c) *Let conditions* b) *be all satisfied except for* (1) *and assume instead that $||Aq|-|\tilde{A}\tilde{q}|| < \delta$ and $||Bq|-|\tilde{B}\tilde{q}|| < \delta$. Then either for the direction $q' \in \Sigma_p$ of any geodesic $pq$ we have $|\measuredangle \tilde{A}\tilde{p}\tilde{q}-|A'q'|| < \varkappa(\delta)$ and $|\measuredangle \tilde{B}\tilde{p}\tilde{q}-|B'q'|| < \varkappa(\delta)$, or $|pq| < \varkappa(\delta)$, or $|pq| > \pi-\varkappa(\delta)$.*

d) *Let conditions* c) *be satisfied and let there be given points $r$ on the geodesic $pq$ and $\tilde{r}$ on $\tilde{p}\tilde{q}$ so that $||pr|-|\tilde{p}\tilde{r}|| < \delta$. Then either $||Ar|-|\tilde{A}\tilde{r}|| < \varkappa(\delta)$ and $||Br|-|\tilde{B}\tilde{r}|| < \varkappa(\delta)$, or $|Ap|+|Aq|+|pq| > 2\pi-\varkappa(\delta)$, or $|Bp|+|Bq|+|pq| > 2\pi-\varkappa(\delta)$.*

The proofs of these assertions are not complicated and are left to the reader.

**9.3. Lemma.** *Let a complete $(n-1)$-dimensional space $M$ with curvature $\geqslant 1$ have an $(n, \delta)$-explosion $(A_i, B_i)$. Then for any point $q \in M$ we have*

$$\left|\sum_{i=1}^{n} \cos^2|A_i q|-1\right| < \varkappa(\delta).$$

*Proof.* We use induction with respect to the dimension. If $n = 2$, the assertion is obvious. To prove the induction step we chose an arbitrary point $p \in A_n$ and construct the points $\tilde{A}_i$, $\tilde{B}_i$ $(1 \leqslant i \leqslant 1)$, $\tilde{p}$, $\tilde{q}$ on the unit sphere $S^{n-1}$ so that $(\tilde{A}_i, \tilde{B}_i)$ form an $(n-1, 0)$-explosion, where $|\tilde{p}\tilde{A}_i| = |\tilde{p}\tilde{B}_i| = \frac{\pi}{2}$ for all $i$, $|\tilde{p}\tilde{q}| = |pq|$, and also $|\measuredangle \tilde{A}_i\tilde{p}\tilde{q}-|A_i'q'|| < \varkappa(\delta)$; here $A_i'$, $q' \subset \Sigma_p$, and $q'$ is the direction of one of the geodesics $pq$. The last requirement can be satisfied since by the inductive hypothesis $\left|\sum_{i=1}^{n-1} \cos^2|A_i'q'|-1\right| < \varkappa(\delta)$, because by Lemma 9.2 a) the sets $(A_i', B_i')$ form an $(n-1, \varkappa(\delta))$-explosion in $\Sigma_p$. Since $||A_i'q'|+|B_i'q'|-\pi| < \varkappa(\delta)$, and $\measuredangle \tilde{A}_i\tilde{p}\tilde{q} + \measuredangle \tilde{B}_i\tilde{p}\tilde{q} = \pi$, we get $|\measuredangle \tilde{B}_i\tilde{p}\tilde{q}-|B_i'q'|| < \varkappa(\delta)$. Now to prove the induction step it is sufficient to apply Lemma 9.2 b) and suppose that

$$\sum_{i=1}^{n-1} \cos^2 |\bar{A}_i \tilde{q}| = 1 - \cos^2 |\bar{p}\tilde{q}|. \blacksquare$$

**9.4. *Theorem.*** *Let $M$ be an $n$-dimensional* SCBB *and let the point $p \in M$ have an $(n, \delta)$-explosion $(a_i, b_i)$. Then the map $f: M \to \mathbb{R}^n$ given by $f(q) = (|a_1 q|, \ldots, |a_n q|)$ maps a small neighbourhood $U$ of the point $p$ almost isometrically onto a domain in $\mathbb{R}^n$.*

The condition that $f$ is almost isometric means that $||f(q)f(r)| - |qr|| < \varkappa(\delta, \delta_1)|qr|$ for any points $q, r \in U$, where

$$\delta_1 = \max_{1 \leqslant i \leqslant n} \{|pa_i|^{-1} \operatorname{diam} U, \quad |pb_i|^{-1} \operatorname{diam} U\}.$$

*Proof.* In view of Theorem 5.4 we need verify only the condition for an almost isometry. Let $q, r \in U$. By Lemma 5.6 we have $|\tilde{\measuredangle} a_i qr - \measuredangle a_i qr| < \varkappa(\delta, \delta_1)$ for any geodesics $qr$, $qa_i$ and consequently

$$\text{(2)} \qquad \left| \frac{|a_i q| - |a_i r|}{|qr|} - \cos \measuredangle a_i qr \right| < \varkappa(\delta, \delta_1).$$

We can apply Lemma 9.3 to the direction $r' \in \Sigma_q$ of the chosen geodesic $qr$ if we suppose that the sets $a_i', b_i' \subset \Sigma_q$ form an $(n, \varkappa(\delta, \delta_1))$-explosion in $\Sigma_q$. Thus $\left|\sum_{i=1}^{n} \cos^2 \measuredangle a_i qr - 1\right| < \varkappa(\delta, \delta_1)$ and hence $\left|\sum_{i=1}^{n} \frac{(|a_i q| - |a_i r|)^2}{|qr|^2} - 1\right| < \varkappa(\delta, \delta_1)$, which is what was required. $\blacksquare$

**9.5. *Theorem.*** *Let $M$ be a complete $(n-1)$-dimensional space with curvature $\geqslant 1$ which has an $(n, \delta)$-explosion $(A_i, B_i)$. Then, for small $\delta > 0$, $M$ is almost isometric to the unit sphere $S^{n-1}$, that is, there exists a homeomorphism $\tilde{f}: M \to S^{n-1}$ such that $||\tilde{f}(r)\tilde{f}(q)| - |rq|| < \varkappa(\delta)|rq|$ is satisfied for any $q, r \in M$.*

*Proof.* We consider the cone $C(M)$ with vertex $p$ and the points $a_i, b_i \in C(M)$ such that the directions of the rays $pa_i$, $pb_i$ lie in $A_i$ and $B_i$ respectively. Clearly $(a_i, b_i)$ form an $(n, \delta)$-explosion at the point $p$. We can choose $a_i, b_i$ so far from $p$ that for the neighbourhood $U = \{q \in C(M) : |pq| < 2\}$ the assertion of the preceding theorem is satisfied with some function $\varkappa(\delta)$ instead of $\varkappa(\delta, \delta_1)$. Consider the map $\tilde{f}: M \to S^{n-1}$, which is a composite of the map $f$ constructed in the previous theorem and the central projection onto the unit sphere with centre $f(p)$. (We suppose, as usual, that $M$ is embedded in $C(M)$ according to the rule $x \to (x, 1)$.) We show that $\tilde{f}$ is almost isometric. Since $f$ is almost isometric, we have $||f(q)f(p)| - 1| < \varkappa(\delta)$ for any point $q \in M$ and consequently the absolute distortion of the distances with $\tilde{f}$ is small, that is, $||f(q)f(r)| - |qr|| < \varkappa(\delta)$ for any $q, r \in M$. It remains to verify the condition of almost isometry for near points $q$, $r$. For this it is sufficient to prove that if $|qr| < \varkappa(\delta)$, then $\left| \measuredangle f(p)f(q)f(r) - \frac{\pi}{2} \right| < \varkappa(\delta)$. Since in our case, as in the previous theorem, the inequalities (2) are satisfied (with $\varkappa(\delta)$ instead of $\varkappa(\delta, \delta_1)$) and the analogous inequalities with $p$ instead of $r$, it is sufficient to verify that $\left| \sum_{i=1}^{n} \cos \measuredangle a_i qr \cdot \cos \measuredangle a_i qp \right| < \varkappa(\delta)$. This inequality is implied by the following assertion. If we are given an $(n, \varkappa(\delta))$-explosion $(a_i', b_i')$ in a complete $(n-1)$-dimensional space $\Sigma_q$ with curvature $\geqslant 1$ and the points $p'$, $r'$ are such that $\left| |p'r'| - \frac{\pi}{2} \right| < \varkappa(\delta)$, then $\left| \sum_{i=1}^{n} \cos|a_i'r'| \cdot \cos|a_i'p'| \right| < \varkappa(\delta)$. The last assertion is implied by the fact that $\Sigma_q$, like $M$, admits a map into the unit sphere $S^{n-1}$ with small absolute distortion of distances. ■

**9.6. *Corollary.*** (a) *A complete $(n-1)$-dimensional space with curvature $\geqslant 1$ which has an $(n, \delta)$-explosion with $\delta > 0$ sufficiently small is a space without boundary.*

(b) *For sufficiently small $\delta > 0$ the $(n, \delta)$-burst points in an $n$-dimensional* SCBB *are interior points.*

*Proof.* Assertion (a) is easily proved by induction with respect to dimension taking into account Theorem 9.5 since, in a space which is almost isometric to the standard sphere of the same dimension, any point has a maximum possible explosion for the given dimension and consequently the space of directions also has a maximal explosion at any point. Assertion (b) follows immediately from (a). ■

**9.7.** We say that the $(m, \delta)$-explosion $(a_i, b_i)$ at the point $p$ is $R$-long if $|a_i p| > \delta^{-1}R$, $|b_i p| > \delta^{-1}R$ for all $i$. The set of points of the space $M$ that have $R$-long $(m, \delta)$-explosion is denoted by $M(m, \delta, R)$.

The rest of this section is devoted to proving the following theorem.

**9.8. *Theorem.*** *Let $M$, $M_1$ be compact $n$-dimensional* SCBB *with the same lower bound for the curvature. Suppose we are given an $\nu$-approximation $h : M_1 \to M$. Then, if the numbers $\delta$, $\nu/\delta^3 R$ are sufficiently small, there exists a map $\bar{h} : M_1(n, \delta, R) \to M$ that is (uniformly) $C\nu$-near to $h$ and such that*

$$\left| 1 - \frac{|\bar{h}(x)\bar{h}(y)|}{|xy|} \right| < \varkappa(\delta, \nu/\delta^3 R),$$

*whenever $|xy| \leqslant \delta^3 R$.*

We recall that a $\nu$-approximation was defined in 7.8.

**9.9. *Corollary.*** *If, under the conditions of the theorem, each point of the space $M$ is an $(n, \delta)$-burst point, then there exists a $\varkappa(\delta, \nu)$-almost isometry $\bar{h} : M_1 \to M$ which is (uniformly) $C\nu$-near to $h$.*

To prove the corollary it is sufficient to note that in this case $M = MN(n, \delta, R)$ for some $R > 0$, and $M_1 = M_1(n, 10\delta, R/10)$.

The map $\bar{h}$ is glued together from almost isometries on balls of radius $\delta R$ using a partition of unity and the construction for the centre of gravity described below (see [21]). After studying the properties of the centre of gravity we give for completeness a formal proof of Theorem 9.8.

**9.10. Centre of gravity.**
Let the point $p$ in an $n$-dimensional SCBB $M$ have an $(n, \delta)$-explosion $a_i$, $b_i$ and let $f : M \to \mathbb{R}^n$ be the corresponding explosion map (see 5.5). We fix a small neighbourhood $U$ of the point $p$ such that $f(U)$ is convex in $\mathbb{R}^n$. Then for any finite collection of points $X = (x_1, \ldots, x_l)$, $x_j \in U$, and collection of weights $W = (w_1, \ldots, w_l)$, $w_j \geqslant 0$, $\sum_{j=1}^{l} w_j = 1$, the centre of gravity is defined to be $X_W = f^{-1}\left(\sum_{j=1}^{l} w_j f(x_j)\right)$.

**9.11. *Lemma.*** *In addition to the assumptions of* 9.10 *let the point $p$ have another $(n, \delta)$-explosion $(s_i, t_i)$ and let*

$$\delta_1 = \max\left\{\frac{\operatorname{diam} U}{\min_i \{|pa_i|, |pb_i|\}}, \frac{\max_i \{|pa_i|, |pb_i|\}}{\min_i \{|ps_i|, |pt_i|\}}\right\}$$

*be small. We assume that the collections of points in $U$, $Q = (q, \ldots, q_l)$ and $R = (r_1, \ldots, r_l)$, satisfy*

$$\max_j \{|q_j r_j|\} \leqslant (1+\delta) \min_j \{|q_j r_j|\},$$

*and for every $i$, $1 \leqslant i \leqslant n$,*

$$|\max_j \tilde{\measuredangle}\, s_i q_j r_j - \min_j \tilde{\measuredangle}\, s_i q_j r_j| \leqslant \delta.$$

*Then for any weights $W^1$ and $W^2$ such that $\| W^1 - W^2 \| < \delta_1$ the centres of gravity $Q_{W^1}$ and $R_{W^2}$ (with respect to the explosion $(a_i, b_i)$) satisfy*

$$\left|1 - \frac{|q_j r_j|}{|Q_{W^1} R_{W^2}|}\right| < \varkappa(\delta, \delta_1), \quad |\tilde{\measuredangle}\, s_i q_j r_j - \tilde{\measuredangle}\, s_i Q_{W^1} R_{W^2}| < \varkappa(\delta, \delta_1)$$

*for all $j = 1, \ldots, l$, $i = 1, \ldots, n$.*

*Proof.* We note that, for any $q, r \in U$, any point $z$ from the collection $\{a_i, b_i, s_i, t_i\}$, and any geodesics $qz$, $qr$ we have (by Lemma 5.6) $|\tilde{\measuredangle} zqr - \measuredangle zqr| < \varkappa(\delta, \delta_1)$. In addition, for any $x$ from the collection $\{a_i, b_i\}$ and $y$ from the collection $\{s_i, t_i\}$ we have $|\tilde{\measuredangle} xqy - \tilde{\measuredangle} xry| < \varkappa(\delta, \delta_1)$ and for any geodesics $qx$, $qy$ $|\tilde{\measuredangle} xqy - \measuredangle xqy| < \varkappa(\delta, \delta_1)$. Hence if we take into account that the $\Sigma_{q_j}$ are almost isometric to $S^{n-1}$, we deduce that $|\max_j \tilde{\measuredangle} xq_j r_j - \min_j \tilde{\measuredangle} xq_j r_j| < \varkappa(\delta, \delta_1)$ for any such point $x$. The last estimate easily implies the assertion of the lemma. ■

**9.12.** *Proof of Theorem* 9.8. We choose, from the cover of the set $\overline{M}_1 = M_1(n, \delta, R)$ by all possible balls of radius $\delta R$ with centres in $\overline{M}_1$, a finite subcover $\{B_{x_i}(\delta R)\}$, $1 \leqslant i \leqslant N_1$. We can suppose that balls a third as small as before cover $\overline{M}_1$ and the multiplicity of the cover is bounded by a number $N$ which depends only on $n$. For each centre $x_i$ we choose an $R$-long $(n, \delta)$-explosion $(s_j^i, t_j^i)$ and a $\delta R$-long $(n, \delta)$-explosion $(a_j^i, b_j^i)$, $1 \leqslant j \leqslant n$, and let $f_i$ denote the explosion map for $(a_j^i, b_j^i)$. Clearly for small $\nu > 0$ the pairs $(h((a_j^i), h(b_j^i))$ form an $(n, 2\delta)$-explosion at the point $h(x_i)$ and we can consider the corresponding explosion map $g_i$.

The composition $h_i = g_i^{-1} \circ f_i$ on $B_{x_i}(\delta R)$ is $\varkappa(\delta, \nu/R\delta)$-almost isometric, $C\nu$-near to $h$. We introduce neighbourhoods $U_i$, $h_i(B_{x_i}(\delta R/2)) \subset U_i \subset \subset h_i(B_{x_i}(\delta R))$ such that the $g_i(U_i)$ are convex in $\mathbb{R}^n$ and we define the functions $\varphi_i : M_1 \to \mathbb{R}$ by

$$\varphi_i(x) = \begin{cases} (1-2|xx_i|/\delta R)^N, & \text{if } x \in B_{x_i}(\delta R/2), \\ 0, & \text{if } x \in M \setminus B_{x_i}(\delta R/2). \end{cases}$$

For an arbitrary point $z \in \overline{M}_1$ let us define the sequence $\{z_i\}$, $1 \leqslant i \leqslant N_1$, inductively:

$z_1 = h_1(z)$ if $z \in B_{x_1}(\delta R)$, and $h(z)$ otherwise,

$z_{i+1}$ is the centre of gravity (with respect to the explosion map $g_{i+1}$) of the points $z_i$ and $h_{i+1}(z)$ with weights $1 - \dfrac{\varphi_{i+1}(z)}{\Sigma_{i+1}}$ and $\varphi_{i+1}(z)/\Sigma_{i+1}$ respectively, where $\Sigma_{i+1}$ denotes $\sum_{j=1}^{i+1} \varphi_j(z)$. (If $\Sigma_{i+1} = 0$, we put $z_{i+1} = h(z)$.) We put $\overline{h}(z) = z_{N_1}$. Obviously $|z_i h(z)| \leqslant C\nu$ for all $i$, consequently $|\overline{h}(z) h(z)| \leqslant C\nu$. We still have to verify the condition of almost isometry for $\overline{h}$ in the small.

We first estimate the local Lipschitz constants of the functions $\varphi_i(x)/\Sigma_i(x)$. Let $\max\limits_{1 \leqslant j \leqslant i} \varphi_j(x) = \varphi_{j_0}(x)$. Then for $y$ sufficiently near to $x$ we have

$$\left| \frac{\varphi_i(x)}{\Sigma_i(x)} - \frac{\varphi_i(y)}{\Sigma_i(y)} \right| \leqslant C \frac{|xy|}{\delta R/2 - |xx_{j_0}|}. \tag{1}$$

In fact

$$\left| \frac{\varphi_i(x)}{\Sigma_i(x)} - \frac{\varphi_i(y)}{\Sigma_i(y)} \right| = \left| \frac{\sum_{j=1}^{i-1} (\varphi_i(x)(\varphi_j(y) - \varphi_j(x)) - \varphi_j(x)(\varphi_i(y) - \varphi_i(x))}{\Sigma_i(x)\Sigma_i(y)} \right| \leqslant$$

$$\leqslant 2N \frac{|\varphi_{j_0}(x) - \varphi_{j_0}(y)|}{\varphi_{j_0}(x)} \leqslant 3N^2 \frac{|xy|}{\delta R/2 - |xx_{j_0}|}.$$

Let $z, y \in \overline{M}_1$ be such that $|zy| < \delta^3 R$ and let $\nu/R < \delta^3$. We find a number $i_0$, $1 \leqslant i_0 \leqslant N_1$, such that

$$\varphi_i(z) \leqslant R\delta^2 |yz|^{-1} \Sigma_{i-1}(z) \tag{2}$$

for all $i > i_0$, but

(3) $$\varphi_{i_0}(z) \geqslant R\delta^2 |yz|^{-1} \Sigma_{i_0-1}(z).$$

Since $\max_j \varphi_j(z) \geqslant (1/3)^N$, then (2) implies that

(4) $$\varphi_{i_0}(z) \geqslant C(\delta^{-2}R^{-1}|yz|)^{N-1}$$

and consequently

(5) $$\delta R/2 - |zx_{i_0}| \geqslant C\delta^{-1}|yz|.$$

From the obvious inequality $\varphi_i(y) \leqslant 2^N(\varphi_i(z) + (2|yz|/\delta R)^N)$ we get

$$\Sigma_{i_0-1}(y) \leqslant C \cdot \Sigma_{i_0-1}(z) + C \cdot (|yz|/\delta R)^N \overset{(3)}{\leqslant} C\delta^{-2}|yz|R^{-1}\varphi_{i_0}(z) + C \cdot (|yz|/\delta R)^N \overset{(4)}{\leqslant}$$
$$\leqslant C\delta^{-2}|yz|R^{-1}\varphi_{i_0}(z)(1+\delta^N) \leqslant C\delta^{-2}|yz|R^{-1}\varphi_{i_0}(z) \leqslant C\delta^{-2}|yz|R^{-1}\varphi_{i_0}(y),$$

because (5) implies immediately that $|1 - \varphi_{i_0}(z)/\varphi_{i_0}(y)| \leqslant C\delta$. Thus we have

$$|1 - \varphi_{i_0}(z)/\Sigma_{i_0}(z)| \leqslant C\delta^{-2}|zy|R^{-1}, \quad |1 - \varphi_{i_0}(y)/\Sigma_{i_0}(y)| \leqslant C\delta^{-2}|zy|R^{-1},$$

and consequently, since $\nu < \delta^3 R$,

(6) $$|h_{i_0}(z)z_{i_0}| < C\delta|yz|, \quad |h_{i_0}(y)y_{i_0}| < C\delta|yz|.$$

In addition, (1) and (5) imply that for $i > i_0$

(7) $$\left|\frac{\varphi_i(z)}{\Sigma_i(z)} - \frac{\varphi_i(y)}{\Sigma_i(y)}\right| < C\delta.$$

Now if we take into account (6), (7), we can prove that $\bar{h}$ is a local almost isometry by using an inductive argument and applying Lemma 9.11 to each operation of taking the centre of gravity, beginning with the one that reduces to the construction of $y_{i_0+1}$, $z_{i_0+1}$ and taking the pairs $(h(s_j^{i_0}), h(t_j^{i_0}))$ as the "long" explosion. It is only necessary to check that for any point $h(s)$ from the collection $h(s_j^{i_0})$ and any $i \geqslant i_0$ for which $\varphi_i(z) > 0$ or $\varphi_i(y) > 0$ we have

$$|\tilde{\measuredangle}\, h(s)h_i(y)h_i(z) - \tilde{\measuredangle}\, h(s)h_{i_0}(y)h_{i_0}(z)| < \varkappa\left(\delta, \frac{\nu}{\delta R}\right),$$
$$|1 - |h_i(y)h_i(z)|/|h_{i_0}(y)h_{i_0}(z)|| < \varkappa\left(\delta, \frac{\nu}{\delta R}\right).$$

The second inequality follows directly from the almost isometry of $h_i$ and $h_{i_0}$ on the corresponding balls. To prove the first inequality it is sufficient to verify that for $j = i$ or $i_0$

$$|\tilde{\measuredangle}\, h(s)h_j(y)h_j(z) - \tilde{\measuredangle}\, syz| < \varkappa\left(\delta, \frac{\nu}{\delta R}\right).$$

This estimate is established in the same way as in the proof of Lemma 9.11, since for any point $u$ from the collection $\{a_i^j, b_i^j\}$ we have

$$|\widetilde{\measuredangle}\, h(s)h_j(y)h_j(u) - \widetilde{\measuredangle}\, syu| < \varkappa\left(\delta, \frac{\nu}{\delta R}\right),$$

$$|\widetilde{\measuredangle}\, h(u)h_j(y)h_j(z) - \widetilde{\measuredangle}\, uyz| < \varkappa\left(\delta, \frac{\nu}{\delta R}\right)$$

and for any point $v$ from the collection $\{s, a_i^j, b_i^j\}$ we have

$$|\widetilde{\measuredangle}\, h(v)h_j(y)h_j(z) - \measuredangle\, h(v)h_j(y)h_j(z)| < \varkappa(\delta),$$

$$|\widetilde{\measuredangle}\, vyz - \measuredangle\, vyz| < \varkappa(\delta). \blacksquare$$

**9.13.** *Remark.* If under the conditions 9.9 the dimension of the space $M_1$ is $m$, where $m > n$, and if each point of it is an $(m, \delta)$-burst point, then the $\nu$-approximation $h : M_1 \to M$ may be approximated by the continuous "almost metric submersion" $\bar{h}$. (In particular, $\bar{h} : M_1 \to M$ is a locally trivial fibration.) This assertion generalizes one of Yamaguchi's results [32]. For the proof it is sufficient to carry out the argument from 9.12 twice—first with respect to the "base" and then with respect to the "fibre". We conjecture that in fact the condition that every point of the space $M_1$ is an $(m, \delta)$-burst point could be omitted.

## §10. Hausdorff measure

**10.1.** In this section we show that the Hausdorff measure ("volume") on a FDSCBB preserves many properties of the volume on Riemannian manifolds with curvature bounded below. We begin with the volume comparison theorem.

**10.2.** ***Theorem.*** *Let $M$ be a complete $n$-dimensional space with curvature $\geqslant k$. Then for any $p \in M$, $R > 0$, the $n$-dimensional Hausdorff measure of the ball $B_p(R)$ does not exceed the volume of the ball $\widetilde{B}_{\bar{p}}(R)$ of radius $R$ in the $n$-dimensional complete simply-connected space of constant curvature $k$. In the case of equality these balls are isometric.*

It is clearly sufficient for the proof (for the moment without the case of equality) to construct an injective non-distorting map

$$f_n : B_p(R) \to \widetilde{B}_{\bar{p}}(R).$$

For any map $\varphi : X \to Y$ of metric spaces we let $(\varphi, \mathrm{id}) : C(X) \to C(Y)$ denote the map of cones over $X$, $Y$ in accordance with the formula $(\varphi, \mathrm{id})(x, a) = (\varphi(x), a)$. Now by induction we put

$$f_n = \exp_{\bar{p}} \circ (f_{n-1}, \mathrm{id}) \circ \exp_p^{-1},$$

where $\exp_p^{-1}$ is the map defined in 7.7. Clearly $f_n$ is injective and non-distorting. It is not difficult to check that in the case when the volumes are equal the map $f_n$ is isometric. ■

We now wish to estimate the Hausdorff dimension of the set of singular points of $M$. For this we need two technical lemmas. As a preliminary we stipulate that in the course of this section $M$ denotes a compact $n$-dimensional SCBB with diameter 1, $N_0(\delta) = \max\limits_X \beta_X(\delta)+1$, where the maximum is taken over all $(n-1)$-dimensional complete spaces with curvature $\geqslant 1$. Since all our arguments are essentially local, we consider the lower bound of the curvature $k$ to be zero.

**10.3. *Lemma.*** *For any natural number $l$ there exists a number $N_1(l)$ such that from any $N_1(l)$ points in $M$ one can choose $l+1$ points $x_0, \ldots, x_l$ so that $|x_0x_i| > 2|x_0x_{i-1}|$, $2 \leqslant i \leqslant l$.*

*Proof.* We carry out induction with respect to $l$. The base $l = 1$ is trivial. Suppose we have $N_1(l+1) = 100lN_1(l)N_0(1/10)$ points in $M$, and let $x$ be one of them. If the desired collection $x_0, x_1, \ldots, x_{l+1}$ with $x_0 = x$ does not exist, then all the points except $x$ may be split into $100l$ classes such that for points $y$, $z$ in the same class we have $0.9 \leqslant |xy|/|xz| \leqslant 1.1$. We choose that class in which there are $\geqslant N_1(l)\cdot N_0(1/10)$ points and find in it a subset $G$ from among the $\geqslant N_1(l)$ elements such that for each point $y \in G$ the direction $y' \in \Sigma_x$ of at least one geodesic $xy$ lies in a fixed ball $B_\xi(1/10)$. Thus for any points $y, z \in G$ we have $|yz| < 1/2 \min\{|xy|, |xz|\}$, and consequently if we choose from $G$ by the inductive hypothesis a collection $x_0, \ldots, x_l$, we can add to it the point $x_{l+1} = x$ to obtain the desired collection. ■

**10.4.** We recall that the $(m, \delta)$-explosion $(a_i, b_i)$ at the point $p$ is said to be $R$-long if $\min\{|a_ip|, |b_ip|\} > \delta^{-1}R$. The set of points in $M$ that admit an $R$-long $(m, \delta)$-explosion is denoted by $M(m, \delta, R)$.

**10.5. *Lemma.*** *Let $D_p$ denote the cylinder $\{x\in B_p(R_1) : ||a_ix|-|a_ip|| \leqslant (0.1)R\}$, where $(a_i, b_i)$ is an $R_1$-long $(m-1, \delta)$-explosion at $p$, $R \leqslant \delta R_1$, $\delta < 1/2n$. Then*

$$\beta_{D_p\setminus M(m,\delta,R)}(\delta^{-1}R) \leqslant N_1(N_0(\delta/2)).$$

*Proof.* Let us assume the contrary and choose in $D_p\setminus M(m, \delta, R)$ a subset of $N_1(N_0(\delta/2))$ points with pairwise distances $\geqslant \delta^{-1}R$. By the preceding lemma we can choose a collection $x_0, x_1, \ldots, x_{N_0(\delta/2)}$ from this subset for which $|x_0x_{i+1}| > 2|x_0x_i|$, $1 \leqslant i < N_0(\delta/2)$. By the definition of the function $N_0$ there are points $x_i$, $x_j$, $0 < i < j < N_0(l/2)$, for which $\tilde{\measuredangle} x_ix_0x_j \leqslant \delta/2$ and consequently $\tilde{\measuredangle} x_0x_ix_j > \pi-\delta$. But then the pairs $(a_i, b_i)$, supplemented by the pair $(x_0, x_j)$, obviously form an $R$-long $(m, \delta)$-explosion at $x_i$. This is a contradiction. ■

**10.6.** Let $M(m, -\delta) = \bigcup_{R>0} M(m, \delta, R)$ denote the set of all $(m, \delta)$-burst points.

***Theorem.*** $\dim_H(M \setminus M(m, \delta)) \leqslant m-1$.

*Proof.* We may suppose that $\delta < 1/2n$. We carry out the induction with respect to $m$; the base $m = 0$ is trivial. To prove the induction step it is sufficient to cover $M(m, \delta)$ by a countable collection of balls of the form $B_x(R_x)$, where $R_x$ is so small that $x$ has an $R_x$-long $(m, \delta)$-explosion, and to verify that for any $x \in M(m, \delta)$ the $m$-dimensional Hausdorff measure $B_x(R_x) \setminus M(m+1, \delta)$ is finite. The last assertion follows from Lemma 10.5. ■

**10.6.1.** *Remark.* Theorem 10.6 and Corollary 12.8 proved below imply that the Hausdorff dimension of the interior singular points of an $n$-dimensional SCBB is no greater than $n-2$.

**10.7.** We pass to the question of the convergence of Hausdorff measures for converging sequences of spaces. We begin with the definition.

Let the compact metric spaces $X_i$ converge (in the sense of Hausdorff) to $X$. We fix a sequence of numbers $\nu_i \to 0$ and a sequence of $\nu_i$-approximations $f_i : X \to X_i$ (see 7.8). We shall say that the measures $\mu_i$ on the spaces $X_i$ *converge weakly* to the measure $\mu$ on $X$ if for any set $A \subset X$ such that $\mu(\partial A) = 0$ we have the convergence $\mu_i(U_{\nu_i}(f_i(A))) \to \mu(A)$, where $U_\varepsilon$ is the $\varepsilon$-neighbourhood.

**10.8.** ***Theorem.*** *Let the compact n-dimensional* SCBB $M_i$, $M$ *have the same lower bound for the curvatures and let* $M_i \to M$ *in the sense of Hausdorff. Then the n-dimensional Hausdorff measures* $\mu_i$ *of the spaces* $M_i$ *converge weakly to the n-dimensional Hausdorff measure* $\mu$ *of the space* $M$.

For the proof we need the following technical assertion.

**10.9.** ***Lemma.*** *Let* $M$ *be a compact n-dimensional* SCBB *with diameter* 1. *Then for sufficiently small* $\delta > 0$ *we have, for any integer* $m$, $0 \leqslant m \leqslant n$, *and any* $R > 0$, $\nu < R$,

$$\nu^n \beta_{M\setminus M(m,\delta,R)}(\nu) < \varkappa[\delta, \hat{\nu}, \hat{M}](R).$$

***Corollary.*** *Under the conditions of the lemma*

$$\mu(M\setminus M(n, \delta, R)) < \varkappa[\delta, \hat{M}](R).$$

*Proof.* We carry out induction with respect to $m$. The base $m = 0$ is trivial. We consider the cover of the set $M(m-1, \delta, R)$ by balls of the form $B_x(R_1)$, where the number $R_1 > \delta^{-1}R$ is chosen later. A finite subcover $B_{x_i}(R_1)$ can be chosen from this cover so that the balls $B_{x_i}(R_1/3)$ do not intersect pairwise. According to 8.4 there are no more than $cR_1^{-n}$ balls in this cover. If we divide the ball $B_x(R_1)$ into cylinders and apply to it Lemma 10.5, we deduce that the set $A = M(m-1, \delta, R_1)\setminus M(m, \delta, R)$ is covered by $c(\delta)R_1^{-n}(R_1/R)^{m-1}$ balls of radius $\delta^{-1}R$, and consequently $\nu^n\beta_A(\nu) \leqslant c(\delta)(R/R_1)^{n-m+1}$.

Since for a suitable choice of $R_1$ we have $\varkappa(R_1)+(R/R_1)^{n-m+1} = \varkappa(R)$, the lemma is proved. ■

**10.10.** *Proof of the theorem.* From the $\nu_i$-approximation $f_i : M \to M_i$ it is not difficult to construct a $10\nu_i$-approximation $g_i : M_i \to M$ so that $|(g_i\circ f_i)(x), x| < 20\nu_i$, and thus $g_i(U_{\nu_i}(f_i(A))) \subset U_{100\nu_i}(A)$. If $\mu(\partial A) = 0$, then $\mu(U_{100\nu_i}(A)) \to \mu(A)$ as $i \to \infty$. It can now be seen that Theorem 10.8 follows from Theorem 9.8 and Lemma 10.9. ■

**10.11. Corollary.** *For any point $p$ in a* FDSCBB *the set of directions $\Sigma_p'$ in which the geodesics emanate from $p$ has total measure in $\Sigma_p$.*

This is implied by 10.8 and the proof of 10.2.

**10.12.** *Remark.* If we use 10.6 and 9.4 it is not difficult to carry across the proof of the "co-area formula" of Kronrod and Federer (see [13]) to the case of a FDSCBB.

**10.13. Corollary.** *Let $A$ be a compact subset in an $n$-dimensional complete space $M$ with curvature $\geqslant k$. Then the ratio $\mu(U_R(A))/\mu(\widetilde{B}(R))$ is a non-increasing function of $R$ for all $R > 0$.* (Here $\widetilde{B}(R)$ is a ball of radius $R$ in the $n$-dimensional complete simply-connected space with curvature $k$.)

The proof is analogous to that in [19], p. 65, taking into account 10.12 and the following observation. If $r, s \in M$, $|Ar| = |As| = R$, and $r_1, s_1$ are points on the geodesics $Ar$, $As$ at a distance $R_1$ from $A$, $R_1 \leqslant R$, then the ratio $|rs|/|r_1s_1|$ does not exceed $R/R_1$ (assuming that $k = 0$).

## §11. Functions that have directional derivatives, the method of succssive approximations, level surfaces of almost regular maps

**11.1.** In this section we describe a class of functions on a FDSCBB and a class of surfaces in a FDSCBB, which serve as more or less successful analogues of smooth functions and smooth submanifolds for Riemannian manifolds. One of the applications of the tools prepared here is the investigation carried out in the next section of points with nearly maximal explosions. A more general situation is considered in §13.

**11.2. Lemma.** *Let $p \in M$ be a point in a* FDSCBB. *Then for any $\delta > 0$ there is a deleted neighbourhood $U$ of the point $p$ so small that:*

a) *for any point $q \in U$ there is a point $q_1 \in M$ such that $|pq| < \delta|pq_1|$ and the pair $(p, q_1)$ forms a $(1, \delta)$-explosion at $q$;*

b) $\measuredangle qpr - \tilde{\measuredangle} qpr < \delta$ *for any triangle $\Delta pqr$ with $q, r \in U$; further:*

c) *for any triangle $\Delta pqr$ with $q, r \in U$ each angle of $\Delta pqr$ differs from the corresponding angle of $\tilde{\Delta} pqr$ by less than $\delta$;*

d) *if* $q \in U$ *is joined to* $p$ *by more than one geodesic, then the angles in any lune* $pqp$ *are less than* $\delta$.

*In addition, if* $A \subset M$ *is a compact set not containing* $p$, *then for any* $\delta > 0$ *there is a deleted neighbourhood* $U$ *of* $p$ *so small that*:

e) $|\tilde{\measuredangle} qpA - |q'A'|| < \delta$, $|\tilde{\measuredangle} q_1 qA - |q'A'|| < \delta$ *is satisfied for any point* $q \in U$ *and* $|\tilde{\measuredangle} q_1 pA - |q'A'|| < \delta$ *for some point* $q_1 \in M$ *such that* $|pq| \leq \delta |pq_1|$ *and* $(p, q)$ *forms a* $(1, \delta)$-*explosion at* $q$. (Here $q', A' \subset \Sigma_p$, and $\tilde{\measuredangle} qpA$ is the angle in the triangle with sides $|qp|$, $|qA|$, $|pA|$ on the $k$-plane.)

We shall not give detailed proofs. They are based on the fact that, as $\Sigma_p$ is compact, we can choose an arbitrarily dense finite net of directions $\xi_i$ in $\Sigma_p$ which are realized by the geodesics $pa_i$, and moreover on account of the proximity of $a_i$ to $p$ we may suppose, by Lemma 7.5, that the angles of the triangle $\Delta a_i p a_j$ differ little from the corresponding angles of $\tilde{\Delta} a_i p a_j$. In addition, it may be supposed that the angles $\tilde{\measuredangle} a_i pA$ differ little from $|a_i' A'|$. If we now put $U = \{q : 0 < |pq| \ll \delta \min\{|pa_i|, |pA|\}$, then assertions a)–e) are easily verified by elementary geometric arguments. ∎

**11.3.** *Definition.* A function $f$ on some domain of a FDSCBB is said to have *directional derivatives* if it is Lipschitz and for any point $p$ in this domain there exists a continuous function $f'_{(p)}$ on $\Sigma_p$ (called the *derivative of* $f$ *at the point* $p$) such that along each geodesic $pa$ the derivative of $f$ (with respect to arc length) at $p$ is equal to $f'_{(p)}(a')$.

We note that for such a function $f$ the differential ratios $\varphi(R, \xi) = \frac{1}{R}(f(\exp_p R\xi) - f(p))$ satisfy a Lipschitz condition in $\xi$ uniformly along $R$, so that $\varphi(R, \xi)$ converges to $f'_{(p)}$ uniformly on $\Sigma_p$ as $R \to 0$ (the fact that $\varphi(R, \xi)$ is not defined for all $\xi \in \Sigma_p$ is immaterial).

**11.4.** *Examples.* The basic examples of functions that have directional derivatives are the distance functions. If $A \subset M$ is a compact set in a FDSCBB, then the function $f(x) = |Ax|$ on $M \setminus A$ satisfies Definition 11.3 with $f'_{(p)}(\xi) = -\cos|A'\xi|$, where $\xi \in \Sigma_p$. This follows immediately from Lemma 11.2 e).

Other examples can be obtained by using the following assertions.

1) Admissible arithmetic operations on functions that have directional derivatives give functions that have directional derivatives, and the derivatives are calculated by the standard rules.

2) The maximum of a finite number of functions that have directional derivatives is a function that has directional derivatives.

3) If $f$ is a function that has directional derivatives and $\varphi$ is a (one-sided) differentiable function on $\mathbb{R}$, then $\varphi \circ f$ is a function that has directional derivatives.

**11.5. The method of successive approximations.**
The following assertion is a formal generalization of the first part of Theorem 5.4 (in the finite-dimensional case).

***Lemma.*** *Let the functions* $f_1, \ldots, f_m$ *that have directional deriatives be given in the domain* $U \subset M$ *of an n-dimensional* SCBB. *Suppose that for some* $\delta$, $\varepsilon$, $0 < n\delta < \varepsilon$, *and for any point* $p \in U$ *there exist directions* $\xi_i^{\pm} \in \Sigma_p$, $1 \leqslant i \leqslant m$, *such that* $|f'_{j(p)}(\xi_i^{\pm})| < \delta$ *when* $i \neq j$ *and* $f'_{i(p)}(\xi_i^{+}) > \varepsilon$, $f'_{j(p)}(\xi_i^{-}) < -\varepsilon$. *Then the map* $f = \{f_1, \ldots, f_m\} : U \to \mathbb{R}^m$ *is* $\frac{1}{2}(\varepsilon - n\delta)$*-open* (see 5.3) *with respect to the metric in* $\mathbb{R}^m$ *given by the norm* $\|x\| = \sum_{i=1}^{m} |x^i|$ *for* $x = (x^1, \ldots, x^m)$.

*Proof.* Let $p \in U$, $A = \{\bar{a} \in \mathbb{R}^m : B_p((\varepsilon - n\delta)^{-1}\|\bar{a}, f(p)\|) \subset U$, and suppose there is an $a \in U$ such that $f(a) = \bar{a}$, $2\|f(p), \bar{a}\| \geqslant (\varepsilon - n\delta)|ap|\}$. Obviously $A$ is closed, $f(p) \in A$. Suppose the assertion of the lemma is not satisfied, and $\bar{a}_0$ is a point that satisfies the condition $B_p(2(\varepsilon - n\delta)^{-1}\|\bar{a}_0, f(p)\|) \subset U$ but does not lie in $A$. Let the point $\bar{a}_1 \in A$ be the nearest point in $A$ to $\bar{a}_0$, and let the point $a_1 \in U$ be such that $f(a_1) = \bar{a}_1$ and $2\|a_1, f(p)\| \geqslant (\varepsilon - n\delta)|pa_1|$. We choose the direction $\xi$ in $\Sigma_{a_1}$ so near to one of the directions $\xi_i^{\pm}$ that $|f'_{i(a_1)}(\xi)| > \varepsilon$, $|f'_{j(a_1)}(\xi)| < \delta$ for $j \neq i$ and $\operatorname{sgn} f'_{i(a_1)}(\xi) = \operatorname{sgn}(\bar{a}_0^i - \bar{a}_1^i)$. Then for sufficiently small $\mu > 0$ we see that $f(\exp_{a_1}(\mu\xi) \in A$ and $\|f(\exp_{a_1}(\mu\xi)), \bar{a}_0\| < \|\bar{a}_1 \bar{a}_0\|$, which contradicts the choice of $\bar{a}_1$. ∎

**11.6.** ***Corollary.*** *Let the point* $p \in U$ *and the number* $\varepsilon_1 > 0$ *be fixed under the conditions of Lemma* 11.5. *Then for sufficiently small* $R$ $(R < c(p, f, \varepsilon_1))$ *there is, for every direction* $\xi \in \Sigma_p$ *such that* $f'_{i(p)}(\xi) = 0$ *for all* $i$, *a point* $q \in U$ *such that* $R < |pq| < R(1+\varepsilon_1)$, $f_i(p) = f_i(q)$ *for all* $i$, *and* $q' \subset \Sigma_p$ *is in a* $\varepsilon_1$*-neighbourhood of* $\xi$.

*In particular, if we are now given functions* $g_1, \ldots, g_l$ *in* $U$ *that have directional derivatives, then one can ensure also that*

$$\left|\frac{g_j(q) - g_j(p)}{|pq|} - g_{j(p)}(\xi)\right| < \varepsilon_1.$$

*Proof.* To construct the point $q$ it is necessary to get a first approximation $q_0$ by shifting the point $p$ along a geodesic with direction $\xi_0 \in \Sigma_p$ very near to $\xi$ and then applying the method of successive approximations (Lemma 11.5) in order to compensate for the alteration in $f_i$. Since $\|f(p) - f(q_0)\| < \varkappa[\xi, |\widehat{pq_0}|](|\xi\xi_0|) \cdot |pq_0|$ and the openness parameter for $f$ depends only on $\varepsilon$, $\delta$, then for sufficiently small $|\xi\xi_0|$ $(|\xi\xi_0| < c[\widehat{\xi}, |\widehat{pq_0}|])$ one can construct the point $q$ where $|qq_0| < \varkappa[\widehat{\xi}, |\widehat{pq_0}|](|\xi\xi_0|) \cdot |pq_0|$. If $|\xi\xi_0|$ and $|pq_0|$ are sufficiently small $(|\xi\xi_0| < c[\widehat{\xi}, |\widehat{pq_0}|]$, $|pq_0| < c[\widehat{\xi}])$, then this last inequality implies that $||pq| - |pq_0|| < \frac{1}{4}\varepsilon_1|pq_0|$ and by Lemma 11.2 b) that $q' \in \Sigma_p$ is in an $\varepsilon_1$-neighbourhood of $\xi$. Finally, if $R$ is sufficiently small ($R < c[\hat{\xi}]$), then there exists in $\Sigma_p$ a sufficiently dense net of directions that can be realized by geodesics of length not less than $R(1+\varepsilon_1)$, consequently with a choice of $\xi_0$ and $q_0$ subject to the condition indicated above that $|\xi\xi_0|$ and $|pq_0|$ be small it is possible to stipulate that the condition $|pq_0| = R(1+\frac{1}{2}\varepsilon_1)$ be satisfied and consequently $R < |pq| < R(1+\varepsilon_1)$. ∎

**11.7. Almost regular maps.**
The definition of almost regular maps, which we give below, is apparently too narrow. However, it turned out to be sufficient in the context of this paper.

*Definition.* A function $f$ on a domain $U \subset M$ of a FDSCBB is said to be $\delta$-*almost regular* if $f$ is the distance function of some compact set $A \subset M$, $A \cap U = \varnothing$, and if for every point $p \in U$ there is a compact set $B_{(p)}$ which does not contain $p$ such that $\tilde{\measuredangle} ApB_{(p)} > \pi - \delta$ (we recall that $\tilde{\measuredangle} ApB_{(p)}$ denotes the angle of the triangle on the $k$-plane with sides $|Ap|$, $|AB_{(p)}|$, $|pB_{(p)}|$).

The map $f = (f_1, \ldots, f_m) : U \to \mathbb{R}^m$ is said to be $\delta$-*almost regular* if its coordinate functions $f_i$ are $\delta$-almost regular and if the compact sets $A_i$, $B_{i(p)}$ figuring in the definition of almost regularity satisfy

$$\tilde{\measuredangle} A_i p B_{j(p)} > \frac{\pi}{2} - \delta, \quad \tilde{\measuredangle} A_i p A_j > \frac{\pi}{2} - \delta,$$

$$\tilde{\measuredangle} B_{i(p)} p B_{j(p)} > \frac{\pi}{2} - \delta \text{ when } i \neq j.$$

(The parameter $\delta > 0$, which we consider to be small, is part of the definition. As a rule our arguments will be independent of $\delta$ if $\delta$ is less than a well-determined "ceiling", depending on the dimension, and also in §12 on the non-degeneracy characteristic of the FDSCBB under consideration (see 8.7).)

Obviously almost regular maps satisfy the conditions of Lemma 11.5 with $\varepsilon = 1 - \delta$.

**11.8.** We now turn to studying the properties of level surfaces of almost regular maps. We begin with dimension estimates.

*Theorem. A level surface $\Pi$ of an almost regular map from an n-dimensional* SCBB *$M$ into $\mathbb{R}^m$ has topological dimension no greater than $n-m$. In addition, the set of points in $\Pi$ that are not $m_1$-burst points in $M$ have topological dimension no greater than $m_1 - 1 - m$. In particular, the set of points in $M$ that are not $m_1$-burst points has dimension no greater than $m_1 - 1$.*

The proof of the theorem is by a simple inverse induction with respect to $m$, in which the induction step is based on the following assertion.

**11.9. *Lemma.*** *Let $\Pi$ be a level surface of a $\varkappa(\delta)$-almost regular map $f = (f_1, \ldots, f_m): M \to \mathbb{R}^m$. We introduce the notation $S_R = \{q \in M, |pq| = R\}$, $\Pi_R = \Pi \cap S_R$. Then for sufficiently small $R > 0$ there is a neighbourhood $U \supset \Pi_R$ in which the map $\bar{f} : M \to \mathbb{R}^{m+1}$, $\bar{f}(q) = (f_1(q), \ldots, f_m(q), |pq|)$, is $\varkappa(\delta)$-almost regular.* (The case $\Pi_R = \varnothing$ is not excluded.)

This lemma follows from Lemma 11.2 e) and 5.6.

**11.10.** *Remark.* In the general case the points in an $n$-dimensional SCBB that are not $n$-burst points may fill a set of topological dimension $n-1$; this, for example, occurs for manifolds with boundary. However, it will be shown in the next section that for an interior point of an $n$-dimensional SCBB an $(n-1)$-burst point is by implication an $n$-burst point, therefore in an $n$-dimensional SCBB without boundary the points that are not $n$-burst points fill a set of topological dimension no greater than $n-2$. Cf. 10.6.1.

**11.11.** Let us now turn to the more refined properties.

***Theorem.*** *On a level surface of an almost regular map on a* FDSCBB *the intrinsic metric, which is equivalent to the extrinsic metric, is induced locally. In particular, such a surface is locally linearly connected.*

More precisely, let $f = (f_1, \ldots, f_m) : U \to \mathbb{R}^m$ be an almost regular map, let $p \in U$, let $A_i$ and $B_i$ be compact sets that figure in the definition of $f$ being almost regular at the point $p$, and let $\Pi = \{q : f_i(q) = f_i(p)$ for all $i\}$ be a level surface, $q, r \in \Pi$, $\delta_1 = \max\{|pq|, |pr|\}/\min\{|pA_i|, |pB_i|, |p, \Pi \cap \partial U|\}$. If $\delta$ (the parameter for $f$ to be almost regular) and $\delta_1$ are sufficiently small ($\delta, \delta_1 > c$), then $q$ and $r$ can be connected in $\Pi$ by a curve of length no greater than $|qr|(1+\varkappa(\delta, \delta_1))$.

*Proof.* We need to verify that a point $s \in \Pi$ can be found arbitrarily near to $r$ such that $|rs| < (1+\varkappa(\delta, \delta_1))\cdot(|qr|-|qs|)$. After this the proof concludes by standard arguments. To construct such a point $s$ it is sufficient, by Corollary 11.6, to find a direction $\xi \in \Sigma_r$ such that $f'_{i(r)}(\xi) = 0$ for all $i$ and $|q'\xi| < \varkappa(\delta, \delta_1)$, $q' \subset \Sigma_r$. Since the sets $A'_i, B'_i \subset \Sigma_r$ form an $(m, \varkappa(\delta, \delta_1))$-explosion in $\Sigma_r$ and $|q'A'_i| > \pi/2-\varkappa(\delta, \delta_1)$, $|q'B'_i| > \pi/2-\varkappa(\delta, \delta_1)$ (see Corollary 5.7), then all that remains is for us to prove the following assertion.

**11.12. *Lemma.*** *Let $M$ be a complete $n$-dimensional space with curvature $\geqslant 1$, let $A_i$, $B_i$ form an $(m, \delta)$-explosion in $M$, and let $Q \subset M$ be a compact set such that $|QA_i| > \pi/2-\delta$, $|QB_i| > \pi/2-\delta$. Then for any point $q \in Q$, if $\delta$ is sufficiently small ($\delta < c$), there is a point $\xi \in M$ such that $|\xi A_i| = \pi/2$ for all $i$ and $|\xi q| < \varkappa(\delta)$.*

*Proof.* Clearly, if $\delta > 0$ is sufficiently small, then the map $f : M \to \mathbb{R}^m$, $f(x) = (|xA_1|, \ldots, |xA_m|)$, is almost regular, with parameter $100\delta^2 n$, in the domain

$$U=\{x\in M\colon \min(|xA_i|, |xB_i|)>\pi/2-10\sqrt{n\delta} \text{ for all } i\}.$$

If furthermore $\delta < 1/100n$, then by Lemma 11.5 the map $f$ will be 1/2-open in $U$ and hence in the $4\sqrt{n}\delta$-neighbourhood of any point $q \in Q$ there is a desired point $\xi$. ■

**11.13.** We now define and study the space of directions and the tangent cone to the level surface of an almost regular map at some point of the surface.

*Definition.* Let $\gamma$ be a curve in a FDSCBB $M$, where $\gamma(0) = p$ and $\gamma(t) \neq p$ for sufficiently small $t > 0$. The *direction of the curve at the point p* is the limit in $\Sigma_p$ of the directions of the geodesics $p\gamma(t)$ as $t \searrow 0$ if this limit exists. The *space of directions* $\Sigma\Pi_p$ of the set $\Pi \subset M$ at the point $p \in \Pi$ is the set of directions at the point $p \in \Pi$ of all possible curves at $\Pi$ beginning at $p$ and having a direction at $p$ (compare 7.2.1).

**11.14.** ***Theorem.*** *Let $p \in \Pi$ be a point on a level surface of an almost regular map $f = (f_1, ..., f_m)$. Then the sets $\Lambda := \{\xi \in \Sigma_p, f_i{}'_{(p)}(\xi) = 0$ for all $i\}$ and $\Sigma\Pi_p$ coincide.*

*Proof.* The inclusion $\Sigma\Pi_p \subset \Lambda$ obviously follows from the fact that all the $f_i$ have directional derivatives. The inverse inclusion easily follows from Theorem 11.11, Lemma 11.2 b), and the following assertion.

**11.15.** ***Lemma.*** *Under the conditions of Theorem* 11.4 *for any $\varepsilon > 0$ and for sufficiently small $R > 0$ ($R < c(p, f, \varepsilon)$) one can find, for any direction $\xi \in \Lambda$ on the surface $\Pi_R = \{q \in \Pi : |pq| = R\}$, a point $q$ such that $q' \subset \Sigma_p$ is contained in the $\varepsilon$-neighbourhood of $\xi$.*

*Proof.* By Corollary 11.6 we can construct for some $R_1 : |R-R_1| < (0.1)\varepsilon R$ a point $q_1 \in \Pi_{R_1}$ for which $q_1' \in \Sigma_p$ is contained in the $(0.1)\varepsilon$-neighbourhood of $\xi$. If $R$ is sufficiently small, then, if we apply the method of successive approximations (Lemma 11.5) to the map $\bar{f}$ (which is almost regular by Lemma 11.9), where $\bar{f}(x) = (f_1(x), ..., f_m(x), |px|)$ with initial approximation $q_1$, we get the desired point $q$. ■

**11.16.** ***Corollary.*** *Under the conditions of Theorem* 11.14 *the spaces $\Pi_R$ with "metrics" $\rho(q, r) = \arccos(1-|qr|^2/2R^2)$ converge in the Hausdorff metric to the space of directions $\Sigma\Pi_p$, in which the metric is defined as the restriction of the metric on $\Sigma_p$.*

(The word "metric" is given in quotation marks, since the triangle inequality may be violated for the function $\rho$.)

**11.17.** ***Corollary.*** *Under the conditions of Theorem* 11.14

a) *$\Sigma\Pi_p$ being non-empty is equivalent to $\Pi_R$ being non-empty for sufficiently small $R > 0$.*

b) *$\Sigma\Pi_p$ being linearly connected is equivalent to $\Pi_R$ being linearly connected for sufficiently small $R > 0$. In addition, the numbers of linearly connected components of $\Sigma\Pi_p$ and $\Pi_R$ are the same.*

*Proof.* Assertion a) is obvious. To prove b) we note that the spaces $\Pi_R$ with "metric" $\rho$ from 11.16 are uniformly locally connected by Theorem 11.11, taking into account Lemma 11.9. On the other hand, the space $\Sigma\Pi_p$ is locally linearly connected by Theorem 11.11, since by Theorem 11.14, $\Sigma\Pi_p$ is the level surface of the map $\Sigma_p \to \mathbb{R}^m$ with coordinate functions arc cos $f'_{i(p)}$, and it is almost regular in a neighbourhood of $\Sigma\Pi_p$. Therefore assertion b) follows from the nearness of $\Pi_R$ and $\Sigma\Pi_p$ in the Hausdorff metric. ■

## §12. Level lines of almost regular maps

**12.1.** Let $M$ be an $n$-dimensional SCBB, let $f = (f_1, \dots, f_m) : M \to \mathbb{R}^m$ be an almost regular map, and let $\Pi$ be a level surface of $f$. This tells us immediately that $m \leqslant n$, and if $m = n$, then $\Pi$ consists of isolated points. In this section we show that if $m = n-1$, then $\Pi$ is a one-dimensional manifold with boundary, and moreover the boundary points of $\Pi$ are precisely the points of the boundary of $M$ lying in $\Pi$. Throughout this section $\varepsilon$ denotes a lower bound of $Vr_{n-1}(\Sigma_p)$ taken over all points $p$ from $M$ (from the compact subset under consideration) (see 8.7). We suppose that the almost regular parameter $\delta$ is less than a well-determined "ceiling" depending on $\varepsilon$ ($\delta < c(\varepsilon)$). This requirement is apparently connected with the method of proof, and most likely the assertions themselves still remain true under the weaker condition[(1)] $\delta < c$. Our first aim is to show that when $m < n$ the surface $\Pi$ cannot have isolated points. For this two lemmas are necessary.

**12.2.** ***Lemma.*** *Let $M$ be a complete $n$-dimensional space with curvature $\geqslant 1$ having an $(m, \delta)$-explosion $A_i$, $B_i$, $m \leqslant n$. Let the point $\xi \in M$ be such that $\sum_{i=1}^{m} \cos^2 |\xi A_i| < 1-\varepsilon_1$. Then, if $\delta$ is sufficiently small ($\delta < c(\varepsilon_1)$), we can find a point $\zeta$ in $M$ such that $|A_i\zeta| = \pi/2$ for all $i$.*

*Proof.* We first construct a point $\eta$ such that $||A_i\eta| - \pi/2| < \varkappa(\delta)$ for all $i$; after this, Lemma 11.12 gives us the construction of the point $\zeta$. To construct the point $\eta$ we may suppose that $A_i$ and $B_i$ are points. We take points $\widetilde{A}_i$, $\widetilde{B}_i$, $\widetilde{\xi}$ on the unit sphere $S^m$ such that $\widetilde{A}_i$, $\widetilde{B}_i$ form an $(m, 0)$-explosion and $|\widetilde{A}_i\widetilde{\xi}| = |A_i\xi|$ for all $i$. We construct a sequence $\widetilde{\xi}_0 = \widetilde{\xi}, \widetilde{\xi}_1, \dots, \widetilde{\xi}_m = \widetilde{\eta}$ inductively by the following rule: $\widetilde{\xi}_{i+1}$ lies on the great semicircle $\widetilde{A}_{i+1}\widetilde{\xi}_i\widetilde{B}_{i+1}$ at a distance $\pi/2$ from $\widetilde{A}_{i+1}$ (and $\widetilde{B}_{i+1}$). It is obvious that $\sum_{i=1}^{m} \cos^2 |\widetilde{\xi}_{j+1}\widetilde{A}_i| \leqslant \sum_{i=1}^{m} \cos^2 |\widetilde{\xi}_j\widetilde{A}_i| < 1-\varepsilon_1$, and consequently the perimeters of the triangles $\Delta\widetilde{A}_j\widetilde{\xi}_{j-1}\widetilde{B}_i$, $\Delta\widetilde{A}_j\widetilde{\xi}_{j-1}\widetilde{A}_i$, $\Delta\widetilde{B}_j\widetilde{\xi}_{j-1}\widetilde{A}_i$, $\Delta\widetilde{B}_j\widetilde{\xi}_{j-1}\widetilde{B}_i$ do not exceed $2\pi - c(\varepsilon_1)$ when $i \neq j$. In addition, it is obvious that $|\widetilde{\xi}_j\widetilde{A}_i| = \pi/2$ when $i \leqslant j$, so that $|\widetilde{\eta}\widetilde{A}_i| = \pi/2$ for all $i$. We carry this construction across to $M$.

(1) This follows from a recent result of A. Petrunin.

We construct a sequence $\xi_0 = \xi, \xi_1, ..., \xi_m = \eta$ inductively by the following rule: $\xi_{i+1}$ lies on the two-link broken line $A_{i+1}\xi_i B_{i+1}$ at a distance $\pi/2$ from $A_{i+1}$. Now, if we apply induction with respect to $j$, it can be proved that $||A_i\xi_j| - |\widetilde{A}_i\widetilde{\xi}_j|| < \varkappa(\delta)$, $||B_i\xi_j| - |\widetilde{B}_i\widetilde{\xi}_j|| < \varkappa(\delta)$ for all $i$. In fact, the base of induction (when $j = 0$) follows from the conditions $|\widetilde{A}_i\widetilde{\xi}| = |A_i\xi|$, $|A_iB_i| > \pi - \delta$, $|\widetilde{A}_i\widetilde{B}_i| = \pi$, and the induction step is ensured by Lemma 9.2 d). Thus $||A_i\eta| - \pi/2| < \varkappa(\delta)$ and the lemma is proved. ■

**12.3. Lemma.** *Let $M$ be a complete $n$-dimensional space with curvature $\geqslant 1$ having an $(m, \delta)$-explosion $A_i$, $B_i$, $m \leqslant n$. Then*

$$Vr_n\left\{\xi \in M: \sum_{i=1}^{m} \cos^2|\xi A_i| > 1 - \delta_1\right\} < \varkappa(\delta, \delta_1).$$

*Proof.* We use induction; the induction step consists of increasing $m$ and $n$ by 1. If $m = 1$, then the assertion follows from Lemma 8.2 and Corollary 8.3.

We consider the general case. We may suppose that $A_m$ is a point. Let $\xi \in M$ be such that $\sum_{i=1}^{m} \cos^2|\xi A_i| > 1 - \delta_1$. We construct points $\widetilde{A}_i$, $\widetilde{B}_i$, $\widetilde{\xi}$ on the unit sphere $S^m$ so that $(\widetilde{A}_i, \widetilde{B}_i)$ form an $(m, 0)$-explosion in $S^m$ and $|\widetilde{A}_i\widetilde{\xi}| = |A_i\xi|$ for all $i$. Clearly either $|\widetilde{A}_m\widetilde{\xi}| < \varkappa(\delta_1)$, or $|\widetilde{A}_m\widetilde{\xi}| > \pi - \varkappa(\delta_1)$, or $\sum_{i=1}^{m-1} \cos^2|\widetilde{A}'_i\widetilde{\xi}'| > 1 - \varkappa(\delta_1)$, where $\widetilde{A}'_i, \widetilde{\xi}' \in \Sigma_{\widetilde{A}_m}$. If we apply Lemma 9.2 c), we deduce that either $|A_m\xi| < \varkappa(\delta, \delta_1)$, or $|A_m\xi| > \pi - \varkappa(\delta, \delta_1)$, or $\sum_{i=1}^{m-1} \cos^2|A'_i\xi'| > 1 - \varkappa(\delta, \delta_1)$, where $A'_i$, $\xi' \in \Sigma_{A_m}$. Now our assertion follows from Lemma 8.2, Corollary 8.3, and the inductive hypothesis applied to $\Sigma_{A_m}$. ■

**12.4. Corollary.** *Let $M$ be a complete $n$-dimensional space with curvature $\geqslant 1$ having an $(m, \delta)$-explosion $A_i$, $B_i$, $m \leqslant n$, where $Vr_n(M) \geqslant \varepsilon$. Then if $\delta > 0$ is sufficiently small ($\delta < c(\varepsilon)$), there is a point $\xi \in M$ such that $|\xi A_i| = \pi/2$ for all $i$.*

This follows from Lemmas 12.2 and 12.3.

**12.5. Corollary.** *When $m < n$ and $\delta$ is sufficiently small* ($\delta < c(\varepsilon)$, see 12.1) *a level surface of a $\delta$-almost regular map from an $n$-dimensional* SCBB *into $\mathbb{R}^m$ does not have isolated points.*

This follows from 12.4 and 11.17 a).

**12.6. Lemma.** *Let $M$ be a complete $n$-dimensional space with curvature $\geqslant 1$ having an $(m, \delta)$-explosion $(A_i, B_i)$, $m \leqslant n-1$, where $Vr_{n-1}(M) \geqslant \varepsilon$. Then if $\delta$ is sufficiently small ($\delta < c(\varepsilon)$), the set $\Lambda = \{\xi \in M : |A_i\xi| = \pi/2$ for all $i\}$ is linearly connected.*

*Proof.* $\Lambda$ is locally linearly connected by Theorem 11.11 because it is a level surface of an almost regular map. Let $\Lambda_1$ and $\Lambda_2$ be the nearest components of $\Lambda$, where $q \in \Lambda_1$ and $r \in \Lambda_2$ are their nearest points. We shall show that $|qr| > \pi - \varkappa(\delta)$. In fact, otherwise the sets $A_i'$, $B_i'$ form an $(m, \varkappa(\delta))$-explosion in the space of directions $\Sigma_q$ and there we would have $|A_i'r'| > \pi/2 - \varkappa(\delta)$, $|B_i'r'| > \pi/2 - \varkappa(\delta)$. This follows from the angle comparison theorem in $M$.

Therefore by Lemma 11.12 a direction $\xi \in \Sigma_q$ can be found such that $|A_i'\xi| > \pi/2$ for all $i$ and $|r'\xi| < \varkappa(\delta)$, and hence by Corollary 11.6 a point $s \in \Lambda$ can be found near $q$ that is nearer to $r$ than $q$, which contradicts the choice of $q$ and $r$. So we deduce that $\Lambda$ has no more than two components and, if there are two of them $\Lambda_1$, $\Lambda_2$, then they are at a distance of at least $\pi - \varkappa(\delta)$. But then by adding $A_{m+1} = \Lambda_1$ and $B_{m+1} = \Lambda_2$ to the existing $A_i$, $B_i$ we get an $(m+1, \varkappa(\delta))$-explosion in $M$. If $\delta$ were chosen sufficiently small, then by Corollary 12.4 a point $\xi \in M$ could be found at a distance $\pi/2$ from all $A_i$, $1 \leqslant i \leqslant m+1$. But then $\xi \in \Lambda \setminus (\Lambda_1 \cup \Lambda_2)$—a contradiction. ■

**12.7. *Theorem.*** a) *Let $m$ be a complete $n$-dimensional space with curvature $\geqslant 1$ having an $(n, \delta)$-explosion $(A_i, B_i)$, where $Vr_n(M) \geqslant \varepsilon$. Then, if $\delta$ is sufficiently small ($\delta < c(\varepsilon)$), the set $\Lambda = \{\xi \in M : |A_i\xi| = \pi/2$ for all $i\}$ consists of a single point if $M$ has a boundary and of two points at a distance $\pi - \varkappa(\delta)$ if $M$ does not have a boundary.*

b) *Let $\Pi$ be a level surface of an $\delta$-almost regular map from an $(n+1)$-dimensional* SCBB *into $\mathbb{R}^n$. Then, if $\delta$ is sufficiently small ($\delta < c(\varepsilon)$,* see 12.1), *$\Pi$ is a one-dimensional manifold, and moreover the boundary points $\Pi$ are precisely those boundary points of the* FDSCBB *that lie in $\Pi$.*

*Proof.* Assertion b) follows from a) when applied to the spaces of directions $\Sigma_p$ at the points $p \in \Pi$, taking into account Theorem 11.14 and its corollaries. In fact, if $p$ is a boundary point, and consequently $\Lambda \subset \Sigma_p$ is a singleton, the "surfaces" $\Pi_R = \{q \in \Pi : |pq| = R\}$, which consist of isolated points since they are level surfaces of an almost regular map into $\mathbb{R}^{n+1}$ (Lemma 11.9), turn out to be singletons by Corollary 11.17 b); in addition, it is obvious that the point $\Pi_R$ depends continuously on $R$. If however $p$ is an interior point and $\Lambda \subset \Sigma_p$ consists of two points at a distance $\pi - \varkappa(\delta)$, then $\Pi_R$ also consists of two points at a distance $2R(1 - \varkappa(\delta))$ (by Corollary 11.16), so that $\Pi$ turns out, in a neighbourhood of $p$, to be an arc containing $p$ as an interior point.

To prove assertion a) we use induction with respect to the dimension. The base $n = 1$ is obvious. In the general case, if $q, r \in \Lambda$, then $|qr| > \pi - \varkappa(\delta)$, since otherwise the mid-point of the geodesic $qr$ would turn out to be an $(n+1)$-burst point with explosion $A_i$, $B_i$ and $q$, $r$. Thus $\Lambda$ contains no more than two points, and if it has two points, then $M$ has an $(n+1, \varkappa(\delta))$-explosion and by Corollary 9.6 a) it does not have a boundary. All that remains is for us to verify that if $M$ has no boundary, then $\Lambda$ cannot be a singleton. By the inductive hypothesis (more precisely by assertion b) that follows from it) the set $\Lambda_n = \{\xi \in M : |\xi A_i| = \pi/2$ when $i \leqslant n-1\}$ is a one-dimensional manifold without boundary, which by Lemma 12.6 is a circle.

Since Lemma 11.12 implies that $|A_n\Lambda_n| < \varkappa(\delta)$ and $|B_n\Lambda_n| < \varkappa(\delta)$, then obviously there are two points on the circle $\Lambda_n$ at a distance $\pi/2$ from $A_n$, which is what was required. ■

**12.8.** ***Corollary.*** *An* $(n-1)$*-burst interior point in an n-dimensional* SCBB *is an n-burst point.*

**12.9. Further results.**

**12.9.1.** A similar approach enables us to show that a small neighbourhood of an $(n-1)$-burst point $p$ in the boundary of an $n$-dimensional SCBB admits an almost isometric map onto the cube in $\mathbb{R}^n$, where the part of the boundary occurring in this neighbourhood maps onto one of the hyperfaces of this cube. Here the explosion map (see 5.5) provides all the coordinate functions of the desired almost isometry except one—the one that is constant on the indicated hyperface. A basic observation which is necessary for the proof is as follows. Near $p$, every level line of the explosion map contains exactly one boundary point, and moreover these points may be characterized as the maximum points of the restriction to the corresponding level line of the distance function from some point $q$. Any point near $p$ lying on the same level line as $p$ can be taken as $q$.

**12.9.2.** Small neighbourhoods of interior $(n-2)$-burst points in an $n$-dimensional SCBB also admit coordinates of the distance type. In more detail, let the point $p$ have an $(n-2)$-explosion $(a_i, b_i)$. Then the corresponding sets of directions $A_i, B_i \subset \Sigma_p$ form a $(n-2)$-explosion in $\Sigma_p$. By Theorem 12.7 b) the level line $\Pi = \{\xi \in \Sigma_p : |A_i\xi| = \pi/2, 1 \leqslant i \leqslant n-2\}$ is a circle, whose diameter, as is easily shown using Lemma 8.6, is not less than $c(\varepsilon)$. We choose four points $\xi_1, \xi_2, \xi_3, \xi_4 \in \Pi$ lying on $\Pi$ in the cyclic order indicated by their numbers at approximately equal distances from each other, and we construct at equal small distances from $p$ the points $q_1, q_2, q_3, q_4$ such that $q_j' \subset \Sigma_p$ are very near to $\xi_j$, $1 \leqslant j \leqslant 4$. Then in a small neighbourhood of $p$ the map $f(q) = (|a_1q|, \ldots, |a_{n-2}q|, |q_1q|-|q_3q|, |q_2q|-|q_4q|)$ is a $c(\varepsilon)$-bi-Lipschitz homeomorphism onto a domain in $\mathbb{R}^n$. We know a proof of this assertion which requires very laborious arguments; it is set out in §14 of the original version of this article, which was distributed as a preprint.

**12.9.3.** The result in 12.9.2 and also 12.8 and 11.2 a) imply that in the two-dimensional case the interior of a SCBB is a two-dimensional manifold with the metric singularities forming a discrete set. A simple additional argument shows that a two-dimensional SCBB in the large is a two-dimensional manifold whose boundary is just the boundary of a FDSCBB in the sense of 7.19. In the three-dimensional case the interior of a SCBB may have non-isolated metric singularites, however 12.9.2 and 11.2 a) imply that the interior topological singularities are isolated. In fact, they have small neighbourhoods which are homeomorphic to the cone over $\mathbb{R}P^2$; see 13.2 b).

## §13. Subsequent results and open questions

All the spaces in this section are assumed to be finite-dimensional and complete.

**13.1.** The theorems formulated in 13.1–13.6 are the result of research by the third author. The proofs are published in a separate paper.

If one considers examples one is naturally led to the conjecture that a small spherical neighbourhood of a point in a FDSCBB is homeomorphic to the tangent cone at this point. This conjecture has been verified for points with an explosion near to being maximal by using almost regular maps. However, for $(n-3)$-burst points in an $n$-dimensional SCBB this method, apparently, cannot give the necessary results. To prove the conjecture in complete generality one has to introduce the more general concept of an admissible map.

A map $f = (f_1, \ldots, f_m) : U \to \mathbb{R}^m$ from a domain $U$ in an $n$-dimensional SCBB $M$ is said to be $(\varepsilon, \delta)$-*admissible* (where $\varepsilon$, $\delta$ are fixed and moreover $\varepsilon \gg \delta > 0$) if for all $x \in U$, $1 \leqslant i \leqslant m$ we have $f_i(x) = |A_i x|$, where $A_i \subset M$ are compact sets that do not intersect $U$, and where for every point $p \in U$ there is a compact set $B_{(p)} \subset M$ such that the inequalities $|A_i' A_j'| > \pi/2-\delta$, $i \neq j$, $|A_i' B_{(p)}'| > \pi/2+\varepsilon$ are satisfied in $\Sigma_p$.

We note that according to the given definition an admissible function is a distance function that does not have "almost critical" points in $U$.

It can be shown that an admissible map is open, and when $m = n$ it is locally homeomorphic. The advantage of admissible maps over almost regular maps, which plays a most important role in the proofs, is as follows.

Let $p \in \Pi$ be a point on a level surface of an almost regular map $f : U \to \mathbb{R}^m$, and let $P$ denote the set of points in $U$ near which $f$ cannot be added to another coordinate function to make an almost regular map into $\mathbb{R}^{m+1}$. Then, if $p \in P$, there are no other points from $P$ on the level surface $\Pi$ near to $p$. However, there may be many points of $P$ near to $p$ on nearby level surfaces; we can only assert that on each such surface they are isolated.

If almost regularity is everywhere replaced by admissibility, then it can be proved that there exists a spherical neighbourhood $B_p(R)$ of the point $p$ in which on each of the nearby level surfaces of the map $f$ there is no more than one point in $P$. In addition, if there is such a point on some level surface, then it can be described in a canonical way, because at this point the average distance from points of the set $\partial B_p(R) \cap \{x \in U : f_i(x) \geqslant f_i(p) \text{ for all } i\}$ takes its greatest value among points on this level surface that are near $p$.

**13.2.** If admissible maps are used, the following assertion can be proved.

***Theorem.*** a) *A proper admissible map is the projection of a locally trivial fibration.*

b) *A small spherical neighbourhood of a point in a* FDSCBB *is homeomorphic to the tangent cone at this point.*

c) *A* FDSCBB *is homeomorphic to a stratified (into topological manifolds) space.*

d) *For a given compact set in an n-dimensional space $M$ with curvature $\geqslant k$ there exists $\nu > 0$ such that every compact n-dimensional space $M_1$ with curvature $\geqslant k$ which is $\nu$-near to $M$ in the sense of Hausdorff must be homeomorphic to $M$. Moreover for each $\nu$-approximation $M_1$ to the space $M$ there exists a $\varkappa(\nu)$-near homeomorphism from $M$ onto $M_1$.*

It has not been established that the homeomorphisms in a)–d) are bi-Lipschitz, although there are apparently no known counterexamples.

The above assertions are all proved in a parallel way by induction. The geometric background of the proof has been given above; the necessary topological technique is borrowed from [31].

**13.3.** In the previous subsection the boundary points of a FDSCBB were in no way distinguished from other singular points. Nevertheless the approach described above enables one to give a more convincing justification for the definition of the boundary of a FDSCBB given in 7.19 and to prove some natural properties of a FDSCBB with a boundary. Namely, the following assertions are true.

***Theorem.*** a) *The property that a point in a* FDSCBB *is a boundary point is determined by the topology of its small conic neighbourhood; the property that a* FDSCBB *has a boundary is determined by its topology.*

b) *The boundary of a* FDSCBB *is closed and homeomorphic to a space stratified into manifolds.*

c) *If $M$ is a* FDSCBB *with a boundary $N$ and if $p \in N$, then $\Sigma N_\nu$ coincides with the boundary of $\Sigma_p$. In addition, an intrinsic metric is induced locally on $N$.*

d) (Doubling theorem). *Let $M$ be a space with curvature $\geqslant k$ and with a boundary $N$. We let $\overline{M}$ denote the result of gluing two copies of $M$ along the common boundary $N$ and furnish this with the natural intrinsic metric. Then $\overline{M}$ is a space with curvature $\geqslant k$ and without boundary.*

Thus, compact FDSCBB with a boundary are the same as convex compact spaces in FDSCBB without boundary.

The proofs of these assertions modulo Theorem 13.2 are not very complicated.

**13.4.** We note two open questions. 1) It is not clear whether the components of the boundary of a space with curvature $\geqslant k$ with the induced intrinsic metric are spaces with curvature $\geqslant k$. In other words, are the components of the boundary of a convex set in a space with curvature $\geqslant k$ spaces with curvature $\geqslant k$?

2) Is the gluing theorem generalizing 13.3 true: if the boundaries of two spaces $M$, $M_1$ with curvature $\geqslant k$ are component-wise isometric, then is the

result of gluing $M$ and $M_1$ with respect to this isometry a space with curvature $\geqslant k$?

**13.5. Equidistant curves.**
Let $M$ be a convex compact subset with non-empty boundary $N$ in a space with curvature $\geqslant 0$. It can be shown that the equidistant curves of the set $M_t = \{x \in M : |xN| \geqslant t\}$ are convex.

From what has been said and from Theorem 13.2 a), b) it follows that the interior of a convex compact set in a space with curvature $\geqslant k > 0$ is homeomorphic to a cone over a compact space with curvature $\geqslant 1$. A similar assertion can be proved also for the whole convex compact set.

The concepts described also open the way to proving an analogue of the well-known theorem of Cheeger and Gromoll [12] for non-regular spaces of non-negative curvature.

**13.6.** A direct generalization of the theorem of Grove and Shiohama [24] follows from Theorem 13.2 a), b). *A space with curvature* $\geqslant 1$ *and with diameter* $> \pi/2$ *is homeomorphic to the suspension over a space with curvature* $\geqslant 1$.

In addition, if we use the results of 13.2, we can get bounds for the Betti numbers of spaces with curvature $\geqslant 0$ by a method analogous to [18].

**13.7.** The next three open questions present difficulties: these are questions about the approximation of FDSCBB by Riemannian manifolds, about deformation of the metric of a FDSCBB, and about their analytic description. Apparently these questions are interlinked.

13.2 d) implies that FDSCBB which are not topological manifolds do not admit approximations by complete Riemannian manifolds of the same dimension and with sectional curvatures uniformly bounded below. This does not exclude the possibility of approximations of arbitrary compact subsets of a FDSCBB $M$ by Riemannian manifolds of larger dimension that collapse onto $M$. For example, a flat Klein bottle, as is known, can collapse to a segment. For other examples see [32].

As yet there are no means of deforming the metric of a FDSCBB while preserving (or at least controllably changing) the lower bound of the curvature. Such deformations are useful, for example, in solving extremal problems. In particular it seems probable, but completely unclear how to prove it, that if there is a point $p$ with diam $\Sigma_p < \pi$ in a FDSCBB $M$, then there exists a deformation of the metric of $M$ which is concentrated about $p$ and which preserves the lower bound for the curvature and decreases the volume.

The metric of a FDSCBB (and even more general so-called manifolds of bounded curvature) may be given by using a "non-regular" linear element which is defined almost everyhwere (see [3], [30]). It is not known whether it is possible to give a more detailed description of a FDSCBB in the case of larger dimension. It is well known [8] that for spaces of two-sided bounded curvature a linear element has fairly high smoothness, at least $C^{1,\alpha}$, $0 < \alpha < 1$. It can be expected that a space with curvature bounded below, in which all the geodesics are locally extendable, can also be represented with the help of a fairly smooth linear element.

## References

[1] A.D. Alexandrov, A theorem on triangles in a metric space and some of its applications, Trudy Mat. Inst. Steklov **38** (1951), 5–23. MR **14**–198.

[2] ———, Uber eine Verallgemeinerung der Riemannschen Geometrie, Schr. Forschungsinst. Math. **1** (1957), 33–84. MR **19**–304.

[3] ——— and V.A. Zalgaller, Two-dimensional manifolds of bounded curvature, Trudy Mat. Inst. Steklov. **63** (1962). MR **27** # 1911.

[4] ———, V.N. Berestovskii, and I.G. Nikolaev, Generalized Riemannian spaces, Uspekhi Mat. Nauk **41**:3 (1986), 3–44. MR **88e**:53103. = Russian Math. Surveys **41**:3 (1986), 1–54.

[5] V.N. Berestovskii, Spaces with bounded curvature and distance geometry, Sibirsk. Mat. Zh. **27**:1 (1986), 11–25. MR **88e**:53110. = Siberian Math. J. **27** (1986), 8–19.

[6] ———, "Submetrics" of three-dimensional forms of non-negative curvature, Sibirsk. Mat. Zh. **28**:4 (1987), 44–56. MR **89h**:53078. = Siberian Math. J. **28** (1987), 552–562.

[7] ———, Homogeneous manifolds with an intrinsic metric. I, II, Sibirsk. Mat. Zh. **29**:6 (1988), 17–29; **30**:2 (1989), 14–28. MR **90c**:53175; **90f**:53085. = Siberian Math. J. **29** (1988), 887–897; **30** (1989), 180–190.

[8] ——— and I.G. Nikolaev, Multi-dimensional generalized Riemannian spaces, Current problems in mathematics. Fundamental directions Vol. 70 (Geometry 4), VINITI, Moscow 1989, pp. 190–272.

[9] S.V. Buyalo, Shortest paths on convex hypersurfaces of a Riemannian space. Studies in topology, Zap. Nauchn. Sem. Leningrad. Otdel. Mat. Inst. Steklov. (LOMI) **66** (1976), 114–132. MR **58** # 30929.

[10] J. Cheeger, Finiteness theorems for Riemannian manifolds, Amer. J. Math. **92** (1970), 61–74. MR **41** # 1055.

[11] ——— and D. Ebin, Comparison theorems in Riemannian geometry, North-Holland Math. Library, Vol. 9, North-Holland Publ. Co., Amsterdam–Oxford 1975. MR **56** # 16538.

[12] ——— and D. Gromoll, On the structure of complete manifolds of non-negative curvature, Ann. of Math. (2) **96** (1972), 413–443. MR **46** # 8121.

[13] H. Federer, Geometric measure theory, Die Grundlehren der mathematischen Wissenschaften, Band 153, Springer-Verlag, Berlin–New York 1969. MR **41** # 1976. Translation: *Geometricheskaya teoriya mery*, Nauka, Moscow 1987. MR **89f**:49034.

[14] K. Fukaya, A boundary of the set of Riemannian manifolds with bounded curvatures and diameters, J. Differential Geom. **28** (1988), 1–21. MR **89h**:54090.

[15] R.E. Greene and Hung Hsi Wu, Lipschitz convergence of Riemannian manifolds, Pacific J. Math. **131** (1988), 119–141. MR **89g**:53063.

[16] D. Gromoll, W. Klingenberg, and W. Meyer, Riemannsche Geometrie im Grossen (Riemannian geometry in the large), Lecture Notes in Math. **55** (1975). MR **51** # 1651.
Translation: *Rimanova geometriya v tselom*, Mir, Moscow 1971.
[17] M. Gromov, Synthetic geometry in Riemannian manifolds, Proc. Internat. Congr. Math. (Helsinki 1978), Acad. Sci. Fennica, Helsinki 1980, pp. 415–419. MR **81g**:53029.
[18] ———, Curvature, diameter and Betti numbers, Comment. Math. Helv. **56** (1981), 179–195. MR **82k**:53062.
[19] ———, Structures métriques pour les variétés riemanniennes (Eds J. Lafontaine and P. Pansu), Textes mathématiques 1, CEDIC, Paris 1981. MR **85e**:53051.
[20] K. Grove, Metric differential geometry, Lecture Notes in Math. **1263** (1987), 171–227. MR **88i**:53075.
[21] ——— and H. Karcher, How to conjugate $C^1$-close group actions, Math. Z. **132** (1973), 11–20. MR **50** # 8575.
[22] ——— and P. Petersen, Manifolds near the boundary of existence, J. Differential Geom. **33** (1991), 379–394.
[23] ———, ———, and Jyh-Yang Wu, Geometric finiteness theorems via controlled topology, Invent. Math. **99** (1990), 205–214. MR **90k**:53075.
Erratum, ibid. **104** (1991), 221–222.
[24] ——— and K. Shiohama, A generalized sphere theorem, Ann. of Math. (2) **106** (1977), 201–211. MR **58** # 18268.
[25] A.D. Milka, Geodesics and shortest lines on convex hypersurfaces, I. Ukrain. Geom. Sb. No. 25 (1982), 95–110. MR **84g**:53075. II, Smoothness of a hypersurface at points of a shortest line, Ukrain. Geom. Sb. No. 26 (1983), 103–110. MR **85f**:53046.
[26] I.G. Nikolaev, Parallel translation of vectors in spaces of curvature that is bilaterally bounded in the sense of A.D. Aleksandrov, Sibirsk. Mat. Zh. **24**:1 (1983), 130–145. MR **84e**:53083.
= Siberian Math. J. **24** (1983), 106–119.
Smoothness of the metric of spaces with bilaterally bounded curvature in the sense of A.D. Aleksandrov, Sibirsk. Math. Zh. **24**:2 (1983), 114–132. MR **84h**:53098.
= Siberian Math. J. **24** (1983), 247–263.
(Announcement: Parallel translation and smoothness of the metric of spaces of bounded curvature, Dokl. Akad. Nauk **250** (1980), 1056–1058. MR **81d**:53052.
= Soviet Math. Dokl. **21** (1980), 263–265.)
[27] Y. Otsu, K. Shiohama, and T. Yamaguchi, A new version of differentiable sphere theorem, Invent. Math. **98** (1989), 219–228. MR **90i**:53049.
[28] S. Peters, Konvergenz Riemannscher Mannigfaltigkeiten, Bonner Mathematische Schriften No. 169, Universität Bonn, Math. Inst., Bonn 1986. MR **87k**:58046.
[29] ———, Convergence of Riemannian manifolds, Compositio Math. **62** (1987), 3–16. MR **88i**:53076.
[30] Yu.G. Reshetnyak, Two-dimensional manifolds of bounded curvature, Modern problems in mathematics. Fundamental directions Vol. 70 (Geometry 4), VINITI, Moscow 1989, pp. 7–189.
[31] L.C. Siebenmann, Deformation of homeomorphisms on stratified sets, Comment. Math. Helv. **47** (1972), 123–163. MR **47** # 7752.

[32] T. Yamaguchi, Collapsing and pinching under a lower curvature bound, Ann. of Math. (2) **133** (1991), 317–357. MR **92b**:53067.

Translated by A. West

St. Petersburg Branch of the Steklov Institute of Mathematics
Institut des Hautes Études Scientifiques

Received by the Editors 4 November 1991

Invent. math. 82, 307–347 (1985)

*Inventiones mathematicae*


# Pseudo holomorphic curves in symplectic manifolds

M. Gromov
Institut des Hautes Etudes Scientifiques, 35, route de Chartres, F-91440 Bures-sur-Yvette, France

## § 0. Introduction

Consider a smooth manifold $V$ with an *almost complex* structure $J$, that is a $C^\infty$-field of *complex linear* structures in the tangent spaces $T_v(V)$, $v \in V$. For example, if $\dim V = 2$, then $(V, J)$ is called a *Riemann surface*. A smooth map between two such manifolds, say $f:(V', J') \to (V, J)$, is called (pseudo) *holomorphic* if the differential $\mathscr{D}_f: T(V') \to T(V)$ is a *complex* linear map for the structures $J'$ and $J$.

*Definitions.* A *parametrized* (pseudo holomorphic) *J-curve* in an almost complex manifold $(V, J)$ is a holomorphic map of a Riemann surface into $V$, say $f:(S, J') \to (V, J)$. The image $C = f(S) \subset V$ is called a (non-parametrized) *J-curve* in $V$. A curve $C \subset V$ is called *closed* if it can be (holomorphically!) parametrized by a *closed* surface $S$. We call $C$ *regular* if there is a parametrization $f: S \to V$ which is a *smooth proper embedding*. A curve is called *rational* if one can choose $S$ diffeomorphic to the sphere $S^2$.

*0.1. Example.* Let $(V, J)$ be the product of two Riemann surfaces, $(V, J) = (S_1 \times S_2, J_1 \oplus J_2)$. Then graphs of holomorphic maps $(S_1, J_1) \to (S_2, J_2)$ clearly are regular $J$-curves in $V$. Conversely, every regular $J$-curve $C$ in $V$ whose projection to $S_1$ is a proper map of degree one (for the orientations induced by the structures $J$ on $C$ and $J_1$ on $C_1$) is the graph of a holomorphic map $S_1 \to S_2$, as a simple argument shows. If, for instance, $S_1$ and $S_2$ are diffeomorphic to $S^2$, then the existence of a holomorphic diffeomorphism $(S^2, J_1) \to (S^2, J_2)$ is insured by the *Riemann mapping theorem* for spheres. This is expressed in our language as *the existence of a regular rational curve* $C \subset V = (S^2 \times S^2, J_1 \oplus J_2)$ *which is homologous to the diagonal in* $S^2 \times S^2$ (i.e. the projections of $C$ to both factors have degree one).

### *0.2. Curves in tame manifolds*

The goal of this paper is an extension of basic facts on curves in complex manifolds to the almost complex case. Our main results concern the existence of such curves in the presence of an auxiliary *symplectic structure* on $(V, J)$.

*Definitions.* An anti-symmetric $\mathbb{R}$-bilinear form $\omega_0$ on $\mathbb{C}^n$ is called *positive* if it is positive on every complex line in $\mathbb{C}^n$, that is $\omega(x, \sqrt{-1}x) > 0$ for all non-zero vectors $x \in \mathbb{C}^2$. Next, an exterior differential 2-form $\omega$ on an almost complex manifold $(V, J)$ is called *J-positive* if $\omega$ is positive on every tangent space $T_v(V)$, $v \in V$, for the implied complex linear structure $J | T_v(V)$. Such a form clearly is positive on every $J$-curve $C$ in $V$.

*Examples.* (a) If $\dim_{\mathbb{R}} V = 2$ then positive forms on $V$ are those area forms whose orientations agree with the orientation induced by $J$.

(b) Let $\omega_1$ and $\omega_2$ be positive forms on Riemann surfaces $(S_1, J_1)$ and $(S_2, J_2)$ correspondingly and let $\tilde{\omega}_1$ and $\tilde{\omega}_2$ be their pull-backs to $V = S_1 \times S_2$ (for the projections $V \to S_1$ and $V \to S_2$). Then the form $\omega = \tilde{\omega}_1 + \tilde{\omega}_2 \overset{\text{def}}{=} \omega_1 \oplus \omega_2$ on $V$ clearly is $J$-positive for $J = J_1 \oplus J_2$.

We say that $J$ is *tamed* by a 2-form $\omega$ on $V$ if $\omega$ is closed (i.e. $d\omega = 0$) and $J$-positive on $V$.

*Remark.* If a 2-form $\omega$ on $V$ is $J$-positive, then, obviously, the top exterior power $\omega^n$ for $2n = \dim V$ nowhere vanishes on $V$. *Closed* 2-forms on $V$ with nonvanishing $\omega^n$ are called *symplectic forms* (or structures) on $V$.

*0.2.A. A generalization of Riemann's mapping theorem to non-split almost complex structures on $S_1 \times S_2$.* Let $S_1$ and $S_2$ be closed connected surfaces with area forms $\omega_1$ and $\omega_2$ of total areas $A_1 = \int_{S_1} \omega_1$ and $A_2 = \int_{S_2} \omega_2$ and denote by $\omega$ the (symplectic!) form $\omega_1 \oplus \omega_2$ on $V = S_1 \times S_2$. Let $J$ be a $C^\infty$-smooth almost complex structure on $V$ tamed by $\omega$.

**Theorem.** (See 2.4.C.). *If $S_1$ is diffeomorphic to the sphere $S^2$ and if $A_2 = kA_1$ for an integer $k \geq 1$, then there is a unique regular rational $J$-curve $C = C_v \subset V$ which contains a given point $v \in V$ and which is homologous to the sphere $S_1 \times s \subset V$, $s \in S_2$. Furthermore, let $S_2$ also be diffeomorphic to $S^2$, and let $A_2 = A_1$. Then there exists a connected regular $J$-curve $C$ in $V$ which represents the homology class $p[S_1] + q[S_2] \in H_2(V, \mathbb{Z})$ for arbitrary non-negative integers $p$ and $q$, and which has $\text{genus}(C) = pq - p - q + 2$. In fact, these curves $C$ form a smooth manifold $M = M_{pq}(J)$ of dimension $2(pq + p + q)$.*

*Remark.* Our version of Riemann's mapping theorem is similar to the one discovered by Schapiro [Sch] in 1941 for mappings in the plane which satisfy a quasi linear system of two elliptic equations. Schapiro's theorem has been greatly generalized since (see [Lav, Be, B-H, Gil, Wen]).

*0.2.A'.* Let us generalize 1.2.A. by considering a closed manifold $V_2$ of dimension $2(n-1) > 2$ with a symplectic form $\omega_2$, such that $\int_{S^2} \omega_2 = kA_1$ for every smoothly mapped sphere $S^2 \to V$, for some *integer* $k = k(S^2 \to V)$, and for the area $A_1$ of the surface $(S^2, \omega_1)$ of 0.2.A.

**Theorem.** (See 2.3.C.). *Let $J$ be a $C^\infty$-smooth almost complex structure on $V = S^2 \times V_2$ tamed by the (symplectic!) form $\omega_1 \oplus \omega_2$. Then there exists a (possibly singular and non-unique) rational $J$-curve $C = C_v \subset V$ which contains a given point $v \in V$ and which is homologous to the sphere $S^2 \times v_2 \subset V$, $v_2 \in V_2$.*

*0.2.B.* Now, let $V = \mathbb{C}P^n$ with the standard symplectic form $\omega$ (which is uniquely characterized up to a scalar factor by being invariant under the obvious $U(n+1)$-action on the complex projective space $\mathbb{C}P^n$) and let $J$ be a $C^\infty$-smooth almost complex structure on $V$ tamed by $\omega$.

**Theorem.** *Any two points $v_1$ and $v_2$ in $V$ lie on some* (possibly non-unique and singular) *rational $J$-curve $C \subset V$ which is homologous to the projective line $\mathbb{C}P^1 \subset \mathbb{C}P^n \approx V$. Moreover, let* $\dim V = 4$ (i.e. $n=2$). *Then the curve $C = C(v_1, v_2)$ is non-singular and it is unique for $v_1 \neq v_2$. Further, any five points in general position* (i.e. no three of them lie on one of the above curves $C$) *in $V$ lie on a unique nonsingular rational curve homologous to $2\mathbb{C}P^1$, and any nine generic point lie on a unique regular curve of* genus one *which is homologous to $3\mathbb{C}P^1$; in general, any $k$-point in general position for $k = \frac{d(d+3)}{2}$ lie on a unique regular curve of degree $d$* (i.e. homologous to $d\mathbb{C}P^1$) *and of genus* $\frac{d^2 - 3d + 2}{2}$.

This is shown in §§ 2.3.C., 2.4.B$_1''$. with a precised notion of the "general position".

*0.2.B′. Example.* (Pointed out by M. Berger.) Let $g$ be an arbitrary $C^\infty$-smooth Riemannian metric on $\mathbb{C}P^2$ and let $\omega = \omega_g$ be a non-zero harmonic 2-form on $\mathbb{C}P^2$ for this metric. Then clearly, this $\omega$ is self dual (i.e. $*\omega = \omega$) as $\operatorname{rank} H^2(\mathbb{C}P^2) = 1$. If $\omega_v = \omega \mid T_v(\mathbb{C}P^2) \neq 0$ for some $v \in \mathbb{C}P^2$, then, since $\dim T_v = 4$, there is a unique complex linear structure on $T_v$ for which

$$\omega_v(x, \sqrt{-1}\, y) = \|\omega_v\|^2 g_v(x, y) \tag{*}$$

for all vectors $x$ and $y$ in $T_v$. Now, let us assume the form $\omega$ to be nowhere zero on $V$ and thus get an almost complex structure, say $J_g$, on $\mathbb{C}P^2$ which is tamed by $\omega$ and which satisfies the (Kähler) identity $\omega(x, \sqrt{-1}\, y) = g'(x, y)$ for the conformal metric $g' = \|\omega\|^2 g$. With this one easily observes (see 2.3.E$_2'$.) that every $J_g$-curve in $V$ is a *minimal surface* for the metric $g'$. Moreover every closed $J_g$-curve is absolutely $g'$-area minimizing in its homology class.

In order to apply the above theorem we additionally assume the existence of a homotopy of metrics, say $g_t$ for $0 \leqq t \leqq 1$, which joins $g = g_1$ with the ($U(3)$-invariant) Fubini-Studi metric $g_0$ on $\mathbb{C}P^2$ and such that the harmonic form $\omega_{g_t}$ does not vanish on $\mathbb{C}P^2$ for all $t \in [0, 1]$. Then, by a theorem of Moser (see [Mo, Wei, Gro$_3$]), there is a self-diffeomorphism of $\mathbb{C}P^2$ which sends the form $\omega_g$ to the (standard!) form $\omega = \omega_{g_0}$. Now, our theorem provides a huge family of absolutely minimizing surfaces for the conformal metric $g'$ on $\mathbb{C}P^2$.

**Corollary.** (See [Gro$_2$], compare [Ber].) *If $g$ is $C^\infty$-close to $g_0$ then there is a smoothly embedded sphere $S^2 \subset \mathbb{C}P^2$ homologous to the projective line $\mathbb{C}P^1 \subset \mathbb{C}P^2$ such that*

$$\left(\frac{\operatorname{Area}_g S^2}{\operatorname{Area}_{g_0} \mathbb{C}P^1}\right)^2 \leqq \left(\frac{\operatorname{Vol}_g \mathbb{C}P^2}{\operatorname{Vol}_{g_0} \mathbb{C}P^2}\right).$$

*Remark.* Every closed surface $S \subset \mathbb{C}P^2$ homologous to $d\mathbb{C}P^1$ satisfies the opposite inequality for the metric $g'$

$$\left(\frac{\mathrm{Area}_{g'} S}{\mathrm{Area}_{g_0} \mathbb{C}P^1}\right)^2 \geqq d^2 \left(\frac{\mathrm{Vol}_{g'} \mathbb{C}P^2}{\mathrm{Vol}_{g_0} \mathbb{C}P^2}\right),$$

with the equality for $J_{g'}$-holomorphic curves in $\mathbb{C}P^2$.

## *0.3. Global invariants of symplectic manifolds*

Let us use the results of 0.2. to distinguish symplectic manifolds, for example, open subsets in $\mathbb{R}^{2n}$ with the symplectic form $\omega_0 = \sum_{i=1}^{n} dx_i \wedge dy_i$.

**0.3.A. Theorem.** *Let $V'$ be an open subset in the manifold $V = S^2 \times V_2$ of 0.2.A'. and let $J'$ be a $C^\infty$-smooth almost complex structure on $V'$ which is tamed by the form $\omega' = \omega | V'$ for $\omega = \omega_1 \oplus \omega_2$* (see 0.2.A'.). *Then, for every point $v' \in V'$, there is a connected proper $J'$-curve ("proper" means that $C'$ is closed in $V'$ as a point-set and that* $\dim_{\mathbb{R}} C' = 2$*) which contains $v'$ and such that $\int_{C'} \omega' \leqq A_1 = \int_{S^2} \omega_1$.*

*Proof.* The structure $J'$ can be easily extended (with a slight perturbation near the boundary $\partial V' \subset V$) to a smooth structure $J$ on $V$ tamed by $\omega$. The Theorem 0.2.A'. provides a $J$-holomorphic curve $C = C_{v'} \subset V$ through $v'$ for which $\int_C \omega = \int_{S^2} \omega_1$. Then one takes the connected component of $v'$ in the intersection $C \cap V'$ for $C'$. (See 2.3.E. for details.)

**Corollary.** *Consider a symplectic diffeomorphism of the open round ball $B(R) \subset \mathbb{R}^{2n}$ onto an open subset $V' \subset \mathbb{R}^{2n}$ which is contained in the $\varepsilon$-neighborhood of the subspace $\mathbb{R}^{2n-2} \subset \mathbb{R}^{2n}$ spanned by the vectors $x_i, y_i$ for $i = 1, \ldots, n-1$. Then the radius $R$ satisfies the inequality $R \leqq \varepsilon$.* (A diffeomorphism $f: B \to V'$ is called symplectic if $f^*(\omega_0) = \omega_0$ for $\omega_0 = \sum_{i=1}^{n} dx_i \wedge dy_i$.)

*Proof.* The subset $V'$, can be symplecticly embedded into $(S^2 \times T^{n-2}, \omega_1 \oplus \omega_2)$ for some form $\omega_1$ on $S^2$ with $A_1 = \int_{S^2} \omega_1 = \pi\varepsilon^2$ and for the form $N\pi\varepsilon^2 \sum_{i=1}^{n-1} dx_i \wedge dy_i$ on the torus $T^{n-2} = \mathbb{R}^{2n-2}/\mathbb{Z}^{2n-2}$ for a large integer $N$. This insures a $J$-curve of $\omega_0$-area $\leqq \pi\varepsilon^2$ through the center of the ball $B = B(R)$ (as $B$ is symplectomorphic to $V'$) for any structure $J$ on $B$ tamed by $\omega_0$. On the other hand, for the standard complex structure on $\mathbb{C}^n = \mathbb{R}^{2n}$, every holomorphic curve through the center has area $\geqq \pi R^2$ (see 2.3.$E_2$.). Hence $R \leqq \varepsilon$. Q.E.D.

*Remarks.* The (linear) symplectic diffeomorphism $(x_i, y_i) \to (\delta x_i, \delta^{-1} y_i)$ sends every ball to an arbitrarily small $\varepsilon$-neighborhood of the subspace $\mathbb{R}^n \subset \mathbb{R}^{2n}$ spanned by $y_1, \ldots, y_n$, provided $\delta > 0$ is sufficiently small.

The above corollary gives an upper bound on the *radius* of a symplectic manifold $V$, which by definition is the upper bound of the radii of the balls $B(R)\subset\mathbb{R}^{2n}$ which admit symplectic imbeddings into $V$. For example, the product of open disks, $V=D(R_1)\times D(R_2)$ for $D(R)\subset(\mathbb{R}^2, dx, \wedge dy)$ has by the above,

$$\operatorname{Rad} V=\min(R_1, R_2).$$

Therefore, the existence of a symplectic diffeomorphism between two such products, say between $V(R_1, R_2)$ and $V(R_1', R_2')$ with $R_1\leqq R_2$ and $R_1'\leqq R_2'$, implies the equality $R_1=R_1'$ and, hence, the equality $R_2=R_2'$ as $\operatorname{Vol} V=\pi^2R_1^2R_2^2$ is symplectictly invariant. A finer argument (see [Gro$_2$]) gives a similar result for products of $n$ disks for all $n=2, 3, \ldots$.

*0.3.B. Packing inequalities.* Let $V_1$ and $V_2$ be disjoint images of symplectic diffeomorphisms of the balls $B(R_1)$ and $B(R_2)$ into $B(R)\subset\mathbb{R}^{2n}$. Then, arguing as earlier with 0.2.B. in place of 0.1.A′. one gets the inequality

$$R_1^2+R_2^2\leqq R^2.$$

Similarly, for disjoint symplectic images $\operatorname{Im} B(R_j)\subset B(R), j=1, \ldots, 5$, one gets now for $n=2$ only,

$$\sum_{i=1}^{5} R_i^2\leqq 2R^2.$$

*0.3.C. Symplectic diffeomorphisms of 4-dimensional manifolds.* Let $V=(S^2\times S^2, \omega_1\oplus\omega_2)$.

**Theorem.** *If* $\int_{S^2}\omega_1=\int_{S^2}\omega_2$ *then the group of symplectic diffeomorphisms of* $V$ *contracts onto the subgroup of isometries of* $S^2\times S^2$ *for the standard Kähler metric on* $S^2\times S^2$ *(this subgroup is the obvious* $\mathbb{Z}/2\mathbb{Z}$ *extension of* $SO(3)\times SO(3)$*).*

This is easily seen with the $J$-curves provided by 0.2.A. for all almost complex structures $J$ tamed by the symplectic form $\omega=\omega_1\oplus\omega_2$ on $V$. The actual argument (see 2.4.A$_1$.) is similar to the usual proof of the contractibility of $\operatorname{Diff} S^2$ onto $O(3)$ based on the Riemann mapping theorem for spheres.

*Warning.* The homotopy type of the symplectic diffeomorphism group of $(S^2\times S^2, \omega_1\oplus\omega_2)$ becomes more complicated (and interesting) if one allows $\int_{S^2}\omega_1\neq\int_{S^2}\omega_2$ (see 2.4.C$_2$.).

*Remark.* The proof indicated above also applies to $\mathbb{C}P^2$ with the standard symplectic form $\omega_0$ and shows the group of symplectic diffeomorphisms of $\mathbb{C}P^2$ to be contractible onto the subgroup of isometries (see 2.4.B$_3'$.). A similar argument (see 2.4.A$_2''$.) yields the following

**Theorem.** *Let an open manifold* $(V, \omega)$ *be symplectically diffeomorphic to* $(\mathbb{R}^4, \omega_0 =dx_1\wedge dy_1+dx_2\wedge dy_2)$ *at infinity. If the Hurewicz homomorphism* $\pi_2(V) \to H_2(V;\mathbb{R})$ *vanishes, then* $(V, \omega)$ *is symplectically diffeomorphic to* $(\mathbb{R}^4, \omega_0)$. *(A "diffeomorphism at infinity" means a diffeomorphism between the complements of compact subsets in the manifolds in question.)*

## 0.4. *Pseudo holomorphic curves with boundaries*

Let $W$ be a *totally real* submanifold in an almost complex manifold $V$, that is $\dim W = \frac{1}{2}\dim V$ and the subbundle $T(W) \subset T(V)$ contains no *complex* line for the implied complex structure in $T(V)$. Our existence theorems for closed curves in $V$ (see § 0.2.) generalize to holomorphic maps $f: S \to V$, where $S$ is a compact Riemann surface with a boundary and $f(\partial S) \subset W$.

*Definitions.* (a) An almost complex manifold $V$ is called *convex at infinity* if there is an exhaustion of $V$ by relatively compact subsets $V_1 \subset V_2 \subset \ldots V_k \subset \ldots \subset V$, such that every holomorphic map $f: D^2 \to V$ with $f(\partial D^2) \subset V_k$ has $f(D^2) \subset V_{k+1}$ for all $k = 1, 2, \ldots$, where $D^2$ is the unit disk, $D^2 = \{z \in \mathbb{C} \mid \|z\| \leqq 1\}$.

(b) A submanifold in a symplectic manifold, say $W \subset (V, \omega)$, is called *Lagrange* if $\dim W = \frac{1}{2}\dim V$ and the form $\omega \mid W$ is everywhere zero on $W$.

(b′) *Example.* Consider the product of ε-disks in $\mathbb{R}^{2n}$,

$$D_\varepsilon^{2n} = \{x_1, y_1, \ldots, x_n, y_n \mid x_i^2 + y_i^2 \leqq \varepsilon^2,\ i = 1, \ldots, n\}.$$

This product contains the product of $n$ ε-circles $\{x_i^2 + y_i^2 = \varepsilon^2\}$, which clearly is Lagrange for the form $\omega_0 = \sum_{i=1}^{n} dx_i \wedge dy_i$ on $\mathbb{R}^{2n}$.

(c) Consider a closed totally real submanifold $W \subset (V, J)$ and let $F: W_0 \times D^2 \to V$, for some manifold $W_0$, be a smooth map with the following five properties.

(i) $F$ diffeomorphically maps $W_0 \times S^1$ onto $W$ for $S^1 = \partial D^2$.

(ii) The map $F_\varepsilon = F \mid W_0 \times S_\varepsilon^1$ for $S_\varepsilon^1 = \{z \in \mathbb{C} \mid \|z\| = \varepsilon\} \subset D^2$ is a totally real embedding for $0 < \varepsilon \leqq 1$, that is $W_\varepsilon = F(W_0 \times S_\varepsilon^1)$ is a totally real submanifold in $V$.

(iii) There exists a symplectic form $\omega_\varepsilon$ on $V$ for every $\varepsilon \in [0, 1]$ which tames $J$ and such that $W_\varepsilon$ is a Lagrange submanifold for $\omega_\varepsilon$. We do not require the continuity of $\omega_\varepsilon$ in $\varepsilon$ but rather the following integrality condition.

(iv) Fix a point $w_0 \in W_0$, let $D_\varepsilon^2 = w_0 \times D_\varepsilon^2 \subset W_0 \times D_\varepsilon^2$, for some $w_0 \in W_0$ be the ε-disk $\{\|z\| \leqq \varepsilon^2\}$ mapped to $V$ by $F$ and denote by $\langle \omega_\varepsilon, D_\varepsilon^2 \rangle$ the integral of $\omega_\varepsilon$ over $D_\varepsilon^2$, that is $\int_{D_\varepsilon^2} F^*(\omega_\varepsilon)$. Then we insist, for *every* smoothly mapped disk $D$ in $V$ with the boundary $\partial D$ in $W_\varepsilon$, on the relation

$$\langle \omega_\varepsilon, D \rangle = k \langle \omega_\varepsilon, D_\varepsilon^2 \rangle \qquad (*)$$

for some *integer* $k = k(\varepsilon)$. (This (*) automatically holds true, for instance, if $\operatorname{rank} H_2(V, W) \leqq 1$.)

(v) There exists a symplectic form $\omega_0$ on $V$ taming $J$ for which the embeddings $F_\varepsilon$ *symplecticly split* for the small $\varepsilon \leqq \varepsilon_0 > 0$. Namely, there is a split neighborhood $U = U' \times \mathbb{R}^2 \subset V$, such that $\omega_0 \mid U = \omega_0' \oplus \omega_0''$ for some symplectic forms on $U'$ and on $\mathbb{R}^2$ respectively. Furthermore, there exist embeddings $F_\varepsilon': W_0 \to U'$ and $F_\varepsilon'': S_\varepsilon^1 \to \mathbb{R}^2$ such that $F_\varepsilon = F_\varepsilon' \times F_\varepsilon''$ for $0 < \varepsilon < \varepsilon_0$.

**0.4.A. Theorem** (see 2.3.D.). *Let $(V, J)$ be an almost complex manifold convex at infinity and let $W \subset V$ be a closed totally real submanifold which admits a map $F: W_0 \times D^2 \to V$ with the properties* (i)–(v). *If the integral $\langle \omega, D_{\varepsilon_0}^2 \rangle$ is positive then* *there exists a J-holomorphic map* $f: D^2 \to V$ *with* $f(\partial D^2) \subset W$, *which is homotopic (as the map between pairs of spaces,* $(D^2, \partial D^2) \to (V, W)$*) to the disk* $D_1^2 = D_{\varepsilon=1}^2 \to V$.

*0.4.A*$_1$. *Examples, remarks, and corollaries.* Our theorem applies to the images $W = \alpha(T^n) \subset \mathbb{C}^n$ of the torus $\{z_1, \ldots, z_n \mid \|z_i\| = 1,\ i = 1, \ldots, n\} \subset \mathbb{C}^n$ under arbitrary symplectic diffeomorphisms $\alpha: (\mathbb{C}^n, \omega_0) \to (\mathbb{C}^n, \omega_0)$ for $\omega_0 = \sum_{i=1}^{n} dx_i \wedge dy_i$, where $z_i = x_i + \sqrt{-1}\, y_i$. Thus we obtain *non-constant* holomorphic maps $(D^2, \partial D^2) \to (\mathbb{C}^n, W)$, (compare [Bi , We, B-G]). In fact our proof (see 2.3.B.) delivers the following more general

**0.4.A$_2$. Theorem.** *For an arbitrary closed* $C^\infty$*-smooth Lagrange submanifold* $W \subset \mathbb{C}^n$ *there exists a non-constant holomorphic map* $f: (D^2, \partial D^2) \to (\mathbb{C}^n, W)$. *It follows, that the relative class* $[\omega_0] \in H^2(\mathbb{C}^n, W; \mathbb{R})$ *is non-zero.*

*Remark.* This theorem partially solves Bennequin's conjecture on the existence of a holomorphic curve in $\mathbb{C}^2$ whose boundary lies in a given *totally real* torus $T^2 \subset \mathbb{C}^2$.

**0.4.A$'_2$. Corollary.** *There exists a symplectic structure* $\omega$ *on* $\mathbb{R}^{2n}$ *for all* $n \geqq 2$ *which admits no symplectic embedding into* $(\mathbb{R}^{2n} = \mathbb{C}^n, \omega_0)$.

Indeed, one easily produces a structure $\omega$ on $\mathbb{R}^{2n}$ such that $0 = [\omega] \in H^2(\mathbb{R}^{2n}, W, \mathbb{R})$, where $W$ is some Lagrange submanifold for this $\omega$ (see 2.3.B$_5$. for details).

We shall also exhibit (see 2.4.D$_1$.) *contact structures* on $\mathbb{R}^{2n+1}$ for all $n \geqq 1$ which do not embed into the standard structure on $\mathbb{R}^{2n+1}$. Such examples for $n = 1$ are due to Bennequin [Ben] and Eliashberg [El$_2$].

*0.4.A*$_3$. We shall use Theorem 0.4.A. in 2.3.B$_4$. to study intersection points between Lagrange submanifolds and thus to prove new fixed point theorems for *exact* symplectic diffeomorphisms. (Compare the geometric method developed by Eliashberg for surfaces [El$_1$] and the new variational approach discovered by Conley and Zehnder, [C-Z]).

*0.4.A*$_4$. The existence theorem for holomorphic maps $f: (D^2, \partial D^2) \to (\mathbb{C}^n, W)$ that are solutions to the equation $\frac{\partial f}{\partial \bar z} = 0$ generalize to the non-homogeneous equation $\frac{\partial f}{\partial \bar z} = g$ for $C^\infty$-maps $g: D^2 \to \mathbb{C}^n$ and for certain *non-compact* Lagrange submanifolds $W \subset \mathbb{C}^n$ (see 2.3.B.). However, if $W$ is compact, then the equation $\frac{\partial f}{\partial \bar z} = \mathrm{const} \in \mathbb{C}^n$ is unsolvable if const is large compared to Diam $W$ (see 2.3.B$_1$.). In fact, one has a "Fredholm alternative" for this equation: Either there is a non-constant solution of $\frac{\partial f}{\partial \bar z} = 0$, or the equation $\frac{\partial f}{\partial \bar z} = g$ is solvable for all $g$, provided the Lagrange submanifold $W \subset \mathbb{C}^n$ in question is sufficiently regular at infinity (see 2.3.B.).

*0.4.A*$_5$. The classical Riemann mapping theorem for holomorphic maps $D^2 \to D' \subset \mathbb{C}$ concerns holomorphic disks in $\mathbb{C}^2$ with boundaries in the Lagrange (!) torus $\partial D^2 \times \partial D' \subset \mathbb{C}^2$. We shall refine (see 2.4.E.) 0.4.A. for $n=2$ (compare 0.2.A.) in order to include Riemanns theorem into our framework.

## *0.5. Historical remarks and references*

The existence of global symplectic geometry was first (to authors' knowledge) conjectured by Arnold in the sixties. However, the evidence coming from subsequent results on symplectic immersions and embeddings (see [Gro$_1$, Gro$_2$] and references therein) seemed to indicate a "symplectic flexibility" with no room for non-trivial geometry. The dilemma was resolved (in favour of geometry) in seventies by Eliashberg who proved *every $C^1$-diffeomorphism, which is a $C^0$-limit of symplectic diffeomorphisms, to be symplectic*, (see [El$_2$]). In fact, this limit theorem is only a tip of the iceberg of (yet unpublished) Eliashberg's symplectic theory. Next crucial step was made by Bennequin who developed a beautiful geometry of *Legendre curves* in contact 3-manifolds (see [Ben]). Both theories (worked out independently by Eliashberg and by Bennequin) employ a direct (and quite complicated) geometric analysis of the standard symplectic and contact structures. A different approach, based on the calculus of variations, was used by Conley and Zehnder [C-Z] in their astounding solution to Arnold's fixed point problem for the symplectic $2n$-torus (this was solved earlier for surfaces by Eliashberg [El$_1$]). Their method has been further developed in [Ch, Si, Fl, H, F-W] (see [Wei, D-D, Ren] for different aspects of the symplectic geometry).

Our study of pseudo-holomorphic curves extends (a small portion of) the Bers-Vekua theory of pseudo-analytic functions to quasi-linear equations. The basic non-linear ingredient is the vanishing cycle phenomenon for pseudo-holomorphic curves (see 1.5.) that was discovered for minimal surfaces by Sacks and Uhlenbeck [S-U]. Besides, the present paper implicitly uses some considerations pertaining to rudimentary theory of plane algebraic curves (these were patiently explained to me by Pierre Deligne) as well as ideas on calibrated geometries (see 2.3.C$''_2$.) communicated to me by Blaine Lawson.

# 1. Compactness properties of holomorphic curves

## *1.1. A bound on the curvature of a holomorphic curve*

Start with a regular holomorphic curve $C_0 \subset \mathbb{C}^n$ through the origin $0 \in \mathbb{C}^n$, take the tangent and the normal space to $C_0$ at $0 \in C_0$, say $T \subset \mathbb{C}^n$ and $N = \mathbb{C}^n \ominus T \subset \mathbb{C}^n$ and let $f: T \to N$ be the map whose graph in $\mathbb{C}^n = T \oplus N$ equals $C_0$ near the origin. Since $C_0$ is holomorphic, the map $f$ also is holomorphic and, in particular, it is a harmonic map. Since the differential $\mathcal{D}_f$ vanishes at $0 \in C_0$ the *mean curvature* of $C_0$ satisfies at $0 \in C_0$,

$$M\,\mathrm{Curv}(C_0, 0) = \Delta f(0) = 0\,. \tag{1}$$

**1.1.A. Corollary.** *Let $J$ be a $C^1$-smooth almost complex structure on $V$, let $\mu$ be a $C^2$-smooth Riemannian metric and let $C \subset V$ be a regular $J$-curve in $V$ through a point $v_0 \in V$. Then the mean curvature of $C$ with respect to $\mu$ satisfies*

$$\|\mathrm{Trace}_{\bar{\mu}} II_{\mu}(C, v_0)\| \leqq \mathrm{const} = \mathrm{const}(J, \mu, v_0) \tag{2}$$

*where $\bar{\mu} = \mu + J\mu$ and $II$ is the second fundamental form.*

*Proof.* The curvature of $C$ changes by a finite amount with the change of $\mu$ according to the chain rule formula for the second derivatives. Thus we may assume that $(V, v_0) = (\mathbb{C}^n, 0)$, $2n = \dim V$, and that the structure $J$ on $T_0(\mathbb{C}^n)$ equals the standard complex structure of $\mathbb{C}^n$. Then, by the chain rule again, there is a holomorphic curve $C_0 \subset \mathbb{C}^n$ tangent to $C$ at the origin whose curvature differs from that of $C$ by a finite amount $\leqq \mathrm{const}(J)$. Q.E.D.

*1.1.B. A bound on the Gauss curvature of $C \subset (V, J)$. If $V$ is compact then the Gauss curvature of the induced metric in $C$ satisfies*

$$G\,\mathrm{Curv}(\mu | C) \leqq \mathrm{const} = \mathrm{const}(J, \mu) \tag{3}$$

*for all Riemannian $C^2$-metrics on $V$ and for all regular $J$-curves $C$ in $V$.*

*Proof.* Use (2) and Gauss' theorema egregium

### *1.2. The Schwarz lemma for conformal maps*

Let $V$ be a surface with a Riemannian metric $\mu$ and let $(V_0, \mu_0)$ be a surface which is rotationally symmetric around a fixed point $v_0 \in V_0$, such that $\mathrm{Area}(V_0, \mu_0) \geqq \mathrm{Area}(V, \mu)$ and such that for any relatively compact domain $V' \subset V$ the ball $V_0' \subset V_0$ around $v_0$ of Area = Area $V'$ satisfies

$$\mathrm{length}\, \partial V_0' \leqq \mathrm{length}\, \partial V'. \tag{4}$$

Let $f: D^2 \to V$ be an injective conformal map and let $f_0: D^2 \to V_0$ be a conformal homeomorphism of the unit disk $D^2 \subset \mathbb{C}$ onto a ball in $V_0$ around $v_0 \in V_0$, such that the center of $D^2$ goes to $v_0 \in V_0$ and such that $\mathrm{Area}\, f_0(D^2) = \mathrm{Area}\, f(D^2)$.

*1.2.A. The areas of the images of the concentric disks $D_\varepsilon \subset D_1 = D^2$ satisfy for all $\varepsilon \in [0, 1]$,*

$$\mathrm{Area}\, f(D_\varepsilon) \leqq \mathrm{Area}\, f_0(D_\varepsilon). \tag{5}$$

*In particular, the differentials of the maps $f$ and $f_0$ at the center $0 \in D^2$ satisfy*

$$\|\mathscr{D}_f\|(0) \leqq \|\mathscr{D}_{f_0}\|(0). \tag{$5_0$}$$

*Proof.* Since $f$ is conformal,

$$\frac{\partial}{\partial \varepsilon} \mathrm{Area}\, f(D_\varepsilon) = \int_{\partial D_\varepsilon} \|\mathscr{D}_f\|^2 \geqq (2\pi\varepsilon)^{-1} \Big( \int_{\partial D_\varepsilon} \|\mathscr{D}_f\| \Big)^2 = (2\pi\varepsilon)^{-1} (\mathrm{length}\, f(\partial D_\varepsilon))^2$$

while the map $f_0$ obviously satisfies

$$\frac{\partial}{\partial \varepsilon} \operatorname{Area} f_0(D_\varepsilon) = (2\pi\varepsilon)^{-1} (\operatorname{length} f_0(\partial D_\varepsilon))^2 .$$

Now, we apply (4) to $V'(\varepsilon) = f(\partial D_\varepsilon) \subset V$ and integrate over $[\varepsilon, 1]$. Q.E.D.

As a corollary we get the ordinary Schwarz lemma,

*1.2.A′. If the Gauss curvature of $(V, \mu)$ satisfies $G\operatorname{Curv}(V, \mu) \leqq -1$, then $\|\mathscr{D}_f\|(0) \leqq 2$ for all conformal maps $f: D^2 \to V$.*

*Proof.* The hyperbolic plane $V_0$ satisfies (4) by a well known isoperimetric inequality (see [B-Z] and references therein). In fact (4) holds true for multiple domains in $V$, which allows non-injective maps $f$. Q.E.D.

Our applications of 1.2.A. need another simple inequality.

*1.2.B.* (See [B-Z].) *Let $G\operatorname{Curv}(V, g) \leqq C$ for $C > 0$, let $S_C^2$ be the sphere of constant curvature $C$ and let $B$ be a round ball in $S_C^2$. Then every simply connected relatively compact domain $V' \subset V$ with* area $V'$ = area $B$ *satisfies*

$$\operatorname{length} \partial V' \geqq \operatorname{length} \partial B . \tag{6}$$

With this we obtain the following generalization of 1.2.A.

**1.2.B′. Corollary.** *Every bijective conformal map $f: B \to V'$ satisfies at the center $b \in B$,*

$$\|\mathscr{D}_f\|(b) \leqq 1 , \tag{6′}$$

*unless $B = S^2$.*

*Remark.* The inequality (6′) holds true for *all* conformal maps $B \to V'$ as the conformal maps $f: B \to V'$ which maximize $\|\mathscr{D}_f\|(b)$ are bijective anyway, since $V'$ is simply connected.

### *1.3. Gradient bounds for pseudo-holomorphic curves*

We want to give an a priori bound for the norm of the differential of a parametrized $J$-curve $f: S \to (V, J)$ with respect to the given Riemannian metrics $\mu_0$ on $S$ and $\mu$ on $V$. We assume here the metric $\mu$ and $\mu_0$ to be *Hermitian* for the implied complex structures in $T(S)$ and $T(V)$. In this case the map $f$ of $S$ on its image is conformal for the induced metric $\mu | f(S)$. The notion of conformality can be made precise at the singular points of $f$ where the differential $\mathscr{D}_f$ vanishes. But we do not need it here, as every map $f$ can be slightly perturbed in an ambient manifold $(V', J') \supset (V, J)$ to a non-singular $J'$-holomorphic map $f': S \to V'$. Thus, in all estimates below we may assume the map $f$ to be an immersion.

*1.3.A. Let $V$ and $S$ be compact and let $J$ be tamed by an exact 2-form $\omega = dl$. Then every $J$-holomorphic map $f: S \to V$ satisfies*

$$\|\mathscr{D}_f\|(s) \leqq \operatorname{const}(\operatorname{dist}(s, \partial S))^{-1} , \tag{7}$$

for all $s \in S$ and for const = const$(V, S)$.

*Proof.* Every compact $J$-curve $S' \subset V$ with boundary satisfies

$$\operatorname{Area} S' \leqq C_1 \int_{S'} \omega = C_1 \int_{\partial S'} l \leqq C_2 \operatorname{length} \partial S'. \tag{8}$$

The induced metric in $S$ also satisfies (6) since the curvature is bounded from above (see 1.1.B.). Now, we assume without loss of generality $S = D^2$ and apply Schwarz' lemma (see 1.2.A., 1.2.B'.) to the composed map $f_s = f \circ \alpha_s : D^2 \to V$ where $\alpha_s$ is a conformal automorphism of $D^2$ sending the center $0 \in D^2$ to $s \in D^2$. This yields

$$\|\mathscr{D}_{f_s}\|(0) \leqq \operatorname{const}(V, \omega), \tag{9}$$

which is equivalent to (7). Q.E.D.

**1.3.A'. Corollary.** *Remove an interior point $s_0 \in S$ and let $f_0 : S \backslash s_0 \to V$ be a $J$-holomorphic map. Then the estimate* (7) *holds true for all $s \neq s_0$. In particular, $f_0$ extends to a Lipschitz map $f : S \to V$.*

*Proof.* Take the Poincaré metric $\mu_0'$ on $S$ which has a cusp at $s_0$, that is $\mu_0' = dr^2 + e^{-r} d\Theta^2$ near $v_0$, for $r \in [r_0, \infty)$ and for the cyclic parameter $\Theta \in [0, 2\pi]$. Then, by 1.3.A. the map $f_0$ is Lipschitz with respect to $\mu_0'$. Therefore length $f_0(\partial \varrho) \to 0$ as $\varrho \to \infty$ for the circles $\partial \varrho = \{r = \varrho\}$ (these are, in the metric $\mu_0$, slightly perturbed small circles around $s_0$). Since $f(S \backslash s_0)$ has bounded mean curvature (see 1.1.A.) and since $V$ is compact this implies the existence of the limit $v_0 = \lim_{s \to s_0} f_0(s)$. Now, we must sharpen the inequality (7) for a surface $S'$ in a small ball in $V$ around $v_0 \in V$, where $S'$ is non-singular and $J$-holomorphic outside $v_0$. We may assume without loss of generality that the symplectic form $\omega$ on $T_{v_0}(V)$ equals the imaginary part of the metric $\mu$, that is $\omega(\tau_1, \tau_2) = \mu(\tau_1, \sqrt{-1}\,\tau_2)$ for all vectors $\tau_1$ and $\tau_2$ in $T_{v_0}(V)$. Recall that by Wirtinger's inequality any pair of *orthonormal* vectors in $T_{v_0}(V) = \mathbb{C}^n$ satisfies

$$\omega(\tau_1, \tau_2) \leqq 1 \tag{10}$$

where the equality holds if and only if $\tau_2 = \sqrt{-1}\,\tau_1$.

**1.3.B. Lemma.** *Let $S'$ lie in the $\varepsilon$-ball $B_\varepsilon \subset V$ around $v_0$ and let $S''$ be another oriented surface in $B_\varepsilon$ with $\partial S'' = \partial S'$. Then*

$$\operatorname{Area} S'' \geqq (1 + C\varepsilon) \operatorname{Area} S' \tag{11}$$

*for some constant $C = C(V)$.*

*Proof.* By the above

$$(1 - C_1 \varepsilon) \operatorname{Area} S' \leqq \int_{S'} \omega = \int_{S''} \omega \leqq (1 + C_1 \varepsilon) \operatorname{Area} S''.$$

**1.3.B'. Corollary.** *The surface $S'$ satisfies*

$$\operatorname{Area} S' \leqq (4\pi)^{-1} (1 + C'\varepsilon) (\operatorname{length} \partial S')^2 \tag{12}$$

Moreover,

$$\operatorname{Area} S' \leqq (4\pi)^{-1} (1 + C'\delta) (\operatorname{length} \partial S')^2 \tag{12'}$$

for $\delta = \min(\varepsilon, \operatorname{length} \partial S')$.

*Proof.* Indeed, (12) obviously holds for minimal surfaces $S''$ with $\partial S''=\partial S'$ and then (11) yields (12) for $S'$. Next, every surface $S'$ whose subdomains satisfy (12) clearly lies in a $\delta'$-neighborhood of $\partial S'$ for $\delta'=\text{const length } \partial S'$ which implies (12′).

Now we show $\|\mathscr{D}_f\|(s)\leqq\text{const}$ for $s\neq s_0$ by assuming $(S,s)=(D^2,0)$ and by applying the proof of 1.2.A. to the continuous map $f$ which is conformal on $D^2\backslash s_0$. Q.E.D.

*Remark.* The above argument (unlike the proof of 1.2.B′.) applies to any Hölder continuous almost complex structure $J$ on $V$.

*1.3.C. Estimates at the boundary.* Let $W$ be a totally real submanifold in $(V,J)$. The above estimates easily extend to holomorphic maps $(S,\partial S)\to(V,W)$ as follows. First we assume the taming form $\omega$ to be equal on $T(V)|W$ to the imaginary part of the metric $\mu$ and we require $\omega$ to vanish on $W$. Then every holomorphic curve $(S',\partial S')\subset(V,W)$ obviously is normal to $W$. Next, we require $W$ to be totally geodesic for $\mu$. Then $\partial S'$ also is a geodesic in $S'$ for the induced metric. Notice that the above choice of $\omega$ and $\mu$ is always possible in a small neighborhood $U\subset V$ of every point $w\in W$. Now, one can estimate the gradient of a holomorphic map $f: S\to S'\subset V$ by using the doubles of $S$ and $S'$ and then by applying the earlier estimates for (closed!) doubled surfaces. Alternatively, one can repeat word for word the argument used for the interior points. Thus one obtains

**1.3.C′. Lemma.** *Fix an* arc *in the boundary of $S$, say $\partial'\subset\partial S$ and let $f: S\backslash s_0\to V$ be a holomorphic map for an interior point $s_0\in\partial'$, such that $f(\partial'\backslash s_0)\subset W$. If $\omega$ is exact then*

$$\|\mathscr{D}_f\|(s)\leqq\text{const}(\text{dist}(s,\partial S\backslash\partial'))^{-1}, \tag{13}$$

*for all $s\in S\backslash s_0$ and for* const $=\text{const}(V,S,\partial')$, *where* (compare 1.3.A.) *$V$ and $S$ are assumed compact and where the boundary $\partial S$ may have corners at the ends of $\partial'$.*

## *1.4. Jets of pseudo-holomorphic maps and sections*

Take an almost complex manifold $(V,J)$ and let $P=PV$ denote the manifold of $J$-complex lines in $T(V)$. Let $\Theta\subset T(P)$ be the $2n$-dimensional subbundle for $2n=\dim V$ whose fiber $\Theta_p$ at a line $p\in P$ equals the inverse image of this line $p\in T(V)$ under the differential of the projection $\pi: P\to V$. The bundle $\Theta$ carries a natural complex linear structure, say $\tilde{J}$ with the following properties

(i) The differential $\mathscr{D}_\pi: T(P)\to T(V)$ is *complex* linear on $\Theta$.

(ii) The induced complex structure on each fiber $\pi^{-1}(v)\approx\mathbb{C}P^{n-1}$, $v\in V$, equals the standard structure on $\mathbb{C}P^{n-1}$.

(iii) The tangential lift $\tilde{C}\subset P$ of every regular $J$-holomorphic curve $C\subset V$ is $\tilde{J}$-holomorphic.

This makes sense as $\tilde{C}$ is everywhere tangent to $\Theta$ by the very definition of $\Theta$.

The existence of $\tilde{J}_0$ is obvious for the standard complex structure $J_0$ on $\mathbb{C}^n$. Next, for a fixed $C_0\subset V$ we identify a small neighborhood $U\subset V$ of some point $v_0\in C_0$ with a small neighborhood in $\mathbb{C}^n$, such that $J_0|T_{v_0}(V)=J|T_{v_0}(V)$ and such that the curve $C_0$ becomes $J_0$-holomorphic as well as $J$-holomorphic. Now, the space of 2-jets of $J$-curves $C \subset V$ tangent to $C_0$ at $v_0$ equals, by the differentiation rules, to that of $J_0$-curves tangent to $C_0$. Hence, the structure $\tilde{J}_0$ also serves as $\tilde{J}$ on $\Theta_p$ for $p = T_{v_0}(C_0)$.

The structure $\tilde{J}$ extends (non-uniquely) to complex linear structure on $T(P)$, say $J'$ on $P$. This $J'$ can be chosen $C^{r-1}$-smooth for a $C^r$-smooth structure $J$.

*1.4.A.* Consider an almost complex manifold $X$ fibered over a surface $S$. A section $S \to X$ is called *holomorphic* if its image is a holomorphic curve in $X$. The space of differentials of holomorphic sections, called $X^{(1)} \to S$, naturally embeds into $PX$ as an open subset.

Now, we apply the above construction to $PX$ and restrict the resulting almost complex structure to $X^{(1)} \subset PX$. For this structure, say $J'$ on $X^{(1)}$, the differential $\mathscr{D}_f^1 : S \to X^{(1)}$ for $\mathscr{D}_f^1(s) = \mathscr{D}_f | T_s(S) : T_s(S) \to T_v(V)$, $v = f(s)$, is a $J'$-holomorphic section for every holomorphic section $S \to X$. Then, by induction, we have a structure on $X^{(r)} = (X^{(r-1)})^{(1)}$ for which the $r^{\text{th}}$-differential $\mathscr{D}_f^r : S \to X^{(r)}$ is holomorphic. An important special case is that of $X = (S, J) \oplus (V, J_0)$, where holomorphic sections correspond to parametrized *$J$-curves in $V$.*

*Thus the $r^{\text{th}}$-order differentials $r = 1, 2, \ldots,$ of holomorphic maps $f$ are holomorphic for some almost complex structure on $X^{(r)}$; therefore, one can estimate all order differentials of $f$ by using the estimates on $\|\mathscr{D}_f\|$ given in* §§ 1.2. *and* 1.3.

*1.4.B. Example.* A continuous extension of a holomorphic map $S \setminus s_0 \to V$ to $S$ (see 1.3.A'.) is $C^\infty$-smooth and hence, holomorphic on $S$.

*Proof.* Consider the fiber $F_0 \approx \mathbb{C}^n$, $2n = \dim V$, of the fibration $X^{(1)} \to X = (S \times V)$ at the point $(s_0, f(s_0)) \in X$. Take a relatively compact open subset $U_0 \subset F_0$ whose small neighborhoods $U'_0 \subset X^{(1)}$ contain the $\mathscr{D}_{f_0}^1$-images of small neighborhoods in $S \setminus s_0$ around $s_0$ (such an $U_0$ does exist as $\|\mathscr{D}_{f_0}\| \leqq \text{const}$ by 1.3.A'.). Then the standard symplectic form $\omega_0$ on $\mathbb{C}^n \approx F_0$ obviously extends to an exact form on $U'_0$ which tames $J'$ on $U'_0$. Now, we apply 1.3.A'. to the ($J'$-holomorphic!) map $\mathscr{D}_{f_0}^1 : S \setminus s_0 \to X^{(1)}$ and thus continuously extend the differential of $f_0$ to $S$. Then, by induction, we get the differentials $\mathscr{D}_f^r$ on $S$ for $r = 2, 3, \ldots$. Q.E.D.

*1.4.C. On the equation $\bar{\partial} f = g$.* Denote by $Y^{(1)} \to S \times V$ the bundle of homomorphisms $T_s(S) \to T_v(V)$. If $S$ and $V$ are given almost complex structures, then we distinguish the subbundle $X^{(1)} \subset Y^{(1)}$ which consists of *complex* linear homomorphisms and we denote $\bar{X}^{(1)} \to S \times V$ the quotient bundle $Y^{(1)}/X^{(1)}$. Now, we assign to each $C^1$-map $f: S \to V$ the section $\bar{\partial} f$ of the bundle $\bar{X}^{(1)}$ over the graph $\Gamma_f \subset S \times V$ by composing the differential of $f$ with the quotient homomorphism $Y^{(1)} \to \bar{X}^{(1)}$. If $g: S \times V \to \bar{X}$ is a smooth section we write $\bar{\partial} f = g$ for the equation $\bar{\partial} f = g | \Gamma_f$.

The following obvious fact brings this equation into the framework of holomorphic maps.

*1.4.C'. There exists a unique almost complex structure $J_g$ on $S \times V$ (which also depends on the given structures in $S$ and in $V$), such that the (germs of) $J_g$-holomorphic sections $f: S \to S \times V$ are exactly and only the solutions of the equation $\bar{\partial} f = g$. Furthermore, the fibers $s \times V \subset S \times V$ are $J_g$-holomorphic (*i.e. *the subbundles $T(s \times V) \subset T(S \times V)$ are $J_g$-complex) and the structure $J_g | s \times V$ equals the original structure on $V = s \times V$.*

*1.4.D. Remarks on curvature equations.* Many geometrically significant second order elliptic systems on surfaces can be expressed by $\bar{\partial}=0$ with auxiliary almost complex structures.

*Examples. 1.4.D$_1$.* Let $S$ be a surface with a metric $g$ of positive curvature and let $X\to S$ be the bundle whose fiber $X_s$, $s\in X$, consists of quadratic forms $h$ on $T_s(S)$, such that

$$\text{Discriminant}_g h=\text{Curv}_s(V,g). \qquad (*)$$

The relation $(*)$ is satisfied by the second fundamental forms $h$ of isometric immersions $(V,g)\to\mathbb{R}^3$. These forms $h$ also satisfy the Codazzi equations which can be equivalently expressed with an almost complex structure $J_g$ on $X$, such that holomorphic sections $S\to X$ are exactly solutions to the Codazzi (as well as Gauss) equations.

*1.4.D$_2$.* Consider a Riemannian manifold $V$ and let $G=Gr_2V$ be the bundle of oriented 2-planes in $T(V)$. Then we define as earlier a subbundle $\Theta\subset T(G)$ of dimension $2m-2$ for $m=\dim V$, such that the tangential lift $\tilde{S}\subset G$ of every surface $S\subset V$ is everywhere tangent to $\Theta$. Take the canonical $(m-2)$-dimensional bundle $B\to G$ and pick a section $\mu: G\to B$.

*1.4.D$'_2$. There is a unique complex linear structure $J_\mu$ in $\Theta$ such that the lift $\tilde{S}\subset G$ is $J_\mu$-holomorphic if and only if the mean curvature vector $M$ of the underlying surface $S\subset V$ satisfies*

$$M_s(S)=\mu(T_s(S))$$

*for all $s\in S$.*

The proof is straightforward.

*1.4.D$_3$.* The cotangent bundle and the twistor bundle of a Riemannian manifold $V$ carry natural almost complex structure. See [Law$_2$, E-S] for a geometric discussion on the corresponding holomorphic curves.

### *1.5. Weak limits of holomorphic curves*

In the course of a deformation the topology of a holomorphic curve $C\subset V$ may jump. For example the quadrics $x^2+y^2=\varepsilon$ degenerate to a pair of lines for $\varepsilon\to 0$. Similarly, graphs $C\subset S^2\times S^2$ of conformal diffeomorphisms $S^2\to S^2$ degenerate to pairs of fibers, $s_1\times S^2\cup S^2\times s_2$. Let us describe how such degeneration happens for general pseudo-holomorphic curves.

*1.5.A. Cusp-curves.* Take a system of disjoint simple closed curves $\gamma_i$ in a closed surface $S$ for $i=1,\dots,k$, and denote by $S^0$ the surface obtained from $S\setminus\bigcup_{i=1}^k\gamma_i$ by the one-point compactification at every end of $S\setminus\bigcup_{i=1}^k\gamma_i$. Denote by $\bar{S}$ the space obtained from $S$ by shrinking every $\gamma_i$ to a single point and observe the obvious map $\alpha: S^0\to\bar{S}$ gluing pairs of points $s'_i$ and $s''_i$ in $S^0$, such that $\bar{s}_i=\alpha(s'_i)=\alpha(s''_i)\in\bar{S}$ are the singular (or cuspidal) points in $\bar{S}$.

An almost complex structure in $\bar{S}$ by definition is that in $S^0$. A continuous map $\beta: \bar{S} \to V$ is called a (parametrized $J$-holomorphic) *cusp-curve* in $V$ if the composed map $\beta \circ \alpha: S^0 \to V$ is holomorphic.

*1.5.A′. Weak convergence.* A sequence of closed $J$-curves $C_j \subset V$ is said to *weakly converge* to a cusp-curve $\bar{C} \subset V$ if the following four conditions are satisfied

(i) all curves $C_j$ are parametrized by a fixed surface $S$ whose almost complex structure depends on $j$, say $C_j = f_j(S)$ for some holomorphic maps

$$f_j: (S, J_j) \to (V, J).$$

(ii) There are disjoint simple closed curves $\gamma_i \in S$, $i = 1, \ldots, k$, such that $\bar{C} = \bar{f}(\bar{S})$ for a map $\bar{f}: \bar{S} \to V$ which is holomorphic for some almost complex structure $\bar{J}$ on $\bar{S}$.

(iii) The structures $J_j$ uniformly $C^\infty$-converge to $\bar{J}$ on compact subsets in $S \setminus \bigcup_{i=1}^{k} \gamma_i$.

(iv) The maps $f_j$ uniformly $C^\infty$-converge to $\bar{f}$ on compact subsets in $S \setminus \bigcup_{i=1}^{k} \gamma_i$. Moreover, $f_j$ uniformly $C^0$-converge on entire $S$ to the composed map $S \to \bar{S} \xrightarrow{\bar{f}} V$. Furthermore,

$$\text{Area}_\mu f_j(S) \to \text{Area}_\mu \bar{f}(\bar{S}) \quad \text{for} \quad \bar{u} \to \infty,$$

where $\mu$ is a Riemannian metric in $V$ and where the area is counted with the geometric multiplicity

**1.5.B. Compactness theorem.** *Let $C_j$ be a sequence of closed $J$-curves of a fixed genus in a compact manifold $(V, J, \mu)$. If the areas of $C_j$ are uniformly bounded,*

$$\text{Area}_\mu C_j \leqq A, \quad j = 1, 2, \ldots,$$

*then some subsequence weakly converges to a cusp-curve $\bar{C}$ in $V$.*

*Proof.* We may assume as earlier the curves $C_j$ to be regular and first we study induced metrics $\mu_j$ in $C_j$. We apply (12′) to small concentric balls $B_\varepsilon \subset (C_j, \mu_j)$ for $0 < \varepsilon \leqq \varepsilon_0$ and conclude by the standard argument to the inequality

$$\text{Area}\, B_\varepsilon \geqq \varepsilon^2, \quad \text{for} \quad \varepsilon \leqq \varepsilon_0, \tag{14}$$

provided the positive number $\varepsilon_0 = \varepsilon_0(V, J, \mu)$ is sufficiently small. Using this we easily find an $\varepsilon$-net $F_j \subset (C_j, \mu_j)$ containing $N$ points for a fixed integer $N = N(V, J, \mu)$, such that every topological annulus $C' \subset C_j - F_j$ satisfies

$$\text{Diam}_\mu C' \leq 10\, \text{length}_\mu \partial C'. \tag{15}$$

Furthermore, let $C'$ be conformally equivalent to the cylinder $S^1 \times [0, l]$ where $S^1$ is the circle of unit length, and let $S_t^1 \subset C'$ be the curve in $C'$ corresponding to the circle $S^1 \times t$ for $t \in [0, l]$. Then obviously

$$\int_a^b (\text{length}\, S_t^1)^2\, dt \leqq \text{Area}\, C' \leqq A, \tag{16}$$

for all $[a,b]\subset[0,l]$. Hence, the annulus $C'_t\subset C'$ between the curves $S^1_t$ and $S^1_{l-t}$ satisfies

$$\operatorname{diam}_\mu C'_t \leqq 20(A/t)^{1/2}, \tag{17}$$

for all $t\in[0,l]$.

Now, let $\mu_j^*$ be a metric of constant curvature $-1$ in $C_j\setminus F_j$ conformally equivalent to $\mu_j$. Then for every $\mu_j^*$-ball $B_\varrho$ in $C_j-F$ of radius $\varrho\leqq 0.1$. there obviously exists an annulus $C'\subset C_j\setminus F_j$, such that $C'_t\supset B_\varrho$ for $t=0.01\,|\log\varrho|$. (In fact, such $C'$ exists for every complete surface whose Gauss curvature $G$ satisfies $|G|\leqq 1$.) This implies with (17) the uniform continuity of the (inclusion) maps $(C_j\setminus F_j,\mu_j^*)\to(V,\mu)$, and hence (see 1.3. and 1.4.) a uniform bound on the $r^{\text{th}}$ order differentials for every $r=1,2,\ldots$.

Next, by the standard (and obvious) properties of hyperbolic surfaces there is a subsequence, which is still denoted by $C_j$, such that

(a) There exist $k$ closed geodesics, say

$$\gamma_i^j\subset(C_j\setminus F_j,\mu_j^*,\qquad i=1,\ldots,k,\ j=1,2,\ldots,$$

whose $\mu_j^*$-length converges to zero as $j\to\infty$, where $k$ is a fixed integer.

(b) There exist $k$ closed curves in a fixed surface, say $\gamma_j$ in $S$, and an almost complex structure $\bar J$ on the corresponding (singular) surface $\bar S$, such that the almost complex structures $J_j$ on $C_j\setminus F_j$ induced from $(V,J)$ $C^\infty$-converge to $\bar J$ outside $\bigcup_{j=1}^k \gamma_j$. Namely, there exist continuous maps $g^j: C_j\to\bar S$ which are homeomorphisms outside the geodesics $\gamma_i^j$, which pinch these geodesics to the corresponding singular points of $\bar S$ (that are the images of $\gamma_i$) and which send $F_j$ to a fixed subset $F$ in the non-singular locus of $\bar S$. Now, the convergence $J_j\to\bar J$ is understood as the uniform $C^\infty$-convergence $g_*^j(J_j)\to\bar J$ on the compact subsets in the non-singular locus of $\bar S$ which is identified with $S\setminus\bigcup_{i=1}^k\gamma_i$.

The limit cusp-curve $\bar C\subset V$, that is a holomorphic map $\bar f:\bar S\to V$, is constructed by first taking the maps

$$\bar f_j=(g^j)^{-1}: S\setminus\bigcup_{i=1}^k\gamma_i\to V.$$

The previous discussion shows the differentials of these maps to be uniformly bounded. Hence, some subsequence, still called $\bar f_j$, converges outside the finite subset $F\subset S$ and the limiting map extends by 1.3.A'. to the required holomorphic map $\bar f$. Finally one slightly perturbes the maps $\bar f_j$ and then extends them in an obvious way to maps $f_j: S\to C_j\to V$ satisfying the properties (ii)–(iv) in 1.5.A.

*1.5.C. Remark.* A similar compactness theorem for minimal surfaces is due to Sacks and Uhlenbeck [S-U].

*1.5.$D_1$. Generalizations.* Theorem 1.5.B. immediately generalizes to maps $f_j: S\to V_j$ for *convergent* sequences of almost complex manifolds. Typically (see 2.3.) one starts with a fibration over a base $X\to B$ with a smooth family of almost complex structures $J_b$ in the (compact) fibers $X_b$ and one applies the compactness theorem to maps $f_j: S\to X_{b_j}$ for $b_j\to b$ as $j\to\infty$.

*1.5.D*$_2$. Take a totally real submanifold $W \subset (V, J)$ and consider compact holomorphic curves $C \subset V$ with boundaries, such that $\partial C \subset W$. Limits of these are *cusp-curves with boundaries*, $(\bar{C}, \partial\bar{C}) \subset (V, W)$, which are, topologically speaking, obtained by shrinking to points some (short) closed loops in $C$ and also some (short) segments in $C$ between boundary points. This is seen by looki g on the double $C \bigcup_{\partial C} C$.

**Corollaries 1.5.E$_1$.** *Take a holomorphic fibration of a compact manifold over a closed surface,* $(X, J) \to (S, J_0)$ *and let* $f_j: S \to X$, $j = 1, 2, \ldots$, *be a sequence of holomorphic sections which diverges in the* $C^\infty$*-topology and such that* Area $f_j(S) \leqq$ Const $< \infty$. *Then there is a non-empty finite subset* $\Gamma \subset S$, *such that some subsequence of* $f_j$ *uniformly* $C^\infty$*-converges on compact subsets in* $S \setminus \Gamma$ *and the limit holomorphicly extends to a section* $f_\infty: S \to X$. *Furthermore, each fiber* $X_\gamma \subset X$, $\gamma \in \Gamma$ *contains a non-constant rational curve* $C_\gamma \subset X_\gamma$ *which meets the image* $f_\infty(S)$ *at* $f_\infty(\gamma)$.

*Proof*. Take the limit cusp-curve $\bar{C} \subset X$ and take the projections of the cuspidal points in $\bar{C}$ for the points $\gamma \in \Gamma$.

*1.5.E*$_2$. Call a homotopy class $\alpha$ of maps $S^2 \to (V, J)$ *J-simple* if every holomorphic cusp-map $f: S^2 \to V$, which pinches some curves $\gamma_i \subset S^2$ to points and which is holomorphic on $S^2 \setminus \bigcup_i \gamma_i$ for an appropriate complex structure on this complement, is non-constant on at most one component of $S^2 \setminus \bigcup_i \gamma_i$. For example, $\alpha$ is simple if it admits no decomposition $\alpha = \sum_{i=1}^{l} \alpha_j$ for $l \geqq 2$, (and for some choice of the base point in $V$) where every $\alpha_j$ can be realized by a non-constant rational $J$-curve in $V$. Take a metric $\mu$ in $V$ fix a constant $A \geqq 0$ and consider the space of $C^\infty$-map $f: S^2 \to V$ in a $J$-simple homotopy class $\alpha$, such that

(i) Area $f(S^2) \leqq A$, where the area is counted with the geometric multiplicity.

(ii) $\operatorname{dist}(f(s_i), f(s_j)) \geqq \delta$ for a fixed triple of points $s_i \in S^2$, $i = 1, 2, 3$, for a fixed $\delta > 0$ and for $i \neq j$.

(iii) $\bar{\partial} f = g$ for a $C^\infty$-section $g$ of the bundle $Y^{(1)}/X^{(1)}$ over $S \times V$ (see 1.4.C.).

Observe that in the presence of a taming form $\omega$, the bound (i) is automatic,

$$\text{Area}\, f(S^2) \leqq A(\|g\|, [f]),$$

where $\|g\|$ is the $C^0$-norm of $g$ and $[f] \in H_2(V, \mathbb{R})$ is the homology class of $f$.

*1.5.E*$'_2$. *There exists an* $\varepsilon > 0$, *such that the* $C^r$*-bound* $\|g\|_r \leqq \varepsilon$ *for some* $r$ *(in fact, one may take* $r = 1$*) implies the compactness of the above space of maps* $f: S^2 \to V$.

*Proof*. The bounds on Area $f(A)$ and on $\|g\|$ give a bound on the areas of the graphs $C_f \subset S \times V$ of the maps $f$. We express the equation $\bar{\partial} f = g$ with the almost complex structures $J_g$ on $S \times V$ indicated in 1.4.C. and then apply 1.5.C. to the $J_g$-curves $C_f \subset S \times V$. If the compactness fails for a given $g$, we get cusp-curves $\bar{C}_f(g) \subset S \times V$, and then with $\varepsilon \to 0$ we get a cusp curve $\bar{C} = \bar{C}(g = 0) \subset (S, J_0) \oplus (V, J)$. Since the homotopy class of the maps $f: S^2 \to V$ is $J$-simple, the curve $\bar{C}$ may have at most two components, say $\bar{C} = \bar{C}_0 \cup \bar{C}_1$, where $\bar{C}_0$ is the graph of a *constant* map

$S^2 \to V$. This map is the uniform limit of our maps $f$ on compact subsets outside a fixed point $\gamma \in S^2$ (see 1.5.E$_1$.). Hence, these $f$ bring arbitrarily close together two out of three points $s_i$, and the proof is concluded by contradiction.

## 2. The index computations for pseudo-holomorphic curves

### *2.1. The reduced operator $\bar{\partial}$*

Fix a $C^\infty$-smooth connection in $T(V)$ compatible with a given complex linear structure $J$ on $T(V)$ and denote by $\Pi(v_1, v_2)$ the corresponding parallel transport $T_{v_1}(V) \to T_{v_2}(V)$ along the shortest geodesic arc between $v_1$ and $v_2$ for a fixed Riemannian metric in $V$. This (complex linear!) map $\Pi(v_1, v_2)$ is well defined and $C^\infty$-smooth in $v_1$ and $v_2$ if the points $v_1$ and $v_2$ in $V$ are sufficiently close. Next we take a $J$-curve $f_0: S \to V$ and we denote by $Y_0 \to S$ the vector bundle $\mathrm{Hom}_{\mathbb{R}}(T(S), f_0^*(T(V))$ and let $X_0 = \mathrm{Hom}_{\mathbb{C}}(T(S), f_0^*(T(V)))$. In other words, $Y_0$ and $X_0$ are the restrictions of the bundles $Y^{(1)}$ and $X^{(1)}$ over $S \times V$ (see 1.4.C.) to the graph of $S$ in $S \times V$. Put $\bar{X}_0 = Y_0 / X_0$ and define the operator $\bar{\partial}_0$ from the space of maps $f: S \to V$ which are $C^0$-close to $f_0$ to the space of sections $S \to \bar{X}_0$ by

$$\bar{\partial}_0 f(s) = \tilde{\Pi}\, \bar{\partial} f, \quad \text{where} \quad \tilde{\Pi} = \tilde{\Pi}(f(s), f_0(s))$$

is the obvious extension of $\Pi$ to the bundle $Y^{(1)}/X^{(1)}$ over $S \times V$. This $\bar{\partial}_0$ is a non-linear (in fact, quasi-linear) $C^\infty$-smooth first order differential operator. Furthermore, $\bar{\partial}_0$ is elliptic. Indeed, the linearization of $\bar{\partial}_0$ at every $f$ has, pointwise, the same principal symbol as the ordinary Cauchy-Riemann operator $\bar{\partial}$. For example, the linearization of $\bar{\partial}_0$ at $f_0$, say $L_0 \bar{\partial}_0$, sends sections of the bundle $f_0^*(T(V))$ to those of $\bar{X}_0$.

*2.1.A. Let $c$ denote the (first) Chern number of $f_0^*(T(V))$, and let $q = \mathrm{genus}\, S$. Then the index of $L_0 \bar{\partial}_0$ is,*

$$\mathrm{Ind}\, L_0 \bar{\partial}_0 = 2(c + n(1-q)) \quad \text{for} \quad 2n = \dim_{\mathbb{R}} V .$$

*Proof.* Apply Atiyah-Singer index theorem. Alternatively, use the classical Riemann-Roch theorem and the homotopy invariance of the index.

*2.1.A$_1$. Examples.* Let $(V, J) = (S, J_0) \oplus (V', J')$ and let $f_0$ be an isomorphism $S \to S \times v' \subset S \times V'$. Then $c = 2 - 2q$ and so Index $= (2n+4)(1-q)$. Observe that the space of holomorphic maps $f: S \to V$ homotopic to $f_0$ has dimension $d = 2n + \sigma - 2$, where $\sigma$ is the dimension of the automorphism group of $S$, that is

$$\sigma = \sigma_q = \begin{cases} 6 & \text{for} \quad q = 0 \\ 2 & \text{for} \quad q = 1 \\ 0 & \text{for} \quad q \geqq 2 . \end{cases}$$

Thus Index $= d$ for $q = 0$. That is the value $g = 0$ is *regular* for the (non-linear) operator $\bar{\partial}_0$ in this case, as holomorphic maps are the solutions to the equation

$\bar{\partial}_0 f=0$. We express this by saying that the *virtual dimension* of the space of curves which is predicted by the index equals the actual dimension.

*2.1.A*$_2$. Take a regular algebraic curve of degree $\delta \geqq 0$ in $\mathbb{C}P^2$. Then

$$c=3\delta, \quad \text{and} \quad q=\tfrac{1}{2}(\delta^2-3\delta+2).$$

Hence,

$$\text{Index}=2\delta(6-\delta).$$

On the other hand, the dimension $d'$ of all curves of degree $\delta$ is $\delta(\delta+3)$, and the dimension $d$ for a *generic* curve $C$ is $d'-d''_q+\sigma_q$ where $d''_q=6q-6+\sigma_q$ is the dimension of the moduli space of curves of the genus $q$, and so the virtual dimension equals the dimension for generic curves of all genera $q$.

*2.1.B. The operator $\bar{\partial}_v$ for regular curves $C_0 \subset V$.* The space of non-parametrized $J$-curves which are $C^1$-close to $C_0$ can be identified with a subspace of sections of the normal bundle $N=T(V)/T(C_0)$. One concocts in an obvious way a non-linear operator $\bar{\partial}_v$, such that the solutions $f$ to $\bar{\partial}_v f=0$ are (non-parametrized) $J$-curves near $C_0$. Then the linearization of $\bar{\partial}_v$ has

$$\text{Index}=2(c_v+(n-1)(1-q))$$

where $c_v=c+2(q-1)$ is the normal Chern number of $C \subset V$. This index for regular curves of degree $\delta$ in $\mathbb{C}P^2$ equals $2(\delta^2+1-q)=\delta(\delta+3)$ which is the actual dimension of the space of curves. The operator $\bar{\partial}_v$ is harder to define for non-regular curves, but the above formula for the index with $c+2q-2$ for $c_v$ still gives the right value for the virtual dimension $d'$ of the non-parametrized curves.

Indeed,

$$\begin{aligned} d' &= \operatorname{Ind} L_0 \bar{\partial}_0+6q-6=2(c+n(1-q))+6q-6 \\ &= 2(c_v-2q+2+n(1-q))+6q-6=2(c_v+(n-1)(1-q)). \end{aligned}$$

(see [Gro$_2$] for further discussion on $\bar{\partial}_v$ for non-regular curves).

*2.1.C. Regular curves in 4-dimensional manifolds.* The linearization $L_v$ of the operator $\bar{\partial}_v$ at a curve $C \subset V$ acts on the bundle $N=T(V)/T(C) \to C$, such that the solutions to the equation $L_v f=0$ are holomorphic sections $f: C \to N$ for some almost complex structure on $N$. The adjoint operator, say $L_v^*$ acts on the dual bundle, $N^*=\operatorname{Hom}_{\mathbb{C}}(N, T(C))$, and again the equation $L_v^* g=0$ distinguishes holomorphic sections for an appropriate almost complex structure.

*2.1.C*$_1$. *If the Chern number $c^*=c(N^*)=2(1-q)-c_v$ is negative, and if* $\dim_{\mathbb{R}} N=2$, *then* $\dim \operatorname{Ker} L_v^*=0$. *Therefore* $\dim \operatorname{Ker} L_v=\operatorname{Ind} L_v$.

*Proof*. Assume the contrary and let $g: C \to N^*$ be a non-zero solution to $L_v^* g=0$. Since $c^*<0$, the index of the intersection of $g(C) \subset N$ with the zero section is negative. But this is well known (see [Be]) to be impossible.

*2.1.C*$_2$. *Remark.* Consider two closed connected $J$-curves, say $C_1$ and $C_2$ in $(V, J)$. Then their intersection is finite, unless $C_1=C_2$. Furthermore, if $\dim V=4$, then the intersection number Int at every intersection point is positive. Moreover, if $C_1$ is non-regular at an intersection point $c \in C_1 \cap C_2$ then $\mathrm{Int}_c$ is at least two. In fact, $\mathrm{Int}_c = 1$ if and only if both curves are regular at $c$ and transversally meet at $c$.

The above facts are obvious for transversal intersection points and any intersection point can be made transversal by a small local perturbation of the curves in question. The existence of "sufficiently many" local holomorphic curves to make such a perturbation possible is immediate in the *real analytic* case by the *Cauchy-Kovalevskaya theorem*. (In fact, for $C^{an}$-curves, one can study the intersection by a direct infinitesimal computation.) If the manifolds and curves in question are $C^\infty$ one gets the above perturbation using rudimentary elliptic theory, which applies to $C^r$-manifolds for $r>1$ (see [N-W]). However, one can easily construct two $J$-curves in a 4-manifold with a *continuous* structure $J$, which meet at a single point with zero intersection number.

*2.1.*$C_2'$. Let $(V, J)$ be an $n$-dimensional $C^\infty$-manifold for $n>4$ and let $V'$ be a complex submanifold in $V$ (i.e. $T(V')$ is a $J$-complex subbundle in $T(V)$) which is closed in $V$ as a subset. Then one sees with the above argument that every $J$-curve $C \subset V$ meets $V'$ at at most finitely many points unless some component of $C$ is contained in $V'$. Furthermore, if $\operatorname{codim} V'=2$, then the intersection number at every point $v \in V' \cap C$ is positive. Moreover, this intersection number is at least 2 unless $C$ is regular at $v$ and meets $V'$ transversally. (It is unclear if any two complex submanifolds in $(V, J)$ of complementary dimensions have positive intersection numbers at isolated intersection points.)

*2.1.D. Holomorphic curves with boundaries.* Let $W$ be a totally real submanifold (see 0.4.) in $(V, J)$. Then the general elliptic theory applies to compact $J$-curves in $V$ whose boundaries lie in $W$. In fact, if the structure $J$ is integrable near $W$ and if $W$ is a $C^{an}$-submanifold, then one can reduce the local study of $C$ to that of the doubled curve $C \bigcup_{\partial C} C$ with the canonical anti-holomorphic involution $I: U \to U$ of a small neighborhood $U \subset V$ of $W$. In fact, any structure $J$ on $V$ can be $\varepsilon$-perturbed near $W$, in order to acquire such an involution and in most geometrically interesting cases one can work with the perturbed structure $J_\varepsilon$. One may return to $J$ by letting $\varepsilon \to 0$ and by applying the estimates in 1.3.

The index of the pertinent operators for curves with boundaries can be computed as earlier with the general index theorem for manifolds with an involution. We only need in this paper the case of discs $f:(D^2, \partial D^2) \to (V, W)$. The induced tangent bundle $f^*(T(V))$ on $D^2$ doubles to a bundle over $S^2 = D^2 \bigcup_{\partial D^2} D^2$, whose first Chern number is denoted by $c'$. Then the linearised operator $\bar\partial_0$ has index $=c'+n$ and the operator $\bar\partial_v$ for *regular* discs has index $=c'+n-3$.

*Example.* Let $V = \mathbb{C}^n$, and let $\mathscr{R}$ denote the manifold of linear totally real subspaces in $\mathbb{C}^n$. This $\mathscr{R}$ is homotopy equivalent to $GL_n\mathbb{C}/GL_n\mathbb{R}$ and $H^1(\mathscr{R}) = \mathbb{Z}$. Moreover, there is a distinguished generator, say $\alpha \in H^1(\mathscr{R})$, such that $\alpha(\beta)=2$ for the 1-cycle $\beta$ in $\mathscr{R}$ coming from the oriented (!) circle $S^1 = U(1) \subset GL_n\mathbb{C}$.

Now, to every map $f:(D^2, \partial D^2) \to (\mathbb{C}^n, W)$ one assigns a map $S^1 = \partial D^2 \to \mathscr{R}$ by $s \mapsto T_{f(s)}(W)$, $s \in S^1$, and then one observes that $c'$ equals the value of $\alpha$ on this circle $S^1 \to \mathscr{R}$. For instance, the embedded disc $D^2 \to \mathbb{C}$ has $c'=2$.

*2.1.D″*. The normal operator $\bar\partial_\nu$ is especially useful for $2n = \dim V = 4$, because the index computation gives the actual dimension of the space of discs in question, provided $c' \geqq 1$. This follows from 2.1.$C_1$. which applies to the sphere $S^2 = D^2 \bigcup_{\partial D^2} D^2$.

## *2.2. Globalization of $\bar\partial$*

Denote by $G$ the space of $C^\infty$-smooth almost complex structures on $S \times V$ for which the projection to $(S, J_0)$ is holomorphic and whose restrictions to the fibers $s \times V \subset S \times V$, $s \in S$, equal a given structure $J$ on $V$. Let $F$ be the space of $C^\infty$-sections $f: S \to S \times V$ and let $H$ be the space of pairs $(f, g) \in F \times G$ for which $f$ is $g$-holomorphic which is equivalent to the equation $\bar\partial f = g$ (see 1.4.C.). Denote by $\Delta: H \to G$ and $\Phi: H \to F$ the projections and observe that $\Phi$ has a natural structure of a locally trivial fibration whose structure group is as follows. Denote by $A$ the group of $C^\infty$-smooth automorphisms of the bundle $(T(V), J)$, that are diffeomorphisms of $V$ lifted to complex linear automorphisms of $T(V)$. Next, let $A(S)$ be the group of $C^\infty$-maps $S \to A$ that is the group of complex linear automorphism of the vertical tangent bundle of the fibration $S \times V \to S$. The group $A(S)$ acts on the spaces $F$, $G$ and $H$ in an obvious way and these actions are transitive. Denote by $A_0 = A_{f_0}(S) \subset A(S)$ the stabilizer of some $f_0 \in F$, consider the principal fibration $P: A(S) \to A(S)/A_0 = F$ with the structure group $A_0$ and observe that the fibration $\Phi$ is naturally associated to $P$.

Now, fix a point $v \in V$ and consider the fibration $P_v: A \to V$ for $P_v(a) = a(v) \in V$. This $P_v$ has a natural class of local sections. Indeed, the action of $A$ on $V$ factors through the action of Diff $V$ on $V$. The fibration Diff $V \to V$ for diff$\mapsto$diff$(v)$ has an obvious class of local sections which then lift to $A$ with the operators $\Pi(v, v')$ of 2.1. Finally, with local sections $V \to A$ we get local sections $F \to A(S)$ which give the desired local splittings of the fibration $\Phi: H \to F$. With these we get a natural infinite dimensional manifold structure on the space $H$.

**2.2.A. Lemma.** *The map $\Delta: H \to G$ is Fredholm for the obvious infinite dimensional manifold structure in $G$.*

*Proof*. Indeed ther kernel and the cokernel of the differential of $\Delta$ at $h_0 = (f_0, g_0)$ are canonically isomorphic to those of the linearization $L_0\bar\partial_0$ of $\bar\partial_0$. Since $L_0\bar\partial_0$ is an elliptic operator it is Fredholm.

*2.2.A′. Remarks.* (1) The only purpose of these $H$, $P$ etc. is to match the current "infinite dimensional manifolds" terminology.

(2) In order to apply to $\Delta$ the standard properties of Fredholm maps we need an appropriate scale of norms (or Banach manifold structures) on $H$ and on $F$. One can use here, as usual, either the Hölder norms $C^{r,\alpha}$ on $G$ and $C^{r+1,\alpha}$ on $F$ for some $\alpha \in [0, 1]$ and $r = 0, 1, 2, \ldots$, or, alternatively, the Sobolev scales $H^r$ and $H^{r+1}$ respectively (see [Nir] for an exposition of the subject).

*2.2.B*. The above discussion reduces the study of solutions $f$ of $\bar\partial f = g$ to that of the pull-back $\Delta^{-1}(g) \subset H$. If $g$ is a *regular* value of $\Delta$, which is the case for generic structures $g$, then $\Delta^{-1}(g)$ is a finite dimensional submanifold in $H$. Furthermore, if the map $\Delta$ is *proper*, then the $\mathbb{Z}_2$-homology class of $[\Delta^{-1}g] \in H_i(H, \mathbb{Z}_2)$ for $i = \operatorname{Index} L\bar{\partial}_0$ is independent of $g$, as the space $G$ is path-connected. (More interesting cobordism invariant of $\Delta^{-1}(g)$ are considered in [Gro$_2$].) The properness of $\Delta$ over $G$, or over some subspace in $G$ amounts to the compactness of families of $g$-holomorphic curves in $(S \times V, g)$ for $g$ running over a compact subset in $G$. Now, we have the following recipe for solving the equation $\bar{\partial} f = g$: *start with a regular value $g_0$ for $\Delta$ such that $[\Delta^{-1} g_0] \neq 0$ and then deform $g_0$ to $g$ such that $\Delta$ is proper over a neighborhood of the path between $g_0$ and $g$.*

We shall be using this recipe in various situations in an informal way without specifying each time the spaces and maps in question, since the only meaningful objects are holomorphic curves which are unsensible to a choice of the infinite dimensional phraseology.

## *2.3. The existence theorems for holomorphic curves*

In what follows $(V, J)$ is a $2n$-dimensional almost complex $C^\infty$-manifold without boundary tamed by some symplectic form $\omega$ on $V$.

*2.3.A. Let $V$ be compact and contain no non-constant rational $J$-curves. (This is obviously so, for instance, if the Hurewitz homomorphism $\pi_2(V) \to H_2(V; \mathbb{R})$ vanishes.) Then the equation $\bar{\partial} f = g$ is solvable with a contractible $C^\infty$-map $f: S^2 \to V$ for all $C^\infty$-sections $g: S^2 \times V \to \bar{X}^{(1)}$* (compare 1.4.C.). *Moreover, there exists a solution $f$, such that $f(s) = v$ for given points $s \in S^2$ and $v \in V$.*

*Proof*. The index of the pertinent operator $\bar{\partial}$ here equals $2n$ and $g = 0$ is a regular value for $\Delta$ (compare 2.2.B.), as the space of solutions $\{f \mid \bar{\partial} f = 0\} = \Delta^{-1}(0) \subset H$ consists of the constant maps $f: S^2 \to V$. The $\mathbb{Z}_2$-homology class $[\Delta^{-1}(0)]$ goes onto the fundamental class $[V] \subset H_{2n}(V, \mathbb{Z}_2)$, for the projection $f \mapsto f(s) \in V$ for a given $s \in S^2$. Next, the map $\Delta$ is proper according to 1.5.E$_1$., since $V$ contains no rational curves, and since the areas of $g$-holomorphic sections $f: S^2 \to (S^2 \times V, g)$ are bounded by const$\|g\|$. This bound is obtained with the symplectic forms $(C\omega_0) \oplus \omega$ on $S^2 \times V$ which tame $g$, where $\omega_0$ is the area form on $S^2$ and $C = C(g)$ is a sufficiently large positive constant. Now, with the recipe in 2.2.B., the class $[\Delta^{-1}(g)]$ goes to $[V]$ for *all generic* $g$ and so the map $f \mapsto f(s) \in V$ sends $\Delta^{-1}(g)$ *onto* $V$ for all (generic or not) $g$. Q.E.D.

*2.3.A'*. The above argument applies to non-compact manifolds $V$ in-so-far as we can control the "size" of the image $f(S)$. For example, let $\mu$ be a *complete* Riemannian metric on $V$ such that the form $\omega$ is *uniformly positive with respect* to $\mu$. That is $\omega(\tau, \sqrt{-1}\,\tau) \geqq \varepsilon > 0$ for all $\mu$-unit vectors $\tau \in T(V)$. Then every contractible map $f: S \to V$ obviously has

$$\operatorname{Area}_\mu f(S) \leqq \text{Const} = \text{const}(\|\bar{\partial} f\|)$$

where the $C^0$-norm $\|\bar{\partial} f\|$ is measured with respect to $\mu$. Now, let us assume the injectivity radius of $\mu$ to be bounded away from zero, $\operatorname{Rad}_v(V, \mu) \geqq \varepsilon > 0$ for all $v \in V$

and let the sectional curvature of $\mu$ to be bounded from above, $K(V,\mu) \leqq \text{const} < \infty$. Then one immediately sees that

$$\operatorname{Diam}_\mu f(S) \leqq \text{const}' \operatorname{Area}_\mu f(S)$$

for const $=$ const$(\|\bar{\partial}\|)$, and so

$$\operatorname{Diam}_\mu f(S) \leqq \text{const}'(\|\bar{\partial} f\|).$$

Now, assuming the existence of the above $\mu$ we conclude:

*If a $C^\infty$-section $g: S^2 \times V \to \bar{X}^{(1)}$ is bounded on $S^2 \times V$ with the norm measured by $\mu$ and a fixed metric in $S^2$, then the equation $\bar{\partial} f = g$ is solvable with a $C^\infty$-map $f: S^2 \to V$ homotopic to zero and such that $f(s) = v$ for given points $s \in S^2$ and $v \in V$, where we assume as earlier the non-existence of rational $J$-curves in $(V, J)$.*

*2.3.B.* Let $V, J, \mu, \omega$, and $g$ be as above and let $W \subset V$ ba an $\omega$-Lagrangian properly embedded (i.e. without boundary and closed in $V$ as a subset) $C^\infty$-submanifold. We are interested in $C^\infty$-maps $f: D^2 \to V$ which send the boundary circle $\partial D^2$ into $W$ and which satisfy the equation $\bar{\partial} f = g$. If the map $f: (D^2, \partial D^2) \to (V, W)$ is contractible, then the index of the corresponding operator $L\bar{\partial}$ equals $n$ and the value $g = 0$ is regular for the map $\Delta$. To apply the previous argument we again need a bound on $\operatorname{Diam} f(D^2)$. This is achieved with the following condition on $W$: the second fundamental form of $W \subset (V, \mu)$ is bounded, $\|II(W)\| \leqq \text{const} < \infty$ and, moreover, there exists positive numbers $\varepsilon > 0$ and $C > 0$, such that every two points $w_1$ and $w_2$ in $W$ for which $\operatorname{dist}_V(w_1, w_2) \leqq \varepsilon$ can be joined by a path in $W$ of length $\leqq C\varepsilon$. This condition on $W$ insures the desired bound on $\operatorname{Diam} f(D^2)$ which gives us the following

**Theorem.** (Compare 2.3.D.) *If every $J$-holomorphic map $(D^2, \partial D^2) \to (V, W)$ is constant (as well as every rational curve in $V$) then there is a contractible $C^\infty$-map $f: (D^2, \partial D^2) \to (V, W)$ which satisfies $\bar{\partial} f = g$ for a given (bounded!) $C^\infty$-section $D^2 \times V \to \bar{X}^{(1)}$ and such that $f(s) = w$ for given points $s \in \partial D^2$ and $w \in W$.*

*Example.* Let $V = \mathbb{C}^n$ and let $W$ be obtained from $\mathbb{R}^n \subset \mathbb{C}^n$ by a symplectic diffeomorphism $\alpha: \left(\mathbb{C}^n, \sum_{i=1}^n dx_i \wedge dy_i\right) \circlearrowleft$, such that the norms of the differentials $D_\alpha$ and $D_\alpha^{-1}$ are bounded on $\mathbb{C}^n$ and the norm of the second differential of $\alpha$ also is bounded. Then the assumptions of the theorem are met by the equation $\frac{\partial f}{\partial \bar{z}} = g$ for an arbitrary $C^\infty$-map $g: D^2 \to \mathbb{C}^n$; hence, this equation is solvable with a $C^\infty$-map $f: D^2 \to \mathbb{C}^n$ which sends the boundary circle $\partial D^2$ into $W$ and such that $f(s) = w$ for given points $s \in \partial D^2$ and $w \in W$.

*2.3.B$_1$.* Let $W$ be a closed Lagrange submanifold in $\mathbb{C}^n$ and let $g_0 \in \mathbb{C}^n$ be a non-zero vector. Since every solution $f$ to the equation $\frac{\partial f}{\partial \bar{z}} = g_0$ is harmonic, it satisfies

$$\|g_0\| \leqq \text{const}_n \operatorname{Diam} f(\partial D^2),$$

and so there is no solution $f: (D^2, \partial D^2) \to (\mathbb{C}^n, W)$ to this equations for large $\|g_0\|$. Hence, by the above theorem there must exist a non-constant holomorphic map

$(D^2, \partial D^2) \to (\mathbb{C}^n, W)$. In fact, this argument applies to closed *quasi-Lagrange* submanifolds $W \subset \mathbb{C}^n$ which by definition are those totally real submanifolds in $\mathbb{C}^n$ which are Lagrange for some symplectic form $\omega$ on $\mathbb{C}^n$ which tames (i.e. is positive on) the complex structure of $\mathbb{C}^n$. Notice, that the quasi-Lagrange submanifolds form an *open* subset in the space of all submanifolds in $\mathbb{C}^n$ and this subset is invariant under holomorphic automorphisms of $\mathbb{C}^n$. Thus we obtain the following

**2.3.B$_1$. Theorem.** *Every closed quasi-Lagrange submanifold $W \subset \mathbb{C}^n$ admits a (non-constant!) holomorphic disk $(D^2, \partial D^2) \to (\mathbb{C}^n, W)$.*

**2.3.B$_2$. Corollary.** *The class $[\omega] \in H^2(\mathbb{C}^n, W; \mathbb{R})$ is non-zero.*

This may be expressed by saying that $\mathbb{C}^n$ contains no closed *exact* Lagrange submanifolds $W$.

*2.3.B$_3$.* Let us generalize the above to closed Lagrange submanifolds $W \subset V = (V', \omega') \oplus (\mathbb{R}^2, dx \wedge dy)$, where the form $\omega'$ either tames an almost complex structure $J'$ on $V'$ which is convex at infinity or there is another structure on $V'$ also called $J'$ which satisfies the conditions in 2.3.B. with some metric $\mu'$ on $V'$. Denote by $J_0$ the standard complex structure on $\mathbb{R}^2 = \mathbb{C}$ and consider $C^\infty$-maps $f: (D^2, \partial D^2) \to (V, W)$, whose projections $f': D^2 \to V'$ are $J'$-holomorphic and the projections $f: D^2 \to \mathbb{C}$ satisfy the equation $\frac{\partial f_0}{\partial \bar{z}} = g_0$ for a constant $g_0 \in \mathbb{C}$. If $|g_0|$ is large, no such map $f$ exists; therefore, we obtain as earlier either a non-constant holomorphic disk $(D^2, \partial D^2) \to (V, W)$ or a non-constant rational curve $S^2 \to V'$. It follows that no $W$ is *weakly exact* in $(V, \omega = \omega' + dx \wedge dy)$ which, by definition means the existence of a $C^\infty$-disk $(D^2, \partial D^2) \to (V, W)$, such that

$$\int_{D^2} \omega \neq 0.$$

*2.3.B$'_3$.* Consider an isotopy of Lagrange submanifolds in $V'$ given by a $C^\infty$-map $F': W' \times [0,1] \to V'$ and let $\tilde{\omega}'$ be the pull-back of the form $\omega'$ to $W' \times [0,1]$. The form $\tilde{\omega}'$ clearly is exact, $\tilde{\omega}' = d\tilde{l}'$, where the 1-form $\tilde{l}'$ on $W' \times [0,1]$ is closed on every submanifold $W' \times t$ for $t \in [0,1]$. Recall that $F'$ is called an *exact isotopy* if the class $[\tilde{l}' | W' \times t] \in H^1(W' = W' \times t; \mathbb{R})$ is constant in $t \in [0,1]$.

**Theorem.** *Let $W'$ be a closed manifold, let $F'$ be an exact isotopy and let the submanifolds $W'_t = F'(W' \times t) \subset (V', \omega')$, $t \in [0,1]$, be weakly exact. Then the intersection $W'_0 \cap W'_1$ is non-empty.*

*Proof.* Fix a positive $\varepsilon < 1$ and take a $C^\infty$-map $\varrho: S^1 \to [0,1]$, where the circle $S^1$ is parametrized by $\Theta \in [-1, +1]$, such that the $\varepsilon$-neighborhood $I_0$ of $0 \in S^1$ goes to $0 \in [0,1]$ and the $\varepsilon$-neighborhood $I_1$ of $\pm 1 \in S^1$ goes to $1 \in [0,1]$. Let $\tilde{l}^*$ be the pull-back of the form $\tilde{l}'$ to $W' \times S^1$ under the map $(w', \Theta) \mapsto (w', \varrho(\Theta))$ and assume without loss of generality the form $\tilde{l}^*$ to vanish on $W' \times (I_0 \cup I_1)$. Then the form $\tilde{l}^* | W' \times \Theta$ is exact for all $\Theta \in S^1$ and so $\tilde{l}^* = d\Psi + \varphi\, d\Theta$ for some $C^\infty$-functions $\Psi$ and $\varphi$ on $W' \times S^1$ where $\varphi$ may be assumed zero on $W' \times (I_0 \cup I_1)$.

Next, consider a map $\alpha$ of the annulus $S^1 \times [\varphi_-, \varphi_+]$ into $\mathbb{R}^2$, where $\varphi_-$ and $\varphi_+$ are the lower and the upper bound of the function $\varphi$ correspondingly, such that

(i) The pull-back under $\alpha$ of the form $dx \wedge dy$ on $\mathbb{R}^2$ equals $-d\varphi \wedge d\Theta$.

(ii) The map $\alpha$ is bijective on $I \times [\varphi_-, \varphi_+]$ where $I \subset S^1$ is some closed subset, such that $I \cup I_0 \cup I_1 = S^1$; furthermore, the origin $0 \in \mathbb{R}^2$ is a unique double point of the map $\alpha$ on $S^1 \times 0$, that is

$$0 = \alpha(0,0) = \alpha(\pm 1, 0),$$

and $\alpha$ is injective on $S^1 = S^1 \times 0$ minus $\{0, \pm 1\}$.

(iii) The curve $S_0^1 = \alpha(S^1 \times 0) \subset \mathbb{R}^2$ "bounds" zero area in $\mathbb{R}^2$, that is $\int_{S_0^1} x\,dy = 0$, for the 1-form $x\,dy$ on $\mathbb{R}^2$.

Now, let $F^*: W' \times S^1 \to V' \times \mathbb{R}^2$ be given by $(w', \Theta) \mapsto (F'(w, \varrho(t)), \alpha(\varphi, \Theta))$. Then

(i)′ The pull-back under $F^*$ of the form $\omega = \omega' + dx \wedge dy$ equals $d\tilde{l}^* - d\varphi \wedge d\Theta = 0$ on $W' \times S^1$.

(ii)′ The set of double points of $F^*$ is $W_0' \cap W_1' \subset V' = V' \times 0 \subset V' \times \mathbb{R}^2$.

(iii)′ If $F^*$ has no double point then the Lagrange submanifold $W = F^*(W' \times S^1) \subset (V' \times \mathbb{R}^2, \omega' + dx \wedge dy)$ is weakly exact if and only if $W_0' \subset V'$ is such.

Finally, according to 2.3.$B_3$., the manifold $W$ cannot be weakly exact; therefore the intersection $W_0 \cap W_1$ is non-empty. Q.E.D.

*2.3.$B_4$. Fixed points of symplectic diffeomorphisms.* Let $\delta: (V, \omega) \to (V, \omega)$ be a symplectic diffeomorphism, that is $\delta^*(\omega) = \omega$ and let $\Gamma_\delta$ be the graph of $\delta$ in the symplectic manifold $(V, \omega) \oplus (V, -\omega)$. Then $\Gamma_\delta$ is Lagrange and the fixed points of $\delta$ correspond to the intersection points of $\Gamma_\delta$ with the diagonal $\Delta \subset V \times V$. Call $\delta$ *exact* if the graph $\Gamma_\delta \subset V \times V$ can be brought to $\Delta$ by an exact isotopy of Lagrange submanifolds in $V \times V$.

Now, we see with 2.3.$B_3$.

*2.3.$B_4'$. Let $(V, \omega)$ be a closed symplectic manifold such that the class $[\omega] \in H^2(V, \mathbb{R})$ vanishes on the image of the Hurewicz homomorphisms $\pi_2(V) \to H_2(V)$. Then every exact symplectic diffeomorphism of $V$ has a fixed point.* (Compare [C-Z, Ch, F-W, F 1, H, L-S, Si]. We shall sharpen our techniques in [$\text{Gro}_2$] in order to remove the condition $[\omega | \pi_2(V)] = 0$ and to give a lower bound on the number of the fixed points.)

*2.3.$B_4''$.* Consider the cotangent bundle $V' = T^*(W_0')$ of a smooth manifold $W_0'$, let $l'$ be the canonical 1-form of $V'$ and let $W' \subset V'$ be a closed Lagrange submanifold for the symplectic form $\omega' = dl'$.

**Theorem.** *If the* (necessarily closed) *form $l' | W'$ is exact, then $W'$ intersects the zero section $W_0' \hookrightarrow V$.*

*Proof*. The manifold $V' = T^*(W_0')$ has a canonical diffeotopy $v' \mapsto tv'$ for all cotangent vector $v' \in T^*(W_0')$ and for $t \in [0, \infty]$. The induced isotopy on $W'$ clearly is Lagrange; it is exact if and only if the form $l' | W'$ is exact. If $W'$ misses the zero section, then the isotoped manifolds $W_t'$ are disjoint from $W'$ for $t \to \infty$ (as well as for $t \to 0$). Next, we take an almost complex structure $J'$ on $V'$ which is tamed by $\omega'$ and such that $J'$ is invariant under the diffeomorphism $v' \mapsto 2v'$ outside some neighborhood $U' \subset V'$ of the zero section $W_0' \subset V'$. Such a $J'$ satisfies the assumptions of 2.3.B$_3$. with some metric $\mu'$ on $V'$ with the similar invariance property outside $U'$ and then 2.3.B$_3$. yields the theorem. (See [Ch] and [H] for a lower bound on $\#(W' \cap W_0')$ for submanifolds $W'$ which are obtained from $W_0'$ by exact Lagrange isotopies.)

*Example*. (Compare [C-Z, Ch, F-W].) Let $(V, \omega)$ be a closed symplectic manifold, let $(V', \omega') = (V, \omega) \oplus (V, -\omega)$ and let $\tilde{V}' \to V'$ be the covering of $V'$ whose fundamental group equals the fundamental group of the diagonal, $\pi_1(\Delta) \subset \pi_1(V')$. Consider a symplectic diffeomorphism $\varphi: V \to V$ which is homotopic to the identity. Then the graph $\Gamma_\varphi \subset V \times V$ lifts to a (closed!) submanifold, say $W'$ in $V'$ which is Lagrange for the lifted form $\tilde{\omega}'$ on $\tilde{V}'$. If the class $[\omega]$ vanishes on the image of $\pi_2(V)$ in $H_2(V)$, then the form $\tilde{\omega}'$ is exact, $\tilde{\omega}' = d\tilde{l}'$, let us assume the (closed!) form $\tilde{l}' | W'$ to be exact. This is so for instance, if $\varphi$ admits an exact symplectic diffeotopy to the identity.

Now, if there is a symplectic diffeomorphism of the cotangent bundle $T^*(\Delta)$ onto $\tilde{V}'$ which brings the zero section of $T^*(\Delta)$ onto the lift $\tilde{\Delta}$ of $\Delta$ to $\tilde{V}'$ (e.g. $V = T^{2n} = \mathbb{R}^{2n}/\mathbb{Z}^{2n}$) then the above theorem insures a fixed point of $\varphi$. More generally, *the proof* of the theorem insures a fixed point of $\varphi$ if there exists a diffeotopy $\delta_t$ of $\tilde{V}'$ for $t \in [0, 1]$, such that

(1) $\delta_1 = \mathrm{Id}$ and for $t \to 0$ the diffeotopy $\delta_t$ $C^\infty$-converges to a map $\tilde{V}' \to \tilde{\Delta}$.

(2) the diffeomorphism $\delta_t$ is *conformally symplectic*, that is $\delta_t(\tilde{\omega}') = \mathrm{const}_t \tilde{\omega}'$ for all $t \in [0, 1]$.

Such a diffeotopy is easier to produce in some cases than the symplectic diffeomorphism $T^*(\tilde{\Delta}) \to \tilde{V}'$.

*2.3.B$_5$*. Let us produce a symplectic structure on $\mathbb{R}^{2n}$ which does not embed into the standard structure. Start with the manifold $V_1 = T^n \times \mathbb{R}^n$ and then "kill" the generators of $\pi_1(T^n) \approx \mathbb{Z}^n$ by attaching $n$ 2-handles $D^2 \times \mathbb{R}^{2n-2}$ to $V_1$. The resulting manifold, say $V_2$, can be easily equipped with a symplectic form $\omega_2$, such that $\omega_2$ vanishes on $T^n$ as well as on the axes $D^2 \times 0$ of the handles. The manifold $V_2$ embeds into $\mathbb{R}^{2n}$ and the form $\omega_2$ can be extended to a symplectic form $\omega$ on $\mathbb{R}^{2n}$ by the $h$-principle for symplectic forms on open manifolds (see [Gro$_1$, Gro$_3$]). Thus we get an *exact* Lagrange torus $T^n \subset (\mathbb{R}^{2n}, \omega)$ (for which the class $[\omega] \in H^2(\mathbb{R}^{2n}, T^n; \mathbb{R})$ vanishes) and so 2.3.B$_2$. allows no embedding $(\mathbb{R}^{2n}, \omega) \to \left(\mathbb{R}^{2n}, \omega_0 = \sum_{i=1}^{n} dx_i \wedge dy_i\right)$.

*2.3.C*. Let the symplectic structure on $V$ split, $(V, \omega) = (\mathbb{C}P^r, \omega_1) \oplus (V_2, \omega_2)$, where $\omega_1$ is the standard structure on the complex projective space $\mathbb{C}P^r$, $r \geq 1$, and where the form $\omega_2$ satisfies the following condition (compare 0.2.A.).

(*) The integral of $\omega_2$ over any smoothly mapped sphere $S^2 \to V_2$ equals $k \int_{\mathbb{C}P^1} \omega_1$ for the projective line $\mathbb{C}P^1 \subset \mathbb{C}P^r$ and for some *integer* $k = k(S^2 \to V_2)$.

Next, take three distinct points $s_i \in S^2$, $i = 1, 2, 3$ and three properly embedded disjoint submanifolds $\Sigma_i$ in $V$, which transversally intersect $\mathbb{C}P^r \times v_2 \subset V$ for a fixed point $v_2 \in V_2$, such that $\Sigma_i \cap \mathbb{C}P^r \times v_2$ consists of a single point for $i = 1$ and $i = 2$, while the submanifold $\Sigma_3 \cap \mathbb{C}P^r \times v_2$ has codimension two in $\mathbb{C}P^r \times v_2 = \mathbb{C}P^r$ and it is not homologous to zero in $H_{2r-2}(\mathbb{C}P^r; \mathbb{Z}_2)$.

Finally, we assume the existence of a metric $\mu$ on $V$, which has (relative to the structures $J$ and $\omega$ on $V$) the properties indicated in 2.3.A′. (If $V_2$ is compact, any metric $\mu$ will do.)

**2.3.C. Theorem.** *There exists a holomorphic map $f: S^2 \to V$ homotopic to the embedding $S^2 \approx \mathbb{C}P^1 \subset \mathbb{C}P^r = \mathbb{C}P^r \times v_2 \subset V$ such that $f(s_i) \in \Sigma_i$ for $i=2$ and 3 while $f(s_1)=\sigma_1$ for a given point $\sigma_1 \in \Sigma_1$.*

*Proof.* Start with a simple algebraic

**2.3.C$_2$. Lemma.** *The space of linear complex structures on $\mathbb{R}^{2n}$ which are tamed by a given 2-form on $\mathbb{R}^{2n}$ is contractible.*

*Proof.* Let $\Omega = GL_{2n}\mathbb{R}/\mathrm{Sp}\,l_{2n}$ be the space of non-singular anti-symmetric bilinear forms on $\mathbb{R}^{2n}$, (where $\mathrm{Sp}\,l_{2n} \subset GL_{2n}\mathbb{R}$ is the automorphism group of the standard form $\sum_{i=1}^{n} x_i \wedge y_i$ on $\mathbb{R}^{2n}$) and let $\mathcal{J} = GL_{2n}\mathbb{R}/GL_n\mathbb{C}$ be the space of complex linear structures on $\mathbb{R}^{2n}$ that are linear automorphisms $J: \mathbb{R}^{2n} \to \mathbb{R}^{2n}$, such that $J^2 = -1$.

Let $\mathcal{T}$ consist of the pairs $(\omega, J) \subset \Omega \times \mathcal{J}$ where $\omega$ tames $J$ and let $\mathcal{T}' \subset \mathcal{T}$ be distinguished by the condition $\omega(Jx, Jy) = \omega(x, y)$ for all $x$ and $y$ in $\mathbb{R}^{2n}$. Clearly $\mathcal{T}' = GL_{2n}\mathbb{R}/U(n)$. The natural projections of $\mathcal{T}$ and $\mathcal{T}'$ to $\mathcal{J}$ are fibrations with convex fibers. Hence, the inclusion $\mathcal{T}' \hookrightarrow \mathcal{T}$ is a homotopy equivalence. Since the inclusion $U(n) \subset \mathrm{Sp}\,l_{2n}$ is a homotopy equivalence, the projection $\mathcal{T}' \to \Omega$ also is such. It follows that the projection $\mathcal{T} \to \Omega$ is a homotopy equivalence, and since it is a fibration, the fibers are contractible.

**2.3.C$'_2$. Corollary.** *The space of almost complex structures on $V$ tamed by a fixed form $\omega$ is contractible. In particular this space is connected.*

*2.3.C$''_2$. Remark.* A form $\omega$ is said to *calibrate* $J$ if $(\omega, J) \subset \mathcal{T}'$ and this definition obviously extends to forms and almost complex structures on manifolds. For example, the 2-form of a Kähler manifold does calibrate the underlying complex structure. In fact, choosing an almost complex structure $J$ which calibrates $\omega$ on $V$ is equivalent to a reduction of the structure group $\mathrm{Sp}\,l_{2n}$ of $(T(V), \omega)$ to $U(n)$. Thus we get with these $\omega$ and $J$ a Riemannian metric on $V$, namely $\mu(\tau, \tau') = \omega(\tau, \sqrt{-1}\,\tau')$, which has the following remarkable (though obvious) property:

*Every $J$-curve $C \subset V$ is absolutely $\mu$-area minimizing in its homology class.*

See [H-L] for amazing examples of calibrated geometries beyond dimension two.

Now, we prove the theorem as follows. We start with a split structure $J_0$ on $V = \mathbb{C}P^r \times V_2$ and we study the equation $\bar{\partial} f = g$ which for $g=0$ gives us $J_0$-holomorphic maps $f: S^2 \to V$.

The index of the pertinent operator here is $2(n+r+1)$. Furthermore, the value $g=0$ is regular for $\Delta$ (compare 2.2.B.) but the manifold $\Delta^{-1}(0)$, which consists of holomorphic maps of $S^2$ to the fibers $S^2 \times v_2$, $v_2 \in V_2$, is by no means compact as it properly contains the conformal automorphism group of each sphere $S^2 \times v_2$. Yet, we regain the compactness by restricting to the subspace $F' \subset F$ of those maps $f: S^2 \to V$ which send $s_i \to \Sigma_i$ for $i=1,2,3$, then by taking $H' = \Phi^{-1}(F') \subset H$ (compare 2.2.A′.) and by letting $\Delta' = \Delta | H': H' \to F'$. We assume, by slightly perturbing the submanifolds $\Sigma_i$ if necessary, the value $g=0$ to be regular for $\Delta'$. Then $(\Delta')^{-1}(0)\subset H'$ is a $2(n-r)$ dimensional submanifold which is compact if and only if $V_2$ (and hence each of $\Sigma_i$) is compact. In the compact case the homology class $[(\Delta')^{-1}(0)]\in H_{2(n-r)}(H';\mathbb{Z}_2)$ goes to $[\Sigma_1]\in H_{2(n-r)}(\Sigma_1;\mathbb{Z}_2)$ under the map $f\mapsto f(s_1)$. Now, the integrality condition on $\omega_2$ makes the homotopy class of the maps $S^2\to V$ in question *simple* (see 1.5.E$_2$.) for any almost complex structure on $V$. Therefore, for each small $g$, the space of solutions to the equation $\bar\partial f=g$ is compact by 1.5.E$'_2$. Hence, the class $[(\Delta')^{-1}(g)]$ goes onto $[\Sigma_1]$ for all small regular $g$ and this property is preserved under any homotopy which brings $J_0$ to $J$ via almost complex structures $J_t$ tamed by $\omega$ with $J_1=J$, where the existence of $J_t$ is insured by 2.3.C$'_2$. Thus the map $(\Delta')^{-1}(g)\to\Sigma_1$ is surjective for small generic $g$ and $J_t=J$ and hence, it is surjective for $g=0$ as well. This concludes the proof for compact manifolds $V$ and the non-compact case follows with the discussion in 2.3.A$'$.

*2.3.C$_3$. Remarks and corollaries.* (a) The Theorems 0.2.A$'$. and 0.2.B. immediately follow from 2.3.C$_1$., except for finer statements for $n=4$; these are discussed in 2.4.

(b) The projective lines in $\mathbb{C}P^r$ could be replaced in the above proof by any family of curves in a given (almost) complex manifold $V_1$, in-so-far as the index consideration gives an adequate description of the actual properties of the family in question. We shall study further examples (and counter examples) of such families in [Gro$_2$].

*2.3.D. The proof of 0.4.A.* The isotopy of submanifolds $F(W_0\times S^1)\subset V$ for $\varepsilon\in\,]0,1]$ can be followed by a homotopy of almost complex structures, $J_\varepsilon$ on $V$ for $J_1=J$ such that

(1) $F(W_0\times S^1_\varepsilon)$ is totally real for $J_\varepsilon$ for all $\varepsilon\in\,]0,1]$;

(2) there is a compact subset in $V$ outside which $J_\varepsilon=J$ for all $\varepsilon\in\,]0,1]$;

(3) the structure $J_\varepsilon$ is split on $U$ for $\varepsilon<\varepsilon_0$ (compare ($v$) in 0.4.) that is $(U,J_\varepsilon)=(U'\times J'_\varepsilon)\oplus|||\mathbb{R}^2,J''_\varepsilon)$.

Next, we take three disjoint submanifolds $\Sigma_i\subset W_\varepsilon=W_0\times S^1_\varepsilon$ which have non-zero intersection indices mod 2 with $S^1=w_0\times S^1\subset V$ and we look at the space of maps $f:(D^2,\partial D^2)\to(V,W_\varepsilon)$ such that $f(s_i)\in F(\Sigma_i)$, $i=1,2,3$ for some fixed distinct points $s_i\in\partial D^2$. We look at the equation $\bar\partial f=g_\varepsilon$ with small $g_\varepsilon$ which is zero for $\varepsilon\leqq\varepsilon_0$. The convexity at infinity of $V$ keeps the solutions within a fixed compact subset in $V$ and so the homology class of the space of solution, say $[\Delta^{-1}g_\varepsilon]\in H_{n-1}(H;\mathbb{Z}_2)$ is invariant under the homotopy. As this class is non-zero for $\varepsilon<\varepsilon_0$ it is non-zero for $\varepsilon=1$ as well and so, as earlier, we obtain a holomorphic map $f:(D^2,\partial D^2)\to(V,W)$ in the required homotopy class, such that $f(s_i)\in\Sigma_i$ for $i=1,2$ and $f(s_3)=w$ for a given point $w\in W$.

*2.3.D$'$. Remark.* The above proof equally applies to non-compact properly embedded Lagrange submanifolds $W$ and to proper maps $F:W_0\times D^2\to V$, provided there exist metrics $\mu_\varepsilon$ on $V$ as in 2.3.A$'$. such that the submanifolds $F(W_\varepsilon)\subset V$, $\varepsilon\in\,]0,1]$, satisfy the conditions of 2.3.B. for the forms $\omega_\varepsilon$. In fact, the existence of $\mu_\varepsilon$ makes the convexity at infinity unnecessary.

*2.3.E. J-curves in open manifolds.* Consider an open subset $V'$ in a symplectic manifold $(V,\omega)$, let $A_i\subset V'$ $i=1,\dots,k$ be compact subsets and let $J'$ be an almost complex structure on $V'$ which is tamed by the form $\omega|V'$. Then one easily obtains with 2.3.$C_2$. a sequence of almost complex structures on $V$ tamed by $\omega$, say $J_j$, $j=1,2,\ldots$, such that for any compact subset $A\subset V'$ the structures $J_j$ for $j>j_0=j_0(A)$ satisfy $J_j|A=J'|A$.

*2.3.$E_1$. Let $f_j: S_j\to V$, $j=1,2,\ldots$ be proper $J_j$-curves, such that the image $f_j(S_j)$ intersects $A_i$ for all i and j and such that $\int_{S_j} f_j^*(\omega)\leqq C_0$ for some constant $C_0<\infty$. Then there exists a proper $J'$-curve $f: S\to V'$ with the same properties: $f(S)\cap A_i\neq 0$, $i=1,\ldots,k$ and $\int_S f^*(\omega)\leqq C_0$.* (Here a curve is called proper if it is a proper non-constant map on every component of the underlying Riemann surface $S$.)

The proof is immediate with the compactness theorems in 1.5.

**2.3.$E_1'$. Corollary.** *Let $V$ be a compact split manifold as in* 2.3.C. *Namely $(V,\omega)=(\mathbb{C}P^r,\omega_1)\oplus(V_2,\omega_2)$ where $\omega_2$ satisfies the integrality condition* (*) *in* 2.3.C. *Let $A_1$ consist of a single point, let $A_2$ support a $(2n-2r)$-dimensional cycle with non-zero* mod 2 *intersection with $\mathbb{C}P^r=\mathbb{C}P^r\times v_2\subset V$ and let $A_3$ contain a $(2n-2)$-cycle whose intersection with $\mathbb{C}P^r$ is non-zero in $H_{2r-2}(\mathbb{C}P^r;\mathbb{Z}_2)$. Then there is a proper $J'$-curve $f: S\to V'$ which meets $A_i$ for $i=1,2,3$, and such that $\int_S f^*(\omega)\leqq \int_{\mathbb{C}P^1}\omega_1$.*

*2.3.$E_1''$. Remark.* The existence of the curves $C'\subset V'$ claimed in 0.3. is obvious with 2.3.$E_1'$.

*2.3.$E_2$.* One can produce using 2.3.$E_1'$. symplectic manifolds $(V',\omega')$ which do not symplecticly embed into the above $(V,\omega)$. For this one must exhibit a particular almost complex structure $J'$ on $V'$ tamed by $\omega'$, for which the integral $\int_C \omega'$ can be estimated from below for all $J'$-curves $C$ in $V'$ which intersect given subsets $A_i$ in $V'$. There are the following methods to obtain such estimates.

*2.3.$E_2'$.* Let the form $\omega'$ *calibrate* (see 2.3.$C_2''$.) a structure $J'$. Then all $J'$-curves $C$ in $V'$ are minimal surfaces for the metric $\omega'(\tau,\sqrt{-1}\,\tau)$ and $\text{area}(C)=\int_C\omega'$. Now, there are various lower bounds on the areas of minimal surfaces in particular Riemannian manifolds (see [$\text{Law}_1$]). For example, every minimal surface $C\subset\mathbb{R}^{2n}$ which passes through the origin and whose boundary lies in the sphere of radius $R$ has $\text{area}(C)\geqq\pi R^2$. (This has been used in 0.3. to rule out symplectic embeddings of the $R$-ball $B(R)\subset\mathbb{R}^{2n}$ into the product $D(\varepsilon)\times\mathbb{R}^{2n-2}$ for the $\varepsilon$-disk $D(\varepsilon)\subset\mathbb{R}^2$ with $\varepsilon<R$.)

*2.3.$E_2''$.* Let $V'=V\setminus V_0$, where $V$ is a complex analytic manifold and $V_0$ is a complex submanifold in $V$. Then the closure of every holomorphic curve $C'\subset V'$ of finite area in $V$ is a one-dimensional complex analytic subvariety, say $C\subset V$ by *Bishop-Stoll Theorem* (see [Bi, St]). Thus, in case $V$ is Kähler with the form $\omega$, we can bound the integral $\int_{C'}\omega$ from below by $\int_C\omega$.

*Example.* Let $V$ split into the product of closed Riemann surfaces,

$$(V,J,\omega)=(S_1,J_1,\omega_1)\oplus\ldots(S_n,J_n,\omega_n).$$

Then, obviously

$$\int_C \omega \geqq \min_i \int_{S_i} \omega_i \quad \text{for} \quad i=1, \ldots, n.$$

Hence, the same bound holds for proper curves $C'$ in any Zariski open subset $V' \subset V$. It follows that there is no symplectic embedding of $V'$ into the product $D(\varepsilon) \times \mathbb{R}^{2n-2}$ for $\varepsilon < \min_{i=1,\ldots,n} \int_{S^i} \omega_i$ (compare 0.3.).

*2.4. J-curves in 4-manifolds*

The $J$-curves insured by the existence theorems in 2.3. are often unique and regular for $2n = \dim V = 4$.

*2.4.A. Examples.* Let an almost complex structure $J$ on $\mathbb{C}P^2$ be tamed by the standard symplectic form $\omega$ on $\mathbb{C}P^2$. Then, according to 2.3.$C_1$., there is a rational $J$-curve $C \subset \mathbb{C}P^2$ through two given points $v$ and $v'$ in $\mathbb{C}P^2$ which is homologous to the projective line $\mathbb{C}P^1 \subset \mathbb{C}P^2$. Since the algebraic intersection number between two such curves equals one, any two of them, say $C$ and $C'$ in $\mathbb{C}P^2$, necessarily meet at a single point, say at $v \in C \cap C'$, unless $C = C'$ (see 2.1.$C_2$.). Furthermore, the curves $C$ and $C'$ are regular at $v$ and meet transversally. Hence, $C$ is regular at all points $v \in C$ and it is uniquely determined by $v$ and $v' \neq v$. Moreover, the curve $C = C(v, v')$ smoothly depends on $(v, v')$.

*2.4.A'.* Take a variety $\mathscr{C}$ of $J$-curves $C$ in some 4-manifold and let $dC$ be some measure on $\mathscr{C}$. Then one defines a (possibly singular) 2-form (or current) $\omega'$ on $V$, whose integral $\int_S \omega'$ is *defined* for all surfaces $S \subset V$ by $\int_S \omega' = \int_{\mathscr{C}} \operatorname{Int}(S, C)\, dC$, where Int stands for the intersection number. This $\omega'$ clearly is $J$-non-negative; if the curves $C \in \mathscr{C}$ are closed then $\omega'$ is closed. Now, if the family $\mathscr{C}$ is "sufficiently large", like the one formed by the curves $C = C(v, v') \subset \mathbb{C}P^2$, and if the measure $dC$ is chosen smooth and positive, then $\omega'$ is an actual smooth $J$-positive form on $V$. Therefore the presence of a taming symplectic structure on $(V, J)$ is necessary for the existence of "many" closed $J$-curves in $V$ for $\dim V = 4$. (In fact, the form $\omega'$ built with the above curves $C(v, v')$ necessarily is isomorphic to the standard form on $\mathbb{C}P^2$, (compare 2.4.$A_1'$. below).

*2.4.$A_1$.* Let $(V, \omega) = (S^2, \omega_0) \oplus (S^2, \omega_0)$ for some area form $\omega_0$ on $S^2$. Then, as earlier, every structure $J$ on $V$ tamed by $\omega$ defines two transversal fibrations of $V$ into rational $J$-curves and these fibrations continuously (even smoothly) depend on $J$. Take one of the two curves through a fixed point $v_0 = (s_0, s_0) \in V$, $s_0 \in S^2$, say $C_1$ in $V$ and choose a $J$-holomorphic diffeomorphism $a_1 = a_1(J|C_1, \omega|C_1)$ of $C_1$ onto $S^2$, such that $a_1(v_0) = s_0$. We construct such an $a_1$ by some (there are many of these) procedure which is canonical modulo rotations of $S^2$ around $s_0 \in S^2$, and we insist on the map $a_1$ to be an isometry (i.e. $a_1^*(\omega_0) = \omega|C_1$) in case the curve $(C_1, J|C_1, \omega|C_1)$ is isometric to the standard sphere $(S^2, J_0, \omega_0)$. A similar map $a_2$ is constructed for the second curve $C_2 \in V$ through $v_0$. Then the pair $(a_1, a_2)$ defines with the above fibrations of $V$ by spheres a continuous (as our "canonical", by definition, means continuous in parameters) map $\alpha$ of the space $\bar{\mathcal{J}}$ of almost complex structures $J$ on $V$ tamed by $\omega$ into the space $\bar{D}=\mathrm{Diff}_0 V/SO(2)\times SO(2)$, where $\mathrm{Diff}_0$ consists of those diffeomorphisms $d$ of $V$ for which $d(v_0)=v_0$ and which fix $H_2(V)$. Observe that for every representative $d\in \mathrm{Diff}_0 V$ of $\alpha(J)\in\bar{D}$ the induced form $d^*(\omega)$ is $J$-positive and so the form $t\omega+(1-t)\,d^*(\omega)$ is non-singular for all $t\in[0,1]$. Then, by Moser's theorem (see [Mo]) there is a *symplectic* diffeomorphism of $V, \omega)$, say $\bar{d}\in\mathrm{Diff}_0$, which continuously depends on $d^*(\omega)$, and such that $\bar{d}=d$ if $d$ is symplectic to start with. Thus we get a map of $\bar{\mathcal{J}}$ to $\bar{\mathcal{S}}_0=\mathcal{S}_0/SO(2)\times SO(2)$ say $\beta:\bar{\mathcal{J}}\to\bar{\mathcal{S}}_0$ for the group $\mathcal{S}_0\subset\mathrm{Diff}_0 V$ of symplectic diffeomorphisms. Now, if $J=J_0\oplus J_0$ for the canonical complex structure on $S^2$ then, by the above construction the induced structure $\sigma^*(J)$ on $V$ satisfies for all $\sigma\in\mathcal{S}_0$,

$$\beta(\sigma^*(J))=\sigma \bmod SO(2)\times SO(2).$$

Since the space $\bar{\mathcal{J}}$ is contractible, (see 2.3.C$'_2$.), we now conclude that $\bar{\mathcal{S}}_0$ is contractible. Hence, (compare 0.3.C.) *the group of symplectic diffeomorphisms contracts to the subgroup of isometries.*

*2.4.A$'_1$.* Let us give a criterion for a closed manifold $(V,\omega)$ to split as $(S^2,\omega_0)\oplus(S^2,\omega_0)$. We assume the existence of two embedded spheres $S_1$ and $S_2$ with trivial normal bundles in $V$, such that the form $\omega$ does not vanish on $S_1$ and on $S_2$ and such that $S_1$ and $S_2$ have a single intersection point where they meet transversally with the intersection number $+1$ for the orientations induced by $\omega|S_1$ and $\omega|S_2$.

*If every smoothly mapped sphere $S\to V$ has $\int_S\omega=k\int_{S_1}\omega=k\int_{S_2}\omega$ for some integer $k=k(S\to V)$, then the manifold $(V,\omega)$ splits, $(V,\omega)=(S^2,\omega_0)\oplus(S^2,\omega_0)$.*

*Proof.* Take an $\omega$-tame almost complex structure $J$ on $V$ for which $S_1$ and $S_2$ are $J$-holomorphic. Then $S_1$ as well as $S_2$ includes into a family of rational $J$-curves which give a topological splitting $V=S^2\times S^2$ with $J$-holomorphic fibers $S^2\times s$ and $s\times S^2$ for all $s\in S^2$. Take any area form $\omega_0$ for which $\int_{S^2}\omega_0=\int_{S_1}\omega=\int_{S_2}\omega$ and observe the form $\omega_0\oplus\omega_0$ to be $J$-positive. This insures a linear symplectic homotopy between $\omega$ and $\omega_0\oplus\omega_0$ on $V$ and then Moser's Theorem yields a symplectic diffeomorphism $(V,\omega)\leftrightarrow(V,\omega_0\oplus\omega_0)$. Q.E.D.

*2.4.A$''_2$. Remark.* Suppose the form $\omega$ has been already split near $S_1\cup S_2\subset V$. Then the above argument allows an extension of this splitting to $V$. Since the complement $V\backslash(S_1\cup S_2)$ is symplectomorphic to the product of two disks in $\mathbb{R}^4$, this shows the standard structure on $\mathbb{R}^4$ to be uniquely determined by its behaviour at infinity (compare 0.3.C.).

*2.4.B.* Let $f:S\to(V,J)$ be a non-constant $J$-curve. Then at every point $s_0\in S$ there is an integer $m=0,1,\ldots$ such that the $i$-th differential $\mathcal{D}^i_f$ vanishes at $s_0$ for $i=1,2,\ldots,m$ while $\mathcal{D}^{m+1}_f(s_0)\neq 0$. This is obvious in the real analytic case (which suffices for most geometric applications) and the $C^\infty$-case is not much harder (see [Gro$_2$]).

**2.4.$B_1$. Lemma.** *Let the map $f: S \to V$ be injective and let a small holomorphic perturbation $f': S \to V$ have $D^1_{f'}(s) \neq 0$ for all s close to $s_0$. Then, for* $\dim V = 4$, *the map $f'$ has at least $m = m(s_0)$ double points near $s_0$.*

This is seen by looking at the local branched covering of order $m+1$ of $V$ around a generic regular $J$-curve in $V$ through the point $v_0 = f(s_0) \in V$ (see [$\mathrm{Gro}_2$] for details).

**2.4.$B'_1$. Corollary.** *Let a sequence of complex structures $J'_j$ on a closed connected surface $S$ $C^\infty$-converge to a structure $J'$ and let the structures $J_j$ on $V$ $C^\infty$-converge to $J$. Let $f_j: (S, J'_j) \to (V, J_j)$ be regular curves which converge to a curve $f: (S, J') \to (V, J)$ for $j \to \infty$. If* $\dim V = 4$, *then either the curve $f$ is regular or all points $f(s) \in V$ are multiple points of $f$.*

This corollary implies the regularity of rational curves in the manifolds $S^2 \times S_2$ (see 0.2.A.) where the existence of these curves is insured by 2.3.$C_1$. Moreover, this corollary shows in many cases the space of *regular* $J$-curves in 4-manifolds to be compact, which allows us to work with the normal operator $\partial_v$ (see 2.1.C.).

*2.4.$B''_1$. Example.* Let $V \approx \mathbb{C}P^2$ and let the structure $J$ be tamed by the standard symplectic form $\omega$ on $CP^2$. Then the virtual dimension (predicted by the index of the linearized $\partial_v$) of regular rational curves $C$ of degree 2 (i.e. homologous to $2\mathbb{C}P^1 \subset \mathbb{C}P^2$) passing through 5 given points in $\mathbb{C}P^2$ equals zero. Moreover, the value zero is regular for $\bar{\partial}_v$ (see 2.1.B.) which makes such a curve stable under deformations of $J$ as long as it does not degenerate to a cusp curve (compare 2.2.B.). One can show that the only possible cusp-curves here are pairs of curves of degree one, which insures a unique regular curve $C$ provided that not three points lie on a degree one curve. (See [$\mathrm{Gro}_2$] for details.) A similar argument applies to regular curves of degree $d$ through $d(d+3)/2$ *generic* points in $V$. Here the genericity condition, which insures the existence of a curve, is essentially the same as earlier: the points should not lie on a cusp-curve of degree $d$. However, one needs (even for the standard $\mathbb{C}P^2$) an additional genericity to rule out non-regular values of the pertinent operator (see [$\mathrm{Gro}_2$]).

*2.4.$B_2$.* Consider a regular rational curve $C_0$ in a 4-dimensional manifold $(V, J)$ and let the normal Chern number be zero. If $V$ is compact and if the class $[C_0]$ is $J$-simple (see 1.5.$E_2$.) then one obtains with 1.5.$E'_2$. and 2.4.$B'_1$. a unique maximal neighborhood $U \subset V$ of $C_0$ which is foliated (or rather fibrated) by holomorphic "translates" of $C_0$, such that the curve $C_u$ through $u \in U$ has $\mathrm{Area}\, C_u \to \infty$ for $u \to \partial U \subset V$. In particular, if the structure $J$ is tamed by some $\omega$, then $V$ is fibrated into rational curves through all points $v \in V$.

*2.4.$B'_2$. Example.* Let $(V, \omega)$ be a closed symplectic 4-manifolds and let $S^2 \subset V$ be an embedded sphere on which the form $\omega$ does not vanish. Then there exists an almost complex structure $J$ on $V$ tamed by $\omega$ for which this sphere is $J$-holomorphic. Now, the $J$-simplicity of $[S^2]$ can be insured by some topological condition. For instance, let the class $[S^2] \in H_2(V; \mathbb{R})$ generate the image of the Hurewicz homomorphism. Then $V$ is foliated by symplectic "translates" of $S^2$. (See [$\mathrm{Gro}_2$] for further topological obstructions on 4-dimensional symplectic manifolds which contain symplectic surfaces $S$ with sufficiently positive (depending on genus $S$) normal bundles.)

*2.4.B$_3$*. The above discussion extends up to certain degree to closed symplectic manifolds $V$ of dimension $2n \geqq 6$. For example, let $(V, \omega)$ admit a symplectic embedding of the space $(\mathbb{C}P^{n-1}, \omega_0)$ for $2n = \dim V$, such that the normal bundle of $\mathbb{C}P^{n-1} \subset V$ is trivial. Consider a $\omega$-tame almost complex structure $J$ on $V$ whose restriction to $\mathbb{C}P^{n-1}$ equals the standard structure on $\mathbb{C}P^{n-1}$. Then, if the class of $[\mathbb{C}P^1]$ is $J$-simple, the family of the projective lines in $\mathbb{C}P^{n-1}$ extends to all of $V$, thus imposing certain restriction on the topology of $V$. For instance, every map of such a $V$ into an aspherical manifold sends the fundamental class $[V] \in H_{2n}(V)$ to zero. (See [Gro$_2$] for the proofs and for further examples.)

*2.4.B$'_3$*. The above extension of $J$-curves from $V_0 \subset V$ to $V$ for codim $V_0 = 2$ can be played in reverse order. For example, take a closed symplectic submanifold $V$ of dimension $2n-2$ in the standard space $(\mathbb{C}P^n, \omega_0)$ such that $V$ is homologous to $\mathbb{C}P^{n-1} \subset \mathbb{C}P^n$. The space $\mathbb{C}P^n$ has "many" $J$-curves for every almost complex structure $J$ tamed by $\omega_0$. In particular this is true for the structures $J$ which make the submanifold $V$ $J$-complex. Every rational $J$-curve in $\mathbb{C}P^n$ homologous to $[\mathbb{C}P^1]$ meets such a $V$ at no more than one point as it follows from 2.1.C$'_2$. Thus, we obtain with 2.3.C$_1$. a rational curve in $V$ through every two points in $V$. If $\dim V = 4$ one then easily shows with these curves (compare 2.4.A$'_1$.) that $V$ is diffeomorphic to $\mathbb{C}P^2$ and that the induced (symplectic!) form $\omega_0 | V$ is isomorphic to the standard one. (See [Gro$_2$] for the proofs and further examples.)

*2.4.C*. Let $(V, \omega) = (S^2, \omega_1) \oplus (S^2, \omega_2)$. If $\int_{S^2} \omega_1 = \int_{S^2} \omega_2$ then $J$-curves in $V$ for every $\omega$-tame structure $J$ display essentially the same properties as $J_0$-curves for a split structure $J_0 = J_1 \oplus J_2$, as seen by an argument similar to that in 2.4. But if $\int_{S^2} \omega_1 \neq \int_{S^2} \omega_2$, then one needs an additional genericity assumption on $J$ to insure a proper behaviour of $J$-curves. If, for instance $\int_{S^2} \omega_1 > k \int_{S^2} \omega_2$, then the homology class $\alpha$ of $[S^2 \times s_0] - k[s_0 \times S^2]$ can be realized by an embedded sphere $S \subset V$ which is the graph of a map $S^2 \to S^2$ of degree $-k$, such that the form $\omega | S$ does not vanish. This sphere is $J$-holomorphic for certain $\omega$-tame structures $J$, which form by the discussion in 2.1. a $2k$-codimensional subvariety in the space of all structures. If such a sphere $S$ is $J$-holomorphic, then the class $p[S^2 \times s_0] + q[s_0 \times S^2]$ contains no regular $J$-curves unless $q - pk \geqq 0$ as it follows from 2.1.C$_2$. In particular, the sphere $S^2 \times s_0$ in $V$ admits no $J$-holomorphic realization. It is not hard to show however (see [Gro$_2$] that this sphere can be realized by a $J$-holomorphic curve for a *generic* $\omega$-tame structure. Moreover, the same remains true for an arbitrary (possibly non-split) symplectic form $\omega'$ on $V$ which can be joined with $\omega$ by a homotopy of symplectic forms $\omega_t$ on $V$.

*2.4.C$_1$*. The "generic" existence theorem generalizes to $2n$-manifolds $V$ for all $n \geqq 2$. This gives some control over deformations of periods of symplectic forms on $V$. For example, if $(V, \omega_0)$ is fibrated (or ruled) by 2-spheres, such that $\omega_0$ restricts to a non-vanishing (area) form on each fiber $S^2 \subset V$, then every symplectic form $\omega$ on $V$ homotopic to $\omega_0$ has $\int_{S^2} \omega \neq 0$ (see [Gro$_2$], [McD]).

*2.4.C₂*. Denote by $\mathscr{J}'=\mathscr{J}'(\omega_1,\omega_2)$ the space of those almost complex structures $J$ on $V$ tamed by $\omega=\omega_1\oplus\omega_2$ for which both spheres $s_0\times S^2$ and $S^2\times s_0$ are realizable by rational $J$-curves in $V$. Then, the argument in 2.4.A$_1$. shows this $\mathscr{J}'$ to be weakly homotopy equivalent to the space $\mathscr{S}/SO(3)\times SO(3)$ for the group $\mathscr{S}=\mathscr{S}(\omega_1,\omega_2)$ of symplectic diffeomorphisms of $V=(S^2,\omega_1)\oplus(S^2,\omega_2)$. If $\int_{S^2}\omega_1\neq\int_{S^2}\omega_2$, then the complement $\mathscr{J}\backslash\mathscr{J}'$ contains a closed codimension two subvariety $\Sigma\subset\mathscr{J}$ which linked to a circle $S^1\subset\mathscr{J}'$, whose class $[S^1]\in H_1(\mathscr{J}';\mathbb{R})$ does not vanish. It follows that the group $\mathscr{S}$ is not contractible to $SO(3)\times SO(3)$. (See [Gro$_2$] for further information.)

*2.4.D*. Let $(V,J)$ be a smooth almost complex manifold with boundary $\partial V$ of $V$. This boundary is called *strictly J-convex* if no holomorphic disk $D$ in $V$ with boundary $\partial D\subset \operatorname{Int} V$ intersects $\partial V$ and if this property is stable under small $C^2$-perturbations of $J$. Thid is equivalent to the usual pseudoconvexity for complex analytic manifolds. In fact one can cast the above definition into analytic language by first defining *strictly J-convex* (corresponding to strictly plurisubharmonic) functions $f\colon V\to\mathbb{R}$ by requiring the 2-form $d(\sqrt{-1}\,df)$ to be positive and then by calling $\partial V$ $J$-convex in case there is a $J$-convex function in a small collar neighborhood $\partial V\times[0,1]\subset V$ for $\partial V\times 0=\partial V$, such that $f|\partial V=1$, $f|\partial V\times 1=0$ and df nowhere vanishes.

Next, we take a totally real submanifold $W\subset\partial V$ and we study $J$-holomorphic curves $C\subset V$ whose boundaries lie in $W$. Since $\partial V$ is convex no $J$-holomorphic deformation of $C$ ever reaches the boundary at the interior points $c\in C$. Hence, the existence theorems 0.4.A. and 2.3.D. insure in the present case holomorphic disks$(D^2,\partial D^2)\to(V,W)$. Furthermore, if $\dim V=4$, then regular $J$-curves are closed in the space of all curves (like closed curves in 2.4.B$_1'$.). That is a limit $f\colon(S,\partial S)\to(V,W)$ of regular $J$-curves $f_j\colon(S,\partial S)\to(V,W)$ is regular unless the map $f$ is nowhere one-to-one on some connected component of $S$. This can be immediately derived from 2.4.B$_1'$. in case the structure $J$ is complex near $W$ and $W$ is real analytic. Indeed, $V$ lies in a slightly greater manifold $V^+\supset V$, such that a small neighborhood $U^+\subset V^+$ of $W$ admits an anti-holomorphic involution $I\colon U^+\to U^+$. Since $\partial V$ is $J$-convex, the intersection $V\cap I(U^+\cap V)$ equals $W$. Thus a small analytic extension of every curve $(C,\partial C)\subset(V,W)$ to $V^+$ by $C^+=C\cup I(C\cap U^+)\subset V^+$ is regular in-so-far as $C$ is regular. Now, by 2.4.B$_1$. the limit curve may acquire no singularity in an interior point and the above extension turns the boundary points of $C$ to interior points of $C^+$. (This argument generalizes to all manifolds $V$, see [Gro$_2$]. But the special case we have treated suffices in most geometrically interesting cases.)

*2.4.D₁*. *Contact structures*. Consider a one-codimensional subbundle $\Theta\subset T(X)$ for a $(2n-1)$-dimensional manifold $X$. Assume the complementary bundle $T(X)/\Theta$ to be orientable and let $l$ be a 1-form on $X$ whose kernel equals $\Theta$. Then the subbundle $\Theta$ is called *contact* (or *a contact structure* on $X$) if the differential $d\Theta$ is non-singular on $\Theta$, that is the $(2n-2)$-form $(d\Theta)^{n-1}$ does not vanish on $\Theta$.

Next, let $(V,J)$ be an almost complex manifold with a strictly $J$-convex boundary $C=\partial V$. Then the subbundle $\Theta=T(X)\cap\sqrt{-1}\,T(X)\subset T(X)$ clearly is contact. Furthermore, if $2n=\dim V=4$, then every totally real submanifold $W\subset X=\partial V\subset V$ is everywhere transversal to $\Theta$. Moreover, for all $n\geqq 2$ the boundary $\partial C\subset\partial V=X$ of every regular $J$-curve $(C,\partial C)\subset(V,\partial V)$ is everywhere transversal to $\Theta$. In particular, if $n=2$, then the boundaries of regular $J$-curves $(C,\partial C)\subset(V,W)$ give us closed transversals to the line field $\Theta\cap T(W)$ for every totally real surface $W\subset\partial V$.

*2.4.D$'_1$. Example.* Let $V$ be the unit ball in $\mathbb{C}^4$ with $\partial V=S^3\subset\mathbb{C}^4$ and let $W$ be the intersection of $S^3$ with the real hyperplane $\operatorname{Im}z_2=0$ for the coordinates $z_1$ and $z_2$ in $\mathbb{C}^2$. This $W$ is *not* totally real. In fact, there are exactly two points in $W$ with complex tangents, namely the points $(z_1=0, z_2=\pm 1)$. Furthermore, there is a one-parameter family of holomorphic disks in $V$ with boundaries in $W$: these are

$$\{\operatorname{Re}z_2=t, \operatorname{Im}z_2=0\} \quad\text{for}\quad t\in\,]-1,1[\,.$$

Let $V'$ be obtained from $V$ by removing the two complex tangency points from $W\subset V$ and let $W'=W\cap V'\subset V'$. This $W'$ clearly is totally real in $V'$. Moreover, one easily constructs a symplectic form $\omega'$ on $V'$ which tames the complex structure of $V'$, for which $W'$ is Lagrange and such that $\omega'$ satisfies the conditions of 2.3.B. with some metric $\mu'$ on $V'$. Now, the above family of disks fits into our framework. In particular, this family is stable under deformations of the almost complex structure on $V$ which do not change the geometry near the two complex tangency points, which keep $\partial V$ convex and $W$ totally real outside these points and which are tamed by some symplectic forms on $V'$. In fact, one may even allow deformations which do change geometry near the complex points but yet keep the complex tangency *elliptic* in the following sense. An isolated complex tangency point $w_0\in W\subset\partial V\subset V$ is called elliptic if it is *focal* for the line field $\Theta\cap T(W)$ on $W\setminus\{w_0\}$. That is for some choice of the orientation of this field all orbits of the resulting vector field on $W$ asymptotically approach $w_0$; moreover, this should remain true for all $C^2$-small perturbations $W'$ of $W$ and (or) of the almost complex structure on $V$. Namely, every such perturbation is required to have a unique complex tangency point $w'_0$ in a fixed small neighborhood of $w_0$, and this $w'_0$ must have the same "focal" behaviour as $w_0$.

*2.4.D$_2$.* Let $(V,J)$ be an almost complex 4-manifold with a $J$-convex boundary and let $W\subset\partial V$ be a smoothly embedded 2-sphere with two elliptic complex tangency points $w_0$ and $w_1$ on $S^2$ and with no complex tangency besides these points.

**Theorem.** *If the manifold $(V,J)$ contains no non-constant rational curve and if $J$ can be tamed by some symplectic form $\omega$ on $V$, then there exists a unique smooth family of embedded $J$-holomorphic discs*

$$(D_t^2,\partial D_t^2)\subset(V,W\setminus\{w_0,w_1\})$$

*for $t\in\,]0,1[$, for which the resulting map $F: D^2\times\,]0,1[\,\to V$ is a smooth embedding, such that $F(x,t)\to w_0$ for $t\to 0$ and $F(x,t)\to w_1$ for $t\to 1$.*

*Proof*. Let us assume for the sake of simplicity the geometry near the points $w_0$ and $w_1$ to be as for the standard sphere $S^2\subset S^3\subset B^4\subset\mathbb{C}^2$. Then we do have our disks near the points $w_0$ and $w_1$. Next, by 2.1.D. and by 1.5.D$_2$. this local family of disks admits the desired global extension, since the areas of the disks are a priori bounded by const area $W$ and since the transversality of $\partial D^2 \subset W$ to $\Theta \cap T(W)$ prevents the disks from degeneration to cusp-curves (see [Gro$_2$] for more general results).

*2.4.D$'_2$.* Examples and corollaries. (a) The above theorem applies to the image $W \subset S^3 \subset B^4 \subset \mathbb{C}^2$ of the standard sphere $S^2 \subset S^3$ under an arbitrary contact transformation of $S^3$ (compare [B-G]).

(b) According to 2.4.D$_2$. the field $\Theta \cap T(W)$ has not closed orbits, as the curves $\partial D_t^2 \subset W$ cover $W \setminus \{w_0, w_1\}$ and they are everywhere transversal to $\Theta \cap T(W)$. In fact, the above proof applies to an embedded closed disk $W \subset \partial V$ with a unique interior elliptic complex tangency point $w_0 \in \operatorname{Int} W$. This shows the existence of a regular holomorphic disk $(D^2, \partial D^2) \subset (V, W)$ whose boundary $\partial D^2 \subset W$ is tangent to the boundary $\partial W$ at some point $w' \in \partial W$. Hence, *the boundary $\partial W$ cannot be a periodic orbit of the field $\Theta \cap T(W)$ on $W$.* This conclusion for disks $W$ in the standard sphere $S^3 \subset \mathbb{C}^4$ is due to Eliashberg [El$_2$] and to Bennequin [Ben]. In fact, Bennequin's analysis of the structure of complex points of surfaces in $S^3 \subset \mathbb{C}^2$ is by far more precise and general.

(c) The above results generalizes to submanifolds $W$ of dimension $n$ in those $(2n-1)$-dimensional contact manifolds $X$ which appear as $J$-convex boundaries of almost complex $2n$-dimensionals manifolds $(V, J)$, where $J$ can be tamed by a symplectic form on $V$. The submanifolds $W \subset X = \partial V$ in question are quite special: The space $\Theta \cap T_w(W)$ must be of dimension $n-1$ for all $w \in W$ outside a codimension two submanifold $W_0 \subset W$ and the hyperplane field $\Theta \cap T(W)$ on $W \setminus W_0$ must be integrable. One shows in certain cases (see [Gro$_2$] the existence of "sufficiently many" $J$-holomorphic disks $(D^2, \partial D^2) \to (V, W \setminus W_0)$. This imposes a non-trivial global condition on the geometry of the foliation on $W \setminus W_0$ tangent to the field $\Theta \cap T(W)$ on $W \setminus W_0$. Then one easily produces examples of submanifolds $W$ in some contact manifolds $X$ diffeomorphic to $\mathbb{R}^{2n-1}$ where this condition is not met; this prevents any contact embedding of such an $X$ into $\mathbb{R}^{2n-1}$ with the standard contact structure (given by the form $\sum_{i=1}^{n-1} x_i\, dy_i + dz$).

*2.4.E. General elliptic equations for surfaces in $W^4$.* Recall that the tangent space $T_x(G)$ of the Grassmann manifold $G = Gr_2\mathbb{R}^4$ of oriented planes in $\mathbb{R}^4$ is naturally isomorphic, for all $x \in G$, to the space of linear maps $G_x \to \mathbb{R}^4/G_x$ for the plane $G_x \subset \mathbb{R}^4$ corresponding to $x \in G$. This defines a unique conformal structure on $G$ whose quadratic form $h$ has signature $(++--)$ and whose zero locus $\{\tau \in T_x(G) \mid h(\tau, \tau) = 0\}$ consists of singular maps $G_x \to \mathbb{R}^4/G_x$. A smooth connected closed embedded surface $E_0 \subset G$ is called *elliptic* if the form $h$ is definite on $T_e(E_0) \subset T_e(G)$ for all $e \in E_0$. For example, the projective line $\mathbb{C}P^1 \subset Gr_2(\mathbb{R}^4 = \mathbb{C}^2)$ is elliptic.

If $E_0$ is elliptic, then, obviously, every line in $\mathbb{R}^4$ is contained in a unique plane belonging to $E_0$. It follows (see [G-W] that $E_0$ can be brought to $\mathbb{C}P^1$ or to $\overline{\mathbb{C}P}^1$ by an isotopy of elliptic surfaces in $Gr_2\mathbb{R}^4$, where $\overline{\mathbb{C}P}^1$ denotes the set of complex lines in $\mathbb{R}^4 = \mathbb{C}^2$ with the reversed orientations.

Next, let $E$ be a smooth 6-dimensional submanifold in the Grassmann bundle $Gr_2V$ of a 4-manifold $V$. Such an $E$ is called *elliptic* if it transversally meets every fiber $Gr_2T_v(V)=Gr_2\mathbb{R}^4$, $v\in V$, in an elliptic surface, say $E_v=E\cap GrT_v(V)$. A smoothly embedded surface $C\subset V$ is called an *E-curve* if the tangential lift $\tilde{C}\subset Gr_2V$ of $C$ lies in $E$. Fix an $E$-curve $C_0\subset V$ and consider infinitesimal deformations of $C_0$ which are given by sections of the normal bundle $N=N(C_0\subset V)$. Among these, we distinguish deformations which (infinitesimally) keep $\tilde{C}_0$ in $E$. Then there is an (obviously constructed) linear differential operator, say $\bar{\partial}_E$ on sections $f: C_0\to N$, such that the (germs of) solutions to the equation $\bar{\partial}_E$ f $=0$ are exactly and only the distinguished deformations. Furthermore, the operator $\bar{\partial}_E$ is (easily seen to be) elliptic for all elliptic submanifolds $E\subset Gr_2V$. (In fact, the ellipticity of $E$ and that of $\bar{\partial}_E$ are equivalent.) Moreover, there is a unique almost complex structure $J=J(E)$ on the space $N$, such that (the 1-jets of) $J$-holomorphic sections $f: C_0\to N$ are exactly and only (the 1-jets of) solutions to $\bar{\partial}_E f=0$. This is well known and the proof is an exercise in linear algebra. In fact, $J$ can be seen geometrically by observing for every elliptic surface $E\subset Gr_2\mathbb{R}^4$ and for every point $e\in E$ the existence of a complex linear structure on $\mathbb{R}^4$ for which the corresponding projective line $\mathbb{C}P^1\subset Gr_2\mathbb{R}^4$ contains $e$ and has $T_e(\mathbb{C}P^1)=T_e(E)$.

Indeed, the natural action of the group $GL_4\mathbb{R}$ on the space of tangent planes $\tau\subset T(Gr_2\mathbb{R}^4)$ has precisely two open orbits corresponding to the type (definite or indefinite) of the form $h|\tau$. This affinity between $E$ and the Cauchy-Riemann equation insures the Schwarz Lemma of 1.2. for $E$-curves. Namely, one fixes a Riemannian metric $\mu$ and then the argument in 1.2. provides a bound on $\|\mathscr{D}_f\|$ for conformal embeddings $f: S\subset V$ whose images $C=f(S)$ are $E$-curves and where $S$ is any Riemann surface. From this point on all facts on $J$-curves proven in the present paper extend to $E$-curves with an obvious change of terminology. Here are instances of that.

*Non-linear Riemann mapping theorem.* A symplectic form $\omega$ on $V$ is said to *tame* an elliptic submanifold $E\subset Gr_2V$ if $\omega$ is positive on all planes $\tau\subset T(V)$ which belong to $E$.

*2.4.$E_1$. Let $(V,\omega)=(S^2,\omega_0)\oplus(S^2,\omega_0)$ and let E be tamed by $\omega$. Then there is a unique $C^\infty$-splitting $V=S_1\times S_2$, such that the fibers $S_1\times s_2$ and $s_1\times S_2$ are regular E-curves in V for all $(s_1,s_2)\in S_1\times S_2$. Furthermore, if no two out of three given points $v_1, v_2$ and $v_3$ in V are contained in one of these fibers, then there is a unique regular E-curve $C\subset S_1\times S_2$ whose projections to $S_1$ and to $S_2$ are diffeomorphisms.*

*Proof*. The group of symplectic automorphisms of the bundle $(T(V),\omega)$ obviously acts on elliptic submanifolds $E\subset G_2V$ tamed by $\omega$. This provides a homotopy $E_t$ of $\omega$-tame submanifolds in $Gr_2V$, such that $E_1=E$ and such that the fibers $S^2\times s$, $S^2\times s$, $s\in S^2$ of the original splitting of $V$ are $E_0$-curves. Now use the homotopy of $E_t$-curves homologous to $S^2\times s$ as in 2.3.C. Thus one obtains for $t=1$ the fibers $S_1\times s_2$ and also the fibers $s_1\times S_2$. Finally, to construct $C=C(v_1,v_2,v_3)$ we start with some $E_0$ which does admit a $E_0$-curve $C_0$ homologous to the diagonal in $S^2\times S^2$. Then we take three distinct points $v_i(0)\in C$, $i=1,2,3$, and let $v_i(t)\in V$, $t\in[0,1]$, be a homotopy, such that $v_i(1)=v_i$ and such that the triple $\{v_i(t)\}$ satisfies the "no two points on a $E_t$-fiber" condition for a homotopy $E_t$ of $E_0$ to $E_1=E$. Then we obtain as earlier a homotopy of $E_t$-curves $C_t=C(E_t,v_i(t))$, where $C_1$ is the required curve $C$.

*2.4.$E_2$*. An involution $I$ of a manifold $V$ with a given $E \subset Gr_2 V$ is called a *conjugation* if $I(\bar{\tau}) \in E$ for all planes $\tau \in E$ where $\bar{\tau}$ denotes the plane with the reversed orientation. For example, if $V$ is a complex algebraic manifold defined over $\mathbb{R}$, then the Galois involution is a conjugation for the submanifold of complex lines in $Gr_2 V$.

Now, we return to $V = S^2 \times S^2$ with the $E$-curves $S_1(v)$ and $S_2(v)$ in $V$ passing through all $v \in V$ and with $C(v_1, v_2, v_3)$ through $v_1$, $v_2$ and $v_3$. Let the map $I: (s_1, s_2) \mapsto (\bar{s}_1, \bar{s}_2)$ be a conjugation where $s \mapsto \bar{s}$ denotes the reflection of $S^2$ in the equator $S^1 \subset S^2$. Observe that the fixed point set of this $I$ is the torus $W = S^1 \times S^1$ and that $W$ necessarily is *totally real* in $(V, E)$, that is no tangent plane to $W$ belongs to $E$.

*2.4.$E_2'$. If $v \in W$, then the $E$-curves $S_1(v)$ and $S_2(v)$ are $I$-invariant and their intersections with $W$ are simple closed (1-dimensional) curves which are homologous in $W$ to the circles $S^1 \times s$ and $s \times S^1$, $s \in S^1$, correspondingly. Furthermore, every $E$-curve $C = C(v_1, v_2, v_3)$ for $v_i \in W$, $i = 1, 2, 3$ also is $I$-invariant and the intersection $C \cap W$ is a simple closed curve in $W$.*

*Proof*. Choose the homotopy $E_t$ such that the (fixed) involution $I$ conjugates $E_t$ for all $t \in [0, 1]$. Then $I$ acts on the space of $E_t$-curves by an involution and this action is $C^\infty$-stable in $t$ like any other smooth action of a compact group on a manifold. Now, the desired property of $I$ for $t = 1$ follows from that for $t = 0$ as earlier.

An alternative argument consists in observing that the curve $I(C)$ with the reversed orientation is an $E$-curve for every $E$-curve $C$ in $E$. If $v \in W$, then the $E$-curves $S_1(v)$ and $I(S_1(v))$ meet at $v$ (as $I(v) = v$), and hence, $S_1(v) = I(S_1(v))$. The rest of the proof is obvious.

*2.4.$E_3$*. Consider a compact manifold $V$ with a boundary and call this boundary *$E$-convex* for a given $E \subset Gr_2 V$ if every connected $E$-curve $C \subset V$ whose boundary $\partial C$ lies in the interior $\operatorname{Int} V$ never meets the boundary $\partial V$. If this convexity is stable under small $C^2$-perturbations of $E$, then $\partial V$ is called *strictly $E$-convex*.

Since $E$ is elliptic, each tangent space $T_v(\partial V)$, $v \in \partial V$ contains a unique plane $\partial_v \subset T_v(\partial V)$ which belongs to $E$ (regardless of the convexity of $E$) and every totally real submanifold $W \subset \partial V$ is everywhere transversal to the subbundle $\Theta = \bigcup_{v \in V} \Theta_v \subset T(\partial V)$. Hence, every closed connected component of $W$ is a 2-torus.

Next, let $\omega$ be a symplectic structure on $V$ which tames $E$ and let $W \subset \partial V$ be a Lagrange torus. Moreover, assume the existence of a symplectic embedding $F_0: (D^2, \omega_0) \oplus (D^2, \omega_0) \to (V, \omega)$, where $D^2$ is the unit disk in $\mathbb{R}^2$ with $\omega_0 = dx \wedge dy$, such that the torus $\partial D^2 \times \partial D^2$ goes onto $W$, and let every smoothly mapped disk $(D, \partial D) \to (V, W)$ have $\int_D \omega = k \int_{D^2} \omega_0 = k\pi$, for some integer $k = k(D^2 \to V)$.

*2.4.$E_3'$. Let the boundary $\partial V$ be strictly $E$-convex. Then there exists an embedding $F: D^2 \times D^2 \to V$, such that*

(i) *$F(\partial D^2 \times \partial D^2) = W$ and the map $F$ is homotopic to $F_0$ by a homotopy of maps $(D^2 \times D^2, \partial D^2 \times \partial D^2) \to (V, W)$;*

(ii) *the images $F(s \times D^2)$ and $F(D^2 \times s)$ are $E$-curves in $V$ for all $s \in D^2$.*

*Furthermore, the image* $F(D^2 \times D^2) \subset V$ *is uniquely determined by the conditions* (i) *and* (ii).

*Proof.* Start with the construction of the $E$-curves $F(D^2 \times s) \subset V$, $s \in \partial D^2$, with boundaries $F(\partial D^2 \times s)$ in $W$. This is done as in 2.4.E$_1$. with a homotopy $E_t$ of $E = E_1$ to an elliptic submanifold $E_0$ for which the original disks $F_0(D^2 \times s)$ are $E_0$-curves and such that the boundary $\partial V$ is strictly $E_t$-convex for all $t \in [0, 1]$. The only remaining point is to check the regularity of the pertinent $E_t$-disks (obtained by a homotopy of the disks $F_0(D^2 \times s)$) at their boundary points. This is especially easy (compare 2.1.D.) under the following assumption on $E$ over a small neighborhood $U \subset V$ of the torus $W$.

(*) There exists an extension of $E|U$ to a $C^\infty$-*smooth* elliptic submanifold $E' \subset Gr_2 U'$ for some manifold $U'$ containing $U$ as an open subset, $U \subset U'$, such that $U'$ admits a $E'$-conjugation $I: U' \to U'$, which keeps $W$ fixed and for which $I(U) \cap U = W$.

If (*) is satisfied, then every $E$-disk $(D^2, \partial D^2) \subset (V, W)$ extends across the boundary to an $E'$-disk in $W' = V \cup U'$. Thus the boundary regularity reduces to the interior regularity of the extended disk. Finally, one achieves (*) for every $E$ by applying a small perturbation to $E|U$, and thus one obtains (with some extra work) embeddings of $D^2 \times \partial D^2$ and of $\partial D^2 \times D^2$ in $V$ which send the disks $D^2 \times s$ and $s \times D^2$, for all $s \in \partial D^2$, to $E$-disks in $V$. Now, these embeddings extend to the required map $F: D^2 \times D^2 \to V$ as follows. Start with an extension $F_1: D^2 \times D^2 \to V$ which satisfies (i) (but may violate (ii) outside the boundary $\partial(D^2 \times D^2) = D^2 \times \partial D^2 \cup \partial D^2 \times D^2$) and consider the pull-back $E_1 = F_1^*(E) \subset Gr_2(D^2 \times D^2)$. Then use the standard embedding $D^2 \times D^2 \subset S^2 \times S^2$ and extend $E_1$ to an elliptic submanifold $E_1' \subset Gr_2(S^2 \times S^2)$. If $E$ satisfies (*), then choose an $E_1'$ which admits a conjugation $I: S^2 \times S^2 \to S^2 \times S^2$ which keeps the torus $F_1^{-1}(W) \subset S^2 \times S^2$ fixed and such that $(D^2 \times D^2) \cap I(D^2 \times D^2) = F_1^{-1}(W)$. Now, the $E$-curves $F(s \times D^2)$ for $s \in \operatorname{Int} D^2$ are obtained by first taking $E'$-curves in $S^2 \times S^2$ homologous to $s \times S^2$, then by intersecting these with $D^2 \times D^2$ and finally by bringing the intersections to $V$ by the map $F_1$. The conclusion of the proof under the assumption (*) is obvious at this point and a simple approximation argument yields the proof in the general case.

2.4.E$_3''$. *Take three (generic) points* $w_1$, $w_2$ *and* $w_3$ *in* $W$ *such that neither the circle* $F(s_1 \times D^2) \subset W$ *nor* $F(\partial D^2 \times s_2) \subset W$ *contains more than one of these points for all* $s_1$ *and* $s_2$ *in* $\partial D^2$. *Then there exists a unique E-disk* $D = D(w_1, w_2, w_3)$ *in* $V$, *such that*

(i) *D is contained in the image* $F(D^2 \times D^2) \subset V$ *and D is transversal to the disks* $F(s \times D^2)$ *and* $F(D^2 \times s)$ *for all* $s \in D^2$;

(ii) *the boundary* $\partial D$ *is contained in W and it contains the points* $w_1$, $w_2$ *and* $w_3$.

*Proof.* Repeat the argument used in 2.4.E$_1$. to produce the $E$-sphere $C = C(v_1, v_2, v_3)$. Alternatively, recall the above $E_1' \subset Gr_2(S^2 \times S^2)$ and take the intersection of the $E_1'$-sphere $C(v_1, v_2, v_3)$ for $v_i = F^{-1}(w_i)$, $i = 1, 2, 3$, with $D^2 \times D^2$ (brought back to $V$ by the map $F_1$) for $D$.

*Remark.* This theorem is due to Lavrentyev [Lav] who assumes the existence of an $E$-splitting of $D^2 \times D^2$ to start with and who puts the emphasis on the map $f: D^2 \to D^2$ rather than on the graph $\Gamma_f \subset D^2 \times D^2$ that is our $E$-disk $D$.

*Acknowledgements.* I am deeply grateful to Dusa McDuff who has painstakingly read the first draft of the paper and located a multitude of inconsistences and errors. I thank Pierre Pansu and Jean-Claude Sikorav for pointing out mistakes in my original treatment of surfaces.

## References

[Be] Bers, L.: An outline of the theory of pseudoanalytic functions. Bull. Am. Math. Soc. **62**, 291–332 (1956)
[Ben] Bennequin, D.: Enlacements et équations de Pfaff. Astérisque **107–108**, 87–162 (1982)
[Ber] Berger, M.: Du côté de chez Pu. Ann. Sci. Ec. Norm. Super., IV. Ser. **4**, 1–44, 241–260 (1972)
[B-G] Bedford, E., Gaveau, B.: Envelopes of holomorphy of certain 2-spheres in $\mathbb{C}^2$. Am. J. Math. **105**, 975–1009 (1983)
[B-H] Begher, H., Hile, G.: Riemann boundary value problem for non-linear elliptic systems. In: Complex Variables, vol. 1, pp. 239–261. U.S.A.: Gordon and Breach 1983
[$Bi_1$] Bishop, E.: Conditions for the analyticity of certain sets. Mich. Math. J. **11**, 289–304 (1964)
[$Bi_2$] Bishop, E.: Differentiable manifolds in complex Euclidean space. Duke Math. J. **32**, 1–22 (1965)
[B-Z] Burago, Yu., Zalgaller, V.: Geometric inequalities (in Russian). Leningrad, Nauka 1980 (an English translation is to appear in Springer-Verlag)
[Ch] Chaperon, M.: Questions de géométrie symplectique. Sem. Sud-Rhodanien IV, Balarue, in Travaux en cours. Paris: Hermann 1984 (To appear)
[C-Z] Conley, C., Zehnder, E.: The Birkhoff-Lewis fixed point theorem and a conjecture of V. I. Arnold. Invent. math. **73**, 33–49 (1983)
[D-D] Dazord, P., Desolneux-Moulis, N. (eds.): Séminaire Sud-Rhodanien de Géométrie. Paris: Hermann 1984
[E-S] Eells, J., Salamon, S.: Twistorial constructions of harmonic maps of surfaces into four-manifolds. Warwick, 1984 (Preprint)
[$El_1$] Eliashberg, Ya.: Estimates on the number of fixed points of area preserving transformations (in Russian) Syktyvkar, 1978 (Preprint)
[$El_2$] Eliashberg, Ya.: Rigidity of symplectic and contact structures (Preprint 1981)
[F l] Floer, A.: Proof of the Arnold conjecture for surfaces and generalizations for certain Kähler manifolds. Bochum (Preprint 1984)
[F-W] Fortune, B., Weinstein, A.: A symplectic fixed point theorem for complex projective spaces. Berkeley (Preprint 1984)
[Gil] Gilbert, R. (ed.): Proc. of the special session on elliptic systems in the plane. 87[th]Ann. Meeting of A.M.S. San Francisco, CA (1981)
[G-W] Gluck, H., Warner, F.: Great circle fibrations of the three sphere. Duke Math. J. **50**, 107–132 (1983)
[$Gro_1$] Gromov, M.: A topological technique for the construction of solutions of differential equations and inequalities. ICM 1970, Nice, **2**, 221–225 (1971)
[$Gro_2$] Gromov, M.: Pseudo-holomorphic curves in symplectic manifolds, II. Berlin-Heidelberg-New York-Tokyo: Springer (In press)
[$Gro_3$] Gromov, M.: Partial differential relations. Berlin-Heidelberg-New York-Tokyo: Springer (In press)
[H-L] Harvey, R., Lawson, B.: Calibrated geometries. Acta Math. **148**, 48–156 (1982)
[H] Hofer, H.: Lagragian embedding and critical point theory. (Preprint 1984)
[Lav] Lavrentiev, M.: A fundamental theorem of the theory of quasi-conformal mappings of plane regions. Izv. Akad. Nauk SSSR **12**, 513–554 (1948)
[$Law_1$] Lawson, B.: Lectures on minimal submanifolds (2e édition). Berkeley: Publish or Perish 1980
[$Law_2$] Lawson, B.: Surfaces minimales et la construction de Calabi-Penrose. Sém. Bourbaki n° 625 (1983/84)
[L-S] Laudenbach, F., Sikorav, J.C.: Persistance d'intersection avec la section nulle au cours d'une isotopie hamiltonienne dans un fibre cotangent. Invent. math. **82**, 349–357 (1985)
[McD] McDuff, D.: (to appear)
[Mo] Moser, J.: On the volume elements on a manifold. Trans. Am. Math. Soc. **120**, 286–294 (1965)
[N-W] Nijehuis, A., Wolf, W.: Some integration problem in almost complex and complex manifolds. Ann. Math. **77**, 424–489 (1963)

[Nir] Nirenberg, L.: Topics in non-linear functional analysis. Lect. Notes by R.A. Artino. Courant Inst. N.Y. Univ., N.Y. 1974
[Ren] IIIe Rencontre de Géométrie du Schnepfenried. Astérisque **107–108** (1982)
[Sch] Schapiro, Z.: Sur l'existence des représentations quasiconformes. C.R. (Doklady) Acad. Sci. URSS **3**, 690–692 (1941)
[St] Stoll, W.: The growth of the area of transcendental analytic set, I and II. Math. Ann. **156**, 47–98, 144–170 (1964)
[Si] Sikorav, J.-C.: Points fixes d'une application symplectique homologue à l'identité. Orsay (Preprint 1984)
[S-U] Sacks, J., Uhlenbeck, K.: The existence of minimal immersions of two-spheres. Ann. Math. **113**, 1–24 (1981)
[Vek] Vekua, I.N.: Generalized analytic functions. London: Pergamon 1962
[We] Wells, R.: Function theory on differential submanifolds. In: Contribution to Analysis, Ahlphors, L. (ed.), pp. 407–437. New York, London: Academic Press 1974
[Wei] Weinstein, A.: Lectures on symplectic manifolds. A.M.S. Conf. Board, Reg. Conf. in Math. **29** (1977)
[Wen] Wendland, W.: Elliptic systems in the plane. London: Pitman 1978

Oblatum 24-I-1985

# GROUPS OF POLYNOMIAL GROWTH AND EXPANDING MAPS

*by* Mikhael GROMOV

## Introduction

Consider a group $\Gamma$ generated by $\gamma_1, \ldots, \gamma_k \in \Gamma$. Each element $\gamma \in \Gamma$ can be represented by a word $\gamma_{i_1}^{p_1} \gamma_{i_2}^{p_2} \ldots \gamma_{i_l}^{p_l}$ and the number $|p_1|+|p_2|+\ldots+|p_l|$ is called *the length* of the word. *The norm* $||\gamma||$ (relative to $\gamma_1, \ldots, \gamma_k$) is defined as the minimal length of the words representing $\gamma$. Notice, that one can have several shortest words representing the same $\gamma \in \Gamma$.

*Examples.* — Let $\Gamma$ be the free Abelian group of rank 2 generated by $\gamma_1, \gamma_2$. Each $\gamma \in \Gamma$ can be represented as $\gamma_1^p \gamma_2^q$, $p, q \in \mathbf{Z}$, and $||\gamma|| = |p| + |q|$. (For the identity element $e \in \Gamma$ we set $||e|| = 0$.)

Let $\Gamma$ be the free (non-Abelian) group of rank 2 generated by $\gamma_1, \gamma_2$. Each $\gamma \neq e$ can be uniquely written as $\gamma_1^{p_1} \gamma_2^{p_2} \gamma_1^{p_3} \ldots \gamma_{i_k}^{p_k}$, or as $\gamma_2^{p_1} \gamma_1^{p_2} \gamma_2^{p_3} \ldots \gamma_{i_k}^{p_k}$, where $i_j = 1, 2$ and $p_1, \ldots, p_k$ are non-zero integers. The norm of such a $\gamma$ is equal to $\sum_{i=1}^{k} |p_j|$.

Let $\Gamma$ be the free cyclic group generated by $\gamma_0$ by let us use the generators $\gamma_1 = \gamma_0^2$, $\gamma_2 = \gamma_0^3$ and $\gamma_3 = \gamma_0^4$. Relative to these $\gamma_1, \gamma_2, \gamma_3$ we obviously have $||\gamma_0|| = 2$, $||\gamma_0^2|| = 1$, $||\gamma_0^6|| = 2$ (because $\gamma_0^6 = \gamma_2^2 = \gamma_1 \gamma_3$), $||\gamma_0^{100}|| = 25$ and so on.

*Elementary properties of the norm.* — For any group one obviously has

$$||\gamma|| = ||\gamma^{-1}||,$$
$$||\gamma\gamma'|| \leq ||\gamma|| + ||\gamma'||.$$

Let $\gamma_1, \ldots, \gamma_k$ and $\delta_1, \ldots, \delta_l$ be two systems of generators in $\Gamma$. The corresponding norms $|| \ ||^{\text{old}}$ and $|| \ ||^{\text{new}}$ are not necessarily equal but there obviously exists a positive constant C such that for each $\gamma \in \Gamma$ one has

$$C||\gamma||^{\text{old}} \geq ||\gamma||^{\text{new}} \geq C^{-1}||\gamma||^{\text{old}}.$$

For a group $\Gamma$ with fixed generators we denote by $B(r) \subset \Gamma$, $r \geq 0$, *the ball of radius r* centered at the identity element *e*. In other words, $B(r)$ consists of all $\gamma \in \Gamma$ with $||\gamma|| \leq r$. We denote by $\# B(r)$ the number of elements in $B(r)$.

For a free Abelian group of rank two generated by $\gamma_1$, $\gamma_2$ one has

$$\# B(r) = 2N^2 + 2N + 1 \quad \text{for} \quad r \in [N, N+1).$$

For the free (non-Abelian) group with two generators one has

$$\# B(r) = 2 \cdot 3^N - 1 \quad \text{for} \quad r \in [N, N+1).$$

*Growth of a group.* — One says that a group $\Gamma$ with generators $\gamma_1, \ldots, \gamma_k$ has *polynomial growth* if there are two positive numbers $d$ and C such that for all balls $B(r)$, $r \geq 1$, one has

$$\# B(r) \leq C r^d.$$

One can easily see that this definition does not depend on the particular choice of the generators and so this notion is correctly defined for the finitely generated groups.

*Examples.* — The finitely generated Abelian groups are easily seen to be of polynomial growth. Also the finitely generated nilpotent groups are of polynomial growth (see [14] and the appendix).

If $\Gamma$ is a finite extension of a group of polynomial growth, then $\Gamma$ itself has polynomial growth. So we conclude:

*If a finitely generated group $\Gamma$ has a nilpotent subgroup of finite index then $\Gamma$ has polynomial growth.*

The free groups with $k \geq 2$ generators do not have polynomial growth. They even have *exponential growth*, i.e.

$$\# B(r) \geq C^r, \quad r \geq 1$$

for some real constant $C > 1$. One can immediately see that this property does not depend on the choice of the generators.

The following theorem settles the growth problem for the solvable groups:

(Milnor-Wolf [8] [14].) — *A finitely generated solvable group $\Gamma$ has exponential growth unless $\Gamma$ contains a nilpotent subgroup of finite index.*

This result together with a theorem of Tits (see [13] and § 4) implies:

(Tits.) — *A finitely generated subgroup $\Gamma$ of a connected Lie group has exponential growth unless $\Gamma$ contains a nilpotent subgroup of finite index.*

In this paper we prove the following.

*Main theorem.* — *If a finitely generated group $\Gamma$ has polynomial growth then $\Gamma$ contains a nilpotent subgroup of finite index.*

The proof if given in § 8.

One can combine this theorem with Shub's criterion (see § 1) and obtains the

*Geometric corollary.* — *An expanding self-map of an arbitrary compact manifold is topologically conjugate to an infra-nil-endomorphism.*

The proof and the definitions are given in § 1.

## 1. Expanding maps

A map $f$ from a metric space X to a metric space Y is called *globally expanding* if for any two points $x_1, x_2 \in X$, $x_1 \neq x_2$, one has

$$\operatorname{dist}(f(x_1), f(x_2)) > \operatorname{dist}(x_1, x_2).$$

We call $f$ *expanding* if each point $x \in X$ has a neighbourhood $U \subset X$ such that the restriction of $f$ to U is globally expanding.

Suppose that X and Y are connected Riemannian manifolds of the same dimension. If X is a complete manifold without boundary, then each expanding map is a covering. In particular, when Y is simply connected such a map is a globally expanding homeomorphism.

Let X be a compact connected Riemannian manifold and let $f: X \to X$ be an expanding map. One can see that X has no boundary, and hence, the map $\tilde{f}: Y \to Y$ induced on the universal covering $Y \to X$ is a globally expanding homeomorphism. The inverse map $\tilde{f}^{-1}: Y \to Y$ is contracting. Moreover, for each $\delta > 0$ there is a positive $\varepsilon$ such that for any two points $y_1, y_2 \in Y$ with $\operatorname{dist}(y_1, y_2) \geq \delta$ one has

$$\operatorname{dist}(\tilde{f}^{-1}(y_1), \tilde{f}^{-1}(y_2) \leq (1-\varepsilon) \operatorname{dist}(y_1, y_2).$$

This is obvious. (Notice, that we use in Y the Riemannian metric induced from X by the covering map $Y \to X$.) It follows that $\tilde{f}^{-1}$ has a unique fixed point and that Y is homeomorphic to the Euclidean space $\mathbf{R}^n$, $n = \dim X$. Now, it is clear that $f: X \to X$ also has a fixed point.

All these facts were established by M. Shub (see [11]). (Notice that the definitions used in [11] are slightly different from ours.)

*Examples.* — Consider the torus $T^n = \mathbf{R}^n/\mathbf{Z}^n$. Each linear map $\mathbf{R}^n \to \mathbf{R}^n$ which sends the lattice $\mathbf{Z}^n \subset \mathbf{R}^n$ into itself induces a map $T^n \to T^n$. This map is expanding if and only if all eigenvalues of the covering linear map $\mathbf{R}^n \to \mathbf{R}^n$ have absolute values greater than one.

*Flat manifolds.* — Let $\Gamma$ be a discrete fixed point free group of motions of $\mathbf{R}^n$ with compact quotient $X = \mathbf{R}^n/\Gamma$. A linear map $\mathbf{R}^n \to \mathbf{R}^n$ which respects $\Gamma$ induces a map $X \to X$ and this map is expanding if and only if the covering linear map $\mathbf{R}^n \to \mathbf{R}^n$

has only eigenvalues of absolute value greater than one. It is known (see [4]) that any flat manifold X has an expanding map of the type we have just described.

*Nil-manifolds.* — Let L be a simply connected nilpotent Lie group with a left invariant Riemannian metric and let $\Gamma \subset \mathrm{L}$ be a discrete subgroup with compact quotient $\mathrm{X} = \mathrm{L}/\Gamma$. (Such an X is called a nil-manifold.) An automorphism $\mathrm{A} : \mathrm{L} \to \mathrm{L}$ which sends $\Gamma$ into itself induces a map $\mathrm{X} \to \mathrm{X}$ and this map is expanding if and only if the linear map $a : \ell \to \ell$ induced by A in the Lie algebra $\ell$ of L has all its eigenvalues greater than one in absolute value. Observe, that not all nilpotent Lie groups admit an expanding automorphism.

*Infra-nil-manifolds.* — Let L be as above and denote by Aff(L) the group of transformations of L generated by the left translations and by all automorphisms $\mathrm{L} \to \mathrm{L}$. Let $\Gamma \subset \mathrm{Aff}(\mathrm{L})$ be a group which acts freely and discretely on L. When the quotient $\mathrm{X} = \mathrm{L}/\Gamma$ is compact it is called an *infra-nil-manifold.* Each expanding automorphism $\mathrm{L} \to \mathrm{L}$ which respects $\Gamma$ induces an expanding map $\mathrm{X} \to \mathrm{X}$. Such maps are called expanding *infra-nil-endomorphisms.*

*Topological conjugacy.* — Two maps $f : \mathrm{X} \to \mathrm{X}$ and $g : \mathrm{Y} \to \mathrm{Y}$ are called topologically conjugate if there exists a homeomorphism $h : \mathrm{X} \to \mathrm{Y}$ such that $h \circ f = g \circ h$, i.e. the following diagram commutes

$$\begin{array}{ccc} \mathrm{X} & \xrightarrow{f} & \mathrm{X} \\ {\scriptstyle h}\downarrow & & \downarrow{\scriptstyle h} \\ \mathrm{Y} & \xrightarrow[g]{} & \mathrm{Y}. \end{array}$$

M. Shub discovered the following remarkable fact (see [11]):

*An expanding self-map of a compact manifold* X *is uniquely determined, up to topological conjugacy, by its action on the fundamental group* $\pi_1(\mathrm{X})$.

The following two results of Shub and Franks (see [11]) are especially important for our paper.

*Shub's criterion.* — *An expanding self-map of a compact manifold* X *is topologically conjugate to an expanding infra-nil-endomorphism if and only if the fundamental group* $\pi_1(\mathrm{X})$ *contains a nilpotent subgroup of finite index.*

*The polynomial growth property* (Franks). — *If a compact manifold* X *admits an expanding self-map then the fundamental group* $\pi_1(\mathrm{X})$ *has polynomial growth.*

We prove this in the next section.

These two facts explain why the geometric corollary is a consequence of the main theorem.

Some partial results on the classification of the expanding maps were obtained earlier by Shub [11] and Hirsch [6]. An idea of Hirsch's paper plays an important role in our approach (see § 5).

## 2. Geometric growth

Consider a Riemannian manifold Y and denote by $\mathrm{Vol}_y(r)$, $y \in Y$, the volume of the ball of radius $r$ around $y$. The growth of Y is defined as the asymptotic behavior of $\mathrm{Vol}_y(r)$ as $r \to \infty$.

This concept is due to Efremovič (see [3]) who pointed out that the growth of a manifold Y, which covers a *compact* manifold X, depends only on the fundamental groups $\pi_1(X)$, $\pi_1(Y)$ and the inclusion $\pi_1(Y) \subset \pi_1(X)$.

The corresponding algebraic notion of the growth was introduced by Švarc and by Milnor who, in particular, proved the following theorem.

(Švarc, Milnor). — *Let* $Y \to X$ *be the universal covering and let us fix a set of generators in the fundamental group* $\Gamma = \pi_1(X)$. *Then there is a constant* $C > 0$ *such that for each* $y \in Y$ *and all* $r \geq 1$ *one has*

$$\mathrm{Vol}_y(Cr + C) \geqslant \# B(r) \geqslant \mathrm{Vol}_y(C^{-1} r),$$

*where* $B(r)$ *denotes the ball in* $\Gamma$ *(see the introduction).*

*Proof.* — Let us identify $\Gamma$ with the orbit of $y \in Y$ under the action of $\Gamma$, so that $y$ corresponds to the identity. Denote by $\widetilde{B}_y(r)$ the intersection of $\Gamma$ with the Riemannian ball of radius $r$ in Y centered at $y$. It is not hard to show (see [8], [12]) that

$$C_1 \,\mathrm{Vol}_y(r) \geq \# \widetilde{B}_y(r) \geq C_1^{-1} \,\mathrm{Vol}_y(r), \quad r \geq 1$$

and

$$\# \widetilde{B}_y(C_2 r) \geqslant \# B(r) \geqslant \# \widetilde{B}_y(C_2^{-1} r), \quad r \geq 0.$$

This implies the theorem.

*Corollary.* — *The fundamental group of a compact manifold* X *has polynomial growth if and only if the universal covering* Y *of* X *has polynomial growth, i.e. if for some* C *and* $d$ *one has*

$$\mathrm{Vol}_y(r) \leq C r^d, \quad r \in [1, \infty).$$

Observe that most (complete, non-compact) manifolds have exponential growth, i.e. $\mathrm{Vol}_y(r) \geq C^r - 1$, $C > 1$, $r \in [1, \infty)$, but there are some interesting instances of polynomial growth.

*Examples of manifolds of polynomial growth:*

(a) Complete manifolds of non-negative Ricci curvature;
(b) Real algebraic submanifolds in $\mathbf{R}^q$;
(c) Nilpotent Lie groups with left invariant metrics;
(d) Leaves of Anosov foliations.

*Remarks.* — Polynomial growth for a manifold of non-negative Ricci curvature follows from Rauch's comparison theorem (see [2]); (b) and (c) are easy exercises; (d) has the same nature as the polynomial growth in the presence of an expanding map: a slight modification (required by a minor discrepancy between the notion of expansion) of the following simple lemma yields both facts.

*Let $f: Y \to Y$ be a totally expanding smooth map. Suppose that the Jacobian of this map is bounded by a constant C and that $f$ is uniformly expanding, i.e. for any two points $x, y \in Y$ with $dist(x,y) \geq 1$ one has*

$$dist(f(x), f(y)) \geq (1+\varepsilon)\, dist(x,y), \quad \varepsilon > 0.$$

*Then* Y *has polynomial growth.*

*Proof.* — Each ball B of radius $r \geq 1$ is sent by $f$ onto a set containing a ball of radius $(1+\varepsilon)r$ and whose volume is at most C Vol(B). It follows that

$$\mathrm{Vol}_y((1+\varepsilon)r) \leq C\, \mathrm{Vol}_y(r), \quad r \geq 1,$$

where $y$ is the fixed point of $f$. This inequality implies polynomial growth.

## 3. Elementary properties of the growth

Let $\Gamma$ be a group with a fixed finite system of generators. The norm $\|\ \|$ (see the introduction) provides $\Gamma$ with a left invariant metric

$$\mathrm{dist}(\alpha, \beta) = \|\alpha^{-1}\beta\|.$$

Consider a subgroup $\Gamma'$ and the left action of $\Gamma'$ on $\Gamma$. Denote by X the corresponding factor space $\Gamma/\Gamma'$ and by $f: \Gamma \to X$ the natural projection. Define $\mathrm{dist}(x,y)$, $x, y \in X$ as $\inf_{\alpha,\beta} \mathrm{dist}(\alpha, \beta)$, $\alpha \in f^{-1}(x)$, $\beta \in f^{-1}(y)$. Since the action of $\Gamma'$ on $\Gamma$ is isometric, the function $\mathrm{dist}(x,y)$ is a metric in X.

*Connectivity.* — The space $X = \Gamma/\Gamma'$ has the following two equivalent properties:

(a) for any two points $x, y \in X$ with $\mathrm{dist}(x,y) = p$, where $p$ is an integer, there exist points $x = x_0, x_1, \ldots, x_p = y$, such that $\mathrm{dist}(x_i, x_{i-1}) = 1$, $i = 1, \ldots, p$;

(b) take a ball $B \subset X$ of radius $p$, where $p$ is an integer, and take its $\varepsilon$-neighbourhood $U_\varepsilon(B) \subset X$ where $\varepsilon$ is also a non-negative integer. Then $U_\varepsilon(B)$ is exactly the ball of radius $p+\varepsilon$ concentric to B.

Both properties (as well as their equivalence) are obvious for $\Gamma$ itself and they are preserved when we pass to X.

As an immediate application we have:

*If* X *is infinite then each ball in* X *of radius $r = 0, 1, \ldots,$ contains at least $r+1$ elements.*

This simple fact provides a useful relation between the growth of a group and its subgroups.

We define growth($\Gamma$) as the lower bound of the numbers $d \geq 0$, such that

$$\# \mathrm{B}(r) \leq \mathrm{const}.\, r^d, \qquad r \geq 1.$$

*Splitting lemma.* — *If* $\Gamma' \subset \Gamma$ *is a finitely generated subgroup of infinite index, then*

$$\mathrm{growth}(\Gamma') \leq (\mathrm{growth}(\Gamma)) - 1$$

*Proof.* — The connectivity properties of $X = \Gamma/\Gamma'$ imply that each ball $\mathrm{B}(r)$, $r = 0, 1, \ldots,$ in $\Gamma$ contains at least $r+1$ elements $\alpha_0, \ldots, \alpha_r \in \Gamma$ such that $f(\alpha_i) \neq f(\alpha_j)$ for $i \neq j$ (recall that $f: \Gamma \to X$ is the factor-map). Consider the intersection $\mathrm{B}' = \mathrm{B}(r) \cap \Gamma'$ and its translates $\mathrm{B}'\alpha_i$, $i = 0, \ldots, r$. These sets are disjoint and they are contained in the ball $\mathrm{B}(2r)$. It follows that $\# \mathrm{B}(2r) \geq (r+1)\,(\# \mathrm{B}')$. This yields the lemma.

*Regular growth.* — All balls in $\Gamma$ of a given radius $r$ have the same number of elements. We denote this number by $b(r) = \# \mathrm{B}(r)$. For a group of growth $d < \infty$ we call a number $r$ *i-regular* $i = 1, 2, \ldots,$ if it satisfies the following two conditions:

(a) $\log(b(2^{-j}r)) \geq \log(b(r)) - j(d+1)\log 2, \quad j = 1, 2, \ldots, i,$

(b) $\log(b(2^{j}r)) \leq \log(b(r)) + \beta_j, \quad j = 1, 2, \ldots, i, \quad$ where $\quad \beta_j = 16^{j+1}(d+1).$

*Regularity lemma.* — *There is a sequence* $(r_i)$ *tending to* $\infty$ *such that each* $r_i$ *is i-regular.*

*Proof.* — Start with the sequence $r'_k = 2^k$. Since growth$(\Gamma) = d$ we have

$$\log(b(r'_k)) \leq \mathrm{C} + k\, d \log 2.$$

This inequality implies that there is an infinite subsequence $r_i = 2^{k_i}$ which satisfies (a), i.e.

$$\log(b(2^{k_i - j})) \geq \log(b(2^{k_i})) - j(d+1)\log 2, \quad j = 1 \ \ldots, i.$$

Let us show that this sequence must also satisfy (b); in other words (a) implies (b) for large $r$.

We first prove the following general inequality which is valid for all finitely generated groups

$$b(5r) \leq \frac{(b(4r))^2}{b(r)}, \quad r = 1, 2, \ldots. \tag{*}$$

*Proof.* — Consider a maximal system of points $\gamma_1, \gamma_2, \ldots \in \mathrm{B}(3r)$ such that the distance between any two of them is at least $2r+1$. It is clear that the balls of radius $r$ centered at $\gamma_1, \gamma_2, \ldots,$ do not intersect and the concentric balls of radius $2r$ cover $\mathrm{B}(3r)$. Using the connectivity property of $\Gamma$ we conclude that the concentric balls of radius $4r$ cover $\mathrm{B}(5r)$. This proves (*), because the original balls of radius $r$ were contained in $\mathrm{B}(4r)$ and the total number of their points could not exceed $b(4r)$.

We simplify the notations by setting $\ell(r)=\log(b(r))$. Then the inequality (*) becomes

$$\ell(5r)\leq 2\ell(4r)-\ell(r).$$

When $r$ is divisible by 4 this implies

$$\ell(6r)\leq \ell\left(5r+\frac{5r}{4}\right)\leq 2\ell(5r)-\ell(r),$$

and so

$$\ell(6r)\leq 4\ell(4r)-3\ell(r).$$

In the same way we get

$$\ell(8r)\leq \ell\left(6.\frac{6r}{4}\right)\leq 16\ell(4r)-15\ell(r).$$

It means that for an $r$ divisible by 16 we have

$$\ell(2r)\leq 16\ell(r)-15\ell\left(\frac{r}{4}\right).$$

Applying this inequality $j$ times we get

$$\ell(2^j r)\leq 16^j\left(\ell(r)-\ell\left(\frac{r}{4}\right)\right)+\ell\left(\frac{r}{4}\right).$$

In our case $\ell(r)-\ell\left(\frac{r}{4}\right)\leq 2(d+1)\log 2$ and so

$$\ell(2^j r)\leq 16^{j+1}(d+1)+\ell\left(\frac{r}{4}\right)\leq 16^{j+1}(d+1)+\ell(r),$$

q.e.d.

## 4. Linear representations

A group $\Gamma$ is called, for brevity, *almost nilpotent (almost solvable)* if it contains a nilpotent (solvable) subgroup of finite index.

This section is devoted to the proof of the following.

*Algebraic lemma. — Let* $\Gamma$ *be a finitely generated group of polynomial growth and let* L *be a Lie group with finitely many connected components. Suppose that for each finitely generated infinite subgroup* $\Gamma' \subset \Gamma$ *there is a subgroup* $\Delta \subset \Gamma'$ *of finite index in* $\Gamma'$ *with the following property: for every* $p=1, 2, \ldots,$ *there is a homomorphism* $\Delta \rightarrow \mathrm{L}$ *such that its image contains at least* $p$ *elements. Then* $\Gamma$ *is almost nilpotent.*

Our proof is based on the following fundamental facts.

(Jordan). — *For each Lie group* L *with finitely many components there is a number* $q$ *such that every finite subgroup in* L *contains an Abelian subgroup of index at most* $q$. (See [10].)

(Tits). — *Let* L *be as above and let* $G \subset L$ *be an arbitrary finitely generated subgroup. Then there are only two possibilities*:

(a) G *contains a free group of rank* 2. *In this case* G *has exponential growth.*
(b) G *is almost solvable. In this case* G *has exponential growth unless it is almost nilpotent.*

(See [13].)

We first prove two simple lemmas.

(a) *Let* L *be as above and let* G *be an arbitrary finitely generated group. Suppose that for every number* $p = 1, 2, \ldots,$ *there is a homomorphism* $G \to L$ *such that its image is finite and has at least* $p$ *elements. Then* G *contains a subgroup* $G' \subset G$ *of finite index such that the commutator group* $[G', G'] \subset G'$ *has infinite index and, consequently,* G' *admits a non-trivial homomorphism in* **Z**.

*Proof.* — Let $q$ be as in Jordan's theorem. Take for $G' \subset G$ the intersection of all subgroups in G of index at most $q$. It is clear that G' satisfies all the requirements.

(b) *Let* $\Gamma$ *be a finitely generated group of polynomial growth. Then the commutator subgroup* $[\Gamma, \Gamma]$ *is also finitely generated.*

*Proof.* — It is sufficient to show that the kernel $\Delta \subset \Gamma$ of any surjective homomorphism $g : \Gamma \to \mathbf{Z}$ is finitely generated.

Take a system of generators $\gamma_0, \gamma_1, \ldots, \gamma_k \in \Gamma$ with the following properties:

$$g(\gamma_0) = z_0 \in \mathbf{Z}, \text{ where } z_0 \text{ denotes the generator in } \mathbf{Z},$$

$$\gamma_i \in \Delta, \quad i = 1, \ldots, k.$$

Denote by $\Delta_m \subset \Delta$ the subgroup generated by

$$\{\gamma_0^j \gamma_i \gamma_0^{-j}\}, \quad i = 1, \ldots, k; \quad j = -m, \ldots, 0, 1, \ldots, m.$$

One obviously has

$$\bigcup_0^\infty \Delta_m = \Delta.$$

If for some number $m$ one has $\Delta_m = \Delta_{m+1}$ then $\Delta_m = \Delta$ and the proof is finished.

Otherwise, there is a sequence $\alpha_m \in \Delta$, $m = 0, 1, \ldots,$ such that each $\alpha_m$ is of the form $\alpha_m = \gamma_0^m \gamma_i \gamma_0^{-m}$ or $\alpha_m = \gamma_0^{-m} \gamma_i \gamma_0^m$, for some $i = 1, \ldots, k$ and $\alpha_m$ is not contained in the group generated by $\alpha_0, \alpha_1, \ldots, \alpha_{m-1}$.

Consider all the products $\beta = \beta(\varepsilon_0, \ldots, \varepsilon_m) = \alpha_0^{\varepsilon_0} \alpha_1^{\varepsilon_1} \ldots \alpha_m^{\varepsilon_m}$ where $\varepsilon_i = 0, 1$. It is clear that the equality $\beta(\varepsilon_0, \ldots, \varepsilon_m) = \beta(\varepsilon_0', \ldots, \varepsilon_m')$ implies $\varepsilon_0 = \varepsilon_0'$, $\varepsilon_1 = \varepsilon_1', \ldots,$ $\varepsilon_m = \varepsilon_m'$. So we have $2^{m+1}$ different $\beta$'s.

On the other hand $||\beta|| \leq ||\alpha_0|| + ||\alpha_1|| + \ldots + ||\alpha_m|| \leq (m+1)(2m+1)$ and for the ball $B((m+1)(2m+1)) \subset \Gamma$ we have

$$\# B((m+1)(2m+1)) \geq 2^{m+1}, \quad m = 1, 2, \ldots$$

This contradicts the polynomial growth, q.e.d.

*Proof of the algebraic lemma.* — According to the splitting lemma (see § 3) we can use induction and assume that all finitely generated subgroups in $\Gamma$ of infinite index are almost nilpotent. Let $\Delta \subset \Gamma$ be a subgroup of finite index which has the required homomorphisms into L. If all these homomorphisms have finite images we use lemma (a) and get a subgroup $\Delta' \subset \Delta$ of finite index such that the commutator subgroup $[\Delta', \Delta'] \subset \Delta'$ has infinite index.

If there is a homomorphism $\Delta \to L$ with infinite image, we apply Tits' theorem to this image and again obtain $\Delta' \subset \Delta$ with the same property.

According to lemma (b) the commutator subgroup $[\Delta', \Delta']$ is finitely generated and by the induction hypothesis it is almost nilpotent. It follows that $\Gamma$ is almost solvable and, by the theorem of Milnor-Wolf, $\Gamma$ is almost nilpotent, q.e.d.

*Corollary. — Let* $\Gamma$ *and* $\Gamma'$ *be as in the algebraic lemma. If each* $\Gamma'$ *has a subgroup* $\Delta$ *of finite index such that either* $\Delta$ *satisfies the condition of the lemma or* $\Delta$ *is Abelian, then* $\Gamma$ *is almost nilpotent.*

This is a trivial consequence of the lemma.

## 5. Topological transformation groups

The following deep theorem plays a crucial role in our proof.

(Montgomery-Zippin). — *Let* Y *be a finite dimensional, locally compact, connected and locally connected metric space. If the group* L *of the isometries of* Y *is transitive (on* Y*) then* L *is a Lie group with finitely many connected components.*

The proof immediately follows from the first corollary in § 6.3 of the book [9].

We shall also need an obvious corollary of this theorem.

*Localization lemma. — Let* Y *be as above, let* $U \subset Y$ *be a non-empty open set and let* $p = 1, 2, \ldots$ *There exists a positive* $\varepsilon$ *with the following property:*

*If* $\ell : Y \to Y$ *is a non-trivial (i.e.* $\ell$ *is not the identity) isometry such that* $\mathrm{dist}(u, \ell(u)) \leq \varepsilon$, $u \in U$, *then* $\ell$ *generates in* L *a subgroup of order at least* $p$.

The idea of applying the theory of Montgomery-Zippin to the classification of expanding maps is due to Hirsch (see [6]). He proceeds as follows.

An expanding map $X \to X$ lifts to a globally expanding homeomorphism $f: Y \to Y$ of the universal covering $Y \to X$ and $f$ respects the action of $\Gamma = \pi_1(X)$ on Y. Hirsch views $\Gamma$ as a subgroup of the group of all homeomorphisms of Y and he constructs subgroups $\Gamma = \Gamma_0 \subset \Gamma_1 \subset \ldots \subset \Gamma_i \subset \ldots$ by setting $\Gamma_i = f^{-i} \Gamma f^i$.

The closure of the union $\bigcup_i \Gamma_i$ is a topological group acting on Y, and Hirsch shows that, in some cases, this group satisfies the requirements of the theory of Montgomery-Zippin (the same corollary in § 6.3 of their book) and thus, he realizes $\Gamma$ as a subgroup of a Lie group.

In our approach we do not use the universal covering but construct Y as a limit of discrete spaces.

## 6. Limits of metric spaces

Consider a space Z with a metric $\delta$ and take two sets $X, Y \subset Z$. *The Hausdorff distance* $H^\delta(X, Y)$ is defined as the lower bound of the numbers $\varepsilon > 0$ such that the $\varepsilon$-neighbourhood of X contains Y and the $\varepsilon$-neighbourhood of Y contains X. The Hausdorff distance can be infinite but it has all properties of a metric.

Consider now two arbitrary metric spaces X and Y and denote by Z their disjoint union. A metric $\delta$ on Z is called *admissible* if its restrictions to X and Y are equal to the original metrics in X and Y respectively.

We define the *Hausdorff distance* $H(X, Y)$ as the lower bound $\inf_\delta H^\delta(X, Y)$ where $\delta$ runs over all admissible metrics on $Z = X \cup Y$.

When X and Y are *compact* spaces the Hausdorff distance enjoys all the properties of a metric. In particular, one has:

$$H(X, Y) = 0 \textit{ if and only if } X \textit{ and } Y \textit{ are isometric.}$$

When the spaces are not compact, it is convenient to have reference points in them and to use the following " modified Hausdorff distance " ([1]). For two metric spaces X, Y with distinguished points $x \in X$ and $y \in Y$, we define $\widetilde{H}((X, x), (Y, y))$ as the infimum of all $\varepsilon > 0$ with the following property: there exists an admissible metric $\delta$ on the disjoint union $X \cup Y$ such that $\delta(x, y) < \varepsilon$, that the ball $B_x(1/\varepsilon)$ of radius $1/\varepsilon$ centered at $x$ in X is contained in the $\varepsilon$-neighborhood of Y (with respect to $\delta$) and, similarly, that $B_y(1/\varepsilon)$ in Y is contained in the $\varepsilon$-neighborhood of X. For three spaces with distinguished points, the function $\widetilde{H}$ satisfies the triangular inequality provided that at least two of the three " distances " involved are small enough $\left(\text{say, } \leqslant \frac{1}{2}\right)$.

*Proper spaces.* — A metric space X is called *proper* if for each point $x_0 \in X$, the distance function $x \to \operatorname{dist}(x_0, x)$ is a proper map $X \to \mathbf{R}$, i.e. if each closed ball (of

([1]) The definition of the modified distance $\widetilde{H}$ is due to O. Gabber who kindly pointed out that another function $\widetilde{H}$ introduced in an earlier version of this paper was inadequate.

finite radius) in X is compact. Observe that a Riemannian manifold is proper if and only if it is complete. This is the theorem of Hopf-Rinow (see [2]).

*Convergence.* — We say that a sequence of spaces $X_j$ with distinguished points $x_j \in X_j$ converges to $(Y, y)$, and we write $(X_j, x_j) \underset{j\to\infty}{\longrightarrow} (Y, y)$ if $\lim \widetilde{H}((X_j, x_j), (Y, y)) = 0$. If the spaces $X_j$ are compact *with uniformly bounded diameter*, this implies that $H(X_j, Y) \to 0$. It is not difficult to see that if the spaces $X_j$ are proper and if there exist arbitrarily large real numbers $r$ such that the sequence $(B_{x_j}(r))$ of balls of radius $r$ in the $X_j$ converges for the Hausdorff distance H, then a subsequence of $(X_j, x_j)$ converges (in the above sense) to a proper space with distinguished point.

*Uniform compactness.* — A family $\{X_j\}$, $j \in J$, of compact metric spaces is called *uniformly compact* if their diameters are uniformly bounded and one of the following two equivalent conditions is satisfied:

(a) for each $\varepsilon > 0$ there is a number $N = N(\varepsilon)$ such that each space $X_j$, $j \in J$, can be covered by N balls of radius $\varepsilon$;

(b) for each $\varepsilon > 0$ there is a number $M = M(\varepsilon)$ such that in each space $X_j$, $j \in J$, one can find at most M disjoint balls of radius $\varepsilon$.

*Compactness criterion.* — *Let* $(X_j, x_j)_{j=1,2,\dots}$, *be a sequence of proper metric spaces with distinguished points. If for each* $r \geq 0$ *the corresponding family of balls* $\{B_j(r)\}_{j=1,2,\dots}$, *is uniformly compact, then there is a subsequence* $(X_{j_k}, x_{j_k})_{k=1,2,\dots}$ with $\lim_{k\to\infty} j_k = \infty$, *which converges to a proper metric space* $(Y, y)$.

*Proof.* — To prove the criterion it is sufficient to find a convergent subsequence of $(B_j(r))$ for an arbitrary but fixed number $r$ and thus we can assume without loss of generality that all $X_j$ are compact and that the family $\{X_j\}_{j=1,2,\dots}$ is uniformly compact.

Take the sequence $\varepsilon_i = 2^{-i}$ and let $N_i$ be natural numbers such that each $X_j$ can be covered by $N_i$ balls of radius $\varepsilon_i$. Denote by $A_i$ the set of all finite sequences of the form $(n_1, n_2, \dots, n_i)$, $1 \leq n_1 \leq N_1$, $1 \leq n_2 \leq N_2$, $\dots$, $1 \leq n_i \leq N_i$, and denote by $p_i : A_{i+1} \to A_i$ the natural projection.

For each space $X_j$, $j = 1, \dots$, there are maps $I_j^i : A_i \to X_j$ with the following properties:

(a) the image of $I_j^i$ forms an $\varepsilon_i$-net in $X_j$, i.e. the $\varepsilon_i$-balls centered at the points of this image cover $X_j$;

(b) for each $a \in A_{i+1}$, $i = 1, 2, \dots$, the point $I_j^{i+1}(a)$ is contained in the $2\varepsilon_i$-ball centered at $I_j^i(p_i(a))$.

These maps are constructed as follows. First we cover $X_j$ by $N_1$ balls of radius $\varepsilon_1$ and we take any bijective map from $A_1$ onto the set of centers of these balls. This is our map $I_j^1$. After that we cover each $\varepsilon_1$-ball by $N_2$ balls of radius $\varepsilon_2$ and map $A_2$ on the set of centers of these $\varepsilon_2$-balls so that $(n_1, n_2)$ goes to the center of a ball which we used to cover the $\varepsilon_1$-ball centered at $I_j^1((n_1))$. This is our $I_j^2$. Then we cover each $\varepsilon_2$-ball by $N_3$ balls of radius $\varepsilon_3$ and map $A_3$ onto the set of centers of these $\varepsilon_3$-balls, so that $(n_1, n_2, n_3) \in A_3$ goes to the center of a ball which was used in covering the $\varepsilon_2$-ball centered at $I_j^2((n_1, n_2))$, and so on.

Denote by A the union $\bigcup_{i=1}^{\infty} A_i$ and by $I_j : A \to X_j$ the map corresponding to all $I_j^i$, $i=1, \ldots$ Denote by F′ the space of all bounded functions $f: A \to \mathbf{R}$ with the norm $\|f\| = \sup_{a \in A} |f(a)|$. Denote by $F \subset F'$ the set which consists of all functions satisfying the following inequalities:

$$\text{if} \quad a \in A_1 \subset A, \quad \text{then} \quad 0 \leq f(a) \leq \sup_j \text{Diam } X_j,$$

$$\text{if} \quad a \in A_i, i > 1, \quad \text{then} \quad |f(a) - f(p_{i-1}(a))| \leq 2\varepsilon_{i-1}.$$

The set F is compact.

Let us define a map $h_j : X_j \to F'$ as follows

$$(h_j(x))(a) = \text{dist}(x, I_j(a)), \quad x \in X_j, \quad a \in A,$$

and " dist " is taken relative to the metric in $X_j$.

The property (a) of $I_j^i$ implies that the map $h_j$ is isometric and the property (b) shows that the image of $h_j$ is contained in F. So we have proved:

*If the family* $\{X_j\}$ *is uniformly compact then there is a compact metric space* F *such that each* $X_j$ *can be isometrically embedded into* F.

To complete the proof of the compactness criterion we invoke the following well-known fact.

*Let* F *be a compact metric space with metric* $\delta$. *Then the space of all compact subsets of* F *is a compact space relative to the Hausdorff distance* $H^\delta$.

We now identify each $X_j$ with its image $h_j(X_j) \subset F$ and take a subsequence $X_{j_k}$ which converges to a compact set $Y \subset F$, i.e. $\lim_{k \to \infty} H^\delta(X_{j_k}, Y) = 0$, where $\delta$ is the metric associated to the norm in $F' \supset F$. It is clear that the distance $H(X_{j_k}, Y)$ also converges to zero as $k \to \infty$, q.e.d.

*Example.* — Let $(X_j)$ be a system of complete $n$-dimensional Riemannian manifolds with Ricci curvature bounded from below by a negative constant. Then the sequence $(X_j, x_j)$ has a subsequence which converges to a metric space $(Y, y)$, but this space is not, in general, a manifold. (See [5] for additional examples.)

*Definite convergence.* — Let $(X_j, x_j)$ be a sequence of spaces which converges to $(Y, y)$. By definition, there exists a system of metrics $\delta_j$ in the disjoint unions $X_j \cup Y$ such that, for any given $r \geq 0$ and $\varepsilon > 0$, the following properties hold for almost all $j$: one has

$\delta_j(x_j, y) \leqslant \varepsilon$, the ball $B_{x_j}(r)$ in $X_j$ is contained in the $\varepsilon$-neighborhood of Y (with respect to $\delta_j$) and the ball $B_y(r)$ in Y is contained in the $\varepsilon$-neighborhood of $X_j$. When such metrics $\delta_j$ are chosen and fixed, we say that there is *definite convergence* and we write

$$(X_j, x_j) \underset{j\to\infty}{\Rightarrow} (Y, y).$$

Now it makes sense to speak about convergence of a sequence $x'_j \in X_j$ to a point $y' \in Y$: this just means that $\lim_{j\to\infty} \delta_j(x'_j, y') = 0$. In particular, the reference points $x_j$ converge to $y \in Y$.

*Convergence of maps.* — Consider a sequence $(X_j, x_j) \underset{j\to\infty}{\Rightarrow} (Y, y)$ and a system of maps $f_j : X_j \to X_j$. We say that the maps $f_j$ *converge to a map* $f : Y \to Y$ if for each $r \geq 0$ and each positive $\varepsilon$ there is a number $\mu = \mu(r, \varepsilon) > 0$ and an integer $N = N(r, \varepsilon)$ such that, for all $j \geq N$, one has:

If the points $x' \in B_{x_j}(r) \subset X_j$ and $y' \in B_y(r) \subset Y$ satisfy $\delta_j(x', y') \leq \mu$ then $\delta_j(f_j(x'), f(y')) \leq \varepsilon$.

*Isometry Lemma. — Let* $(X_j, x_j) \underset{j\to\infty}{\Rightarrow} (Y, y)$ *and let* Y *be a proper space. Let* $f_j : X_j \to X_j$ *be isometries such that* $dist_j(x_j, f_j(x_j)) \leq C$ *(where* C *is a constant which does not depend on* $j$ *and* $dist_j$ *denotes the metric in* $X_j$*). Then there is a subsequence* $(X_{j_k}, x_{j_k})$ *such that the maps* $f_{j_k}$ *converge to an isometry* $Y \to Y$.

*Proof.* — Take a sequence $(\varepsilon_i)$, with $\lim_{i\to\infty} \varepsilon_i = 0$ and $\varepsilon_i \leqslant 1/4$, and a sequence $(r_i)$ such that $r_{i+1} \geqslant r_i + C + 1$. Upon passing to a subsequence, we may—and shall—assume that, for all $j$, $\delta_j(x_j, y) \leqslant \varepsilon_j$, that the ball $B_y(r_j)$ is contained in the $\varepsilon_j$-neighborhood of $X_j$ and that the ball $B_{x_j}\left(r_j + C + \frac{1}{2}\right)$ is contained in the $\varepsilon_j$-neighborhood of Y. For all $i$, choose a finite $\varepsilon_i$-net in $B_y(r_i)$. Now, construct a system of maps $g_{ij} : E_i \to E_{i+1}$ as follows. For $e \in E_i$, choose a point $x \in X_j$ such that $\delta_j(e, x) \leqslant \varepsilon_j$. Thus, $x \in B_{x_j}\left(r_i + \frac{1}{2}\right)$ hence $f_j(x) \in B_{x_j}\left(r_i + C + \frac{1}{2}\right)$. Now choose $x' \in Y$ such that $\delta_j(x', x) \leqslant \varepsilon_j$. Then, $x' \in B_y(r_{i+1})$. Finally, take for $g_{ij}(e)$ any point $e'$ of $E_{i+1}$ such that $dist(x', e) \leqslant \varepsilon_{i+1}$.

There exists a sequence $j_1, \ldots, j_k, \ldots$, such that for each $i = 1, 2, \ldots$, the map $g_{ij_k}$ does not depend on $k$ for $k \geqslant i$, i.e. for any two sufficiently large $k$ and $\ell$ we have $g_{ij_k} = g_{ij_\ell}$. It is clear that the corresponding sequence $f_{j_k}$ converges to an isometry $g : Y \to Y$.

*Corollary. — If each space* $X_j$ *is homogeneous (i.e. the group of all isometries of* $X_j$ *is transitive) then* Y *is also homogeneous.*

*Proof.* — Let us construct an isometry $Y \to Y$ which sends $y$ to $y' \in Y$. Take an arbitrary sequence $(x'_j)$ with $x'_j \in X_j$ which converges to $y'$ and take a system of isometries $f_j : X_j \to X_j$ such that $f_j(x_j) = x'_j$. According to the lemma we can assume that $(f_j)$ converges to an isometry $g : Y \to Y$, and $g(y) = y'$.

## 7. Limits of discrete groups

We start with a general construction. Let X be a metric space with metric "dist". We denote by $\lambda X$, $\lambda > 0$, the same X but with a new metric

$$(\text{dist})^{\text{new}} = \lambda(\text{dist}).$$

*Examples.* — When $X = \mathbf{R}^n$ then all spaces $\lambda X$, $\lambda \in (0, \infty)$, are isometric.

When X is a manifold of constant curvature K then $\lambda X$ has curvature $\lambda^{-2}K$.

When X is compact and has diameter D then $\lambda X$ has diameter $\lambda D$.

Let X be an $n$-dimensional manifold of dimension $n$ and let $x \in X$. If $\lim_{i\to\infty} \lambda_i = \infty$ then the sequence $(X_i, x_i)$, $X_i = \lambda_i X$, $x_i = x$, converges to $\mathbf{R}^n$.

When X is a metric space such that each $\varepsilon$-ball, $0 \leq \varepsilon \leq 1$ centered at $x \in X$ can be covered by $p$ balls of radius $\varepsilon/2$, where $p$ is an arbitrary but fixed number, then there is a sequence $\lambda_i \to \infty$ such that $(\lambda_i X, x)$ converges to a proper space $(Y, y)$ which can be regarded as a tangent cone of X at $x$. (The proof follows from the compactness criterion of § 6.)

We are now concerned with the limits of $\lambda_i X$ when $\lambda_i \to 0$. The following examples serve only as illustrations and the proofs (quite simple) are left to the reader.

When X is compact and $\lambda_i \to 0$, then the sequence $\lambda_i X$ converges to the one-point space.

When X is a complete manifold of non-negative Ricci curvature then for some sequence $\lambda_i \to 0$ there is a limit $(Y, y)$ of $(\lambda_i X, x)$, but Y is not always a manifold.

Let X be the free Abelian group of rank two with two fixed generators. If we use in X the metric associated to the norm as in § 3 then the sequence $(\lambda_i X, x)$, $\lambda_i \to 0$, converges to the plane $\mathbf{R}^2$ with the following Minkowski metric

$$\text{dist}((a, b), (a', b')) = |a - a'| + |b - b'|.$$

Take a non-Abelian nilpotent simply connected Lie group X of dimension 3 (notice, that there is only one such group) with a left invariant Riemannian metric. When $\lambda_i \to 0$, then the sequence $(\lambda_i X, x)$ converges to a space $(Y, y)$ which is homeomorphic to X, but the limit metric in Y is not Riemannian. This metric can be described as follows. When we divide the Lie group by its center (which is isomorphic to $\mathbf{R}$) we get a Riemannian submersion $X \to \mathbf{R}^2$ with one dimensional fibers. Take two points $x_1, x_2 \in X$ and consider all smooth curves which are normal to the fibers and which join $x_1$ with $x_2$. Define $\text{dist}(x_1, x_2)$ as the lower bound of the lengths of these curves. This is exactly the limit metric in Y (which is homeomorphic to X). Notice that for each $\lambda > 0$ the space $\lambda Y$ is isometric to Y. Notice also that the Hausdorff dimension of Y is 4 rather than 3. (The definition of the Hausdorff dimension can be found in Ch. VII of [7].)

Let us return to our major topic.

*Main construction.* — Let $\Gamma$ be a group of polynomial growth with a fixed system of generators and with the corresponding metric dist. Let $\{r_i\}, i=1, 2, \ldots$ be a sequence of $i$-regular numbers such that $\lim_{i\to\infty} r_i = \infty$ (see the regularity lemma of § 3). We denote by $e \in \Gamma$ the identity element and we consider the sequence $(\Gamma_i, e_i)$, $\Gamma_i = r_i^{-1}\Gamma$, $e_i = e$. It follows from the definition of the regularity (see inequalities $(a)$ and $(b)$) that the family of $r$-balls in $(\Gamma_i)_{i=1,2,\ldots}$ is uniformly compact (each $\Gamma_i$ is, obviously, a proper space) and, by the compactness criterion (see § 6), there is a convergent subsequence. To avoid double indices we assume that the sequence $(\Gamma_i, e_i)$ itself converges to a space $(Y, y)$.

*Properties of the limit space* Y

(1) Y is a locally compact space, because it is proper.

(2) Y is connected and locally connected. Moreover each ball in Y is connected and even path-connected.

*Proof.* — The connectivity property of $\Gamma$ (see § 3) implies that for any two points $\alpha, \beta \in \Gamma_i$ there is a point $\gamma \in \Gamma_i$ such that

$$\operatorname{dist}_i(\alpha, \gamma) \leq \frac{1}{2}\operatorname{dist}_i(\alpha, \beta) + r_i^{-1},$$

$$\operatorname{dist}_i(\gamma, \beta) \leq \frac{1}{2}\operatorname{dist}_i(\alpha, \beta) + r_i^{-1}.$$

It follows that for any two points $y_1, y_2 \in Y$ there is an $x \in Y$ such that

$$\operatorname{dist}(y_1, x) = \operatorname{dist}(x, y_2) = \frac{1}{2}\operatorname{dist}(y_1, y_2)$$

(" $\operatorname{dist}_i$ " denotes the metric in $\Gamma_i$ and " dist " is the limit metric in Y). This property not only implies the required connectivity of Y, but also shows that any two points $y_1, y_2 \in Y$ can be joined by a curve with the length equal to $\operatorname{dist}(y_1, y_2)$.

(3) The group L of all isometries of Y is transitive on Y.

This follows from the corollary to the isometry lemma (§ 6).

(4) Y is finite dimensional.

*Proof.* — The regularity condition (see inequality $(a)$ in § 3) implies that for $j \leq i$ each ball in $\Gamma_i$ of radius $\frac{1}{2}$ can be covered by $N_j$ balls of radius $2^{-j+1}$ where $N_j = 2^{j(d+1)}$ and $d$ denotes the growth of $\Gamma$. It follows that for *each* $j = 1, 2, \ldots,$ one can cover every $\frac{1}{2}$-ball in Y by $N_j$ balls of radius $2^{-j+1}$. This shows that the Hausdorff dimension of Y is at most $d+1$ and hence (see chapter VII in [7]) Y is finite dimensional.

*Main conclusion. — The group* L *of all isometries of* Y *is a Lie group with finitely many components.*

*Proof.* — Use the theorem of Montgomery-Zippin (see § 5).

## 8. Proof of the main theorem

Take an arbitrary group $\Gamma$ with a fixed finite system of generators and the associated metric. Define

$$D(\gamma, r) = \sup_{\beta} \operatorname{dist}(\gamma\beta, \beta)$$

where $\gamma \in \Gamma$, $r \in [0, \infty)$ and $\beta$ runs over the $r$-ball in $\Gamma$ centered at the identity.

Take now a subgroup $\Gamma' \subset \Gamma$ generated by $\gamma_1, \ldots, \gamma_k$ and set (with an abuse of notations)

$$D(\Gamma', r) = \sup_{1 \leq j \leq k} D(\gamma_j, r).$$

*If the function* $D(\Gamma', r)$, $r \in [0, \infty)$, *is bounded, then* $\Gamma'$ *contains an Abelian subgroup of finite index.*

*Proof.* — If $D(\gamma, r)$ is bounded when $r \to \infty$, then the conjugacy class $\{\beta^{-1}\gamma\beta\}$, $\beta \in \Gamma$, of $\gamma$ is finite and the centralizer of $\gamma$ has finite index in $\Gamma$, q.e.d.

Suppose that $D(\Gamma', r)$ is unbounded but for a divergent sequence $r_i$ the ratio $r_i^{-1} D(\Gamma', r_i)$ converges to zero.

*Displacement lemma. — For each* $\varepsilon > 0$ *there is a sequence* $\alpha_i$, $i = 1, 2, \ldots$, *such that*

$$\lim_{i \to \infty} r_i^{-1} D(\alpha_i^{-1} \Gamma' \alpha_i, r_i) = \varepsilon,$$

*where* $D(\alpha^{-1}\Gamma'\alpha, r) = \sup_{1 \leq j \leq k} D(\alpha^{-1}\gamma_j\alpha, r)$.

*Proof.* — The connectivity property of $\Gamma$ (see § 3) implies that for an arbitrary integer $m$ one has

$$D(\Gamma', r+m) \leq D(\Gamma', r) + 2m.$$

On the other hand it is clear that for any $\alpha \in \Gamma$ and $r \geq 0$

$$D(\alpha^{-1}\Gamma'\alpha, r) \leq D(\Gamma', r + \|\alpha\|).$$

So we have

$$D(\alpha^{-1}\Gamma'\alpha, r) \leq D(\Gamma', r) + 2\|\alpha\|. \qquad (*)$$

Since $D(\Gamma', r)$ is unbounded as a function of $r$, the function $D(\alpha^{-1}\Gamma'\alpha, r)$ is unbounded as a function of $\alpha \in \Gamma$, when $r$ is kept fixed.

When $r_i$ is sufficiently large our assumption $r_i^{-1}D(\Gamma', r_i) \to 0$ implies that

$$D(\Gamma', r_i) < \varepsilon r_i.$$

On the other hand for some $\mu \in \Gamma$ we have

$$D(\mu^{-1}\Gamma'\mu, r_i) > \varepsilon r_i.$$

Using (*) and the connectivity again we conclude that there is an $\alpha_i \in \Gamma$ such that

$$|D(\alpha_i^{-1}\Gamma'\alpha_i, r_i) - \varepsilon r_i| \leq 2,$$

q.e.d.

Let $\Gamma_i$ be as in the main construction (see § 7). The group $\Gamma$ acts isometrically on each $\Gamma_i$ and if $\gamma_i \in \Gamma$ satisfy $r_i^{-1}||\gamma_i|| \leq C < \infty$, $i = 1, 2, \ldots$, then the corresponding isometries $\tilde{\gamma}_i : \Gamma_i \to \Gamma_i$ satisfy the condition of the isometry lemma (see § 6) and we can assume (using a subsequence when it is necessary) that these isometries converge to an isometry $\ell : Y \to Y$.

Let $\Gamma' \subset \Gamma$ be an arbitrary subgroup and $\gamma \in \Gamma'$. By taking $\gamma_i = \gamma$ we get an isometry $\ell = \ell_\gamma : Y \to Y$ and so we get a map $\Gamma' \to L$, where L is the isometry group of Y. (Because $\Gamma'$ is countable we can assume that the convergence takes place for all $\gamma \in \Gamma'$.) This map is, obviously, a homomorphism. (To be precise we must fix a definite convergence $\Gamma_i \Rightarrow Y$ as in § 6 and only then our consideration becomes meaningful.)

The kernel of this homomorphism $\gamma \to \ell_\gamma$ consists of all $\gamma$ in $\Gamma'$ such that

$$\lim_{i\to\infty} r_i^{-1}D(\gamma, r_i) \to 0.$$

This limit exists because we have the convergence of the isometries $\gamma_i = \gamma : \Gamma_i \to \Gamma_i$ to $\ell$.

We are ready to prove the main theorem. Take a subgroup $\Gamma' \subset \Gamma$ generated by $\gamma_1', \ldots, \gamma_j', \ldots, \gamma_k'$. According to the algebraic lemma (§ 4) and the main conclusion of § 7 we must only find a subgroup $\Delta \subset \Gamma'$ of finite index in $\Gamma'$ such that either $\Delta$ is commutative or $\Delta$ has as many homomorphisms into L as is required by the algebraic lemma. (See the corollary to the algebraic lemma.) If the homomorphism $\gamma \to \ell_\gamma$ we constructed above has infinite image, the conditions of the algebraic lemma are satisfied and the proof is finished.

Suppose that the kernel $\Gamma'' \subset \Gamma'$ of the homomorphism $\gamma \to \ell_\gamma$ has finite index in $\Gamma'$. For the group $\Gamma''$ we have

$$\lim_{i\to\infty} r_i^{-1}(D(\gamma, r_i)) = 0, \quad \gamma \in \Gamma'',$$

and we can shorten the notations by assuming $\Gamma' = \Gamma''$.

If the function $D(\Gamma', r)$, $r \in [0, \infty)$, is bounded we have an Abelian $\Delta \subset \Gamma'$ and the proof is finished. Now we come to the last case.

The function $D(\Gamma', r)$ is unbounded but

$$\lim_{i\to\infty} r_i^{-1}D(\Gamma', r_i) = 0.$$

Fix an $\varepsilon > 0$ and construct $\alpha_i = \alpha_i(\varepsilon) \in \Gamma$ as in the displacement lemma. We obtain a new isometric action of $\Gamma'$ on $\Gamma_i$ as follows. First we send $\gamma$ to $\gamma_i = \alpha_i^{-1}\gamma\alpha_i$ and then we act by $\gamma_i$ on $\Gamma_i$ as usual by the left translation. For each generator $\gamma'_1, \ldots, \gamma'_j, \ldots, \gamma'_k \in \Gamma'$ the translations $\gamma_{ji} = \alpha_i^{-1}\gamma'_j\alpha_i : \Gamma_i \to \Gamma_i$ satisfy

$$\lim_{i\to\infty} \sup \operatorname{dist}_i(\gamma_{ji}(e_i), e_i) \leq \varepsilon,$$

where $\operatorname{dist}_i$ denotes the distance in $\Gamma_i$, $e_i \in \Gamma_i$ denotes the reference point in $\Gamma_i$ corresponding to $e \in \Gamma$ and $j = 1, \ldots, k$.

The isometry lemma (§ 6) allows us to assume that for each $j$ the sequence $\gamma_{ji}$, $i \to \infty$, converges to an isometry $\ell_j : Y \to Y$ and so we obtained a homomorphism $A = A(\varepsilon) : \Gamma' \to L$.

Let us show that when $\varepsilon > 0$ is small then the image of $A(\varepsilon)$ is large. The properties of $\alpha_i = \alpha_i(\varepsilon)$ guarantee (see the displacement lemma) that for some $\gamma'_j$, say for $\gamma'_1$, one has

$$\lim_{i\to\infty} r_i^{-1} D(\alpha_i^{-1}\gamma'_1\alpha_i, r_i) = \varepsilon.$$

But

$$r_i^{-1} D(\alpha_i^{-1}\gamma'_1\alpha_i, r) = \sup_x \operatorname{dist}_i(\gamma_{1i}(x), x),$$

where $x$ runs over the unit ball $B_{e_i}(1) \subset \Gamma_i$. So for the limit $\ell_1 = A(\gamma'_1)$ we also have

$$\sup_{y'} \operatorname{dist}(\ell_1(y'), y') = \varepsilon,$$

where $y'$ runs over the ball $B_y(1) \subset Y$.

We apply now the localization lemma (§ 5) and finish the proof of the main theorem.

*Final remarks.* — Let $\Gamma$ be a finitely generated almost nilpotent group *without torsion*. By a result of Auslander-Schenkman (see [1]) one can easily show that $\Gamma$ contains a nilpotent subgroup $\Delta \subset \Gamma$ of index $q$ with $\log\log q \leq 2^d$, where $d = \operatorname{growth}(\Gamma)$.

This allows us to apply the main theorem to infinitely generated groups.

*If $\Gamma$ has no torsion and each finitely generated subgroup $\Gamma' \subset \Gamma$ has growth at most $d$, then $\Gamma$ contains a nilpotent subgroup of finite index.*

Let us now give a more effective version of the Main Theorem.

*For any positive integers $d$ and $k$ there exist positive integers $R$, $N$ and $q$ with the following properties. If a group $\Gamma$ with a fixed system of generators satisfies the inequality $\# B(r) \leq kr^d$ for $r = 1, 2, \ldots, R$, then $\Gamma$ contains a nilpotent subgroup $\Gamma'$ of index at most $q$ and whose degree of nilpotency is at most $N$.*

*Proof.* — We start with a definition. Let $\Delta$ and $\Gamma^i$, $i = 1, 2, \ldots,$ be a system of groups endowed with a fixed system of $k$ generators. We say that the sequence $(\Gamma^i)$ converges to $\Delta$ if there is a sequence of balls $B_i = B(r_i) \subset \Delta$ with $r_i \to \infty$ for $i \to \infty$, and a sequence of bijective maps $f$ of each ball $B_i$ onto the $r_i$-ball in $\Gamma^i$ centered at the identity such that $f(\delta_1 \delta_2^{-1}) = f(\delta_1)\,(f(\delta_2))^{-1}$ for any two elements $\delta_1$, $\delta_2$ in $\Delta$ satisfying $\|\delta_1\| + \|\delta_2\| \leq r_i$.

An arbitrary sequence of groups $(\Gamma^i)$, each with $k$ given generators, always has a convergent subsequence. If the limit group $\Delta$ has a nilpotent subgroup of index $\leq R$ and of degree of nilpotency $\leq N$, then such subgroups exist in the groups $\Gamma^i$ for all sufficiently large $i$'s.

Now, if we suppose that our theorem is false, we get a sequence of groups $\Gamma^j$, $j = 1, \ldots,$ such that the balls in each group $\Gamma^j$ satisfy the inequality $B(r) \leq kr^d$ for $r = 1, \ldots, j$, but none of the groups $\Gamma^j$ has a nilpotent subgroup of index $\leq j$ and of degree of nilpotency $\leq j$. Taking a convergent subsequence and passing to the limit we get a group $\Delta$ such that all balls in $\Delta$ satisfy $B(r) \leq kr^d$, $r = 1, 2, \ldots,$ and such that $\Delta$ contains no nilpotent subgroup of finite index. This contradicts the Main Theorem.

*Question.* — What is the dependence of the numbers R, N and $q$ on $d$ and $k$? In particular, does there exist an *effective* estimate of these numbers in terms of $d$ and $k$?

*A geometric application.* — Let $V = V^n$ be a complete Riemannian manifold such that the values of the Ricci tensor on all unit tangent vectors of V are bounded from below by $-(n-1)K$, $K \geq 0$. Let $\Gamma$ be the group generated by some isometries $\gamma_1, \ldots, \gamma_k$ of V. Suppose that for a point $v \in V$ we have the following inequalities:

$$\operatorname{dist}(v, \gamma(v)) \geq \varepsilon > 0, \qquad \text{for all} \quad \gamma \in \Gamma \backslash e;$$
$$\operatorname{dist}(v, \gamma_i(v)) \leq C\varepsilon, \qquad \text{for} \quad C \geq 1 \quad \text{and} \quad i = 1, \ldots, k.$$

In this case, the geometric growth theorem of Milnor (see [8]) reduces to the inequality

$$\# B(r) \leq 4^n C^n r^n \exp(2nCr\varepsilon\sqrt{K}), \qquad r = 1, 2, \ldots,$$

where $B(r)$ denotes the $r$-ball in $\Gamma$ relative to the word metric associated to $\{\gamma_i\}$. This gives the following

*Geometric theorem.* — *If* $\varepsilon$ *is sufficiently small compared to* $n$, C *and* K, *that is* $\varepsilon \leq \mu = \mu(n, C, K) > 0$, *then the group* $\Gamma$ *is almost nilpotent.*

*Acknowledgements.* — I want to thank H. Bass and J. Tits for their interest in this paper and their many helpful comments and suggestions. I am particularly thankful to J. Tits who kindly agreed to present his short proof of the theorem of Milnor-Wolf in an appendix.

REFERENCES

[1] L. Auslander, E. Schenkman, Free groups and Hirsch-Plotkin radicals, *Proc. A.M.S.*, **16** (1965), 784-788.
[2] J. Cheeger, D. Ebin, *Comparison theorems in Riemannian geometry*, North-Holland, 1975.
[3] V. Efremovič, The proximity geometry of Riemannian manifolds, *Uspehi Mat. Nauk*, **8** (1953), 189 (Russian).

[4] D. Epstein, M. Shub, Expanding endomorphisms of flat manifolds, *Topology*, **7** (1968), 13-141.
[5] M. Gromov, Synthetic geometry of Riemannian manifolds, *Proc. I.C.M. Helsinki*, 1978, vol. I, 415-419 (1979).
[6] M. Hirsch, Expanding maps and transformation groups, *Proc. Symp. Pure Math.*, XIV (1970), 125-133.
[7] W. Hurewicz, H. Wallman, *Dimension Theory*, Princeton U. Press, 1948.
[8] J. Milnor, A note on curvature and fundamental group, *J. Diff. Geom.*, **2** (1968), 1-7.
[9] D. Montgomery, L. Zippin, *Topological transformation groups*, Interscience, N. Y., 1955.
[10] M. Raghunathan, *Discrete subgroups of Lie groups*, Springer, Berlin, 1972.
[11] M. Shub, Expanding maps, *Proc. Symp. Pure Math.* XIV (1970), 273-277.
[12] A. Švarc, A volume invariant of coverings, *Dokl. Akad. Nauk SSSR*, **105** (1955), 32-34.
[13] J. Tits, Free subgroups in linear groups, *J. Algebra*, **20** (1972), 250-270.
[14] J. Wolf, Growth of finitely generated solvable groups and curvature of Riemannian manifolds, *J. Diff. Geom.*, **2** (1968), 421-446.

Institut des Hautes Études scientifiques,
35, route de Chartres,
91440 Bures-sur-Yvette.

*Manuscrit reçu le 11 mars 1980.*

# APPENDIX

*by* Jacques TITS

This appendix to M. Gromov's paper " Groups of polynomial growth and expanding maps "—hereafter referred to as [G]—arose from an attempt to understand J. Wolf's article [10] (1). Our main purpose is to provide a short and self-contained account of the results on nilpotent and solvable groups which are needed for the proof of Gromov's main theorem and its converse. In A1 we show—in a slightly more general context—that a finitely generated nilpotent group has polynomial growth, and, as a by-product of the proof, we obtain a formula of H. Bass [*Proc. Lond. Math. Soc.* (3), 25 (1972), 603-614] giving the degree of that growth (proposition 2). In A2, we observe that one can, at little cost, make Gromov's proof of his main theorem independent of the special case of that theorem for solvable groups, thus providing a new proof for that special case as well (the statement, part of the result of Milnor-Wolf, that the growth of a finitely generated solvable group is either polynomial or exponential, is not included). The present text is made up of excerpts of a conference at the Séminaire Bourbaki (February 1981, exposé n° 572), slightly expanded and adapted.

## A 1. Growth of filtered groups

Let $\Gamma$ be a nilpotent group and let $(\Gamma_i)_{i\in\mathbf{N}^*}$ be a system of subgroups such that $\Gamma=\Gamma_1$, $[\Gamma_i, \Gamma_j]\subset\Gamma_{i+j}$ and $\Gamma_i=\{1\}$ for almost all $i$. By an *f-generating set* of $\Gamma$ (relative to the filtration $(\Gamma_i)$), we mean a subset E of $\Gamma$ such that $E_i=E\cap\Gamma_i$ generates $\Gamma_i$ for all $i$. Set $E'_i=E-E_{i+1}$. We define the *f-length* of a word in the elements of $E\cup E^{-1}$ as the increasing sequence $(n_1, n_2, \ldots)$, where $n_i$ is the length (in the usual sense) of the contribution of $E'_i\cup E'^{-1}_i$ to the word. An element of $\Gamma$ is said to be *of f-length* $\leqslant(r_1, r_2, \ldots)$, with $r_i\in\mathbf{R}_+$, if it can be expressed as a word of f-length $(n_1, n_2, \ldots)$, with $n_i\leqslant r_i$. Assuming E finite, we denote by ${}_fc(r_1, r_2, \ldots)$ the number of such elements. If ${}_fc'$ is the function defined in the same way as ${}_fc$ but starting from another finite f-generating set, there exist nonvanishing constants $a, b\in\mathbf{R}_+$ such that

$$ {}_fc(ar_1, ar_2, \ldots)\leqslant{}_fc'(r_1, r_2, \ldots)\leqslant{}_fc(br_1, br_2, \ldots) $$

(1) Numbers between brackets refer to the bibliography of [G].

for all sequences $(r_i)$. This legitimates the following definition: we say that the group $\Gamma$ has *polynomial f-growth of degree* $\leqslant d$ if there is a constant $c \in \mathbf{R}_+$ such that, for all $r \in \mathbf{R}_+$, one has

$$ {}_{f}c(r, r^2, r^3, \ldots) \leqslant cr^d + 1. $$

*Proposition* 1. — *If $d_i$ denotes the rank of the abelian group $\Gamma_i/\Gamma_{i+1}$, then the group $\Gamma$ has polynomial f-growth of degree $\leqslant id_i$.*

We proceed by double induction: a descending induction on $a = \sup\{i \mid \Gamma = \Gamma_i\}$ and (given $a$) an ordinary induction on the minimum $m$ of the cardinalities of all generating sets of $\Gamma_a/\Gamma_{a+1}$. We choose the f-generating set E so that $\mathrm{Card}\, E'_a = m$, that $[E \cup E^{-1}, E \cup E^{-1}] \subset E$, and that, if an element of $E'_a$ has a nontrivial power inside $\Gamma_{a+1}$, then its power with the smallest strictly positive exponent having that property also belongs to E. Let us choose $y \in E'_a$ and denote by $\Gamma'$ the subgroup of $\Gamma$ generated by $E'_a - \{y\}$. We first prove the following assertion, by induction on $q$:

(*) if $w$ is a word of f-length $(n_1, n_2, \ldots)$ in the elements of $E \cup E^{-1}$ and if $(y_1, y_2, \ldots, y_p)$ is the contribution of $\{y, y^{-1}\}$ to $w$ (thus, $y_i = y$ or $y^{-1}$ for all $i$, and $p \leqslant n_a$), then, for $0 \leqslant q \leqslant p$, there is a word $w_q$ representing the same element of $\Gamma$ as $w$, with the same contribution of $\{y, y^{-1}\}$, starting by $y_1 y_2 \ldots y_q$, and whose f-length $(n'_1, n'_2, \ldots)$ satisfies the relations

$$ n'_i \leqslant n_i + q n_{i-a} + \binom{q}{2} n_{i-2a} + \ldots $$

The assertion is obvious for $q = 0$. The induction hypothesis provides a word $w_{q-1}$ starting with $y_1 \ldots y_{q-1}$ and of f-length $(n''_1, n''_2, \ldots)$ satisfying

$$ n''_i \leqslant n_i + (q-1) n_{i-a} + \binom{q-1}{2} n_{i-2a} + \ldots $$

Now move $y_q$ to the left by successive commutation. Its jumping over an element of $E_i$ introduces a new element belonging to $E_{i+a}$ (possibly the identity). Therefore, we eventually get a word $w_q$ starting with $y_1 \ldots y_q$ and of f-length $(n'_1, n'_2, \ldots)$ such that $n'_i \leqslant n''_i + n''_{i+a}$, hence (*).

Assuming $n_i \leqslant r^i$ (for some $r \in \mathbf{R}_+$), making $q = p \leqslant r^a$, majorizing $\binom{q}{j}$ by $q^j$ $(\leqslant r^{aj})$ and denoting by $e$ the smallest value of $i$ such that $\Gamma_i = \{1\}$, we deduce from (*) that

(**) every element $g \in \Gamma$ of f-length $\leqslant (r, r^2, \ldots)$ can be written as $g = y^s g'$, where $|s| \leqslant r^a$ and $g'$ is an element of $\Gamma'$ of f-length $\leqslant (er - |s|, er^2 - |s|, \ldots)$.

If $\Gamma/\Gamma'$ is infinite, the induction hypothesis (on $(a, m)$) applied to $\Gamma'$ implies the existence of a constant $c' \in \mathbf{R}_+$ such that the number of possible choices for $g'$ is majorized by

$$ c' r^{\Sigma(id_i) - a} + 1. $$

Since there are less than $2r^a+1$ admissible values for $s$, the proposition ensues. If $\Gamma/\Gamma'$ has order $t<\infty$, we can rewrite $g$ as $y^{s_1}g'_1$, where $0\leqslant s_1<t$, $g'_1\in\Gamma'$ and the f-length of $g'_1$ is $\leqslant(er, er^2, \ldots)$. This time, the induction hypothesis enables us to majorize the number of possible elements $g'$ by $c'r^{\Sigma id_i}+1$ (for a suitable $c'$) and the proposition again follows, since $s_1$ takes only finitely many values.

*Lemma* 1. — *Let* $\Gamma$ *be a finitely generated nilpotent group of class* $e$. *Denote by* Z *the last nontrivial term of its descending central series, by* E *a finite generating subset of* $\Gamma$ *and by* $z$ *an element of* Z. *Then, there exists a constant* $c\in\mathrm{R}_+$ *such that, for all* $n\in\mathrm{N}$, $z^n$ *can be expressed as a word of length* $c\sqrt[e]{n}$ *in the elements of* $\mathrm{E}\cup\mathrm{E}^{-1}$.

The proof will be by induction on $e$. It is clearly sufficient to consider the case where $z$ is the commutator $[x,y]$ of an element $x$ of E and an element $y$ of the penultimate nontrivial term of the descending central series. Let $n\in\mathrm{N}$, let $n_1$ be the smallest integer which is strictly larger than $\sqrt[e]{n}$ and let $a_1$, $a_2$ be the integers defined by

$$n=a_1n_1^{e-1}+a_2,\quad a_1<n_1,\quad a_2\leqslant n_1^{e-1}.$$

The induction hypothesis applied to $\Gamma/\mathrm{Z}$ and $y \bmod \mathrm{Z}$ implies the existence of a constant $c'\in\mathbf{R}_+$ (independent of $n$) and of two elements $y_1$ and $y_2$ of length $\leqslant c'n_1$, respectively congruent to $y^{n_1^{e-1}}$ and $y^{a_2} \bmod \mathrm{Z}$. Now, the assertion follows from the fact that the length of

$$z^n=[x^{a_1},y^{n_1^{e-1}}].[x,y^{a_2}]=[x^{a_1},y_1].[x,y_2]$$

is $\leqslant 2a_1+4c'n_1+2$.

*Proposition* 2 (Bass-Wolf). — *Let* $\Gamma$ *be a finitely generated nilpotent group and let* $d_i$ *be the rank of the i-th quotient* $\Gamma_i/\Gamma_{i+1}$ *of its descending central series* $(\Gamma_i)$. *Set* $d=\Sigma id_i$. *Choose arbitrarily a finite generating set* E *of* $\Gamma$ *and, for* $r\in\mathbf{R}_+$, *let* $c(r)$ *represent the number of elements of* $\Gamma$ *which can be expressed as a word of length* $\leqslant r$ *in the elements of* $\mathrm{E}\cup\mathrm{E}^{-1}$. *Then, there exist constants* $c_1, c_2\in\mathbf{R}_+$ *such that*

$$c_1r^d\leqslant c(r)\leqslant c_2r^d+1$$

*for all* $r$; *in particular,* $\Gamma$ *has polynomial growth of degree* $d$.

The existence of $c_2$ is an immediate consequence of proposition 1 applied to the descending central series. We prove the existence of $c_1$ by induction on the class $e$ of $\Gamma$. Set $d'=d-ed_e$. The induction hypothesis implies the existence of a constant $c'_1\in\mathbf{R}_+$ such that, for $r\in\mathbf{R}_+$, there exists a subset of $\Gamma$ of cardinality $\geqslant c'_1r^{d'}$ consisting of elements of length $\leqslant\frac{r}{2}$, pairwise non congruent mod $\Gamma_e$. On the other hand, by lemma 1, there exists $c''_1\in\mathbf{R}_+$ such that $\Gamma_e$ has more than $c''_1r^{ed_e}$ distinct elements of length $\leqslant\frac{r}{2}$. Hence the claim.

(N.B. — The above proof of the existence of $c_1$, including lemma 1, essentially follows the line of the proof given by J. Wolf [10], who also obtains an upper bound for $c(r)$, sufficient to establish the polynomial growth but coarser than the upper bound of proposition 2, which is due to H. Bass [*loc. cit.*]. Our proof of the existence of $c_2$, obtained without prior knowledge of Bass' result, is different from his, at least formally: the introduction of the f-growth makes it less computational and, as it seems, somewhat shorter.)

## A 2. A special case of the theorem of Milnor-Wolf

In Gromov's proof of his main theorem, the only reference to the theorem of Milnor-Wolf (stated in the introduction of [G]) is at the end of [G], § 4 (proof of the " algebraic lemma "), and the reader will easily convince himself that only the following very special case of the theorem is needed there (take for L the group $\Delta'$ of [G]):

*Lemma* 2. — *Let* L *be a group. Suppose that there exists a homomorphism* $\xi$ *of* L *onto* $\mathbf{Z}$ *whose kernel is finitely generated and almost nilpotent. Then* L *itself is almost nilpotent or its growth is exponential.*

The following argument is extracted from [10] (where it is however somewhat hidden). We first prove:

*Lemma* 3. — *Let* $\Lambda$ *be a free abelian group and let* $\alpha : \Lambda \to \Lambda$ *be an automorphism.*

(i) *If* $\alpha$ *(extended to* $\Lambda\otimes\mathbf{C}$*) is semi-simple and if all its eigenvalues have absolute value one, then* $\alpha$ *has finite order.*

(ii) *If* $\alpha$ *has an eigenvalue of absolute value* $\geqslant 2$, *there exists* $\lambda\in\Lambda$ *such that the elements*

$$\varepsilon_0\lambda+\varepsilon_1\alpha(\lambda)+\varepsilon_2\alpha^2(\lambda)+\dots \qquad (\varepsilon_i=0 \text{ or } 1, \text{ and } =0 \text{ for almost all } i)$$

*are pairwise distinct.*

(i) It suffices to observe that the orbits of $\{\alpha^z \mid z\in Z\}$ in $\Lambda\otimes\mathbf{C}$ have compact closures, from which follows that the orbits of that same group in $\Lambda$ are finite.

(ii) If $\beta : \Lambda\to\mathbf{C}$ is a linear form such that $\beta\circ\alpha=\rho\beta$, with $|\rho|\geqslant 2$, then the assertion is true for every $\lambda\in\Lambda$ such that $\beta(\lambda)\neq 0$; indeed, one has

$$\beta\Big(\sum_{i=0}^{\infty}\varepsilon_i\alpha^i(\lambda)\Big)=\Big(\sum_{i=0}^{\infty}\varepsilon_i\rho^i\Big).\beta(\lambda)$$

and, in view of the inequality satisfied by $\rho$, the numbers $\sum_{i=0}^{\infty}\varepsilon_i\rho^i$ are pairwise distinct.

*Proof of lemma* 2. — Let $z$ be an element of L such that $\xi(z)=1$, and let $L_1'$ be the greatest nilpotent normal subgroup of $L_1$; upon substituting $L_1'$ for $L_1$ and the group generated by $L_1'$ and $z$ for L, we may—and shall—assume that $L_1$ is nilpotent. Let

$(L_i)_{1 \leqslant i \leqslant e}$ be a central series of $L_1$ normalized by $z$, and let $\alpha_i$ denote the automorphism of $L_i/L_{i+1}$ induced by int $z$. We suppose the series $(L_i)$ chosen in such a way that, whenever $L_i/L_{i+1}$ is infinite, it is a free abelian group of which $\alpha_i$ is a semi-simple automorphism : any central series normalized by $z$ clearly has a refinement satisfying that condition. If all $\alpha_i$ have finite order, there exists an integer $s \geqslant 1$ such that $z^s$ centralizes each quotient $L_i/L_{i+1}$; then, the group generated by $L_1$ and $z^s$ has finite index in L and is nilpotent, and the lemma is proved. Let us therefore assume that, for some $j$, $\alpha_j$ has infinite order. By lemma 3 (i), there exists an integer $t$ such that $\alpha_i^t$ has an eigenvalue of absolute value $\geqslant 2$ and, by lemma 3 (ii), there exists $x \in L_j$ such that the elements

$$x^{\varepsilon_0}.(z^t x z^{-t})^{\varepsilon_1}.(z^{2t} x z^{-2t})^{\varepsilon_2}. \ldots \qquad (\varepsilon_i = 0 \text{ or } 1 \text{ and } = 0 \text{ for almost all } i)$$

are pairwise distinct. This implies that L has exponential growth, and the proof is complete.

(Note the similarity of the last argument with the proof of lemma (b) of [G], § 4.)

Collège de France,
11, place Marcelin-Berthelot,
75231 Paris Cedex 05.

*Manuscrit reçu le 7 février 1981.*

## Curriculum Vitae

Gregory A. Margulis

Date of Birth: February 24, 1946, Moscow, Russia

**Education:**

1962 Moscow High School N 721
1970 Ph.D., Moscow State University

**Employment History:**

Institute for Problems in Information Transmission
1970–1974 Junior Research Fellow, Moscow University
1975–1986 Senior Research Fellow, Moscow University
1987–1991 Leading Research Fellow, Moscow University
1991– Professor, Yale University

**Honors and Awards:**

1968 Prize of Moscow Mathematical Society for young mathematicians
1978 Fields Medal
1991 Medal of College de France (in connection with the visiting professorship)
1991 Foreign Honorary member of the American Academy of Arts and Sciences
1995 Humboldt Prize
1996 Honorary Fellow of Tata Institute of Fundamental Research
1997 Lobachevsky Prize
1999 Honorary Doctoral Degree from the University of Bielefeld
2001 Member of the National Academy of Sciences
2003 European Academy of Sciences
2004 Connecticut Academy of Science & Engineering
2005 Wolf Prize
2009 AMS Colloquium Lectures
2010 Honorary Doctoral Degree from Ecole Normale Superiere, Paris (ENS)
2011 Honorary Doctoral Degree from the University of York
2011 Dobrushin International Award

## List of Publications

1. Homogeneous flows, moduli spaces and arithmetic, in *Proc. of the Clay Mathematics Institute*, Summer School held at the Centro di Ricerca Matematica Ennio De Giorgi, Pisa, June 11–July 6, 2007. Edited by M. L. Einsiedler, D. A. Ellwood, A. Eskin, D. Kleinbock, E. Lindenstrauss, G. Margulis, S. Marmi and J.-C. Yoccoz. Clay Mathematics Proceedings, Vol. 10 (Amer. Math. Soc. and Clay Mathematics Institute, 2010).
2. D. Kleinbock, G. Margulis and J. Wang, Metric Diophantine approximation for systems of linear forms via dynamics, *Int. J. Number Theory* **6** (2010), No. 5, 1139–1168.
3. W. M. Goldman, F. Labourie and G. Margulis, Proper affine actions and geodesic flows of hyperbolic surfaces, *Ann. of Math.* (2) **170** (2009), No. 3, 1051–1083.
4. J. S. Athreya and G. A. Margulis, Logarithm laws for unipotent flows, I, *J. Mod. Dyn.* **3** (2009), No. 3, 359–378.
5. D. Fisher and G. Margulis, Local rigidity of affine actions of higher rank groups and lattices, *Ann. of Math.* (2) **170** (2009), No. 1, 67–122.
6. M. Einsiedler, G. Margulis and A. Venkatesh, Effective equidistribution for closed orbits of semisimple groups on homogeneous spaces, *Invent. Math.* **177** (2009), No. 1, 137–212.
7. T. Gelander, A. Karlsson and G. A. Margulis, Superrigidity, generalized harmonic maps and uniformly convex spaces, *Geom. Funct. Anal.* **17** (2008), No. 5, 1524–1550.
8. D. Fisher and G. Margulis, Almost isometric actions, property $(T)$, and local rigidity, *Invent. Math.* **162** (2005), No. 1, 19–80.
9. H. Abels, G. A. Margulis and G. A. Soifer, The Auslander conjecture for groups leaving a form of signature $(n-2,2)$ invariant, *Israel J. Math.* **148** (2005), 11–21.
10. A. Eskin, G. Margulis and S. Mozes, Quadratic forms of signature $(2,2)$ and eigenvalue spacings on rectangular 2-tori. *Ann. of Math.* (2) **161** (2005), No. 2, 679–725.
11. G. Margulis, Random walks on the space of lattices and the finiteness of covolumes of arithmetic subgroups, in *Algebraic Groups and Arithmetic*, pp. 409–425 (Tata Inst. Fund. Res., 2004).
12. H. Abels and G. Margulis, Coarsely geodesic metrics on reductive groups, in *Modern Dynamical Systems and Applications*, pp. 163–183 (Cambridge Univ. Press, 2004).

13. A. Eskin and G. Margulis, Recurrence properties of random walks on finite volume homogeneous manifolds, in *Random Walks and Geometry*, pp. 431–444 (Walter de Gruyter, 2004).
14. G. A. Margulis, *On Some Aspects of the Theory of Anosov Systems*. With a survey by Richard Sharp: Periodic orbits of hyperbolic flows. Translated from the Russian by Valentina Vladimirovna Szulikowska. Springer Monographs in Mathematics (Springer-Verlag, 2004).
15. D. Fisher and G. A. Margulis, Local rigidity for cocycles, in *Surveys in Differential Geometry*, Vol. VIII (Boston, MA, 2002), pp. 191–234, Surv. Differ. Geom., Vol. VIII (Int. Press, 2003).
16. G. A. Margulis, Diophantine approximation, lattices and flows on homogeneous spaces, in *A Panorama of Number Theory or the View from Baker's Garden* (Zürich, 1999), pp. 280–310 (Cambridge Univ. Press, 2002).
17. V. V. Beresnevich, V. I. Bernik, D. Y. Kleinbock and G. A. Margulis, Metric Diophantine approximation: The Khintchine–Groshev theorem for nondegenerate manifolds, dedicated to Yuri I. Manin on the occasion of his 65th birthday, *Mosc. Math. J.* **2** (2002), No. 2, 203–225.
18. H. Abels, G. A. Margulis and G. A. Soifer, On the Zariski closure of the linear part of a properly discontinuous group of affine transformations, *J. Differential Geom.* **60** (2002), No. 2, 315–344.
19. V. Bernik, D. Kleinbock and G. A. Margulis, Khintchine-type theorems on manifolds: The convergence case for standard and multiplicative versions, *Internat. Math. Res. Notices* **2001** (2001), No. 9, 453–486.
20. G. A. Margulis and N. Qian, Rigidity of weakly hyperbolic actions of higher real rank semisimple Lie groups and their lattices, *Ergodic Theory Dynam. Syst.* **21** (2001), No. 1, 121–164.
21. G. Margulis, Free subgroups of the homeomorphism group of the circle, *C. R. Acad. Sci. Paris Sér. I Math.* **331** (2000), No. 9, 669–674.
22. W. M. Goldman and G. A. Margulis, Flat Lorentz 3-manifolds and cocompact Fuchsian groups, in *Crystallographic Groups and Their Generalizations* (Kortrijk, 1999), pp. 135–145, Contemp. Math., Vol. 262 (Amer. Math. Soc., 2000).
23. S. I. Adyan, E. I. Zel'manov, G. A. Margulis, S. P. Novikov, A. S. Rapinchuk, L. D. Faddeev and V. I. Yanchevskiĭ, Vladimir Petrovich Platonov (on the occasion of his sixtieth birthday) (in Russian), *Usp. Mat. Nauk* **55** (2000), No. 3(333), 197–204; translation in *Russ. Math. Surv.* **55** (2000), No. 3, 601–610.
24. G. A. Margulis and G. D. Mostow, Some remarks on the definition of tangent cones in a Carnot–Carathéodory space, *J. Anal. Math.* **80** (2000), 299–317.

25. G. Margulis, Problems and conjectures in rigidity theory, in *Mathematics: Frontiers and Perspectives*, pp. 161–174 (Amer. Math. Soc., 2000).
26. G. A. Margulis, A. Nevo and E. M. Stein, Analogs of Wiener's ergodic theorems for semisimple Lie groups, II, *Duke Math. J.* **103** (2000), No. 2, 233–259.
27. G. A. Margulis and È. B. Vinberg, Some linear groups virtually having a free quotient, *J. Lie Theory* **10** (2000), No. 1, 171–180.
28. A. Karlsson and G. A. Margulis, A multiplicative ergodic theorem and nonpositively curved spaces, *Comm. Math. Phys.* **208** (1999), No. 1, 107–123.
29. D. Y. Kleinbock and G. A. Margulis, Logarithm laws for flows on homogeneous spaces, *Invent. Math.* **138** (1999), No. 3, 451–494.
30. G. A. Margulis and S. Mozes, Aperiodic tilings of the hyperbolic plane by convex polygons, *Israel J. Math.* **107** (1998), 319–325.
31. D. Y. Kleinbock and G. A. Margulis, Flows on homogeneous spaces and Diophantine approximation on manifolds, *Ann. of Math.* (2) **148** (1998), No. 1, 339–360.
32. A. Eskin, G. Margulis and S. Mozes, Upper bounds and asymptotics in a quantitative version of the Oppenheim conjecture, *Ann. of Math.* (2) **147** (1998), No. 1, 93–141.
33. G. A. Margulis, Oppenheim conjecture, in *Fields Medallists' Lectures*, pp. 272–327, World Sci. Ser. 20th Century Math., Vol. 5 (World Scientific, 1997).
34. G. Margulis, Existence of compact quotients of homogeneous spaces, measurably proper actions, and decay of matrix coefficients, *Bull. Soc. Math. France* **125** (1997), No. 3, 447–456.
35. H. Abels, G. A. Margulis and G. A. Soifer, Properly discontinuous groups of affine transformations with orthogonal linear part, *C. R. Acad. Sci. Paris Sér. I Math.* **324** (1997), No. 3, 253–258.
36. G. A. Margulis and G. M. Tomanov, Measure rigidity for almost linear groups and its applications, *J. Anal. Math.* **69** (1996), 25–54.
37. S. P. Novikov, L. A. Bunimovich, A. M. Vershik, B. M. Gurevich, E. I. Dinaburg, G. A. Margulis, V. I. Oseledets, S. A. Pirogov, K. M. Khanin and N. N. Chentsova, Yakov Grigor'evich Sinaĭ (on the occasion of his sixtieth birthday) (in Russian), *Usp. Mat. Nauk* **51** (1996), No. 4(310), 179–191; translation in *Russ. Math. Surv.* **51** (1996), No. 4, 765–778.
38. D. Y. Kleinbock and G. A. Margulis, Bounded orbits of nonquasiunipotent flows on homogeneous spaces, in *Sinaĭ's Moscow Seminar on Dynamical Systems*, pp. 141–172, Amer. Math. Soc. Transl. Ser. 2, Vol. 171 (Amer. Math. Soc., 1996).

39. A. Eskin, G. Margulis and S. Mozes, On a quantitative version of the Oppenheim conjecture, *Electron. Res. Announc. Amer. Math. Soc.* **1** (1995), No. 3, 124–130 (electronic).
40. H. Abels, G. A. Margulis and G. A. Soĭfer, Semigroups containing proximal linear maps, *Israel J. Math.* **91** (1995), No. 1–3, 1–30.
41. G. A. Margulis and G. D. Mostow, The differential of a quasi-conformal mapping of a Carnot–Carathéodory space, *Geom. Funct. Anal.* **5** (1995), No. 2, 402–433.
42. G. A. Margulis and G. M. Tomanov, Invariant measures for actions of unipotent groups over local fields on homogeneous spaces, *Invent. Math.* **116** (1994), No. 1–3, 347–392.
43. S. G. Dani and G. A. Margulis, Limit distributions of orbits of unipotent flows and values of quadratic forms, in *I. M. Gel'fand Seminar*, pp. 91–137, Adv. Soviet Math., Vol. 16, Part 1 (Amer. Math. Soc., 1993).
44. G. A. Margulis and G. M. Tomanov, Measure rigidity for algebraic groups over local fields, *C. R. Acad. Sci. Paris Sér. I Math.* **315** (1992), No. 12, 1221–1226.
45. S. G. Dani and G. A. Margulis, On the limit distributions of orbits of unipotent flows and integral solutions of quadratic inequalities, *C. R. Acad. Sci. Paris Sér. I* **314** (1992), 699–704.
46. G. A. Margulis, Dynamical and ergodic properties of subgroup actions on homogeneous spaces with applications to number theory, in *Proc. of the International Congress of Mathematicians*, Vols. I, II (Kyoto, 1990), pp. 193–215 (Math. Soc. Japan, 1991).
47. G. A. Margulis, Lie groups and ergodic theory, in *Algebra and Analysis* (Kemerovo, 1988), pp. 73–86, Amer. Math. Soc. Transl. Ser. 2, Vol. 148 (Amer. Math. Soc., 1991).
48. S. G. Dani and G. A. Margulis, Asymptotic behaviour of trajectories of unipotent flows on homogeneous spaces, *Proc. Indian Acad. Sci. Math. Sci.* **101** (1991), No. 1, 1–17.
49. G. A. Margulis, Compactness of minimal closed invariant sets of actions of unipotent groups, *Geom. Dedicata* **37** (1991), No. 1, 1–7.
50. G. A. Margulis, Discrete subgroups of semisimple Lie groups, in *Ergebnisse der Mathematik und ihrer Grenzgebiete* (3) [*Results in Mathematics and Related Areas* (3)], (Springer-Verlag, 1991).
51. G. A. Margulis, Orbits of group actions and values of quadratic forms at integral points, in *Festschrift in honor of I. I. Piatetski-Shapiro on the occasion of his sixtieth birthday, Part II* (Ramat Aviv, 1989), pp. 127–150, Israel Math. Conf. Proc., Vol. 3 (Weizmann, 1990).

52. G. A. Margulis, Dynamical and ergodic properties of subgroup actions on homogeneous spaces with applications to number theory, a plenary address presented at the *International Congress of Mathematicians* held in Kyoto, August 1990. ICM-90 (Math. Soc. Japan, 1990).
53. S. G. Dani and G. A. Margulis, Values of quadratic forms at integral points: An elementary approach, *Enseign. Math.* (2) **36** (1990), No. 1–2, 143–174.
54. S. G. Dani and G. A. Margulis, Orbit closures of generic unipotent flows on homogeneous spaces of SL(3, **R**), *Math. Ann.* **286** (1990), No. 1–3, 101–128.
55. G. A. Margulis, Indefinite quadratic forms and unipotent flows on homogeneous spaces, in *Dynamical Systems and Ergodic Theory* (Warsaw, 1986), pp. 399–409, Banach Center Publ., Vol. 23 (PWN, 1989).
56. I. Ya. Gol'dsheĭd and G. A. Margulis, Lyapunov exponents of a product of random matrices (in Russian), *Usp. Mat. Nauk* **44** (1989), No. 5(269), 13–60; translation in *Russ. Math. Surv.* **44** (1989), No. 5, 11–71.
57. S. G. Dani and G. A. Margulis, Values of quadratic forms at primitive integral points, *Invent. Math.* **98** (1989), No. 2, 405–424.
58. G. A. Margulis, Discrete subgroups and ergodic theory, in *Number Theory, Trace Formulas and Discrete Groups* (Oslo, 1987), pp. 377–398 (Academic Press, 1989).
59. S. G. Dani and G. A. Margulis, Values of quadratic forms at primitive integral points, *C. R. Acad. Sci. Paris Sér. I Math.* **308** (1989), No. 7, 199–203.
60. F. Grunewald and G. Margulis, Transitive and quasitransitive actions of affine groups preserving a generalized Lorentz-structure, *J. Geom. Phys.* **5** (1988), No. 4, 493–531.
61. G. A. Margulis, Lie groups and ergodic theory, in *Algebra — Some Current Trends* (Varna, 1986), pp. 130–146, Lecture Notes in Math., Vol. 1352 (Springer, 1988).
62. G. A. Margulis, Explicit group-theoretic constructions of combinatorial schemes and their applications in the construction of expanders and concentrators (in Russian), *Problemy Peredachi Informatsii* **24** (1988), No. 1, 51–60; translation in *Problems Inform. Transmission* **24** (1988), No. 1, 39–46.
63. I. Ya. Gol'dsheĭd and G. A. Margulis, The condition of simplicity for the spectrum of Lyapunov exponents (in Russian), *Dokl. Akad. Nauk SSSR* **293** (1987), No. 2, 297–301.
64. G. A. Margulis, Formes quadratriques indéfinies et flots unipotents sur les espaces homogènes [Indefinite quadratic forms and unipotent flows

on homogeneous spaces], *C. R. Acad. Sci. Paris Sér. I Math.* **304** (1987), No. 10, 249–253.

65. G. A. Margulis and J. Rohlfs, On the proportionality of covolumes of discrete subgroups, *Math. Ann.* **275** (1986), No. 2, 197–205.
66. G. A. Margulis and A. S. Omel'chenko, Deformation of plane curves (in Russian), *Problemy Peredachi Informatsii* **20** (1984), No. 4, 31–40.
67. G. A. Margulis, Complete affine locally flat manifolds with a free fundamental group (in Russian), Automorphic functions and number theory, II, *Zap. Nauchn. Sem. Leningrad. Otdel. Mat. Inst. Steklov. (LOMI)* **134** (1984), 190–205.
68. G. A. Margulis, Arithmeticity of the irreducible lattices in the semisimple groups of rank greater than 1, *Invent. Math.* **76** (1984), No. 1, 93–120.
69. A. T. Huckleberry and G. A. Margulis, Invariant analytic hypersurfaces, *Invent. Math.* **71** (1983), No. 1, 235–240.
70. G. A. Margulis, Free totally discontinuous groups of affine transformations (in Russian), *Dokl. Akad. Nauk SSSR* **272** (1983), No. 4, 785–788.
71. G. A. Margulis, Finitely-additive invariant measures on Euclidean spaces, *Ergodic Theory Dynam. Syst.* **2** (1982), No. 3–4, 383–396.
72. G. A. Margulis, Explicit constructions of graphs without short cycles and low density codes, *Combinatorica* **2** (1982), No. 1, 71–78.
73. G. A. Margulis, On the decomposition of discrete subgroups into amalgams. Selected translations, *Selecta Math. Soviet.* **1** (1981), No. 2, 197–213.
74. G. A. Margulis and G. A. Soĭfer, Maximal subgroups of infinite index in finitely generated linear groups, *J. Algebra* **69** (1981), No. 1, 1–23.
75. G. A. Margulis, Some remarks on invariant means, *Monatsh. Math.* **90** (1980), No. 3, 233–235.
76. G. A. Margulis, Multiplicative groups of a quaternion algebra over a global field (in Russian), *Dokl. Akad. Nauk SSSR* **252** (1980), No. 3, 542–546.
77. G. A. Margulis and G. A. Soĭfer, Nonfree maximal subgroups of infinite index of the group $\mathrm{SL}_n(\mathbf{Z})$ (in Russian), *Usp. Mat. Nauk* **34** (1979), No. 4(208), 203–204.
78. G. A. Margulis, Finiteness of quotient groups of discrete subgroups (in Russian), *Functional. Anal. i Prilozhen.* **13** (1979), No. 3, 28–39.
79. G. A. Margulis, Factor groups of discrete subgroups and measure theory (in Russian), *Functional. Anal. i Prilozhen.* **12** (1978), No. 4, 64–76.
80. G. A. Margulis, Factor-groups of discrete subgroups (in Russian), *Dokl. Akad. Nauk SSSR* **242** (1978), No. 3, 533–536.
81. G. A. Margulis and G. A. Soĭfer, A criterion for the existence of maximal subgroups of infinite index in a finitely generated linear group (in Russian), *Dokl. Akad. Nauk SSSR* **234** (1977), No. 6, 1261–1264.

82. G. A. Margulis, Cobounded subgroups in algebraic groups over local fields (in Russian), *Functional. Anal. i Prilozhen.* **11** (1977), No. 2, 45–57; 95.
83. G. A. Margulis, Discrete groups of motions of manifolds of nonpositive curvature (in Russian), in *Proceedings of the International Congress of Mathematicians* (Vancouver, B.C., 1974), Vol. 2, pp. 21–34 (Canad. Math. Congress, 1975).
84. G. A. Margulis, On the action of unipotent groups in the space of lattices, in *Lie Groups and Their Representations* (Proc. Summer School, Bolyai, János Math. Soc., Budapest, 1971), pp. 365–370 (Halsted, 1975).
85. G. A. Margulis, Arithmeticity of nonuniform lattices in weakly noncompact groups (in Russian), *Functional. Anal. i Prilozhen.* **9** (1975), No. 1, 35–44.
86. G. A. Margulis, Non-uniform lattices in semisimple algebraic groups, in *Lie Groups and Their Representations* (Proc. Summer School on Group Representations of the Bolyai János Math. Soc., Budapest, 1971), pp. 371–553 (Halsted, 1975).
87. G. A. Margulis, Probabilistic characteristics of graphs with large connectivity (in Russian), *Problemy Peredači Informacii* **10** (1974), No. 2, 101–108.
88. G. A. Margulis, Arithmetic properties of discrete subgroups (in Russian), *Usp. Mat. Nauk* **29** (1974), No. 1(175), 49–98.
89. G. A. Margulis, Arithmeticity and finite-dimensional representations of uniform lattices (in Russian), *Functional. Anal. i Prilozhen.* **8** (1974), No. 3, 77–78.
90. G. A. Margulis, Explicit constructions of expanders (in Russian), *Problemy Peredači Informacii* **9** (1973), No. 4, 71–80.
91. G. A. Margulis, Arithmeticity of nonuniform lattices (in Russian), *Functional. Anal. i Prilozhen.* **7** (1973), No. 3, 88–89.
92. G. A. Margulis, Metric questions in the theory of $C$-systems (in Russian), *Ninth Mathematical Summer School* (Kaciveli, 1971) (in Russian), pp. 342–348 (Izdanie Inst. Mat. Akad. Nauk Ukrain., 1972).
93. G. A. Margulis, The action of unipotent groups in a lattice space (in Russian), *Mat. Sb. (N.S.)* **86**(128) (1971), 552–556.
94. G. A. Margulis, Certain measures that are connected with $U$-flows on compact manifolds (in Russian), *Functional. Anal. i Prilozhen.* **4** (1970), No. 1, 62–76.
95. G. A. Margulis, The isometry of closed manifolds of constant negative curvature with the same fundamental group (in Russian), *Dokl. Akad. Nauk SSSR* **192** (1970), 736–737.

96. G. A. Margulis, On the arithmeticity of discrete groups, *Sov. Math. Dokl.* **10** (1969), 900–902; translated from *Dokl. Akad. Nauk SSSR* **187** (1969), 518–520 (Russian).
97. G. A. Margulis, Certain applications of ergodic theory to the investigation of manifolds of negative curvature (in Russian), *Functional. Anal. i Prilozhen.* **3** (1969), No. 4, 89–90.
98. G. A. Margulis, Discrete subgroups of real semi-simple Lie groups (in Russian), *Mat. Sb. (N.S.)* **80**(122) (1969), 600–615.
99. D. A. Každan and G. A. Margulis, A proof of Selberg's hypothesis (in Russian) *Mat. Sb. (N.S.)* **75**(117) (1968), 163–168.
100. G. A. Margulis, Positive harmonic functions on nilpotent groups, *Dokl. Akad. Nauk SSSR* **166** (1966), 1054–1057 (Russian); translated as *Sov. Math. Dokl.* **7** (1966), 241–244.

# Preface

I am very thankful to Springer-Verlag for publishing a translation of my thesis, which was defended at Moscow State University in 1970. There have been only two publications related to the thesis. The first one is a short announcement [4], and the second one is [17] which more or less coincides with sections 3 and 4 of the thesis. In the early seventies, I was planning to publish the thesis in its entirety, but those plans were postponed eventually for more than 30 years. The main reason for the postponement was that I became more involved in other projects.

About three years ago I wrote to Joachim Heinze with the suggestion to publish an English translation of my thesis. He agreed, but under the condition that the publication will come "with a competent commentary on the development of the field during the last 30 years since the thesis was written". I am very glad that this translation is published with an excellent survey "Periodic orbits of hyperbolic flows" by Richard Sharp.

I have updated the terminology from the original. In particular, when I wrote my thesis, Anosov systems were called in Russian "$U$-sistemy", where $U$ is the first letter of "Uslovie" (condition). I also made some small changes at the beginning of the proof of Lemma 6.2 in order to make the argument using the Brauer theorem clearer.

I am very thankful to Steve Miller who carefully read this translation and made numerous corrections.

My thesis advisor was Ya. G. Sinai. His influence on my formation as a mathematician is hard to overestimate. Now we are good friends and I would like to take this opportunity to express him my deepest gratitude.

New Haven, CT
October, 2003

*G. A. Margulis*

# On Some Aspects of the Theory of Anosov Systems

*G. A. Margulis*

# Introduction

The theory of Anosov systems is a result of the generalization of certain properties, which hold on geodesic flows on manifolds of negative curvature. It turned out that these properties alone are sufficient to ensure ergodicity, mixing, and, moreover, existence of $K$-partitions. All above-mentioned properties are connected with the asymptotical behavior of variational equations along the trajectories of Anosov systems. Therefore, it would be appropriate to propose that other asymptotical properties of geodesic flows on manifolds of negative curvature hold for the class of Anosov systems, too. However, it would be more rational to consider not all of the Anosov flows, but the class $L$ of Anosov flows that preserve some integral invariant and have no continuous eigenfunctions.

Let $\{T^t\}$ be an Anosov flow from $L$ on a Riemannian manifold $W^n$. Then there exist four foliations $\mathfrak{S}^{l+1}$, $\mathfrak{S}^{k+1}$, $\mathfrak{S}^l$, and $\mathfrak{S}^k$ on $W^n$ that are invariant under $\{T^t\}$. The leaves of these foliations are called the expanding leaves, the contracting leaves, the expanding horospheres, and the contracting horospheres, respectively. Now, a natural concept of $\epsilon$-equivalency is introduced for sets belonging to different expanding leaves. Roughly speaking, two sets belonging to expanding leaves are $\epsilon$-equivalent if one of them may be continuously transformed into the other so that, during the transformation, each point describes a curve belonging entirely to some contracting horosphere, the curve's length being less than $\epsilon$ (for more details, see Definition 2.1).

In § 2 of the present paper the asymptotical behavior of a Lebesgue measure on expanding leaves under the action of an Anosov flow is investigated. It turns out that

$$\lim_{t\to+\infty} \frac{\mu_{\mathfrak{S}^{l+1}}(T^t M_1)}{\mu_{\mathfrak{S}^{l+1}}(T^t M_2)} = 1, \tag{1}$$

where $M_1$ and $M_2$ are two $\epsilon$-equivalent sets.

In § 2 we consider the class $T$ of functions on $W^n$ whose support is a compact subset of some expanding leaf. (The leaves may be different for different functions.) Every function is continuous on its leaf. In the same way, we introduce the concept of $\epsilon$-equivalency for these functions and prove that for any two $\epsilon$-equivalent functions $f_1$ and $f_2$,

$$\lim_{t\to+\infty} \frac{\int T^t f_1 \, d\mu_{\mathfrak{S}^{l+1}}}{\int T^t f_2 \, d\mu_{\mathfrak{S}^{l+1}}} = 1, \tag{2}$$

where $\int f \, d\mu_{\mathfrak{S}^{l+1}}$ is the integral of $f$ over the leaf where $f$ does not vanish. Further, we prove that for any non-negative and non-zero functions $f_1$ and $f_2$ from $T$,

$$0 < c_1 < \varliminf_{t\to+\infty} \frac{\int T^t f_1 \, d\mu_{\mathfrak{S}^{l+1}}}{\int T^t f_2 \, d\mu_{\mathfrak{S}^{l+1}}} \le \varlimsup_{t\to+\infty} \frac{\int T^t f_1 \, d\mu_{\mathfrak{S}^{l+1}}}{\int T^t f_2 \, d\mu_{\mathfrak{S}^{l+1}}} < c_2 < \infty. \tag{3}$$

In § 3 we define a family of functionals $l_t$ $(t \ge 0)$ on $T$ as follows: if $f \in T$, then

$$l_t(f) = \int T^t f \, d\mu_{\mathfrak{S}^{l+1}}. \tag{4}$$

By (2) and (3), the closure of the convex hull of $l_t$ $(t \ge 0)$ is compact with respect to some topology. Hence, by the Tychonoff theorem on a fixed point we deduce that there exists a functional $\widetilde{l}$ such that

1) if $f \in T$, then
$$\widetilde{l}(T^t f) = d^t_{\mathfrak{S}^{l+1}} \widetilde{l}(f), \tag{5}$$
2) if $f_1$ and $f_2$ are $\epsilon$-equivalent for some $\epsilon$, then $\widetilde{l}(f_1) = \widetilde{l}(f_2)$, and
3) $\widetilde{l}$ is positive on the set of non-negative and non-zero functions.

After that, by the Riesz–Markov theorem we prove that there exists a class of countably additive measures $\widetilde{\mu}_{\mathfrak{S}^{l+1}}$ on expanding leaves such that

1) $\widetilde{\mu}_{\mathfrak{S}^{l+1}}(T^t M) = d^t_{\mathfrak{S}^{l+1}} \widetilde{\mu}_{\mathfrak{S}^{l+1}}(M)$,
2) if $M_1$ and $M_2$ are $\epsilon$-equivalent, then $\widetilde{\mu}_{\mathfrak{S}^{l+1}}(M_1) = \widetilde{\mu}_{\mathfrak{S}^{l+1}}(M_2)$, and
3) for any (nonempty) open $M$, $0 < \widetilde{\mu}_{\mathfrak{S}^{l+1}}(M) < \infty$.

In the case of Anosov diffeomorphisms, similar measures were constructed by Ya. G. Sinai in [7], with the help of his theory of Markov partitions.

By using the measures $\widetilde{\mu}_{\mathfrak{S}^{l+1}}$ and some other measures constructed in § 3, we construct in § 4 a special measure $\widetilde{\mu}$ defined on the whole of $W^n$ and invariant under $\{T^t\}$. In the same section we formulate Theorems 2 and 3, which show that the flow $\{T^t\}$ with $\widetilde{\mu}$ is a $K$-flow and the foliations connected with the Anosov flow are metrically transitive with respect to $\widetilde{\mu}$. Theorems 2 and 3 are proved in appendix (§ 8).

The constant $d_{\mathfrak{S}^{l+1}}$ is very important. For example, it is proved in § 8 that $\log_2 d_{\mathfrak{S}^{l+1}}$ is equal to the topological entropy of $\{T^t\}$. The constant $d_{\mathfrak{S}^{l+1}}$ is essential when studying the asymptotical behavior of the number of periodic trajectories, or the volume of a ball with radius $R$ as $R \to \infty$ (for manifolds with negative curvature), etc.

The closed trajectories of $\{T^t\}$ are studied in § 6. As it is known, the problem of the closed geodesics on a compact manifold $M$ of negative curvatuve is equivalent to the problem of the periodic trajectories of a geodesic flow defined on the manifold of linear elements of $M$. In fact, any closed geodesic is naturally associated with a closed trajectory, and the length of the closed geodesic is equal to the period of the corresponding trajectory. (Here we mean that any closed geodesic is given together with its direction. In the case of closed geodesics without the given direction, all the estimations made below are to be divided by two.)

By the well-known Birkhoff theorem we get that for any non-trivial free homotopy class on $M$ there exists a unique closed geodesic[1]. So, any free homotopy class on $M$ corresponds to a conjugacy class in $\pi_1(M)$. Therefore we may say that the asymptotics of the closed geodesics defines the asymptotics of the conjugacy classes in $\pi_1(M)$ (except for the identity class).

It follows from results of Hadamard [11] and Morse [12] that for any two-dimensional compact manifold $M$ with negative curvature, the vectors tangent to $M$ and tangent to closed geodesics are dense in the space of all tangent vectors. Anosov has proved a similar theorem for the $n$-dimensional case. More precisely, he has proved that periodic trajectories of an arbitrary flow from $L$ are dense everywhere. This means that an Anosov flow has, in any case, an infinite number of periodic trajectories. On the other hand, any free homotopy class of $M$ has a unique closed geodesic; therefore, the number of closed geodesics on $M$ is at most countable. Moreover, if we denote by $\nu(R)$ the number of closed geodesics having length less than $R$, then $\nu(R)$ is finite for any $R$. It is easy to see that this assumption remains true for arbitrary Anosov flows.

Since $\nu(R)$ is finite, we can investigate the asymptotical behavior of $\nu(R)$ as $R$ tends to infinity. Ya. G. Sinai proved in [13] that there exist positive constants $K_1$ and $K_2$ such that

$$(n-1)K_2 \le \underline{\lim}_{t\to\infty} \frac{\log \nu(t)}{t} \le \overline{\lim}_{t\to\infty} \frac{\log \nu(t)}{t} \le (n-1)K_1 \tag{6}$$

and if $\{T^t\}$ is a geodesic flow on a compact manifold with negative curvature, then $-K_1^2$ is a lower bound and $-K_2^2$ is an upper bound of the curvature over all two-dimensional directions. (Here $\nu(t)$ stands for the number of periodic trajectories of the Anosov flow $\{T^t\}$ with period that is less than $t$.)

In § 6 we prove that

$$\nu(R) \sim \frac{d^R_{\mathfrak{S}^{l+1}}}{R \cdot \log d_{\mathfrak{S}^{l+1}}}, \tag{7}$$

where $d_{\mathfrak{S}^{l+1}}$ is the constant defined in § 3, or, putting $c = \log d_{\mathfrak{S}^{l+1}}$,

$$\nu(R) \sim \frac{e^{cR}}{cR}. \tag{8}$$

[1] Existence of a closed geodesic for any non-trivial free homotopy class remains true for any compact manifold. As for uniqueness, it is a special property of manifolds with negative curvature (see [11]).

Formula (8) is equivalent to the following formula:

$$\lim_{R\to\infty} \left|\log \nu(R) - (cR - \log R - \log c)\right| = 0. \tag{9}$$

Formula (9) not only gives the first term in the asymptotics of $\log \nu(R)$, but the two following terms as well.

If $M$ is equal to an $n$-dimensional manifold of constant negative curvature $-K^2$, the constant $c$ coincides with $K(n-1)$, and (8) becomes

$$\nu(R) \sim \frac{e^{R(n-1)K}}{R(n-1)K}. \tag{10}$$

This formula for the case of two-dimensional manifolds of constant negative curvature was proved earlier by Ye. G. Litinsky (unpublished).

In § 6 we prove Theorem 4, which implies that periodic trajectories of $\{T^t\}$ may be considered as uniformly distributed over $W^n$.

In § 7 we discuss a problem, similar to the problem of the asymptotical behavior of the number of integer points in a Euclidean ball as its radius tends to infinity. Namely, let $M^s$ be a compact $s$-dimensional manifold of negative curvature and $\overline{M}^s$ its universal covering. Denote by $\pi$ the natural projection $\overline{M}^s \xrightarrow{\pi} M^s$. If $m_1, m_2 \in \overline{M}^s$, then by $a_R(m_1, m_2)$ denote the number of points from $\pi^{-1}(\pi(m_2))$, which belong to the open ball with radius $R$ and the center in $m_1$. If $M^s$ were not a manifold of negative curvature but a torus, and if $\overline{M}^s$ were an $s$-dimensional Euclidean space, then $a_R(m_1, m_2)$ would denote the number of points with integer coordinates belonging to a ball (or, more generally, to an ellipsoid) with radius $R$ and the center in some point. Therefore, the problem of the asymptotical behavior of $a_R(m_1, m_2)$ is similar to the problem of the asymptotical behavior of the number of integer points in a ball.

Denote by $b_R(m)$ the volume of the ball with radius $R$ and center $m$. For the Euclidean case (when $M^s$ is a torus), Gauss proved that

$$\lim_{R\to\infty} \frac{a_R(m_1, m_2)v(M^s)}{b_R(m)} = 1 \tag{11}$$

for any $m_1$, $m_2$, and $m$. (Here $v(M^s)$ is the volume of $M^s$.) In the case of manifolds of negative curvature, Gauss' method gives only

$$\begin{aligned} 0 < c_1 < \varliminf_{R\to\infty} &\frac{a_R(m_1, m_2)v(M^s)}{b_R(m)} \\ \le \varlimsup_{R\to\infty} &\frac{a_R(m_1, m_2)v(M^s)}{b_R(m)} < c_2 < \infty. \end{aligned} \tag{12}$$

The above is related to the fact that the area of a sphere with radius $R$ is commensurable with the volume of a ball of the same radius.

In § 7 we prove the following

**Theorem 8.** *There exists a constant $c(m_1, m_2)$ such that*

$$a_R(m_1, m_2) \sim c(m_1, m_2) \cdot \frac{d^R_{\mathfrak{S}^{l+1}}}{\log d_{\mathfrak{S}^{l+1}}}, \tag{13}$$

*where $d_{\mathfrak{S}^{l+1}}$ is the constant from (7).*

In the case of a manifold $M^s$ with constant curvature $-K^2$, Theorem 8 may be improved to

$$a_R(m_1, m_2) \sim \frac{v(s)e^{R(s-1)K}}{v(M^s) \cdot 2^{s-1}K(s-1)}, \tag{14}$$

where $v(s)$ is the volume of a ball with radius 1 in $s$-dimensional Euclidean space and $v(M^s)$ is the volume of $M^s$.

Finally, in § 7 we consider the asymptotical behavior of the volume of a ball on $\overline{M}^s$ with radius $R$ and center in a point $m \in \overline{M}^s$. More precisely, we prove

**Theorem 9.** *There exists a constant $b(m)$ such that*

$$b_R(m) \sim b(m) \frac{d^R_{\mathfrak{S}^{l+1}}}{\log d_{\mathfrak{S}^{l+1}}} \tag{15}$$

*and $b(m) = \int_{M^s} c(m, \widetilde{m}) d\widetilde{m}$.*

It has already been mentioned that Theorems 2 and 3 are proved in § 8. The proofs use a method similar to the method used in proving the corresponding theorems in the case of Lebesgue measure. Only the way to construct the partition may be considered as a novelty, though it is similar to the method of the Markov partitions used in [14] and [15]. In § 8, it is also proved that $\log_2 d_{\mathfrak{S}^{l+1}}$ coincides with the metrical entropy of the flow $\{T^t\}$ with measure $\widetilde{\mu}$.

In conclusion, let us note the following:

1) almost all the results of the present paper were announced by the author in [4],
2) the content of § 2, 3, and 4 coincides in many respects with the content of [17],
3) the proofs of (10) and (14) given in the present paper may be applied almost entirely to the case of manifolds with constant negative curvature and finite volume. (Their compactness is not necessary.)

The author is grateful to Professor Ya. G. Sinai for useful discussions and constant attention to this work.

# References


1. Anosov, D. V. (1969): *Geodesic flows on closed Riemann manifolds with negative curvature.* Proceedings of the Steklov Institute of Mathematics, No 90, 1967. Translated from the Russian by S. Feder, American Mathematical Society, Providence, R. I., iv+235. (The original Russian edition: Anosov, D. V. (1967): *Geodesic flows on closed Riemannian manifolds of negative curvature.* (Russian) Trudy Mat Inst Steklov 90:209.)
2. Anosov, D. V., Sinaĭ, Ja. G. (1967): *Certain smooth ergodic systems.* (Russian) Uspehi Mat Nauk 22, No 5: 107–172
3. Hille, Einar, Phillips, Ralph S. (1974): *Functional analysis and semi-groups.* Third printing of the revised edition of 1957. American Mathematical Society Colloquium Publications, Vol. XXXI. American Mathematical Society, Providence, R. I., xii+808
4. Margulis, G. A. (1969): *Certain applications of ergodic theory to the investigation of manifolds of negative curvature.* (Russian) Funkcional Anal i Priložen 3, No 4: 89–90
5. Rohlin, V. A. (1967): *Lectures on the entropy theory of transformations with invariant measure.* (Russian) Uspehi Mat Nauk 22, No 5 (137): 3–56
6. Sinaĭ, Ya. G. (1966): *Classical dynamic systems with countably-multiple Lebesgue spectrum. II.* (Russian) Izv Akad Nauk SSSR Ser Mat 30: 15–68
7. Sinaĭ, Ya. G. (1968): *Markov partitions and U-diffeomorphisms.* (Russian) Funkcional Anal i Priložen 2, No 1: 64–89
8. Dinaburg, E. I. (1970): *A correlation between topological entropy and metric entropy.* (Russian) Dokl Akad Nauk SSSR 190: 19–22
9. Rohlin, V. A. (1952): *On the fundamental ideas of measure theory.* Amer. Math. Soc. Translation 1952, No 71: 55 (The original Russian edition: Rohlin, V. A. (1949): *On the fundamental ideas of measure theory.* (Russian) Mat Sbornik N S 25 (67): 107–150)
10. Veršik, A. M. (1965): *A measurable realization of continuous groups of automorphisms of a unitary ring.* (Russian) Izv Akad Nauk SSSR Ser Mat 29: 127–136
11. Hadamard J. (1898): *Les surfaces courlure et leur lignes geodesiques.* J. Math. pures et appl. 5 (4): 27–73

12. Morse M. (1924): *A fundamental class of geodesics on any closed surface of genus greater than one.* Trans. Amer. Math. Soc. 26: 25–60
13. Sinaĭ, Ya. G. (1966): *Asymptotic behavior of closed geodesics on compact manifolds with negative curvature.* (Russian) Izv Akad Nauk SSSR Ser Mat 30: 1275–1296
14. Sinaĭ, Ya. G. (1968): *Construction of Markov partitions.* (Russian) Funkcional Anal i Priložen 2, No 3: 70–80
15. Ratner, M. E. (1969): *The invariant measure for an U-flow on a three-dimensional manifold.* (Russian) Dokl Akad Nauk SSSR 186: 261–263
16. Hopf, E. (1949): *Statistics of geodesic lines on manifolds of negative curvature.* (Russian) Uspehi Matem Nauk (N.S.) 4, No 2(30): 129–170
17. Margulis, G. A. (1970): *Certain measures that are related to Anosov flows.* (Russian) Funkcional Anal i Priložen 4, No 1: 62–76

# Table of Contents

# Introduction

## 1. Statement of Main Results

Let $A$ be a nonempty finite set. Suppose that for each $\alpha \in A$ a local (i.e. non-discrete locally compact) field $k_\alpha$ and a connected semisimple $k_\alpha$-group $\mathbf{G}_\alpha$ have been chosen. Denote by $\mathbf{G}_\alpha(k_\alpha)$ the group of $k_\alpha$-rational points of $\mathbf{G}_\alpha$ and let $G = \prod_{\alpha \in A} \mathbf{G}_\alpha(k_\alpha)$.

In this book the main objects of study are lattices (i.e. discrete subgroups with finite covolume) in the group $G$. We mainly consider the so called irreducible lattices (a precise definition of this concept will be given in 5.9 of Chapter III; here we only note that if for every $\alpha \in A$ the group $\mathbf{G}_\alpha$ has no nontrivial $k_\alpha$-anisotropic factors, then a lattice $\Gamma \subset G$ is irreducible if and only if no subgroup of finite index in $\Gamma$ can be represented in the form of a direct product of two infinite subgroups).

We now provide an important example of an irreducible lattice. Let $K$ be a global field, let $\mathscr{R}$ be the set of all (inequivalent) valuations of the field $K$, and let $\mathscr{R}_\infty \subset \mathscr{R}$ be the set of archimedean valuations of $K$. Denote by $K_v$ the completion of the field $K$ with respect to the valuation $v \in \mathscr{R}$. If $S \subset \mathscr{R}$, then the ring $\{x \in K \mid |x|_v \leq 1 \text{ for all } v \in \mathscr{R} - \mathscr{R}_\infty - X\}$, where $|x|_v$ is the value of the valuation $v$ at $x$, of $S$-integral elements of the field $K$, will be denoted by $K(S)$. Now let $\mathbf{H}$ be a connected non-commutative absolutely almost simple $K$-subgroup of the group $\mathbf{SL}_n$ of unimodular $n \times n$ matrices. We set $\mathscr{T} = \mathscr{T}(\mathbf{H}) = \{v \in \mathscr{R} \mid$ the group $\mathbf{H}$ is anisotropic over $K_v$, or equivalently, the group $\mathbf{H}(K_v)$ is compact$\}$. Let $S \subset \mathscr{R}$ be a finite set such that $S \supset \mathscr{R}_\infty - \mathscr{T}$ and let $\mathbf{H}(K(S))$ denote the subgroup of $\mathbf{H}$ consisting of matrices with entries in $K(S)$.

We set $H_S = \prod_{v \in S} \mathbf{H}(K_v)$ and identify the group $\mathbf{H}(K(S))$ with its image under the diagonal embedding in $H_S$. By the Borel-Harish-Chandra-Behr-Harder reduction theorem $\mathbf{H}(K(S))$ is a lattice in $H_S$. It can easily be verified that the lattice $\mathbf{H}(K(S))$ is irreducible. Let $\varphi$: $H_S \to G$ be a continuous isomorphism. Then $\varphi(\mathbf{H}(K(S)))$ is an irreducible lattice in $G$. The irreducible lattices obtained in this way and all the lattices commensurable with them will be called *arithmetic*. This definition is the final one in the case where for each $\alpha \in A$ the group $\mathbf{G}_\alpha$ has no nontrivial $k_\alpha$-anisotropic factors and the groups $\mathbf{G}_\alpha$ are either all simply connected or all adjoint. In the general case one should take as $\varphi$ not only isomorphisms but also continuous homomorphisms from a certain more general class (for details we refer to 1.3 and 1.4 of Chapter IX). If $G = \mathbf{G}(\mathbb{R})$, where $\mathbf{G}$ is a connected semisimple $\mathbb{R}$-group, then an irreducible lattice $\Gamma \subset G$ is arithmetic if and only if there exist a connected non-commutative almost $\mathbb{Q}$-simple $\mathbb{Q}$-group $\mathbf{F}$ and an $\mathbb{R}$-epimorphism $\tau\colon \mathbf{F} \to \mathbf{G}$ such that the Lie group $(\operatorname{Ker}\tau)(\mathbb{R})$ is compact and the subgroups $\tau(\mathbf{F}(\mathbb{Z}))$ and $\Gamma$ are commensurable.

We can now state the arithmeticity theorem (although not in full generality). First, recall that $\operatorname{char} k$ denotes the characteristic of the field $k$ and $\operatorname{rank}_k \mathbf{F}$ denotes the $k$-rank (i.e. the dimension of a maximal $k$-split torus) of the $k$-group $\mathbf{F}$. Let us set $\operatorname{rank} G = \sum_{\alpha \in A} \operatorname{rank}_{k_\alpha} \mathbf{G}_\alpha$.

**(1) Theorem.** (See Theorems A and B and Remark 2 in 1.9 of Chapter IX.) *Let $\Gamma$ be an irreducible lattice in $G$. Suppose that either* $\operatorname{rank} G \geq 2$ *or $\Gamma$ is finitely generated and is of infinite index in the commensurability subgroup* $\operatorname{Comm}_G(\Gamma) \stackrel{\text{def}}{=} \{g \in G \mid$ *the subgroups $g\Gamma g^{-1}$ and $\Gamma$ are commensurable*$\}$. *Assume in addition that for no $\alpha \in A$ the group $\mathbf{G}_\alpha$ has a nontrivial $k_\alpha$-anisotropic factor. Then the lattice $\Gamma$ is arithmetic.*

It should be noted that an irreducible lattice $\Gamma \subset G$ is finitely generated if at least one of the following conditions is satisfied:

(a) the lattice $\Gamma$ is cocompact (i.e. the quotient space $\Gamma \setminus G$ is compact);
(b) there exists $\alpha \in A$ such that $\operatorname{char} k_\alpha = 0$;
(c) $\operatorname{rank} G \geq 2$.

The proof of Theorem 1 makes fundamental use of the superrigidity theorems, more precisely Theorem 2 below. Before stating this theorem we set $G^+ = \prod_{\alpha \in A} \mathbf{G}_\alpha(k_\alpha)^+$, where $\mathbf{G}_\alpha(k_\alpha)^+$ denotes the subgroup of $\mathbf{G}_\alpha(k_\alpha)$ generated by the sets of $k_\alpha$-rational points of unipotent radicals of parabolic $k_\alpha$-subgroups of the group $\mathbf{G}_\alpha$. Note that

1) if for each $\alpha \in A$ the group $\mathbf{G}_\alpha$ is simply connected and has no nontrivial $k_\alpha$-anisotropic factors, then $G^+ = G$;
2) if the groups $\mathbf{G}_\alpha$ have no nontrivial $k_\alpha$-anisotropic factors and the fields $k_\alpha$ are archimedean, then $G^+$ coincides with the identity component of the Lie group $G$.

**(2) Theorem.** (See Theorems 5.4 and 5.6 of Chapter VII.) *Let $\Gamma$ be a lattice in $G$, let $\Lambda$ be a subgroup of* $\operatorname{Comm}_G(\Gamma)$ *containing $\Gamma$, $k$ a local field, $\mathbf{H}$ a connected adjoint $k$-simple $k$-group, and let $\tau\colon \Lambda \to \mathbf{H}(k)$ be a homomorphism. Suppose that for no $\alpha \in A$ the group $\mathbf{G}_\alpha$ has a nontrivial $k_\alpha$-anisotropic factor and the following conditions are satisfied:*

*(i) either* $\operatorname{rank} G \geq 2$ *and the lattice $\Gamma$ is irreducible or the closure of the subgroup $\Lambda$ contains $G^+$;*
*(ii) the subgroup $\tau(\Gamma)$ is Zariski dense in $\mathbf{H}$ but is not relatively compact in $\mathbf{H}(k)$ (in the topology induced from that of the field $k$).*

*Then $\tau$ extends, uniquely, to a continuous homomorphism $\tilde{\tau}\colon G \to \mathbf{H}(k)$.*

In Theorem 2 we actually consider homomorphisms $\tau$ of the lattice $\Gamma$ to algebraic groups over local fields $k$ such that the Zariski closure of the subgroup $\tau(\Gamma)$ is a $k$-simple $k$-group. Nevertheless, we may consider homomorphisms for which this condition is not satisfied and prove for them assertions similar to that of Theorem 2. Making use of these assertions we shall prove the following.

**(3) Theorem.** (See Theorem 5.8 and Corollary 5.9 of Chapter IX.) *Let $\Gamma \subset G$ be an irreducible lattice, $l$ a field, $\mathbf{F}$ an algebraic $l$-group, and let $\delta\colon \Gamma \to \mathbf{F}(l)$ be a homomorphism. Suppose that* $\operatorname{rank} G \geq 2$ *and for no $\alpha \in A$ the group $\mathbf{G}_\alpha$ has a nontrivial $k_\alpha$-anisotropic factor. Then*

*(i) in case* $\operatorname{char} l = 0$, *the Zariski closure of the group $\delta(\Gamma)$ is a semisimple $l$-group;*

*(ii) in case* $\operatorname{char} l \neq \operatorname{char} k_\alpha$ *for all $\alpha \in A$, the group $\delta(\Gamma)$ is finite;*

*(iii) the 1-dimensional cohomology group $H^1(\Gamma, \rho)$ is trivial for every representation $\rho$ of the group $\Gamma$ on a finite-dimensional vector space over a field of characteristic* 0.

In connection with (iii) above, note that if $\rho$ is a continuous non-trivial absolutely irreducible representation of the group $G$ on a finite-dimensional vector space over a local field $k$ of arbitrary characteristic, then, under certain assumptions on $G$ and $\Gamma$, the restriction map of the cohomology groups $H^1_{\text{cont}}(G, \rho) \to H^1(\Gamma, \rho)$ is an isomorphism (see Corollary 5.21 of Chapter VII).

Another group of results is related to the finiteness of factor groups of lattices $\Gamma \subset G$. These results can be stated as follows:

**(4) Theorem.** (See Proposition 5.3 and Theorem 5.4 of Chapter IX.) *Let $\Gamma \subset G$ be an irreducible lattice, let* $\operatorname{rank} G \geq 2$, *and assume that for no $\alpha \in A$ the group $\mathbf{G}_\alpha$ has a nontrivial $k_\alpha$-anisotropic factor. Then*

*(i) for every normal subgroup $N$ of $\Gamma$, either $N$ is contained in the center $\mathscr{Z}(G)$ of $G$ or the group $\Gamma/N$ is finite;*

*(ii) the factor group of the group $\Gamma$ modulo its commutator subgroup is finite.*

In case $k_\alpha$ is archimedean for every $\alpha \in A$, Theorems 1–4 can be restated in terms of Lie group theory. Before doing this we shall give some definitions.

Let $H$ be a connected semisimple Lie group. A connected commutative subgroup $T \subset H$ is said to be an $\mathbb{R}$-*split torus* if for each $x \in T$, $\operatorname{Ad} x$ is semisimple and all its eigenvalues are real. Here Ad denotes as usual the adjoint representation. The dimension of a maximal $\mathbb{R}$-split torus of $H$ is called the *rank* of $H$ and denoted by $\operatorname{rank} H$. We remark that if $H$ is the connected component of the identity in the group of isometries of a symmetric space $X$ of non-positive curvature, then $\operatorname{rank} H$ coincides with the rank of the space $X$, i.e. with the dimension of a maximal totally geodesic flat submanifold $M \subset X$.

We call a lattice $\Gamma \subset H$ *reducible* if there exist two connected infinite normal subgroups $H'$ and $H''$ in $H$ such that $H = H' \cdot H''$, $H' \cap H'' \subset \mathscr{Z}(H)$ and the subgroup $(\Gamma \cap H')\cdot(\Gamma \cap H'')$ is of finite index in $\Gamma$; otherwise $\Gamma$ is called *irreducible*.

Now we can present the Lie-theoretic restatements.

**(1') Theorem.** (See Theorem 6.5 of Chapter IX.) *Let $H$ be a connected semisimple Lie group with trivial center and no non-trivial compact factor groups, and let $\Gamma$ be an irreducible lattice in $H$. Assume that either* $\operatorname{rank} H \geq 2$ *or $\Gamma$ is of infinite index in* $\operatorname{Comm}_H(\Gamma)$. *Then the lattice $\Gamma$ is arithmetic, i.e., there exist a connected non-commutative $\mathbb{Q}$-simple $\mathbb{Q}$-group $\mathbf{F}$ and a continuous epimorphism $\sigma\colon \mathbf{F}(\mathbb{R})^0 \to H$ such that the kernel* $\operatorname{Ker}\sigma$ *is compact and the subgroups $\sigma(\mathbf{F}(\mathbb{Z}) \cap \mathbf{F}(\mathbb{R})^0)$ and $\Gamma$ are commensurable, where $\mathbf{F}(\mathbb{R})^0$ denotes the connected component of the identity in the Lie group $\mathbf{F}(\mathbb{R})$.*

**(2') Theorem.** (See Theorem 6.16 of Chapter IX.) *Let $H$ be a connected semisimple Lie group without non-trivial compact factor groups. Let $\Gamma \subset H$ be a lattice, $\Lambda \supset \Gamma$ a subgroup of* $\operatorname{Comm}_H(\Gamma)$, *$k$ a local field, $\mathbf{F}$ a connected semisimple $k$-group, and $\delta\colon \Lambda \to \mathbf{F}(k)$ a homomorphism such that the subgroup $\delta(\Gamma)$ is Zariski dense in $\mathbf{F}$. Assume that either* $\operatorname{rank} H \geq 2$ *and the lattice $\Gamma$ is irreducible or the subgroup $\Lambda$ is dense in $H$. Then,*

*(a) for $k$ isomorphic neither to $\mathbb{R}$ nor to $\mathbb{C}$, i.e. for non-archimedean $k$, the subgroup $\delta(\Gamma)$ is relatively compact in $\mathbf{F}(k)$;*
*(b) for archimedean $k$, if the subgroup $\delta(\Gamma)$ is not relatively compact in $\mathbf{F}(k)$ and the group $\mathbf{F}$ is adjoint and $k$-simple, $\delta$ extends, uniquely, to a continuous homomorphism $\tilde{\delta}\colon H \to \mathbf{F}(k)$;*
*(c) for $k = \mathbb{R}$, if the group $\mathbf{F}$ is adjoint and has no nontrivial $\mathbb{R}$-anisotropic factors, $\delta$ extends, uniquely, to a continuous homomorphism $\tilde{\delta}\colon H \to \mathbf{F}(\mathbb{R})$.*

**(3') Theorem.** (See Theorem 6.15 of Chapter IX.) *Let $H$ be a connected semisimple Lie group with finite center, and no non-trivial compact factor groups, let $\Gamma$ be an irreducible lattice in $H$, $l$ a field, $\mathbf{F}$ an algebraic $l$-group, and let $\delta\colon \Gamma \to \mathbf{F}(l)$ be a homomorphism. Assume* $\operatorname{rank} H \geq 2$. *Then*

*(i) if* $\operatorname{char} l = 0$, *then the Zariski closure of the subgroup $\delta(\Gamma)$ is a semisimple $l$-group;*
*(ii) if* $\operatorname{char} l \neq 0$, *then the group $\delta(\Gamma)$ is finite;*
*(iii) the one-dimensional cohomology group $H^1(\Gamma, \rho)$ is trivial for every representation $\rho$ of the group $\Gamma$ on a finite-dimensional vector space over a field of characteristic* 0.

**(4') Theorem.** (See Theorem 6.14 of Chapter IX.) *With $H$ and $\Gamma$ as in Theorem $3'$, assume* $\operatorname{rank} H \geq 2$. *Then*

*(i) for every normal subgroup $N$ of the group $\Gamma$, either $N$ is contained in the center of the group $H$ or the factor group $\Gamma/N$ is finite;*
*(ii) the factor group of the group $\Gamma$ modulo its commutator subgroup is finite.*

In Sections 4–7 of Chapter IX we present a number of consequences of the results mentioned above. Among them are Theorems 7.20 and 7.23 of Chapter IX which may be viewed as a strengthening of Mostow's theorem on the strong rigidity of locally symmetric spaces (for spaces of rank $\geq 2$).

We now turn to some applications of the superrigidity theorems and the theorems on finiteness of factor groups of discrete subgroups to the study of algebraic groups over global fields. With $K$, $\mathscr{R}$, $\mathscr{R}_\infty$, $K_v$, $K(S)$, $\mathbf{H}$, $\mathscr{T} = \mathscr{T}(\mathbf{H})$ and $\mathbf{H}(K(S))$ as in the above definition of an arithmetic lattice, we fix a subset $S \subset \mathscr{R}$ such that $S \supset \mathscr{R}_\infty - \mathscr{T}$. We call a subgroup of $\mathbf{H}$ *S-arithmetic* if it is commensurable with $\mathbf{H}(K(S))$. We set $\operatorname{rank}_S \mathbf{H} = \sum_{v \in S} \operatorname{rank}_{K_v} \mathbf{H}$.

Given an $l$-group $\mathbf{F}$ and a field homomorphism $\sigma\colon l \to l'$, we denote by ${}^\sigma\mathbf{F}$ the $l'$-group obtained from $\mathbf{F}$ via $\sigma$, and by $\sigma^0$ the homomorphism of the group $\mathbf{F}(l)$ into ${}^\sigma\mathbf{F}(l')$ induced from $\sigma$. (If $\mathbf{F}$ is a matrix $l$-group, then ${}^\sigma\mathbf{F}$ is obtained by applying the homomorphism $\sigma$ to the equations defining $\mathbf{F}$; furthermore, if $x = (x_{ij}) \in \mathbf{F}(l)$, then $\sigma^0(x) = (\sigma(x_{ij}))$.)

**(5) Theorem.** (See Theorem C of Chapter VIII.) *Let $\Lambda$ be an S-arithmetic subgroup of the group $\mathbf{H}$. Let $l$ be a field, $\mathbf{F}$ a connected non-commutative absolutely almost simple $l$-group, and let $\delta\colon \Lambda \to \mathbf{F}(l)$ be a homomorphism such that the subgroup $\delta(\Lambda)$ is Zariski dense in $\mathbf{F}$. Assume that* $\operatorname{rank}_S \mathbf{H} \geq 2$ *and that either $\mathbf{H}$ is simply connected or $\mathbf{F}$ is adjoint. Then there exist a homomorphism $\sigma\colon K \to l$, an $l$-epimorphism $\eta\colon {}^\sigma\mathbf{H} \to \mathbf{F}$, and a homomorphism $\nu\colon \Lambda \to \mathscr{Z}(\mathbf{F})$ such that $\delta(\lambda) = \nu(\lambda) \cdot \eta(\sigma^0(\lambda))$ for all $\lambda \in \Lambda$.*

**(6) Theorem.** (See Theorem B of Chapter VIII.) *Let $\Lambda$ be an S-arithmetic subgroup of the group $\mathbf{H}$. Let $l$ be a field, $\mathbf{F}$ an algebraic $l$-group, and $\delta\colon \Lambda \to \mathbf{F}(l)$ a homomorphism. Assume that* $\operatorname{char} K = 0$ *and* $\operatorname{rank}_S \mathbf{H} \geq 2$. *Then*

*(i) for* $\operatorname{char} l \neq 0$ *and $\mathbf{H}$ simply connected, the group $\delta(\Lambda)$ is finite;*

*(ii) for* $\operatorname{char} l = 0$, *the Zariski closure of the group $\delta(\Lambda)$ is a semisimple $l$-group;*

*(iii) for* $\operatorname{char} l = 0$ *(i.e. $l \supset \mathbb{Q}$) and $\mathbf{H}$ simply connected, there exist a uniquely determined $l$-morphism $\varphi\colon R_{K/\mathbb{Q}}\mathbf{H} \to \mathbf{F}$ and a homomorphism $\nu\colon \Lambda \to \mathbf{F}$ such that the subgroup $\nu(\Lambda)$ is finite and commutes with $\varphi(R_{K/\mathbb{Q}}\mathbf{H})$ and $\lambda(\gamma) = \nu(\lambda) \cdot \varphi(R^0_{K/\mathbb{Q}}(\lambda))$ for each $\lambda \in \Lambda$, where $R_{K/\mathbb{Q}}$ denotes the restriction of scalars functor and $R^0_{K/\mathbb{Q}}\colon \mathbf{H}(K) \to (R_{K/\mathbb{Q}}\mathbf{H})(\mathbb{Q})$ denotes the natural isomorphism (for precise definitions see 1.7 of Chapter I).*

**(7) Theorem.** (See Theorem A of Chapter VIII.) *Let $\Lambda$ be an S-arithmetic subgroup of the group $\mathbf{H}$ and let $N$ be a normal subgroup in $\Lambda$. Assume that* $\operatorname{rank}_S \mathbf{H} \geq 2$ *and that either $\mathbf{H}$ is simply connected or $S$ is finite. Then either $N$ is contained in the center of $\mathbf{H}$ or $\Lambda/N$ is finite.*

It should be noted that Theorems 5–7 are deduced from Theorems 2–4 (more precisely, from the weak versions of these theorems obtained in Chapters IV and VII) via the Borel-Harish-Chandra-Behr-Harder reduction theorem.

## 2. Synopsis of the Chapters

Chapter I is preliminary. This chapter contains the results from different areas of mathematics used in the succeeding chapters. Here we state without proofs almost all the results from other books and papers which are used in Chapters II–IX.

The results presented in Chapter II can be divided into three parts:

1) density theorems for subgroups with property $(S)$ and, in particular, a version of the Borel-Wang theorem on Zariski density of certain subgroups in algebraic groups;
2) density theorems for projections of discrete subgroups onto direct factors;
3) ergodicity theorems for actions on quotient spaces.

Furthermore we prove a series of assertions concerning the theory of unitary representations and iterations of linear transformations. We make use of these assertions in Chapter II and in Chapters III and V.

Chapter III is devoted to property $(T)$. For a locally compact group this means that the trivial one-dimensional representation is isolated in the space of irreducible unitary representations. It turns out that property $(T)$ for discrete groups is closely connected with their structure. In particular, if a discrete group $\Gamma$ has property $(T)$, then $\Gamma$ is finitely generated and the factor group of the group $\Gamma$ modulo its commutator subgroup is finite. Applications of property $(T)$ in the theory of discrete subgroups are based on the fact that lattices in a sufficiently large class of semisimple groups have this property.

In Chapter IV we prove a weak version of above mentioned Theorem 4 on finiteness of factor groups of discrete subgroups. The proofs are largely based on the results of Chapter III and on the study of invariant algebras of measurable sets.

In Chapters V and VI the theorems on existence of equivariant measurable maps are proved. These are used in Chapter VII in the proof of the superrigidity theorems. In Chapter V we present an approach based on the multiplicative ergodic theorem and in Chapter VI an approach based on boundary theory. Chapter VI, or more precisely its first three sections, may be considered as an introduction to the Furstenberg boundary theory.

In Chapter VII the superrigidity theorems for discrete subgroups are established. In particular we prove Theorem 2 stated above. A series of consequences of these theorems are obtained as well. At the end of Chapter VII, namely in Section 8, certain results on rigidity of ergodic actions of semisimple groups are stated.

In Chapter VIII a series of assertions on normal subgroups and "abstract" homomorphisms of algebraic groups over global fields are established with the help of the results of Chapters VI and VIII. In particular, we prove Theorems 5–7 stated above.

In Chapter IX the arithmeticity theorems are stated in full generality and proved. A series of consequences of these theorems are obtained and, in particular, it is shown how with the help of these theorems certain results of Chapters IV and VII can be strengthened. Sections 6 and 7 are largely devoted to the restatement of some previous results in terms of Lie group theory, symmetric space theory, and the theory of complex manifolds.

In Appendices A and B we prove two theorems which are used in Chapter V. In Appendix C we provide some examples of non-arithmetic lattices in real rank 1 semisimple Lie groups.

Finally, we would like to indicate how one might use this book. The reader who is only interested in the arithmeticity theorems may pass, after Chapter I, immediately to Chapter VI, then to Chapter VII till 5.17, and finally to Chapter IX. Doing this the reader would occasionally have to consult other chapters, in particular Sections 2, 4 and 7 of Chapter II. The reader who is mainly interested in the statements of the results should consult Sections 2, 4, 6 and 7 of Chapter II, Section 5 of Chapter III, Sections 5–8 of Chapter VII, Sections 2, 3 of Chapter VIII, and Sections 1, 3–7 of Chapter IX. It should be noted that in Sections 1, 2 of Chapter II, 1–3 of Chapter III, 1, 3 of Chapter IV, 1–3 of Chapter V, 1–4 of Chapter VII and in Chapter VI the machinery of algebraic group theory has almost never been used.

## 3. Remarks on the Structure of the Book, References and Notation

In this book chapters are divided into sections and sections into paragraphs. In Chapter I, §§1–5, we use a triple spaced numeration, in other chapters only a double spaced one. Cross references to results in another chapter are preceded by the chapter number denoted by a Roman numeral, e.g. Lemma II.3.5, while cross references to results within the same chapter are given without indication of chapter number, e.g. Lemma 3.5

The notation and the terminology used throughout the book are introduced in Section 0 of Chapter I. For the most part, they agree with generally accepted usage. The notation used within a chapter appears at the beginning of that chapter before Section 1. The notation used within a section is usually introduced either at the beginning of that section or before the paragraph with the last number 1 or in the paragraph with the last number 0, e.g. the notation used in Section 3 appears either before paragraph 3.1 or in 3.0.

# Historical and Bibliographical Notes

The Borel-Wang density theorem presented in Chapter II was established by Borel in [Bo 1] for all archimedean $k$ and by Wang in [WS 8] in the general case. At present, there are various generalizations and proofs of this theorem (see [Da 3], [Fu 7], [Mosk], and [Rag 6]). The approach to the proof of the density theorems presented in Sections 2 and 4 of Chapter II is close to the one proposed in [Da 3] and [Fu 7].

Lemma 3.2 of Chapter II is actually a modification of Mautner's lemma proved by Prasad (see [Pr 2], Lemma 2). In the case that the fields $k_\alpha$ are archimedean, Proposition II.3.3 is a particular case of the Moore-Mautner theorem (see [Moo 1], Theorem 1) and in case $\operatorname{char} A = 1$ was in fact established by Prasad in the process of proving Proposition 2 in [Pr 2]. The proof given in this book is a modification of the proofs given in [Moo 1] and [Pr 2]. Lemma II.3.4 is due to Kirillov (see [D-K], the proof of Lemma 5, or [G-G-P], the proof of the lemma in §2 of the appendix of Chapter II). Proposition II.3.9 and Corollary II.3.10 were established in §2.3 of [Mar 8].

Theorem II.5.1 was proved in [Mar 8] and [Pr 5] (the proof given in this book is a version of the one given in [Mar 8]). For the results of Section 6 in Chapter II (see [Mar 8], [Pr 5], and Section 1 in [Pr 6] (for the strong approximation theorem given in 6.8, see also [Kn 3], [Pl 2], [Pl 3], and [Pl 4]). In the case where the fields $k_\alpha$ are archimedean, Theorem II.7.2 is actually a special case of the Mostow-Moore-Mautner theorem (see [Most 5], Lemma 4) and, in case $\operatorname{card} A = 1$ and $\operatorname{char} k_\alpha = 0$ for all $\alpha \in A$, was essentially proved by Dani (see [Da 1], Corollary 3.2). The proof of this theorem given in this book is a modification of the proofs in [Da 1] and [Most 5]. We refer the reader who is interested in the ergodicity problem of flows on the homogeneous spaces of arbitrary real Lie groups to [B-M]. Questions involving the interrelationships between the theory of discrete subgroups and ergodic theory have been studied in [Da 1,2,4-8], [Rat 1-3], [Sul 1], and many other papers.

As was remarked at the beginning of Chapter III, property $(T)$ was introduced by Kazhdan in [Kaz 1]. The main results concerning this property were given in that paper. However the exposition there is very concise. More detailed expositions have been given in [D-K] and [WS 4]. The case of semisimple algebraic groups over fields of characteristic 2 has been considered in [WS 18], where several other results on property $(T)$ of nonsemisimple groups have also been obtained. The implication (ii) $\Rightarrow$ (i) in Theorem III.2.12 was established in [WS 11]. The proof of Watatani's theorem given in Section 3 of Chapter III, asserting that the groups with property $(T)$ are not amalgams, is actually the same as in [Wat]. In III.5.12 we present the results of the papers [Mar 14] and [Sul 2]. These two papers differ one from another by the choice of dense subgroups with property $(T)$.

It should be noted that the forthcoming book [H-V] is specially devoted to property $(T)$.

In Chapter IV we have presented the substance of the papers [Mar 11] and [Mar 12]. The proof of Theorem IV.1.3 is a modification of that of Theorem III.1.2.3 in [D-S]. Theorem IV.1.5 is deduced from Theorem IV.1.3 in the same way as in Section 7 of [Br], the usual density point theorem is deduced from the Vitali covering theorem. Proposition IV.2.4 was in fact established in [Li].

The approach to the proof of the existence of equivariant measurable maps based on an application of the multiplicative ergodic theorem presented in Chapter V was proposed for the first time in [Mar 7]. Another approach (based on boundary theory) presented in Chapter VI was offered by Furstenberg (see [Fu 6] and [Fu 8]). For other approaches, see VII.5.26.

An indication of the proof of Theorem VII.5.9 in case $\Lambda' = \Gamma'$ was given in [Mar 7] and complete details offered in [Mar 9]. The method developed in [Mar 7] and [Mar 9] can immediately be applied to the proof of Theorem VII.5.6 in case $\operatorname{char} k_\alpha = 0$ for all $\alpha \in A$. The approach to the proof of the superrigidity theorems presented in Chapter VII includes some ideas and arguments due to E. Vishik. Repeating the statement made in [Mar 7] the author would particularly like to mention that while proving the superrigidity theorems he was under the influence of Mostow's remarkable ideas on the extension of isomorphisms of lattices to compactifications of symmetric spaces.

Theorem VII.7.5, in the case where the lattice $\Gamma'$ is cocompact, was first established by Mostow in [Most 1]. Developing Mostow's method, Prasad proved Theorem VII.7.5 for non-cocompact lattices $\Gamma'$ of $\mathbb{Q}$-rank 1 (see [Pr 1]) and Theorem VII.7.1 for $\Gamma'$ cocompact (see [Pr 6]). Theorem VII.7.5 in case $\Gamma'$ is non-cocompact was established by the author in the paper entitled "Non-cocompact lattices in semisimple algebraic groups" and published in the same transactions as [Mar 6]. The proof given in this paper is based on the study of unipotent elements in lattices and therefore can not be carried over the case in which $\Gamma'$ is cocompact.

Theorem VIII (A) was first established in 2.4 of [Mar 12]. The majority of the results of VIII.2, though in somewhat weaker form, were established in that paper. Theorem VIII (B) in a slightly different form was stated in [Mar 7]. In case $\operatorname{char} k = 0$, Propositions VIII.3.20 and VIII.3.22 were established in [Vi 4]; several other results on definition rings of linear groups were also obtained in this paper.

The main results of Chapter IX (i.e. Theorems A and B in 1.9) in case $\operatorname{char} k_\alpha = 0$ for all $\alpha \in A$ and $\mathbf{G} - \mathbf{G}^{\mathrm{is}}$ were stated (in a slightly different form) in [Mar 7].

Theorem IX.1.16 under the assumption "$\operatorname{rank}_{\mathbb{R}} \mathbf{G} \geq 2$ and $\mathbf{G} = \mathbf{G}^{\mathrm{is}}$" was first established in [Mar 7] and [Mar 9] and had been conjectured by Selberg in case $\Gamma_0$ is non-cocompact (see [Sel 2], [Sel 3]) and by Piatetski-Shapiro in the general case (see [Pi 2], [Pi 3]). Various results, essentially for the case of non-cocompact lattices, have been obtained by these two authors while aiming at the proof of this conjecture (see [Pi 1], [Pi 2], [Pi 3], [Sel 1], [Sel 3]). For $\Gamma_0$ non-cocompact, Theorem IX.1.16 under the assumption mentioned above was established in [Mar 3] and [Mar 5] by a method based on a study of unipotent elements of $\Gamma_0$ and which is therefore inapplicable to cocompact lattices.

A slightly weaker result on arithmeticity of non-cocompact lattices in real semisimple groups was announced by the author in the paper "On the arithmeticity of discrete subgroups" (Soviet Math. Dokl., 10 (1969), 900–902). Following this, a substantial progress towards a proof of Theorem IX.1.16, for non-cocompact $\Gamma_0$, was also made by Raghunathan (see [Rag 6]), whose method is based again on a study of unipotent elements of $\Gamma_0$.

For certain groups $G$, under the assumption $\operatorname{char} k_\alpha = 0$ for all $\alpha \in A$, Theorem $A$ of IX.1.9 was conjectured (though in somewhat inaccurate form) by Piatetski-Shapiro (see [Pi 2]). For the case of $\mathbb{R}$-groups Theorem B of IX.1.9 was conjectured, at least essentially, in [P-S] and [Pi 2] and so was Proposition IX.7.7. Theorem IX.1.9 B was proved by Kazhdan in [Kaz 2] for non-cocompact lattices in $\mathbf{SL}_2(\mathbb{R})$ consisting of matrices with algebraic integral entries.

Corollary IX.4.17 for the case $\mathbf{G} = \mathbf{G}^{is}$ was conjectured in various forms by Selberg (see [Sel 1] and [Sel 3]). Corollary IX.7.2 yields a positive answer to Selberg's question of whether there exist symmetric spaces without irreducible discrete groups of motions of finite covolume.

# References


[A-M] Alon, N.; Milman, V.D.: $\lambda_1$, isoperimetric inequalities for graphs and superconcentrators. J. Combinatorial Theory. **38** (1985) 73–88

[An 1] Andreev, E.M.: Convex polyhedra in Lobachevsky space. Math. USSR Sb. **10** (1970) 413–440

[An 2] Andreev, E.M.: Convex polyhedra of finite volume in Lobachevsky space. Math. USSR Sb. **12** (1970) 255–259

[A-W] Akemann, C.A.; Walter, M.E.: Unbounded negative definite functions. Can. J. Math. **33** (1981) 862–871

[Be 1] Behr, H.: Zur starken Approximation in algebraischen Gruppen über globalen Körpern. J. Reine Angew. Math. **229** (1968) 107–116

[Be 2] Behr, H.: Endliche Erzeugbarkeit arithmetischer Gruppen über Funktionenkörpern. Invent. math. **7** (1969) 1–32

[Be 3] Behr, H.: Chevalley groups of rank 2 over $\mathbb{F}_q(t)$ are not finitely presentable. Proc. Symp. "Homological and combinatorial techniques in group theory" (Durham 1977). London Math. Soc., Lecture Notes Ser. 36, 213–224

[Be 4] Behr, H.: Finite presentability of arithmetic groups over global function fields. Proc. Edinburgh Math. Soc. **30** (1987) 23–39

[Bi] Billingsley, P.: Ergodic theory and information. Wiley, New York 1965

[B-K] Bernstein, I.N.; Kazhdan, D.A.: The one-dimensional cohomology of discrete subgroups. Funct. Anal. Appl. **4** (1970) 1–4

[B-M] Brezin, J.; Moore, C.C.: Flows on homogeneous spaces: a new look. Amer. J. Math. **103** (1981) 571–613

[B-M-S] Bass, H.; Milnor, J.; Serre, J.-P.: Solution of the congruence subgroup problem for $\mathbf{SL}_n$ and $\mathbf{Sp}_{2n}$. Publ. Math. IHES **33** (1967) 59–137

[Bo 1] Borel, A.: Density properties for certain subgroups of semisimple groups without compact components. Ann. Math. **72** (1960) 179–188

[Bo 2] Borel, A.: Compact Clifford-Klein forms of symmetric spaces. Topology **2** (1963) 111–122

[Bo 3] Borel, A.: Some finiteness properties of adèle groups over number fields. Publ. Math. IHES **16** (1963) 1–30

[Bo 4] Borel, A.: Density and maximality of arithmetic subgroups. J. Reine Angew. Math. **224** (1966) 78–89

[Bo 5] Borel, A.: Ensembles fondamentaux pour les groups arithmétiques et formes automorphes. Faculté des Sciences de Paris, cours mimeographié, 1967

[Bo 6] Borel, A.: Linear algebraic groups. Benjamin, New York 1969

[Bo 7] Borel, A.: Some metric properties of arithmetic quotients of symmetric spaces and an extension theorem. J. Differential Geometry, **6** (1972) 543–560

[Bo 8] Borel, A.: Cohomologie de certains groupes discrets et laplacien $p$-adique. Seminaire Bourbaki, Exp. 437, 1973/74, Springer Lecture Notes, No. 431

[Bo 9] Borel, A.: Cohomology of arithmetic groups. Proc. Int. Congress Math. (Vancouver 1974), vol. 1 (1975), pp. 435–442

[Bo 10] Borel, A.: Cohomologie de sous-groupes discrets et représentations de groupes semi-simples. Astérisque **32–33** (1976) 73–112

[Bo 11] Borel, A.: Stable and $L^2$-cohomology of arithmetic groups. Bull. Amer. Math. Soc., New Ser. **3** (1980) 1025–1027

[Bo 12] Borel, A.: Commensurability classes and volumes of hyperbolic 3-manifolds. Ann. Scuola Norm. Sup. Pisa Cl. Sci (4) **8** (1981) 1–23

[Bo-Hard] Borel, A.; Harder, G.: Existence of discrete cocompact subgroups of reductive groups over local fields. J. Reine Angew. Math. **298** (1978) 53–64

[Bo-Hari] Borel, A.; Harish-Chandra: Arithmetic subgroups of algebraic groups. Ann. Math. **75** (1962) 485–535

[Boro 1] Borovoj, M.V.: Abstract simplicity for some anisotropic algebraic groups over number fields. Dokl. Akad. Nauk SSSR **283** (1985) 794–797 (in Russian)

[Boro 2] Borovoj, M.V.: Abstract simplicity of groups of type $\mathbb{D}_n$ over number fields. Usp. Mat. Nauk, V. 43, **5** (263) 179–180 (in Russian)

[Bo-Se 1] Borel, A.; Serre, J.-P.: Corners and arithmetic groups. Comment. Math. Helv. **48** (1973) 436–491

[Bo-Se 2] Borel, A.; Serre, J.-P.: Cohomologie d'immeubles et de groupes $S$-arithmétiques. Topology **15** (1967) 211–232

[Bo-Sp] Borel, A.; Springer, T.A.: Rationality properties of linear algebraic groups II. Tohoku Math. J. **20** (1968) 443–497

[Bo-T 1] Borel, A.; Tits, J.: Groupes réductifs. Publ. Math. IHES **27** (1965) 55–150

[Bo-T 2] Borel, A.; Tits, J.: Eléments unipotents et sous-groupes paraboliques de groupes réductifs. I. Invent. math. **12** (1971) 95–104

[Bo-T 3] Borel, A.; Tits, J.: Compléments à l'article "Groups réductifs". Publ. Math. IHES **41** (1972) 253–276

[Bo-T 4] Borel, A.; Tits, J.: Homomorphismes "abstraits" de groupes algébriques simples. Ann. Math. **97** (1973) 499–571

[Bou 1] Bourbaki, N.: Topologie Générale. Hermann, Paris: Ch. I–II (1951), Ch. III–IV (1951), Ch. V–VIII (1955)

[Bou 2] Bourbaki, N.: Espaces Vectorielles Topologiques. Hermann, Paris: Ch. I–II (1953), Ch. III–V (1955), Fascicule de résultats 1955

[Bou 3] Bourbaki, N.: Intégration. Ch. VI–VIII, Hermann, Paris 1960

[Bou 4] Bourbaki, N.: Variétés Différentielles et Analytiques. Fascicule des résultats, Hermann, Paris: Paragr. 1–7 (1967), Paragr. 8–15 (1971)

[Bou 5] Bourbaki, N.: Groupes et Algèbres de Lie. Hermann, Paris: Ch. I (1971), Ch. II–III (1972), Ch. IV–VI (1968)

[Bo-W] Borel, A.; Wallach, N.: Continuous cohomology, discrete subgroups and representations of reductive groups. Ann. Math. Studies, No. 94, Princeton Univ. Press, Princeton 1980

[Br] Brudno, A.A.: Theory of functions of real variables. Nauka, Moscow 1971 (in Russian)

[B-T 1] Bruhat, F.; Tits, J.: Groupes réductifs sur un corps local, I. Publ. Math. IHES **41** (1972) 5–252

[B-T 2] Bruhat, F.; Tits, J.: Groupes réductifs sur un corps local, II. Publ. Math. IHES **60** (1984) 5–184

[B-T 3] Bruhat, F.; Tits, J.: Groupes algébriques simples sur un corps local: cohomologie galoisienne, décompositions d"Iwasawa et de Cartan. C.R. Acad. Sci. Paris. Série A, **263** (1966) 867–869

[Bu] Bugayenko, V.O.: Groups of automorphisms of unimodular hyperbolic quadratic forms over the ring $\mathbb{Z}[(\sqrt{5}+1)/2]$. Moscow Univ. Math. Bull. **39** (1984) 6–14

[C 1] Chevalley, C.: Theory of Lie groups, I. Princeton University Press, Princeton 1946

[C 2] Chevalley, C.: Théorie des groupes de Lie, II: groupes algébriques. Hermann, Paris 1951

[C 3] Chevalley, C.: Théorie des groupes de Lie, III: théorèmes généraux sur les algèbres de Lie. Hermann, Paris 1955

[Ch 1] Chernousov, V.I.: On the projective simplicity of algebraic groups split over a quadratic extension of a number field. Dokl. Akad. Nauk SSSR **296** (1987) 1301–1405 (in Russian)

[Ch 2] Chernousov, V.I.: On the structure of the groups of rational points of algebraic groups of type $\mathbb{D}_n$. Dokl. Akad. Nauk SSSR **31** (1987) 593–596 (in Russian)

[Ch 3] Chernousov, V.I.: On the projective simplicity for some groups of rational points over number fields. Izv. Akad. Nauk SSSR, Ser. math. **53** (1989) 398–410 (in Russian)

[Ch-W] Choda, M.; Watatani, Y.: Fixed point algebra and property (T). Math. Japonica **27** (1982) 263–266

[Cox] Coxeter, H.S.M.: Discrete groups generated by reflections. Ann. Math. **35** (1934) 588–621

[Co-W] Connes, A.; Weiss, B.: Property (T) and asymptotically invariant sequences. Israel. J. Math. **37** (1980) 209–210

[Da 1] Dani, S.G.: Kolmogorov automorphisms on homogeneous spaces. Amer. J. Math. **98** (1976) 119–163

[Da 2] Dani, S.G.: On invariant measures, minimal sets, and a lemma of Margulis. Invent. math. **51** (1979) 239–260

[Da 3] Dani, S.G.: A simple proof of Borel's density theorem. Math. Zeit. **174** (1980) 81–94

[Da 4] Dani, S.G.: Invariant measures and minimal sets of horospherical flows. Invent. math. **64** (1981) 357–385

[Da 5] Dani, S.G.: Continuous equivariant images of lattice actions on boundaries. Ann. Math. **119** (1984) 111–119

[Da 6] Dani, S.G.: On orbits of unipotent flows on homogeneous spaces. Ergod. Theor. and Dynam. Sys. **4** (1984) 25–34

[Da 7] Dani, S.G.: On orbits of unipotent flows on homogeneous spaces, II. Ergod. Theor. and Dynam. Sys. **6** (1986) 167–182

[Da 8] Dani, S.G.: Orbits of horospherical flows. Duke Math. Z. **53** (1986) 177–188

[De] Deligne, P.: Extensions centrales non-résiduellement finies de groupes arithmétiques. C.R. Acad. Sci. Paris. Série A, **287** (1978) 203–208

[Die] Dieudonné, J.: La géométrie des groupes classiques. Springer, New York 1971

[Dix] Dixmier, J.: Les $C^*$-algèbres et leurs représentations. Gauthier-Villars, Paris 1969

[D-K] Delaroche, C.; Kirillov, A.: Sur les relations entre l'espace dual d'un groupe et la structure de ses sous-groupes fermés. Séminaire Bourbaki, vol. 1967/68, exposé 343, Benjamin, New York 1969

[D-M] Deligne, P.; Mostow, G.D.: Monodromy of hypergeometric functions and non-lattice integral monodromy. Publ. Math. IHES **63** (1986) 5–89

[Dr] Drinfel'd, V.G.: Finitely additive measures on $S^2$ and $S^3$ invariant with respect to rotations. Funct. Anal. Appl. **18** (1984) 245–246

[D-R] Del Junco, A.; Rosenblatt, J.: Counterexamples in ergodic theory and number theory. Math. Ann. **245** (1979) 185–197

[D-S] Dunford, N.; Schwartz, T.: Linear operators, Part I: general theory. Interscience publishers, New York 1958

[Fa] Faddeev, L.D.: The eigen-function expansion of Laplace's operator on the fundamental domain of a discrete group on Lobachevsky plane. Trudy Moskov. Mat. Obšč. **17** (1967) 323–350 (in Russian)

[Fe] Fell, J.M.G.: Weak containment and induced representations of groups. Can. J. Math. **14** (1962) 237–268

[F-G] Feldman, J.; Greenleaf, F.P.: Existence of Borel transversals in groups. Pacific J. Math. **25** (1968) 455–461

[F-K] Furstenberg, H.; Kesten, H.: Products of random matrices. Ann. Math. Statistics **31** (1960) 457–469

[Fu 1] Furstenberg, H.: Non-commuting random products. Trans. Amer. Math. Soc. **108** (1963) 377–428

[Fu 2] Furstenberg, H.: Poisson boundaries and envelopes of discrete groups. Bull. AMS **73** (1967) 350–356

[Fu 3] Furstenberg, H.: The unique ergodicity of the horocycle flow, in: Recent advances in topological dynamics. Springer Lecture Notes No. 318

[Fu 4] Furstenberg, H.: Random walks and discrete subgroups of Lie groups, in: Advances in Probability and related topics, v. 1, Marcel Dekker, New York 1973

[Fu 5] Furstenberg, H.: Boundary theory and stochastic processes on homogeneous spaces, in: Harmonic analysis on homogeneous spaces. (Symposia on Pure and Applied Math.), Williamstown, Mass. 1972, Proceedings, vol. **26** (1973) 193–229

[Fu 6] Furstenberg, H.: Equivariant measurable maps and representations of lattice subgroups. Hebrew University, Jerusalem 1975, preprint

[Fu 7] Furstenberg, H.: A note on Borel's density theorem. Proc. AMS **55** (1976) 209–212

[Fu 8] Furstenberg, H.: Rigidity and cocycles for ergodic actions of semisimple groups (after G.A. Margulis and R. Zimmer). Seminaire Bourbaki, No. 559, 1979/80

[Ga 1] Garland, H.: $p$-adic curvature and the cohomology of discrete subgroups of $p$-adic groups. Ann. Math. (2) **97** (1973) 375–423

[Ga 2] Garland, H.: On the cohomology of discrete subgroups of $p$-adic groups. Proc. In. Congress Math. Vancouver 1974, v. **1** (1975) 449–453

[G-G-P] Gel'fand, I.M.; Graev, M.I.; Piatetski-Shapiro, I.I.: Representation theory and automorphic functions. Fiz. Mat. Giz., Moscow 1966 (in Russian)

[Go] Godement, R.: Les fonctions de type positif et la théorie des groupes. Trans. AMS **63** (1948) 1–84

[G-R] Garland, H.; Raghunathan, M.S.: Fundamental domains for lattices in $R$-rank 1 semisimple Lie groups. Ann. Math. **92** (1970) 279–326

[Gre] Greenleaf, F.: Invariant means on topological groups. van Nostrand, New York 1969

[Gro] Gromov, M.: Hyperbolic manifolds, groups, and actions, in: Riemann surfaces and related topics, ed. I. Kra and B. Maskit, Annals of Math. Studies, No. 97. Princeton University Press 1981

[Gro-P] Gromov, M.; Piatetski-Shapiro, I.: Non-arithmetic groups in Lobachevsky spaces. Publ. Math. IHES (1988)

[G-S 1] Grunewald, F.; Schwermer, J.: A non-vanishing theorem for the cuspidal cohomology of $\mathbf{SL}_2$ over imaginary quadratic integers. Math. Ann. **258** (1981) 183–200

[G-S 2] Grunewald, F.; Schwermer, J.: Free non-abelian quotients of $\mathbf{SL}_2$ over orders of imaginary quadratic number fields. J. Algebra **69** (1981) 298–304

[G-S 3] Grunewald, F.; Schwermer, J.: Arithmetic quotients of hyperbolic 3-space, cusp forms and link complements. Duke Math. J. **48**, No. 2 (1981) 351–358

[Gu] Guivarc'h, Y.: Quelques propriétés asymptotiques des produits de matrices aléatoires. Springer Lecture Notes, No. **774** (1980) 178–250

[G-V] Gel'fand, I.M.; Vilenkin, N.J.: Some applications of harmonic analysis. Equipped Hilbert spaces (Generalized functions No. 4), Gosud. Izdat. Fiz. Mat. Lit., Moscow 1961 (in Russian)

[Hal 1] Halmos, P.R.: Measure Theory. New York 1950

[Hal 2] Halmos, P.R.: Lectures on ergodic theory. Math. Soc. of Japan, Tokyo 1956

[Har 1] Harder, G.: Eine Bemerkung zum schwachen Approximationsatz. Arch. d. Math. **19** (1968) 465–471

[Har 2] Harder, G.: Minkowskische Reduktionstheorie über Funktionenkörpern. Invent. math. **7** (1969) 33–54

[Har 3] Harder, G.: A Gauss-Bonnet formula for discrete arithmetically defined groups. An. Sc. d. Ec. Norm. Sup. 4 Serie t. **4** (1971) 409–455

[Har 4] Harder, G.: On the cohomology of discrete arithmetically defined groups, in: Proc. Int. Colloq. on discrete subgroups of Lie groups and applications to moduli (Bombay 1973). Oxford University Press 1975, 129–160

[Har 5] Harder, G.: Über die Galoiskohomologie halbeinfacher algebraischer Gruppen, III. J. Reine Angew. Math. **274/275** (1975) 125–138

[Har 6] Harder, G.: Die Kohomologie $S$-arithmetischer Gruppen über Funktionenkörpern. Invent. math. **42** (1977) 135–175

[Har 7] Harder, G.: Eisenstein cohomology of arithmetic groups. The case $\mathbf{GL}_2$. Invent. math. **89** (1987) 37–118

[He] Helgason, S.: Differential Geometry and Symmetric Spaces. Academic Press, New York 1962

[Ho] Hopf, E.: Ergodentheorie. Ergebnisse der Mathematik, vol. 2. J. Springer, Berlin 1937

[Hov] Hovanskiĭ, A.G.: Hyperplane sections of polyhedra, toroidal manifolds, and discrete groups in Lobachevsky space. Funct. Anal. Appl. (1986) 50–61 (in Russian)

[Hö] Höfer, T.: Ballquotienten als verzweigte Überlegungen der projektiven Ebene. Diss. Universität Bonn, Bonn 1985

[Hul] Hulanicki, A.: Means and Følner conditions on locally compact groups. Studia Math. **27** (1966) 87–104

[Hum 1] Humphreys, J.E.: Linear Algebraic Groups. Springer, New York 1975

[Hum 2] Humphreys, J.E.: Arithmetic groups. Springer Lecture Notes, No. **789** (1980)

[H-V] Harpe, P.; Valette, A.: La propriété (T) de Kazhdan pour les groupes localement compacts. To appear

[J-M] Jonson, D.; Millson, J.: Deformation spaces associated to compact hyperbolic manifolds. Preprint

[J-W-W] James, D.; Waterhouse, W.; Weisfeiler, B.: Abstract homomorphisms of algebraic groups: problems and bibliography. Communications in Algebra **9** (1981) 95–114

[Kaz 1] Kazhdan, D.A.: On the connection between the dual space of a group and the structure of its closed subgroups. Funct. Anal. Appl. **1** (1967) 63–65

[Kaz 2] Kazhdan, D.A.: Construction of $\Gamma$-rational groups for certain discrete subgroups $\Gamma$ of the group $\mathbf{SL}(2,\mathbb{R})$. Funct. Anal. Appl. **2** (1968) 34–37

[Kaz 3] Kazhdan, D.A.: On arithmetic varieties. Proc. of the Summer School of the Bolyai Janos Math. Soc. on Lie groups and their representations, Budapest 1971. Akademia Kiado, Budapest 1975, pp. 151–217

[Kaz 4] Kazhdan, D.A.: Some applications of the Weil representation, Mimeographed notes.

[Kar] Karpelevich, F.I.: Transitivity surfaces of a semisimple subgroup of the motion group of a symmetric space. Dokl. Akad. Nauk SSSR **93** (1953) 401–404 (in Russian)

[Ki] Kirillov, A.A.: Basic representation theory. Nauka, Moscow 1972 (in Russian)

[Kn 1] Kneser, M.: Orthogonale Gruppen über algebraischen Zahlkörpern. J. Reine Angew. Math. **196** (1956) 213–220

[Kn 2] Kneser, M.: Schwache Approximation in algebraischen Gruppen. Colloque sur la théorie des groupes algébriques. CBRM Brussels (1962) 41–52

[Kn 3] Kneser, M.: Strong approximation I, II. Proc. Symp. Pure Math. **9** (1966) 99–103, 187–196

[Kn 4] Kneser, M.: Normal subgroups of integral orthogonal groups. Algebraic $K$-theory and its geometric applications. Springer Lecture-Notes No. **108** (1969) 67–71

[Kn 5] Kneser, M.: Normalteiler ganzzahliger Spingruppen. J. Reine Angew. Math. 311/312 (1979) 191–214

[Ko] Kostant, B.: On the existence and irreducibility of certain series of representations. Bull. AMS **75** (1969) 627–642

[K-S-F] Kornfeld, I.P.; Sinai, J.G.; Fomin, S.V.: Ergodic Theory. Gosud. Izd. Fiz. Mat. Lit., Moscow 1980 (in Russian)

[Ku] Kurosh, A.G.: Group Theory. Nauka, Moscow 1967 (in Russian)

[La] Lachaud, G.: Spectral analysis of automorphic forms on rank one groups by perturbation methods, in: Harmonic Analysis of Homogeneous spaces. Proc. Symp. Pure Math. **26** (1973) 441–450

[Lan 1] Lang, S.: Algebra. Addison-Wesley, Reading, Mass. 1965

[Lan 2] Lang, S.: $\mathbf{SL}_2(\mathbb{R})$. Addison-Wesley, Reading, Mass. 1975

[Li] Lind, D.A.: Translation invariant sigma algebras of groups. Proc. AMS **42** (1974) 218–221

[Lo] Loève, M.: Probability Theory. D. Van Nostrand Comp. Inc., Princeton New Jersey 1960

[L-R] Losert, V.; Rindler, H.: Almost invariant sets. Bull. London Math. Soc. **13** (1981) 145–148

[L-S] Lee, R.; Schwermer, J.: Cohomology of arithmetic subgroups of $\mathbf{SL}_3$ at infinity. J. Reine Angew. Math. **330** (1982) 100–131

[Lu 1] Lubotzky, A.: Group presentation, $p$-adic analytic groups and lattices in $\mathbf{SL}_2(\mathbb{C})$. Ann. Math. (2) **118** (1983) 115–130

[Lu 2] Lubotzky, A.: Trees and discrete subgroups of Lie groups over local fields. Bull. AMS **20** (1989) 23–30

[Macd] Macdonald, I.G.: Spherical functions on a group of $p$-adic type. Univ. of Madras, India, 1971

[Mack] Mackey, G.W.: Induced representations of locally compact groups. Ann. Math. **55** (1952) 101–139

[Mak 1] Makarov, V.S.: A class of decompositions of the Lobachevsky space. Soviet Math. Dokl. **6** (1965) 400–401

[Mak 2] Makarov, V.S.: On a certain class of discrete groups of Lobachevsky space having an infinite fundamental region of finite measure. Soviet Math. Dokl. **7** (1966) 328–331

[Mak 3] Makarov, V.S.: On a certain class of 2-dimensional Fedorov groups. Math. USSR Izv. **1** (1968) 515–524

[Mal] Mal'cev, A.I.: On a class of homogeneous spaces, AMS Transl. **39** (1951)

[Mar 1] Margulis, G.A.: The isometry of closed manifolds of constant negative curvature with the same fundamental group. Soviet Math. Dokl. **11** (1970) 722–723

[Mar 2] Margulis, G.A.: Explicit construction of expanders. Probl. Inform. Transmission (1975) 325–332 (in Russian)

[Mar 3] Margulis, G.A.: Arithmeticity of discrete subgroups. Russ. Math. Surv. **29** (1974) 107–156

[Mar 4] Margulis, G.A.: Arithmeticity and finite-dimensional representations of uniform lattices. Funct. Anal. Appl. **8** (1974) 258–259

[Mar 5] Margulis, G.A.: Arithmeticity of non-uniform lattices in weakly noncompact groups. Funct. Anal. Appl. **9** (1975) 31–38

[Mar 6] Margulis, G.A.: On the action of unipotent groups in the space of lattices, in: Proc. of the Summer School of the Bolyai Janos Math. Soc. on Lie groups and their representations, Budapest 1971, Akademia Kiado, Budapest 1975, 365–370

[Mar 7] Margulis, G.A.: Discrete groups of motions of manifolds of non-positive curvature. Proc. Int. Congress Math. (Vancouver 1974), AMS Transl. **109** (1977) 33–45

[Mar 8] Margulis, G.A.: Cobounded subgroups of algebraic groups over local fields. Funct. Anal. Appl. **11** (1977) 119–128

[Mar 9] Margulis, G.A.: Arithemticity of irreducible lattices in semisimple groups of rank greater than 1, appendix to Russian translation of M. Raghunathan, Discrete subgroups of Lie groups. Mir, Moscow 1977 (in Russian). Invent. math. **76** (1984) 93–120

[Mar 10] Margulis, G.A.: Factor groups of discrete subgroups. Soviet Math. Dokl. **19** (1978) 1145–1149

[Mar 11] Margulis, G.A.: Quotient groups of discrete subgroups and measure theory. Funct. Anal. Appl. **12** (1978) 295–305

[Mar 12] Margulis, G.A.: Finiteness of quotient groups of discrete subgroups. Funct. Anal. Appl. **13** (1979) 178–187

[Mar 13] Margulis, G.A.: On the multiplicative group of a quaternion algebra over a global field. Soviet Math. Dokl. **21** (1980) 780–784

[Mar 14] Margulis, G.A.: Some remarks on invariant means. Monatshefte für Math. **90** (1980) 233–235

[Mar 15] Margulis, G.A.: On the decomposition of discrete subgroups into amalgams. Selecta Mathematica Sovetica **1** (1981) 197–213

[Mar 16] Margulis, G.A.: Finitely-additive invariant measures on Euclidean spaces. Ergodeor. Th. and Dynam. Sys. **2** (1982) 383–396

[Mar 17] Margulis, G.A.: Free totally discontinuous groups of affine transformations. Soviet Math. Dokl. **28** (1983) 435–439

[Mar 18] Margulis, G.A.: Complete affine locally flat manifolds with free fundamental group. Zapiski Naučn. Sem. Leningrad Otd. Mat. Inst. Steklov **134** (1984) 190–205 (in Russian)

[Ma-R] Margulis, G.A.; Rohlfs, J.: On the proportionality of covolumes of discrete subgroups. Math. Ann. **275** (1986) 197–205

[Mat] Matsumoto, H.: Sur les sous-groupes arithmétiques des groupes semi-simples deployés. Ann. Scient. E. Norm. Sup., 4-th Series, **2** (1969) 1–62

[Mel] Mezzljakov, Ju.I.: A survey of latest results on automorphisms of classical groups, in: Collection "Isomorphisms of classical groups", Mir, Moscow 1976, 250–259 (in Russian)

[Me2] Mezzljakov, Ju.I.: Theory of isomorphisms of classical groups in 1976–1980, in: Collection "Isomorphisms of classical groups over integral rings", Mir, Moscow 1980, 252–258 (in Russian)

[Milli] Millionshchikov, V.M.: A criterion for the stability of the probable spectrum of linear system of differential equations with recurrent coefficients, and a criterion for the almost reducibility of systems with almost periodic coefficients. Math. USSR Sb. **7** (1969) 171–193

[Mills 1] Millson, J.: On the first Betti number of a constant negatively curved manifold. Ann. Math. **104** (1976) 235–247

[Mills 2] Millson, J.: Real vector bundles with discrete structural group. Topology **18** (1979) 83–89

[Mills 3] Millson, J.: Cycles and harmonic forms on locally symmetric spaces. Preprint

[Mi-R] Millson, J.; Raghunathan, M.S.: Geometric construction of cohomology for arithmetic groups I. Papers dedicated to the memory of V.K. Patodi Indian Acad. Sci. Bangalore 1980

[Moo 1] Moore, C.C.: Ergodicity of flows on homogeneous spaces. Amer. J. Math. **88** (1966) 154–178

[Moo 2] Moore, C.C.: The Mautner phenomenon for general unitary representations. Pacific J. Math. **86** (1980) 155–169

[Mosk] Moskowitz, M.: On the density theorem of Borel and Furstenberg. Arkch. Math. **16** (1978) 11–27

[Most 1] Mostow, G.D.: Self-adjoint groups. Ann. Math. **62** (1955) 44–55

[Most 2] Mostow, G.D.: Some new decomposition theorems for semi-simple groups. Memoirs AMS **14** (1955) 31–54

[Most 3] Mostow, G.D.: Fundamental groups of homogeneous spaces. Ann. Math. **66** (1957) 249–255

[Most 4] Mostow, G.D.: Quasi-conformal mappings in $n$-space and the rigidity of hyperbolic space forms. Publ. Math. IHES **34** (1968) 53–104

[Most 5] Mostow, G.D.: Intersections of discrete subgroups with Cartan subgroups. J. Indian Math. Soc. **34** (1970) 203–214

[Most 6] Mostow, G.D.: The rigidity of locally symmetric spaces. Proc. Int. Congr. Math. **2** (1970) 187–197

[Most 7] Mostow, G.D.: Strong rigidity of locally symmetric spaces. Ann. Math. Studies, No. 78, Princeton University Press, Princeton, N.J. 1973

[Most 8] Mostow, G.D.: Discrete subgroups of Lie groups. Adv. Math. **16** (1975) 112–123

[Most 9] Mostow, G.D.: Existence of a non-arithmetic lattice in $\mathbf{SU}(2,1)$. Proc. Nat. Acad. Sci. USA **75** (1978) 3029–3033

[Most 10] Mostow, G.D.: On a remarkable class of polyhedra in complex hyperbolic space. Pacific J. Math. **86** (1980) 171–276

[Most 11] Mostow, G.D.: Existence of nonarithmetic monodromy groups. Proc. Nat. Acad. Sci. USA **78** (1981) 5948–5950

[M-S] Mostow, G.D.; Siu, Y.T.: A compact Kaehler surface of negative curvature not covered by the ball. Ann. Math. **112** (1980) 321–360

[M-T] Mostow, G.D.; Tamagawa, T.: On the compactness of the arithmetically defined homogeneous spaces. Ann. Math. **76** (1962) 440–463

[N 1] Nikulin, V.V.: On factor groups of the automorphism groups of hyperbolic forms modulo subgroups generated by 2-reflections. Soviet Math. Dokl. **20** (1979) 1156–1158

[N 2] Nikulin, V.V.: On arithmetic groups generated by reflections in Lobachevsky spaces. Math. USSR Izv. **16** (1981) 573–601

[N 3] Nikulin, V.V.: On the classification of arithmetic groups generated by reflections in Lobachevsky spaces. Math. USSR Izv. **18** (1982) 99–123

[N 4] Nikulin, V.V.: On the factor groups of the automorphism groups of hyperbolic forms by the subgroups generated by 2-reflections. Applications to algebraic geometry, in Sovremennye problemy matematiki 18, Moscow VINITI 1981, 3–114 (in Russian)

[O] Oseledec, V.I.: A multiplicative ergodic theorem. Trans. Moscow Math. Soc. **19** (1968) 197–231

[Pi 1] Piatetski-Shapiro, I.I.: Discrete groups of analitic automorphisms of a polycilinder and automorphic functions. Dokl. Akad. Nauk SSSR **124** (1959) 760–763 (in Russian)

[Pi 2] Piatetski-Shapiro, I.I.: Automorphic forms and arithmetic groups. Proc. Int. Congr. Math. (Moscow 1966), Mir, Moscow 1968, 232–247 (in Russian), AMS Transl. (2) **70** (1968) 185–201

[Pi 3] Piatetski-Shapiro, I.I.: Discrete subgroups of Lie groups. Trans. Moscow Math. Soc. **18** (1968) 1–18

[Pl 1] Platonov, V.P.: Linear algebraic group theory and torsion groups. Izv. Akad. Nauk SSSR, Ser. math. **30** (1966) 573–620 (in Russian)

[Pl 2] Platonov, V.P.: The problem of strong approximation and the Kneser-Tits hypothesis for algebraic groups. Math. USSR Izv. **3** (1970) 1139–1147; ibid. **4** (1970) 784–786

[Pl 3] Platonov, V.P.: Arithmetic and structural problems in linear algebraic groups. Proc. Int. Congress Math. (Vancouver 1974), AMS Transl. **109** (1977)

[Pl 4] Platonov, V.P.: Arithmetic theory of algebraic groups. Usp. Mat. Nauk. v. 37, **3** (255) (1982) 3–54 (in Russian)

[Pl-J] Platonov, V.P.; Janchevskiĭ, V.I.: On Harder's conjecture. Soviet Math. Dokl. **16** (1975) 424–427

[Pl-R 1] Platonov, V.P.; Rapinchuk, A.S.: On the group of rational points of 3-dimensional groups. Dokl. Akad. Nauk SSSR **247** (1979) 279–282 (in Russian)

[Pl-R 2] Platonov, V.P.; Rapinchuk, A.S.: The multiplicative structure of division rings over number fields and the Hasse norm principle. Proc. Steklov Inst. Math. **165** (1985) 187–205

[Pr 1] Prasad, G.: Strong rigidity of $\mathbb{Q}$-rank 1 lattices. Invent. math. **21** (1973) 255–286

[Pr 2] Prasad, G.: Triviality of certain automorphisms of semi-simple groups over local fields. Math. Ann. **218** (1975) 219–227

[Pr 3] Prasad, G.: Discrete subgroups isomorphic to lattices in Lie groups. Amer. J. Math., v. **98** (1976) 241–261, 853–863

[Pr 4] Prasad, G.: Non-vanishing of the first cohomology. Bull. Soc. Math. France **105** (1977) 415–418

[Pr 5] Prasad, G.: Strong approximation for semi-simple groups over function fields. Ann. Math. **105** (1977) 553–572

[Pr 6] Prasad, G.: Lattices in semi-simple groups over local fields. Studies in algebra and number theory. Adv. Math. Suppl. Studies **6** (1979)

[Pr 7] Prasad, G.: Elementary proof of a theorem of Bruhat-Tits-Rousseau and a theorem of Tits. Bull. Soc. Math. France **110** (1982) 197–202

[Pro] Prokhorov, M.N.: The non-existence of discrete groups of reflections with non-compact fundamental polyhedron of finite volume in Lobachevsky spaces of higher dimension. Izv. AN SSSR Ser. math. **50** (1986) 320–332 (in Russian)

[Pr-R 1] Prasad, G.; Raghunathan, M.S.: Cartan subgroups and lattices in semi-simple groups. Ann. Math. **96** (1972) 296–317

[Pr-R 2] Prasad, G.; Raghunathan, M.S.: On the congruence subgroup problem: determination of the "metaplectic kernel". Invent. math. **71** (1983) 21–42

[Pr-R 3] Prasad, G.; Raghunathan, M.S.: Topological central extension of semi-simple groups over local fields. Ann. Math. **119** (1984) 141–201, 203–268

[Pr-R 4] Prasad, G.; Raghunathan, M.S.: On the Kneser-Tits problem. Comment. Math. Helv. **60** (1985) 107–121

[P-S] Piatetski-Shapiro, I.I.; Šafarevič, I.R.: Galois theory of transcendental extensions and uniformization. AMS Transl. (2) **69** (1968) 111–145

[Rag 1] Raghunathan, M.S.: On the first cohomology of discrete subgroups of semisimple Lie groups. Amer. J. Math. **87** (1965) 102–139

[Rag 2] Raghunathan, M.S.: Vanishing theorems for cohomology groups associated to discrete subgroups of semi-simple Lie groups. Osaka J. Math. **3** (1966) 243–256. Corrections, Osaka J. Math. **16** (1979) 295–299

[Rag 3] Raghunathan, M.S.: Cohomology of arithmetic subgroups of algebraic groups, I. Ann. Math. **86** (1967) 409–424

[Rag 4] Raghunathan, M.S.: Cohomology of arithmetic subgroups of algebraic groups, II. Ann. Math. **87** (1968) 279–304

[Rag 5] Raghunathan, M.S.: Discrete subgroups of Lie groups. Springer, New York 1972

[Rag 6] Raghunathan, M.S.: Discrete groups and $\mathbb{Q}$-structures on semisimple Lie groups. Proc. Int. Coll. on Discrete subgroups of Lie groups and appl. to moduli, Oxford University Press 1975, 225–321

[Rag 7] Raghunathan, M.S.: On the congruence subgroup problem. Publ. Math. IHES **46** (1976) 107–161

[Rag 8] Raghunathan, M.S.: A proof of Oseledec's multiplicative ergodic theorem. Israel J. Math. **32** (1979) 356–362

[Rag 9] Raghunathan, M.S.: Arithmetic lattices in semisimple groups. Proc. Indian Acad. Sci. (Math. Sci.) **91** (1982) 133–138

[Rag 10] Raghunathan, M.S.: Torsion in cocompact lattices in $\mathbf{SO}(2,n)$. Math. Ann. **266** (1984) 403–419

[Rag 11] Raghunathan, M.S.: On the group of norm 1 elements in a division algebra. Math. Ann. **279** (1988) 457–484

[Rag 12] Raghunathan, M.S.: On the congruence subgroup problem, II. Invent. math. **85** (1986) 73–117

[Rag 13] Raghunathan, M.S.: Discrete subgroups of algebraic groups over local fields of positive characteristics. To appear in Proceedings of Indian Academy of Sciences

[Rap 1] Rapinchuk, A. S.: Multiplicative arithmetic of division algebras over number fields and the metaplectic problem. Izv. Akad. Nauk SSSR, Ser. math. **51** (1987) 1033–1065 (in Russian)

[Rap 2] Rapinchuk, A.S.: The congruence subgroup problem for algebraic groups and strong approximation in affine manifolds. Dokl. Akad. Nauk SSSR **32** (1988) 581–584 (in Russian)

[Rat 1] Ratner, M.: Rigidity of horocycle flows. Ann. Math. **115** (1983) 597–614

[Rat 2] Ratner, M.: Horocycle flows, joinings, and rigidity of products. Ann. Math. **118** (1983) 277–313

[Rat 3] Ratner, M.: Ergodic theory in hyperbolic space. Preprint

[Rohlf] Rohlfs, J.: On the cuspidal cohomology of the Bianchi modular groups. Preprint

[Rohli 1] Rohlin, V.A.: On the fundamental ideas of measure theory. AMS Transl. (1) **10** (1962) 1–54

[Rohli 2] Rohlin, V.A.: Lectures on entropic transformation theory with invariant measure. Usp. Mat. Nauk **22** (1967) 3–56 (in Russian)

[Rosenb] Rosenblatt, J.: Uniqueness of invariant means for measure preserving transformations. Trans. AMS **265** (1981) 623–636

[Rosenl] Rosenlicht, M.: Questions of rationality for solvable algebraic groups over non-perfect fields. Annali di Mat. IV **61** (1963) 97–120

[R-S] Rohlfs, J.; Speh, B.: Representations with cohomology in the direct spectrum of subgroups of $\mathbf{SO}(n,1)(\mathbb{Z})$ and Lefschetz numbers. Ann. Sci. Ec. Norm. Super., IV, Sér. **20** (1987) 89–136

[Ru] Ruelle, D.: Ergodic theory of differential dynamical systems. Preprint IHES 1978

[Schm] Schmidt, K.: Amenability, Kazhdan's property T, strong ergodicity and invariant means for ergodic group actions. Ergod. Theor. and Dynam. Syst. **1** (1981) 223–226

[Scho] Schoenberg, I.J.: Metric spaces and positive definite functions. Trans. AMS **44** (1938) 522–536

[Schw 1] Schwartzmann, O.V.: On discrete arithmetic subgroups of complex Lie groups. Mat. Sb. **77** (1968) 541–544 (in Russian)

[Schw 2] Schwartzmann, O.V.: Discrete groups of reflections in complex ball. Funct. Anal. Appl. **18** (1984) 88–89 (in Russian)

[Schw 3] Schwartzmann, O.V.: Discrete groups of reflections in complex ball. Questions of group theory and homological algebra. Proc. Jaroslavl' (1985) 61–77 (in Russian)

[Schwe 1] Schwermer, J.: Sur la cohomologie de $\mathbf{SL}_n(\mathbb{Z})$ a l'infini et les séries d'Eisenstein. C.R. Acad. Sci. Paris **289** (1979) 413–415

[Schwe 2] Schwermer, J.: Kohomologie arithmetisch definierter Gruppen und Eisensteinreihen. Springer Lecture Notes, No. 988, 1983

[Sel 1] Selberg, A.: On discontinuous groups in higher dimensional symmetric spaces. Contributions to Function Theory (Int. Colloq. Bombay, 1960). Tata Inst. Fund. Res., Bombay 1960, 147–164

[Sel 2] Selberg, A.: Discontinuous groups and harmonic analysis. Proc. Int. Congr. Math. (Stockholm 1962), Inst. Mittag-Leffler, Ojursholm, 1963, 177–189

[Sel 3] Selberg, A.: Recent development in the theory of discontinuous groups of motions of symmetric spaces. Proc. 15th Scandinavian Math. Congr. (Oslo 1968), Springer Lecture Notes No. 118 (1970) 99–120

[Ser 1] Serre, J.-P.: Le problème de groupes de congruence pour $\mathbf{SL}_2$. Ann. Math. **92** (1970) 489–527

[Ser 2] Serre, J.-P.: Cohomologie des groupes discrets. Prospects in Math., Ann. Math. Studies **70** (1971) 77–169

[Ser 3] Serre, J.-P.: Arbres, amalgames, $\mathbf{SL}_2$. Soc. Math. France, Asterisque **46** (1977)

[Sha] Shafarevich, I.R.: Foundations of Algebraic Geometry. Nauka, Moscow 1973 (in Russian)

[Shil] Shilov, G.E.: Mathematical analysis. Fiz. Mat. Giz., Moscow 1960 (in Russian)

[Shir] Shirakawa, H.: An example of infinite measure preserving ergodic geodesic flow on a surface with constant negative curvature. Comment. Math. Univ. Sacti Pauli **31** (1982) 163–182

[So] Soulé, C.: The cohomology of $\mathbf{SL}_3(\mathbb{Z})$. Topology **17** (1978) 1–22

[Spa] Spatzier, R.J.: On lattices acting on boundaries of semisimple groups. Ergod. Theor. and Dynam. Sys. **1** (1981) 489–494

[Spr] Springer, T.A.: Reductive groups, in: Automorphic forms, representations and $L$-functions. Proc. Symp. Pure Math. v. 33, Part 1, 3–27

[Ste] Steinberg, R.: Lectures on Chevalley groups. Notes prepared by John Faulkner and Robert Wilson, Yale University 1967

[Stu] Stuhler, U.: Homological properties of certain arithmetic groups in function field case. Invent. math. **57** (1980) 269–281

[Sul 1] Sullivan, D.: On the ergodic theory at infinity of an arbitrary discrete group of hyperbolic motions, in Proc. of the Stony Brook Conf. on Kleinian groups and Riemann surfaces, New York 1978

[Sul 2] Sullivan, D.: For $n > 3$ there is only one finitely additive rotationally invariant measure on the $n$-sphere defined on all Lebesgue measurable subsets. Bull. AMS **4** (1981) 121–123

[Sup] Suprunenko, D.A.: Matrix groups. Nauka, Moscow 1972 (in Russian)

[Th] Thurston, W.: The geometry and topology of 3-manifolds. Mimeographed notes, Princeton University

[Ti 1] Tits, J.: Algebraic and abstract simple groups. Ann. Math. **80** (1964) 313–329

[Ti 2] Tits, J.: Classification of algebraic semisimple groups, algebraic groups, and discontinuous groups. Symposium Colorado, Boulder 1965, AMS Proc. Symp. Pure Math. IX

[Ti 3] Tits, J.: Free subgroups in linear groups. J. Algebra **20** (1972) 250–270

[Ti 4] Tits, J.: Travaux de Margulis sur les sous-groupes discretes de groupes de Lie. Springer Lecture Notes, No. 577 (Seminaire Bourbaki 1975/76), Springer 1977, exposè 482, 114–130

[To 1] Tomanov, G.M.: Sur la structure des groupes algébriques simples des type $\mathbb{D}_n$ definis sur des corps de nombres. C.R. Acad. Sci. Paris **306** (1988) 647–650

[To 2] Tomanov, G.M.: On the congruence-subgroup problem for groups of type $\mathbb{G}_2$. C.R. de l'Ac. Bulg. des Sciences **42** (1989) No. 6

[To 3] Tomanov, G.M.: Projective simplicity of groups of rational points of simply connected algebraic groups over number fields. Banach Center Publ. (Topics in Algebra) **26** (1989)

[To 4] Tomanov, G.M.: On the congruence-subgroup problem for some anisotropic algebraic groups over number fields. J. Reine Angew. Math. **402** (1989) 138–152

[To 5] Tomanov, G.M.: Remarques sur la structure des groupes algébriques définis sur des corps de nombres. C.R. Acad. Sci. Paris **310** (1990) 33–36

[T-W] Tong, G.L.; Wang, S.P.: Harmonic forms dual to geodesic cycles in quotients of $\mathbf{SU}(p, 1)$. Math. Ann. **258** (1982) 289–318

[Va 1] Vaserštein, L.N.: On the group $\mathbf{SL}_2$ over Dedekind rings of arithmetic type. Math. USSR Sb. **18** (1972) 321–332

[Va 2] Vaserštein, L.N.: Structure of classical arithmetic groups of rank greater than 1. Matem. Sb. **91** (1973) 445–470 (in Russian)

[Ve] Venkataramana, T.N.: On superrigidity and arithmeticity of lattices in semisimple groups over local fields of arbitrary characteristic. Invent. math. **92**, no. 2 (1988) 255–306

[Vi 1] Vinberg, E.B.: Discrete groups generated by reflections in Lobachevsky spaces. Math. USSR Sb. **1** (1967) 429–444

[Vi 2] Vinberg, E.B.: Compact Lie groups. Moscow Univ. Press, Moscow 1967 (in Russian)

[Vi 3] Vinberg, E.B.: Some examples of crystallographic groups in Lobachevsky spaces. Math. USSR Sb. **7** (1969) 617–622

[Vi 4] Vinberg, E.B.: Rings of definition of dense subgroups of semisimple linear groups. Math. USSR Izv. **5** (1971) 45–55

[Vi 5] Vinberg, E.B.: Discrete linear groups generated by reflections. Math. USSR Izv. **5** (1971) 1083–1119

[Vi 6] Vinberg, E.B.: Some arithmetic discrete groups in Lobachevsky spaces, in Proc. Int. Coll. on Discrete subgrops of Lie groups and appl. to moduli (Bombay 1973). Oxford University Press (1975) 323–348

[Vi 7] Vinberg, E.B.: Properties of the root decomposition of a semisimple Lie algebra over an algebraically nonclosed field. Funct. Anal. Appl. **9** (1975) 17–21

[Vi 8] Vinberg, E.B.: Absence of crystallographic groups of reflections in Lobachevsky spaces of large dimension. Funct. Anal. Appl. **15** (1981) 128–130

[Vi 9] Vinberg, E.B.: Two almost algebraic K3 surfaces. Math. Ann. **265** (1983) 1–21

[Vi 10] Vinberg, E.B.: Absence of crystallographic groups of reflections in Lobachevsky spaces of large dimension. Trans. Moscow Math. Soc. **47** (1984) 68–102 (in Russian)

[Vi 11] Vinberg, E.B.: Discrete groups of reflections in Lobachevsky spaces. Proc. Int. Congr. Math., Warsaw 1983, 585–592

[Vi 12] Vinberg, E.B.: Hyperbolic reflection groups. Russ. Math. Surv. **40** (1985) 31–75

[V-K] Vinberg, E.B.; Kaplinskaja, I.M.: On the groups $O_{18,1}(\mathbb{Z})$ and $O_{19,1}(\mathbb{Z})$. Soviet Math. Dokl. **19** (1978) 194–197

[V-O] Vinberg, E.B.; Oniščik, A.L.: Seminaire on algebraic groups and Lie groups. Moscow Univ. Press, Moscow 1969 (in Russian)

[Wae 1] van der Waerden, B.L.: Algebra I. Springer, New York 1971

[Wae 2] van der Waerden, B.L.: Algebra II. Springer, New York 1967

[Wat] Watatani, Y.: Property (T) of Kazhdan implies property (FA) of Serre. Math. Jap. **27** (1982) 97–103

[Weil 1] Weil, A.: On discrete subgroups of Lie groups I, II. Ann. Math. (2) **27** (1960) 369–384; ibid. (2) **75** (1962) 578–602

[Weil 2] Weil, A.: Adeles and algebraic groups. Inst. for Advanced Study, Princeton, N.J. 1961

[Weil 3] Weil, A.: Remarks on the cohomology of groups. Ann. Math. **80** (1964) 149–157

[Weil 4] Weil, A.: Basic number theory. Springer, New York 1967

[Weis 1] Weisfeiler, B.: On abstract monomorphisms of $k$-forms of **PGL**(2). J. Algebra **57** (1979) 522–543

[Weis 2] Weisfeiler, B.: Abstract homomorphisms of big subgroups of algebraic groups. Notre Dame Math. Lect. University of Notre Dame Press **10** (1982) 135–181

[Wh] Whitney, H.: Elementary structure of real algebraic varieties. Ann. Math. **66** (1957) 545–556

[WH 1] Wang, H.C.: On a maximality property of discrete subgroups with fundamental domain of finite measure. Amer. J. Math. **89** (1967) 124–132

[WH 2] Wang, H.C.: in: Topics on totally discontinuous groups in symmetric spaces. W. Boothly (ed.) M. Dekker 1972, pp. 460–487

[WS 1] Wang, S.P.: Limit of lattices in a Lie groups. Trans. AMS **133** (1968) 519–526

[WS 2] Wang, S.P.: On a theorem of representation of lattices. Proc. AMS **23** (1969) 583–587

[WS 3] Wang, S.P.: On a conjecture of Chabauty. Proc. AMS **23** (1969) 569–572

[WS 4] Wang, S.P.: The dual space of semi-simple Lie groups. Amer. J. Math. **91** (1969) 921–937

[WS 5] Wang, S.P.: On the centralizer of a lattice. Proc. AMS **21** (1969) 21–23

[WS 6] Wang, S.P.: On the limit of subgroups in a group. Amer. J. Math. **92** (1970) 708–724

[WS 7] Wang, S.P.: Some properties of lattices in a Lie group. Illinois J. Math. **14** (1970) 35–39

[WS 8] Wang, S.P.: On density properties of $S$-subgroups of locally compact groups. Ann. Math. **94** (1971) 325–329

[WS 9] Wang, S.P.: On a lemma of Mahler. Amer. J. Math. 95, No. **4** (1973) 703–712

[WS 10] Wang, S.P.: On the first cohomology group of discrete groups with property (T). Proc. AMS **42** (1974) 621–624

[WS 11] Wang, S.P.: On isolated points in the dual spaces of locally compact groups. Math. Ann. **218** (1975) 19–34

[WS 12] Wang, S.P.: On subgroups with property (P) and maximal discrete subgroups. Amer. J. Math. **97** (1975) 404–414

[WS 13] Wang, S.P.: Homogeneous spaces with finite invariant measure. Amer. J. Math. **98** (1976) 311–324

[WS 14] Wang, S.P.: On density properties of certain subgroups of locally compact groups. Duke Math. J. **43** (1976) 561–578; Correction Duke Math. J. **45** (1978) 953

[WS 15] Wang, S.P.: On $L$-subgroups of locally compact groups. Adv. Math. **28** (1978) 89–100

[WS 16] Wang, S.P.: On density properties of certain subgroups with boundedness conditions. Monatsh. Math. **89** (1980) 141–162

[WS 17] Wang, S.P.: A note on free subgroups in linear groups. J. Algebra **71** (1981) 232–234

[WS 18] Wang, S.P.: On the Mautner phenomenon and groups with property (T). Amer. J. Math. **104** (1982) 1191–1210

[Zi 1] Zimmer, R.L.: Strong rigidity for ergodic actions of semisimple Lie groups. Annals of Math. **112** (1980) 511–529

[Zi 2] Zimmer, R.L.: Orbit equivalence and rigidity for ergodic actions of Lie groups. Ergod. Theor. and Dynam. Sys. **1** (1981) 237–253

[Zi 3] Zimmer, R.L.: Equivariant images of projective space under the action of $\mathbf{SL}(n,\mathbb{Z})$. Ergod. Theor. and Dynam. Sys. **1** (1981) 519–622

[Zi 4] Zimmer, R.L.: Ergodic theory, group representations, and rigidity. Bull. AMS **6** (1982) 383–416

[Zi 5] Zimmer, R.L.: On the Mostow rigidity theorem and measurable foliations by hyperbolic space. Israel J. Math. **43** (1982) 281–290

[Zi 6] Zimmer, R.L.: Volume preserving actions of lattices in semisimple groups on compact manifolds. Publ. Math. IHES **59** (1984) 5–33

[Zi 7] Zimmer, R.L.: Kazhdan groups acting on compact manifolds. Invent. math. **75** (1984) 425–436

[Zi 8] Zimmer, R.L.: Ergodic theory and semisimple groups. Birkhäuser, Boston 1985

[Z-S] Zariski, O.; Samuel, P.: Commutative Algebra I, II. D. Van Nostrand Comp. Inc. Princeton, New Jersey 1958, 1960

Докл. Акад. Наук СССР
Том 272 (1983), № 4

Soviet Math. Dokl.
Vol. 28 (1983), No. 2

# FREE TOTALLY DISCONTINUOUS GROUPS OF AFFINE TRANSFORMATIONS

UDC 512.817

G. A. MARGULIS

Let $A(n)$ be the group of affine transformations of $n$-dimensional real space $\mathbf{R}^n$. The group $A(n)$ is a semidirect product $GL(n, \mathbf{R}) \cdot V(n)$, where $V(n)$ is the group of parallel translations. We will say that a subgroup $\Gamma$ of $A(n)$ acts *totally discontinuously* on $\mathbf{R}^n$ if for any compactum $K \subset \mathbf{R}^n$ the set $\{\gamma \in \Gamma \mid \gamma K \cap K \neq \emptyset\}$ is finite. Milnor [1] conjectured that any subgroup $\Gamma$ of $A(n)$ that acts totally discontinuously on $\mathbf{R}^n$ is almost polycyclic (i.e. contains a polycyclic subgroup of finite index). We should mention that in [1] this conjecture was stated in a different, but equivalent, form: the fundamental group of any complete affine variety is almost polycyclic. It is easy to prove Milnor's conjecture for $n = 2$. However, for $n = 3$ the answer to this conjecture is negative. More precisely, as we will show below (see Theorems 2 and 3), there exist (noncommutative) free subgroups of $A(3)$ that act totally discontinuously on $\mathbf{R}^3$.

**1.** The Euclidean norm of a vector $x \in \mathbf{R}^3$ will be denoted by $\|x\|$. The symbol $\angle(A, B)$ will denote the angle (nonnegative and at most $\pi/2$) between $A$ and $B$, where $A$ and $B$ are planes or straight lines in $\mathbf{R}^3$. For the distance in the Euclidean metric between points, between points and sets, and between sets in $\mathbf{R}^3$ we will use the notation $\rho(\cdot, \cdot)$.

**2.** We define a bilinear form $B(x, y) = x_1y_1 + x_2y_2 - x_3y_3$, where $x = (x_1, x_2, x_3) \in \mathbf{R}^3$ and $y = (y_1, y_2, y_3) \in \mathbf{R}^3$. We denote the cone $\{x \in \mathbf{R}^3 \mid B(x, x) = 0\}$ by $W$. We denote one of the connected components of the set $W - \{0\}$ by $W^+$, the other by $W^-$.

**3.** Let $H$ denote the connected identity component of the group $O(2, 1)$ of linear transformations preserving the bilinear form $B(x, y)$. Put $G = H \cdot V(3)$. Let $\varphi: G \to H$ be the natural epimorphism ($\varphi(hv) = h$, $h \in H$, $v \in V(3)$). We will say that $g \in G$ is *hyperbolic* if $\varphi(g)$ has a real eigenvalue different from 1. It is easy to show that $\varphi(g)$ is diagonalizable and has positive eigenvalues. One of these eigenvalues is equal to 1, a second, which we denote by $\lambda(g)$, is less than 1, and the third is $\lambda(g)^{-1}$.

Suppose $g \in G$ is hyperbolic. Put

$$A^0(g) = \{x \in \mathbf{R}^3 \mid \varphi(g)x = x\}, \quad A^-(g) = \{x \in \mathbf{R}^3 \mid \varphi(g)x = \lambda(g)x\},$$
$$A^+(g) = \{x \in \mathbf{R}^3 \mid \varphi(g)x = \lambda(g)^{-1}x\}.$$

It is easy to see that $A^-(g) \subset W$, $A^+(g) \subset W$, and $B(x, x) > 0$ for any $x \in A^0(g) - \{0\}$. Having chosen an orientation in $\mathbf{R}^3$, we associate with $g$ a uniquely determined positively oriented basis $(x^0(g), x^-(g), x^+(g))$, where $x^0(g) \in A^0(g)$, $x^-(g) \in A^-(g) \cap W^+$, $x^+(g) \in A^+(g) \cap W^+$ and $B(x^0(g), x^0(g)) = \|x^-(g)\| = \|x^+(g)\| = 1$. There exists in $\mathbf{R}^3$ exactly one straight line $C_g$ parallel to $A^0(g)$ and invariant under $g$. The restriction of the transformation $g$ to $C_g$ is a parallel translation by a vector $d_g \in A^0(g)$. Put $\alpha(g) = B(d_g, x^0(g))$. It is easy to show that

(1) $\alpha(g^n) = |n|\alpha(g)$ for $n \neq 0$, and
(2) $\alpha(g) = B(gx - x, x^0(g))$ for any $x \in \mathbf{R}^3$.

1980 *Mathematics Subject Classification*. Primary 22E40.

Note that in the right-hand side of (1) there appears as the coefficient of $\alpha(g)$ not $n$, as we might expect, but $|n|$. If $\alpha(g) \neq 0$ or, equivalently, if $g$ has no fixed points, then we attach a sign to $g$ by taking $g$ to be positive if $\alpha(g) > 0$ and negative if $\alpha(g) < 0$. It is important to note that, in view of (1), $g$ and $g^{-1}$ have the same sign. Put $D^-(g) = A^0(g) + A^-(g)$ and $D^+(g) = A^0(g) + A^+(g)$. We have

$$D^-(g)\{x \in \mathbf{R}^3 \mid B(x, x^-(g)) = 0\}, \quad D^+(g) = \{x \in \mathbf{R}^3 \mid B(x, x^+(g)) = 0\},$$
$$\mathbf{R}^3 = D^-(g) \oplus A^+(g) = D^+(g) \oplus A^-(g).$$

The planes $D^-(g)$ and $D^+(g)$ are tangent to the cone $W$. Let $E^-(g)$ and $E^+(g)$ be the planes containing $C_g$ and parallel to $D^-(g)$ and $D^+(g)$ respectively. For $x \in \mathbf{R}^3$ we denote by $B_g^-(x)$ and $B_g^+(x)$ the straight lines passing through $x$ and parallel to $A^-(g)$ and $A^+(g)$ respectively.

**4.** An element $g$ of the group $G$ is called *$\varepsilon$-hyperbolic* if it is hyperbolic and

$$\rho(x^-(g), x^+(g)) > \varepsilon.$$

Hyperbolic elements $g, h \in G$ are called *$\varepsilon$-transverse* if the distance from each of the vectors $x^-(g)$, $x^+(g)$ to each of the vectors $x^-(h)$, $x^+(h)$ is greater than $\varepsilon$. If $g$ and $h$ are $\varepsilon$-transverse for some $\varepsilon > 0$ (i.e. if $\{x^-(g), x^+(g)\} \cap \{x^-(h), x^+(h)\} = \emptyset$), we say that $g$ and $h$ are *transverse*.

Suppose $k \geq 1$ and there are given two nonnegative functions $f_1$ and $f_2$ on some subset $E$ of the direct product of $k$ copies of $G$. We will say that $f_1$ *is of order at most* $f_2$, and write $f_1 \ll f_2$, if for any $\varepsilon > 0$ there exists $p(\varepsilon) > 0$ such that $f_1(g_1, \dots, g_k) \leq p(\varepsilon) f_2(g_1, \dots, g_k)$ whenever $(g_1, \dots, g_k) \in E$ and the elements $g_i$, $1 \leq i \leq k$, are $\varepsilon$-hyperbolic and pairwise $\varepsilon$-transverse. If $f_1 \ll f_2$ and $f_2 \ll f_1$, we will write $f_1 \sim f_2$ and say that $f_1$ is *equivalent* to $f_2$. It is easy to show that

$$\|x^0(g)\| \sim 1, \qquad |\alpha(g)| \sim \|d_g\|. \tag{3}$$

**5.** LEMMA 1. *For any $\varepsilon > 0$ there exists $a(\varepsilon) > 1$ such that if elements $h$ and $g$ of $G$ are $\varepsilon$-hyperbolic and $\varepsilon$-transverse, $\lambda(h) < a(\varepsilon)^{-1}$, and $\lambda(g) < a(\varepsilon)^{-1}$, then:*

1) *$gh$ is hyperbolic,*
2) $a(\varepsilon)^{-1}\lambda(g)\lambda(h) < \lambda(gh) < a(\varepsilon)\lambda(g)\lambda(h)$,
3) $\rho(x^+(gh), x^+(g)) < a(\varepsilon)\lambda(g)$,
4) $\rho(x^-(gh), x^-(h)) < a(\varepsilon)\lambda(h)$,
5) $\angle(E_{gh}^+, E_g^+) < a(\varepsilon)\lambda(g)$,
6) $\angle(E_{gh}^-, E_h^-) < a(\varepsilon)\lambda(h)$,
7) $\rho(x^0(gh), B_g^+(x^0(g))) < a(\varepsilon)\lambda(g)$,
8) $\rho(x^0(gh), B_h^-(x^0(h))) < a(\varepsilon)\lambda(h)$,
9) *the group generated by $g$ and $h$ is free (and noncommutative).*

Put $U(x, \varepsilon) = \{y \in \mathbf{R}^3 \mid \rho(x, y) \leq \varepsilon\}$. Assertions 1) and 3) follow from the fact that for some function $a(g, h)$ equivalent to $\lambda(g)$ the set $U(x^+(g), a(g, h))$ is mapped into itself under the action of the transformation $x \to gh(x)/\|gh(x)\|$. Replacing $g$ by $h^{-1}$ and $h$ by $g^{-1}$, we deduce 4) from 3). Assertions 2) and 5)–8) follow easily from 1), 3) and 4) (it may be necessary to alter $a(\varepsilon)$). Assertion 9) can be proved by an argument analogous to one used in [2].

**6.** Since $\rho(g^{-1}x, E_g^+) = \lambda(g)^{-1}\rho(x, E_g^+)$ for any hyperbolic $g \in G$ and any $x \in \mathbf{R}^3$, it follows that

$$\rho(x, E_g^+) \leq \lambda(g)\rho(x, g^{-1}x)/(1 - \lambda(g)). \tag{4}$$

In order to guarantee that the transformation $gh$ is $\varepsilon/2$-hyperbolic, we will assume in the following lemma that $\lambda(h) < a(\varepsilon)^{-1}\varepsilon/4$ and $\lambda(g) < a(\varepsilon)^{-1}\varepsilon/4$, where $a(\varepsilon)$ is defined in Lemma 1.

LEMMA 2. *For any $\varepsilon > 0$ there exists $c(\varepsilon) > 1$ such that if elements $h$ and $g$ of $G$ are $\varepsilon$-hyperbolic and $\varepsilon$-transverse and if $q \in \mathbf{R}^3$, then*

$$\rho(q, E^+_{gh}) \le \rho(q, E^+_g) + c(\varepsilon)\lambda(g)[|\alpha(g)| + \rho(q, C_g) + \rho(q, C_h)] + c(\varepsilon)\lambda(gh)|\alpha(h)|; \tag{5}$$

$$\begin{aligned}\rho(q, C_{gh}) \le & c(\varepsilon)(\lambda(g) + \lambda(h))\rho(q, C_g) \\ & + c(\varepsilon)[\rho(q, C_h) + \rho(q, E^+_g) + \lambda(g)|\alpha(g)| + \lambda(h)|\alpha(h)|];\end{aligned} \tag{6}$$

$$|\alpha(gh) - \alpha(g) - \alpha(h)| \le c(\varepsilon)[\rho(C_g, C_h) + \lambda(g)|\alpha(g)| + \lambda(h)|\alpha(h)|]. \tag{7}$$

We will show briefly how to prove Lemma 2. Suppose $x \in C_g$ is such that $\rho(q, x) = \rho(q, C_g)$. Put $y = B^-_g(g^{-1}x) \cap E^-_{h^{-1}}$. It is easy to show that $\rho(h^{-1}y, gy) \ll \|d_g\| + \|d_h\| + \rho(x, C_h)$. It then follows from (4) that

$$\rho(gy, E^+_{gh}) \ll \lambda(gh)[\|d_g\| + \|d_h\| + \rho(x, C_h)].$$

On the other hand, it is rather easy to show that

$$\rho(gy, E^+_g) \ll \lambda(g)[\|d_g\| + \rho(x, C_h)], \quad \rho(q, gy) \ll \rho(q, C_g) + \lambda(g)[\|d_g\| + \rho(x, C_h)].$$

Now to prove (5) we use (3) and observe that, in view of Lemma 1, $\angle(E^+_g, E^+_{gh}) \ll \lambda(g)$ and $\lambda(g) \ll \lambda(gh)$. Next, since $gh$ is $\varepsilon/2$-hyperbolic, it follows easily that $\rho(q, C_{gh}) \ll \rho(q, E^+_{gh}) + \rho(q, E^-_{gh})$. This together with (5) easily implies (6). Now suppose $u \in C_g$ is such that $\rho(u, C_h) = \rho(C_g, C_h)$. Put $v = B^-_g(u) \cap E^+_h$. It is easy to prove that

$$\rho(gv - v, d_g) \ll \rho(C_g, C_h) \quad \text{and} \quad \rho(v - h^{-1}v, d_h) \ll \rho(C_g, C_h).$$

Therefore, $\rho(gv - h^{-1}v, d_g + d_h) \ll \rho(C_g, C_h)$. Using (2) and (3) we easily obtain

$$|\alpha(gh) - B(d_g + d_h, x^0(gh))| \ll \rho(C_g, C_h).$$

From this inequality, using (3), assertions 7) and 8) of Lemma 1, and the equalities $B(d_g, B^+_g(x^0(g))) = \alpha(g)$ and $B(d_h, B^-_n(x^0(h))) = \alpha(h)$, we obtain (7).

REMARK. The main part of Lemma 2 is inequality (8).*

7. THEOREM 1. *Suppose $h_1, h_2 \in G$ are hyperbolic and transverse. Assume either that at least one of the numbers $\alpha(h_1)$ and $\alpha(h_2)$ is equal to zero, or that $h_1$ and $h_2$ have different signs.*

*Then the group $\Gamma$ generated by $h_1$ and $h_2$ does not act totally discontinuously on $\mathbf{R}^3$.*

PROOF. In view of assertion 9) of Lemma 1, we can assume, replacing $h_1$ and $h_2$ by sufficiently large powers, that $\Gamma$ is free. By the conditions imposed on $h_1$ and $h_2$, there exists a sequence $\{m_i, n_i\}$ of distinct pairs of nonnegative integers $m_i$ and $n_i$ such that

$$\sup_i |m_i\alpha(h_1) + n_i\alpha(h_2)| < \infty.$$

Put $h(i) = h_1^{m_i}h_2^{n_i}$. Using (6) and (7), we obtain that

$$L(q) \stackrel{\text{def}}{=} \sup_i \rho(q, C_{h(i)}) < \infty \quad \text{for any } q \in \mathbf{R}^3,$$

$$M \stackrel{\text{def}}{=} \sup_i |\alpha(h(i))| < \infty.$$

---

**Editor's note.* The inequality number in this remark is reproduced as it appears in the original.

But $U(q, L(q) + M) \cap (h(i)U(q, L(q))) \neq \emptyset$ for any $i$. Since $\Gamma$ is free, this gives us the theorem.

**8.** LEMMA 3. *Suppose $\varepsilon > 0$ is sufficiently small and $h_1, h_2 \in G$ are $2\varepsilon$-hyperbolic and $2\varepsilon$-transverse, act on $\mathbf{R}^3$ without fixed points, and have the same sign. Put $L = \min(|\alpha(h_1)|, |\alpha(h_2)|)$ and $A = \rho(C_{h_1}, C_{h_2})$. Assume that $\lambda(h_i)^{-1} > 20(c^2(\varepsilon) + a(\varepsilon))$ for $i = 1, 2$ and that $L \geq 60c^2(\varepsilon)A$, where $a(\varepsilon)$ and $c(\varepsilon)$ are defined in Lemmas 1 and 2.*

*Then there exists $g \in G$ such that if $h$ can be expressed in terms of $h_1$ and $h_2$ as a word of length $j$, then $gh$ is $\varepsilon$-hyperbolic and*

$$|\alpha(gh)| \geq jL/8. \tag{8}$$

*Outline of the proof of Lemma* 3. Suppose $q \in C_{h_1}$ is such that $\rho(q, C_{h_2}) = A$. For a sufficiently small $\varepsilon > 0$ there always exists a $2\varepsilon$-hyperbolic element $g \in G$ that is $2\varepsilon$-transverse to $h_1$ and $h_2$ such that $\alpha(g) = 0$ and $q \in C_g$. In addition, we may assume that $\lambda(g)$ is as small as we like. Now suppose there are given sequences $\{i_l\} \subset \{1, 2\}$ and $\{k_l\} \subset \mathbf{Z} - \{0\}$, where $i_{l+1} \neq i_l$ ($\mathbf{Z}$ is the set of integers). Put $g_0 = g$, $g_l = g_{l-1}h_{i_l}^{k_l}$, $K_0 = 1$ and $K_l = K_{l-1} + |k_l|$. Using Lemma 1, we can easily show that if $\lambda(g)$ is sufficiently small, then for any $l \geq 0$ we have

$$\lambda(g_{l+1}) < 2^{-K_{l+1}}\lambda(g), \tag{9}$$

(A) $g_l$ and $h_{i_{l+1}}$ are $\varepsilon$-hyperbolic and $\varepsilon$-transverse.

Put $t_l = \alpha(g_l)$, $f_l = \alpha(h_{i_l})$, and $T = \max(|\alpha(h_1)|, |\alpha(h_2)|)$. Using Lemma 2, along with (9) and (A), we can prove that for a sufficiently small $\lambda(g)$

$$t_{l+1} \geq t_l + (3/4)|k_{l+1}|f_{l+1} - (T/4). \tag{10}$$

Changing the orientation of $\mathbf{R}^3$, if necessary, we may assume that $\alpha(h_1) > 0$ and $\alpha(h_2) > 0$. Then $f_l > 0$ and, since $i_{l+1} \neq i_{l+2}$, we have $f_{l+1} + f_{l+2} = T + L$. Therefore, in view of (10), $t_{l+2} \geq t_l + [(|k_{l+1}| + |k_{l+2}|)L/8]$. On the other hand, it follows easily from (7) that $t_1 \geq |k_1|L/2$. Therefore, $\alpha(g_l) = t_l \geq K_l L/8$, which proves (8).

THEOREM 2. *Suppose $\varepsilon > 0$ and $h_1, h_2 \in G$ are such that the conditions of Lemma 3 are satisfied and $L > 0$.*

*Then the group $\Gamma$ generated by $h_1$ and $h_2$ is free and acts totally discontinuously on $\mathbf{R}^3$.*

PROOF. Let $g$ be the element constructed in Lemma 3. Suppose $U \subset \mathbf{R}^3$ is a compactum. It follows from Lemma 3, with the aid of (2) and (3), that the set

$$\{h \in \Gamma \mid ghU \cap gU \neq \emptyset\} = \{h \in \Gamma \mid hU \cap U \neq \emptyset\}$$

is finite. In view of assertion 9) of Lemma 1, this proves the theorem.

A special case of Theorem 2 is

THEOREM 3. *Suppose $\varepsilon > 0$ is sufficiently small and $h_1, h_2 \in G$ are $\varepsilon$-hyperbolic and $\varepsilon$-transverse, act on $\mathbf{R}^3$ without fixed points, and have the same sign. Assume that $C_{h_1} \cap C_{h_2} \neq \emptyset$ and $\lambda(h_i)^{-1} > 20(c^2(\varepsilon) + a(\varepsilon))$, $i = 1, 2$.*

*Then the group $\Gamma$ generated by $h_1$ and $h_2$ is free and acts totally discontinuously on $\mathbf{R}^3$.*

REMARK. It would be interesting to know for which finitely generated free subgroups $\Lambda$ of $H$ there exists a subgroup $\Gamma$ of $G$ acting totally discontinuously on $\mathbf{R}^3$ such that $\varphi(\Gamma) = \Lambda$. It follows from a theorem of Auslander (see [3], §8.24) that such groups $\Lambda$

must be discrete. Is this a sufficient condition? Apparently it is not. It is more likely that the following condition is sufficient: $\Lambda$ is discrete and contains no unipotent elements.

Institute of Problems of Information Transmission
Academy of Sciences of the USSR
Received 26/JAN/83

BIBLIOGRAPHY

1. John Milnor, Advances in Math. **25** (1977), 178–187.
2. J. Tits, J. Algebra **20** (1972), 250–270.
3. M. S. Raghunathan, *Discrete subgroups of Lie groups*, Springer-Verlag, 1972.

Translated by G. A. KANDALL

DYNAMICAL SYSTEMS AND ERGODIC THEORY
BANACH CENTER PUBLICATIONS, VOLUME 23
PWN – POLISH SCIENTIFIC PUBLISHERS
WARSZAWA 1989

# INDEFINITE QUADRATIC FORMS AND UNIPOTENT FLOWS ON HOMOGENEOUS SPACES

G. A. MARGULIS

*Institute of Information Transmission, Academy of Sciences of the U.S.S.R.
Moscow, U.S.S.R.*

**1.** Let $B$ be a real nondegenerate indefinite quadratic form in $n$ variables. It is well known that if $n \geqslant 5$ and the coefficients of $B$ are rational then $B$ represents zero nontrivially, i.e. there exist integers $x_1, \ldots, x_n$ not all equal to 0, such that $B(x_1, \ldots, x_n) = 0$. Theorem 1 stated below can be considered as an analogue of this assertion in the case when $B$ is not proportional to a form with rational coefficients. Note that in Theorem 1 the condition $n \geqslant 5$ is replaced by a weaker condition $n \geqslant 3$.

THEOREM 1. *Suppose that $n \geqslant 3$ and that $B$ is not proportional to a form with rational coefficients or, it is the same, the ratio of some two coefficients of $B$ is irrational. Then for any $\varepsilon > 0$, there exists integers $x_1, \ldots, x_n$ not all equal to 0, such that $|B(x_1, \ldots, x_n)| < \varepsilon$.*

One can easily inderstand that if Theorem 1 is proved for some $n_0$ then it is proved for all $n \geqslant n_0$. So it is enough to prove this theorem for $n = 3$. Let us note if $n = 2$ then the analogous assertion is not true; to see this one can consider the form $x_1^2 - \lambda x_2^2$ where $\lambda$ is an irrational positive number such that $\sqrt{\lambda}$ has a continued fraction development with bounded partial quotients.

Theorem 1 gives the answer to Davenport's conjecture (see [4] [1]). It has been earlier proved in the following cases: (a) $n \geqslant 21$ (see [5]); (b) $n = 5$ and $B$ is of the type

$$B(x_1, \ldots, x_5) = \lambda_1 x_1^2 + \ldots + \lambda_5 x_5^2$$

(see [4]). Proofs given in [4] and [5] are based on the use of methods of analytic number theory.

(1) For the case $n \geqslant 5$ this conjecture is due to A. Oppenheim (see *The minima of indefinite ternary quadratic forms*, Ann. of Math. 32 (1931), 271–288).

In Section 2, Theorem 1 will be deduced from a result on the action of the group preserving $B$ on the space of lattices. It should be noted this result (Theorem 2) is essentially equivalent to Theorem 1.

**2.** Let as usual, $\boldsymbol{C}$, $\boldsymbol{R}$, $\boldsymbol{Q}$ and $\boldsymbol{Z}$ denote the sets of complex, real, rational and integer numbers respectively and let $\mathrm{SL}(n, \boldsymbol{R})$ (resp. $\mathrm{SL}(n, \boldsymbol{Z})$) denotes the group of unimodular matrices of order $n$, with real (resp. integer coefficients). Let $G = \mathrm{SL}(3, \boldsymbol{R})$ and $\Gamma = \mathrm{SL}(3, \boldsymbol{Z})$. Let us denote by $\Omega$ the space of lattices in $\boldsymbol{R}^3$ having determinant 1. The quotient space $G/\Gamma$ can be naturally identified with $\Omega$ (under this indentification the coset $g\Gamma$ goes to the lattice $g\boldsymbol{Z}^3$). For $y \in \Omega = G/\Gamma$ let $G_y$ denotes the stationary subgroup $\{g \in G | \; gy = y\}$. Note that if $y = g\boldsymbol{Z}^3$ then $G_y = g\Gamma g^{-1}$.

Let us denote by $H$ the group of elements of $G$ preserving the form $2x_1 x_3 - x_2^2$. The group $H$ is locally isomorphic to $\mathrm{SL}(2, \boldsymbol{R})$ and the number of its connected components is equal to 2 (besides $H$ is connected in the Zariski topology).

THEOREM 2. *If $z \in \Omega = G/\Gamma$ and the orbit $Hz$ is relatively compact in $\Omega$ then the quotient space $H/H \cap G_z$ is compact.*

*Remark.* Let $P \subset G$ be a closed subgroup and $z \in G/\Gamma$. Then the quotient space $P/P \cap G_z$ is compact iff the orbit $Pz$ is compact.

Theorem 2 will be proved in Section 6 with the help of some assertions from Sections 3, 4 and 5. Now we give the reduction of Theorem 1 to Theorem 2. As it was noted in Section 1, it is enough to prove Theorem 1 for $n = 3$. Let $H_B$ denotes the group of elements of $G$ preserving $B$. Since $n = 3$ and the form is indefinite, we have, in some basis of $\boldsymbol{R}^3$, the form $B$ has the type $\lambda(x_1 x_3 - x_2^2)$ where $\lambda = \pm 1$. So $H = g_B H_B g_B^{-1}$ for some $g_B \in G$.

Suppose now the assertion of Theorem 1 is not true, i.e. $|B(x)| > \varepsilon$ for some $\varepsilon > 0$ and all $x \in \boldsymbol{Z}^3$, $x \neq 0$. Then, as $H_B$ preserves $B$, we have $|B(x)| > \varepsilon$ for any $h \in H$ and $x \in h\boldsymbol{Z}^3$. In view of Mahler compactness criterion, this implies the set $H_B \boldsymbol{Z}^3$ ist relatively compact in $\Omega$. Now we apply Theorem 2 for $z = g_B \boldsymbol{Z}^3$ and get that the quotient space $H/H \cap G_z$ and, consequently, the quotient space $H_B/H_B \cap \Gamma$ are compact. Then, in view of Borel's density theorem (see [1]), $H_B \cap \Gamma$ is Zariski dense in $H$. On the other hand, $\Gamma$ consists of matrices with integer coefficients and (see [2] chapter AG, p. 14.4) if in an affine manifold the set of $\boldsymbol{Q}$-rational points is Zariski dense then this manifold is defined over $\boldsymbol{Q}$. So $H_B$ is a $\boldsymbol{Q}$-subgroup of $G$. Hence, $H_B = H_{\sigma B}$ for any automorphism $\sigma$ of $\boldsymbol{C}$ over $\boldsymbol{Q}$ where $\sigma B$ is obtained from $B$ by applying of $\sigma$ to coefficients of $B$. It follows forms $\sigma B$ and $B$ are proportional for any $\sigma$. But this contradicts to the assumption "the ratio of some two coefficients of $B$ is irrational".

**3.** In this section, $G$ is an arbitrary second countable locally compact group and $\Omega$ is a homogeneous space of $G$. Let us denote by $\bar{A}$ the closure of a subset $A$ of a topological space, by $\mathcal{N}_G(F)$ the normalizer of a subgroup $F \subset G$ in $G$, by $e$ the identity element of $G$ and by $N^+$ the set of positive integer numbers.

LEMMA 1. *Let $F$, $P$ and $P'$ ($F \subset P$, $F \subset P'$) be closed subgroups of $G$ and let $Y$ and $Y'$ be closed subsets of $\Omega$ and let $M \subset G$. Suppose that*

(I) *$PY = Y$ and $P'Y' = Y'$;*

(II) *$mY \cap Y' \neq \emptyset$ for any $m \in M$;*

(III) *$Y$ is a compact minimal $F$-invariant subset (minimality means that $Fy$ is dense in $Y$ for any $y \in Y$).*

*Then $hY \subset Y'$ for any $h \in \mathcal{N}_G(F) \cap \overline{P'MP}$.*

*Proof.* Set $S = \{g \in G | \; gY \cap Y' \neq \emptyset\}$. Conditions (I) and (II) imply that $S \supset P'MP$. On the other hand, as $Y$ is compact and $Y'$ is closed, the set $S$ is closed in $G$. So $S \supset \overline{P'MP}$. Hence, $hY \cap Y \neq \emptyset$, i.e. $hy = y'$ for some $y \in Y$ and $y' \in Y'$. But $h \in \mathcal{N}_G(F)$, $FY = Y$ and $FY' = Y'$. Therefore, $hgy = (hgh^{-1})hy \in Fy' \subset FY' = Y'$ for any $g \in F$. In other words, $hFy \subset Y'$. But the orbit $Fy$ is dense in $Y$ and $Y'$ is closed. So $hY \subset Y'$.

LEMMA 2. *Let $F$ be a closed subgroup of $G$ and let $Y$ be a closed minimal $F$-invariant subset and let $g \in \mathcal{N}_G(F)$. If $gY \cap Y \neq \emptyset$ then $gY = Y$.*

*Proof.* As $g \in \mathcal{N}_G(F)$ and $FY = Y$ we have $FgY = gY$ and, consequently, $F(gY \cap Y) = gY \cap Y$. But $gY \cap Y \neq \emptyset$ and $Y$ is a minimal closed $F$-invariant subset. So $gY \cap Y = Y$. Hence, $gY = Y$.

Lemmas 1 and 2 immediately imply

LEMMA 3. *Let $F$ and $P$ ($F \subset P$) be closed subgroups of $G$ and let $Y \subset \Omega$ and $M \subset G$. Suppose that (I) $PY = Y$; (II) $mY \cap Y \neq \emptyset$ for any $m \in M$; (III) $Y$ is a compact minimal $F$-invariant subset. Then $hY = Y$ for any $h \in \mathcal{N}_G(F) \cap \overline{PMP}$.*

LEMMA 4. *Let $F$ be a closed subgroup of $G$ and let $y \in \Omega$. Suppose that the quotient space $F/F \cap G_y$ is not compact (equivalently, the orbit $Fy$ is not compact) and that $Fy$ is a compact minimal $F$-invariant subset. Then the closure of the subset $\{g \in G - F | \; gy \in Fy\}$ contains $e$.*

*Proof.* Suppose the contrary. Then there exists a relatively compact neighbourhood $U \subset G$ of the identity such that $Uy \cap Fy = (U \cap F)y$. Let us represent $F$ as the union of increasing sequence of compact subsets $K_n$, $n \in N^+$. Since $F/F \cap G_y$ is not compact, for any $n \in N^+$ there exists $z_n \in Fy$ such that $K_n z_n \cap (U \cap F)y = \emptyset$. But $Uy \cap Fy = (U \cap F)y$ and $K_n z_n \subset Fy$. So $K_n z_n \cap Uy = \emptyset$. This fact and the fact that the set $Uy$ is open in $\Omega$ imply that the closure of the set $\Psi \overset{\text{def}}{=} \bigcup_{n \in N^+} K_n z_n$ does not contain $y$. Further as $\overline{Fy}$ is compact, we can assume (replacing $\{z_n\}$ by a subsequence) that the sequence $\{z_n\}$ tends to $z \in \overline{Fy}$ as $n \to \infty$. Then as $K_n z_n \subset \psi$, $K_n \subset K_{n+1}$ and $F = \bigcup_{n \in N^+} K_n$ we have $Fz \subset \bar{\Psi}$. But $y \notin \bar{\Psi}$. So $y \notin \overline{Fz}$. In view of the inclusion $z \in \overline{Fy}$, this contradicts to the fact that $\overline{Fy}$ is a minimal closed $F$-invariant subset.

**4.** Till the end of the paper, $G$, $\Gamma$, $\Omega$, $H$ and $G_y$ denote the same as in Section 2. In this section, we shall formulate Lemmas 5–7 about closures of subsets of the type $P'MP$ where $P'$ and $P$ are subgroups of $G$ and $M$ is a subset of $G$, satisfying the condition $e \in \bar{M}$. These lemmas will be proved in Section 8. We have to fix some notations. Let

$$d(t) = \begin{bmatrix} t & 0 & 0 \\ 0 & 1 & 0 \\ 0 & 0 & t^{-1} \end{bmatrix},$$

$$v_1(t) = \begin{bmatrix} 1 & t & t^2/2 \\ 0 & 1 & t \\ 0 & 0 & 1 \end{bmatrix} = \exp t \begin{bmatrix} 0 & 1 & 0 \\ 0 & 0 & 1 \\ 0 & 0 & 0 \end{bmatrix},$$

$$v_2(t) = \begin{bmatrix} 1 & 0 & t \\ 0 & 1 & 0 \\ 0 & 0 & 1 \end{bmatrix} = \exp t \begin{bmatrix} 0 & 0 & 1 \\ 0 & 0 & 0 \\ 0 & 0 & 0 \end{bmatrix}.$$

Let $D = \{d(t) |\ t > 0\}$, $V_1 = \{v_1(t) |\ t \in R\}$, $V_2 = \{v_2(t) |\ t \in R\}$, $V_2^+ = \{v_2(t) |\ t > 0\}$ and $V_2^- = \{v_2(t) |\ t < 0\}$.

Further, we set

$$V = V_1 \cdot V_2 = \left\{ \begin{bmatrix} 1 & v_1 & v_2 \\ 0 & 1 & v_1 \\ 0 & 0 & 1 \end{bmatrix} \middle| \ v_1, v_2 \in R \right\}$$

and denote by $W \subset G$ the group of unipotent upper triangular matrices. It is clear that $D$ normalizes each of subgroups $V_1$, $V_2$, $V$ and $W$ and that $V$ is commutative. Let us also note $D \subset H$ and $V_1 = H \cap W$.

LEMMA 5. (I) *Let $U$ be a connected unipotent subgroup of $G$ and let $M \subset G$. Suppose that $e \in \bar{M} - M$.*

(a) *If $M \subset G - \mathcal{N}_G(U)$ then the connected component of $e$ of the set $\mathcal{N}_G(U) \cap \overline{UMU}$ differs from $U$.*

(b) *If $M \subset G - U$ then the closure of the subgroup generated by the set $\mathcal{N}_G(U) \cap \overline{UMU}$ contains a one-parameter subgroup which is not contained in $U$.*

(II) *If* $M \subset G - V_1$ *and* $e \in \bar{M} - M$ *then the closure of the subgroup generated by the set* $\mathcal{N}_G(V_1) \cap \overline{V_1 M V_1}$ *contains either* $V$ *or a subgroup of the type* $vDV_1v^{-1}$ *where* $v \in V$.

LEMMA 6. *If* $M \subset G - \mathcal{N}_G(V)$ *and* $e \in \bar{M}$ *then* $W \cap \overline{VMV}$ *generates* $W$.

LEMMA 7. *If* $M \subset G - \dot{H}$ *and* $e \in \bar{M}$ *then* $\overline{HMDV_1}$ *contains either* $V_2^+$ *or* $V_2^-$.

5. As in [3], by the horospherical subgroup corresponding to an element $g \in G$ we mean the subgroup

$$\{u \in G| \ g^j u g^{-j} \to e \text{ as } j \to \infty\}.$$

This subgroup will be denoted by $U_g$. It is well known $U_h$ is a closed unipotent subgroup and $g$ normalizes $U_g$. If $U_g \neq \{e\}$, $F$ denotes the subgroup generated by $U_g$ and $g$ and $\Lambda$ is a lattice in $G$ then the orbit $Fz$ is dense in $G/\Lambda$ for any $z \in G/\Lambda$ (see Propositions 1.1 and 1.2 in [3]); recall that a lattice in $G$ is a discrete subgroup such that the corresponding quotient space has a finite volume with respect to Haar measure). But $\Gamma = \mathrm{SL}(3, \boldsymbol{Z})$ is a lattice in $G = \mathrm{SL}(3, \boldsymbol{R})$ (see [6], Corollary 10.5) and, as can be easily checked, $W = U_{d(t)}$ for any $0 < t < 1$. So we have

LEMMA 8. *For any* $y \in \Omega = G/\Gamma$ *the orbit* $DWy$ *is dense in* $\Omega$ *and, consequently, is not relatively compact in* $\Omega$.

As it was noted, $W = U_{d(t)}$ for any $0 < t < 1$. On the other hand (see [6]) Theorem 1.12), if $y \in \Omega = G/\Gamma$, $\gamma \in G_y$, $\gamma \neq e$, $\{g_n| \ n \in N^+\} \subset G$ and $g_n \gamma g_n^{-1} \to e$ as $n \to \infty$ then the set $\{g_n y| \ n \in N^+\}$ is not relatively compact in $\Omega$. So we have

LEMMA 9. *Let* $y \in \Omega$. *Suppose the orbit* $Dy$ *is relatively compact in* $\Omega$. *Then* $W \cap G_y = \{e\}$ *and consequently the quotient space* $U/U \cap G_y$ *is not compact for any nontrivial closed subgroup* $U$ *of* $W$.

LEMMA 10. *For any* $y \subset \Omega$, *the orbit* $DVy$ *is not relatively compact in* $\Omega$.

*Proof.* Suppose the contrary i.e. that $\overline{DVy}$ is compact. Then the $DV$-invariant subset $\overline{DVy}$ contains a minimal closed $V$-invariant subset $Y$. One can assume that $y \in Y$. Let us denote the subgroup $\{g \in \mathcal{N}_G(V)| \ gY = Y\}$ by $F$. As $Y$ is closed, $F$ is closed. Let us denote the connected component of identity of the group $F$ by $F^0$. One can easily check that the connected component of the identity of the group $\mathcal{N}_G(V)$ coincides with $DW$. This easily implies that any closed unimodular connected subgroup of $\mathcal{N}_G(V)$ containing $V$ is itself contained in $W$. Thus either $F^0 \subset W$ or $F^0$ is nonunimodular. But a nonunimodular group does not contain a lattice. In view of Lemma 9, this implies that the quotient space $F^0/F^0 \cap G_y$ and, consequently, the quotient space $F/F \cap G_y$ are noncompact. Now we use Lemma 4 and see that the closure of $\Phi \stackrel{\text{def}}{=} \{g \in G-F| \ gY \cap Y \neq \emptyset\}$ contains $e$. On the other hand, it follows from Lemma 2 that $P \cap \overline{\mathcal{N}_G(V)} = \emptyset$. So the closure of the set $M \stackrel{\text{def}}{=} \{g \in G-\mathcal{N}_G(V)| \ gY \cap Y \neq \emptyset\}$ contains $e$. Then in view of Lemma 6, $\overline{VMV} \cap W$ generates $W$. This fact, Lemma 3 and the inclusion $W \subset \mathcal{N}_G(V)$ imply $WY = Y$. Now in view of Lemma 8, the set $\overline{DVy} \supset \overline{DY} = \overline{DWY}$ is not compact. Thus we arrive at a contradiction which shows that the lemma is true.

LEMMA 11. *For any $y \in \Omega$, the sets $DV_1 V_2^+ y$ and $DV_1 V_2^- y$ are not relatively compact in $\Omega$.*

*Proof.* If $n \in \boldsymbol{N}^+$, $t \in \boldsymbol{R}$ and $|t| \leqslant n$ then $v_2(t) v_2(n) y \in V_2^+ y$. This easily implies that if $z$ is a limit point of $\{v_2(n) y| \ n \in \boldsymbol{N}^+\}$ then $V_2 z \subset \overline{V_2^+ y}$ and consequently $DVz = DV_1 V_2 z \subset \overline{DV_1 V_2^+ y}$. But if $V_2^+ y$ is relatively compact in $\Omega$ then the set of limit points of $\{v_2(n) y| \ n \in \boldsymbol{N}^+\}$ is not empty. Therefore and in view of Lemma 10, $DV_1 V_2^- y$ is not relatively compact in $\Omega$. The fact $DV_1 V_2^- y$ is not relatively compact one can prove analogously.

**6. Proof of Theorem 2.** Since the set $\overline{Hz}$ is compact and $H$-invariant, it contains a minimal closed $H$-invariant set $X$. Then as $H \supset V_1$ and $X$ is compact, the set $X$ contains a minimal closed $V_1$-invariant set $Y$. Let us choose $y \in Y$. As $Hy \subset X$ is relatively compact and $D \subset H$, Lemma 9 implies that $V_1/V_1 \cap G_y$ is not compact. Therefore and in view of Lemma 4, the closure of $M_1 \stackrel{\text{def}}{=} \{g \in G-V_1| \ gY \cap Y \neq 0\}$ contains $e$. Let us denote by $\Psi$ the closure of the subgroup generated by $\mathcal{N}_G(V_1) \cap \overline{V_1 M_1 V_1}$. Lemma 3 implies

(1) $\Psi Y = Y$.

As $HX = X$, $Y \subset X$ and $DV_1 \subset H$ we have $DV_1 Y \subset X$. So and in view of (1), $DV_1 \Psi Y \subset X$. Using this fact, the compactness of $X$ and Lemma 11 we get

(2) $DV_1 \Psi \not\supset V_1 V_2^+$ and $DV_1 \Psi \not\supset V_1 V_2^-$.

According to Lemma 5 (II), $\Psi$ contains either $V$ or $vDV_1 v^{-1}$ where $v \in V$. But if $v \in V - V_1$ one can easily check that $DV_1 (vDV_1 v^{-1})$ contains either $DV_1 V_2^+$ or $DV_1 V_2^-$. Therefore and in view of (2), $\Psi \supset DV_1$. This fact and (1) imply

(3) $DV_1 Y = Y$.

Set

(4) $M = \{g \in G-H| \ gy \in \overline{Hz}\}$.

Suppose that $e \in M$. Then in view of Lemma 7, $\overline{HMDV_1}$ contains either $V_2^+$ or $V_2^-$. On the other hand in view of (3) and Lemma 1, $gY \subset \overline{Hz}$ for any $g \in \mathcal{N}_G(V_1) \cap \overline{HMDV_1}$. So either $V_2^+ Y \subset \overline{Hz}$ or $V_2^- Y \subset \overline{Hz}$. This fact, the equalities $DV_1 V_2^+ = V_2^+ DV_1$ and $DV_1 V_2^- = V_2^- DV_1$ and equality (3) imply that $Hz$ contains either $DV_1 V_2^+ Y$ or $DV_1 V_2^- Y$. In view of Lemma 11, this contradicts to the compactness of $\overline{Hz}$. This we have shown

(5) $e \notin \bar{M}$.

We have $y \in Y \subset X \subset \overline{Hz}$ and $X$ is a minimal closed $H$-invariant subset. So and in view of Lemma 4, (4) and (5) imply that $H/H \cap G_y$ is compact. On the other hand, as $y \in \overline{Hz}$ and in view of (4) and (5), $y \in Hz$. So $H/H \cap G_z$ is compact.

7. In the proofs of Lemmas 5–7, we shall need some assertions about unipotent groups of linear transformations. Let us formalate them in the form of the following lemma; we only recall beforehand that (see [7]) all orbits are closed for linear actions of connected unipotent groups.

LEMMA 12. *Let $U$ be a connected unipotent group of linear transformations of $\boldsymbol{R}^n$ and let $Y \subset \boldsymbol{R}^n$. Let $L = \{x \in \boldsymbol{R}^n | \; Ux = x\}$.*

*(I) If $p \in L \cap \bar{Y}$ and $L \cap Y = \emptyset$ then the connected component of $p$ in the set $\overline{UY} \cap L$ is not compact.*

*(II) Let $X \subset \boldsymbol{R}^n$ be a closed subset. Suppose for any $x \in (\overline{UY} \cap X) - L$ the connected component of $x$ in the set $Ux \cap X$ is not compact. If $p \in L \cap \overline{Y \cap X}$ and $L \cap Y = \emptyset$ then the connected component of $p$ in the set $\overline{UY} \cap X \cap L$ is not compact.*

*Proof.* The connected unipotent group $U$ is isomorphic to its Lie algebra $\mathfrak{G}$ as an algebraic manifold (this isomorphism cah be realized by the logarithmical mapping ln: $U \to \mathfrak{G}$. On the other hand, (a) for any $x \in \boldsymbol{R}^n$, the coordinates of ux are regular functions of $u \in U$; (b) the set of values of any nonconstant regular function on a linear space is noncompact. Therefore, for any $x \in \boldsymbol{R}^n - L$, the connected closed set $Ux$ is not compact. Hence, (II) implies (I).

Let us prove (II) using induction on $n$. Since $U$ is unipotent and in view of Lie–Kolchin theorem, there exists a $U$-invariant linear subspace $S$ in $\boldsymbol{R}^n$ of codimension 1 such that $S \supset L$ and $U$ acts on $\boldsymbol{R}^n/S$ trivially. Further, let us choose a sequence $\{y_i \in Y \cap X| \; i \in N^+\}$ tending to $p$. Let us denote by $\Psi$ the upper topological limit of sets $Uy_i \cap X$, i.e. $\Psi$ is the set of limit points of all sequences of the type $\{z_i \in Uy_i \cap X| \; i \in N^+\}$. Since $p \in L \subset S$, $y_i \to p$ as $i \to \infty$ and $U$ acts on $\boldsymbol{R}^n/S$ trivially we have $\Psi \subset S$. On the other hand, since $y_i \to p$ as $i \to \infty$ and the connected component of $y_i$ of the set $Uy_i \cap X$ is not compact for any $i \in N^+$ it follows that the connected component of $p$ in the set $\Psi$ is not compact. So, and in view of the inclusion $\Psi \subset \overline{UY} \cap X$ the connected component of $p$ in the set $\overline{UY} \cap X \cap S$ is not compact. Then either the connected component of $p$ in the set $\overline{UY} \cap X \cap L$ is not compact or $p$

contains a point $q$ which belongs to the closure of the set $(\overline{UY} \cap X \cap S) - L$. But $\overline{UA} \cap X \cap S \subset \overline{UY} \cap X \cap S$ for any $A \subset \overline{UY}$. So we can replace $X$ by $X \cap S$ and $Y$ by $\overline{UY} \cap X \cap S - L$ and assume that $X \subset S$ and $Y \subset S$. This gives the possibility to produce the desired induction.

**8. Proof of Lemma 5.** (I) (a) Let $\pi$: $G \to G/U$ be the natural projection and $T = \{x \in G/U \mid Ux = x\}$. Since $\mathcal{N}_G(U) = \{g \in G \mid UgU = gU\}$, we have $\pi(\mathcal{N}_G(U)) = T$. So (I) (a) follows from

(x) the connected component of $\pi(e)$ of the set $T \cap \overline{U\pi(M)}$ is not compact and consequently differs from $\pi(e)$.

The connected unipotent subgroup $U$ is algebraic and has no rational characters. So, and in view of a theorem of Chevalley (see [2], Theorem 5.1), there exist $m \in \mathbf{N}^+$, a faithful rational representation $\alpha$: $G \to \mathrm{GL}(m, \mathbf{R})$ and $x_0 \in \mathbf{R}^m$ such that $U = \{g \in G \mid \alpha(g)x_0 = x_0\}$. According to the lemma on the orbit closure (see [2], Proposition 1.8), $\alpha(G)x_0$ is a smooth manifold which is open in its closure. This implies that the map $gU \to \alpha(g)x_0$ is a homeomorphism of $G/U$ onto $\alpha(G)x_0$. Now to prove (x), it remains to apply Lemma 12 (I) for $Y = \alpha(M)x_0$ and $p = x_0$; we should only note that (1) as $e \in \bar{M}$, $M \subset G - \mathcal{N}_G(U)$ and $T = \pi(\mathcal{N}_G(U))$ we have $\pi(e) \in T \cap \overline{\pi(M)}$ and $T \cap \pi(M) = \emptyset$; (2) as $U$ is connected and unipotent and the representation $\alpha$ is rational, $\alpha(U)$ is connected and unipotent.

(b) Let us denote by $\Psi$ the closure of the subgroup generated by $\mathcal{N}_G(U) \cap \overline{UMU}$ and by $\Psi^0$ the connected component of identity of the Lie group $\Psi$. It is enough to prove $\Psi^0 \neq U$. In view of (a), one can assume $M \subset \mathcal{N}_G(U)$. Then $\Psi \supset M$. But $e \in \bar{M} - M$. So $\Psi^0 \cap M \neq \emptyset$ and as $M \subset G - U$ we have $\Psi^0 \neq U$.

(II) One can directly check that the connected component of identity of the group $\mathcal{N}_G(V_1)$ coincides with $DV$. This easily implies that, for any one-parameter subgroup $S \subset \mathcal{N}_G(V_1)$, $S \not\subset V_1$, the subgroup $SV_1$ contains either $V$ or a subgroup of the type $vDV_1v^{-1}$ where $v \in V$. Now it remains to use the assertion (I) (b).

*Proof of Lemma* 6. Let us denote by $a_{ij}(g)$ the coefficient of a matrix $g$ standing on the intersection of $i$th row and $j$th column. Set $W^+ = \{w \in W \mid a_{12}(w) \geqslant a_{23}(w)\}$ and $W^- = \{w \in W \mid a_{12}(w) \leqslant a_{23}(w)\}$. One can easily see that each of the sets $W^+$ and $W^-$ generates $W$. So it is enough to prove

(A1) $\overline{VMV}$ contains either $W^+$ or $W^-$.

The Lie algebra $\mathfrak{G}$ of $G$ is naturally identified with the space of real matrices of order 3 having the trace zero. Let $E_{ij}$ denotes the matrix such that $a_{ij}(E_{ij}) = 1$ and $a_{kl}(E_{ij}) = 0$ if $k \neq i$ or $l \neq j$. Set $x_0 = E_{12} + E_{23} \in \mathfrak{G}$. Let us denote by Ad the adjoint representation of $G$ and note that $(\mathrm{Ad}\, g)x = gxg^{-1}$ for any $g \in G$ and $x \in \mathfrak{G}$. According to the lemma on the orbit closure, $(\mathrm{Ad}\, G)x_0$ is a smooth manifold which is open in its closure. This implies the map sending $gV$ onto $(\mathrm{Ad}\, g)x_0$ is a homeomorphism of $G/V$ onto $(\mathrm{Ad}\, G)x_0$. So (A1) is equivalent to

(A2) $\overline{(\mathrm{Ad}\, V)(\mathrm{Ad}\, M)x_0}$ contains either $(\mathrm{Ad}\, W^+)x_0$ on $(\mathrm{Ad}\, W^-)x_0$.

Let $\mathfrak{V}$, $\mathfrak{V}_1$ and $\mathfrak{V}_2$ denote the Lie algebras of $V$, $V_1$ and $V_2$ respectively. Note $\mathfrak{V}_1 = \{tx_0 |\ t \in \boldsymbol{R}\}$, $\mathfrak{V}_2 = \{tE_{13} |\ t \in \boldsymbol{R}\}$ and $\mathfrak{V} = \mathfrak{V}_1 + \mathfrak{V}_2$. Set $\mathfrak{V}_2^+ = \{tE_{13} |\ t > 0\}$ and $V_2^- = \{tE_{13} |\ t < 0\}$. Direct calculations show that $(\mathrm{Ad}\, W^+)x_0 = x_0 + \mathfrak{V}_2^+$ and $(\mathrm{Ad}\, W^-)x_0 = x_0 + \mathfrak{V}_2^-$. Further, one can easily check that $\mathscr{N}_G(V) = \{g \in G |\ (\mathrm{Ad}\, g)x_0 \in \mathfrak{V}\}$. But $M \subset G - \mathscr{N}_G(V)$. So $V \cap (\mathrm{Ad}\, M)x_0 = \emptyset$. As $e \in \bar{M}$ we have $x_0 \in \overline{(\mathrm{Ad}\, M)x_0}$. In view of aforesaid, (A2) is a particular case of

(B1) if $Y \subset (\mathrm{Ad}\, G)x_0$, $x_0 \in \bar{Y}$ and $V \cap Y = \emptyset$ then $\overline{(\mathrm{Ad}\, V)Y}$ contains either $x_0 + \mathfrak{V}_2^+$ or $x_0 + \mathfrak{V}_2^-$.

Set $X = \{x \in \mathfrak{G} |\ a_{12}(x) = 1\}$ and denote by $N$ the set of nilpotent elements of $\mathfrak{G}$. We have $x_0 \in X \cap N$, $(\mathrm{Ad}\, G)x_0 \subset N$ and $y/a_{12}(y) \in X \cap N$ if $y \in N$ and $a_{12}(y) \neq 0$. So (B1) can easily be deduced from

(B2) if $Y \subset X \cap N$, $x_0 \in \bar{Y}$ and $\mathfrak{V} \cap Y = \emptyset$ then $\overline{(\mathrm{Ad}\, V)Y}$ contains either $x_0 + \mathfrak{V}_2^+$ or $x_0 + \mathfrak{V}_2^-$.

Since $N$ is closed and invariant under $\mathrm{Ad}\, G$, we have $\overline{(\mathrm{Ad}\, V)Y} \subset N$ for any $Y \subset N$. On the other hand, one can easily check that (a) $X \cap \mathfrak{V} = x_0 + \mathfrak{V}_2$ and if $S \subset x_0 + \mathfrak{V}_2$ is a connected closed noncompact subset and $x_0 \in S$ then either $S \supset x_0 + \mathfrak{V}_2^+$ or $S \supset x_0 + \mathfrak{V}_2^-$; (b) $U = \{x \in \mathfrak{G} |\ (\mathrm{Ad}\, V)x = x\}$; (c) the group $V$ and consequently the group $\mathrm{Ad}\, V$ are connected and unipotent. Therefore and in view of Lemma 12 (II), (B2) follows.

(C) for any $x \in (X \cap N) - V$ the connected component of $x$ in the set $(\mathrm{Ad}\, V)x \cap X$ is not compact.

Let us denote by $\mathscr{T} \subset G$ the space of upper triangular matrices with the trace zero. If $x \in (\mathscr{T} \cap X \cap N) - V$ then $x = x_0 + sE_{23} + tE_{13}$ where $s \neq 0$. In this case direct calculations show that $(\mathrm{Ad}\, V)x = x_0 + \mathfrak{V}_2 \subset X$. So one can assume $x \notin \mathscr{T}$. One can easily check that the isotropy subgroup $V_y = \{v \in V |\ (\mathrm{Ad}\, v)y = y\}$ is trivial for any $y \in \mathfrak{G} - \mathscr{T}$. In particular $V_x = \{e\}$. On the other hand, since (see [7]) all orbits are closed for linear actions of connected unipotent groups, the set $(\mathrm{Ad}\, V)x$ is closed. So the orbit mapping $v \to (\mathrm{Ad}\, v)x$, $v \in V$, is proper. Now to prove (C), it remains to show that the set

$$\Phi \overset{\mathrm{def}}{=} \{v \in V |\ a_{12}(vxv^{-1}) = 1\}$$

is not compact. It is directly checked the polynomial

$$P(s,t) \overset{\mathrm{def}}{=} a_{12}\left[\begin{bmatrix} 1 & s & t \\ 0 & 1 & s \\ 0 & 0 & 1 \end{bmatrix} x \begin{bmatrix} 1 & s & t \\ 0 & 1 & s \\ 0 & 0 & 1 \end{bmatrix}^{-1}\right]$$

is linear with respect to $t$ for any $s$. But $P(0, 0) = 1$. So either $P(0, t) = 1$ for any $t \in \boldsymbol{R}$ or, for any but a finite number $s \in \boldsymbol{R}$, there exists $t(s) \in \boldsymbol{R}$ such that $P(s, t(s)) = 1$. Hence, $\Phi$ is not compact.

*Proof of Lemma* 7. Let $a_{ij}(g)$, $\mathfrak{G}$, Ad, $\mathfrak{B}$, $\mathfrak{B}_1$, $\mathfrak{B}_2$, $\mathfrak{B}_2^+$ and $\mathfrak{B}_2^-$ be as in the proof of Lemma 6. Further, let us denote by $\mathfrak{H} \subset \mathfrak{G}$ the Lie algebra of $H$ and by $P \subset \mathfrak{G}$ the orthogonal complement to $\mathfrak{H}$ with respect to the Killing form. One can easily check

(1) $P = \{x \in \mathfrak{G} | \ a_{11}(x) = a_{33}(x),\ a_{12}(x) = -a_{23}(x) \text{ and } a_{21}(x) = -a_{32}(x)\}$.

We have $\mathfrak{H} + P = \mathfrak{G}$ and $\mathfrak{H} \cap P = \{0\}$, and hence $e$ (resp. 0) has a neighbourhood $A$ in $\mathfrak{H}$ (resp. $B$ in $P$) such that the map $(a, b) \to a \cdot \exp b$ is a homeomorphism of $A \times B$ onto a neighbourhood of $e$ in $G$. On the other hand, the conclusion is not altered if we replace $M$ by a set of elements $h(m) \cdot m$ $(m \in M,\ h(m) \in H)$. It follows that it suffices to consider the case where $M \subset \exp B$. Let us consider two cases: (a) $M \cap \overline{V_2} = \emptyset$; (b) $M \cap V_2 \neq \emptyset$.

(a) Set $M_0 = \log M$. As $e \in \bar{M}$ and $e \notin M$, we have

$$0 \in \bar{M}_0 \quad \text{and} \quad 0 \notin M_0. \tag{2}$$

It is immediately checked that $\mathfrak{B} = \{x \in \mathfrak{G} | \ (\operatorname{Ad} V_1) x = x\}$. But $\mathfrak{B}_2 = \mathfrak{B} \cap P$. Thus

$$\mathfrak{B}_2 = \{x \in P | \ (\operatorname{Ad} V_1) x = x\}. \tag{3}$$

The subalgebra $\mathfrak{H}$ and the Killing form are invariant under $\operatorname{Ad} H$. Therefore, $P$ is invariant under $\operatorname{Ad} H$. In particular, $(\operatorname{Ad} V_1) P = P$. Now applying Lemma 12 (I) and using (2), (3) and the equality $\overline{M \cap V_2} = \emptyset$ we see that the connected component of zero in the set $\overline{(\operatorname{Ad} V_1) M_0} \cap V_2$ is not compact. Consequently, $\overline{(\operatorname{Ad} V_1) M_0}$ contains either $\mathfrak{B}_2^+$ or $\mathfrak{B}_2^-$. But $\exp \mathfrak{B}_2^+ = V_2^+$, $\exp \mathfrak{B}_2^- = V_2^-$ and

$$HMDV_1 \supset V_1 M V_1 \supset \bigcup_{v \in V_1} (vMv^{-1}) = \exp(\operatorname{Ad} V_1) M_0.$$

So $\overline{HMDV_1}$ contains either $V_2^+$ or $V_2^-$.

(b) Let $v_2(t_0) \in M \cap V_2$. Then

$$HMDV_1 \supset DMD \supset Dv_2(t_0) D \supset \{d(t) v_2(t_0) d(t)^{-1} | \ t > 0\}$$
$$= \{v_2(t^2 t_0) | \ t > 0\}.$$

But $t_0 \neq 0$ (because $e \notin M$). Thus $HMDV_1$ contains either $V_2^+$ or $V_2^-$.

**9. Concluding remarks.** The study of closed subsets of invariant under unipotent subgroups plays the main role in the proof of Theorem 2. In this connection we note that Raghunathan has formulated the following.

CONJECTURE. Let $F$ be a connected semisimple Lie group, $\Lambda$ be a lattice in $F$ and $U \subset F$ be a closed subgroup. Suppose that $U$ is unipotent i.e. that Adu is unipotent for any $u \in U$. Then, for any $x \in F/\Lambda$, there exists a subgroup $P \subset F$ containing $U$ such that the closure of the orbit $Ux$ coincides with $Px$.

Raghunathan also noted the connection of his conjecture with Davenport's conjecture (which was proved at the present paper).

Raghunathan's conjecture was proved by Dani (see [3]) in the case when $U$ is a horospherical subgroup. Using methods of the present paper, it is possible to prove this conjecture in the case where $F = \mathrm{SL}(3, \boldsymbol{R})$ and $Ux$ is relatively compact in $F/\Lambda$.

Let us also note that Raghunathan conjecture can be generalized if we omit the assumption that $F$ is semisimple and replace the condition "$U$ is unipotent" by the condition "$U$ is generated by unipotent elements".

The results of the present paper were announced in [8].

## References

[1] A. Borel, *Density properties of certain subgroups of semisimple groups*, Ann. of Math. 72 (1960), 179–188.

[2] –, *Linear algebraic groups*, Benjamin, New York 1969.

[3] S. G. Dani, *Orbits of horospherical flows*, Duke Math. J. 53 (1986), 177–188.

[4] H. Davenport, H. Heilbronn, *On indefinite quadratic forms in five variables*, J. London Math. Soc. 21 (1946), 185–193.

[5] H. Davenport, H. Ridout, *Indefinite quadratic forms*, Proc. London Math. Soc. (3) 9 (1959), 544–555.

[6] M. S. Raghunathan, *Discrete subgroups of Lie groups*, Springer, Berlin–Heidelberg–New York 1972.

[7] M. Rosenlicht, *On quotient varieties and the affine embeddings of certain homogeneous spaces*, Trans. Amer. Math. Soc. 101 (1961), 211–223.

[8] G. A. Margulis, *Formes quadratiques indéfinies et flots unipotents sur les espaces homogènes*, C. R. Acad. Sc. Paris, t. 304, Série I, n° 10, 1987.

C. R. Acad. Sci. Paris, t. 314, Série I, p. 699-704, 1992 699

Théorie des groupes/*Group Theory*

# On the limit distributions of orbits of unipotent flows and integral solutions of quadratic inequalities

Shrikrishna Gopalrao DANI and Gregori Aleksandrovitch MARGULIS

***Abstract*** – We give some results on the convergence of averages along orbit-segments of unipotent flows to their limit distributions, being uniform on certain compact sets of initial points and defining unipotent one-parameter subgroups. As an application we deduce a result giving lower estimates for the number of integral solutions of systems of inequalities of the form $|Q(x)|<\varepsilon$ and $\|x\|\leqq r$, where Q is a non-degenerate indefinite quadratic form on $\mathbf{R}^n$, $n\geqq 3$; the estimates are also uniform on certain compact sets of quadratic forms.

## Sur les distributions limites des orbites de flots unipotents et les solutions entières d'inégalités quadratiques

***Résumé*** – *Nous donnons quelques résultats assurant l'uniformité de la convergence des « moyennes sur des morceaux d'orbites » de flots unipotents, sur certains ensembles compacts de points initiaux et de sous-groupes à un paramètre, définissant les flots. Comme application nous donnons une minoration du nombre des solutions entières des systèmes d'inéquations de la forme* $|Q(x)|<\varepsilon$ *et* $\|x\|\leqq r$, *où* Q *est une forme quadratique non dégénérée indéfinie sur* $\mathbf{R}^n$, $n\geqq 3$ ; *les minorations sont aussi uniformes sur certains ensembles compacts des formes quadratiques.*

***Version française abrégée*** – Soient G un groupe de Lie connexe et Γ un réseau dans G; soit μ la mesure de probabilité invariante sur G/Γ. Soit $U=\{u_t\}$ un sous-groupe à un paramètre unipotent de G. On s'intéresse aux adhérences et aux « distributions » des $\{u_t\}$-orbites sur G/Γ et aux applications à des problèmes diophantiens.

Récemment Marina Ratner a démontré, en réponse aux conjectures de Raghunathan, que pour chaque $x\in G/\Gamma$ l'adhérence de l'orbite de $x$ est un sous-espace homogène de la forme $Hx$, où H est un sous-groupe fermé de G, contenant U, et que l'orbite est distribuée uniformément par rapport à une mesure de probabilité H-invariante sur $Hx$ (*voir* [7] et [8]). Soit $\mathscr{G}(U)$ l'ensemble des $x$ qui ne sont pas contenus dans des sous-espaces homogènes propres comme ci-dessus. Soit $C_b(G/\Gamma)$ l'espace des fonctions réelles continues bornées sur G/Γ. On a en particulier

$$\frac{1}{T}\int_0^T \varphi(u_t x)\,dt \to \int_{G/\Gamma}\varphi\,d\mu \qquad \text{pour} \quad \varphi\in C_b(G/\Gamma) \quad \text{et} \quad x\in\mathscr{G}(U).$$

Nous énonçons maintenant quelques résultats concernant l'uniformité de la convergence comme ci-dessus sur certains sous-ensembles compacts et donnons une application à l'approximation diophantienne.

A. THÉORÈME. – *Soient* G *un groupe de Lie connexe et* Γ *un réseau dans* G. *Soit* $\{u_t^{(i)}\}$ *une suite de sous-groupes à un paramètre unipotents de* G *convergeant vers un sous-groupe unipotent* $\{u_t\}$. *Soient* $\{x_i\}$ *une suite dans* G/Γ *convergeant vers un point* $x\in\mathscr{G}(\{u_t\})$ *et* $\{T_i\}$ *une suite dans* $\mathbf{R}^+$ *telle que* $T_i\to\infty$. *Alors, pour tout* $\varphi\in C_b(G/\Gamma)$,

$$\frac{1}{T_i}\int_0^{T_i}\varphi(u_t^{(i)}x_i)\,dt \to \int_{G/\Gamma}\varphi\,d\mu.$$

B. THÉORÈME. – *Soient* G *un groupe de Lie connexe et* Γ *un réseau dans* G. *Soient* $U=\{u_t\}$ *un sous-groupe à un paramètre unipotent*, $\varphi\in C_b(G/\Gamma)$, K *un sous-ensemble compact de* G *et*

Note présentée par Jacques TITS.

700 C. R. Acad. Sci. Paris, t. 314, Série I, p. 699-704, 1992

$\varepsilon>0$. *Alors il existe des sous-groupes propres fermés* $H_1, \ldots, H_k$ *et des sous-ensembles compacts* $C_i$ *de* $\{g\in G \mid g^{-1}Ug\subseteq H_i\}$, $i=1, \ldots, k$, *tels que* 1) *pour chaque* $i$, $H_i \cap \Gamma$ *est un réseau dans* $H_i$ *et* 2) *pour tout sous-ensemble compact* F *de* $K-\bigcup C_i$ *il existe* $T_0\geqq 0$ *tel que, pour tout* $x\in F\Gamma/\Gamma$ *et* $T\geqq T_0$,

$$\left|\frac{1}{T}\int_0^T \varphi(u_t x)\,dt-\int_{G/\Gamma}\varphi\,d\mu\right|<\varepsilon.$$

La démonstration de ces théorèmes depend d'idées analogues à celles utilisées dans [4] et du théorème de Ratner sur la classification des mesures invariantes par des flots unipotents [7]. Notons que le théorème A implique le théorème de « rigidité de distribution » de Ratner [8]; on a ici une autre façon de le déduire de sa classification des mesures invariantes. Le théorème donne aussi un résultat de M. Burger et P. Sarnak.

C. Corollaire. — *Soient* $\mathcal{Q}(p)$ *l'espace de formes quadratiques sur* $\mathbf{R}^n$ *de discriminant* $\pm 1$ *et signature* $(p, n-p)$, *où* $n\geqq 3$ *et* $1\leqq p<n$. *Soit* $\mathcal{K}$ *un sous-ensemble compact de* $\mathcal{Q}(p)$. *Alors pour tous* $0<a<b$ *et* $\theta>0$ *il existe un sous-ensemble fini* $\mathcal{S}$ *de* $\mathcal{K}$ *tel que chaque* $Q\in\mathcal{S}$ *est proportionnelle à une forme rationnelle et pour chaque sous-ensemble compact* $\mathcal{C}$ *de* $\mathcal{K}-\mathcal{S}$ *il existe* $r_0\geqq 0$ *tel que pour tout* $Q\in\mathcal{C}$ *et* $r\geqq r_0$ *le cardinal de* $\{x\in\mathbf{Z}^n \mid a<Q(x)<b \text{ et } \|x\|\leqq r\}$ *est plus grand que* $(1-\theta)\lambda(\{v\in\mathbf{R}^n \mid a<Q(v)<b \text{ et } \|v\|\leqq r\})$, *où* $\lambda$ *est la mesure de Lebesgue sur* $\mathbf{R}^n$. *En outre, si* $n\geqq 5$, *pour tout* $\varepsilon>0$ *il existe* $c>0$ *et* $r\geqq r_0$ *tels que pour tous* $Q\in\mathcal{K}$ *et* $r\geqq r_0$ *le cardinal de* $\{x\in\mathbf{Z}^n \mid 0\leqq Q(x)<\varepsilon \text{ et } \|x\|\leqq r\}$ *est plus grand que*

$$c\lambda(\{v\in\mathbf{R}^n \mid 0\leqq Q(v)<\varepsilon \text{ et } \|v\|\leqq r\}).$$

---

Introduction. — Study of orbits of unipotent flows on homogeneous spaces leads to various results on diophantine approximation. In recent years, in particular, it was applied to obtain a proof of the Oppenheim conjecture on values of quadratic forms at integral points and also certain strengthenings of that result (*cf.* [5], [3] and [4]); *see* also [6], for a more general perspective). Major progress has now been made in the understanding of the dynamics of the unipotent flows, through the work of Marina Ratner, in response to a conjecture of Raghunathan, where she obtained a classification of ergodic invariant measures of the flows [7] and deduced that any orbit under such a flow is uniformly distributed on a homogeneous subset [8].

In various contexts it is also of interest to know how the convergences involved in the uniform distribution depend on the starting point and the unipotent one-parameter subgroup inducing the flow; in particular, whether they are uniform on suitable compacts sets. It turns out that certain ideas involved in [4] and developed further in [9], in both of which partial results were obtained towards Raghunathan's conjecture, can be applied together with Ratner's classification of invariant measures to obtain certain answers in this regard. In this Note we announce some results on this theme and an application to diophantine approximation and give an outline of the proofs.

Let G be a connected Lie group and let $\Gamma$ be a lattice in G, namely a discrete subgroup such that the quotient $G/\Gamma$ has a (unique) G-invariant probability measure, say $\mu$. A one-parameter subgroup $U=\{u_t\}$ of G is referred to as unipotent if $\operatorname{Ad} u_t$ is unipotent for all $t\in\mathbf{R}$. For any subgroup S of G let $\mathcal{S}(S)$ denote the set of points $x\in G/\Gamma$ such that $Hx$ is closed and has finite H-invariant measure for some proper closed subgroup H containing S and let $\mathcal{G}(S)=G/\Gamma-\mathcal{S}(S)$. Let $l$ denote the Lebesgue measure. One

C. R. Acad. Sci. Paris, t. 314, Série I, p. 699-704, 1992 

of our main results can then be formulated as follows:

1. THEOREM. — *Let* G *and* Γ *be as above. Let* S *be any subgroup which is generated by the unipotent one-parameter subgroups contained in it. Let* F *be a compact subset contained in* $\mathscr{G}(S)$. *Then for any* $\varepsilon>0$ *there exists a neighbourhood* Ω *of* $\mathscr{S}(S)$ *such that for any unipotent one-parameter subgroup* $\{u_t\}$ *of* G, *any* $x\in F$ *and* $T\geqq 0$,

$$l(\{t\in[0, T]\,|\,u_t x\in\Omega\})\leqq\varepsilon T.$$

Our proof of this is independent of Ratner's work and goes along the lines of some results in [4] and [9]; *see* Sec. 2 for an outline of the proof. We note that in the particular case when $\{u_t\}=S$ and F is a singleton set the assertion can also be deduced from Theorem B of [8]. Together with Ratner's classification of invariant measures of unipotent flows and Theorem 3.5 of [2], by a routine argument, Theorem 1 yields the following:

2. THEOREM. — *Let* $\{u_t^{(i)}\}$ *be a sequence of unipotent one-parameter subgroups converging to a unipotent one-parameter subgroup* $\{u_t\}$; *that is,* $u_t^{(i)}\to u_t$ *for all t. Let* $\{x_i\}$ *be a sequence in* G/Γ *converging to some* $x\in\mathscr{G}(\{u_t\})$ *and let* $\{T_i\}$ *be a sequence in* $\mathbf{R}^+$, *tending to infinity. Then for any bounded continuous function* φ *on* G/Γ,

$$\frac{1}{T_i}\int_0^{T_i}\varphi(u_t^{(i)}x_i)\,dt\to\int_{G/\Gamma}\varphi\,d\mu.$$

It may be noted that Theorem 2 in particular implies Ratner's distribution rigidity theorem for unipotent flows (*cf.* [8], Theorem B); our method gives another way of deducing it from her classification of the invariant measures. A result of M. Burger and P. Sarnak on the limit distributions of closed orbits of certain subgroups of G on G/Γ, which they applied to deduce uniform distribution of Hecke points (*cf.* [1], "Theorem 5.2"), also readily follows from Theorem 2.

For any closed subgroups H and U of G let $X(H, U)=\{g\in G\,|\,U\subset gHg^{-1}\}$. We note that if H is a proper subgroup and $H\cap\Gamma$ is a lattice in H then $X(H, U)\Gamma/\Gamma$ is contained in $\mathscr{S}(U)$. Via a variation of Proposition 2.3 below (in which the set F is not required to be disjoint from XΓ but the conclusion is only for all large T) we also prove the following variation of Theorem 2.

3. THEOREM. — *Let* $U=\{u_t\}$ *be a unipotent one-parameter subgroup of* G *and let* φ *be a bounded continuous function on* G/Γ. *Let* K *be a compact subset of* G/Γ *and let* $\varepsilon>0$ *be given. Then there exist finitely many proper closed subgroups* $H_1, \ldots, H_k$ *such that* $H_i\Gamma$ *is closed and* $H_i\cap\Gamma$ *is a lattice in* $H_i$ *for all i, and compact subsets* $C_1, \ldots, C_k$ *of* $X(H_1, U), \ldots, X(H_k, U)$ *respectively, for which the following holds: For any compact subset* F *of* $K-\bigcup C_i\Gamma/\Gamma$ *there exists a* $T_0\geqq 0$ *such that for all* $x\in F$ *and* $T\geqq T_0$,

$$\left|\frac{1}{T}\int_0^{T}\varphi(u_t x)\,dt-\int_{G/\Gamma}\varphi\,d\mu\right|<\varepsilon.$$

We now describe an application, giving lower estimates for the number of integral solutions of quadratic inequalities. Let $n\geqq 3$ be fixed and for $1\leqq p<n$ let $\mathscr{Q}(p)$ denote the space of quadratic forms on $\mathbf{R}^n$ with discriminant $\pm 1$ and signature $(p, n-p)$, equipped with its natural topology. Let $\|\,\|$ and λ respectively denote the Hilbert norm and the Lebesgue measure on $\mathbf{R}^n$ and for any set E let $|E|$ denote the cardinality of E. For any $Q\in\mathscr{Q}(p)$ and $0\leqq a<b$ let $V_Q(a, b)=\{v\in\mathbf{R}^n\,|\,a\leqq Q(v)<b\}$.

4. COROLLARY. — *Let* $\mathscr{K}$ *be a compact subset of* $\mathscr{Q}(p)$. *Let* $\nu$ *be a continuous positive function on the sphere* $\{v\in\mathbf{R}^n \mid \|v\|=1\}$ *and let* $\Omega=\{v\in\mathbf{R}^n \mid \|v\|<\nu(v/\|v\|)\}$. *Then we have the following:*

(i) *for any* $a, b\in\mathbf{R}$, $0<a<b$, *and* $\theta>0$, *there exists a finite subset* $\mathscr{S}$ *of* $\mathscr{K}$ *such that each* $Q\in\mathscr{S}$ *is a multiple of a rational quadratic form and for any compact subset* $\mathscr{C}$ *of* $\mathscr{K}-\mathscr{S}$ *there exists* $r_0\geqq 0$ *such that for all* Q *in* $\mathscr{C}$ *and* $r\geqq r_0$,

$$|V_Q(a, b)\cap r\Omega\cap \mathbf{Z}^n|\geqq(1-\theta)\lambda(V_Q(a, b)\cap r\Omega).$$

(ii) *if* $n\geqq 5$, *for any* $\varepsilon>0$ *there exist* $c>0$ *and* $r_0\geqq 0$ *such that for all* $Q\in\mathscr{K}$ *and* $r\geqq r_0$, $|V_Q(0, \varepsilon)\cap r\Omega\cap \mathbf{Z}^n|\geqq c\lambda(V_Q(0, \varepsilon)\cap r\Omega)$.

It is easy to see that when $\Omega$ is the unit sphere the above lower estimates are asymptotically proportional to $r^{n-2}$. We mention here that similar estimates, for a single Q, amounting to the same exponent of growth in terms of $r$ but smaller constant, were obtained earlier by the first named author jointly with Shahar Mozes.

2. SKETCH OF THE PROOFS. — As in [4] and [9] the argument depends on an elementary property of polynomials, which we now formulate as follows:

2.1. LEMMA. — *Let* $n\in\mathbf{N}$ *and* $\varepsilon>0$ *be given. Let* $\delta=\varepsilon^n/n^n(n+1)$. *Let* $k\geqq 1$ *and* $\varphi_1, \ldots, \varphi_k$ *be polynomials of degree at most n such that* $\sup|\varphi_i(0)|\geqq 1$. *Then for any* $T\geqq 0$,

$$l(\{t\in[0, T]|, |\varphi_i(t)|<\delta, \forall i\})\leqq\varepsilon l(\{t\in[0, T]|, |\varphi_i(t)|<1, \forall i\}).$$

The lemma follows easily from the Lagrange interpolation formula applied over each connected component of the set on the right-hand side.

2.2. PROPOSITION. — *Let* V *be a finite-dimensional real vector space and let* A *be an algebraic subvariety of* V. *Then for any compact subset* C *of* A *and* $\varepsilon>0$ *there exists a compact subset* D *of* A *such that the following holds: for any neighbourhood* $\Omega$ *of* D *in* V *there exists a neighbourhood* $\Psi$ *of* C *in* V *such that for any unipotent one-parameter subgroup* $\{u_t\}$ *of* GL(V), *any* $v\in V-\Omega$ *and any* $T\geqq 0$,

$$l(\{t\in[0, T]\mid u_t v\in\Psi\})\leqq\varepsilon l(\{t\in[0, T]\mid u_t v\in\Omega\}).$$

If $\psi$ is a polynomial on V such that $A=\{v\in V\mid\psi(v)=0\}$, the Proposition can be deduced by applying Lemma 2.1 to $\|u_t v\|^2$ and $\psi(u_t v)$, which are polynomials in $t$ of bounded degree for $\{u_t\}$ as in the hypothesis and $v\in V$.

2.3. PROPOSITION. — *Let* G *be a Lie group and let* $\Gamma$ *be a discrete subgroup of* G. *Let* $\rho: G\to GL(V)$ *be a finite-dimensional representation over a real vector space* V. *Let* $p\in V$ *be such that* $\rho(\Gamma)p$ *is closed and let* $\eta: G\to V$ *be the orbit map defined by* $\eta(g)=\rho(g)p$ *for all* $g\in G$. *Let* A *be an algebraic subvariety of* V *and let* $X=\eta^{-1}(A)$. *Let* $Y=\{g\in X\mid\exists\gamma\in\Gamma$ *such that* $\rho(\gamma)p\neq\pm p$ *and* $g\gamma\in X\}$. *Let* C *be a compact subset of* A, K *be a compact subset of* $(G-Y\Gamma)/\Gamma$ *and* F *be a compact subset of* $G-X\Gamma$. *Let* $\varepsilon>0$ *be given. Then there exists a neighbourhood* $\Psi$ *of* C *in* V *such that for any one-parameter subgroup* $\{u_t\}$ *such that* $\{\rho(u_t)\}$ *is unipotent, any* $g\in F$ *and* $T\geqq 0$,

$$l(\{t\in[0, T]\mid\eta(u_t g\gamma)\in\Psi \textit{ for some } \gamma\in\Gamma \textit{ and } u_t g\Gamma\in K\})\leqq\varepsilon T.$$

The proof goes through several steps, of which we shall describe the main aspects. Let D be a compact subset of A such that the conclusion of Proposition 2.2 holds for C as above and $\varepsilon/4$ instead of $\varepsilon$ there. Using the fact that $\rho(\Gamma)p$ is closed we can find a relatively compact neighbourhood $\Omega$ of D in V, disjoint from $\eta(F\Gamma)$ and such that if $g$,

C. R. Acad. Sci. Paris, t. 314, Série I, p. 699-704, 1992 

$g' \in G$, $g\Gamma = g'\Gamma \in K$ and $\eta(g)$, $\eta(g') \in \bar{\Omega}$ then $\eta(g) = \pm\eta(g')$. Let $\Psi$ be a neighbourhood of C as in the conclusion of Proposition 2.2, for this $\Omega$.

Now let $\{u_t\}$ be a one-parameter subgroup as in the hypothesis. We set $T_0(g) = 2\varepsilon^{-1}\inf\{t \geq 0 \mid \text{for any } \gamma \in \Gamma\ \exists s \in [0, t] \text{ such that } \eta(u_s g\gamma) \notin \Omega\}$. Clearly $T_0(g) = 0$ for all $g \in F$. We shall uphold the assertion in the Proposition for all $g \in G$ and $T \geqq T_0(g)$; such an extension is of relevance to the variation that we alluded to earlier.

Now let $g \in G$ and $T \geqq T_0(g)$ be given. For each $q \in \eta(\Gamma) = \rho(\Gamma)p$ let

$$I(q) = \{t \in (0, T) \mid \rho(u_t g)q \in \Omega\}$$

and

$$J(q) = \{t \in (\varepsilon T/2, T) \mid \rho(u_t g)q \in \Psi \text{ and } u_t g\Gamma \in K\}.$$

To prove the proposition it is enough to show that $l(\bigcup_{q \in \eta(\Gamma)} J(q)) \leqq \varepsilon T/2$. For each $q \in \eta(\Gamma)$ we find a subset $I'(q)$ such that $J(q) \subset I'(q) \subset I(q)$, $l(J(q)) \leqq \varepsilon\, l(I'(q))/4$ for all $q \in \eta(\Gamma)$ and $I'(q_1)$, $I'(q_2)$ are disjoint unless $q_1 = \pm q_2$. The desired conclusion follows immediately from this. The set $I'(q)$ is formed of truncated parts of connected components of $I(q)$ by taking from any connected component $(r, s)$ with $s > T_0(g)$, the part $(r, s')$ where $s' = \sup\{t \in (r, s) \mid u_t g\Gamma \in K\}$. The inequality $l(J(q)) \leqq \varepsilon\, l(I'(q))/4$ then follows from Proposition 2.2 applied in each connected component $(r, s)$ of $I(q)$. The disjointness of $I'(q_1)$, $I'(q_2)$ when $q_1 \neq \pm q_2$ follows from the condition involved in the choice of $\Omega$. This completes the proof.

We now outline a proof of Theorem 1 in the case when G is the connected component of the identity in a real algebraic group and $\Gamma$ is an arithmetic lattice in G.

Let a compact subset F of $G/\Gamma$ and an $\varepsilon > 0$ be given. There exists a compact subset K of $G/\Gamma$ such that for any unipotent one-parameter subgroup $\{u_t\}$ of G, any $x \in F$ and $T \geqq 0$, $l(\{t \in [0, T] \mid u_t x \in K\}) > (1 - \varepsilon/2)T$ (*cf.* [2]). Now let H be any algebraic subgroup of G such that $H \cap \Gamma$ is a Zariski-dense lattice in H. We apply Proposition 2.3 with $V = \Lambda^h \mathscr{G}$ and $\rho = \Lambda^h \mathrm{Ad}: G \to GL(V)$, where $\mathscr{G}$ is the Lie algebra of G and $h$ is the dimension of H. We choose $p$ to be a non-zero vector in the line corresponding to the Lie subalgebra of H (rationality considerations show that $\rho(\Gamma)p$ is closed) and A to be the algebraic hull of $\rho(X(H, S))p$ in V. Together with an inductive hypothesis Proposition 2.3 implies that for any $\theta > 0$, $X(H, S)\Gamma/\Gamma$ has a neighbourhood $\Omega$ such that for any unipotent one-parameter subgroup $\{u_t\}$ of G, any $x \in F$ and $T \geqq 0$, $l(\{t \in [0, T] \mid u_t x \in \Omega \cap K\}) \leqq \theta T$. The choice of K and the countability of the subgroups H as above then yields the Theorem (*see* the proof of Theorem 6.1 of [9] for an idea of the argument).

We mention here that while dealing with a general lattice we prove that the corresponding $\Gamma$-orbit in the representation space as above is closed.

We now indicate how Corollary 4 (i) is deduced from Theorem 2 in the case $n = 3$, $p = 1$ and $\mathscr{K}$ a singleton set, say consisting of a form Q; a closer analysis along the same lines, using Theorem 3, yields the general result. Corollary 4 (ii) follows from a similar argument, together with Meyer's theorem. Let $G = SL(3, \mathbf{R})$ and $\Gamma = SL(3, \mathbf{Z})$. Let $Q_0$ be the quadratic form defined by $Q_0(v_1, v_2, v_3) = 2v_1 v_3 - v_2^2$. Let H be the special orthogonal group of $Q_0$, $\{u_t\}$ be the one-parameter subgroup of H consisting of all upper triangular unipotent matrices in H ($\{v_1(t)\}$ as in [4]), M be the compact subgroup of H consisting of all orthogonal matrices in H and let $m$ be the normalized Haar measure on M.

704 C. R. Acad. Sci. Paris, t. 314, Série I, p. 699-704, 1992

Now, let $\delta>0$ be a small number and let $\sigma=(1-\delta)$. Let $a\leqq\alpha<\beta\leqq b$ be such that $\alpha/\beta>\sigma$. Let $\Delta$ be a small neighbourhood of $\{s(e_1+e_3)/\sqrt{2}\mid s\in(\alpha,\beta)\}$ in $V_{Q_0}(\alpha,\beta)$, intersecting the trajectories of $\{u_t\}$ in segments of equal length and let $\chi$ be the characteristic function of $\Delta$. Let $\psi$ be the function on $G/\Gamma$ defined by $\psi(g\Gamma)=\sum_{x\in g\mathbf{Z}^n}\chi(x)$. Then $\psi$ is an integrable function on $G/\Gamma$ and its integral is $\lambda(\Delta)$. Clearly, for any $g\in G$, any Borel subset B of M and $T>0$.

$$T^{-1}\int_T^{2T}\int_B\sum_{x\in g\mathbf{Z}^n}\chi(u_t mx)\,dm\,dt=T^{-1}\int_T^{2T}\int_B\psi(u_t mg\Gamma)\,dm\,dt.$$

There exists a $g\in G$ such that $Q(v)=Q_0(gv)$, for all $v\in\mathbf{R}^3$. When Q is not a multiple of a rational form, Theorem 2 yields that for such a $g$ and all large T the right-hand side is at least $\sigma\lambda(\Delta)m(B)$. On the other hand for all large T the left hand side above is bounded by $\sigma^{-1}\lambda(\Delta)m(B)\lambda(E(T))^{-1}|E(T)\cap g\mathbf{Z}^3|$, where $E(T)=\{v\in\mathbf{R}^3\mid u_t mv\in\Delta$ for some $T\leqq t\leqq 2T$ and $m\in B\}$. This yields a lower estimate for $|E(T)\cap g\mathbf{Z}^3|$. The desired estimate for $|V_Q(a,b)\cap r\Omega\cap\mathbf{Z}^3|$, which is the same as $|V_{Q_0}(a,b)\cap r(g\Omega)\cap g\mathbf{Z}^3|$, is obtained by filling the sets $V_{Q_0}(a,b)\cap r(g\Omega)$ closely with mutually disjoint sets of the form $E(T)$ and using the above estimates.

The authors would like to thank the Sonderforschungsbereich-170, Göttingen, and the Mathematical Sciences Research Institute, Berkeley, for hospitality while this work was done. Thanks are also due to Nimish Shah and Shahar Mozes for useful comments.

Note remise le 15 janvier 1992, acceptée le 19 mars 1992.

References

[1] M. Burger and P. Sarnak, Ramanujan duals II, *Invent. Math.*, 106, 1991, pp. 1-11.

[2] S. G. Dani, On orbits of unipotent flows on homogeneous spaces II, *Ergod. Th. Dynam. Sys.*, 6, 1986, pp. 167-182.

[3] S. G. Dani and G. A. Margulis, Values of quadratic forms at primitive integral points, *Invent. Math.*, 98, 1989, pp. 405-424.

[4] S. G. Dani and G. A. Margulis, Orbit closures of generic unipotent flows on homogeneous space of SL(3, **R**), *Math. Ann.*, 286, 1990, pp. 101-128.

[5] G. A. Margulis, Discrete subgroups and ergodic theory, in: *Number Theory, Trace Formulas and Discrete Subgroups*, Academic Press, New York-London, 1989, pp. 377-398.

[6] G. A. Margulis, *Dynamical and ergodic properties of subgroup actions on homogeneous spaces with applications to number theory*, ICM-1990 Plenary address, Preprint.

[7] M. Ratner, On Raghunathan's measure conjecture, *Ann. Math.*, 134, 1991, pp. 545-607.

[8] M. Ratner, Raghunathan's topological conjecture and distributions of unipotent flows, *Duke Math. J.*, 63, 1991, pp. 235-280.

[9] N. A. Shah, Uniformly distributed orbits of certain flows on homogeneous spaces, *Math. Ann.*, 289, 1991, pp. 315-334.

S. G. D. : *Mathematical Sciences Research Institute*, 1000 *Centennial Drive, Berkeley, California* 94720, *U.S.A.*;

G. A. M. : *Department of Mathematics, Yale university*, Box 2155 *Yale station, New Haven, Connecticut* 06520, *U.S.A.*

ELECTRONIC RESEARCH ANNOUNCEMENTS
OF THE AMERICAN MATHEMATICAL SOCIETY
Volume 1, Issue 3, 1995

# ON A QUANTITATIVE VERSION OF THE OPPENHEIM CONJECTURE

ALEX ESKIN, GREGORY MARGULIS, AND SHAHAR MOZES

ABSTRACT. The Oppenheim conjecture, proved by Margulis in 1986, states that the set of values at integral points of an indefinite quadratic form in three or more variables is dense, provided the form is not proportional to a rational form. In this paper we study the distribution of values of such a form. We show that if the signature of the form is not $(2,1)$ or $(2,2)$, then the values are uniformly distributed on the real line, provided the form is not proportional to a rational form. In the cases where the signature is $(2,1)$ or $(2,2)$ we show that no such universal formula exists, and give asymptotic upper bounds which are in general best possible.

Let $Q$ be an indefinite nondegenerate quadratic form in $n$ variables. Let $\mathcal{L}_Q = Q(\mathbb{Z}^n)$ denote the set of values of $Q$ at integral points. The Oppenheim conjecture, proved by Margulis (cf. [Mar]) states that if $n \geq 3$, and $Q$ is not proportional to a form with rational coefficients, then $\mathcal{L}_Q$ is dense. The Oppenheim conjecture enjoyed attention and many studies since it was conjectured in 1929 mostly using analytic number theory methods. In this paper[1] we study some finer questions related to the distribution the values of $Q$ at integral points.

## 1.

Let $\nu$ be a continuous positive function on the sphere $\{v \in \mathbb{R}^n \mid \|v\| = 1\}$, and let $\Omega = \{v \in \mathbb{R}^n \mid \|v\| < \nu(v/\|v\|)\}$. We denote by $T\Omega$ the dilate of $\Omega$ by $T$. Define the following set:

$$V^Q_{(a,b)}(\mathbb{R}) = \{x \in \mathbb{R}^n \mid a < Q(x) < b\}.$$

We shall use $V_{(a,b)} = V^Q_{(a,b)}$ when there is no confusion about the form $Q$. Also let $V_{(a,b)}(\mathbb{Z}) = V^Q_{(a,b)}(\mathbb{Z}) = \{x \in \mathbb{Z}^n \mid a < Q(x) < b\}$. The set $T\Omega \cap \mathbb{Z}^n$ consists of $O(T^n)$ points, $Q(T\Omega \cap \mathbb{Z}^n)$ is contained in an interval of the form $[-\mu T^2, \mu T^2]$, where $\mu > 0$ is a constant depending on $Q$ and $\Omega$. Thus one might expect that for any interval $[a,b]$, as $T \to \infty$,

$$|V_{(a,b)}(\mathbb{Z}) \cap T\Omega| \sim c_{Q,\Omega}(b-a)T^{n-2}, \tag{1.1}$$

---

Received by the editors December 6, 1995.
1991 *Mathematics Subject Classification.* Primary 11J25, 22E40.
Research of the first author partially supported by an NSF postdoctoral fellowship and by BSF grant 94-00060/1.
Research of the second author partially supported by NSF grants DMS-9204270 and DMS-9424613.
Research of the third author partially supported by the Israel Science foundation and by BSF grant 94-00060/1.

[1]The full version is available electronically at `http://www.math.uchicago.edu/~eskin`

where $c_{Q,\Omega}$ is a constant depending on $Q$ and $\Omega$. This may be interpreted as "uniform distribution" of the sets $Q(\mathbb{Z}^n \cap T\Omega)$ in the real line. Our main result is that (1.1) holds if $Q$ is not proportional to a rational form, and has signature $(p,q)$ with $p \geq 3$, $q \geq 1$. We also determine the constant $c_{Q,\Omega}$.

If $Q$ is an indefinite quadratic form in $n$ variables, $\Omega$ is as above and $(a,b)$ is an interval, we show that there exists a constant $\lambda = \lambda_{Q,\Omega}$ so that as $T \to \infty$,

$$\text{Vol}(V_{(a,b)}(\mathbb{R}) \cap T\Omega) \sim \lambda_{Q,\Omega}(b-a)T^{n-2}. \tag{1.2}$$

Our main result is the following:

**Theorem 1.** *Let $Q$ be an indefinite quadratic form of signature $(p,q)$, with $p \geq 3$ and $q \geq 1$. Suppose $Q$ is not proportional to a rational form. Then for any interval $(a,b)$, as $T \to \infty$,*

$$|V_{(a,b)}(\mathbb{Z}) \cap T\Omega| \sim \lambda_{Q,\Omega}(b-a)T^{n-2},$$

*where $n = p+q$, and $\lambda_{Q,\Omega}$ is as in (1.2).*

Only the upper bound in this formula is new: the asymptotically exact lower bound was proved in [DM]. Also a lower bound with a smaller constant was obtained independently by M. Ratner, and by S. G. Dani jointly with S. Mozes (both unpublished).

If the signature of $Q$ is $(2,1)$ or $(2,2)$, then no universal formula like (1.1) holds. In fact, we have the following theorem:

**Theorem 2.** *Let $\Omega_0$ be the unit ball, and let $q = 1$ or $2$. Then for every $\epsilon > 0$ and every interval $(a,b)$ there exists a quadratic form $Q$ of signature $(2,q)$ not proportional to a rational form, and a constant $c > 0$ such that for an infinite sequence $T_j \to \infty$,*

$$|V_{(a,b)}(\mathbb{Z}) \cap T\Omega_0| > cT_j^q(\log T_j)^{1-\epsilon}.$$

The case $q = 1$, $b \leq 0$ of Theorem 2 was noticed by P. Sarnak and worked out in detail in [Bre]. The quadratic forms constructed are of the form $x_1^2 + x_2^2 - \alpha x_3^2$, or $x_1^2 + x_2^2 - \alpha(x_3^2 + x_4^2)$, where $\alpha$ is extremely well approximated by squares of rational numbers.

However in the $(2,1)$ and $(2,2)$ cases, we can still establish an upper bound of the form $cT^q \log T$. This upper bound is effective, and is uniform over compact sets in the set of quadratic forms. We also give an effective uniform upper bound for the case $p \geq 3$.

**Theorem 3.** *Let $\mathcal{O}(p,q)$ denote the space of quadratic forms of signature $(p,q)$ and discriminant $\pm 1$, let $n = p+q$, $(a,b)$ be an interval, and let $\mathcal{D}$ be a compact subset of $\mathcal{O}(p,q)$. Let $\nu$ be a continuous positive function on the unit sphere and let $\Omega = \{v \in \mathbb{R}^n \mid \|v\| < \nu(v/\|v\|)\}$. Then, if $p \geq 3$ there exists a constant $c$ depending only on $\mathcal{D}$, $(a,b)$ and $\Omega$ such that for any $Q \in \mathcal{D}$ and all $T > 1$,*

$$|V_{(a,b)}(\mathbb{Z}) \cap T\Omega| < cT^{n-2}.$$

*If $p = 2$ and $q = 1$ or $q = 2$, then there exists a constant $c > 0$ depending only on $\mathcal{D}$, $(a,b)$ and $\Omega$ such that for any $Q \in \mathcal{D}$ and all $T > 2$,*

$$|V_{(a,b)} \cap T\Omega \cap \mathbb{Z}^n| < cT^{n-2}\log T.$$

Also, for the $(2,1)$ and $(2,2)$ cases, we have the following "almost everywhere" result:

**Theorem 4.** *For almost all quadratic forms $Q$ of signature $(p,q) = (2,1)$ or $(2,2)$*

$$|V_{(a,b)}(\mathbb{Z}) \cap T\Omega| \sim \lambda_{Q,\Omega}(b-a)T^{n-2},$$

*where $n = p+q$, and $\lambda_{Q,\Omega}$ is as in* (1.2).

Theorem 4 may be proved using a recent general result of Nevo and Stein [NS]; we present a self-contained argument which suffices for the application in the full paper. In [Sar], P. Sarnak proved that an analogous asymptotic formula holds for almost all forms within a specific two-parameter family; this family arises in problems related to Quantum Chaos.

Finally, following [DM] we give a "uniform" version of Theorem 1:

**Theorem 5.** *Let $\mathcal{D}$ be a compact subset of $\mathcal{O}(p,q)$, with $p \geq 3$. Let $n = p+q$, and let $\Omega$ be as in Theorem 3. Then for every interval $[a,b]$ and every $\theta > 0$, there exists a finite subset $\mathcal{P}$ of $\mathcal{D}$ such that each $Q \in \mathcal{P}$ is a scalar multiple of a rational form and for any compact subset $\mathcal{F}$ of $\mathcal{D} - \mathcal{P}$ there exists $T_0$ such that for all $Q$ in $\mathcal{F}$ and $T \geq T_0$,*

$$(1-\theta)\lambda_{Q,\Omega}(b-a)T^{n-2} \leq |V_{(a,b)}(\mathbb{Z}) \cap T\Omega| \leq (1+\theta)\lambda_{Q,\Omega}(b-a)T^{n-2},$$

*where $\lambda_{Q,\Omega}$ is as in* (1.2).

As in Theorem 1 only the upper bound is new; the asymptotically exact lower bound, which holds even for $SO(2,1)$ and $SO(2,2)$, was proved in [DM].

**Remark 1.** If we consider $|V_{(a,b)}(\mathbb{R}) \cap T\Omega \cap \mathcal{P}(\mathbb{Z}^n)|$ instead of $|V_{(a,b)}(\mathbb{Z}) \cap T\Omega|$ (where $\mathcal{P}(\mathbb{Z}^n)$ denotes the set of primitive lattice points), then Theorem 1 and Theorem 5 hold provided one replaces $\lambda_{Q,\Omega}$ by $\lambda'_{Q,\Omega} = \lambda_{Q,\Omega}/\zeta(n)$, where $\zeta$ is the Riemann zeta function.

**Remark 2.** Theorem 1 and Theorem 5, as well as lower bounds in [DM] and even the proof of the Oppenheim conjecture in [Mar], are not effective. It seems to be a very difficult problem to give effective versions of these results.

**Passage to the space of lattices.** We now relate the counting problem of Theorem 1 to a certain integral expression involving the orthogonal group of the quadratic form and the space of lattices $SL(n,\mathbb{R})/SL(n,\mathbb{Z})$. Roughly this is done as follows. Let $f$ be a bounded function on $\mathbb{R}^n - \{0\}$ vanishing outside a compact subset. For $g \in SL(n,\mathbb{R})$ let

$$\tilde{f}(g) = \sum_{v \in \mathbb{Z}^n} f(gv). \tag{1.3}$$

The proof is based on the identity of the form

$$\int_K \tilde{f}(a_t k)\, dk = \sum_{v \in \mathbb{Z}^n} \int_K f(a_t k v)\, dk \tag{1.4}$$

obtained by integrating (1.3). In (1.4) $\{a_t\}$ is a certain diagonal subgroup of the orthogonal group of $Q$, and $K$ is a maximal compact subgroup of the orthogonal group of $Q$. Then for an appropriate function $f$, the right hand side is related to the number of lattice points $v \in [e^t/2, e^t]\partial\Omega$ with $a < Q(v) < b$. We then establish the asymptotics of the left-hand side using the ergodic theory of unipotent flows and some other techniques.

**Lattices.** Let $\Delta$ be a lattice in $\mathbb{R}^n$. We say that a subspace $L$ of $\mathbb{R}^n$ is $\Delta$-*rational* if $L \cap \Delta$ is a lattice in $L$. For any $\Delta$-rational subspace $L$, we denote by $d_\Delta(L)$ or simply by $d(L)$ the volume of $L/(L\cap\Delta)$. Let us note that $d(L)$ is equal to the norm of $e_1 \wedge \cdots \wedge e_\ell$ in the exterior power $\bigwedge^\ell(\mathbb{R}^n)$ where $\ell = \dim L$ and $(e_1, \cdots, e_\ell)$ is a basis over $\mathbb{Z}$ of $L \cap \Delta$. If $L = \{0\}$ we write $d(L) = 1$. A lattice is $\Delta$ *unimodular* if $d_\Delta(\mathbb{R}^n) = 1$. The space of unimodular lattices is isomorphic to $SL(n,\mathbb{R})/SL(n,\mathbb{Z})$.

Let us introduce the following notation:

$$\alpha_i(\Delta) = \sup\left\{\frac{1}{d(L)} \,\middle|\, L \text{ is a } \Delta\text{-rational subspace of dimension } i\right\}, \quad 0 \le i \le n,$$
$$\alpha(\Delta) = \max_{0\le i\le n} \alpha_i(\Delta).$$

The following lemma is known as the "Lipschitz Principle":

**Lemma** ([Sch, Lemma 2]). *Let $f$ be a bounded function on $\mathbb{R}^n$ vanishing outside a compact subset. Then there exists a positive constant $c = c(f)$ such that*

$$\tilde{f}(\Delta) < c\alpha(\Delta) \tag{1.5}$$

*for any lattice $\Delta$ in $\mathbb{R}^n$. Here $\tilde{f}$ is the function on the space of lattices defined in (1.3).*

**Quadratic Forms.** Let $n \ge 3$, and let $p \ge 2$. We denote $n - p$ by $q$, and assume $q > 0$. Let $\{e_1, e_2, \ldots, e_n\}$ be the standard basis of $\mathbb{R}^n$. Let $Q_0$ be the quadratic form defined by

$$Q_0\left(\sum_{i=1}^n v_i e_i\right) = 2v_1v_n + \sum_{i=2}^p v_i^2 - \sum_{i=p+1}^{n-1} v_i^2 \quad \text{for all } v_1, \ldots, v_n \in \mathbb{R}.$$

It is straightforward to verify that $Q_0$ has signature $(p,q)$. Let $G = SL(n,\mathbb{R})$, the group of $n\times n$ matrices of determinant 1. For each quadratic form $Q$ and $g \in G$, let $Q^g$ denote the quadratic form defined by $Q^g(v) = Q(gv)$ for all $v \in \mathbb{R}^n$. By the well known classification of quadratic forms over $\mathbb{R}$, for each $Q \in \mathcal{O}(p,q)$ there exists $g \in G$ such that $Q = Q_0^g$. For any quadratic form $Q$ let $SO(Q)$ denote the special orthogonal group corresponding to $Q$; namely $\{g \in G \mid Q^g = Q\}$. Let $H = SO(Q_0)$. Then the map $H\backslash G \to \mathcal{O}(p,q)$ given by $Hg \to Q_0^g$ is a homeomorphism.

For $t \in \mathbb{R}$, let $a_t$ be the linear map so that $a_te_1 = e^{-t}e_1$, $a_te_n = e^te_n$, and $a_te_i = e_i$, $2 \le i \le n-1$. Then the one-parameter group $\{a_t\}$ is contained in $H$. Let $\hat{K}$ be the subgroup of $G$ consisting of orthogonal matrices, and let $K = H \cap \hat{K}$. It is easy to check that $K$ is a maximal compact subgroup of $H$, and consists of all $h \in H$ leaving invariant the subspace spanned by $\{e_1 + e_n, e_2, \ldots, e_p\}$. We denote by $m$ the normalized Haar measure on $K$.

**The main theorems.** To prove Theorem 1 one may use the following theorem:

**Theorem 6.** *Suppose $p \ge 3$, $q \ge 1$. Let $\phi$ be a continuous function on $G/\Gamma \approx$ the space of lattices in $\mathbb{R}^n$ with determinant 1. Assume that for some $s$, $0 < s < 2$, and some $C > 0$,*

$$|\phi(\Delta)| < C\alpha(\Delta)^s, \quad \textit{for all } \Delta \in G/\Gamma.$$

*Let $x_0 \in G/\Gamma$ be a unimodular lattice such that $Hx_0$ is not closed. Let $\nu$ be any continuous function on $K$. Then*

$$\lim_{t \to +\infty} \int_K \phi(a_t k x_0)\nu(k)\, dm(k) = \int_K \nu\, dm \int_{G/\Gamma} \phi(y)\, d\mu(y).$$

To prove Theorem 5 we use the following generalization:

**Theorem 7.** *Suppose $p \geq 3$, $q \geq 1$. Let $\phi$, $\nu$ be as in Theorem 6, and let $\mathcal{C}$ be any compact set in $G/\Gamma$. Then for any $\epsilon > 0$ there exist finitely many points $x_1, \dots, x_\ell \in G/\Gamma$ such that*

(i) *The orbits $Hx_1, \dots, Hx_\ell$ are closed and have finite $H$-invariant measure.*
(ii) *For any compact subset $F$ of $\mathcal{C} - \bigcup_{1 \leq i \leq \ell} Hx_i$, there exists $t_0 > 0$, so that for all $x \in F$ and $t > t_0$,*

$$\left| \int_K \phi(a_t k x)\nu(k)\, dm(k) - \int_{G/\Gamma} \phi\, d\mu \int_K \nu\, dm \right| \leq \epsilon.$$

If the function $\phi$ is bounded, then Theorem 6 and Theorem 7 follow easily from [DM, Theorem 3]). This theorem is a refined version of Ratner's uniform distribution theorem [Rat4]; the proof uses Ratner's measure classification theorem (see [Rat1, Rat2, Rat3]), Dani's theorem on the behavior of unipotent orbits at infinity [Dan1, Dan2], and "linearization" techniques.

Both [DM, Theorem 3] and Ratner's uniform distribution theorem hold for bounded continuous functions, but not for arbitrary continuous functions from $L^1(G/\Gamma)$. However, for a non-negative bounded continuous function $f$ on $\mathbb{R}^n$, the function $\tilde{f}$ defined in (1.3) is non-negative, continuous, and $L^1$ but unbounded (it is in $L^s(G/\Gamma)$ for $1 \leq s < n$, where $G = SL(n, \mathbb{R})$, and $\Gamma = SL(n, \mathbb{Z})$). As it was done in [DM] it is possible to obtain asymptotically exact lower bounds by considering bounded continuous functions $\phi \leq \tilde{f}$. However, to carry out the integrals in (1.4) and prove the upper bounds in the theorems stated above we need to examine carefully the situation at the "cusp" of $G/\Gamma$, i.e. outside of compact sets.

By (1.5) the function $\tilde{f}(g)$ on the space of unimodular lattices $G/\Gamma$ is majorized by the function $\alpha(g)$. The function $\alpha$ is more convenient since it is invariant under the left action of the maximal compact subgroup $\hat{K}$ of $G$, and its growth rate at infinity is known explicitly. Theorems 1 and 5 are proved by combining [DM, Theorem 3] with the following integrability estimate:

**Theorem 8.** *If $p \geq 3$, $q \geq 1$ and $0 < s < 2$, or if $p = 2$, $q \geq 1$ and $0 < s < 1$, then for any lattice $\Delta$ in $\mathbb{R}^n$*

$$\sup_{t > 0} \int_K \alpha(a_t k \Delta)^s\, dm(k) < \infty.$$

*The upper bound is uniform as $\Delta$ varies over compact sets in the space of lattices.*

This result can be interpreted as follows. For a lattice $\Delta$ in $G/\Gamma$ and for $h \in H$, let $f(h) = \alpha(h\Delta)$. Since $\alpha$ is left-$\hat{K}$ invariant, $f$ is a function on the symmetric space $X = K \backslash H$. Theorem 8 is the statement that if $p \geq 3$, then the averages of $f^s$, $0 < s < 2$ over the sets $K a_t K$ in $X$ remain bounded as $t \to \infty$, and the bound is uniform as one varies the base point $\Delta$ over compact sets. We remark that in the case $q = 1$, the rank of $X$ is 1, and the sets $Ka_tK$ are metric spheres of radius $t$, centered at the origin.

If $(p, q) = (2, 1)$ or $(2, 2)$, Theorem 8 does not hold even for $s = 1$. The following result is, in general, best possible:

**Theorem 9.** *If $p = 2$ and $q = 2$, or if $p = 2$ and $q = 1$, then for any lattice $\Delta$ in $\mathbb{R}^n$,*

$$\sup_{t>1} \frac{1}{t} \int_K \alpha(a_t k \Delta)\, dm(k) < \infty.$$

*The upper bound is uniform as $\Delta$ varies over compact sets in the space of lattices.*

## 2.

We now outline the proof of Theorems 8 and 9. From its definition, the function $\alpha(g)$ is the maximum over $1 \leq i \leq n$ of left-$\hat{K}$ invariant functions $\alpha_i(g)$. The main idea of the proof is to show that the $\alpha_i$ satisfy a system of integral inequalities which imply the desired bound.

If $p \geq 3$ and $0 < s < 2$, or if $(p, q) = (2, 1)$ or $(2, 2)$ and $0 < s < 1$, we show that for any $c > 0$ there exist $t > 0$, and $\omega > 1$ so that the functions $\alpha_i^s$ satisfy the following system of integral inequalities in the space of lattices:

$$A_t \alpha_i^s \leq c_i \alpha_i^s + \omega^2 \max_{0 \leq j \leq n-i, i} \sqrt{\alpha_{i+j}^s \alpha_{i-j}^s}, \tag{2.1}$$

where $A_t$ is the averaging operator $(A_t f)(\Delta) = \int_K f(a_t k \Delta)$, and $c_i \leq c$. If $(p, q) = (2, 1)$ or $(2, 2)$ and $s = 1$, then (2.1) also holds (for suitably modified functions $\alpha_i$), but some of the constants $c_i$ cannot be made smaller than 1.

Let $f_i(h) = \alpha_i(h\Delta)$, so that each $f_i$ is a function on the symmetric space $X$. When one restricts to an orbit of $H$, (2.1) becomes:

$$A_t f_i^s \leq c_i f_i^s + \omega^2 \max_{0 \leq j \leq n-i, i} \sqrt{f_{i+j}^s f_{i-j}^s}. \tag{2.2}$$

If $\operatorname{rank} X = 1$, then $(A_t f)(h)$ can be interpreted as the average of $f$ over the sphere of radius $t$ in $X$, centered at $h$. We show that if the $f_i$ satisfy (2.2), then for any $\epsilon > 0$, the function $f = f_{\epsilon,s} = \sum_{0 \leq i \leq n} \epsilon^{i(n-i)} f_i^s$ satisfies the scalar inequality:

$$A_t f \leq cf + b, \tag{2.3}$$

where $t$, $c$ and $b$ are constants. We show that if $c$ is sufficiently small, then (2.3) for a fixed $t$ together with the uniform continuity of $\log f$ imply that $(A_r f)(1)$ is bounded as a function of $r$, which is the conclusion of Theorem 8. If $c = 1$, which will occur in the $SO(2, 1)$ and $SO(2, 2)$ cases, then (2.3) implies that $(A_r f)(1)$ is growing at most linearly with the radius, which is the conclusion of Theorem 9.

Throughout the proof we consider the functions $\alpha(g)^s$ for $0 < s < 2$ even though for the application to quadratic forms we only need $s = 1 + \delta$. This yields a better integrability result, and is also necessary for the proofs of Theorem 6 and Theorem 7.

**Acknowledgments:** The authors would like to thank Peter Sarnak for useful conversations.

## References

[Bre] T. Brennan, Princeton University undergraduate thesis, 1994.

[Dan1] S.G. Dani, On orbits of unipotent flows on homogeneous spaces, *Ergod. Theor. Dynam. Syst.* **4**(1984), 25–34. MR **86b:**58068

[Dan2] S.G. Dani, On orbits of unipotent flows on homogenous spaces II, *Ergod. Theor. Dynam. Syst* **6**(1986), 167–182. MR **88e:**58052

[DM] S.G. Dani and G.A. Margulis, Limit distributions of orbits of unipotent flows and values of quadratic forms, *Advances in Soviet Math.* **16**(1993), 91–137. MR **95b:**22024

[Mar] G. A. Margulis, Discrete Subgroups and Ergodic Theory, In *Number theory, trace formulas and discrete subgroups (a symposium in honor of A. Selberg)*, pages 377–398, Academic Press, Boston, MA, 1989. MR **90k:**22013a

[NS] A. Nevo and E. Stein, A generalization of Wiener's pointwise ergodic theorem. Preprint.

[Rat1] M. Ratner, Strict measure rigidity for nilpotent subgroups of solvable groups, *Invent. Math.* **101**(1990), 449–482. MR **92h:**22015

[Rat2] M. Ratner, On measure rigidity of unipotent subgroups of semisimple groups, *Acta. Math.* **165**(1990), 229–309. MR **91m:**57031

[Rat3] M. Ratner, On Raghunathan's measure conjecture, *Annals of Math.* **134**(1991), 545–607. MR **93a:**22009

[Rat4] M. Ratner, Raghunathan's topological conjecture and distributions of unipotent flows, *Duke Math. J.* **63**(1991), 235–290. MR **93f:**22012

[Sch] W. Schmidt, Asymptotic formulae for point lattices of bounded determinant and subspaces of bounded height, *Duke Math. J.* **35**(1968), 327–339. MR **37:**161

[Sar] P. Sarnak, Values at integers of binary quadratic forms. In preparation.

Department of Mathematics, University of Chicago, Chicago, IL 60637, USA
*E-mail address*: eskin@math.uchicago.edu

Department of Mathematics, Yale University, New Haven, CT, USA
*E-mail address*: margulis@math.yale.edu

Institute of Mathematics, Hebrew University, Jerusalem 91904, ISRAEL
*E-mail address*: mozes@math.huji.ac.il

Annals of Mathematics, **148** (1998), 339–360

# Flows on homogeneous spaces and Diophantine approximation on manifolds

By D. Y. Kleinbock and G. A. Margulis*


## Abstract

We present a new approach to metric Diophantine approximation on manifolds based on the correspondence between approximation properties of numbers and orbit properties of certain flows on homogeneous spaces. This approach yields a new proof of a conjecture of Mahler, originally settled by V. G. Sprindžuk in 1964. We also prove several related hypotheses of Baker and Sprindžuk formulated in 1970s. The core of the proof is a theorem which generalizes and sharpens earlier results on nondivergence of unipotent flows on the space of lattices.


## 1. Introduction

We start by recalling several basic facts from the theory of simultaneous Diophantine approximation. For $\mathbf{x}, \mathbf{y} \in \mathbb{R}^n$ we let

$$\mathbf{x} \cdot \mathbf{y} = \sum_{i=1}^{n} x_i y_i, \quad \|\mathbf{x}\| = \max_{1 \le i \le n} |x_i|,$$

$$\Pi(\mathbf{x}) = \prod_{i=1}^{n} |x_i| \quad \text{and} \quad \Pi_+(\mathbf{x}) = \prod_{i=1}^{n} |x_i|_+,$$

where $|x|_+$ stands for $\max(|x|, 1)$. One says that a vector $\mathbf{y} \in \mathbb{R}^n$ is *very well approximable* (cf. [S2]), to be abbreviated as VWA, if the following two equivalent conditions are satisfied:

(i) for some $\varepsilon > 0$ there are infinitely many $\mathbf{q} \in \mathbb{Z}^n$ such that

$$|\mathbf{q} \cdot \mathbf{y} + p| \cdot \|\mathbf{q}\|^n \le \|\mathbf{q}\|^{-n\varepsilon} \tag{1.1}$$

for some $p \in \mathbb{Z}$;

*1991 *Mathematics Subject Classification.* Primary 11J13, 11J83; Secondary 22E99, 57S25. The work of the first author was supported in part by NSF Grant DMS-9304580 and that of the second author by NSF Grant DMS-9424613.

(ii) for some $\varepsilon > 0$ there are infinitely many $q \in \mathbb{Z}$ such that

$$\|q\mathbf{y} + \mathbf{p}\|^n \cdot |q| \leq |q|^{-\varepsilon} \tag{1.2}$$

for some $\mathbf{p} \in \mathbb{Z}^n$.

The equivalence of (1.1) and (1.2) follows from Khintchine's Transference Principle; see e.g. [C, Chapter V]. It is well known (see [C, Chapter VII]) that almost every $\mathbf{y} \in \mathbb{R}^n$ is not VWA. A more difficult question arises if one considers almost all points $\mathbf{y}$ on a submanifold $M$ of $\mathbb{R}^n$ (in the sense of the natural measure class on $M$). In 1932, K. Mahler [Mah] conjectured that almost all points on the curve

$$M = \{(x, x^2, \ldots, x^n) \mid x \in \mathbb{R}\} \tag{1.3}$$

are not very well approximable. V. Sprindžuk's proof of this conjecture (see [Sp1], [Sp2]) has eventually led to the development of a new branch in approximation theory, usually referred to as "Diophantine approximation with dependent quantities" or "Diophantine approximation on manifolds." According to Sprindžuk's terminology, a submanifold $M \subset \mathbb{R}^n$ is called *extremal* if almost all $\mathbf{y} \in M$ are not VWA. Since 1964, extensive classes of extremal manifolds have been found by Sprindžuk and his followers; see [Sp3], [Sp4] for a review. However, the following conjecture made by Sprindžuk in 1980 has remained unproved:

CONJECTURE $\mathrm{H}_1$ ([Sp4]). *Let $f_1, \ldots, f_n$ be real analytic functions in $\mathbf{x} \in U$, $U$ a domain in $\mathbb{R}^d$, which together with 1 are linearly independent over $\mathbb{R}$. Then the manifold $M = \{\mathbf{f}(\mathbf{x}) \mid \mathbf{x} \in U\}$ is extremal.*

The case $n = 2$ was settled by Schmidt in 1964 [S1], and very recently V. Beresnevich and V. Bernik [BeBe] proved the above result for $n = 3$.

Note that the validity of the above conjecture for polynomials $f_1, \ldots, f_n \in \mathbb{Q}[x]$ with $\deg(f_i) \leq n$ follows from Sprindžuk's original proof of Mahler's Conjecture. However, there exists a strengthening of Conjecture $\mathrm{H}_1$ which has not been proved even for the curve (1.3). Namely, one says that $\mathbf{y} \in \mathbb{R}^n$ is *very well multiplicatively approximable* (to be abbreviated as VWMA) if the following two equivalent conditions are satisfied:

(i) for some $\varepsilon > 0$ there are infinitely many $\mathbf{q} \in \mathbb{Z}^n$ such that

$$|\mathbf{q} \cdot \mathbf{y} + p| \cdot \Pi_+(\mathbf{q}) \leq \Pi_+(\mathbf{q})^{-\varepsilon} \tag{1.1M}$$

for some $p \in \mathbb{Z}$;

(ii) for some $\varepsilon > 0$ there are infinitely many $q \in \mathbb{Z}$ such that

$$\Pi(q\mathbf{y} + \mathbf{p}) \cdot |q| \leq |q|^{-\varepsilon} \tag{1.2M}$$

for some $\mathbf{p} \in \mathbb{Z}^n$.

In other words, instead of taking the $n^{\text{th}}$ power of the maximum of the "differences" $|qy_i+p_i|$ in (1.2) one simply multiplies them, and instead of taking the $n^{\text{th}}$ power of the maximum of the numbers $|q_i|$ in (1.1) one multiplies those which are different from zero. The equivalence of (1.1M) and (1.2M) can be shown by modification of the argument needed to prove the standard form of Khintchine's Transference Principle; see a remark in [Sp3, p. 69]. A manifold $M \subset \mathbb{R}^n$ is said to be *strongly extremal* if almost all $\mathbf{y} \in M$ are not VWMA. Clearly (1.$i$) implies (1.$i$M), $i = 1, 2$; therefore any strongly extremal manifold is extremal. In his book [B], A. Baker raised the question of confirming the strong extremality of the curve (1.3). Later, in his review [Sp4], Sprindžuk also stated

CONJECTURE $\mathrm{H}_2$. *Any manifold $M \subset \mathbb{R}^n$ satisfying the assumptions of Conjecture* $\mathrm{H}_1$ *is strongly extremal.*

We remark that the validity of the above conjecture for polynomials with $n \leq 4$ is the recent result of V. Bernik and V. Borbat [BeBo]. See also [VC] and references therein for other special cases.

In this paper we present a proof of the above conjectures as well as some of their generalizations. More precisely, consider a $d$-dimensional submanifold $M = \{\mathbf{f}(\mathbf{x}) \mid \mathbf{x} \in U\}$ of $\mathbb{R}^n$, where $U$ is an open subset of $\mathbb{R}^d$ and $\mathbf{f} = (f_1, \dots, f_n)$ is a $C^m$ imbedding of $U$ into $\mathbb{R}^n$. For $l \leq m$, say that $\mathbf{y} = \mathbf{f}(\mathbf{x})$ is an *$l$-nondegenerate point* of $M$ if the space $\mathbb{R}^n$ is spanned by partial derivatives of $\mathbf{f}$ at $\mathbf{x}$ of order up to $l$. We will say that $\mathbf{y}$ is *nondegenerate* if it is $l$-nondegenerate for some $l$. One can view this condition as an infinitesimal version of not lying in any proper affine hyperplane, i.e. of the linear independence of $1, f_1, \dots, f_n$ over $\mathbb{R}$. Indeed, if the functions $f_i$ are analytic, it is easy to see that the linear independence of $1, f_1, \dots, f_n$ over $\mathbb{R}$ in a domain $U$ is equivalent to all points of $M = \mathbf{f}(U)$ being nondegenerate. Thus Conjecture $\mathrm{H}_2$ would follow from

THEOREM A. *Let $f_1, \dots, f_n \in C^m(U)$, $U$ an open subset of $\mathbb{R}^d$, be such that almost every point of $M = \{\mathbf{f}(\mathbf{x}) \mid \mathbf{x} \in U\}$ is nondegenerate. Then $M$ is strongly extremal.*

Note that if $n = 2$ and $d = 1$, the above assumption says that the curve $\{(f_1(x), f_2(x)) \mid x \in U\}$, $U \subset \mathbb{R}$, has nonzero curvature almost everywhere. This was exactly the assumption used by Schmidt [S1], and the strong extremality of such a curve is the subject of Theorem 14 from [Sp3, Ch. 2].

Our proof of Theorem A is based on the correspondence (cf. [D1], [K1], [K2]) between approximation properties of numbers $\mathbf{y} \in \mathbb{R}^n$ and behavior of certain orbits in the space of unimodular lattices in $\mathbb{R}^{n+1}$. More precisely, one acts on the lattice

$$\Lambda_{\mathbf{y}} = \begin{pmatrix} 1 & \mathbf{y}^T \\ 0 & I_n \end{pmatrix} \mathbb{Z}^{n+1} \tag{1.4}$$

by semisimple elements of the form

$$g_{\mathbf{t}} = \operatorname{diag}(e^t, e^{-t_1}, \ldots, e^{-t_n}),\ \mathbf{t} = (t_1, \ldots, t_n),\ t_i \geq 0,\ t = \sum_{i=1}^{n} t_i \tag{1.5}$$

(the latter notation will be used throughout the paper, so that whenever $t$ and $\mathbf{t}$ appear in the same formula, $t$ will stand for $\sum_{i=1}^n t_i$).

Define a function $\delta$ on the space of lattices by

$$\delta(\Lambda) \stackrel{\text{def}}{=} \inf_{\mathbf{v} \in \Lambda \setminus \{0\}} \|\mathbf{v}\| \tag{1.6}$$

(note that the ratio of $1+\log\big(1/\delta(\Lambda)\big)$ and $1+\operatorname{dist}(\Lambda, \mathbb{Z}^{n+1})$ is bounded between two positive constants for any right invariant Riemannian metric "dist" on the space of lattices). We prove in Section 2 that for any very well multiplicatively approximable $\mathbf{y} \in \mathbb{R}^n$ there exists $\gamma > 0$ and infinitely many $\mathbf{t} \in \mathbb{Z}_+^n$ such that

$$\delta(g_{\mathbf{t}} \Lambda_{\mathbf{y}}) \leq e^{-\gamma t}; \tag{1.7}$$

in other words, if $\operatorname{dist}(g_{\mathbf{t}}\Lambda_{\mathbf{y}}, \mathbb{Z}^{n+1})$ grows sublinearly as a function of $\mathbf{t}$, then $\mathbf{y}$ is not VWMA. Thus to prove the strong extremality of $M$ as in Theorem A it is enough to show that for any nondegenerate point $\mathbf{y}_0 = \mathbf{f}(\mathbf{x}_0)$ there is a neighborhood $B$ of $\mathbf{x}_0$ in $U$ such that for almost all $\mathbf{y} \in \mathbf{f}(B)$ and any $\gamma > 0$, there are at most finitely many $\mathbf{t} \in \mathbb{Z}_+^n$ such that (1.7) holds.

In view of Borel-Cantelli, the latter can be done by estimating the measure of the sets

$$E_{\mathbf{t}} \stackrel{\text{def}}{=} \{\mathbf{x} \in B \mid \delta(g_{\mathbf{t}} \Lambda_{\mathbf{f}(\mathbf{x})}) \leq e^{-\gamma t}\} \tag{1.8}$$

for any fixed $\mathbf{t} \in \mathbb{Z}_+^n$, so that

$$\sum_{\mathbf{t} \in \mathbb{Z}_+^n} |E_{\mathbf{t}}| < \infty \tag{1.9}$$

(here and hereafter $|\cdot|$ stands for the Lebesgue measure). Such estimates are obtained in Section 5 by modifying proofs of earlier results on nondivergence of unipotent flows in the space of lattices. According to the theorem of Dani [D2] strengthening an earlier result of Margulis [Mar], for any $c > 0$ and any lattice $\Lambda$ in $\mathbb{R}^k$ there exists $\varepsilon > 0$ such that for any unipotent subgroup $\{u_x | x \in \mathbb{R}\}$ of $\mathrm{SL}_k(\mathbb{R})$ one has

$$\big|\{x \in [0, T] \mid \delta(u_x \Lambda) < \varepsilon\}\big| \leq cT\,. \tag{1.10}$$

Similar estimates are known for any polynomial map from $\mathbb{R}^d$ to $\mathrm{GL}_k(\mathbb{R})$ instead of $x \to u_x$, cf. [Sh] (or [EMS] for a bigger class of maps). In Section 3 (see

Proposition 3.4 for a precise statement) we show that if $\mathbf{y}_0 = \mathbf{f}(\mathbf{x}_0)$ is nondegenerate, $\mathbf{x}_0$ has a neighborhood on which linear combinations of $1, f_1, \ldots, f_n$ behave like polynomials of uniformly bounded degree. Then in Sections 4 and 5 we modify the argument of Margulis and Dani in order to get a quantitative relation between $c$ and $\varepsilon$ in the analogue of (1.10) (see Proposition 2.3) which will guarantee convergence in (1.9). The last section of the paper deals with several possible extensions of the main result, as well as some open questions.

REFERENCES

[B] A. BAKER, *Transcendental Number Theory*, Cambridge Univ. Press, Cambridge, 1975.

[Be] V. BERNIK, A proof of Baker's conjecture in the metric theory of transcendental numbers, Dokl. Akad. Nauk SSSR **277** (1984), 1036–1039 (Russian).

[BeBe] V. BERESNEVICH and V. BERNIK, On a metrical theorem of W. Schmidt, Acta Arith. **75** (1996), 219–233.

[BeBo] V. BERNIK and V. BORBAT, Polynomials with differences in values of coefficients and a conjecture of A. Baker, Vestsī Akad. Navuk Belarusī Ser. Fīz.-Mat. Navuk (1997), no. 3, 5–8 (Russian).

[C] J. W. S. CASSELS, *An Introduction to Diophantine Approximation*, Cambridge Tracts in Math. and Math. Physics No. 45, Cambridge Univ. Press, Cambridge, 1957.

[D1] S. G. DANI, Divergent trajectories of flows on homogeneous spaces and Diophantine approximation, J. Reine Angew. Math. **359** (1985), 55–89.

[D2] ———, On orbits of unipotent flows on homogeneous spaces, II, Ergodic Theory Dynamical Systems **6** (1986), 167–182.

[DM] S. G. DANI and G. A. MARGULIS, Limit distributions of orbits of unipotent flows and values of quadratic forms, Adv. Soviet Math. vol. 16, Amer. Math. Soc., Providence, RI, 1993, 91–137.

[DRV1] M. M. DODSON, B. P. RYNNE and J. A. G. VICKERS, Metric Diophantine approximation and Hausdorff dimension on manifolds, Math. Proc. Cambridge Philos. Soc. **105** (1989), 547–558.

[DRV2] ———, Khintchine-type theorems on manifolds, Acta Arith. **57** (1991), 115–130.

[DRV3] ———, Simultaneous Diophantine approximation and asymptotic formulae on manifolds, J. Number Theory **58** (1996), 298–316.

[EMS] A. ESKIN, S. MOZES and N. SHAH, Non-divergence of translates of certain algebraic measures, Geom. Funct. Anal. **7** (1997), 48–80.

[K1] D. KLEINBOCK, Flows on homogeneous spaces and Diophantine properties of matrices, Duke Math. J. (to appear).

[K2] ———, Nondense orbits of nonquasiunipotent flows and applications to Diophantine approximation, Ph.D. Thesis, Yale University, 1996.

[Ko1] E. KOVALEVSKAYA, Simultaneously extremal manifolds, Dokl. Akad. Nauk BSSR **31** (1987), 405–408 (Russian).

[Ko2] ———, Strongly jointly extremal manifolds, Vestsī Akad. Navuk BSSR Ser. Fīz.-Mat. Navuk (1987), no. 6, 16–19 (Russian).

[Mah] K. MAHLER, Über das Mass der Menge aller *S*-Zahlen, Math. Ann. **106** (1932), 131–139.

[Mar] G. A. MARGULIS, On the action of unipotent groups in the space of lattices, Proc. Summer School on group representations, (Budapest 1971), Académiai Kiado, Budapest, 1975, pp. 365–370.

[Mat] P. MATTILA, *Geometry of Sets and Measures in Euclidean Spaces. Fractals and Rectifiability*, Cambridge Studies in Advanced Mathematics, 44, Cambridge University Press, Cambridge, 1995.

[R] M. S. RAGHUNATHAN, *Discrete Subgroups of Lie Groups*, Springer-Verlag, New York, 1972.

[S1] W. SCHMIDT, Metrische Sätze über simultane Approximation abhänginger Grössen, Monatsh. Math. **68** (1964), 154–166.

[S2] ———, Diophantine approximation, LNM 785, Springer-Verlag, Berlin and New York. 1980.

[Sh] N. SHAH, Limit distributions of polynomial trajectories on homogeneous spaces, Duke Math. J. **75** (1994), 711–732.

[Sp1] V. SPRINDŽUK, More on Mahler's conjecture, Doklady Akad. Nauk SSSR **155** (1964), 54–56 (Russian); English transl. in Soviet Math. Dokl **5** (1964), 361–363.

[Sp2] ———, Mahler's problem in metric number theory, Translations of Mathematical Monographs, vol. 25, Amer. Math. Soc., Providence, RI, 1969.

[Sp3] ———, *Metric Theory of Diophantine Approximations*, John Wiley & Sons, New York, 1979.

[Sp4] ———, Achievements and problems in Diophantine approximation theory, Russian Math. Surveys **35** (1980), 1–80.

[VC] A. I. VINOGRADOV and G. V. CHUDNOVSKY, The proof of extremality of certain manifolds, Contributions to the theory of transcendental numbers, Math. Surveys Monographs, 19, Amer. Math. Soc., Providence, R.I.,1984, 421–447.

(Received January 23, 1997)

Annals of Mathematics, **170** (2009), 67–122

# Local rigidity of affine actions of higher rank groups and lattices

By DAVID FISHER and GREGORY MARGULIS


## Abstract

Let $J$ be a semisimple Lie group with all simple factors of real rank at least two. Let $\Gamma < J$ be a lattice. We prove a very general local rigidity result about actions of $J$ or $\Gamma$. This shows that almost all so-called "standard actions" are locally rigid. As a special case, we see that any action of $\Gamma$ by toral automorphisms is locally rigid. More generally, given a manifold $M$ on which $\Gamma$ acts isometrically and a torus $\mathbb{T}^n$ on which it acts by automorphisms, we show that the diagonal action on $\mathbb{T}^n \times M$ is locally rigid.

This paper is the culmination of a series of papers and depends heavily on our work in two recent articles. The reader willing to accept the main results of those papers as "black boxes" should be able to read the present paper without referring to them.


## 1. Introduction

Throughout this paper $J$ is a (connected) semisimple Lie group with no compact factors and all simple factors of real rank at least two, and $\Gamma < J$ is a lattice. The purpose of this paper is to prove the following:

THEOREM 1.1. *Let $\rho$ be a quasi-affine action of $J$ or $\Gamma$ on a compact manifold $X$. Then the action is $C^{\infty,\infty}$ and $C^{3,0}$ locally rigid. Furthermore, there exists an integer $k_0$, such that the action is $C^{k,k-n}$ locally rigid for all $k > k_0$, where $n = \frac{1}{2}\dim X + 3$.*

First author partially supported by NSF grant DMS-0226121 and DMS-0643546. Second author partially supported by NSF grants DMS-0244406 and DMS-0801195. The authors would also like to thank the FIM at ETHZ for hospitality and support.

*Remarks on regularity.* The number $k_0 = \max(k_1, n)$ where $k_1$ is determined by properties of certain foliations associated to the dynamics of $\rho(g_i)$, for a specific finite set of choices of $g_1, \ldots, g_k$ in $J$ or $\Gamma$. If $k$ is even, we can let $n = \frac{1}{2}\dim X + 2$ instead.[1]

We now proceed to define the terms in the theorem. We say $H$ is a *connected real algebraic group* if it is the connected component of the real points $H(\mathbb{R})^0$ of an algebraic group defined over $\mathbb{R}$.

*Definition* 1.2. a) Let $H$ be a connected real algebraic group, $\Lambda < H$ a cocompact lattice. Assume a topological group $G$ acts continuously on $H/\Lambda$. We say that the $G$ action on $H/\Lambda$ is *affine* if every element of $G$ acts via an affine diffeomorphism.

b) More generally, let $M$ be a compact manifold. Assume a group $G$ acts affinely on $H/\Lambda$. Choose a Riemannian metric on $M$ and a cocycle over the $G$ action $\iota : G \times H/\Lambda \to \mathrm{Isom}(M)$. We call the *skew product action* of $G$ on $H/\Lambda \times M$ defined by $d \cdot (x, m) = (d \cdot x, \iota(d, x) \cdot m)$ a *quasi-affine action.*

We always write $X = H/\Lambda \times M$. Recall that an *affine diffeomorphism* $d$ of $H/\Lambda$ is one covered by a diffeomorphism $\tilde{d}$ of $H$ where $\tilde{d} = A \circ T_h$ where $A$ is an automorphism of $H$ such that $A(\Lambda) = \Lambda$ and $T_h$ is left translation by $h \in H$. The full group of affine diffeomorphisms of $H/\Lambda$ is a finite dimensional Lie group which we write as $\mathrm{Aff}(H/\Lambda)$. The definition of acting affinely given above is equivalent to saying the action is given by a homomorphism $\pi : G \to \mathrm{Aff}(H/\Lambda)$. See [FM03, §6.1] for a description of $\mathrm{Aff}(H/\Lambda)$ and a classification of affine actions of $J$ or $\Gamma$ as above. Note also that the case of quasi-affine actions as defined here includes products of affine actions with trivial actions. Another class of examples give the following:

COROLLARY 1.3. *Let $J$ be as above and $\Gamma < J$ a lattice. Then any action of $\Gamma$ by automorphisms of $\mathbb{T}^m$ is $C^{\infty,\infty}$ and $C^{3,0}$ locally rigid. Furthermore there exists a positive integer $k_0 \geq 3$, depending on the action, such that the action is $C^{k,k-\frac{1}{2}m-3}$ locally rigid for all $k \geq \min(\frac{1}{2}m + 3, k_0)$.*

We now formally define local rigidity in this context.

*Definition* 1.4. Given a topological group $G$ and a continuous $C^\infty$ action, $\rho : G \times X \to X$, by diffeomorphisms on a manifold $X$, we say that the *action is $C^{k,r}$ locally rigid*, where $r \leq k$, if any continuous action $\rho'$ by $C^k$ diffeomorphisms, that is sufficiently $C^k$ close to $\rho$ is *conjugate* to $\rho$ by a small $C^r$ diffeomorphism. We say

[1] While this paper was under review, we learned that results in [RT05] combined with certain results from [FM05] allow us to achieve remarkably less loss of regularity, yielding $C^{k,k-3}$ local rigidity in place of $C^{k,k-n}$ local rigidity. We explain this briefly at the end of the paper.

that a continuous action $\rho$ is $C^{\infty,\infty}$ *locally rigid* if any continuous $C^{\infty}$ action which is sufficiently $C^{\infty}$ close to $\rho$ is conjugate to $\rho$ by a small $C^{\infty}$ diffeomorphism.

The special case of $C^{k,k}$ local rigidity says exactly that the homomorphism $\rho : G \to \mathrm{Diff}^k(X)$ is locally rigid. In other words that any homomorphism close to $\rho$ is conjugate to $\rho$ by a small element of $\mathrm{Diff}^k(X)$. Since the $C^{\infty}$ topology is defined as the inverse limit of the $C^k$ topologies, two $C^{\infty}$ diffeomorphisms are $C^{\infty}$ close if they are $C^k$ close for some large $k$. Our proof shows explicitly that a $C^{\infty}$ perturbation $\rho'$ of $\rho$ which is $C^k$ close to $\rho$ is conjugate to $\rho$ by a $C^{\infty}$ diffeomorphism which is $C^{k-n}$ close to the identity where $n$ is as in Theorem 1.1. The topology we take on $\mathrm{Hom}(G, \mathrm{Diff}^k(X))$ to define *close* above is the compact-open topology.

Gromov in [Gro88] and Zimmer in [Zim87] suggested that one might be able to "essentially classify" all volume-preserving smooth actions of higher rank semisimple groups and their lattices on compact manifolds. This would be, in a sense, a "nonlinear" analogue of the second author's superrigidity theorems, since one of the consequences of the superrigidity theorems is a classification of all finite dimensional linear representations of higher rank lattices (modulo issues concerning finite image representations). In [Zim87], Zimmer also proposed the study of local rigidity of known actions of higher rank lattices on compact manifolds, as a "nonlinear" analogue of the classical local rigidity theorems of Calabi-Vesentini, Selberg and Weil. These show that any cocompact lattice $\Gamma$ in any simple Lie group $J$ is locally rigid, as long as $J$ is not locally isomorphic to $\mathrm{SL}_2(\mathbb{R})$ [CV60], [Sel], [Wei62]. That is, any embedding of $\Gamma$ in $J$ close to the defining one $i : \Gamma \to J$ is simply a conjugate of $i$ by a small element of $J$. Since $J$ acts transitively on $J/\Gamma$, our theorem can be taken to be a generalization of Weil's result in the case when $J$ is a higher rank simple group. A perturbation $\Gamma'$ of $\Gamma$ in $J$ defines a perturbation of the original $J$ action on $J/\Gamma$ since $J/\Gamma$ and $J/\Gamma'$ are diffeomorphic. The conjugacy between these actions can easily be seen to give a conjugacy between $\Gamma$ and $\Gamma'$.

Many results have been proven concerning local rigidity of affine actions of higher rank lattices and Lie groups, particularly when the action is assumed to satisfy some strong hyperbolicity condition. The first results of this kind are due to Hurder [Hur92]. He proved that the standard action on $\mathbb{T}^n$ of any finite index subgroup in $\mathrm{SL}_n(\mathbb{Z})$ is deformation rigid for $n \geq 3$. (This involves assuming a path of nearby actions and obtaining a path of conjugacies.) The same actions were shown to be locally rigid in [KL91] and [KLZ96]. Many other results along these lines were obtained by many authors; we refer to the introduction of [MQ01] for a more detailed discussion. Here we mention that all standard Anosov actions on tori and nilmanifolds were proven to be locally rigid in [KS97] and all so-called weakly hyperbolic actions were proven to be locally rigid in [MQ01].

For isometric actions, there are also results. In [Ben00], Benveniste shows that any isometric action of any cocompact lattice in a group $J$ as above is $C^{\infty,\infty}$ locally rigid. The interested reader should refer to the introduction to [FM05] for a discussion of earlier, weaker results by Zimmer concerning (certain) perturbations of isometric actions of groups with property (T). In our previous paper [FM05], we have proven:

THEOREM 1.5. *Let $G$ be a locally compact, compactly generated group with property (T). Let $(X, g)$ be a compact Riemannian manifold, and let $\rho$ be an action of $G$ on $X$ by isometries. Then the action $C^{\infty,\infty}$ is locally rigid and is $C^{k,k-\kappa}$ locally rigid for any $k \geq 2$ and any $\kappa > 0$.*

We remark that Theorem 1.5 holds for a much broader class of groups than Theorem 1.1.

The proof of Theorem 1.1 uses a foliated generalization of Theorem 1.5 also proven in [FM05]. This result is recalled below in Section 4 where it is applied in the course of our proof.

For actions which are neither weakly hyperbolic nor isometric for all previous results, due to Nitica and Torok, concern affine actions which are products of Anosov actions and trivial actions [NT95], [NT01], [Tör03]. For example, take the standard action of $\mathrm{SL}_n(\mathbb{Z})$ on $\mathbb{T}^n$ and let $\rho$ denote the action obtained on $\mathbb{T}^n \times S^1$ by taking $\gamma(t, s) = (\gamma t, s)$. Then Nitica and Torok show that, given $k > 0$, any $C^\infty$ action $\rho'$ that is sufficiently $C^2$ close to $\rho$ is conjugate to $\rho$ by a $C^0$ small, $C^k$ diffeomorphism. (This result does not imply $C^\infty$ local rigidity because the size of the perturbation must be made smaller to obtain more derivatives in the conjugacy.) Their full result is more general, allowing one to replace the standard action of $\mathrm{SL}_n(\mathbb{Z})$ on $\mathbb{T}^n$ by any so-called TNS action of a higher rank lattice on a torus. They also prove some more general results for deformation rigidity, but always for products of TNS and trivial actions.

We briefly note two of the more major differences between the cases considered by Nitica and Torok and the general case considered here. First, in the general case, the *central foliation* for the group action is not necessarily by compact leaves. Secondly, in the general case, the action along the central foliation is isometric but not necessarily trivial. (For a definition of the central foliation, see § 2.1 below.)

We note here that nonlocally rigid volume-preserving actions of higher rank semisimple groups and their lattices on compact manifolds have been constructed, first in [KL96] and later and more generally in [Ben96]. Those in [Ben96] are even shown to have smooth volume-preserving deformations. See also [Fis08] for a more general construction and another proof that the deformations are nontrivial. A weaker result is shown in [KL96].

*Acknowledgements.* The authors would like to thank Amie Wilkinson for many helpful conversations concerning the work of Hirsch, Pugh and Shub in [HPS77]. They would also like to thank Ben Schmidt for bringing the work of Rauch and Taylor in [RT05] to their attention.

## References

[Aus63] L. AUSLANDER, On radicals of discrete subgroups of Lie groups, *Amer. J. Math.* **85** (1963), 145–150. MR 27 #2583 Zbl 0217.37002

[Ben96] E. J. BENVENISTE, Exotic actions of semisimple groups and their deformations, unpublished manuscript and University of Chicago Ph.D. dissertation, 1996.

[Ben00] ———, Rigidity of isometric lattice actions on compact Riemannian manifolds, *Geom. Funct. Anal.* **10** (2000), 516–542. MR 2001j:37055 Zbl 0970.22012

[BM81] J. BREZIN and C. C. MOORE, Flows on homogeneous spaces: a new look, *Amer. J. Math.* **103** (1981), 571–613. MR 83e:22009 Zbl 0506.22008

[CV60] E. CALABI and E. VESENTINI, On compact, locally symmetric Kähler manifolds, *Ann. of Math.* **71** (1960), 472–507. MR 22 #1922b

[CC00] A. CANDEL and L. CONLON, *Foliations*, vol. I, *Grad. Studies Math.* **23**, A.M.S., Providence, RI, 2000.

[Fis02] D. FISHER, On the arithmetic structure of lattice actions on compact spaces, *Ergodic Theory Dynam. Systems* **22** (2002), 1141–1168. MR 2004j:37004 Zbl 1026.37001

[Fis07] ———, Local rigidity of group actions: past, present, future, in *Dynamics, Ergodic Theory, and Geometry*, *Math. Sci. Res. Inst. Publ.* **54**, Cambridge Univ. Press, Cambridge, 2007, pp. 45–97. MR 2369442 Zbl 1144.22008

[Fis08] ———, Deformations of group actions, *Trans. Amer. Math. Soc.* **360** (2008), 491–505. MR 2008m:37046 Zbl 1126.37012

[FM03] D. FISHER and G. A. MARGULIS, Local rigidity for cocycles, in *Surveys in Differential Geometry* (Boston, 2002), vol. 8, Internat. Press, Somerville, MA, 2003, pp. 191–234. MR 2004m:22032 Zbl 1062.22044

[FM05] ———, Almost isometric actions, property (T), and local rigidity, *Invent. Math.* **162** (2005), 19–80. MR 2006m:53063 Zbl 1076.22008

[Gro88] M. GROMOV, Rigid transformations groups, in *Géométrie Différentielle* (Paris, 1986), *Travaux en Cours* **33**, Hermann, Paris, 1988, pp. 65–139. MR 90d:58173 Zbl 0652.53023

[Guy02] M. GUYSINSKY, The theory of non-stationary normal forms, *Ergodic Theory Dynam. Systems* **22** (2002), 845–862. MR 2003e:37028 Zbl 1010.37026

[GK98] M. GUYSINSKY and A. KATOK, Normal forms and invariant geometric structures for dynamical systems with invariant contracting foliations, *Math. Res. Lett.* **5** (1998), 149–163. MR 99b:58215 Zbl 0988.37063

[HPS77] M. W. HIRSCH, C. C. PUGH, and M. SHUB, *Invariant Manifolds*, *Lecture Notes in Math.* **583**, Springer-Verlag, New York, 1977. MR 58 #18595 Zbl 0355.58009

[Hur92] S. HURDER, Rigidity for Anosov actions of higher rank lattices, *Ann. of Math.* **135** (1992), 361–410. MR 93c:22018 Zbl 0754.58029

[KL91] A. KATOK and J. LEWIS, Local rigidity for certain groups of toral automorphisms, *Israel J. Math.* **75** (1991), 203–241. MR 93g:58076 Zbl 0785.22012

[KL96] ———, Global rigidity results for lattice actions on tori and new examples of volume-preserving actions, *Israel J. Math.* **93** (1996), 253–280. MR 96k:22021 Zbl 0857.57038

[KLZ96] A. KATOK, J. LEWIS, and R. ZIMMER, Cocycle superrigidity and rigidity for lattice actions on tori, *Topology* **35** (1996), 27–38. MR 97e:22009 Zbl 0857.57037

[KS94] A. KATOK and R. J. SPATZIER, Subelliptic estimates of polynomial differential operators and applications to rigidity of abelian actions, *Math. Res. Lett.* **1** (1994), 193–202. MR 95b:35042 Zbl 0836.57026

[KS97] ———, Differential rigidity of Anosov actions of higher rank abelian groups and algebraic lattice actions, *Tr. Mat. Inst. Steklova* **216** (1997), 292–319. MR 99i:58118 Zbl 0938.37010

[Mar77] G. A. MARGULIS, Discrete groups of motions of manifolds of non-positive curvature, in *Proc. Internat. Congress Math., Vancouver* 1974, *AMS* Translations, vol. 109, 1977, pp. 33–45.

[Mar91] ———, *Discrete Subgroups of Semisimple Lie Groups*, Springer-Verlag, New York, 1991. MR 92h:22021 Zbl 0732.22008

[MQ01] G. A. MARGULIS and N. QIAN, Rigidity of weakly hyperbolic actions of higher real rank semisimple Lie groups and their lattices, *Ergodic Theory Dynam. Systems* **21** (2001), 121–164. MR 2003a:22019 Zbl 0976.22009

[MZ55] D. MONTGOMERY and L. ZIPPIN, *Topological Transformation Groups*, Interscience Publishers, New York, 1955. MR 17,383b Zbl 0068.01904

[MS88] C. C. MOORE and C. SCHOCHET, *Global Analysis on Foliated Spaces*, *Math. Sci. Res. Inst. Publ.* **9**, Springer-Verlag, New York, 1988. MR 89h:58184 Zbl 0648.58034

[NT95] V. NIŢICĂ and A. TÖRÖK, Cohomology of dynamical systems and rigidity of partially hyperbolic actions of higher-rank lattices, *Duke Math. J.* **79** (1995), 751–810. MR 96k:58168 Zbl 0849.58049

[NT01] ———, Local rigidity of certain partially hyperbolic actions of product type, *Ergodic Theory Dynam. Systems* **21** (2001), 1213–1237. MR 2002k:37041 Zbl 0987.37015

[PR01] G. PRASAD and A. S. RAPINCHUK, Irreducible tori in semisimple groups, *Internat. Math. Res. Notices* **23** (2001), 1229–1242. MR 2003f:20083a Zbl 1057.22025

[RT05] J. RAUCH and M. TAYLOR, Regularity of functions smooth along foliations, and elliptic regularity, *J. Funct. Anal.* **225** (2005), 74–93. MR 2006e:35065 Zbl 1073.35077

[Sel] A. SELBERG, On discontinuous groups in higher-dimensional symmetric spaces, in *Contributions to Function Theory*, *Internat. Colloq. Function Theory* (Bombay, 1960), Tata Institute of Fundamental Research, pp. 147–164. MR 24 #A188 Zbl 0201.36603

[Tör03] A. TÖRÖK, Rigidity of partially hyperbolic actions of property (T) groups, *Discrete Contin. Dyn. Syst.* **9** (2003), 193–208. MR 2003i:37019 Zbl 1017.37015

[Wei62] A. WEIL, On discrete subgroups of Lie groups. II, *Ann. of Math.* **75** (1962), 578–602. MR 25 #1242 Zbl 0131.26602

[Zim76] R. J. ZIMMER, Extensions of ergodic group actions, *Illinois J. Math.* **20** (1976), 373–409. MR 53 #13522 Zbl 0334.28015

[Zim87] ______, Actions of semisimple groups and discrete subgroups, in *Proc. Internat. Congress of Mathematicians* (Berkeley, Calif., 1986), vol. 2, Amer. Math. Soc., Providence, RI, 1987, pp. 1247–1258. MR 89j:22024 Zbl 0671.57028

[Zim90] ______, *Essential Results of Functional Analysis*, *Chicago Lectures in Math.*, University of Chicago Press, Chicago, 1990. MR 91h:46002 Zbl 0708.46001

(Received August 30, 2004)

*E-mail address*: fisherdm@indiana.edu
INDIANA UNIVERSITY, DEPARTMENT OF MATHEMATICS, RAWLES HALL, BLOOMINGTON, IN 47401, UNITED STATES

*E-mail address*: gregory.margulis@yale.edu
YALE UNIVERSITY, DEPARTMENT OF MATHEMATICS, P.O. BOX 208283, NEW HAVEN, CT 06520, UNITED STATES